Operations Research Pr

Operations Research Pr

Diethard Klatte • Hans-Jakob Lüthi
Karl Schmedders
Editors

Operations Research Proceedings 2011

Selected Papers of the International Conference on Operations Research (OR 2011), August 30–September 2, 2011, Zurich, Switzerland

 Springer

Editors
Diethard Klatte
Karl Schmedders
Institut für Betriebswirtschaftslehre
Universität Zürich
Zürich
Switzerland

Hans-Jakob Lüthi
Institute of Operations Research
ETH Zürich
Zürich
Switzerland

Colored tables and figures may be viewed online on the book webpage at www.springer.com

ISSN 0721-5924
ISBN 978-3-642-29209-5   ISBN 978-3-642-29210-1 (eBook)
DOI 10.1007/978-3-642-29210-1
Springer Heidelberg New York Dordrecht London

Library of Congress Control Number: 2012939852

Printed on acid-free paper

Springer is part of Springer Science+Business Media (www.springer.com)

Preface

The German Operations Research Society, GOR, has a long-standing tradition to report on the progress in the field of Operations Research, as evidenced by presentations in their annual conference in the form of conference proceedings. About every four years the annual conference is hosted by one of the other two German-speaking OR societies: the Austrian society, ÖGOR, or the Swiss community, SVOR. The *International Conference on Operations Research (OR 2011)* took place at the University of Zurich from August 30 to September 2, 2011. This volume contains a selection of papers presented at this conference.

More than 840 scientists and students from over 50 different countries attended OR 2011 and presented 620 papers in 16 parallel topical streams, as well as special award sessions. These proceedings are organized around these streams and award sessions. The stream co-chairs organized the review process of submitted papers from their respective streams. The editors made the final selection based on recommendations by referees as well as the stream co-chairs. The final acceptance rate of submitted papers was about 50 percent.

Instead of a catchy mission statement or slogan for OR 2011, we, the organizers, committed ourselves to realizing a "Sustainable OR 2011". Efficiency of resource usage lies at the heart of Operations Research, and environmental protection has become an active academic topic at OR conferences. For example, at OR 2011 the stream "Energy, Environment and Climate" was prominently visible with 8 sessions dedicated to this topic. However, scientific and technological advances alone are not sufficient to achieve a sustainable way of living.

With the project "Sustainable OR 2011" we intended to increase the awareness of the conference's carbon footprint in the areas of travel, accommodation, catering, conference materials and infrastructure. Part of the conference's footprint has been compensated by supporting carbon offset projects with "myclimate", a nonprofit spinoff foundation of the ETH Zurich. We invite the conference participants and readers of these proceedings to join these efforts and to "do the best and offset the rest"!

The conference was designed according to our shared understanding of Operations Research as an interdisciplinary science focusing on modeling complex sociotechnical systems to gain insight into behavior under interventions by decision makers. Dealing with "organized complexity" lies at the core of OR and designing useful support systems to master the challenge of system management in complex environment is the ultimate goal of our professional societies. To this end, two fundamental competences need to be steadily improved, algorithmic techniques and systemic modeling. The editors hope that both aspects are as well-balanced in these proceedings as they were in the conference.

We thank the program committee and, in particular, the stream co-chairs for helping us during the review process of more than 800 papers submitted for presentation at OR 2011. Many thanks to all the anonymous reviewers who evaluated the papers submitted for these proceedings. We owe much gratitude to our colleagues Marco Laumanns, Michele Marcionelli and Marc Wiedmer who implemented and maintained the paper submission system and greatly helped us in preparing the LATEXdocuments and optimizing the layout process. Thank you very much!

We are grateful to Christian Rauscher from Springer-Verlag for his support in publishing this proceedings volume.

Zurich, January 2012   *Diethard Klatte*
                                                             Hans-Jakob Lüthi
                                                             Karl Schmedders

Contents

Stream II
Discrete Optimization, Graphs and Networks

Stream III
Decision Analysis, Decision Support

Stream IV
Energy, Environment and Climate

Stream V
Financial Modeling, Risk Management, Banking

Stream IX
Metaheuristics and Biologically Inspired Approaches

Stream X
Network Industries and Regulation

Stream XI
OR in Industry, Software Applications, Modeling Languages

Stream XII
Production Management, Supply Chain Management

Stream XIII
Scheduling, Time Tabling and Project Management

Stream XIV
Stochastic Programming, Stochastic Modeling and Simulation

Stream I
Continuous Optimization and Control

Stephan Dempe, TU Freiberg, GOR Chair
Gustav Feichtinger, TU Vienna, ÖGOR Chair
Diethard Klatte, University of Zurich, SVOR Chair

On the controlling of liquid metal solidification in foundry practice

Andrey Albu and Vladimir Zubov

Abstract The optimal control problem of metal solidification in casting is considered. The process is modeled by a three dimensional two phase initial-boundary value problem of the Stefan type. The optimal control problem was solved numerically using the gradient method. The gradient of the cost function was found with the help of the conjugate problem. The discreet conjugate problem was posed with the help of the Fast Automatic Differentiation technique.

1 Introduction

An important stage in metal casting is cooling and solidification of melted metal in the mold. Quality of the obtained detail greatly depends on how the crystallization process proceeds and thus this process has to be controlled.

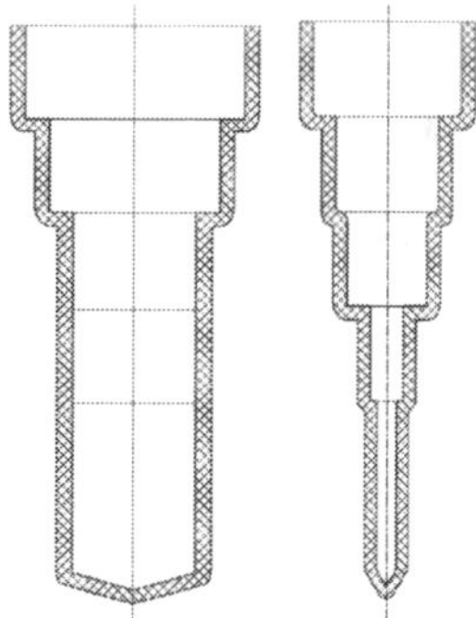

Fig. 1

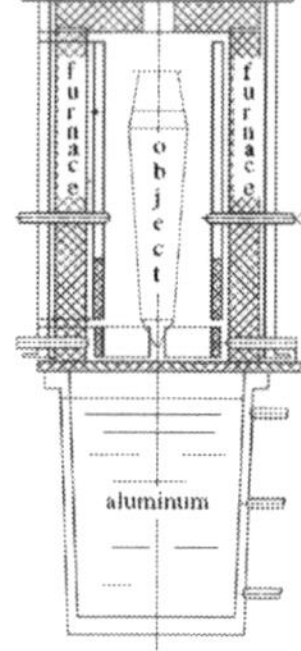

Fig. 2

A mold with specified outer and inner boundaries (see Fig. 1) is filled with liquid metal (the hatched area in Fig. 1 depicts the mold and the internal unhatched area

Andrey Albu
Computing Center RAS, Vavilov st. 40, 119333 Moscow, Russia, e-mail: andrey.albu@gmail.com

Vladimir Zubov
Computing Center RAS, Vavilov st. 40, 119333 Moscow, Russia, e-mail: zubov@ccas.ru

D. Klatte et al. (eds.), *Operations Research Proceedings 2011*, Operations Research Proceedings,
DOI 10.1007/978-3-642-29210-1_1, © Springer-Verlag Berlin Heidelberg 2012

shows the inside space filled with liquid metal). The mold and the metal inside it are heated up to prescribed temperatures T_{form} and T_{met} respectively. Next, the mold filled with metal (which is hereafter referred to as the object) begins to cool gradually under varying surrounding conditions. The different parts of its outer boundary are under different thermal conditions. The setup for metal casting is shown in Fig. 2. The upper part of this setup consists of a furnace and a mold moving inside it. The lower part is the coolant representing a large tank filled with liquid aluminum whose temperature is slightly higher than the aluminum melting point.

The cooling of the liquid metal in the furnace proceeds as follows. On the one hand, the object is slowly immersed in the low temperature liquid aluminum, which causes the solidification of the metal. On the other hand, the object gains heat from the furnace, which prevents the solidification process from proceeding too fast.

One essential feature of this problem is the moving interface between liquid and solid phases. The law of motion of this interface is unknown in advance and has to be determined. Problems of this class (Stefan problems) are noticeably more complicated than those not involving phase transitions.

2 Mathematical formulation of the problem

The process of cooling down of the object is described by the heat-type equation:

$$\frac{\partial H}{\partial t} = \frac{\partial}{\partial x}\left(K\frac{\partial T}{\partial x}\right) + \frac{\partial}{\partial y}\left(K\frac{\partial T}{\partial y}\right) + \frac{\partial}{\partial z}\left(K\frac{\partial T}{\partial z}\right), \qquad (x,y,z) \in Q. \qquad (1)$$

Here x, y, and z are the coordinates of the point; t is time; Q is the domain (object) with a piecewise smooth boundary Γ; $T(x,y,z,t)$ is the substance temperature at the point with coordinates (x,y,z) at time t. The heat content function $H(T(x,y,z,t))$ is defined as

$$H(T(x,y,z,t)) = \begin{cases} H_1(T), & (x,y,z) \in metal, \\ H_2(T), & (x,y,z) \in mold, \end{cases}$$

$$H_1(T) = \begin{cases} \rho_S c_S T, & T < T_1, \\ \rho_S c_S T + \dfrac{\rho_S \gamma (T - T_1)}{T_2 - T_1}, & T_1 \le T < T_2, \\ \rho_L c_L (T - T_2) + \rho_S c_S T_2 + \rho_S \gamma, & T \ge T_2, \end{cases} \qquad H_2(T) = \rho_\Phi c_\Phi T,$$

γ is the specific heat of melting.

The thermal conductivity K is also different for the metal and the mold:

$$K(T) = \begin{cases} K_1(T), & (x,y,z) \in metal, \\ k_\Phi, & (x,y,z) \in mold, \end{cases}$$

$$K_1(T) = \begin{cases} k_S, & T < T_1, \\ \dfrac{k_L - k_S}{T_2 - T_1}T + \dfrac{k_S T_2 - k_L T_1}{T_2 - T_1}, & T_1 \leq T < T_2, \\ k_L, & T \geq T_2. \end{cases}$$

The constants c_S, c_L, c_Φ, ρ_S, ρ_L, ρ_Φ, k_S, k_L, k_Φ, T_1, and T_2 are assumed to be known. The domain separating the phases is determined by a narrow range of temperatures $[T_1, T_2]$, in which the thermodynamic coefficients change rapidly.

The mold and the metal are cooled via their interaction with the surroundings. The individual parts of the outer boundary of the object are in different thermal conditions. The basic types of thermal conditions at a point of the outer boundary of the object can be described as follows (see [1]).

1) The point is in the liquid aluminum.
In this case, the following processes have to be taken into account:
(I) the heat lost by the object due to its own radiation;
(II) the heat gained from the surrounding liquid aluminum due to its radiation;
(III) the heat transfer due to thermal conduction between the coolant and the object.
2) The point is outside the liquid aluminum.
In this case, the following processes have to be taken into account:
(I) the heat lost by the object due to its own radiation;
(II) the heat gained from the emitting walls of the furnace;
(III) the heat gained from the emitting surface of the liquid aluminum.

The conditions of heat transfer with the surrounding medium are set on the boundary Γ of Q. These conditions depend on surface point and time and can be written in the following general form:

$$\tilde{\alpha}T + \tilde{\beta}T_{\mathbf{n}} = \tilde{\gamma}. \tag{2}$$

Here $\tilde{\alpha}$, $\tilde{\beta}$, and $\tilde{\gamma}$ are given functions of the coordinates (x, y, z) of a point on Γ and temperature $T(x, y, z, t)$, and $T_{\mathbf{n}} = \frac{\partial T}{\partial \mathbf{n}}$ is the derivative of T in the $\mathbf{n}$-direction (the external normal to the surface Γ).

It should be noted that the thermodynamic coefficients have a jump on the metal-mold interface. Two conditions are set on this surface: the temperature and the heat flux must be continuous. Thus, to solve the direct problem is to determine a function $T(x, y, z, t)$ that satisfies Eq. (1) in Q, conditions (2) on the outer boundary Γ of Q, two conditions on the metal-mold interface (continuity of heat flux and temperature) and two conditions on the phase interface (the jump of heat flux is proportional to the speed of the phase interface and temperature is continuous).

Different technological requirements can be imposed on the evolution of the phase boundary to obtain a detail of the desired quality. In this work certain ones are considered: the shape of the phase boundary should be close to a plane and its law of motion should be close to a preset one. One of the consequences of satisfying these requirements would be absence of "bubbles" of liquid metal. These "bubbles" are essentially areas of liquid metal that are surrounded by already solidified metal. If such areas are present during the solidification process, the quality of the detail would be unacceptable for sure.

The evolution of the solidification front is affected by numerous parameters that could be controlled, but in practice most often the speed $u(t)$ at which the mold moves relative to the furnace is used as the control. To find a control $u(t)$ that will satisfy the mentioned technological requirements, we state the optimal control problem and write down the following cost functional:

$$I(u) = \frac{1}{t_2(u) - t_1(u)} \int_{t_1(u)}^{t_2(u)} \iint_S [Z_{pl}(x,y,t,u) - z_*(t)]^2 dxdydt.$$

Here t_1 is the time when the crystallization front arises; t_2 is the time when the crystallization of metal completes; $(x,y,Z_{pl}(x,y,t,u))$ are the real coordinates of the phase interface at the time t; $(x,y,z_*(t))$ are the desired coordinates of the phase interface at the time t (see Fig. 3); S is the largest cross section of the mold that is filled with metal.

3 Numerical algorithm for solving the optimal control problem

The object being investigated is approximated by a body, which consists of a finite number of rectangular parallelepipeds. The approximated body is placed mentally into a certain large parallelepiped. In this large parallelepiped a basic rectangular grid is introduced in such a way that all external surfaces of the approximated body and the surfaces, which divide metal and form, coincide with the grid surfaces. Besides the basic grid, the auxiliary grid is built, whose surfaces are parallel to the surfaces of the basic grid and are displaced relative to it with a half-step in all directions. As a result the entire object being investigated is broken by the surfaces of auxiliary grid into elementary volumes. The use of such grids allows us to accurately calculate the heat fluxes.

A numerical algorithm was developed for solving the initial-boundary value problem (1)-(2) (see [2]). The algorithm that solves the direct problem is based on the heat balance law:

$$\iiint_V \left[H\left(T(x,y,z,t^j + \tau)\right) - H\left(T(x,y,z,t^j)\right) \right] dV = \int_{t^j}^{t^j+\tau} \iint_S K(T)T_{\mathbf{n}}dsdt,$$

which means that the increment in heat content within volume V over a fixed time interval τ is equal to the amount of heat transferred through the surface S of V over the same time interval.

For each elementary volume the equation of heat balance is written in terms of heat content (see [1]). The finite-difference approximation of the heat balance equation is based on the Peaceman-Rachford scheme. In the computation of the direct problem primary attention was given to the evolution of the solidification front and to how it is affected by the parameters of the problem.

The optimal control problem was solved by reducing the original problem to a nonlinear programming problem. The control function was approximated by a piecewise constant function. The received nonlinear programming problem was solved by the gradient method. The efficiency of the gradient methods depends essentially on the accuracy which the gradient of the cost functional is calculated with. The approximate finite-difference evaluation of the cost functional gradient in the given optimal control problem is associated with huge difficulties (see [3]). In [4] an effective method is proposed for the evaluation of the cost functional gradient in this optimal control problem. The method is based on the Fast Automatic Differentiation (FAD) technique (see [5]) and produces the exact value of the gradient for the chosen approximation of the direct problem and functional.

The generalized FAD technique can be described as follows. The goal of any optimal control problem is to optimize a cost functional that depends on controls and state variables. The controls and state variables are connected by certain constraints (for example, for given controls, the state variables are determined by solving a boundary value problem for a system of partial differential equations). The first step in the generalized FAD technique involves the discretization of the functional and the constraints. As a result, the cost functional is associated with a function of a finite number of variables, while the constraints are associated with a set of algebraic equations. Thus, we have to optimize a function of a finite number of variables that are related by a set of algebraic equations. The second step is to evaluate the gradient of the discrete cost function that is subjected to the constraints. The FAD technique delivers a unique finite-difference scheme for the conjugate problem and a unique formula for determining the gradient of the cost functional. The value of the gradient of the cost function, calculated according to formulas of the FAD-methodology, is precise for the chosen approximation of the optimal control problem. Let us especially note that the machine time needed for calculation of the gradient components in the considered problem is not more than half of the machine time needed to solve the direct problem. This fact is supported by numerous computations and agrees with the conclusions drawn in [5].

Figure 4 shows the optimal law of motion of the mold relative to the furnace for parameters provided in [1]. Figures 5-8 illustrate isotherms at different times in two cross sections through the object's vertical axis of symmetry parallel to the object facets. Since the object is symmetric about the vertical axis, the figures present only halves of the cross sections. The light vertical and horizontal lines inside the object separate the metal and the mold. The light curves show lines of constant temperature and the heavy curve depicts the interface between two phases in the metal. Figures 5, 6 (the first experiment) illustrate the process of metal solidification in a mold moving relative to the furnace at a constant speed $u(t) = 2.0\ mm/min$. Figures 7, 8 (the second experiment) correspond to the optimal law of motion.

Figures 5-8 vividly show the advantages of the optimal process of metal crystallization. In the second experiment the behavior of the phase boundary is close to the

desired one, whereas "bubbles" of liquid metal form and collapse inside the object
in the first experiment, which results in a casting of poor quality.

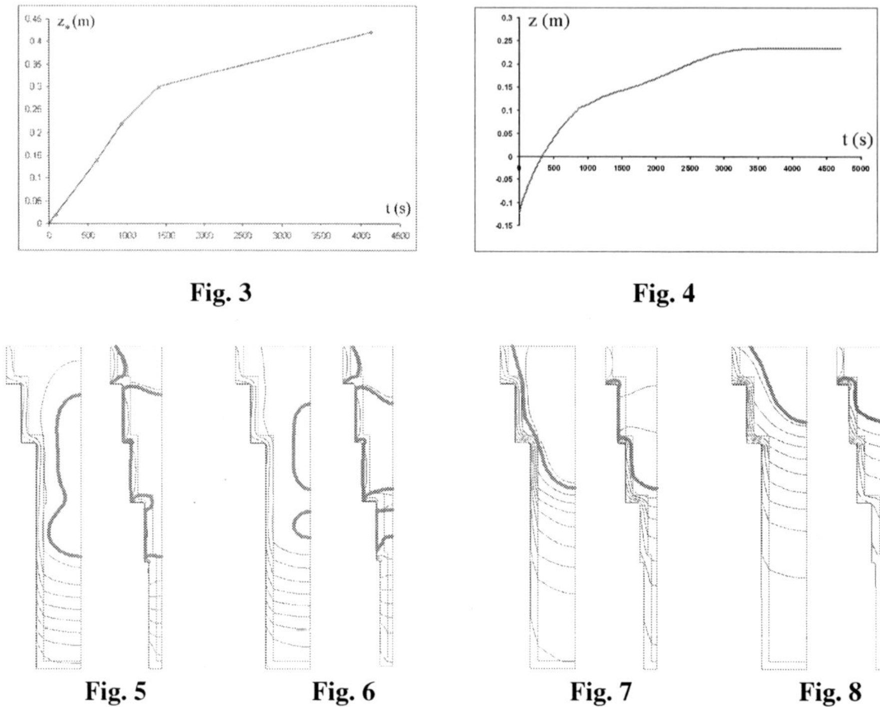

Fig. 3   Fig. 4

Fig. 5   Fig. 6   Fig. 7   Fig. 8

Acknowledgements This work was supported by the Program for Fundamental Research of Presidium of RAS P17, by RFBR (No. 11-01-12136-ofi-m-2011) and by the Program Leading Scientific Schools (NSh-4096.2010.1).

References

1. Albu, A.F., Zubov, V.I.: Mathematical Modeling and Study of the Process of Solidification in Metal Casting. Computational Mathematics and Mathematical Physics **47**, 843–862 (2007)
2. Albu, A.F., Zubov, V.I.: Functional Gradient Evaluation in an Optimal Control Problem Related to Metal Solidification. Computational Mathematics and Mathematical Physics **49**, 47–70 (2009)
3. Albu, A.V., Albu, A.F., Zubov, V.I.: Functional Gradient Evaluation in the Optimal Control of a Complex Dynamical System. Computational Mathematics and Mathematical Physics **51**, 762–780 (2011)
4. Albu, A.V., Zubov, V.I.: Choosing a Cost Functional and a Difference Scheme in the Optimal Control of Metal Solidification. Computational Mathematics and Mathematical Physics **51**, 21–34 (2011)
5. Evtushenko, Y.G.: Computation of Exact Gradients in Distributed Dynamic Systems. Optimizat. Methods and Software **9**, 45–75 (1998)

Optimal dividend and risk control in diffusion models with linear costs[*]

Hiroaki Morimoto

Abstract We consider the optimization problem of dividends and risk exposures of a firm in the diffusion model with linear costs. The variational inequality associated with this problem is given by the nonlinear form of elliptic type. Using the viscosity solutions technique, we solve the corresponding penalty equation and show the existence of a classical solution to the variational inequality. The optimal policy of dividend payment and risk exposure is shown to exist.

1 Introduction

We consider the optimal dividend and risk control problem of a firm in the diffusion model with linear costs. Let R_t be the risk process refered to the reserve of the firm at time $t \geq 0$. The risk process evolves according to the stochastic differential equation:

$$dR_t = \mu dt + \sigma dB_t, \quad R_0 = x > 0,$$

on a complete probability space $(\Omega, \mathscr{F}, P)$, carrying a one-dimensional standard Brownian motion $\{B_t\}$, endowed with the natural filtration $\mathscr{F}_t$ generated by $\sigma(B_s, s \leq t)$ for $t \geq 0$, where $\mu > 0$ denotes the profit per unit time and $\sigma \neq 0$ is a diffusion coefficient.

A control policy (a, L) is described by a pair of the risk exposure $a = \{a_t\}$ and the flow $L = \{L_t\}$ of dividend payments. The portion $1 - a_t$ of the reserve is paid for reinsurance and L_t denotes the total amount of dividend paid out up to time t. The policy (a, L) is said to be admissible if $\{a_t\}$ is an $\{\mathscr{F}_t\}$-progressively measurable process such that

$$0 \leq a_t \leq 1, \quad t \geq 0,$$

and $\{L_t\}$ is a nonnegative, nondecreasing, continuous $\{\mathscr{F}_t\}$-adapted process with $x - L_0 > 0$. We respectively denote by $\mathscr{A}$ and $\mathscr{L}$ the class of all admissible risk exposures a and dividends L. Given $(a, L) \in \mathscr{A} \times \mathscr{L}$, the dynamics of the risk process $\{R_t\}$ is given by

Department of Mathematics, Ehime University
Matsuyama 790-0826, Japan
e-mail: morimoto@mserv.sci.ehime-u.ac.jp

[*] This work was supported by JSPS KAKENHI 21540188.

D. Klatte et al. (eds.), *Operations Research Proceedings 2011*, Operations Research Proceedings, 9
DOI 10.1007/978-3-642-29210-1_2, © Springer-Verlag Berlin Heidelberg 2012

$$dR_t = a_t(\mu dt + \sigma dB_t) - vR_t dt - dL_t, \quad R_0 = x - L_0 > 0, \tag{1}$$

where vx represents the linear cost function of the reserve x or the debt payment for $v \geq 0$.

The objective is to find an optimal policy $(a^*, L^*) = \{(a_t^*, L_t^*)\}$ so as to maximize the expected present value of dividends up to bankruptcy:

$$J(a,L) = E\left[\int_0^\vartheta e^{-\alpha t} dL_t\right], \tag{2}$$

over all $(a,L) \in \mathscr{A} \times \mathscr{L}$, where $\vartheta = \vartheta(R) := \inf\{t \geq 0 : R_t = 0\}$ and $\alpha > 0$ is a discount factor. In case of $v = 0$, this problem has been studied by Højgaard and Taksar [4] and Taksar [7]. We refer to Morimoto [5, 6] for the viscosity solutions technique in the stochastic optimization problem.

Our approach consists in finding a classical solution v of the following variational inequality associated with the problem:

$$v'(x) \geq 1, \quad x > 0, \tag{3}$$

$$-\alpha v + \max_{0 \leq a \leq 1}\left(\frac{1}{2}a^2\sigma^2 v'' + a\mu v'\right) - vxv' \leq 0, \quad x > 0, \tag{4}$$

$$\left\{-\alpha v + \max_{0 \leq a \leq 1}\left(\frac{1}{2}a^2\sigma^2 v'' + a\mu v'\right) - vxv'\right\}(v' - 1)^+ = 0, \quad x > 0, \tag{5}$$

$$v(0) = 0. \tag{6}$$

In order to solve the variational inequality (3) - (6), we study the penalty equation of the form:

$$-\alpha u + \frac{1}{2}\varepsilon^2 x^2 u'' + \mathscr{M}u - vxu' + \frac{1}{\varepsilon}(u' - 1)^- = 0, \quad x > 0, \tag{7}$$

$$u(0) = 0, \tag{8}$$

where $\mathscr{M}u = \max_{0 \leq a \leq 1}(\frac{1}{2}a^2\sigma^2 u'' + a\mu u')$ and $\varepsilon \in (0,1)$. We show the existence of a solution $u \in C^2(0,\infty) \cap C[0,\infty)$ to the penalty equation (7), (8). By using the penalization method, we prove the convergence of u to a concave viscosity solution $v \in C^2(0,\infty) \cap C[0,\infty)$ of (3) - (6) as $\varepsilon \to 0$. Furthermore, we present the optimal policy (a^*, L^*) with the reflecting barrier at the free boundary x^* for v.

2 The penalized problem

For each $(a,c) \in \mathscr{C} \times \mathscr{C}$, there exists a unique nonnegative solution $\{X_t\}$ of

$$dX_t = 1_{\{t \leq \vartheta(X)\}}\left\{a_t(\mu dt + \sigma dB_t) + \varepsilon X_t d\bar{B}_t - vX_t dt - \frac{c_t}{\varepsilon}dt\right\}, \quad X_0 = x \geq 0, \tag{9}$$

where $\{\bar{B}_t\}$ is a Brownian motion, mutually independent of $\{B_t\}$, and $\mathscr{C}$ denotes the class $\mathscr{A}$ for $\{(B_t,\bar{B}_t)\}$. Since $(u'-1)^- = \max_{0\le c\le 1}(1-u')c$, we observe that the penalty equation (7) is the Hamilton-Jacobi-Bellman equation associated with the maximization problem:

$$u(x) := \sup_{(a,c)\in\mathscr{C}\times\mathscr{C}} E\left[\int_0^\theta e^{-\alpha t}\frac{c_t}{\varepsilon}dt\right], \tag{10}$$

subject to (9), where $\theta := \vartheta(X)$ and the supremum is taken over all systems $(\Omega,\mathscr{F},P,\{\mathscr{F}_t\};\{(B_t,\bar{B}_t)\},\{a_t\},\{c_t\})$.

Lemma 1. *There exists a concave supersolution* $f \in C^1(0,\infty)\cap C^2((0,\infty)\setminus\{m\})$ *of (7), (8), independent of* ε, *for some* $m > 0$.

Proof. Define
$$f(x) = \begin{cases} k(x), & x \le m, \\ x-m+k(m), & x \ge m, \end{cases}$$

where $k(x) = Kx^\lambda$ and $m = \sigma^2(1-\lambda)/\mu$. For a suitable choice of $0 < \lambda < 1 < K$, we see that f is a supersolution, i.e.,

$$-\alpha f + \frac{1}{2}\varepsilon^2 x^2 f'' + \mathscr{M}f - vxf' + \frac{1}{\varepsilon}(f'-1)^- \le 0, \quad x > 0, x \ne m.$$

Theorem 1. *We have*

$$0 \le u(x) \le f(x), \quad x \ge 0,$$
$$|u(x) - u(y)| \le f(|x-y|), \quad x,y \ge 0.$$

Proof. By (9) and (10), we see $\theta \ge \vartheta(Y)$ and $u(x) \ge u(y)$ if $x \ge y$. Applying the generalized Ito formula for convex functions to f, we can show the assertions.

Definition 1. Let $\zeta \in C[0,\infty)$ satisfy (8). Then ζ is called a viscosity subsolution (resp., supersolution) of (7), (8) if, whenever for $\phi \in C^2$, $\zeta - \phi$ attains its local maximum (resp., minimum) at $z > 0$, then

$$-\alpha\zeta + \frac{1}{2}\varepsilon^2 x^2 \phi'' + \mathscr{M}\phi - vx\phi' + \frac{1}{\varepsilon}(\phi'-1)^-\bigg|_{x=z} \ge 0 \quad (resp., \le 0).$$

We call ζ a viscosity solution of (7), (8) if it is both a viscosity sub- and supersolution of (7), (8).

Theorem 2. *u is a viscosity solution of (7), (8).*

Proof. By Theorem 1, we can see that the dynamic programming principle holds for u. Therefore, we obtain the viscosity property of u.

Theorem 3. *We have*
$$u \in C^2(0,\infty)\cap C[0,\infty).$$

Proof. For any $0 < p < q$, we consider the boundary value problem:

$$-\alpha w + \frac{1}{2}\varepsilon^2 x^2 w'' + \mathcal{M}w - \nu x w' + \frac{1}{\varepsilon}(w'-1)^- = 0, \quad x \in (p,q),$$

$$w(p) = u(p), \quad w(q) = u(q). \tag{11}$$

By uniformly ellipticity, the theory of fully nonlinear elliptic equations [3] yields that there exists a unique solution $w \in C^2(p,q) \cap C[p,q]$ of (11). By the uniqueness result on viscosity solutions, we have $w = u$ and u is smooth.

Theorem 4. *u is concave on $[0,\infty)$.*

3 Variational inequalities

3.1 Viscosity solutions

Definition 2. Let $\zeta \in C[0,\infty)$ satisfy (6). Then ζ is called a viscosity solution of (3) - (6), if the following assertions are satisfied:

(a) For any $\phi \in C^2(0,\infty)$ and any local minimum point $\bar{z} > 0$ of $\zeta - \phi$,

$$\phi'(\bar{z}) \geq 1, \quad -\alpha\zeta + \mathcal{M}\phi - \nu x \phi'\Big|_{x=\bar{z}} \leq 0,$$

(b) For any $\phi \in C^2(0,\infty)$ and any local maximum point $z > 0$ of $\zeta - \phi$,

$$\{-\alpha\zeta + \mathcal{M}\phi - \nu x \phi'\}(\phi'-1)^+\Big|_{x=z} \geq 0.$$

Theorem 5. *There exists a subsequence $\{u_{\varepsilon_n}\}$ such that*

$$u_{\varepsilon_n} \to v \in C[0,\infty) \quad \text{locally uniformly in } (0,\infty) \text{ as } \varepsilon_n \to 0. \tag{12}$$

Furthermore, v is a viscosity solution of (3) - (6).

Proof. Let $0 < p < q$ be arbitrary. By concavity and Theorem 1, we get

$$0 \leq u'_\varepsilon(x)x \leq u_\varepsilon(x) - u_\varepsilon(0) \leq \|f\|_{C[p,q]}, \quad x \in [p,q].$$

Hence

$$\sup_\varepsilon \|u'_\varepsilon\|_{C[p,q]} < \infty. \tag{13}$$

Thus, by the Ascoli-Arzelà theorem, there exists a subsequence $\{u_{\varepsilon_n}\}$ satisfying (12). By Theorem 2, passing to the limit, we obtain the viscosity property of v.

3.2 Regularity and the free boundary

Theorem 6. *We have*

$$u'_{\varepsilon_n}(x) \geq 1 \quad \text{for} \quad x > 0.$$

Proof. By Theorems 4 and 5, we note that v is concave and twice differentiable almost everywhere. We recall that $\partial v(q) = \{v'(q)\}$ at a differentiable point $q > 0$ of v. Then, by the viscosity property of v, we have the assertion.

Now, let $A_n(x)$ denote the maximizer of $\max_{0 \leq a \leq 1}(\frac{1}{2}a^2\sigma^2 u''_{\varepsilon_n} + a\mu u'_{\varepsilon_n})$, i.e.,

$$A_n(x) = G(M_n(x)), \quad M_n(x) = -\mu u'_{\varepsilon_n}/\sigma^2 u''_{\varepsilon_n},$$

where

$$G(x) = \begin{cases} x & \text{if } 0 \leq x \leq 1, \\ 1 & \text{if } 1 < x. \end{cases}$$

Lemma 2. *For any* $0 < p < q$, *we have*

$$M_n(x) \geq (2\alpha p/\mu) \wedge 1, \quad x \in [p,q], \tag{14}$$

$$\sup_n \sup\{|A_n(x) - A_n(y)|/|x-y| : x,y \in [p,q], x \neq y\} < \infty. \tag{15}$$

Proof. By (7), concavity and Theorem 6, we have

$$\alpha u_{\varepsilon_n} + vx u'_{\varepsilon_n} = \frac{1}{2}\varepsilon_n^2 x^2 u''_{\varepsilon_n} + \max_{0 \leq a \leq 1}(\frac{1}{2}a^2\sigma^2 u''_{\varepsilon_n} + a\mu u'_{\varepsilon_n}) \leq \frac{\mu}{2}u'_{\varepsilon_n} M_n(x).$$

By concavity, $x u'_{\varepsilon_n} \leq u_{\varepsilon_n}$. Thus we get $\alpha x \leq (\mu/2)M_n(x)$, which implies (14).

Next, let $A_n(x) = M_n(x)$ on $(p_1,q_1) \subset [p,q]$. By (14) and (13), we get

$$0 \leq \sup_{x \in [p_1,q_1]} -\sigma^2 u''_{\varepsilon_n}(x) < \infty, \quad \sup_n \|A'_n\|_{C[p_1,q_1]} < \infty,$$

which implies (15).

Theorem 7. *We have*

$$v \in C^2(0,\infty), \quad v' \geq 1 \quad on \ (0,\infty).$$

Proof. For any $0 < p < q$, we set $(\bar{p}, \bar{q}) = (p/2, q + p/2)$. Consider the boundary value problem:

$$\frac{1}{2}\varepsilon_n^2 x^2 \zeta'' + \frac{1}{2}A_n(x)^2\sigma^2 \zeta'' + \{A_n(x)\mu - vx\}\zeta' = \alpha u_{\varepsilon_n}, \quad x \in (\bar{p}, \bar{q}), \tag{16}$$

$$\zeta(\bar{p}) = u_{\varepsilon_n}(\bar{p}), \quad \zeta(\bar{q}) = u_{\varepsilon_n}(\bar{q}).$$

By Theorems 3 and 6, we see that u_{ε_n} solves (16). By Lemma 2 and the interior Schauder estimates for (16), we have $\sup_n \|u_{\varepsilon_n}\|_{C^{2,\gamma}[p,q]} < \infty, 0 < \gamma < 1$, which completes the proof. We remark that v is a classical solution of (3) - (6).

Theorem 8. *There exists the free boundary* $x^* \in (0,\infty)$ *for* v, *which fulfills*

$$x^* = \sup\{x > 0 : v'(x) > 1\}.$$

Proof. By the contradiction arguments, we can see that $\{\cdot\}$ is non-empty and $x^* < \infty$.

4 Optimal policies

Consider the SDE with reflecting barrier conditions for the free boundary x^*:

$$dR_t^* = \bar{A}(R_t^*)(\mu dt + \sigma dB_t) - vR_t^* dt - dL_t^*, \quad R_0^* = x, \tag{17}$$

$$L_t^* = \int_0^t 1_{\{R_s^* = x^*\}} dL_s^*, \tag{18}$$

$$L_t^* \text{ is continous and nondecreasing}, \tag{19}$$

$$R_t^* \leq x^*, \quad t \geq 0, \tag{20}$$

$$\int_0^t 1_{\{R_s^* = x^*\}} ds = 0, \quad t \geq 0, \tag{21}$$

where $\bar{A}(x)$ is the continuous extension of $A(x) := G(-\mu v'/\sigma^2 v'')$ for $x > 0$ and $\bar{A}(x) = 0$ for $x \leq 0$.

Lemma 3. *We have* $\lim_{x \to 0+} A(x) = 0$ *and* $\bar{A}$ *is Lipschitz on* $(-\infty, x^*]$.

Theorem 9. *We assume* $0 < x \leq x^*$. *Then the optimal policy* (a^*, L^*) *for (2) subject to (1) is given by* $a_t^* = \bar{A}(R_t^*)$ *and* $\{L_t^*\}$ *of (17) - (21).*

Proof. According to [1], by Lemma 3, there exists a unique solution $\{(R_t^*, L_t^*)\}$ of (17) - (21). Applying Ito's formula to (3) - (6), we can obtain the optimality.

5 Conclusion

The optimal policy with the reflecting barrier at the free boundary is shown to exist.

References

1. Bensoussan, A., Lions, J.L.: Contrôl Impulsionnel et Inéquations Quasi-Variationnelles. Gauthier-Villars, Paris (1982)
2. Crandall, M.G., Ishii, H., Lions, P.L.: User's guide to viscosity solutions of second order partial differential equations. Bull. Amer. Math. Soc. 27, 1-67 (1992)
3. Gilbarg, D., Trudinger, N.S.: Elliptic Partial Differential Equations of Second Order. Springer-Verlag, Berlin (1983)
4. Højgaard, B., Taksar, M.: Controlling risk exposure and dividends payout schemes: insurance company example. Math. Finance 9, 153-182 (1999)
5. Morimoto, H.: Optimal dividend payments in the stochastic Ramsey model. Stochastic Process. Appl. 120, 427-441 (2010)
6. Morimoto, H.: Stochastic Control and Mathematical Modeling: Applications in Economics. Cambridge University Press, New York (2010)
7. Taksar, M.I. : Optimal risk and dividend distribution control models for an insurance company. Math. Methods Oper. Res. 51, 1-42 (2000)

An optimal control problem for a tri-trophic chain system with diffusion

N. C. Apreutesei

Abstract An optimal control problem is studied for a nonlinear reaction-diffusion system that describes the behavior of a trophic chain composed by a predator, a pest and a plant species. A pesticide is introduced in the ecosystem. It is regarded as a control variable. The purpose is to minimize the pest density and to maximize the plant density.

1 Introduction

We study an optimal control problem associated to a reaction-diffusion system which models a tri- trophic food chain consisting of a predator (whose density is y^2), a pest (of density y^1) and a plant (of density y^3). The plant can be a vegetable species or a cereal from a cultivable land for agriculture, while the parasite is an insect that destroyes the plant species. The predator can be a bird species. Their habitat is modelled by a bounded domain $\Omega \subset \mathbb{R}^d$, $d \leq 3$. The densities y^1, y^2, y^3 depend both on the time $t \in [0, T]$ and on the spatial position $x \in \Omega$.

In order to reduce the negative action of the pest, one spreads a pesticide in the habitat. Suppose that its effect on the pest is proportional to its density y^1. Denote by $u(t,x)$ the proportionality factor at time t and spatial position x. We regard it as a control variable. The dynamics of the ecosystem can be described by the partial differential system

$$\begin{cases} y_t^1 = \alpha_1 \Delta y^1 + y^1 \left(a_1 - b_1 y^2 + c_1 y^3 - u \right) \\ \quad y_t^2 = \alpha_2 \Delta y^2 + y^2 \left(-a_2 + b_2 y^1 \right) \\ \quad \quad y_t^3 = y^3 \left(a_3 - b_3 y^1 \right) \end{cases} , \quad (t,x) \in Q \qquad (1.1)$$

$$\left(y^1 \right)_v = \left(y^2 \right)_v = 0, \ (t,x) \in \Sigma, \qquad (1.2)$$

N. C. Apreutesei
Department of Mathematics, Technical University "Gh. Asachi" Iasi, Romania, e-mail: napreut@gmail.com

D. Klatte et al. (eds.), *Operations Research Proceedings 2011*, Operations Research Proceedings, 15
DOI 10.1007/978-3-642-29210-1_3, © Springer-Verlag Berlin Heidelberg 2012

$$y^i(0,x) = y_0^i(x) > 0, \ x \in \Omega, \ i = 1,2,3, \tag{1.3}$$

where $Q = [0,T] \times \Omega$, $\Sigma = [0,T] \times \partial\Omega$ and $(y^i)_v$ is the outward normal derivative of y^i. All the involved coefficients are supposed to be positive.

Our purpose is to minimize the pest density and to maximize the harvest of cereals (or vegetables) at the end of a time interval $[0,T]$. More exactly, we solve the control problem

$$Min \ \Phi(y,u), \ \Phi(y,u) = \int_{\Omega} \left(y^1 - y^3\right)(T,x)\,dx, \ y = \left(y^1,y^2,y^3\right), \tag{1.4}$$

$$u \in \mathcal{U} = \left\{u \in L^2(Q), \ 0 \le u(t,x) \le 1, \ (t,x) \in Q\right\}.$$

This optimal control is quite standard due to the absence of any state constraints. Though known results are not applicable here. The mathematical result is not a routine consequence of [5]. The structure of the paper is the following. Section 2 is dedicated to the existence of a global solution for the system $(1.1) - (1.3)$. The existence of an optimal solution and and some necessary optimality conditions are found in Section 3. Other related control problems can be found in [1], [3] - [7].

2 The existence of a global solution

In order to prove the existence of a strong solution for problem $(1.1) - (1.3)$, we write it in the form of an abstract Cauchy problem in the Hilbert space $H = L^2(\Omega)^3$, namely

$$\begin{cases} y'(t) = Ay(t) + f(t,y(t)), \ t \in [0,T] \\ \qquad\quad y(0) = y_0, \end{cases} \tag{2.1}$$

where $y_0 = (y_0^1,y_0^2,y_0^3)$, $A : D(A) \subseteq H \to H$ is the linear operator

$$Ay = \left(\alpha_1 \Delta y^1, \alpha_2 \Delta y^2, 0\right), \ (\forall) y = \left(y^1,y^2,y^3\right) \in D(A),$$

$$D(A) = \left\{y = (y^1,y^2,y^3) \in H^2(\Omega)^2 \times L^\infty(\Omega), \ (y^1)_v = (y^2)_v = 0, \ x \in \partial\Omega\right\},$$

and $f(t,y) = (f_1(t,y), f_2(t,y), f_3(t,y))$ is the nonlinear term from (1.1) :

$$\begin{cases} f_1(t,y(t)) = y^1\left(a_1 - b_1 y^2 + c_1 y^3 - u\right) \\ \quad f_2(t,y(t)) = y^2\left(-a_2 + b_2 y^1\right) \\ \quad f_3(t,y(t)) = y^3\left(a_3 - b_3 y^1\right) \end{cases}, \ t \in [0,T], \tag{2.2}$$

$$y \in D(f) = \{y \in H, \ f(t,y(t)) \in H, \ (\forall) t \in [0,T]\}.$$

To prove the existence of a solution for problem (2.1), we use Proposition 1.2, p. 175, [2]. Operator A is the infinitesimal generator of a C_0-semigroup of contractions on H, it is self-adjoint and dissipative. But function f is not Lipschitz in $y \in H$, uniformly with respect to $t \in [0, T]$, so the above result cannot be applied directly. Using a truncation procedure, we will associate an auxiliary problem to which we can apply the mentioned result.

Let $N > 0$ be sufficiently large and $f^N = \left(f_1^N, f_2^N, f_3^N\right)$ be obtained from $f = (f_1, f_2, f_3)$ in the following way. If $\left|y^1\right| \le N$, then y^1 in $f\left(t, y^1, y^2, y^3\right)$ remains unchanged. If $y^1 > N$, y^1 from (2.2) is replaced by N. If $y^1 < -N$, then y^1 is replaced by $-N$. We do the same for y^2 and y^3. Thus function f^N becomes Lipschitz continuous in y uniformly with respect to $t \in [0, T]$. Supposing that $y_0 \in D(A)$ and $y_0^i > 0$ on Ω, $i = 1, 2, 3$, we conclude with the aid of Proposition 1.2 from [2] that the truncated problem

$$y_N'(t) = Ay_N(t) + f^N(t, y_N(t)), \ t \in [0, T], \ y_N(0) = y_0, \tag{2.3}$$

has a unique strong solution $y_N = \left(y_N^1, y_N^2, y_N^3\right) \in W^{1,2}(0, T; H)$ with $y_N^1, y_N^2 \in L^2\left(0, T; H^2(\Omega)\right) \cap L^\infty\left(0, T; H^1(\Omega)\right)$.

Now we want to derive from this result the existence of a global solution for problem $(1.1) - (1.3)$. Since the proof is standard, we will sketch here only the main steps, without any detail. First we show that y_N is bounded on Q. Next one proves that $y_N^i(t, x) > 0$, $(\forall)(t, x) \in Q$, $i = 1, 2, 3$. Indeed, one multiplies (2.3) by $\left(y_N^1\right)^-(t, x) = -\inf\left\{y_N^1(t, x), 0\right\}$ and use the writing $y_N^1 = \left(y_N^1\right)^+ - \left(y_N^1\right)^-$. One integrates over $[0, t] \times \Omega$ and employes Green's formula, followed by Gronwall's inequality, to obtain the desired estimate.

Now we can show that for $N > 0$ large enough, $\left|y_N^i(t, x)\right| \le N$, $(\forall)(t, x) \in (0, \theta) \times \Omega$, $i = 1, 2, 3$, for some $\theta \in (0, T)$. Thus f^N coincides with f for $t \in (0, \theta)$, so $y_N = \left(y_N^1, y_N^2, y_N^3\right)$ is a solution of problem $(1.1) - (1.3)$ defined at least on $(0, \theta) \times \Omega$. Finally, one proves that each y_N^i is bounded on $(0, \theta) \times \Omega$, so y_N^i is in fact defined on the whole set Q.

Therefore, we have proved the following global existence result for the boundary value problem $(1.1) - (1.3)$.

Theorem 1. *Let Ω be a bounded domain from R^d, $d \le 3$, with the boundary of class $C^{2+\alpha}$, $\alpha > 0$. If $\alpha_1, \alpha_2, a_i, b_i, c_1 > 0$ $(i = 1, 2, 3)$, $u \in U$ and $y_0 = \left(y_0^1, y_0^2, y_0^3\right) \in D(A)$, $y_0^i > 0$ on Ω, $i = 1, 2, 3$, then problem $(1.1) - (1.3)$ has a unique strong solution $y = \left(y^1, y^2, y^3\right) \in W^{1,2}(0, T; H)$ such that $y^1, y^2 \in L^2\left(0, T; H^2(\Omega)\right) \cap L^\infty\left(0, T; H^1(\Omega)\right)$ and $y^i \in L^\infty(Q)$, $y^i > 0$ on Q for $i = 1, 2, 3$.*

3 The optimal control problem

This section is devoted to the study of the optimal control problem $(1.1) - (1.4)$. We begin with the existence of an optimal solution. The proof is standard and we omit it.

Theorem 2. *If α_1, α_2, a_i, b_i, $c_1 > 0$ $(i = 1,2,3)$ and $y_0 \in D(A)$, $y_0^i > 0$ on Ω, $i = 1,2,3$, then the optimal control problem $(1.1)-(1.4)$ admits an optimal solution (y^*, u^*), $y^* = \left(y^{1*}, y^{2*}, y^{3*}\right)$.*

We deduce now some necessary conditions of optimality. Let A^* be the adjoint of the operator A, f_y^* be the adjoint of the Jacobian matrix f_y and $l(y(T)) = y^1(T,\cdot) - y^3(T,\cdot)$ be the function under the integral from (1.4). Let u^* be an optimal control and $y^* = \left(y^{1*}, y^{2*}, y^{3*}\right)$ be the corresponding optimal state. The adjoint system associated to our problem is

$$\begin{cases} p'(t) + A^* p(t) = -f_y^*(t,y^*)\,p(t),\ \text{a. e. on } (0,T) \\ \qquad\qquad p(T) = -\nabla l(y^*(T)), \end{cases}$$

where $p = \left(p^1, p^2, p^3\right)$ is the adjoint variable. In detail, it becomes

$$\begin{cases} p_t^1 = -\alpha_1 \Delta p^1 - (a_1 - u^*)\,p^1 - y^{2*}\left(-b_1 p^1 + b_2 p^2\right) - y^{3*}\left(c_1 p^1 - b_3 p^3\right) \\ p_t^2 = -\alpha_2 \Delta p^2 + a_2 p^2 - y^{1*}\left(-b_1 p^1 + b_2 p^2\right) \\ p_t^3 = -a_3 p^3 - y^{1*}\left(c_1 p^1 - b_3 p^3\right),\ (t,x) \in Q \end{cases} \tag{3.1}$$

$$\left(p^1\right)_v = \left(p^2\right)_v = 0 \text{ on } \Sigma, \tag{3.2}$$

$$p^1(T,x) = -1,\ p^2(T,x) = 0,\ p^3(T,x) = 1 \text{ on } \Omega. \tag{3.3}$$

One can easily prove that this problem admits a unique strong solution $p = \left(p^1, p^2, p^3\right) \in W^{1,2}(0,T;H)$ with $p \in L^\infty(Q)^3$.

Now we establish an optimality condition for our problem.

Theorem 3. *If (y^*, u^*) is an optimal solution of problem $(1.1)-(1.4)$ and $p = \left(p^1, p^2, p^3\right)$ is the solution of adjoint system $(3.1)-(3.3)$, then $\displaystyle\int_Q y^{1*} p^1 (u^* - \widetilde{u})\,dtdx \le 0$, for any $\widetilde{u} \in U$.*

Proof. Let (y^*, u^*), $(y^\varepsilon, u^\varepsilon)$ be an optimal pair and an arbitrary pair respectively, where $u^\varepsilon = u^* + \varepsilon u_0 \in \mathcal{U}$, with $u_0 \in L^2(Q)$ $(\varepsilon > 0)$, while $y^\varepsilon = \left(y^{1\varepsilon}, y^{2\varepsilon}, y^{3\varepsilon}\right)$ is the corresponding solution of $(1.1)-(1.3)$. By a comparison theorem, we can easily prove that $y^{i\varepsilon}$ are bounded on Q, uniformly with respect to ε $(i = 1,2,3)$.

If we denote by $z^{i\varepsilon}$ the ratio $z^{i\varepsilon} = \left(y^{i\varepsilon} - y^{i*}\right)/\varepsilon$, $i = 1,2,3$, from system $(1.1)-(1.3)$ for (y^*, u^*) and for $(y^\varepsilon, u^\varepsilon)$, one obtains

$$\begin{cases} z_t^{1\varepsilon} = \alpha_1 \Delta z^{1\varepsilon} + \left(a_1 - b_1 y^{2\varepsilon} + c_1 y^{3\varepsilon} - u^\varepsilon\right) z^{1\varepsilon} - b_1 y^{1*} z^{2\varepsilon} + c_1 y^{1*} z^{3\varepsilon} - y^{1*} u_0 \\ z_t^{2\varepsilon} = \alpha_2 \Delta z^{2\varepsilon} + b_2 y^{2\varepsilon} z^{1\varepsilon} + \left(-a_2 + b_2 y^{1*}\right) z^{2\varepsilon} \\ z_t^{3\varepsilon} = -b_3 y^{3\varepsilon} z^{1\varepsilon} + \left(a_3 - b_3 y^{1*}\right) z^{3\varepsilon},\ (t,x) \in Q \end{cases}$$
$$\tag{3.4}$$

$$\left(z^{1\varepsilon}\right)_v = \left(z^{2\varepsilon}\right)_v = 0 \text{ on } \Sigma, \tag{3.5}$$

$$z^{1\varepsilon}(0,x) = z^{2\varepsilon}(0,x) = z^{3\varepsilon}(0,x) = 0 \text{ on } \Omega. \tag{3.6}$$

This can be written in the form

$$z_t^\varepsilon = Az^\varepsilon + B^\varepsilon z^\varepsilon + D, \ t \in (0,T), \ z^\varepsilon(0) = 0,$$

where

$$B^\varepsilon(t) = \begin{pmatrix} \left(a_1 - b_1 y^{2\varepsilon} + c_1 y^{3\varepsilon} - u^\varepsilon\right) & -b_1 y^{1*} & c_1 y^{1*} \\ b_2 y^{2\varepsilon} & \left(-a_2 + b_2 y^{1*}\right) & 0 \\ -b_3 y^{3\varepsilon} & 0 & \left(a_3 - b_3 y^{1*}\right) \end{pmatrix},$$

$$D(t) = \begin{pmatrix} -y^{1*} u_0 \\ 0 \\ 0 \end{pmatrix}.$$

The solution of this problem is given by

$$z^\varepsilon(t) = \int_0^t S(t-s) B^\varepsilon(s) z^\varepsilon(s)\, ds + \int_0^t S(t-s) D(s)\, ds, \ t \in [0,T], \tag{3.7}$$

where $S(t), t \geq 0$, is the C_0−semigroup of contractions generated by A. Gronwall's Lemma implies that $||z^{i\varepsilon}||_{L^2(Q)}$ are bounded with respect to ε and therefore $||y^{i\varepsilon} - y^{i*}||_{L^2(Q)} = \varepsilon ||z^{i\varepsilon}||_{L^2(Q)} \to 0$ as $\varepsilon \to 0$. In other words, $y^{i\varepsilon} \to y^{i*}$ in $L^2(Q)$ as $\varepsilon \to 0$, for $i = 1,2,3$.

Passing to the limit formally as $\varepsilon \to 0$ in $(3.4) - (3.6)$, we arrive at

$$\begin{cases} z_t^1 = \alpha_1 \Delta z^1 + \left(a_1 - b_1 y^{2*} + c_1 y^{3*} - u^*\right) z^1 - b_1 y^{1*} z^2 + c_1 y^{1*} z^3 - y^{1*} u_0 \\ z_t^2 = \alpha_2 \Delta z^2 + b_2 y^{2*} z^1 + \left(-a_2 + b_2 y^{1*}\right) z^2 \\ z_t^3 = -b_3 y^{3*} z^1 + \left(a_3 - b_3 y^{1*}\right) z^3, \ (t,x) \in Q \end{cases} \tag{3.8}$$

$$\left(z^1\right)_v = \left(z^2\right)_v = 0 \text{ on } \Sigma, \tag{3.9}$$

$$z^1(0,x) = z^2(0,x) = z^3(0,x) = 0 \text{ on } \Omega \tag{3.10}$$

or equivalently,

$$z_t = Az + Bz + D, \ t \in (0,T), \ z(0) = 0,$$

where B is the matrix whose elements are the limits of the elements of B^ε. As previously, this problem admits a unique bounded on Q solution $z = \left(z^1, z^2, z^3\right) \in W^{1,2}(0,T;H)$ and

$$z(t) = \int_0^t S(t-s) B(s) z(s)\, ds + \int_0^t S(t-s) D(s)\, ds, \ t \in [0,T]. \tag{3.11}$$

By (3.7) and (3.11), we deduce with the aid of Gronwall's Lemma, that $z^{i\varepsilon} \to z^i$ in $L^2(Q)$, for $i = 1,2,3$.

Since (y^*, u^*) is a minimal pair, we have $\Phi(y^*, u^*) \leq \Phi(y^\varepsilon, u^\varepsilon)$ for all $\varepsilon > 0$. Dividing by $\varepsilon > 0$ and passing to the limit as $\varepsilon \to 0$, one arrives at

$$\int_\Omega \left(z^1 - z^3\right)(T,x)\, dx \geq 0. \tag{3.12}$$

By $(3.1) - (3.3)$ and $(3.8) - (3.10)$ we can write

$$\sum_{i=1}^{3} \left(p^i z_t^i + z^i p_t^i \right) = \sum_{i=1}^{2} \alpha_i \left(p^i \Delta z^i - z^i \Delta p^i \right) - y^{1*} u_0 p^1.$$

Integrating over Q, with the aid of Green's formula and of the boundary and initial conditions, one obtains that $\int_{\Omega} \left(z^1 - z^3 \right) (T, x)\, dx = \int_{Q} y^{1*} u_0 p^1 dt dx$. This, together with (3.12), implies that $\int_{Q} y^{1*} u_0 p^1 dt dx \geq 0$. The claim follows taking $u_0 = -u^* + \widetilde{u}$, where $\widetilde{u} \in \mathcal{U}$ is arbitrarily chosen. $\quad\square$

Remark 1. When the optimal control $u^* = 0$, there is no action of the pesticide in the habitat and the system coincides with the initial (uncontrolled) one. If $u^* = 1$, then the pesticide acts at its maximum capacity.

Acknowledgements This work was supported by CNCS- UEFISCDI (Romania) project number PN II-IDEI 342/2008.

References

1. Apreutesei, N.: Necessary optimality conditions for a Lotka-Volterra three species system. Math. Model. Nat. Phenom. **1**, 123-135 (2006)
2. Barbu, V.: Mathematical Methods in Optimization of Differential Systems. Kluwer Academic Publishers, Dordrecht (1994)
3. Feichtinger, G., Veliov, V.: On a distributed control problem arising in dynamic optimization of a fixed-size population. SIAM J. Optim. **18**, no. 3, 980–1003 (2007)
4. Haimovici, A.: A control problem for a Volterra three species system. Mathematica - Revue d'Analyse Numérique et de Théorie de l'Approx. **23** (46), no. 1, 35–41 (1980)
5. J. P. Raymond, H. Zidani, Hamiltonian Pontryagin's principles for control problems governed by semilinear parabolic equations. Appl. Math. Optim. **39**, no. 2, 143–177 (1999)
6. Wu, X. J., Li, J., Upadhyay, R. K.: Chaos control and synchronization of a three-species food chain model via Holling functional response. Int. J. Comput. Math. **87**, no. 1, 199–214 (2010)
7. Yunusov, M.: Optimal control of an ecosystem of three trophic levels (Russian). Dokl. Akad. Nauk Tadzh. SSR **30**, no.5, 277–281 (1987)

Necessary Optimality Conditions for Improper Infinite-Horizon Control Problems

Sergey M. Aseev and Vladimir M. Veliov

Abstract The paper revisits the issue of necessary optimality conditions for infinite-horizon optimal control problems. It is proved that the normal form maximum principle holds with an explicitly specified adjoint variable if an appropriate relation between the discount rate, the growth rate of the solution and the growth rate of the objective function is satisfied. The main novelty is that the result applies to general non-stationary systems and the optimal objective value needs not be finite (in which case the concept of overtaking optimality is employed).

1 Introduction

Infinite-horizon optimal control problems arise in many fields of economics, in particular in models of economic growth. Typically, the utility functional to be maximized is defined as an improper integral of the discounted instantaneous utility on the time interval $[0, \infty)$. The last circumstance gives rise to specific mathematical features of the problems and different pathologies (see [4, 7, 9, 11, 13]).

The contribution of the present paper is twofold. First we extend the version of the Pontryagin maximum principle for infinite-horizon optimal control problems with dominating discount established in [3, 4] to a more general class of non-autonomous problems and relax the assumptions. Second, we adopt the classical needle variations technique [12] to the case of infinite-horizon problems. Thus, the approach in the present paper essentially differs from the ones used in [3, 4, 6]. The needle variations technique is a standard tool in the optimal control theory. The advantage of this technique is that as a rule it produces (if applicable) the most general versions of the Pontryagin maximum principle. Nevertheless, application of needle

Sergey M. Aseev
International Institute for Applied Systems Analysis, Schlossplatz 1, A-2361 Laxenburg, Austria,
and Steklov Mathematical Institute, Gubkina 8, 119991 Moscow, Russia, e-mail: aseev@mi.ras.ru

Vladimir M. Veliov
Institute of Mathematical Methods in Economics, Vienna University of Technology, Argentinierstr.
8/E105-4, A-1040 Vienna, Austria, e-mail: veliov@tuwien.ac.at

D. Klatte et al. (eds.), *Operations Research Proceedings 2011*, Operations Research Proceedings,
DOI 10.1007/978-3-642-29210-1_4, © Springer-Verlag Berlin Heidelberg 2012

variations technique is not so straightforward in the case of infinite-horizon problems.

Another important feature of our main result is that it is applicable also for problems where the objective value may be infinite. In this case the notion of overtaking optimality is adapted (see [7]). In contrast to the known results, the maximum principle that we obtain has a normal form, that is, the multiplier of the objective function in the associated Hamiltonian can be taken equal to one.

2 Statement of the problem and assumptions

Let G be a nonempty open convex subset of R^n and U be an arbitrary nonempty set in R^m. Let $f : [0,\infty) \times G \times U \mapsto R^n$ and $g : [0,\infty) \times G \times U \mapsto R^1$.

Consider the following optimal control problem (P):

$$J(x(\cdot),u(\cdot)) = \int_0^\infty e^{-\rho t} g(t,x(t),u(t))\,dt \to \max, \tag{1}$$

$$\dot{x}(t) = f(t,x(t),u(t)), \qquad x(0) = x_0, \tag{2}$$

$$u(t) \in U. \tag{3}$$

Here $x_0 \in G$ is a given initial state of the system and $\rho \in R^1$ is a "discount" rate (which could be even negative).

Assumption (A1): *The functions* $f : [0,\infty) \times G \times U \mapsto R^n$ *and* $g : [0,\infty) \times G \times U \mapsto R^1$ *together with their partial derivatives* $f_x(\cdot,\cdot,\cdot)$ *and* $g_x(\cdot,\cdot,\cdot)$ *are continuous in* (x,u) *on* $G \times U$ *for any fixed* $t \in [0,\infty)$, *and measurable and locally bounded in* t, *uniformly in* (x,u) *in any bounded set.* [1]

In what follows we assume that the class of *admissible controls* in problem (P) consists of all measurable locally bounded functions $u : [0,\infty) \mapsto U$. Then for any initial state $x_0 \in G$ and any admissible control $u(\cdot)$ plugged in the right-hand side of the control system (2) we obtain the following Cauchy problem:

$$\dot{x}(t) = f(t,x(t),u(t)), \qquad x(0) = x_0. \tag{4}$$

Due to assumption $(A1)$ this problem has a unique solution $x(\cdot)$ (in the sense of Carathéodory) which is defined on some time interval $[0,\tau]$ with $\tau > 0$ and takes values in G (see e.g. [9]). This solution is uniquely extendible to a maximal interval of existence in G and is called *admissible trajectory* corresponding to the admissible control $u(\cdot)$.

If $u(\cdot)$ is an admissible control and the corresponding admissible trajectory $x(\cdot)$ exists on $[0,T]$ in G, then the integral

---

[1] The local boundedness of these functions of t, x and u (take $\phi(\cdot,\cdot,\cdot)$ as a representative) means that for every $T > 0$ and for every bounded set $Z \subset G \times U$ there exists M such that $\|\phi(t,x,u)\| \leq M$ for every $t \in [0,T]$ and $(x,u) \in Z$.

$$J_T(x(\cdot), u(\cdot)) := \int_0^T e^{-\rho t} g(t, x(t), u(t)) \, dt$$

is finite. This follows from (A1), the definition of admissible control and the existence of $x(\cdot)$ on $[0, T]$.

We will use the following modification of the notion of weakly overtaking optimal control (see [7]).

Definition 1: *An admissible control $u_*(\cdot)$ for which the corresponding trajectory $x_*(\cdot)$ exists on $[0, \infty)$ is locally weakly overtaking optimal (LWOO) if there exists $\delta > 0$ such that for any admissible control $u(\cdot)$ satisfying*

$$\text{meas}\, \{t \geq 0 : u(t) \neq u_*(t)\} \leq \delta$$

and for every $\varepsilon > 0$ and $T > 0$ one can find $T' \geq T$ such that the corresponding admissible trajectory $x(\cdot)$ is either non-extendible to $[0, T']$ in G or

$$J_{T'}(x_*(\cdot), u_*(\cdot)) \geq J_{T'}(x(\cdot), u(\cdot)) - \varepsilon.$$

Notice that the expression $d(u(\cdot), u_*(\cdot)) = \text{meas}\, \{t \in [0, T] : u(t) \neq u_*(t)\}$ generates a metric in the space of the measurable functions on $[0, T]$, $T > 0$, which is suitable to use in the framework of the needle variations technique (see [2]).

In the sequel we denote by $u_*(\cdot)$ an LWOO control and by $x_*(\cdot)$ – the corresponding trajectory.

Assumption (A2): *There exist numbers $\mu \geq 0$, $r \geq 0$, $\kappa \geq 0$, $\beta > 0$ and $c_1 \geq 0$ such that for every $t \geq 0$*
(i) $\|x_*(t)\| \leq c_1 e^{\mu t}$;
(ii) for every admissible control $u(\cdot)$ for which $d(u(\cdot), u_*(\cdot)) \leq \beta$ the corresponding trajectory $x(\cdot)$ exists on $[0, \infty)$ in G and it holds that

$$\|g_x(t, y, u_*(t))\| \leq \kappa (1 + \|y\|^r) \quad \text{for every } y \in \text{co}\, \{x(t), x_*(t)\}.$$

Assumption (A3): *There are numbers $\lambda \in R^1$, $\gamma > 0$ and $c_2 \geq 0$ such that for every $\zeta \in G$ with $\|\zeta - x_0\| < \gamma$ equation (4) with $u(\cdot) = u_*(\cdot)$ and initial condition $x(0) = \zeta$ (instead of $x(0) = x_0$) has a solution $x(\zeta; \cdot)$ on $[0, \infty)$ in G and*

$$\|x(\zeta; t) - x_*(t)\| \leq c_2 \|\zeta - x_0\| e^{\lambda t}.$$

The last two assumptions can be viewed as definitions of the constants μ, r and λ, which appear in the key assumption below, called *dominating discount* condition.

Assumption (A4):
$$\rho > \lambda + r \max\{\lambda, \mu\}.$$

For an arbitrary $\tau \geq 0$ consider the following linear differential equation (the linearization of (4) along $(x_*(\cdot), u_*(\cdot))$:

$$\dot{y}(t) = f_x(t, x_*(t), u_*(t))y(t), \quad t \geq 0 \tag{5}$$

with initial condition

$$y(\tau) = y_0. \tag{6}$$

In the next section we present necessary optimality conditions in the form of Pontryagin's maximum principle.

3 Main result

Due to assumption $(A1)$ the partial derivative $f_x(\cdot, x_*(\cdot), u_*(\cdot))$ is measurable and locally bounded. Hence, there is a unique (Carathéodory) solution $y_*(\cdot)$ of the Cauchy problem (8), (6) which is defined on the whole time interval $[0, \infty)$. Moreover,

$$y_*(t) = K_*(t, \tau)y_*(\tau), \quad t \geq 0, \tag{7}$$

where $K_*(\cdot, \cdot)$ is the Cauchy matrix of differential system (8) (see [10]). Recall that

$$K_*(t, \tau) = Y_*(t)Y_*^{-1}(\tau), \quad t, \tau \geq 0,$$

where $Y_*(\cdot)$ is the fundamental matrix solution of (8) normalized at $t = 0$. This means that the columns $y_i(\cdot)$, $i = 1, \ldots, n$, of the $n \times n$ matrix function $Y_*(\cdot)$ are (linearly independent) solutions of (8) on $[0, \infty)$ that satisfy the initial conditions

$$y_i^j(0) = \delta_{i,j}, \quad i, j = 1, \ldots, n,$$

where

$$\delta_{i,i} = 1, \quad i = 1, \ldots, n, \quad \text{and} \quad \delta_{i,j} = 0, \quad i \neq j, \quad i, j = 1, \ldots, n.$$

Analogously, consider the fundamental matrix solution $Z_*(\cdot)$ (normalized at $t = 0$) of the linear adjoint equation

$$\dot{z}(t) = -[f_x(t, x_*(t), u_*(t))]^* z(t). \tag{8}$$

Then $Z_*^{-1}(t) = [Y_*(t)]^*, t \geq 0$.

Define the normal-form Hamilton-Pontryagin function $\mathscr{H} : [0, \infty) \times G \times U \times R^n \mapsto R^1$ for problem (P) in the usual way:

$$\mathscr{H}(t, x, u, \psi) = e^{\rho t}g(t, x, u) + \langle f(t, x, u), \psi \rangle, \quad t \in [0, \infty), \ x \in G, \ u \in U, \ \psi \in R^n.$$

The following theorem presents the main result of the paper – a version of the Pontryagin maximum principle for non-autonomous infinite-horizon problems with dominating discount.

Theorem 1. *Assume that $(A1)$–$(A4)$ hold. Let $u_*(\cdot)$ be an admissible LWOO control and let $x_*(\cdot)$ be the corresponding trajectory. Then*

(i) *For any $t \geq 0$ the integral*

$$I_*(t) = \int_t^\infty e^{-\rho s}[Z_*(s)]^{-1} g_x(s, x_*(s), u_*(s))\, ds \tag{9}$$

converges absolutely.

(ii) *The vector function $\psi : [0, \infty) \mapsto R^n$ defined by*

$$\psi(t) = Z_*(t) I_*(t), \quad t \geq 0 \tag{10}$$

is (locally) absolutely continuous and satisfies the conditions of the normal form maximum principle, i.e. $\psi(\cdot)$ is a solution of the adjoint system

$$\dot{\psi}(t) = -\mathscr{H}_x(t, x_*(t), u_*(t), \psi(t)) \tag{11}$$

and the maximum condition holds:

$$\mathscr{H}(t, x_*(t), u_*(t), \psi(t)) \overset{a.e.}{=} \sup_{u \in U} \mathscr{H}(t, x_*(t), u, \psi(t)). \tag{12}$$

The proof of the above theorem employs approximations with finite-horizon problems in which the solution of the adjoint equation can be defined explicitly. The key point is to use the dominating discount condition (A4) to show locally uniform convergence of the finite-horizon adjoint functions to an infinite-horizon adjoint function for which the normal form maximum principle holds. The proof is too long to be presented in this note, but it is available at http://orcos.tuwien.ac.at/research/research_reports as Research Report 2011-06.

4 Comments

1. The dominating discount condition (A4) is easy to check for the so-called one-sided Lipschitz systems. Namely, assume that in addition to (A1), the following condition is satisfied for some real number λ:

$$\langle f(t,x,u) - f(t,y,u), x - y \rangle \leq \lambda \|x - y\|^2 \qquad \forall\, t \geq 0,\ x, y \in G,\ u \in U. \tag{13}$$

If we assume (for simplification) that the derivative $\|g_x(t,x,u)\|$ is bounded when $t \geq 0$, $x \in G$ and $u \in U$, then the dominating discount condition (A4) reduces to

$$\rho > \lambda.$$

Notice that λ can be negative in (13), in which case the last inequality may even be satisfied for a negative number ρ. A model with $\lambda < \rho < 0$ arising in capital growth theory, to which our result is applicable will presented in a full-size publication.

2. Another way of verifying the dominating discount condition (A4) is presented in [3, 4]. It involves calculation (or estimation from above) of the maximal element of the spectrum of the homogeneous part of the linearized dynamics, provided that the latter is regular (see e.g. [6] or [3, 4] for the above terms).

In both cases discussed in points 1 and 2 in this section it is possible to prove that, assuming boundedness of g_x as in Point 1, the normal form of the maximum principle is fulfilled with the unique bounded solution $\psi(t)$ of the adjoint equation. The precise formulations and proofs will be given in a full size paper.

5 Acknowledgements

This research was financed by the Austrian Science Foundation (FWF) under grant No I476-N13. The first author was partly supported by the Russian Foundation for Basic Research (grant No 09-01-00624-a).

References

1. A. V. Arutyunov, The Pontryagin Maximum Principle and Sufficient Optimality Conditions for Nonlinear Problems, *Differential Equations*, **39** (2003) 1671-1679.
2. S. M. Aseev and A. V. Kryazhimskii, The Pontryagin Maximum Principle and Transversality Conditions for a Class of Optimal Control Problems with Infinite Time Horizons, *SIAM J. Control Optim.*, **43** (2004) 1094-1119.
3. S. M. Aseev and A. V. Kryazhimskii, The Pontryagin Maximum Principle and Optimal Economic Growth Problems, *Proc. Steklov Inst. Math.*, **257** (2007) 1-255.
4. J.-P. Aubin and F. H. Clarke, Shadow Prices and Duality for a Class of Optimal Control Problems, *SIAM J. Control Optim.*, **17**, 567-586 (1979).
5. D. A. Carlson, A. B. Haurie, A. Leizarowitz, Infinite Horizon Optimal Control. Deterministic and Stochastic Systems, Springer, Berlin, 1991.
6. L. Cesari, Asymptotic Behavior and Stability Problems in Ordinary Differential Equations, Springer, Berlin, 1959.
7. G. Feichtinger and V.M. Veliov, On a distributed control problem arising in dynamic optimization of a fixed-size population, *SIAM Journal on Control and Optimization*, 18 (2007), 980 - 1003.
8. A. F. Filippov, Differential Equations with Discontinuous Right-Hand Sides, Nauka, Moscow, 1985 (Kluwer, Dordrecht, 1988).
9. H. Halkin, Necessary Conditions for Optimal Control Problems with Infinite Horizons, *Econometrica*, **42** (1974) 267-272.
10. P. Hartman, Ordinary Differential Equations, J. Wiley & Sons, New York, 1964.
11. P. Michel, On the Transversality Condition in Infinite Horizon Optimal Problems, *Econometrica*, **50** (1982) 975-985.
12. L. S. Pontryagin, V. G. Boltyanskij, R. V. Gamkrelidze, E. F. Mishchenko, The Mathematical Theory of Optimal Processes, Fizmatgiz, Moscow, 1961 (Pergamon, Oxford, 1964).
13. K. Shell, Applications of Pontryagin's Maximum Principle to Economics, Mathematical Systems Theory and Economics 1, Springer, Berlin, 1969, 241-292 (Lect. Notes Oper. Res. Math. Econ. **11**).

Efficient Serial and Parallel Coordinate Descent Methods for Huge-Scale Truss Topology Design

Peter Richtárik and Martin Takáč

Abstract In this work we propose solving *huge-scale* instances of the truss topology design problem with coordinate descent methods. We develop four efficient codes: *serial* and *parallel* implementations of *randomized* and *greedy* rules for the selection of the variable(s) (potential bar(s)) to be updated in the next iteration. Both serial methods enjoy an $O(n/k)$ iteration complexity guarantee, where n is the number of potential bars and k the iteration counter. Our parallel implementations, written in CUDA and running on a graphical processing unit (GPU), are capable of speedups of up to two orders of magnitude when compared to their serial counterparts. Numerical experiments were performed on instances with up to 30 million potential bars.

1 Introduction

In the process of designing mechanical structures such as railroad bridges, airplanes or buildings, one faces the problem of designing a truss—a system of elastic bars of varying volumes with endpoints at nodes, which are usually given by a 2D/3D grid, and are either fixed (and cannot move) or free—capable of withstanding a specified vector of forces applied to the nodes in an "optimal way". After the forces are applied, the structure deforms (bars are stretched, free nodes move) to an equilibrium position, storing potential energy (compliance). Trusses of smaller compliance are more rigid. The goal of Truss Topology Design (TTD) is for a given grid structure of nodes and a vector of forces acting on them to construct a truss of at most a given total volume having *minimum compliance*.

In recent years, optimization methods have been applied to solving various formulations of TTD. For instance, a single load TTD can be cast as a linear program [5],

Peter Richtárik
School of Mathematics, University of Edinburgh, UK, e-mail: Peter.Richtarik@ed.ac.uk

Martin Takáč
School of Mathematics, University of Edinburgh, UK, e-mail: Takac.MT@gmail.com

D. Klatte et al. (eds.), *Operations Research Proceedings 2011*, Operations Research Proceedings, 27
DOI 10.1007/978-3-642-29210-1_5, © Springer-Verlag Berlin Heidelberg 2012

robust TTD as a semidefinite program [1], TTD with integer variables as an integer linear semidefinite program [3]. We refer to [9] for a comprehensive survey.

Our work is motivated by the need to solve *huge-scale* TTD problems using methods with *provable iteration complexity guarantees*. These are problems on which second order methods fail due to memory limitations and for which even the evaluation of the gradient is time-consuming. We argue that coordinate descent methods (CDM) are well suited for this task; these are some of the reasons:

1. CDMs in general have *low per-iteration memory* and *work requirements*, which is an essential prerequisite for any method aspiring to solve huge-scale problems.
2. This effect is *extreme* for TTD problems under formulation (1), which we propose in this paper, due to *inherent super-sparsity* of matrix A caused by each bar being related to two 2D forces. Each column of A will have at most 4 nonzeros and one iteration of a (non-greedy) CDM will hence only need $O(1)$ *memory* and will perform $O(1)$ *work*, independently of the dimension of the problem.
3. CMDs can be easily *parallelized*, achieving *huge speedups* when compared to serial implementations.

For alternative heuristic approaches to huge-scale TTD problem, see [2, 10].

2 Problem formulation

Consider an $r \times c$ grid of nodes, m being free (although due to space limitations we only work with 2D trusses in this paper, the results are applicable to the 3D case as well). We allow, potentially, all pairs of nodes to be joined by a bar, except we require that no two potential bars intersect at more than one point. By n we denote the number of potential bars, $w_i \geq 0$ is the weight of bar i, the total weight of all bars cannot exceed 1. The collection of 2D forces (load) acting at the free nodes is represented by $d \in \mathbb{R}^{2m}$; the collection of node displacement associated with bar i after the load is applied is denoted by $a_i \in \mathbb{R}^{2m}$.

The compliance of a truss with weights w under the load d is equal to $\frac{1}{2}d^T v$, where v is any solution of the equilibrium system $\sum_i w_i a_i a_i^T v = d$. The problem of *minimizing compliance* subject to $\sum_i w_i = 1$ can be equivalently written as the linear program $\max_v \{d^T v \,:\, |a_i^T v| \leq 1, \, i = 1, \ldots, n\}$, the dual of which is equivalent to $\min_x \{\|x\|_1 \,:\, Ax = d\}$, where $A = [a_1, \ldots, a_n]$. For more detail about the construction above we refer to Section 1.3.5 in [5]. In [7], a gradient method is described for solving all the above problems simultaneously[1], in relative scale. In this paper we work with a penalized (and scaled) version of the last problem:

$$\min_{x \in \mathbb{R}^n} \{\tfrac{1}{2}\|Ax - d\|_2^2 + \lambda \|x\|_1\}, \qquad \lambda > 0. \tag{1}$$

Although this problem is unconstrained, our methods can deal with simple lower and upper bounds on the variables x.

---

[1] A single iterative procedure is run on the input $\{A, d\}$, having a natural interpretation for each of the problems, returning solutions of *all* of them.

3 Iteration complexity of the serial methods

Consider now the problem

$$F^* \stackrel{\text{def}}{=} \min_{x \in \mathbb{R}^n} \{ F(x) \equiv f(x) + g_1(x^{(1)}) + \cdots + g_n(x^{(n)}) \}, \qquad (2)$$

where f and $\{g_i\}$ satisfy these assumptions: (A1) $f : \mathbb{R}^n \to \mathbb{R}$ is convex and has coordinate-wise Lipschitz gradient, uniformly in x, with constants $L_1, \ldots, L_n > 0$, i.e., $|\nabla_i f(x) - \nabla_i f(x + t e_i)| \leq L_i |t|$ for all i and $x, x + t e_i \in$ dom F, (A2) functions $g_i : \mathbb{R} \to \mathbb{R} \cup \{+\infty\}$ are convex and closed. Problem (1) is a special case of (2) with

$$f(x) = \tfrac{1}{2}\|Ax - d\|_2^2, \qquad g_i(x^{(i)}) = \lambda |x^{(i)}|, \qquad L_i = \|a_i\|_2^2, \qquad (3)$$

in which case Step 4 of Algorithm 1 can be computed in closed form and is known as *soft-thresholding*.

Algorithm 1 Serial Coordinate Descent

1: Choose initial point $x_0 \in \mathbb{R}^n$
2: **for** $k = 0, 1, \ldots$ **repeat**
3:    Choose $i \in \{1, 2, \ldots, n\}$ *randomly* or *greedily*
4:    $t_i^* \leftarrow \arg\min_{t \in \mathbb{R}} \nabla_i f(x_k) t + \tfrac{L_i}{2} t^2 + g_i(x_k^{(i)} + t)$
5:    $x_{k+1} \leftarrow x_k + t_i^* e_i$

The following result gives iteration complexity guarantees for randomized and greedy version of Algorithm 1. The greedy selection rule in part (ii) coincides with the one in [4]; the authors do not, however, provide any complexity bounds.

Theorem 1. *Let f and $\{g_i\}$ satisfy assumptions (A1), (A2) and choose $x_0 \in$ dom F and $0 < \varepsilon < F(x_0) - F^*$. Further let $C = \max\{R_L^2(x_0), F(x_0) - F^*\}$, where $R_L(x_0) = \max_x \max_{x^* \in X^*} \{\|x - x^*\|_L : F(x) \leq F(x_0)\}$, $\|x\|_L = (\sum_{i=1}^n L_i (x^{(i)})^2)^{\frac{1}{2}}$ and X^* is the set of minimizers of (2).*

(i) **Random:** *Choose $0 < \rho < 1$ and consider Algorithm 1, where in Step 3 each coordinate is chosen with probability $\frac{1}{n}$. Then after $K \geq \frac{2nC}{\varepsilon}(1 + \log\frac{1}{\rho}) + 2 - \frac{2nC}{F(x_0) - F^*}$ iterations we have $\text{Prob}[F(x_K) - F(x^*) \leq \varepsilon] \geq 1 - \rho$.*

(ii) **Greedy:** *Let f and $\{g_i\}$ be as in (3) and consider Algorithm 1, where in Step 3 coordinate i is chosen greedily as $i = \arg\max_{j \in \{1,\ldots,n\}} \alpha_k(j)$, where*

$$\alpha_k(j) = \begin{cases} \tfrac{L_j}{2}(t_j^*)^2 + \lambda x_k^{(j)}(sign(x_k^{(j)}) - sign(x_k^{(j)} + t_j^*)), & \text{if } x_k^{(j)} + t_j^* \neq 0, \\ \lambda |x_k^{(j)}| - \nabla_j f(x_k) t_j^* - \tfrac{L_j}{2} t_j^{*2}, & \text{otherwise.} \end{cases}$$

Then after $K \geq \frac{2nC}{\varepsilon} - \frac{2nC}{F(x_0) - F^*}$ iterations we get $F(x_K) - F(x^*) \leq \varepsilon$.

Proof. (Rough Sketch) Statement (i) is identical to part (i) of Theorem 4 in [8]. Part (ii) can be proved in an analogous way using the fact that $\alpha_k(j) = F(x_k) - F(x_k + t_j^* e_j)$ and $F(x_k + t_i^* e_i) = \min_{t,i} F(x_k + t e_i) \leq \tfrac{1}{n} \sum_j F(x_k + t_j^* e_j)$. $\square$

4 Numerical experiments

In this final section we compare numerically our serial methods (SR = Serial Random, SG = Serial Greedy), described in Section 3, with their GPU-accelerated variants (PR = Parallel Random, PG = Parallel Greedy). All algorithms were run on TTD instances of the form (1) with $\lambda = 10^{-4}$. Due to space limitations we need to limit our exposition to a sketch of two experiments only.

Implementation details. Our serial codes (SR, SG) were written in C++, the parallel ones (PR, PG) using a CUDA C/C++ compiler from NVIDIA. All experiments were performed on a system with Intel Xeon CPU X5650@2.67GHz (we have only used one out of the 6 cores) and 48GB DDR2 PC3-1066 6.4GT/s memory. We have used NVIDIA Tesla C2050 GPU device with 448 cores, peak performance of 1.03Tflops for single precision and 3GB GDDR5 RAM. PG was implemented using CUSPARSE and CUBLAS libraries; all linear algebra and the greedy selection rule were implemented in parallel. In case of PR we choose independently for each thread a random coordinate and perform a single iteration of Algorithm 1. Race conditions were avoided using atomic operations.

Experiment 1. In Table 1 the three it/sec columns illustrate how the number of iterations per second decreases with increasing problem size for methods SG, SR and NEST (Nesterov's accelerated full-gradient method [6]).

$r \times c$	$2m$	n	$\|A\|_0 = 4n$	time (it) SeDuMi	it/sec SG	accel PG	it/sec SR	accel PR	it/sec NEST
20×20	800	48,934	195,736	61 (40)	5,101	0.6×	2.4M	28.3×	53.05
30×30	1,800	246,690	986,760	658 (10) NP	1,195	2.7×	2.2M	26.1×	9.01
40×40	3,200	779,074	3,116,296	8.6k (32) NP	380	8.0×	1.9M	24.9×	3.43
50×50	5,000	1,901,930	7,607,720	48.4k (4) NP	159	16.0×	1.5M	27.2×	1.41
80×80	12,800	12,454,678	49,818,712	X	26	42.3×	1.5M	34.0×	0.15
100×100	20,000	30,398,894	121,595,576	X	11	52.2×	1.2M	30.9×	0.05

Table 1 Comparison of performance of SeDuMi, SG, PG, SR, PR and NEST on 6 TTD instances ($A \in \mathbb{R}^{2m \times n}$, NP=failure due to numerical problems, X=no iteration after 10 hours).

For instance, while n increases by a factor of 621 when going from the 20×20 to the 100×100 problem, the per-iteration speed of SR is merely halved (cf. with point 2 in the introduction). The accel columns indicate the acceleration achieved by parallelization. Note that the speedup increases with problem size for the greedy method (to 52× for the largest problem) and stays virtually constant ($\approx 30\times$) for the randomized method. It is not possible to expect the speedup to be equal to the number of cores because i) GPU cores ale slower than a CPU and ii) communication overhead slows GPU implementations down. Iterations of any second order method (we have used SeDuMi (v1.21)) become prohibitively expensive as n increases. Indeed, for the 80×80 problem SeDuMi is not able to perform a single iteration in 10 hours while PR does 34×1.5 million iterations per second. Note that a similar, albeit much less pronounced, effect holds for NEST.

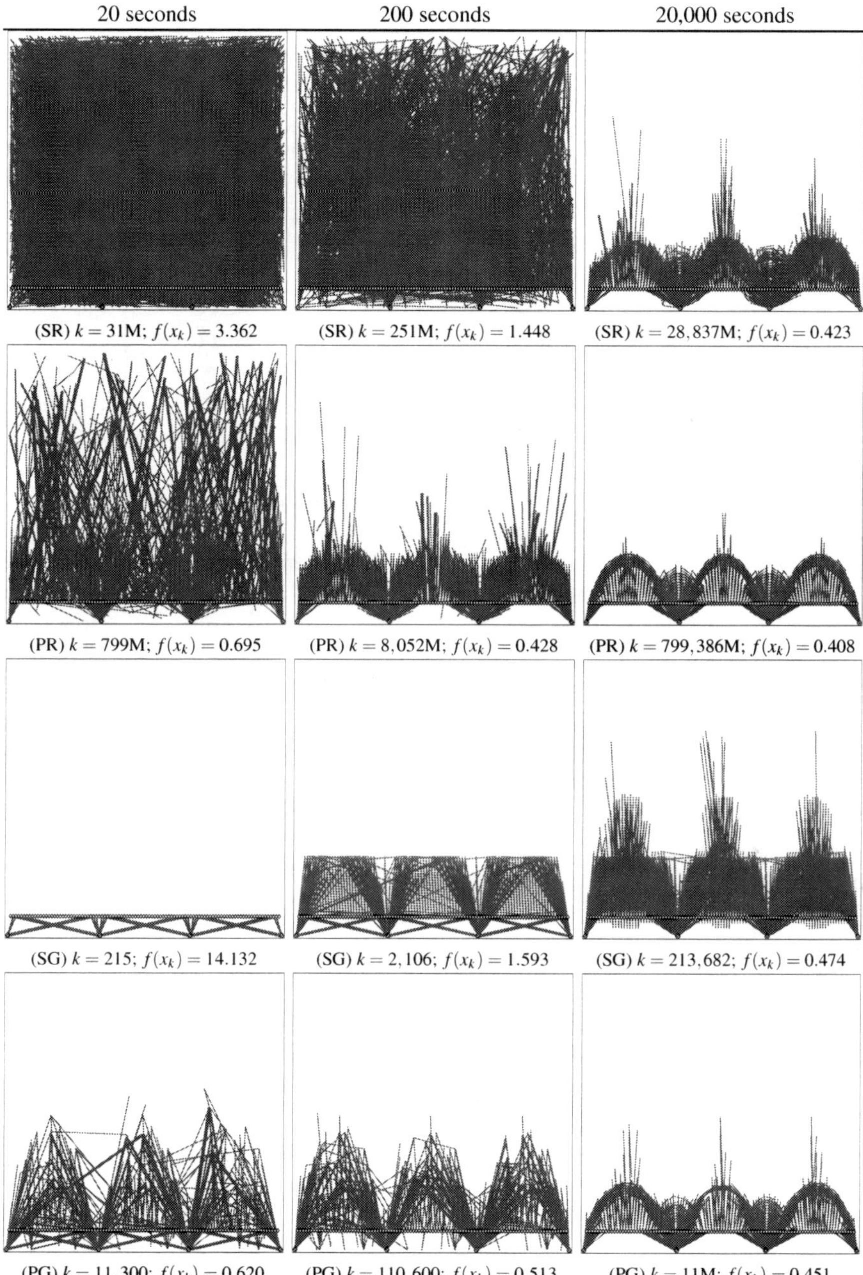

Fig. 1 "Bridge" truss after 20s, 200s ($\approx$3.34 minutes) and 20,000s ($\approx$5.56 hours) of computation time for algorithms SR, PR, SG and PG (rows in this order). Number of iterations ("M" = millions) and objective value is shown under each plot.

Experiment 2. In Figure 1 we run methods SR, PR, SG and PG on a 100×100 "bridge" instance (with more than 30 million variables/potential bars) for 20, 200 and 20,000 seconds. There are 4 fixed nodes evenly spaced at the bottom (large red diamonds), and a unit downward force is applied at every node at height 7 (out of 100) above the "ground" (at the small green diamonds). We have used $x_0 = 0$ for which $f(x_0) = 49$. Note that the parallel methods have managed to push the objective function down from 49 to 0.695 (PR) and 0.620 (PG) after 20 seconds already; despite the size of the problem. These trusses do not resemble a bridge yet, but after 5.56 hours the parallel methods do produce visibly bridge-like structures.

Vanilla methods vs heuristics. Note that while we have tried our best to implement our methods as efficiently as possible, further significant speedups are likely possible by introducing *heuristics*. For instance, note that in an optimal truss most of the potential bars will have zero weight. However, our methods, in the "vanilla" form appearing in this paper (this form enables us to prove rigorous iteration complexity bounds), will unnecessarily consider these zero weight bars in each iteration. Therefore, for instance in the case of SR and PR, decreasing the probability of selecting zero-weight bars will introduce a speedup (see *q-shrinking* strategy in [8]). We have not implemented this or any other heuristics in this work as our goal was to demonstrate that even vanilla methods are able to solve huge-scale problems. However, in solving any practical large TTD problem, we recommend introducing acceleration heuristics.

Acknowledgements The work of the first author was supported in part by EPSRC grant "Mathematics for vast digital resources" (EP/I017127/1); both authors were also supported in part by the Centre for Numerical Algorithms and Intelligent Software (funded by EPSRC grant EP/G036136/1 and the Scottish Funding Council).

References

1. Ben-Tal, A., Nemirovski, A.: Robust truss topology design via semidefinite programming, SIAM Journal on Optimization 7, 991–1016 (1997)
2. Gilbert M., Tyas A.: Layout optimization of large-scale pin-jointed frames, Engineering Computations **20** (8), 1044–1064 (2003)
3. Kočvara, M.: Truss topology design with integer variables made easy, Opt. Online (2010)
4. Li, Y., Osher, S.: Coordinate descent optimization for l_1 minimization with application to compressed sensing; a greedy algorithm, Inverse Problems and Imaging **3**, 487–503 (2009)
5. Nemirovski, A., Ben-Tal, A.: Lectures on Modern Convex Optimization: Analysis, Algorithms, and Engineering Applications. SIAM, Philadelphia, PA, USA (2001)
6. Nesterov, Yu.: Gradient methods for minimizing composite objective function. CORE Discussion Paper #2007/76 (2007)
7. Richtárik, P.: Simultaneously solving seven optimization problems in relative scale. Optimization Online (2009)
8. Richtárik, P., Takáč, M.: Iteration complexity of randomized block-coordinate descent methods for minimizing a composite function. Technical Report ERGO 11-011 (2011)
9. Rozvany, G., Bendsøe, M.P., Kirsch, U.: Layout optimization of structures. Applied Mechanics Reviews, **48** (2), 41–119 (1995)
10. Sokół, T.: Topology optimization of large-scale trusses using ground structure approach with selective subsets of active bars. Extended Abstract, Computer Methods in Mechanics (2011)

About error bounds in metric spaces

Marian J. Fabian, René Henrion, Alexander Y. Kruger and Jiří V. Outrata

Abstract The paper presents a general primal space classification scheme of necessary and sufficient criteria for the error bound property incorporating the existing conditions. Several primal space derivative-like objects – slopes – are used to characterize the error bound property of extended-real-valued functions on metric sapces.

1 Introduction

In this paper f is an extended-real-valued function on a metric space X, $f(\bar{x}) = 0$, $S_f := \{x \in X \mid f(x) \le 0\}$, and $f_+(x) := \max(f(x), 0)$. We are looking for characterizations of the *error bound property*.

Definition 1. f has a local error bound at $\bar{x}$ if there exists a $c > 0$ such that

$$d(x, S_f) \le c f_+(x) \quad \text{for all } x \text{ near } \bar{x}. \tag{1}$$

For the summary of the theory of error bounds and its various applications, the reader is referred to the survey papers [2, 5, 9, 10], as well as the book [1]. Recent extensions to vector-valued functions can be found in [3].

M. Fabian
Mathematical Institute, Academy of Sciences of the Czech Republic, Žitná 25, 11567 Prague 1, Czech Republic; e-mail: fabian@math.cas.cz

R. Henrion
Weierstrass Institute for Applied Analysis and Stochastics, 10117 Berlin, Germany; e-mail: henrion@wias-berlin.de

A. Kruger
School of Information Technology and Mathematical Sciences, Centre for Informatics and Applied Optimization, University of Ballarat, POB 663, Ballarat, Vic, 3350, Australia; e-mail: a.kruger@ballarat.edu.au

J. Outrata
Institute of Information Theory and Automation, Academy of Sciences of the Czech Republic, 18208 Prague, Czech Republic; e-mail: outrata@utia.cas.cz

D. Klatte et al. (eds.), *Operations Research Proceedings 2011*, Operations Research Proceedings, 33
DOI 10.1007/978-3-642-29210-1_6, © Springer-Verlag Berlin Heidelberg 2012

Property (1) can be equivalently defined in terms of the *error bound modulus* [5]:

$$\mathrm{Er}\,(\bar{x}) := \liminf_{\substack{x \to \bar{x} \\ f(x)>0}} \frac{f(x)}{d(x,S_f)}, \tag{2}$$

namely, f has a local error bound at $\bar{x}$ if and only if $\mathrm{Er}\,f(\bar{x}) > 0$. Constant (2) provides a quantitative characterization of this property.

2 Slopes

Primal space characterizations of error bounds can be formulated in terms of slopes. Recall that the (strong) *slope* [4] of f at x $(|f(x)| < \infty)$ is defined as

$$|\nabla f|(x) := \limsup_{u \to x} \frac{(f(x) - f(u))_+}{d(u,x)}. \tag{3}$$

The following modifications of (3) can be convenient for characterizing the error bound property:

$$|\nabla f|^0(\bar{x}) = \liminf_{x \to \bar{x}} \frac{f(x) - f(\bar{x})}{\|x - \bar{x}\|}, \tag{4}$$

$$\overline{|\nabla f|}(\bar{x}) = \liminf_{x \to \bar{x},\, f(x) \to f(\bar{x})} |\nabla f|(x), \tag{5}$$

$$\overline{|\nabla f|}^>(\bar{x}) = \liminf_{x \to \bar{x},\, f(x) \downarrow f(\bar{x})} |\nabla f|(x), \tag{6}$$

$$\overline{|\nabla f|}^\circ(\bar{x}) := \liminf_{x \to \bar{x},\, f(x) \downarrow f(\bar{x})} \sup_{u \neq x} \frac{(f(x) - f_+(u))_+}{d(u,x)}. \tag{7}$$

Constants (4)–(7) are called the *internal slope*, *strict slope*, *strict outer slope*, and *uniform strict slope* of f at $\bar{x}$ respectively.

The relationships between the constants are straightforward.

Proposition 1. *(i)* $\overline{|\nabla f|}(\bar{x}) \leq \overline{|\nabla f|}^>(\bar{x}) \leq \overline{|\nabla f|}^\circ(\bar{x}).$
(ii) $\overline{|\nabla f|}(\bar{x}) = (-|\nabla f|^0(\bar{x}))_+.$
(iii) $|\nabla f|^0(\bar{x}) \leq \overline{|\nabla f|}^\circ(\bar{x}).$
(iv) If $|\nabla f|^0(\bar{x}) > 0$ then $|\nabla f|^0(\bar{x}) = \mathrm{Er}\,f(\bar{x}).$

The inequalities in Proposition 1 can be strict.

Example 1. Let $f : \mathbb{R} \to \mathbb{R}$ be defined as follows:

$$f(x) = \begin{cases} 0 & \text{if } x < 0, \\ x & \text{if } x \geq 0. \end{cases}$$

Obviously $\overline{|\nabla f|}(0) = |\nabla f|(0) = |\nabla f|^0(0) = 0$. At the same time, $\overline{|\nabla f|}^>(0) = \overline{|\nabla f|}^\circ(0) = 1$.

The next example is a modification of the corresponding one in [6].

Example 2. Let $f : \mathbb{R} \to \mathbb{R}$ be defined as follows:

$$f(x) = \begin{cases} -x & \text{if } x \leq 0, \\ \dfrac{1}{i} & \text{if } \dfrac{1}{i+1} < x \leq \dfrac{1}{i}, \ i = 1,2,\ldots, \\ x & \text{if } x > 1. \end{cases}$$

Obviously $\overline{|\nabla f|}^{>}(0) = |\nabla f|(0) = 0$. At the same time, $\overline{|\nabla f|}^{\diamond}(0) = |\nabla f|^{0}(0) = 1$.

The function in the above example is discontinuous. However, the second inequality in Proposition 1 (i) can be strict for continuous and even Lipschitz continuous functions. The function in the next example is piecewise linear and Clarke regular at 0 (that is, directionally differentiable, and its Clarke generalized directional derivative coincides with the usual one).

Example 3. Let $f : \mathbb{R} \to \mathbb{R}$ be defined as follows:

$$f(x) = \begin{cases} -x & \text{if } x \leq 0, \\ x\left(1 + \dfrac{1}{i}\right) - \dfrac{1}{i(i+1)} & \text{if } \dfrac{1}{i+1} < x \leq \dfrac{1}{i+1} + \dfrac{1}{(i+1)^2}, \ i = 1,2,\ldots, \\ \dfrac{1}{i} & \text{if } \dfrac{1}{i+1} + \dfrac{1}{(i+1)^2} < x \leq \dfrac{1}{i}, \ i = 1,2,\ldots, \\ x & \text{if } x > 1. \end{cases}$$

f is everywhere Fréchet differentiable except for a countable number of points. One can find a point $x > 0$ arbitrarily close to 0 with $|\nabla f|(x) = 0$ (on a horizontal part of the graph). The slopes of non-horizontal parts of the graph decrease monotonously to 1 as $x \downarrow 0$. It is not difficult to check that $\overline{|\nabla f|}^{>}(0) = |\nabla f|(0) = 0$ while $\overline{|\nabla f|}^{\diamond}(0) = |\nabla f|^{0}(0) = 1$.

If f is convex then the second inequality in Proposition 1 (i) holds as equality.

For the function f in Example 1, it holds $|\nabla f|(0) < \overline{|\nabla f|}^{>}(0)$. In the nonconvex case one can also have the opposite inequality.

Example 4. Let $f : \mathbb{R} \to \mathbb{R}$ be defined as follows:

$$f(x) = \begin{cases} x & \text{if } x < 0, \\ x^2 & \text{if } x \geq 0. \end{cases}$$

Obviously $|\nabla f|(0) = 1$ while $\overline{|\nabla f|}^{>}(0) = 0$. Note that despite slope $|\nabla f|(0)$ being positive, the function in this example does not have a local error bound at 0. Hence, condition $|\nabla f|(\bar{x}) > 0$ is not in general sufficient for the error bound property to hold at $\bar{x}$.

3 Error bound criteria

The next theorem generalizes and strengthens [5, Theorem 2].

Theorem 1. *(i)* $\operatorname{Er} f(\bar{x}) \leq \overline{|\nabla f|}^{\circ}(\bar{x})$.
(ii) If X is complete and f_{+} is lower semicontinuous near $\bar{x}$, then $\operatorname{Er} f(\bar{x}) = \overline{|\nabla f|}^{\circ}(\bar{x})$.

Proof. (i) If $\underline{\operatorname{Er}} f(\bar{x}) = 0$ or $\overline{|\nabla f|}^{\circ}(\bar{x}) = \infty$, the conclusion is trivial. Let $0 < \gamma <$ $\operatorname{Er} f(\bar{x})$ and $\overline{|\nabla f|}^{\circ}(\bar{x}) < \infty$. We are going to show that $\overline{|\nabla f|}^{\circ}(\bar{x}) \geq \gamma$. By (2), there is a $\delta > 0$ such that

$$\frac{f(x)}{d(x, S_f)} > \gamma. \tag{8}$$

for any $x \in B_{\delta}(\bar{x})$ with $f(x) > f(\bar{x})$. Take any $x \in B_{\delta}(\bar{x})$ with $f(\bar{x}) < f(x) \leq f(\bar{x}) + \delta$ (Such points x exist since $\overline{|\nabla f|}^{\circ}(\bar{x}) < \infty$.) By (8), one can find a $w \in S_f$ such that

$$\frac{f(x)}{d(x, w)} > \gamma.$$

It follows that $\overline{|\nabla f|}^{\circ}(\bar{x}) \geq \gamma$.

(ii) Let X be complete and f_{+} be lower semicontinuous near $\bar{x}$. Thanks to (i), we only need to prove that $\operatorname{Er} f(\bar{x}) \geq \overline{|\nabla f|}^{\circ}(\bar{x})$. If $\operatorname{Er} f(\bar{x}) = \infty$, the inequality is trivial. Let $\operatorname{Er} f(\bar{x}) < \gamma < \infty$. Chose a $\delta > 0$ such that f_{+} is lower semicontinuous on $B_{(\gamma^{-1}+1)\delta}(\bar{x})$. Then by (2), there is an $x \in B_{\delta \min(1/2, \gamma^{-1})}(\bar{x})$ such that

$$0 < f(x) < \gamma d(x, S_f).$$

Put $\varepsilon = f(x)$. Then $f_{+}(x) \leq \inf f_{+} + \varepsilon$. Applying to f_{+} the Ekeland variational principle with an arbitrary $\lambda \in (\gamma^{-1}\varepsilon, d(x, S_f))$, one can find a w such that $f(w) \leq f(x)$, $d(w, x) \leq \lambda$ and

$$f_{+}(u) + (\varepsilon/\lambda)d(u, w) \geq f_{+}(w), \qquad \forall u \in B_{(\gamma^{-1}+1)\delta}(\bar{x}). \tag{9}$$

Obviously,

$$d(w, x) < d(x, S_f) \leq d(x, \bar{x}), \tag{10}$$
$$d(w, \bar{x}) \leq d(w, x) + d(x, \bar{x}) < 2d(x, \bar{x}) \leq \delta,$$
$$f(w) \leq f(x) < \gamma d(x, \bar{x}) \leq \delta.$$

Besides, $f(w) > 0$ due to the first inequality in (10). It follows from (9) that

$$f(w) \leq f_{+}(u) + (\varepsilon/\lambda)d(u, w) \leq \gamma d(u, w)$$

for all $u \in B_{(\gamma^{-1}+1)\delta}(\bar{x})$. If $u \notin B_{(\gamma^{-1}+1)\delta}(\bar{x})$, then $d(u, w) > (\gamma^{-1}+1)\delta - d(w, \bar{x}) > \gamma^{-1}\delta$, and consequently

$$f(w) < \delta < \gamma d(u, w)$$

Thus, in both cases

$$\sup_{u \neq w} \frac{f(w) - f_+(u)}{d(u,w)} \leq \gamma.$$

This implies the inequality $\overline{|\nabla f|}^{\diamond}(\bar{x}) \leq \mathrm{Er}\, f(\bar{x})$. $\quad\square$

Without lower semicontinuity, the inequality in Theorem 1 (i) can be strict.

Example 5. Let $f : \mathbb{R} \to \mathbb{R}$ be defined as follows:

$$f(x) = \begin{cases} -3x & \text{if } x \leq 0, \\ 3x - \dfrac{1}{2^i} & \text{if } \dfrac{1}{2^{i+1}} < x \leq \dfrac{1}{2^i},\ i = 0, 1, \ldots, \\ 2x & \text{if } x > 1. \end{cases}$$

Obviously, $\mathrm{Er}\, f(0) = 1$ while $\overline{|\nabla f|}^{\diamond}(0) = 3$.

Example 6. Let $f : \mathbb{R}^2 \to \mathbb{R}$ be defined as follows:

$$f(x_1, x_2) = \begin{cases} x_1 + x_2 & \text{if } x_1 > 0, x_2 > 0, \\ -x_1 & \text{if } x_1 > 0, x_2 \leq 0, \\ -x_2 & \text{if } x_2 > 0, x_1 \leq 0, \\ 0 & \text{otherwise}, \end{cases}$$

and let $\mathbb{R}^2$ be equipped with the Euclidean norm. The function is discontinuous on the set $\{(t,0) \in \mathbb{R}^2 : t > 0\} \cup \{(0,t) \in \mathbb{R}^2 : t > 0\}$. Then $\mathrm{Er}\, f(0) = 2$ and $\overline{|\nabla f|}^{\diamond}(0) = 3$.

In view of Theorem 1, inequality $\overline{|\nabla f|}^{\diamond}(\bar{x}) > 0$ provides a necessary and sufficient error bound criterion for lower semicontinuous functions on complete metric spaces. In a slightly different form, a similar condition for the calmness [q] property of level set maps first appeared in [8, Proposition 3.4]; see also [7, Corollary 4.3].

Taking into account Proposition 1, inequalities

$$|\nabla f|^0(\bar{x}) > 0, \quad \overline{|\nabla f|}(\bar{x}) > 0 \quad \text{and} \quad \overline{|\nabla f|}^{>}(\bar{x}) > 0$$

provide sufficient error bound criteria

The relationships among the primal space error bound criteria are illustrated in Fig. 1 (X is complete and f_+ is lower semicontinuous near $\bar{x}$).

In Banach spaces, it is possible to formulate corresponding dual space error bound criteria in terms of subdifferential slopes [5].

Acknowledgements The research was partially supported by the Institutional Research Plan of the Academy of Sciences of Czech Republic, AVOZ 101 905 03; the Grant Agency of the Czech Republic, Projects 201/09/1957 and 201/11/0345; the DFG Research Center MATHEON "Mathematics for key technologies" in Berlin; and the Australian Research Council, Project DP110102011.

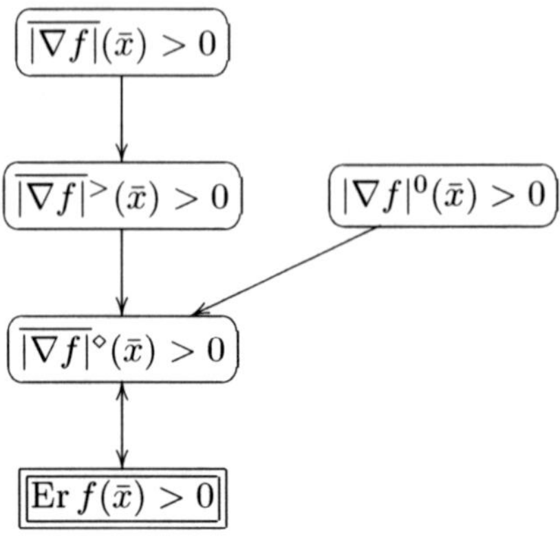

Fig. 1 Primal space criteria

References

1. Auslender, A., Teboulle, M.: Asymptotic Cones and Functions in Optimization and Variational Inequalities. Springer Monographs in Mathematics. Springer-Verlag, New York (2003)
2. Azé, D.: A survey on error bounds for lower semicontinuous functions. In: Proceedings of 2003 MODE-SMAI Conference, *ESAIM Proc.*, vol. 13, pp. 1–17. EDP Sci., Les Ulis (2003)
3. Bednarczuk, E.M., Kruger, A.Y.: Error bounds for vector-valued functions: necessary and sufficient conditions. Nonlinear Anal. (2011). DOI 10.1016/j.na.2011.05.098
4. De Giorgi, E., Marino, A., Tosques, M.: Problems of evolution in metric spaces and maximal decreasing curve. Atti Accad. Naz. Lincei Rend. Cl. Sci. Fis. Mat. Natur. (8) **68**(3), 180–187 (1980)
5. Fabian, M.J., Henrion, R., Kruger, A.Y., Outrata, J.V.: Error bounds: necessary and sufficient conditions. Set-Valued Var. Anal. **18**(2), 121–149 (2010).
6. Ioffe, A.D., Outrata, J.V.: On metric and calmness qualification conditions in subdifferential calculus. Set-Valued Anal. **16**(2–3), 199–227 (2008)
7. Klatte, D., Kruger, A.Y., Kummer, B.: From convergence principles to stability and optimality conditions. J. Convex Anal. To be published
8. Kummer, B.: Inclusions in general spaces: Hoelder stability, solution schemes and Ekeland's principle. J. Math. Anal. Appl. **358**(2), 327–344 (2009).
9. Lewis, A.S., Pang, J.S.: Error bounds for convex inequality systems. In: Generalized Convexity, Generalized Monotonicity: Recent Results (Luminy, 1996), *Nonconvex Optim. Appl.*, vol. 27, pp. 75–110. Kluwer Acad. Publ., Dordrecht (1998)
10. Pang, J.S.: Error bounds in mathematical programming. Math. Programming, Ser. B **79**(1-3), 299–332 (1997). Lectures on Mathematical Programming (ISMP97) (Lausanne, 1997)

Electricity Price Modelling for Turkey

Miray Hanım Yıldırım, Ayşe Özmen, Özlem Türker Bayrak, Gerhard Wilhelm Weber

Abstract This paper presents customized models to predict next-day's electricity price in short-term periods for Turkey's electricity market. Turkey's electricity market is evolving from a centralized approach to a competitive market. Fluctuations in the electricity consumption show that there are three periods; day, peak, and night. The approach proposed here is based on robust and continuous optimization techniques, which ensures achieving the optimum electricity price to minimize error in periodic price prediction. Commonly, next-day's electricity prices are forecasted by using time series models, specifically dynamic regression model. Therefore electricity price prediction performance was compared with dynamic regression. Numerical results show that CMARS and RCMARS predicts the prices with 30% less error compared to dynamic regression.

1 Introduction

Turkish electricity market had been controlled by the government. However, due to rapid increase in population, industrialization, and urbanization, legal reforms were made in the last ten years to establish a competitive market. As a result, the market is evolving from monopolistic publicly owned production and distribution into a private market. This evolution started with dissociation of electricity generation, transmission, and distribution. Now, it continues with privatization of public seg-

Miray Hanım Yıldırım
Inst. of Appl. Math., Middle East Technical Univ., Ankara, Turkey,
Dept. of Industrial Engineering, Çankaya Univ., Ankara, Turkey, e-mail: maslan@cankaya.edu.tr

Ayşe Özmen
Inst. of Appl. Math., Middle East Technical Univ., Ankara, Turkey,
e-mail: ayseozmen19@gmail.com

Özlem Türker Bayrak
Dept. of Industrial Engineering, Çankaya Univ., Ankara, Turkey, e-mail: ozlem@cankaya.edu.tr

Gerhard Wilhelm Weber
Inst. of Appl. Math., Middle East Technical Univ., Ankara, Turkey, e-mail: gweber@metu.edu.tr

D. Klatte et al. (eds.), *Operations Research Proceedings 2011*, Operations Research Proceedings,   39
DOI 10.1007/978-3-642-29210-1_7, © Springer-Verlag Berlin Heidelberg 2012

ments. In 2001, a new market law is introduced to create a liberalized market. From then to 2010, although the majority is traded through two-sided contracts, 15% of the market operates under free conditions. This indicates demand and supply are balanced in 15% of the market. It was planned to complete this evolution into a spot market by the end of 2010. However, due to legal consequences and infrastructure works, this transition will be completed by the end of 2011.

In electricity market, since there are fluctuations in the demand, the prices can change in short term periods, even in a day. In Turkey, these periods are planned as day (06:00-17:00), peak (17:00-22:00) and night (22:00-06:00). Here, we propose a customized electricity price modelling of Turkey to make short term projections for the competitive market, where only day ahead prices are forecasted.

Several models are proposed in the literature for competitive electricity markets. Here, there are three main approaches; game theory models, time series models, and production cost models [3]. On the other hand, short term predictions of prices for a competitive market can be made by using data-driven approaches, which are covered by time series models. Both stationary and non-stationary time series models are used for short term predictions [1, 8]. Among these models, one traditional model, which is dynamic regression, and two customized approaches, namely CMARS and RCMARS, are used in this study.

In the following section, a traditional approach, namely the dynamic regression is presented. Additionally,CMARS and RCMARS models are proposed. Finally, numerical results, conclusions, and future work are presented in the last two sections.

2 The Electricity Price Modelling Approach

Today, since the market is evolving into a competitive form, electricity price forecasting models are being developed in Turkey. In these studies, main drivers of daily electricity price are determined by using traditional price models. Traditional price models generally require a large amount of data. However, since the data for the new market will be scarce, it is important to have a model that predicts the day-ahead price with less data, while minimizing the error. As a result, robust optimization techniques are selected for price forecasting, since they do not require much data. Here, CMARS and RCMARS models are used for this purpose. Proposed models are compared with dynamic regression model.

2.1 Dynamic Regression Approach

Since the variables change in time, an effective method for price modelling is to use a dynamic procedure. (1) presents a dynamic regression model that consists of electricity price p_{t+1} at time $t+1$ explained by past prices at time $t, t-1, t-2, \ldots, t-n$, and the values of demand at time $t, t-1, t-2, \ldots, t-n$.

$$p_{t+1} = \beta_0 d_t + \beta_1 d_{t-1} + \ldots + \beta_n d_{t-n} + \delta_0 p_t + \delta_1 p_{t-1} + \ldots + \delta_n p_{t-n} + \varepsilon_t. \tag{1}$$

Here β_i and δ_i represent the coefficients, and ε_t stands for the error term. This error term is assumed to be independent and normally distributed with mean zero and constant variance. This method is used in order to overcome the serial correlation in error [4,5].

Since the efficiency of the method depends on the selection of explanatory variables, the appropriate model for Turkish electricity market is developed by using real data set of March 2011. The resulting model is

$$p_{t+1} = \beta_0 d_t + \beta_1 d_{t-7} + \delta_0 p_t + \varepsilon_t. \tag{2}$$

The model relates next day's price to current day's demand and price as well as the demand of the same day of the previous week.

2.2 CMARS

CMARS is developed as an alternative to Multivariate Adaptive Regression Splines (MARS) [6]. The aim of this method is not only to minimize the estimation error but also to maximize efficiency as much as possible. Therefore, the method relies on Penalized Residual Sum of Squares (PRSS) that performs as a Tikhonov regularization problem and this problem can be solved by using a conic quadratic programming (CQP) approach in order to minimize the error [7]. The general model is given by $Y = f(x) + \varepsilon$, where y represents the response variable, x represents the vector of explanatory variables, and ε is the error term.

The first step is to create a large model that has maximum number of basis functions (BF) by using MARS. The model, which involves BFs, can be stated as:

$$y = \alpha_0 + \sum_{n=1}^{N} \alpha_n \psi_n(x) + \varepsilon, \tag{3}$$

where $\psi(n = 1, 2, \ldots, N)$ are the set of BFs, and α_n is the unknown coefficient for the nth BF.

As the next step, the MARS models are obtained for each subset by using the Salford MARS System. Then the maximum number of BFs (M_{max}) and the highest degree of interactions are determined. As a result, the largest model for the first period, i.e. the day, is found to be

$$\begin{aligned}
y_d = {}& \alpha_0 + \alpha_1 \max\{0, x_3 - 0.63\} + \alpha_2 \max\{0, 0.63 - x_3\} + \alpha_3 \max\{0, x_2 + 2.04\} \cdot \\
& \max\{0, 0.63 - x_3\} + \alpha_4 \max\{0, x_1 + 2.7\} \cdot \max\{0, x_3 - 0.63\} + \alpha_5 \\
& \max\{0, x_1 + 2.7\} + \alpha_6 \max\{0, x_1 - 0.51\} + \alpha_7 \max\{0, 0.51 - x_1\} + \\
& \alpha_8 \max\{0, x_1 + 0.28\} + \alpha_9 \max\{0, -0.28 - x_1\} + \varepsilon.
\end{aligned}$$

Similarly, the largest models are found for the other periods. Afterwards, PRSS is constructed with the maximum number of BFs and discretized to handle the multidimensional integrals [7]. As a result, the approximation of PRSS can be defined as

$$PRSS = \|y - \psi(\tilde{b})\alpha\|_2^2 + \lambda\|L\alpha\|_2^2, \tag{4}$$

where $\lambda = \lambda_m(\varphi^2)$ represents the penalty parameter for some $\varphi > 0, \varphi \in \mathbb{R}$. Here, L stands for the diagonal $(M_{max} + 1) \times (M_{max} + 1)$-matrix. As the last step, PRSS is formulated as a CQP problem:

$$\min_{t,\alpha} t$$
$$\text{subject to}$$
$$\|\psi(\tilde{b})\alpha - y\|_2 \leq t$$
$$\|L\alpha\|_2 \leq \sqrt{M}$$
$$t \geq 0,$$

where $\sqrt{M}$ is an appropriate bound. CMARS is used to find minimum PRSS for various values of $\sqrt{M}$. The model is solved in MATLAB environment for three explanatory variables and the results are given in section 3.

2.3 RCMARS

Electricity price models involve uncertain parameters. For instance, small perturbations in electricity price and demand may cause unstabilities in solution. In order to avoid unstable solutions, all input and output variables are assumed as random variables, opposite to the dynamic regression, where only output (dependent variable) is assumed as random variable. This robustification is applied to BFs obtained from MARS. Thus, variables are transformed into standard normal form. Using standard normal form, uncertainty sets are obtained for the input parameters. Hence, RCMARS model takes its general form as $Y = f(\check{x}) + \varepsilon$, where y is the response variable, $\check{x}$ is the vector of noisy explanatory variables, and ε is the error term [2, 7].

Uncertainty matrices and vectors for the input and output parameters are constructed based on polyhedral uncertainty sets that are represented by standard confidence intervals. To solve the problem, PRSS in (4) is reformulated as a CQP with an uncertainty

$$PRSS = \|\check{y} - \psi(\check{b})\alpha\|_2^2 + \lambda\|L\alpha\|_2^2. \tag{5}$$

Here, PRSS is the TR problem that is solved through CQP [2]. Furthermore, polyhedral uncertainty sets are used to overcome the complexity of the optimization problem. Robust optimization model is obtained as

$$\min_{\varpi,\alpha} \varpi$$

subject to

$$\|W^j\alpha - z^i\|_2 \leq \varpi \qquad (i = 1,\ldots,2^N;\ j = 1,\ldots,2^{N \times M_{max}})$$

$$\|L\alpha\|_2 \leq \sqrt{M},$$

where $M \geq 0$.

3 Numerical Results and Comparison

Proposed models and the traditional model are used to predict day-ahead electricity prices of Turkey. One month is selected and daily periodic data is used to predict the electricity prices. Numerical results are presented in Table 1 for one period. The models give similar results for the other two periods.

Table 1 Comparison of electricity price models in terms of MAE, RMSE, R^2_{adj}, Var and r

Day	Dynamic Regression	CMARS	RCMARS1	RCMARS2	RCMARS3
MAE	0.75	0.53	0.82	0.57	0.54
RMSE	0.99	0.86	1.19	0.88	0.86
R^2_{adj}	0.13	0.26	0.42	0.23	0.26
Var	0.33	0.34	0.007	0.25	0.32
r	0.33	0.73	0.35	0.74	0.73

Three different performance measures, namely the mean absolute error (MAE), the root mean square error (RMSE), the adjusted R^2, the variance (Var), and the correlation coefficient (r) are used to assess the prediction performance of the methods. In RCMARS, parameters are calculated for nine different uncertainty scenarios using the values under polyhedral uncertainty sets and the results with minimum (RCMARS1), medium (RCMARS2) and maximum (RCMARS3) performance measures are presented in the table. Since uncertainty matrices and vectors for the parameters are constructed based on polyhedral uncertainty sets, which are represented by standard confidence intervals, RCMARS performs better when the length of confidence intervals is decreased. Better predictions are obtained when CMARS and RCMARS are used in Turkish electricity market.

4 Conclusion and Future Work

In this paper, we constructed an electricity price model for Turkish electricity market. One traditional and two new approaches are proposed and then analyzed considering three different types of period in a day. The results show that CMARS and

RCMARS performs better than the dynamic regression. Although not appropriate for small-sized data sets, dynamic regression is used to compare the traditional approach and the customized approaches.

As the future research, the future price prediction performances of three methods can be compared by using predicted residual sum of squares (PRSS) values. Furthermore, a scenario analysis can be made by using RCMARS for electricity market participants, whose bidding strategies are based on price thresholds.

References

1. Aggarwal, S., Saini, L. M., Kumar, A.: Electricity price forecasting in Ontario electricity market using wavelet transform in artificial neural network based model. Int. Journal of Control, Automation, and Systems. **6**, 639–650 (2008)
2. Ben-Tal, A., Nemirovski, A.: Lectures on Modern Convex Optimization: Analysis, Algorithms and Engineering Applications, MPPR-SIAM Series on Optimization. SIAM,Philadelphia (2001)
3. Gonzalez, A.M., Roque, A.M.S., Garcia-Gonzalez, J.: Modelling and forecasting electricity prices with input/output hidden Markov models. IEEE Trans. Power Syst. **20**, 13–24 (2005)
4. Gujarati, D. N., Porter, D. C.: Basic Econometrics. McGraw-Hill, Boston (2009)
5. Nogales, F. J., Contreras, J., Conejo, A. J., Espinola, R.: Forecasting next-day electricity prices by time series models. IEEE Trans. Power Syst. **17**, 342–348 (2002)
6. Weber, G. W., Batmaz, İ., Köksal, G., Taylan, P., Yerlikaya, F.: C-MARS: A new contribution to nonparametric regression with multivariate adaptive regression splines supported by continuous optimization. to appear in Inverse Problems in Science and Engineering (IPSE)
7. Özmen, A.: Robust Conic Quadratic Programming Applied to Quality Improvement- A Robustification of CMARS. Master Thesis. METU, Ankara, Turkey (2010)
8. Zareipour, H., Janjani, A., Leung, H., Motamedi, A., Schellenberg, A.: Classification of future electricity market prices. IEEE Trans. Power Syst. **26**, 165–173 (2011)

Stream II
Discrete Optimization, Graphs and Networks

Friedrich Eisenbrand, EPF Lausanne, SVOR Chair
Bettina Klinz, TU Graz, ÖGOR Chair
Alexander Martin, University of Erlangen-Nuremberg, GOR Chair

The stable set polytope of claw-free graphs with stability number greater than three

Anna Galluccio, Claudio Gentile, and Paolo Ventura

Abstract In 1965 Edmonds gave the first complete polyhedral description for a combinatorial optimization problem: the Matching polytope. Many researchers tried to generalize his result by considering the Stable Set polytope of claw-free graphs. However this is still an open problem. Here we solve it for the class of claw-free graphs with stability number greater than 3 and without 1-joins.

1 Introduction

Given a graph $G = (V, E)$ and a vector $w \in \mathbb{Q}_+^V$ of node weights, the *stable set problem* is to find a set of pairwise nonadjacent nodes *(stable set)* of maximum weight. The *stability number* $\alpha(G)$ of G denotes the maximum cardinality of a stable set of G. The *stable set polytope*, denoted by $STAB(G)$, is the convex hull of the incidence vectors of the stable sets of G.

The study of the stable set polytope of claw-free graphs attracts the attention of the scientific community since early seventies when the pioneering work of Edmonds on the matching polytope [3] was translated for the stable set polytope of line graphs (a line graph $L(G)$ of a graph G is obtained by considering the edges of G as nodes of $L(G)$ and two nodes of $L(G)$ are adjacent if and only if the corresponding edges of G have a common endnode). At that time it seemed natural to look for a linear description of the stable set polytope for classes of graphs that properly contain line graphs such as the *claw-free graphs* (i.e., graphs such that the neighborhood of each node has stability number at most two), but very soon it turned out that the goal was not so easy to achieve [11].

The recent result of Chudnovsky and Seymour [1] on the structure of claw-free graphs gave a new impulse to study the stable set polytope of these graphs. In [1], they proved that every claw-free graph with stability number at least 4 is obtained by iteratively composing graphs belonging to one of a few well defined families with

Anna Galluccio, Claudio Gentile, and Paolo Ventura
Istituto di Analisi dei Sistemi ed Informatica "A. Ruberti", Consiglio Nazionale delle Ricerche,
Viale Manzoni, 30 - 00185 Roma, Italia e-mail: {galluccio,gentile,ventura}@iasi.cnr.it

D. Klatte et al. (eds.), *Operations Research Proceedings 2011*, Operations Research Proceedings,   47
DOI 10.1007/978-3-642-29210-1_8, © Springer-Verlag Berlin Heidelberg 2012

two kinds of operations: *1-join* and *strip composition*. We focus on those graphs that are obtained by composing with strip composition only five types of graphs, called *fuzzy Z_i-strips*, $i = 1,2,3,4,5$, that are explicitly described in [1]. We call *striped graphs* the graphs obtained in this way.

The striped graphs form a large subclass of claw-free graphs. Indeed, they constitute "almost all" claw-free graphs with large stability number as proved by Chudnovsky and Seymour in the following:

Theorem 1. *[1] Every connected claw-free graph that does not admit a 1-join satisfies one of the following conditions: either has stability number at most 3, or is fuzzy circular interval, or is striped.*

When G is fuzzy circular interval, Eisenbrand et al. [4] gave a linear description of $STAB(G)$ consisting of nonnegativity inequalities, clique inequalities, and a class of nonrank inequalities called the *clique-family inequalities* [12]. In this paper we present a linear description of $STAB(G)$ when G is a striped graph, thus completing the case of claw-free graphs with $\alpha(G) \geq 4$ and without 1-join. The result is based on the polyhedral properties of the *2-clique-bond composition*, a graph composition introduced in [7]. For polyhedral definitions we refer to [14].

Given a graph $G = (V,E)$ and two sets $A,B \subseteq V$, we say that A is B-complete (B-anticomplete) if every node in A is (is not, respectively) adjacent to every node in B. A node x is simplicial if its neighbourhood $N(x)$ is a clique. An edge $xy \in E$ is simplicial if $N(x) \setminus \{y\}$ and $N(y) \setminus \{x\}$ are cliques. A claw-free graph admits a 1-join if its node set can be partitioned into four nonempty sets V_i for $i = 1,..,4$, such that $V_2 \cup V_3$ is a clique, V_1 is $V_3 \cup V_4$-anticomplete, and V_4 is $V_1 \cup V_2$-anticomplete.

Given two graphs G_1 and G_2, let $a_0^i b_0^i$ be a simplicial edge of G_i and let $A_i = N(a_0^i) \setminus \{b_0^i\}$ and $B_i = N(b_0^i) \setminus \{a_0^i\}$, $i = 1,2$. The *2-clique-bond* of G_1 and G_2 along the edges $a_0^1 b_0^1$ and $a_0^2 b_0^2$ is the graph G obtained by deleting the nodes a_0^i and b_0^i, for $i = 1,2$, and joining every node in A_1 with every node in A_2 and every node of B_1 with every node of B_2. We denote by $G_i / a_0^i b_0^i$ the graph obtained by contracting the edge $a_0^i b_0^i$ into a single node z_0^i, for $i = 1,2$.

Definition 1. Let G be the 2-clique-bond of G_1 and G_2 along the simplicial edges $a_0^1 b_0^1$ and $a_0^2 b_0^2$. An inequality of $STAB(G)$ is said to be an *even-odd combination* of inequalities if it has the following form:

$$\sum_{v \in V(G_i \setminus \{a_0^i, b_0^i\})} \beta_v^i x_v + \sum_{v \in V(G_j \setminus \{a_0^j, b_0^j\})} \beta_v^j x_v \leq \beta_0^i + \beta_0^j - 1, \qquad (1)$$

where $\beta^i x \leq \beta_0^i$ is valid for $STAB(G_i)$ and different from $x_{a_0^i} + x_{b_0^i} \leq 1$, $\beta^j x \leq \beta_0^j$ is valid for $STAB(G_j / a_0^j b_0^j)$ and different from the clique inequalities supported by $A_j \cup \{z_0^j\}$ or $B_j \cup \{z_0^j\}$, and $\beta_{a_0^i}^i = \beta_{b_0^i}^i = \beta_{z_0^j}^j = 1$, for $i,j = 1,2$ and $i \neq j$.

The following fundamental result holds for the stable set polytope of graphs resulting from 2-clique-bond compositions.

Theorem 2. *[7] Let G_i be a graph with a simplicial edge $a_0^i b_0^i$, $i = 1, 2$, and let G be the 2-clique-bond of G_1 and G_2. Then $STAB(G)$ is described by: (i) non-negativity inequalities; (ii) clique inequalities supported by $A_1 \cup A_2$ and $B_1 \cup B_2$; (iii) facet defining inequalities of $STAB(G_i)$ with zero coefficients on a_0^i and b_0^i for $i = 1, 2$; (iv) even-odd combinations of facet defining inequalities of $STAB(G_i)$ and $STAB(G_j / a_0^j b_0^j)$ for each $i, j = 1, 2$ and $i \neq j$.*

Note that the above result can also be applied iteratively to classes of graphs obtained by consecutively applying 2-clique-bond compositions along a fixed set of pairwise disjoint simplicial edges.

2 The Chudnovsky-Seymour construction of claw-free graphs

Chudnovsky and Seymour [1] describe explicitly how to construct striped graphs:

- Start with a graph G_0 that is a disjoint union of cliques. Take a partition of $V(G_0)$ into stable sets $X_1, \ldots, X_k$ such that each X_i satisfies $|X_i| = 2$.
- For $1 \leq i \leq k$, take a graph G_i (where $G_0, G_1, \ldots, G_k$ are pairwise node-disjoint) and a subset $Y_i \subseteq V(G_i)$ with $|Y_i| = 2$. When we apply this, the pairs (G_i, Y_i) will be taken from an explicit list $\mathscr{L}$ of allowed pairs. In particular, G_i is claw-free, Y_i is stable in G_i, and for each node in Y_i, its neighbour set in G_i is a clique.
- For $1 \leq i \leq k$, take a bijection between X_i and Y_i; for each $x \in X_i$ and its mate $y \in Y_i$, make the $N(x)$ complete to the $N(y)$, and then delete x, y.

Hereafter we refer to this construction as the Chudnovsky-Seymour's construction. When $k = 2$ we speak of *strip composition*. The list $\mathscr{L}$ consists of 5 members $\mathscr{L}_1, \ldots, \mathscr{L}_5$ (see [1] at pg. 1388-1389 for more details) and are defined as follows:

Definition 2. A pair $(G, \{a_0, b_0\})$ is a *linear interval strip*, or a Z_1-*strip*, if its node set is $V = \{a_0 = v_1, v_2, \ldots, v_{n-1}, v_n = b_0\}$ where: (i) for $1 \leq i < j < k \leq n$ if $v_i v_k \in E$ then $v_i v_j \in E$ and $v_j v_k \in E$, (ii) a_0, b_0 are simplicial.

Definition 3. A pair $(G, \{a_0, b_0\})$ is a Z_2-*strip* if its node set is $V = A \cup B \cup C \setminus X$ where: (i) $A = \{a_0, a_1, \ldots, a_n\}$, $B = \{b_0, b_1, \ldots, b_n\}$, and $C = \{c_1, \ldots, c_n\}$ are pairwise disjoint cliques; (ii) for $1 \leq i, j \leq n$, a_i and b_j are adjacent if and only if $i = j$; (iii) for $1 \leq i \leq n$ and $1 \leq j \leq n$, c_i is adjacent to a_j, b_j if and only if $i \neq j$; (iv) $X \subseteq A \cup B \cup C \setminus \{a_0, b_0\}$ such that $|C \setminus X| \geq 2$.

Definition 4. A pair $(G, \{a_0, b_0\})$ is a Z_3-*strip* if its node set is $V = A \cup B \cup C \cup \{z_1, z_2\} \setminus X$ where: (i) $A = \{a_0, a_1, a_2, \ldots, a_n\}$, $B = \{b_0, b_1, b_2, \ldots, b_n\}$, and $C = \{c_1, c_2 \ldots, c_n\}$ are pairwise disjoint cliques; (ii) for $1 \leq i, j \leq n$, $a_i b_j$, $a_i c_j$, $b_i c_j \in E$ if and only if $i = j$; (iii) z_1 is $(A \cup C)$-complete and B-anticomplete; z_2 is $(B \cup C)$-complete and A-anticomplete; (iv) $X \subseteq A \cup B \cup C \setminus \{a_0, b_0\}$.

Definition 5. A pair $(G, \{a_0, b_0\})$ is a Z_4-*strip* if its node set is $V = A \cup B \cup C$ where: (i) $A = \{a_0, a_1, a_2\}$, $B = \{b_0, b_1, b_2, b_3\}$, and $C = \{c_1, c_2\}$ are pairwise disjoint cliques; (ii) $\{a_2, c_1, c_2\}$, $\{a_1, b_1, c_2\}$, $\{b_2, c_1\}$ are cliques.

Definition 6. A pair $(G, \{a_0, b_0\})$ is a Z_5-*strip* if its node set is $V = V(\Gamma) \setminus X$ where: (i) Γ is the graph with nodes $\{v_1, \ldots, v_{13}\}$ and with adjacency as follows: $(v_1, \ldots, v_6)$ is a hole of Γ of length 6; v_7 is adjacent to v_1, v_2; v_8 is adjacent to v_4, v_5; v_9 is adjacent to v_6, v_1, v_2, v_3; v_{10} is adjacent to v_3, v_4, v_5, v_6, v_9; v_{11} is adjacent to $v_3, v_4, v_6, v_1, v_9, v_{10}$; v_{12} is adjacent to $v_2, v_3, v_5, v_6, v_9, v_{10}$; v_{13} is adjacent to $v_1, v_2, v_4, v_5, v_7, v_8$. (ii) $X \subseteq \{v_{11}, v_{12}, v_{13}\}$ and $a_0 = v_7$, $b_0 = v_8$.

A pair $\{u, v\} \subset V$ is *fuzzy* if either $uv \in E$ and $(V, E \setminus \{uv\})$ is claw-free or $uv \notin E$ and $(V, E \cup \{uv\})$ is claw-free. The following definition of thickening slightly extends the original definition in [1] to include some extremal cases.

Definition 7. Given a graph H and a set F (eventually empty) of unordered fuzzy pairs of nodes of $V(H)$, a *thickening* of H on F is a graph G satisfying the following:

- for every $v \in V(H)$ there is a nonempty clique $K_v \subseteq V(G)$, all pairwise disjoint and with union $V(G)$;
- if $uv \notin E(H)$ and $\{u, v\} \notin F$, then K_u is K_v-anticomplete in G;
- if $uv \in E(H)$ and $\{u, v\} \notin F$, then K_u is K_v-complete in G;
- if $\{u, v\} \in F$, then *either* K_u is neither K_v-complete nor K_v-anticomplete in G *or* $K_u = \{u\}$, $K_v = \{v\}$, and $uv \in E(G)$ if and only if $uv \notin E(H)$.

A *fuzzy Z_i-strip* is a thickening of a Z_i-strip and the family $\mathscr{Z}_i$ consists of fuzzy Z_i-strips, for $i = 1, \ldots, 5$.

We say that an edge $a_0^i b_0^i$ is *super simplicial* if it is simplicial and moreover it satisfies $N(A_i \cap B_i) = N(a_0^i) \cup N(b_0^i)$. In order to use the polyhedral properties of the 2-clique-bond composition and in particular Theorem 2, we now add a simplicial edge between nodes a_0 and b_0 of fuzzy strips, thus speaking of *closed fuzzy strips* $(G, a_0 b_0)$. It is not difficult to check that the 2-clique-bond composition applied on claw-free graphs along super simplicial edges produces the same graphs as the strip composition.

Let $G_0, G_1, \ldots, G_k$ be the graphs in the Chudnovsky-Seymour's construction. Denote by G_i^+ the closed fuzzy strip obtained from the pair (G_i, Y_i) by adding an edge e_i connecting the two nodes of Y_i for $i = 1, \ldots, k$. Moreover, let G_0^+ be the graph obtained from G_0 by adding edges f_i between the nodes in the pairs X_i for $i = 1, \ldots, k$. By definition, all these new edges are super simplicial and pairwise nonadjacent in G_0^+. Moreover, G_0^+ is a line graph.

Thus Chudnovsky-Seymour's construction can be reformulated in terms of 2-clique-bond composition as follows:

Theorem 3. *Every striped graph G can be obtained from a line graph G_0^+ by iteratively performing 2-clique-bond compositions of closed fuzzy strips belonging to $\mathscr{Z}_1$, $\mathscr{Z}_2$, $\mathscr{Z}_3$, $\mathscr{Z}_4$, and $\mathscr{Z}_5$ along a set of pairwise nonadjacent super simplicial edges of G_0^+.*

As a consequence of Theorem 3 and Theorem 2, in order to understand the structure of $STAB(G)$ when G is striped, we need to know the inequalities defining the stable set polytopes of all closed fuzzy strips and *contracted closed fuzzy strips* (i.e., the graphs obtained by contracting the simplicial edges involved in the composition).

3 The stable set polytope of striped graphs

Chudnovsky and Seymour stated the following (whose proof can be found in [16]):

Theorem 4. *Let G be a striped graph that is obtained from a line graph G_0^+ by iteratively performing 2-clique-bond compositions of closed fuzzy strips belonging to $\mathcal{Z}_1$. Then $STAB(G)$ is defined by nonnegativity and rank inequalities.*

Let us consider the other types of strips. Starting from the characterization of $STAB(G)$ when $\alpha(G) = 2$ due to Cook [15], we derive:

Theorem 5. *[8] Let (G,a_0b_0) be a closed fuzzy Z_i-strip with $i \in \{2,3,4\}$. Then $STAB(G)$ and $STAB(G/a_0b_0)$ are described by nonnegativity, rank, and lifted 5-wheel inequalities.*

By exploiting the properties of super simplicial edges, we prove that, when composing closed fuzzy Z_1-, Z_2-, Z_3-, and Z_4-strips, the inequalities used in even-odd combinations are only rank inequalities. By iteratively using Theorem 2 and Theorem 5 we are able to describe the stable set polytope of an important subclass of striped graphs.

Theorem 6. *[8] Let G be a graph obtained from a line graph G_0^+ by iteratively performing 2-clique-bond compositions of closed fuzzy strips belonging to $\mathcal{Z}_1$, $\mathcal{Z}_2$, $\mathcal{Z}_3$, and $\mathcal{Z}_4$ along a set of nonadjacent super simplicial edges of G_0^+. Then $STAB(G)$ is described by nonnegativity, rank, and lifted 5-wheel inequalities.*

If fuzzy Z_5-strips are allowed in the construction, then new inequalities become necessary to describe $STAB(G)$. In fact, the following nonrank inequality, named *gear inequality*, is proved to be facet defining [5] for the stable set polytope of a contracted closed fuzzy Z_5-strip G/a_0b_0:

$$\sum_{v\in V(G/a_0b_0)\setminus\{v_9,v_{10}\}} x_v + 2(x_{v_9} + x_{v_{10}}) \le 3. \tag{2}$$

This inequality occurs also in the definition of $STAB(G)$ when $v_{13} \in V(G)$.

Theorem 7. *[9] Let (G,a_0b_0) be a closed fuzzy Z_5-strip. Then $STAB(G)$ and $STAB(G/a_0b_0)$ are described by nonnegativity, rank, lifted 5-wheel, and lifted gear inequalities.*

When fuzzy Z_5-strips are reapetedly used in 2-clique-bond compositions (as it happens in Theorem 3), new inequalities result from iterative even-odd combinations of gear and rank inequalities. More formally,

Definition 8. [6] An inequality $\alpha x \le \beta$ is a *multiple geared rank inequality* if it is (i) either a rank inequality; (ii) or a gear inequality; (iii) or an even-odd combination of two multiple geared rank inequalities.

Roughly speaking, a multiple geared rank inequality has coefficients 1 on every node of G apart from some pairs of nodes that are hubs of the 5-wheels in fuzzy Z_5-strips and have coefficients 2. We can now state our main result:

Theorem 8. *[9] Let G be a striped graph. Then $STAB(G)$ is described by nonnegativity, rank, lifted 5-wheel, and lifted multiple geared rank inequalities.*

The above theorem together with the characterization of rank facet defining inequalities in [10] provides the minimal defining linear system for the stable set polytope of striped graphs. It also proves that the coefficients of the linear system are $\{0,1,2\}$-valued for striped graphs, contrarily to what happens for claw-free graphs with stability number three [13] or for fuzzy circular interval graphs [4].

Together with the result of Eisenbrand et al. [4], Theorem 8 provides the complete linear description of the stable set polytope of claw-free graphs with stability number at least four that do not admit a 1-join. As the polyhedral description of the stable set polytope for graphs G with $\alpha(G) = 2$ is already known [15], in order to answer the longstanding open question of providing a linear description for the stable set polytope of claw-free graphs, only the case $\alpha(G) = 3$ is left open.

References

1. M. Chudnovsky and P. Seymour. Claw-free graphs V: Global structure. *J. Comb. Th. B*, 98:1373–1410, 2008.
2. V. Chvátal. On certain polytopes associated with graphs. *J. Comb. Th. B*, 18:138–154, 1975.
3. J. Edmonds. Maximum matching and a polyhedron with 0, 1 vertices. *J. Res. of Nat. Bureau of Stand. B*, 69B:125–130, 1965.
4. F. Eisenbrand, G. Oriolo, G. Stauffer, and P. Ventura. The stable set polytope of quasi-line graphs. *Combinatorica*, 28(1):45–67, 2008.
5. A. Galluccio, C. Gentile, and P. Ventura. Gear composition and the stable set polytope. *Operations Research Letters*, 36:419–423, 2008.
6. — Gear composition of stable set polytopes and $\mathscr{G}$-perfection. *Mathematics of Operations Research*, 34:813–836, 2009.
7. — 2-clique-bond of stable set polyhedra. Tech. Rep. 09-10, IASI - CNR, 2009. Submitted.
8. — The stable set polytope of claw-free graphs with large stability number I: Fuzzy antihat graphs are $\mathscr{W}$-perfect. Tech. Rep. 10-06, IASI - CNR, 2010. Submitted.
9. — The stable set polytope of claw-free graphs with large stability number II: Striped graphs are $\mathscr{G}$-perfect. Tech. Rep. 10-07, IASI - CNR, 2010. Submitted.
10. A. Galluccio and A. Sassano. The rank facets of the stable set polytope for claw-free graphs. *J. Comb. Th. B*, 69:1–38, 1997.
11. R. Giles and L.E. Trotter. On stable set polyhedra for $K_{1,3}$-free graphs. *J. Comb. Th. B*, 31:313–326, 1981.
12. G. Oriolo. Clique family inequalities for the stable set polytope for quasi-line graphs. *Discrete Applied Mathematics*, 132:185–201, 2003.
13. A. Pêcher, P. Pesneau, and A. Wagler. On facets of stable set polytope of claw-free graphs with maximum stable set size three. *Elec. Notes in Discrete Mathematics*, 28:185–190, 2006.
14. A. Schrijver. *Combinatorial optimization*. Springer Verlag, Berlin, 2003.
15. B. Shepherd. Near-perfect matrices. *Mathematical Programming*, 64:295–323, 1994.
16. G. Stauffer. *On the stable set polytope of claw-free graphs*. PhD thesis, EPF Lausanne, 2005.

Packing Euler graphs with traces

Peter Recht and Eva-Maria Sprengel

Abstract For a graph $G = (V, E)$ and a vertex $v \in V$, let $T(v)$ be a *local trace* at v, i.e. $T(v)$ is an Eulerian subgraph of G such that every walk $W(v)$, with start vertex v can be extended to an Eulerian tour in $T(v)$. In general, local traces are not unique. We prove that if G is Eulerian every maximum edge-disjoint cycle packing $\mathscr{Z}^*$ of G induces maximum local traces $T(v)$ at v for every $v \in V$. In the opposite, if the total size $\sum_{v \in V} |E(T(v))|$ is minimal then the set of related edge-disjoint cycles in G must be maximum.

1 Introduction

Packing edge-disjoint cycles in graphs is a classical graph-theoretical problem. There is a large amount of literature concerning conditions that are sufficient for the existence of some number of disjoint cycles which may satisfy some further restrictions. A selection of related references is given in [6]. The algorithmic problems concerning edge-disjoint cycle packings are typically hard (e.g. see [3], [4], [8]) There are papers in which practical applications of such packings are mentioned ([2], [5], [7]) .

Starting point for the results of the paper is the attempt to find an optimal packing of a graph by the determination of such packings for different subgraphs. In [6] such an approach was studied when the subgraphs were induced by vertex cuts. In the present paper we study the behavior of such packings if the subgraphs are traces.

We consider a finite and undirected graph G with vertex set $V(G)$ and edge set $E(G)$ that contains no loops. For a finite sequence $v_{i_1}, e_1, v_{i_2}, e_2, \ldots, e_{r-1}, v_{i_r}$ of vertices v_{i_j} and pairwise distinct edges $e_j = (v_{i_j}, v_{i_{j+1}})$ of G the subgraph W of G with vertices $V(W) = \{v_{i_1}, v_{i_2}, \ldots, v_{i_r}\}$ and edges $E(W) = \{e_1, e_2, \ldots, e_{r-1}\}$ is called a

Peter Recht
Operations Research und Wirtschaftsinformatik, TU Dortmund, Vogelpothsweg 87, D 44221 Dortmund, e-mail: peter.recht@tu-dortmund.de

Eva-Maria Sprengel
Operations Research und Wirtschaftsinformatik, TU Dortmund, Vogelpothsweg 87, D 44221 Dortmund, e-mail: eva-maria.sprengel@tu-dortmund.de

D. Klatte et al. (eds.), *Operations Research Proceedings 2011*, Operations Research Proceedings, 53
DOI 10.1007/978-3-642-29210-1_9, © Springer-Verlag Berlin Heidelberg 2012

walk with *start vertex* v_{i_1} and *end vertex* v_{i_r}. If W is closed (i.e. $v_{i_1} = v_{i_r}$) we call it a *circuit* in G. A *path* is a walk in which all vertices v have degree $d_W(v) \leq 2$. A closed path will be called a *cycle*. A connected graph in which all vertices v have even degree $d_G(v)$ is called *Eulerian*. For an Eulerian graph G a circuit W with $E(W) = E(G)$ is called an *Eulerian tour*.

For $1 \leq i \leq k$ let $G_i \subset G$ be subgraphs of G. We say G is *induced by* $\{G_1, G_2, \ldots, G_k\}$ if $V(G) = V(G_1) \cup V(G_2) \cup \ldots V(G_k)$ and $E(G) = E(G_1) \cup E(G_2) \cup \ldots E(G_k)$. Two subgraphs $G' = (V', E')$, $G'' = (V'', E'') \subset G$ are called *edge-disjoint* if $E' \cap E'' = \emptyset$.

A *packing* $\mathscr{L}(G) = \{G_1, \ldots, G_q\}$ of G is a collection of subgraphs G_i of G ($i = 1, \ldots, q$) such that all G_i are mutually edge-disjoint and G is induced by $\{G_1, \ldots, G_q\}$. A packing $\mathscr{L}(G) = \{G_1, \ldots, G_q\}$ is called a circuit packing of cardinality $s \geq 0$, if exactly s of the G_i are circuits. If exactly s of the G_i are cycles $\mathscr{L}(G)$ is called a cycle packing of cardinality s. The set of all circuit packings and cycle packings of cardinality s are denoted by $\bar{\mathscr{C}}_s(G)$ and $\mathscr{C}_s(G)$, respectively. If there is no confusion possible we will write $\mathscr{L}$ instead of $\mathscr{L}(G)$. If the cardinality of a circuit packing $\mathscr{L}$ is maximum, all circuits in $\mathscr{L}$ must be cycles. The packing number of G is the maximum cardinality of a cycle packing of G and is denoted by $\nu(G)$. A corresponding cycle packing is called a *maximum cycle packing*.

Section 2 relates maximum cycle packings of G and local traces in the case that G is Eulerian.

In section 3 a "min-max"-type theorem is proved. If the local traces are induced by a set of edge-disjoint cycles and the total size of the traces is minimal, then the set of cycles are a maximum cycle packing of G.

2 Local traces and maximum cycles packings

Let $H = (V, E)$ be an Eulerian Graph. A vertex $v \in V$ is called *proper*, if every walk W, starting at v can be extended to an Euler-tour in H. An Eulerian graph that contains a proper vertex is called a *trace*. Traces were first considered by Ore [9] and Baebler [1].

Using results in [1] and [9], we immediately get a characterization of traces by their packing number.

Proposition 1. *Let* $H = (V, E)$ *be Eulerian and* $\Delta = max\{d_H(u) | u \in V\}$. *Then* $\nu(H) = \frac{1}{2}\Delta = \frac{1}{2}d_H(v)$ *if and only if* v *is a proper vertex of* H.

Proof. Note, that $\nu(H) \geq \frac{1}{2}\Delta$. For $\nu(H) = \frac{1}{2}\Delta$, let $v \in V$ such that $d_H(v) = \Delta$. Assume that there is a cycle $C \subseteq H$ with $v \notin C$. Obviously, each of the components $H'_1, H'_2, \ldots, H'_k$ of $H \setminus E(C)$ is Eulerian. Let H'_i be that component that contains v. Then $d|_{H'_i}(v) = d_H(v) = \Delta$. But then, $\nu(H) \geq 1 + \sum_{j=1}^{k} \nu(H'_j) > \nu(H'_i) \geq \frac{1}{2}d_{H'_i}(v) = \frac{1}{2}d_H(v)$, a contradiction.
Now, let v be a proper vertex. If $\{C_1, C_2, \ldots, C_{\nu(H)}\}$ is a maximum cycle packing of H, then $d_H(v) = 2\nu(H) \geq \Delta$. Since $d_H(v) \leq \Delta$, $\nu(H) = \frac{1}{2}\Delta = \frac{1}{2}d_H(v)$ follows. $\qquad\square$

Now, we transfer the concept of a trace locally to an arbitrary graph $G = (V, E)$. For $v \in V$ an Eulerian subgraph $T(v) = (V(T(v)), E(T(v))) \neq \emptyset$ of G is called *a trace (at v)* , if $v \in V(T(v))$ and v is proper with respect to $T(v)$. A trace $T(v)$ is called *saturated* (at v), if there is no Eulerian subgraph $H \subset G$ such that $T(v) \subsetneqq H$ and v is proper with respect to H.

The number $|E(T(v))|$ is called the *size of the trace* (at v).

A trace $T(v)$ at v is called *maximum*, if $T(v)$ is induced by $k(v)$ edge-disjoint cycles $\{C_1, C_2, \ldots, C_{k(v)}\}$ and $k(v)$ is maximum.

The next theorem shows that maximum traces $T(v)$ can be obtained if a maximum cycle packing of G is known.

Theorem 1. *Let $H = (V, E)$ be Eulerian and $\mathscr{Z}^*$ a maximum packing of H. For $v \in V$ let $\mathscr{Z}^*(v) := \{C_i \in \mathscr{Z}^* | v \in V(C_i)\}$.*
Then $\mathscr{Z}^*(v)$ induces a maximum trace $T(v)$ at v.

Proof. Let $T(v)$ be the subgraph of H induced by the $\frac{d_H(v)}{2}$ cycles in $\mathscr{Z}^*(v)$. Obviously, $T(v)$ is Eulerian, $v \in V(T(v))$ and $v(T(v)) = \frac{d_H(v)}{2}$. We will show, that an arbitrary cycle $C \in T(v)$ passes v. Assume, that is not the case, i.e. there is a cycle C^* such that $v \notin C^*$. Then consider the graph $H^* = T(v) \setminus C^*$.
Obviously, $v(H^*) < v(T(v))$, but since each component of H^* is Eulerian and $d_{H^*}(v) = d_{T(v)}(v) = d_H(v)$, $v(H^*) = \frac{d_H(v)}{2} = v(T(v))$, a contradiction. $\qquad \square$

Note, that the fact that H is Eulerian is crucial, i.e. for an arbitrary graph a maximum cycle packing must not induce a maximum trace at v, even it must not induce a saturated trace.

The opposite direction in Theorem 1 is not true in general. I.e. there is an Eulerian graph $H = (V, E)$ and a cycle packing $\mathscr{Z}_1 = \{C_1, C_2, \ldots, C_s\}$ of cardinality $s < v(H)$ such that H is induced by $\mathscr{Z}_1$ and for every $v \in V$ the subgraph $T(v)$ of H, induced by $\mathscr{Z}_1(v)$ is a maximum trace. I.e. maximum traces of H can be induces by cycle packings of H that are not maximum. Fig.1 illustrates such an example. Observe, that the maximum packing $\mathscr{Z}_2$ in this example is actually unique.

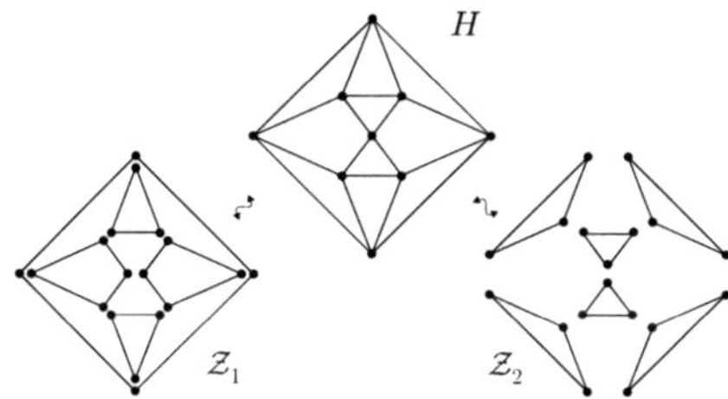

Fig. 1 $|\mathscr{Z}_1| = 5, F(\mathscr{Z}_1) = 66$ whereas $|\mathscr{Z}_2| = 6 = v(H), F(\mathscr{Z}_2) = 54$

3 A "min-size – max-packing"–Theorem

In order to characterize a maximum cycle packing of H by properties of local traces $T(v)$, let us consider a specific subset $\mathscr{C}_s^*$ of $\mathscr{C}((H))$. A packing $\mathscr{Z}$ belongs to $\mathscr{C}_s^*$ if it is a cycle packing of cardinality $s \leq v(H)$, G is induced by $\mathscr{Z}$ and for $v \in V$ the subgraph $T(v)$ of G induced by the cycles in $\mathscr{Z}(v)$ is a trace at v. Let $F(\mathscr{Z}) = \sum_{v \in V} |E(T(v))|$, the total size of the induced local traces. We will prove the following

Theorem 2. *Every $\mathscr{Z}^*$ that minimizes F on $\bigcup_{s=1}^{v(H)} \mathscr{C}_s^*$ is a maximum cycle packing of H, i.e. $\mathscr{Z}^* \in \mathscr{C}_{v(H)}^*$.*

To show the theorem we first prove Lemma 1, which is true not only for Eulerian graphs H. Let $G = (V(G), E(G))$ be a graph and consider a circuit packing $\mathscr{Z}_s(G)$ of order $s \leq v(G)$, i.e. $\mathscr{Z}_s(G) = \{C_1, C_2, \ldots, C_s, \tilde{G}\} \in \bar{\mathscr{C}}_s(G)$, where the C_i's denote the circuits in $\mathscr{Z}_s(G)$. Note, that $E(\tilde{G}) = \emptyset$ if and only if G is Eulerian. In this case $\bar{\mathscr{C}}_0(G) = \emptyset$.
Let $l_i = |E(C_i)|$. For $s \geq 1$, define

$$\bar{L}(\mathscr{Z}_s(G)) = \sum_{i=1}^{s} l_i^2 + |E(\tilde{G})|^2.$$

For $s = 0$, set $\bar{L}(\mathscr{Z}_0(G)) := |E(G)|^2$.

Lemma 1. *Let G be a graph of order $|V(G)|$ and size $|E(G)|$.
For $s \in \{0, 1, 2, \ldots, v(G)\}$ let*

$$\bar{m}_s(G) := \min\{\bar{L}(\mathscr{Z}) \,|\, \mathscr{Z} \in \bar{\mathscr{C}}_s(G)\}$$

and

$$\bar{M}_s(G) := \max\{\bar{L}(\mathscr{Z}) \,|\, \mathscr{Z} \in \bar{\mathscr{C}}_s(G)\}.$$

Then

1. If $\bar{\mathscr{C}}_0(G) \neq \emptyset$, then

$$\bar{m}_{s-1}(G) > \bar{m}_s(G), \text{ and } \bar{M}_{s-1}(G) > \bar{M}_s(G) \quad s = 1, 2, \ldots, v(G)$$

2. If $\bar{\mathscr{C}}_0(G) = \emptyset$, then

$$\bar{m}_{s-1}(G) > \bar{m}_s(G), \text{ and } \bar{M}_{s-1}(G) > \bar{M}_s(G) \quad s = 2, 3, \ldots, v(G)$$

Proof. We will use induction on $r = v(G)$.
Case 1 $\bar{\mathscr{C}}_0(G) \neq \emptyset$:
Obviously $\bar{m}_0(G) = \bar{M}_0(G) = |E(G)|^2$.
Let $v(G) = r = 1$. Let $\mathscr{Z}_1 \in \bar{\mathscr{C}}_1(G)$, i.e. $\mathscr{Z}_1 = \{C_1, \tilde{G}\}$ and $l_1 = |E(C_1)|$.
If $|E(\tilde{G}| \neq \emptyset$, we immediately get $\bar{L}(\mathscr{Z}_1) := l_1^2 + (|E(G)| - l_1)^2 < |E(G)|^2$.

Hence, $\bar{m}_0(G) > \bar{m}_1(G)$ and $\bar{M}_0(G) > \bar{M}_1(G)$.
Case 2 $\bar{\mathscr{C}}_0(G) = \emptyset$**:**
Obviously $\bar{m}_1(G) = \bar{M}_1(G) = |E(G)|^2$.
Let $v(G) = r = 2$. Let $\mathscr{L}_2 \in \bar{\mathscr{C}}_2(G)$, i.e. $\mathscr{L}_2 = \{C_1, C_2\}$ and $l_i = |E(C_i)|$. We immediately get $\bar{L}(\mathscr{L}_2) := l_1^2 + (|E(G)| - l_1)^2 < |E(G)|^2$. Hence, $\bar{m}_1(G) > \bar{m}_2(G)$ and $\bar{M}_1(G) > \bar{M}_2(G)$.
Now, let $r \geq 1$ (if $\bar{\mathscr{C}}_0(G) \neq \emptyset$) or $r \geq 2$ (if $\bar{\mathscr{C}}_0(G) = \emptyset$). Moreover, let us assume that for all graphs G such that $v(G) = r' \leq r$ the relations $\bar{m}_{r'-1}(G) > \bar{m}_{r'}(G)$ and $\bar{M}_{r'-1}(G) > \bar{M}_{r'}(G)$ hold.
Let G be a graph such that $v(G) = r + 1$.

1. *Take $\mathscr{L}_r(G) \in \bar{\mathscr{C}}_r(G)$ such that $\bar{L}(\mathscr{L}_r(G)) = \bar{m}_r(G)$. Take the circuit $C_1 \in \mathscr{L}_r(G)$ of length l_1 and consider the graph $G \setminus C_1$. Obviously, $\mathscr{L} := (\mathscr{L}_r(G) \setminus \{C_1\}) \in \bar{\mathscr{C}}_{r-1}(G \setminus C_1)$. Moreover, $\bar{L}(\mathscr{L}) = \min\{\bar{L}(\mathscr{L}_{r-1}(G \setminus C_1)) | \mathscr{L}_{r-1} \in \bar{\mathscr{C}}_{r-1}(G \setminus C_1)\}$ must hold, otherwise $\mathscr{L}_r(G)$ was not a minimizer in $\bar{\mathscr{C}}_r(G)$. Using the assumption, we get $\bar{L}(\mathscr{L}) = \bar{m}_{r-1}(G \setminus C_1) > \bar{m}_r(G \setminus C_1)$ and, by this, $\bar{m}_r(G) = \bar{m}_{r-1}(G \setminus C_1) + l_1^2 > \bar{m}_r(G \setminus C_1) + l_1^2 \geq \bar{m}_{r+1}(G)$.*
2. *The proof is just the same as in (i.) but instead of taking out a circuit $C_1 \in \mathscr{L}_r(G)$ one takes it out from $\mathscr{L}_{r+1}(G)$.* $\square$

This lemma immediately implies the following

Corollary 1. *Let $v(G) \geq 1$. Every $\mathscr{L}^*(G)$ that minimizes $\bar{L}$ on $\bar{\mathscr{C}}(G) = \bigcup_s \bar{\mathscr{C}}_s(G)$ is a maximum cycle packing of G, i.e. $\mathscr{L}^*(G) \in \bar{\mathscr{C}}_{v(G)}$.* $\square$

Now, we can proceed with the proof of Theorem 1

Proof. We first observe, that for all $v \in V$ and for all $C_i \in \mathscr{L} = \{C_1, C_2, \ldots, C_s\} \subset \mathscr{C}_s^*$ the following is true: $v \in V(C_i) \Leftrightarrow C_i \in \mathscr{L}(v)$.
Therefore, we get $F(\mathscr{L}) = \sum_{v \in V} |E(T(v))| = \sum_{v \in V} \sum_{C_i \in \mathscr{L}(v)} |E(C_i)| =$
$= \sum_{C_i \in \mathscr{L}} \sum_{v \in V(C_i)} |E(C_i)| = \sum_{C_i \in \mathscr{L}} |V(C_i)||E(C_i)| = \sum_{i=1}^{s} |E(C_i)|^2 = \bar{L}(\mathscr{L})$.
Let $\mathscr{L}^*$ be a minimizer of F in $\mathscr{C}_{v(G)}^*$, and assume that there is $\bar{\mathscr{L}}^* \in \mathscr{C}_s^*$, such that $F(\bar{\mathscr{L}}^*) = F(\mathscr{L}^*)$, but $s < v(G)$. We then get $F(\bar{\mathscr{L}}^*) \geq \min\{\bar{L}(\mathscr{L}') | \mathscr{L}' \in \bar{\mathscr{C}}_{s(G)}\} = \bar{m}_{s(G)} > \bar{m}_{v(G)}(G) = F(\mathscr{L}^*)$, a contradiction. $\square$

References

1. Baebler, F.: Über eine spezielle Klasse Euler'scher Graphen , Comment. Math. Helv., vol 27(1), 81-100 (1953)
2. Bafna, V., Pevzner, P.A. : Genome Rearrangement and sorting by reversals, SIAM Journal on Computing Vol 25 (2) 272-289 (1996)
3. Caprara, A.: Sorting Permutations by Reversals and Eulerian Cycle Decompositions, SIAM Journal on Discrete Mathematics 12 (1) 91 - 110 (1999)
4. Caprara, A., Panconesi, A., Rizzi, R.: Packing Cycles in Undirected Graphs, Journal of Algorithms 48 (1) 239 - 256 (2003)

5. Fertin, G., Labarre, A., Rusu, I., Tannier, E., Vialette, S.: Combinatorics of Genome Rearrangement, MIT Press, Cambridge, Ma. (2009)
6. Harant, J., Rautenbach, D., Recht, P., Schiermeyer, I., Sprengel, E.M.: Packing disjoint cycles over vertex cuts, Discrete Mathematics 310 1974-1978(2010)
7. Kececioglu, J., Sankoff, D.: Exact and Approximation Algorithms for Sorting by Reversals with Application to genome rearrangement Algorithmica 13 180 - 210 (1997)
8. Krivelevich, M., Nutov, Z., Salvatpour, M.R., Yuster, J., Yuster, R.: Approximation algorithms and hardness results for cycle packing problems, ACM Trans. Algorithm 3 Article No. 48 (2007)
9. Ore, O.: A problem regarding the tracing of graphs, Elem. Math. 6, 49-53 (1951)

Minimum Cost Hyperassignments with Applications to ICE/IC Rotation Planning

Olga Heismann and Ralf Borndörfer

Abstract Vehicle rotation planning is a fundamental problem in rail transport. It decides how the railcars, locomotives, and carriages are operated in order to implement the trips of the timetable. One important planning requirement is operational regularity, i. e., using the rolling stock in the same way on every day of operation. We propose to take regularity into account by modeling the vehicle rotation planning problem as a minimum cost hyperassignment problem (HAP). Hyperassignments are generalizations of assignments from directed graphs to directed hypergraphs. Finding a minimum cost hyperassignment is $\mathcal{NP}$-hard. Most instances arising from regular vehicle rotation planning, however, can be solved well in practice. We show that, in particular, clique inequalities strengthen the canonical LP relaxation substantially.

1 Introduction

Vehicle rotation planning, also known as rolling stock roster planning, deals with the allocation of vehicles to trips in a timetable, see [4]. We focus in this article on a basic version of the problem that deals with the construction of a cyclic schedule for a standard week, disregarding maintenance, train composition, and some other side constraints. In this setting, we are looking for an assignment of each trip to a follow-on trip which will be serviced by the same vehicle.

Such an assignment is considered operationally *regular*, if many timetabled trips are followed by the same timetabled trips on as many days of the standard week as possible, i. e., if trip 4711 is followed by trip 4712 on Monday, this should also be the case on Tuesday, Wednesday, etc. (provided that these trips exist on these days). In practice, most trips appear on almost every day of operation. In other words, the

Olga Heismann
Zuse Institute Berlin, Takustraße 7, 14195 Berlin, Germany, e-mail: heismann@zib.de

Ralf Borndörfer
Zuse Institute Berlin, Takustraße 7, 14195 Berlin, Germany, e-mail: borndoerfer@zib.de

D. Klatte et al. (eds.), *Operations Research Proceedings 2011*, Operations Research Proceedings, 59
DOI 10.1007/978-3-642-29210-1_10, © Springer-Verlag Berlin Heidelberg 2012

weekly timetable is largely regular, such that there is a good chance to also construct a regular vehicle rotation plan.

Regular vehicle rotation plans are easier to communicate and understand than non-regular ones. They standardize operations, increase robustness, and facilitate real-time scheduling. It is therefore essential to include regularity in vehicle rotation planning models. This can be achieved by considering a suitable concept of hyperassignments, as we will show in the following sections.

2 Notions on Directed Hypergraphs

We start by recalling some basic notions on directed hypergraphs, see also [1] for an introduction (with slightly different requirements for hyperarcs).

Definition 1 (Directed Hypergraph, Directed Graph). A *directed hypergraph D* is a pair (V,A) consisting of a *vertex* set V and a set $A \subseteq 2^V \times 2^V$ of *hyperarcs* (T_a, H_a) such that $T_a, H_a \neq \emptyset$. We call T_a the *tail* of the hyperarc $a \in A$ and H_a the *head* of a. A hyperarc a is called an *arc* if $|H_a| = |T_a| = 1$. If all hyperarcs are arcs, we call D a *directed graph*.

Definition 2 (Outgoing and Ingoing Hyperarcs). Let $D = (V,A)$ be a directed hypergraph. For $W \subseteq V$, $B \subseteq A$ we define

$$\delta_B^{\text{out}}(W) := \{a \in B : T_a \cap W \neq \emptyset\} \quad \text{and} \quad \delta_B^{\text{in}}(W) := \{a \in B : H_a \cap W \neq \emptyset\}$$

to be the *outgoing* and *ingoing hyperarcs* of W, respectively. For $\delta_B^{\text{out}}(\{v\})$ and $\delta_B^{\text{in}}(\{v\})$ we simply write $\delta_B^{\text{out}}(v)$ and $\delta_B^{\text{in}}(v)$, respectively. If no set B is given in the index of δ^{out} or δ^{in}, B is assumed to be the full hyperarc set of the hypergraph in question.

Definition 3 (Cost Function). Given a set S, a *cost function* is a function $c_S : S \to \mathbb{R}$. For $T \subseteq S$ we define
$$c_S(T) := \sum_{s \in T} c_S(s).$$

We are now ready to propose hyperassignments.

Definition 4 (Circulation, Hyperassignment). Let $D = (V,A)$ be a directed hypergraph. $Z \subseteq A$ is called *circulation* in D if $|\delta_Z^{\text{out}}(v)| = |\delta_Z^{\text{in}}(v)|$ for every $v \in V$. A circulation $H \subseteq A$ in D is called a *hyperassignment* if for each $v \in V$ $|\delta_H^{\text{out}}(v)| = 1$.

Problem 1 (Hyperassignment Problem (HAP)).

Input: A pair (D, c_A) consisting of a directed hypergraph $D = (V,A)$ and a cost function $c_A : A \to \mathbb{R}$.

Output: A minimum cost hyperassignment in D w.r.t. c_A, i.e., a hyperassignment H^* in D such that $c_A(H^*) = \min\{c_A(H) : H$ is a hyperassignment in $D\}$, or the information that no hyperassignment in D exists if this is the case.

An ILP formulation of the hyperassignment problem is as follows:

$$\text{minimize} \qquad \sum_{a \in A} c_A(a) x_a \qquad\qquad \text{(HAP)}$$

$$\text{subject to} \quad \sum_{a \in \delta^{\text{in}}(v)} x_a - \sum_{a \in \delta^{\text{out}}(v)} x_a = 0 \qquad \forall v \in V \qquad\qquad \text{(i)}$$

$$\sum_{a \in \delta^{\text{out}}(v)} x_a = 1 \qquad \forall v \in V \qquad\qquad \text{(ii)}$$

$$x \geq 0 \qquad\qquad \text{(iii)}$$

$$x \in \mathbb{Z}^A \qquad\qquad \text{(iv)}$$

3 Complexity of the Hyperassignment Problem (HAP)

We study in this section the complexity of the HAP. Despite its simple form, the problem turns out to be $\mathcal{NP}$-hard even for directed hypergraphs with head and tail cardinality 2. In fact, already the LP-relaxation can be numerically complex.

Theorem 1. *Given a directed hypergraph $D = (V, A)$ satisfying $|T_a| = |H_a| \leq 2$ for all $a \in A$ and a cost function $c_A : A \to \mathbb{R}$, HAP with input (D, c_A) is $\mathcal{NP}$-hard.*

Proof. The 3-dimensional matching problem is $\mathcal{NP}$-complete (see [2], page 46). Given an undirected hypergraph $U = (N \cup O \cup P, E)$, $|N| = |O| = |P|$, $|e \cap N| = |e \cap O| = |e \cap P| = 1 \quad \forall e \in E$, it asks whether a partitioning of U into a subset of elements of E exists.

Construct a directed hypergraph $D = (V, A)$ with $V = N \cup O \cup \{\{p\} \times \{0, 1\} : p \in P\}$ and

$$A = A_1 \cup A_2,$$

$$A_1 = \Big\{ \big((e \cap N) \cup (e \cap O), \{(e \cap P, 0), (e \cap P, 1)\}\big) : e \in E \Big\},$$

$$A_2 = \Big\{ \big(\{(e \cap P, 0)\}, e \cap N\big), \big(\{(e \cap P, 1)\}, e \cap O\big) : e \in E \Big\}.$$

This can be done in polynomial time and the resulting hypergraph satisfies $|T_a| = |H_a| \leq 2$ for all $a \in A$. Choose $c_A : A \to \mathbb{R}, c_A \equiv 0$. Then HAP with input (D, c_A) returns a hyperassignment with cost 0 if and only if a partitioning of U exists. The chosen hyperarcs from A_1 correspond to the edges $e \in E$ in the partitioning of U. This proves the theorem. $\quad\square$

The determinant of submatrices of the coefficient matrix of an ILP is a complexity indicator. For example, if the coefficient matrix is totally unimodular, the LP relaxation is integral. In general, by Cramer's rule, the denominator of the variable values in a basic solution of an LP is at most the determinant of the basis matrix (if the numerators and denominators are relatively prime). For the LP relaxation of the (HAP), the denominators, and therefore also the determinants of basis matrices, can be arbitrarily large. This is the case even if one allows only hyperarcs with head and tail cardinality at most two. The following example illustrates this fact.

Let s be a positive integer and consider the following directed and head cardinality at most two. We want $V = \{u, v_i, w_i, : i \in \{0, \dots, s-1\}\}$ and $A = A_1 \cup A_2$ with

$$A_1 = \{(\{v_i, w_i\}, \{v_i, w_i\}) : i \in \{0, \dots, s-1\}\},$$

$$A_2 = \{(\{u, v_i\}, \{w_{(i+1)} \mod s, u\}), (\{w_i\}, \{u\}), (\{u\}, \{v_i\}) : i \in \{0, \dots, s-1\}\}.$$

The only feasible solution of the LP relaxation of (HAP) is $x_a = \frac{2s-1}{2s}$ for all $a \in A_1$ and $x_a = \frac{1}{2s}$ for all $a \in A_2$. Thus the determinant of the basis matrix is at least $2s$.

An upper bound on the modulus of the determinant is $\prod_{a \in A} |T_a|$ if the hypergraph $D = (V, A)$ can be extended to a graph based hypergraph by adding arcs (this is also true for the hypergraph of the example). This is the case if head and tail cardinalities are equal for each hyperarc. Since every column of the basis matrix can be represented as the sum of columns for the corresponding arcs and basis matrices of (HAP) for directed graphs are totally unimodular, i. e., have determinant with modulus 0 or 1, we can apply the multilinearity of the determinant until we get only such matrices and obtain the bound.

We remark that one can also prove that the gap between the optimum solution of the LP relaxation of (HAP) and the minimum cost hyperassignment can be arbitrarily large. An example of such a HAP instance on a hypergraph with only 6 vertices is given in [3]. Moreover, the reduction from the 3-dimensional matching problem implies that HAP is APX-complete.

4 Hyperassignment Model for Regular Vehicle Rotation Planning

Considering regularity in vehicle rotation planning leads to the minimum cost hyperassignment problem, as we will show now.

Suppose we are given a weekly repeating schedule for long distance trains with all trips that a railway company wants to operate. A *trip* is characterized by its departure day, departure time, departure location, arrival location, and its duration.

Every trip has to be serviced by a vehicle. Between arrival and departure there may be several intermediate stops, but the vehicle must not change during the trip.

After servicing a trip, a vehicle does a *deadhead trip* (possibly of distance zero) from the arrival location of the trip to the departure location of the next trip it services. This deadhead trip has some duration. Afterwards, when the weekday and departure time of the next trip has come, the vehicle services this next trip.

A *vehicle rotation plan* is an assignment of each trip to another follow-on trip. This assignment tells every vehicle which trip it has to service next. Since the schedule is periodic, the sequence of trips for every vehicle is periodic, too. The period is a positive integral multiple of a week.

The *cost* of a pair of trips in the assignment depends on the duration and distance of the associated deadhead trip and on the duration of the breaks before and after the deadhead. Clearly, the longer the deadheads and the breaks are, the more vehicles the railway company needs.

The aim is to find an assignment of minimum cost. It is apparent that vehicle rotation planning as explained so far can be formulated in terms of an assignment problem in a directed graph $D = (V,A)$, where the vertices V are the trips and there is an arc $a \in A$ from every trip to every possible follow-on trip.

Now we include operational regularity. We associate with each trip $v \in V$ a departure weekday $d_v \in \{\mathrm{Mo}, \ldots, \mathrm{Su}\}$. We group all trips that differ only in the departure weekday and call such a set a *train*.

Given an assignment, we can count for each train the number of *non-regular deadheads* in the trips assigned to the trips of the train. Two deadheads are non-regular if the trains the next trips belong to are different or the breaks have a different length. Otherwise they are *regular*. The less the number of unequal deadheads in the vehicle rotation plan, the higher the operational regularity.

This criterion can be modeled in terms of a hyperassignment problem. To this purpose, we define hyperarcs as follows. For each possible set of deadhead trips $a_1, \ldots, a_d \in A$ with the same length between the trips of two trains we introduce a hyperarc $a \in A$ where $T_a = \bigcup_{i=1}^{d} T_{a_i}$ are the timetabled trips of the first train and $H_a = \bigcup_{i=1}^{d} H_{a_i}$ are the timetabled trips of the second train between which the deadhead trips take place. To reward regularity, we set $c_a < \sum_{i=1}^{d} c_{a_i}$. Then we can choose the hyperarc a for the hyperassignment if we would use $a_1, \ldots, a_d$ in the assignment to get lower costs. The difference $c_a - \sum_{i=1}^{d} c_{a_i}$ is the bonus for operational regularity.

5 Computational Results

Solving practical instances of HAP using the ILP (HAP) showed that the separation of clique inequalities associated with this formulation is very important. Our computational results (on an Intel Core i7-870) are summarized in Table 1. It can be seen that the duality gap between the LP solution and the ILP solution is very small if one adds enough clique inequalities, and that it can be large without them. The LP bound improved up to 20 % by adding clique inequalities and the LP-IP gap was reduced in many cases to less than 1 %.

All instances stem from a project with DB Fernverkehr AG, which deals with the optimization of long distance passenger railway transport in Germany. More precisely, they arise from cyclic weekly schedules of ICE 1 trains. Our results show that the HAP is computationally well-behaved. This model therefore provides an excellent basis for incorporating regularity requirements into more complex large-scale real-world vehicle rotation planning models.

Table 1 Computational results with real-world vehicle rotation planning problems using (HAP) and CPLEX 12.1.0. The LP-IP gap is given by $1 - \frac{L}{I}$, where L is the optimum value of the LP relaxation and I is the best integral solution found by CPLEX or Gurobi 3.0.0. The root gap is $1 - \frac{R}{I}$, where R is the optimum value of the LP relaxation before branching but after applying the cuts described in the seventh and eighth column of the table. The root improvement is $\frac{R}{L} - 1$. (*) means that the calculation was aborted.

# rows $(2 \cdot |V|)$	# columns $(|A|)$	nonzeros	LP-IP gap	root gap	root improvement	# clique cuts	# other cuts	root run time (sec.)
534	52056	140081	11.16 %	6.81 %	4.90 %	160	14	8
620	80477	236020	8.72 %	0.00 %	9.54 %	120	2	29
812	102375	216566	0.38 %	0.18 %	0.20 %	24	16	40
1128	267542	732134	4.59 %	0.26 %	4.55 %	263	0	160
1310	363513	1006024	7.85 %	0.22 %	8.28 %	378	2	270
1496	469932	1369224	18.70 %	1.86 %	20.71 %	809	0	971
1696	618348	1787078	5.17 %	0.16 %	5.28 %	925	0	1705
1746	649525	1859898	7.52 %	4.88 %	2.86 %	563	0	1129
1798	647650	1822718	13.60 %	0.95 %	14.65 %	537	0	1099
1798	647650	1822718	13.35 %	0.62 %	14.69 %	604	0	873
2006	855153	2491372	5.76 %	0.68 %	5.39 %	1025	0	2490
2260	1079535	3138752	9.89 %	2.03 %	8.73 %	954	0	5483
2502	1290750	3680124	7.06 %	0.76 %	6.79 %	801	0	4583
2620	1432355	4187296	9.05 %	1.15 %	8.68 %	1068	0	7910
2624	1439453	4087042	14.17 %	5.23 %	10.41 %	951	0	(*) 14400

References

1. Gallo, G., Longo, G., Pallottino, S., Nguyen, S.: Directed hypergraphs and applications. Discrete Applied Mathematics (1993) doi: 10.1016/0166-218X(93)90045-P
2. Garey, M. R., Johnson, D. S.: Computers and Intractability: A Guide to the Theory of NP-Completeness (Series of Books in the Mathematical Sciences). W. H. Freeman (1979)
3. Heismann, O.: Minimum cost hyperassignments. Master's thesis, TU Berlin (2010)
4. Maróti, G.: Operations research models for railway rolling stock planning. Eindhoven University of Technology (2006)

Primal Heuristics for Branch-and-Price Algorithms

Marco Lübbecke and Christian Puchert

Abstract In this paper, we present several primal heuristics which we implemented in the branch-and-price solver GCG based on the SCIP framework. This involves new heuristics as well as heuristics from the literature that make use of the reformulation yielded by the Dantzig-Wolfe decomposition. We give short descriptions of those heuristics and briefly discuss computational results. Furthermore, we give an outlook on current and further development.

1 Introduction

A wealth of practical decision and planning problems in operations research can be modeled as mixed integer programs (MIPs) with a special structure; this structure has been recently successfully exploited in *branch-and-price* algorithms [6] which are based on *Dantzig-Wolfe decomposition* [5]; we assume the reader to be familiar with both concepts. *Primal heuristics* are a very important aspect of MIP solving: In practice, the decision maker is often satisfied with a "good" solution to her problem; besides, such solutions can help accelerate the solution process. We developed and tested heuristics specially tailored for branch-and-price that exploit the fact that two different formulations of the MIP are available via the Dantzig-Wolfe decomposition: the original MIP, and the extended reformulation.

We implemented our heuristics in the branch-and-price solver GCG [8] which is based on the well-known SCIP framework [1]. Whereas today's branch-and-price algorithms are almost always tailored to the application and are thus "reserved to experts," the goal of the GCG project is to provide a more generally accessible solver

Marco Lübbecke
RWTH Aachen University, Operations Research, Templergraben 64, 52056 Aachen, Germany
e-mail: marco.luebbecke@rwth-aachen.de

Christian Puchert
RWTH Aachen University, Operations Research, Templergraben 64, 52056 Aachen, Germany
e-mail: puchert@or.rwth-aachen.de

D. Klatte et al. (eds.), *Operations Research Proceedings 2011*, Operations Research Proceedings, 65
DOI 10.1007/978-3-642-29210-1_11, © Springer-Verlag Berlin Heidelberg 2012

that is available to a much broader range of users. Our work contributes towards this goal by providing primal heuristics for branch-and-price.

In the remainder, we will present three types of such heuristics—called extreme point, restricted master, and column selection heuristics—and discuss computational results. This work is based on the master's thesis of the second author [11].

2 Preliminaries

We want to solve MIPs of the form

$$\begin{aligned}
\min\ & c^T x \\
\text{s. t.}\ & Ax \geq b \\
& x \in X\ ,
\end{aligned} \tag{1}$$

with $X = \{x \in \mathbb{R}^n : Dx \geq d, x_i \in \mathbb{Z} \text{ for all } i \in I \subseteq \{1,\dots,n\}\}$. Typically, $Ax \geq b$ are some "complicated" or "linking" constraints, whereas the set X often has a block structure $X = X_1 \times \cdots \times X_K$ with $X_k = \{x^k \in \mathbb{R}^{n_k} : D^k x^k \geq d^k, x_i^k \in \mathbb{Z} \text{ for all } i \in I_k\}$, $k = 1,\dots,K$, $I_k \subseteq \{1,\dots n_k\}$, $\sum_{k=1}^{K} n_k = n$.

Via Dantzig-Wolfe decomposition, (1) is reformulated to a *master problem*

$$\begin{aligned}
\min\quad & \sum_{k=1}^{K} \sum_{p \in P_k} c^p \lambda_{kp} \\
\text{s. t.}\quad & \sum_{k=1}^{K} \sum_{p \in P_k} a^p \lambda_{kp} \geq b \\
& \sum_{p \in P_k} \lambda_{kp} = 1 \qquad \text{for all } k \\
& \lambda_k \in \mathbb{R}_+^{|P_k|} \quad \text{for all } k\ ,
\end{aligned} \tag{2}$$

where integrality of the x_i is enforced by constraints $\sum_{p \in P_k} \lambda_{kp} x^{kp} = x^k$ for all k and $x_i^k \in \mathbb{Z}$ for all $i \in I_k$ (*convexification*) or $\lambda_k \in \mathbb{Z}_+^{|P_k|}$ for all k (*discretization*). The master problem (2) bears too many variables to explicitly work with and column generation is typically applied. The pricing subproblems to generate variables as needed read as

$$\begin{aligned}
\min\ & (c^k)^T x^k - (\pi^k)^T A^k x^k - \pi_0 \\
\text{s. t.}\ & x^k \in X_k\ ,
\end{aligned} \tag{3}$$

one for each block k.

3 Extreme Points Crossover

Each component $\tilde{x}^k$ of the solution $\tilde{x}$ yielded by solving the master problem (2) is a convex combination $\tilde{x}^k = \sum_{p \in P_k} \tilde{\lambda}_{kp} x^{kp}$, with P_k being the (index) set of extreme points obtained by solving the k-th pricing problem. These are integer feasible and satisfy the constraints $D^k x^k \geq d^k$, while $\tilde{x}^k$ satisfies all linear constraints but is usually not integer feasible. This leads to the question: Can we exploit the integer feasibility of the x^{kp} to obtain a solution $\bar{x}$ satisfying *all* constraints? This is what extreme point based heuristics try to achieve. The extreme points we are particularly interested in are those with $\tilde{\lambda}_{kp} > 0$.

It may happen that there is a number of coordinates in which these extreme points are identical. Especially when their values are also shared by $\tilde{x}$, it seems worth exploring whether there are more integer feasible or feasible points taking the same values. Therefore, *Extreme Points Crossover* performs a neighborhood search on the extreme points by *crossing* them: For each block, it takes those x^{xp} with the highest $\tilde{\lambda}_{kp}$, fixes the variables in which the selected points are identical and solves a sub-MIP.

As its name indicates, the heuristic can be compared to the Crossover heuristic discussed in [3, 12] and implemented in SCIP, which considers a number of already known feasible solution for crossing. The main advantage of our heuristic against this one is that it does not need any previously found feasible solution.

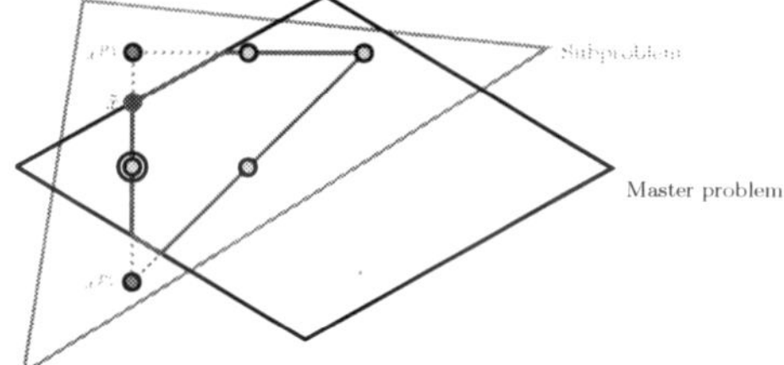

Fig. 1 Extreme Points Crossover: An LP feasible solution $\tilde{x}$ is a convex combination of $x^{p_1}, x^{p_2} \in conv(X)$. All the three points have a coordinate in common which is also shared by the encircled feasible point.

4 Restricted Master

When we solve the master problem, we dynamically add variables that improve the current solution. As the total number of master variables may be exponentially high compared to the number of original variables, this means that the master problem may soon reach a size where searching for feasible solutions would become very time-consuming. This is what the *Restricted Master* heuristic (suggested by Joncour et al. [9]) tries to overcome. It searches for an integer feasible master solution by restricting the formulation again to a subset of promising variables, hence regarding a problem which is of considerably smaller size than the current master formulation.

That is, we solve a sub-MIP of the master formulation (without adding any new variables to it) where all variables not contained in this subset are fixed to zero. As our subset, we chose those master variables which are part of the master LP solution, i.e., the λ_p with $\tilde{\lambda}_p > 0$.

5 Column Selection

Column Selection—also suggested by Joncour et al. [9]—is a constructive approach to build feasible solutions from scratch, with only the knowledge of extreme points previously generated by the pricing problems. In the discretization approach, integrality is enforced by integrality constraints on the master variables. This in particular means that choosing one x^{kp_k}, $p_k \in P_k$, for each block k and building the Cartesian product will lead to a point $\bar{x} = x^{1p_1} \times \cdots \times x^{Kp_K}$ that is feasible w.r.t. the integrality and the subproblem constraints; only the master constraints may be violated.

The heuristic constructs a product as above by successively choosing x^{kp_k} previously generated by the subproblems, while trying to avoid master infeasibility. Technically, this happens by increasing the corresponding master variable λ_{kp_k} in the master problem by one.

There are different approaches trying to achieve that, we implemented the following two:

- *Greedy Column Selection*: Starting with $\bar{\lambda} = 0$, increase a master variable such that the number of currently violated master constraints is decreased as much as possible. As a tie breaker, one can optionally use the objective coefficients of the λ_{kp_k} variables;
- *Relaxation Based Column Selection*: Starting with a rounded down master LP solution $\lfloor \tilde{\lambda} \rfloor$, we increase master variables, preferring those which actually have been rounded (i.e., the λ_{kp_k} with $\tilde{\lambda}_{kp_k} > 0$); this method can also be viewed as a rounding heuristic on the master variables.

6 Results

We ran GCG with the above mentioned heuristics on four types of combinatorial optimization problems: bin packing, vertex coloring, capacitated p-median and resource allocation. These problems bear a bordered block-diagonal structure or at least can be brought into one which can be easily exploited by GCG. Apart from that, we also tested how several of the default SCIP heuristics [3]—such as rounding heuristics, RENS, diving heuristics, and improvement heuristics—perform within the branch-and-price scenario (that is, when applied to the original x variables).

Previously, no primal heuristics had been implemented in GCG; only the default SCIP heuristics on the master variables were used. Now, the new heuristics as well as the SCIP heuristics on the original variable space are available to GCG.

The heuristics did unfortunately not so much improve the solution time of GCG. This is actually no big surprise since it already performed well on especially the bin packing and the smaller coloring instances before the heuristics had been implemented. However, solutions could be found and the number of branch-and-bound nodes could be reduced on a large number of the problems on which the heuristics were tested; in particular, they were successful on some instances where GCG previously failed.

		Restricted Master	Column Selection	Extreme Point
BIN	Solutions found [% inst.]	81.48	74.07	50.00
	Best solution [% inst.]	29.63	0.00	11.11
	Running time [sec, total]	4	17	3226
	Running time [sec, geom. mean]	1.00	1.04	16.48
COLEASY	Solutions found [% inst.]	76.67	60.00	60.00
	Best solution [% inst.]	10.00	3.33	46.67
	Running time [sec, total]	6	14	329
	Running time [sec, geom. mean]	1.02	1.09	3.86
COLHARD	Solutions found [% inst.]	43.48	52.17	4.35
	Best solution [% inst.]	30.43	39.13	4.35
	Running time [sec, total]	40	47	1168
	Running time [sec, geom. mean]	1.48	1.66	5.39
CPMP	Solutions found [% inst.]	0.00	0.00	71.43
	Best solution [% inst.]	0.00	0.00	7.79
	Running time [sec, total]	10	324	681
	Running time [sec, geom. mean]	1.00	2.37	4.75
RAP	Solutions found [% inst.]	92.86	95.24	92.86
	Best solution [% inst.]	16.67	2.38	42.85
	Running time [sec, total]	3	0	67
	Running time [sec, geom. mean]	1.00	1.00	1.52

Table 1 Results: For each heuristic and each test set, we indicate in how many percent of the instances the heuristics found a solution, how often it found the best known solution and how much time it spent over all instances in total and on average, respectively.

Besides, we also had test runs on some MIPLIB2003 [2] and MIPLIB2010 [5] instances; recently, there has been made progress in detecting structures in such general MIPs even if it is not apparently visible (see [3] for details). Our results there indicate that the heuristics are also of use in that general setting; in particular, we recently succeeded with applying diving heuristics to MIPLIB2003 instances. The GCG versions of *Coefficient Diving* and *Fractional Diving*—which use the master LP instead of the original LP—were able to reduce the computational effort for the instances 10teams, modglob, p2756, pp08aCUTS, and rout.

7 Conclusion & Outlook

The heuristics presented here are able to generate feasible solutions for quite a number of problem instances. Yet, there still lies some potential in them. Currently, we are improving the *Extreme Points Crossover* heuristic and the GCG diving heuristics.

In addition, we are developing a column generation based variant of the *Feasibility Pump* heuristic by Fischetti et al. [7]. While the original *Feasibility Pump* decomposes the problem into linear and integer constraints, our heuristic distinguishes—like the Dantzig-Wolfe decomposition—between master and subproblem constraints. That is, by alternately solving modified master problems and subproblems, we try to construct two sequences of points, hopefully converging towards each other and yielding a feasible solution.

Often, like in stochastic programs or other multi-stage decision problems, a matrix has a staircase structure, i.e., we have *linking variables* belonging to multiple blocks. The presence of such variables is undesirable since in order to apply a Dantzig-Wolfe decomposition, one needs to copy these variables and introduce additional master constraints. A way to overcome this issue is to heuristically fix them, which is what *fifty-fifty* heuristics do. Another possibility is letting diving heuristics branch on those variables, hoping that they will get fixed by doing so.

The results show that our heuristics have now become an important part of GCG; a number of them are now included by default. This motivates us to further pursue our research in this important area of computational integer programming.

References

1. Achterberg, T.: SCIP: Solving constraint integer programs. Mathematical Programming Computation **1**(1), 1–41 (2009). http://mpc.zib.de/index.php/MPC/article/view/4
2. Achterberg, T., Koch, T., Martin, A.: MIPLIB 2003. Operations Research Letters **34**(4), 361–372 (2006). http://miplib.zib.de
3. Bergner, M., Caprara, A., Furini, F., Lübbecke, M., Malaguti, E., Traversi, E.: Partial convexification of general MIPs by Dantzig-Wolfe reformulation. In: Günlük, O., Woeginger, G.J. (eds.), Integer Programming and Combinatorial Optimization (IPCO 2011), LNCS, 6655, pp. 39-51. Springer, Berlin (2011)
4. Berthold, T.: Primal Heuristics for Mixed Integer Programs. Master's thesis. Technische Universität Berlin (2006)
5. Dantzig, G.B., Wolfe, P.: Decomposition Principle for Linear Programs. Operations Research **8**(1), 101–111 (1960)
6. Desrosiers, J., Lübbecke, M.: A Primer in Column Generation. In: Desaulniers, G., Desrosiers, J., Solomon, M.M. (eds.), Column Generation, pp. 1-32. Springer, US (2005)
7. Fischetti, M., Glover, F., Lodi, A.: The feasibility pump. Mathematical Programming **104**, 91–104 (2005)
8. Gamrath, G., Lübbecke, M.: Experiments with a Generic Dantzig-Wolfe Decomposition for Integer Programs. In: Festa, P. (ed.), Experimental Algorithms, LNCS, 6049, pp. 239-252. Springer, Berlin (2010)
9. Joncour, C., Michel, S., Sadykov, R., Sverdlov, D., and Vanderbeck, F.: Column Generation based Primal Heuristics. Electronic Notes in Discrete Mathematics **3**, 695–702 (2010). ISCO 2010 – International Symposium on Combinatorial Optimization.
10. T. Koch, T. Achterberg, E. Andersen, O. Bastert, T. Berthold, R. E. Bixby, E. Danna, G. Gamrath, A. M. Gleixner, S. Heinz, A. Lodi, H. Mittelmann, T. Ralphs, D. Salvagnin, D. E. Steffy, and K. Wolter. MIPLIB 2010. *Mathematical Programming Computation*, 3(2):103–163, 2011.
11. Puchert, C.: Primal Heuristics for Branch-and-Price Algorithms. Master's thesis. Technische Universität Darmstadt (2011)
12. Rothberg, E.: An Evolutionary Algorithm for Polishing Mixed Integer Programming Solutions. ILOG Inc. (2005)

Rounding and Propagation Heuristics for Mixed Integer Programming

Tobias Achterberg, Timo Berthold*, and Gregor Hendel

Abstract Primal heuristics are an important component of state-of-the-art codes for mixed integer programming. In this paper, we focus on primal heuristics that only employ computationally inexpensive procedures such as rounding and logical deductions (propagation). We give an overview of eight different approaches. To assess the impact of these primal heuristics on the ability to find feasible solutions, in particular early during search, we introduce a new performance measure, the *primal integral*. Computational experiments evaluate this and other measures on MIPLIB 2010 benchmark instances.

1 Introduction: primal heuristics for MIP

Mixed integer programming (MIP) is to solve the optimization problem

$$\tilde{x}_{\mathrm{opt}} = \mathrm{argmin}\{c^T x \mid Ax \leq b, l \leq x \leq u, x_j \in \mathbb{Z} \text{ for all } j \in J\}, \tag{1}$$

with $A \in \mathbb{R}^{m \times n}$, $b \in \mathbb{R}^m$, $c \in \mathbb{R}^n$, $l, u \in \hat{\mathbb{R}}^n$ (with $\hat{\mathbb{R}} := \mathbb{R} \cup \{\pm\infty\}$), and $J \subseteq \{1, \ldots, n\}$.

In state-of-the-art MIP solvers, primal heuristics play a major role in finding and improving integer feasible solutions at an early stage of the solution process. Knowing good solutions early during optimization helps to prune the search tree and to simplify the problem via dual reductions. Further, it proves the feasibility of a problem and a practitioner might be satisfied with a solution that is proven to be within a certain gap to optimality.

Tobias Achterberg
IBM Deutschland, Germany, e-mail: achterberg@de.ibm.com

Timo Berthold
Zuse Institute Berlin, Takustr. 7, 14195 Berlin, Germany, e-mail: berthold@zib.de

Gregor Hendel
Zuse Institute Berlin, Takustr. 7, 14195 Berlin, Germany, e-mail: hendel@zib.de

* Supported by the DFG Research Center MATHEON *Mathematics for key technologies* in Berlin.

This article gives an overview about rounding and propagation heuristics for MIP that are integrated into SCIP [1], which is a state-of-the-art non-commercial solver and framework for mixed integer programming.

The ZI Round heuristic has been introduced by Wallace [6], more details on the other primal heuristics can be found in [1,3,4].

2 Rounding heuristics

The goal of rounding heuristics is to convert a fractional solution $\bar{x}$ of the system $Ax \leq b, l \leq x \leq u$ into an integral solution, i.e., $x_j \in \mathbb{Z}\, \forall j \in J$. All rounding heuristics described in this section use the notion of *up-* and *down-locks*. For a MIP (1), we call the number of positive coefficients $\xi_j^+ := \left|\{i\colon a_{ij} > 0\}\right|$ the *up-locks* of the variable x_j; the number of negative coefficients is called the *down-locks* ξ_j^- of x_j.

Simple Rounding is a very cheap heuristic that iterates over the set of fractional variables of some LP-feasible point. It only performs roundings, which guarantee to keep all linear constraints satisfied. Consider an integer variable $x_j, j \in J$, with fractional LP solution $\bar{x}_j$. If $\xi_j^- = 0$, we can safely set $\tilde{x}_j := \lfloor \bar{x}_j \rfloor$ without violating any linear constraint. Analogously, if $\xi_j^+ = 0$, we can set $\tilde{x}_j := \lceil \bar{x}_j \rceil$. If all integer variables with fractional $\bar{x}_j$ can be rounded that way, then $\tilde{x}$ will be a feasible MIP solution.

In contrast to Simple Rounding, *Rounding* also performs roundings which potentially lead to a violation of some linear constraints, trying to recover from this infeasibility by further roundings later on. The solutions that can be found by Rounding are a superset of the ones that can be found by Simple Rounding. Like Simple Rounding, the Rounding heuristic takes up- and down-locks of an integer variable with fractional LP value $\bar{x}_j$ into account. As long as no linear constraint is violated, the algorithm iterates over the fractional variables and applies a rounding into the direction of fewer locks, updating the *activities* $A_i\tilde{x}$ of the LP rows after each step, A_i being the i-th row of A. If there is a violated linear constraint, hence $A_i\tilde{x} > b_i$ for some i, the heuristic will try to find a fractional variable that can be rounded in a direction such that the violation of the constraint is decreased, using the number of up- and down-locks as a tie breaker. If no rounding can decrease the violation of the constraint, the procedure is aborted.

ZI Round [6] reduces the *integer infeasibility* of an LP solution step-by-step by shifting fractional values towards integrality, but not necessarily rounding them. For each integer variable x_j with fractional solution value $\bar{x}_j$, the heuristic calculates bounds for both possible rounding directions of $\bar{x}_j$ such that the obtained solution stays LP-feasible. $\bar{x}_j$ is shifted by the corresponding bound into the direction which reduces the *fractionality* $\min\{\bar{x}_j - \lfloor \bar{x}_j \rfloor, \lceil \bar{x}_j \rceil - \bar{x}_j\}$ most. The set of fractional variables might be processed several times by ZI Round. It either terminates with a MIP solution $\tilde{x}$ or aborts if the integer infeasibility could not be decreased anymore or if a predefined iteration limit has been reached.

3 Propagation heuristics

The goal of propagation heuristics is to construct a feasible MIP solution $\tilde{x}$ from scratch or from a start solution, while applying domain propagation to exclude variable values that would lead to an infeasibility or to a solution that is inferior to the incumbent. In contrast to diving or LNS heuristics (see, e.g., [3]), propagation heuristics do not solve any LP during search.

The *Shifting* heuristic is similar to Rounding, but it tries to continue in the case that no rounding can decrease the violation of a linear constraint. In this case, the value of a continuous variable or an integer variable with integral value will be shifted in order to decrease the violation of the constraint. To avoid cycling, the procedure terminates after a certain number of *non-improving shifts*. A shift is called *non-improving*, if it neither reduces the number of fractional variables nor the number of violated rows.

Shift-and-Propagate tries to find a MIP solution by alternately fixing variables and propagating these fixings. Starting with an initial, typically infeasible, solution in which each variable is assumed to be at one of its bounds, it iterates over all integer variables $x_j, j \in J$ in nondecreasing order of their impact on the activity of the linear constraints. In each step, the heuristic fixes a variable x_j to a value $\tilde{x}_j$ such that the overall infeasibility gets maximally reduced. The fixing $x_j = \tilde{x}_j$ is then propagated to reduce further variable domains. In case that the domain propagation detects the infeasibility of the current partial solution, the domain of x_j is reset to its previous state and the variable is postponed. The procedure is aborted when a predefined iteration limit is exceeded.

4 Improvement heuristics

Improvement heuristics consider the incumbent solution as a starting point and try to construct an improved solution with better objective value.

Oneopt is a straightforward improvement heuristic: given a feasible MIP solution $\tilde{x}$, the value of an integer variable $x_j, j \in J$, can be decreased for $c_j > 0$ or increased for $c_j < 0$ if the resulting solution is still feasible. If more than one variable can be shifted, they are sorted by non-decreasing impact $|c_j \delta_j|$ on the objective and sequentially shifted until no more improvements can be obtained. Here, $\delta_j \in \mathbb{Z}$ denotes how far the variable can be shifted into the desired direction without losing feasibility. Oneopt often succeeds in improving solutions which were found by the rounding heuristics described in Section 2, since their defensive approach to round into the direction of fewer locks tends to over-fulfill linear constraints, sacrificing solution quality.

The *Twoopt* heuristic attempts to improve a feasible MIP solution $\tilde{x}$ by altering the solution values of pairs of variables. Only variables which share a pre-defined ratio of LP rows are considered as pairs. Each step of the heuristic consists of improving the objective value by shifting one variable, and then compensating the resulting infeasibilities by shifting a second variable, without completely losing the objective improvement. Similarly to **Oneopt**, pairs are processed in non-decreasing order of their impact on the objective.

5 Computational experiments

In MIP solving, the running time to optimality and the number of branch-and-bound nodes are typical measures for comparison. For primal heuristics, the time needed to find a first feasible solution, an optimal solution, or a solution within a certain gap to optimality are favorable measures that concentrate on the primal part of the solution process. Nevertheless, the trade-off between speed and solution quality is not well covered by any of them.

We suggest a new performance measure that takes into account the overall solution process. The goal is to measure the progress of convergence towards the optimal solution over the entire solving time. Let $\tilde{x}$ be a solution for a MIP, and $\tilde{x}_{\mathrm{opt}}$ be an optimal (or best known) solution for that MIP. We define the *primal gap* $\gamma \in [0,1]$ of $\tilde{x}$ as:

$$\gamma(\tilde{x}) := \begin{cases} 0, & \text{if } |c^t \tilde{x}_{\mathrm{opt}}| = |c^t \tilde{x}| = 0, \\ 1, & \text{if } c^t \tilde{x}_{\mathrm{opt}} \cdot c^t \tilde{x} < 0, \\ \dfrac{|c^t \tilde{x}_{\mathrm{opt}} - c^t \tilde{x}|}{\max\{|c^t \tilde{x}_{\mathrm{opt}}|, |c^t \tilde{x}|\}}, & \text{else.} \end{cases} \tag{2}$$

Let $t_{\max} \in \mathbb{R}_{\geq 0}$ be a limit on the solution time of a MIP solver. Considering a log file of a MIP solver for a certain problem instance within a fixed computational environment, we define its *primal gap function* $p \colon [0, t_{\max}] \mapsto [0, 1]$:

$$p(t) := \begin{cases} 1, & \text{if no incumbent found until point } t, \\ \gamma(\tilde{x}(t)), & \text{with } \tilde{x}(t) \text{ being the incumbent solution at point } t. \end{cases} \tag{3}$$

$p(t)$ is a step function that changes whenever a new incumbent is found. It is zero from the point on at which the optimal solution is found. We define the *primal integral $P(T)$* of a run until a point in time $T \in [0, t_{\max}]$ as:

$$P(T) := \int_{t=0}^{T} p(t)\, dt = \sum_{i=1}^{I} p(t_{i-1}) \cdot (t_i - t_{i-1}), \tag{4}$$

where $t_i \in [0, T]$ for $i \in 1, \dots, I-1$ are the points in time when a new incumbent solution is found, $t_0 = 0$, $t_I = T$.

We suggest to use $P(t_{\max})$ for measuring the quality of primal heuristics. It favors finding good solutions early. The fraction $\frac{P(t_{\max})}{t_{\max}}$ can be seen as the average solution quality during the search process. Spoken differently, the smaller $P(t_{\max})$ is, the

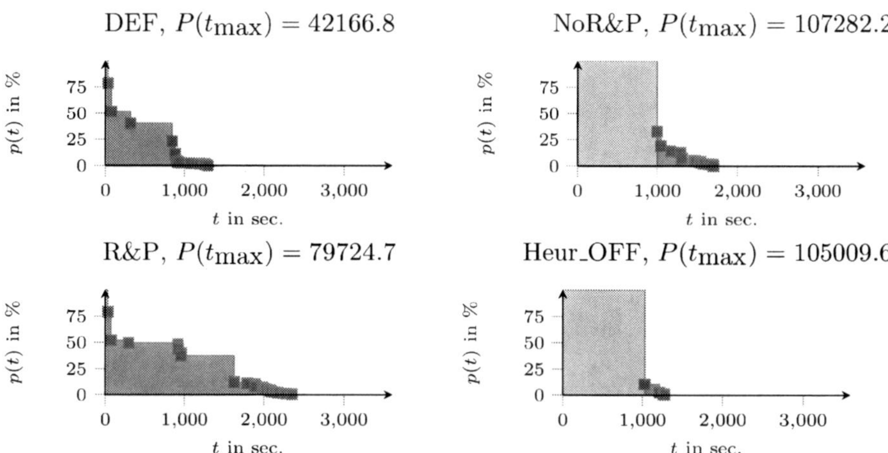

Fig. 1 The solving process depicted for the instance n3seq24

better is the expected quality of the incumbent solution if we stop the solver at an arbitrary point in time.

We used the benchmark set of the MIPLIB 2010 [5] as test set for our experiments. Since we are interested in the primal part of the solution process, we excluded the four infeasible instances triptim1, enlight14, ns1766074, and ash608gpia-3col; further, mspp16 was excluded since it terminated for memory reasons during presolving for all our tests. Thus, 82 test instances remained. We performed four different runs: SCIP without any primal heuristics (Heur_OFF), SCIP using only the rounding and propagation heuristics which are described in this paper (R&P), SCIP using all default heuristics except the ones described in this paper (NoR&P), and SCIP with default settings (DEF). All experiments were conducted with a time limit of one hour, a memory limit of 4 GB on a 3.00 GHz Intel® Core™2 Extreme CPU X9650 with 6144 KB Cache and 8 GB RAM.

Figure 1 exemplarily shows the primal gap function $p(t)$ for the four settings applied to the instance n3seq24. A square shows when a new primal solution is found and its quality. It can be seen that the two settings that use rounding and propagation heuristics find solutions earlier and hence have a smaller primal integral $P(t_{\max})$. For all four settings, SCIP found the optimal solution within an hour, but timed out without proving optimality. Interestingly, with disabled heuristics (Heur_OFF), the optimal solution was found in the smallest amount of time. Nevertheless, the behavior of the default settings (DEF) seems favorable since primal solutions of reasonable quality are found much earlier.

Table 1 shows aggregated results of our experiments. The first row shows the evaluation of the *normalized* primal integral over the entire testset. For each instance, the corresponding integral $P(t_{\max})$ is divided by the integral obtained with the setting Heur_OFF. This reference value is then used to compute the geometric mean $\phi(P)$ for each setting. The remainder of the table shows the geometric means

	DEF	NoR&P	R&P	Heur_OFF
$\phi(P)$	0.44	0.47	0.61	1.0
$\phi(t_1)$	8.54	17.48	10.88	57.66
$\phi(t_{\rm opt})$	215.63	218.18	236.92	263.46
$\phi(t_{\rm solved})$	712.90	676.10	746.48	838.60

Table 1 Computed mean values for all four settings

of the time until the first (t_1) and the best ($t_{\rm opt}$) primal solution were found and the mean solving time ($t_{\rm solved}$).

The advantage of rounding and propagation heuristics can be best seen in the time to first solution t_1: they are valuable for finding start solutions. The impact on the time to the best solution and the overall solving time is much smaller, the latter even showing a degradation. The primal integral $P(t_{\max})$ implies the following ranking: To get good primal solutions, using cheap and expensive heuristics together (DEF) is slightly better than using only expensive heuristics (noR&P), which is considerably better than using only cheap ones (R&P), which is much better than using no heuristics (Heur_OFF). This behavior is only partially expressed by the other measures.

Detailed computational results can be found in the online supplement [2].

References

1. T. Achterberg. *Constraint Integer Programming*. PhD thesis, Technische Universität Berlin, 2007.
2. T. Achterberg, T. Berthold, and G. Hendel. Rounding and propagation heuristics for mixed integer programming. ZIB-Report 11-29, Zuse Institute Berlin, 2011. http://vs24.kobv.de/opus4-zib/frontdoor/index/index/docId/1325/.
3. T. Berthold. Primal Heuristics for Mixed Integer Programs. Master's thesis, Technische Universität Berlin, 2006.
4. G. Hendel. New rounding and propagation heuristics for mixed integer programming. Bachelor's thesis, Technische Universität Berlin, 2011.
5. T. Koch, T. Achterberg, E. Andersen, O. Bastert, T. Berthold, R. E. Bixby, E. Danna, G. Gamrath, A. M. Gleixner, S. Heinz, A. Lodi, H. Mittelmann, T. Ralphs, D. Salvagnin, D. E. Steffy, and K. Wolter. MIPLIB 2010. *Mathematical Programming Computation*, 3(2):103–163, 2011.
6. C. Wallace. ZI round, a MIP rounding heuristic. *Journal of Heuristics*, 16:715–722, 2009.

Inverse Shortest Path Models Based on Fundamental Cycle Bases

Mikael Call and Kaj Holmberg

Abstract The inverse shortest path problem has received considerable attention in the literature since the seminal paper by Burton and Toint from 1992. Given a graph and a set of paths the problem is to find arc costs such that all specified paths are shortest paths. The quality of the arc costs is measured by the deviation from some ideal arc costs.

Our contribution is a novel modeling technique for this problem based on fundamental cycle bases. For 'LP compatible' norms we present a cycle basis model equivalent to the LP dual. The LP dual of our cycle basis model is a path based model that only requires a polynomial number of path constraints. This model is valid also for 'LP incompatible' norms. This yields the first polynomial sized *path* formulation of the original problem.

1 Introduction and Problem Formulation

The inverse shortest path problem (ISP) is the first inverse problem encountered in the OR literature. Burton and Toint [6] motivate the problem by applications in traffic modeling and seismic tomography. Since then, ISP has received considerable attention, e.g. [5, 12]. More recently, closely related variants have also been considered, e.g. [1–3, 7, 9, 10].

Problem Formulation Let $G = (N, A)$ be a biconnected digraph and $\mathscr{P} = \{P_k\}_1^K$ a set of paths. Denote the source and destination of P_k by (s_k, t_k) and all (s_k, t_k)-paths by $\mathscr{P}_k$. Given 'ideal' costs $\bar{c} \in \mathbb{Q}_+^A$, the problem is to find costs $c \in \mathbb{Q}_+^A$ such that all paths $P_1, \ldots, P_k$ are shortest paths and $||c - \bar{c}||$ is minimized for some norm. Using variables $c_{ij} \geq 0$ for all arcs, $(i, j) \in A$, yields a well known model of ISP.

Mikael Call
Linköping University, Department of Mathematics, SE-581 83 Linköping, e-mail: mikael.call@liu.se

Kaj Holmberg
Linköping University, Department of Mathematics, SE-581 83 Linköping, e-mail: kaj.holmberg@liu.se

D. Klatte et al. (eds.), *Operations Research Proceedings 2011*, Operations Research Proceedings, 77
DOI 10.1007/978-3-642-29210-1_13, © Springer-Verlag Berlin Heidelberg 2012

$$\min \ ||c - \bar{c}||$$
$$s.t. \ \sum_{(i,j)\in Q} c_{ij} - \sum_{(i,j)\in P_k} c_{ij} \geq 0, \qquad Q \in \mathscr{P}_k, \ k = 1,\ldots,K, \tag{1}$$
$$c_{ij} \geq 0, \qquad (i,j) \in A.$$

The number of constraints in (1) may be exponential. From a computational point of view, this can be handled by solving shortest path separation problems. In the cases, see below, where (1) can be transformed into a pure LP, this is equivalent to applying column generation to its dual. This approach is used for the L^1 norm in [12]. An alternative that also works for other norms is to use an equivalent extended formulation based on arc reduced cost constraints, cf. Section 5 in [6]. We present a similar model below.

Let A_t be the set of all arcs in a shortest path from some source to destination t, and $\bar{A}_t = A \setminus A_t$ its complement, i.e.

$$A_t = \{(i,j) \in A \mid (i,j) \in P_k \text{ and } t = t_k \text{ for some } k\}. \tag{2}$$

Denote by S_t the nodes spanned by A_t. Further, let $T \subseteq N$ be the set of destinations, i.e., $t \in T$ if and only if $A_t \neq \emptyset$.

For each $t \in T$ and $s \notin S_t$ we know that there must exist some unspecified shortest path from s to t. An auxiliary arc, (s,t), is introduced to represent this fact. Let D_t be the set of all auxiliary arcs with destination t and let $D = \bigcup_{t\in T} D_t$.

In effect, when G is augmented by the arcs in D, it becomes a multi-graph, i.e. there may be two arcs between a node pair $(i,j) \in N \times N$. We denote the induced multi-graph by $\tilde{G} = (N, \tilde{A})$, where $\tilde{A} = A \uplus D$ is a multi-arc set. Here, $\uplus$ is used to emphasize that arc multiplicities are taken into account in the union.

Observe that the arc set A_t induces an ingraph to t that spans all nodes in S_t and that D_t induces an instar to node t that spans all nodes in $A \setminus S_t$. Therefore, the arcs in $A_t \uplus D_t$ form an ingraph that necessarily contains a reverse spanning arborescence rooted at t. Intuitively, this is a shortest path arborescence to destination t.

The fundamental idea in existing polynomial formulations of ISP is that a path, P_k, is a shortest path if and only if all its arcs have reduced cost zero, w.r.t. a feasible node potential, π, associated with its destination, i.e. $c_{ij} + \pi_i - \pi_j = 0$ for $(i,j) \in P_k$. Since paths in $\mathscr{P}$ can have different destinations, one node potential, π^t, is required for each destination, $t \in T$. A node potential is feasible if no reduced cost is negative.

We also use a distance variable d_{st} for each auxiliary arc $(s,t) \in D_t$. This yields a polynomial ISP model.

$$\min \ ||c - \bar{c}||$$
$$\begin{aligned}
s.t. \ \ c_{ij} + \pi_i^t - \pi_j^t &= 0, \qquad & (i,j) \in A_t, \ t \in T, \\
c_{ij} + \pi_i^t - \pi_j^t &\geq 0, \qquad & (i,j) \in \bar{A}_t, \ t \in T, \\
d_{st} + \pi_s^t - \pi_t^t &= 0, \qquad & (s,t) \in D_t, \ t \in T, \\
d_{ij} + \pi_i^t - \pi_j^t &\geq 0, \qquad & (i,j) \in D_t, \ t \in T, \\
c_{ij} &\geq 0, \qquad & (i,j) \in A, \\
d_{st} &\geq 0, \qquad & (s,t) \in D.
\end{aligned} \tag{3}$$

An advantage with (3) over (1) is that it has a polynomial number of constraints. Note that (3) is also more general than (1) if the sets A_t are not restricted to connected ingraphs as in (2). This generalization is important if ISP is a pricing problem in a branch and price scheme. However, disconnectedness can sometimes cause severe optimality condition issues, cf. [8].

The only difference between (3) and other arc reduced cost models in the literature is the presence of the auxiliary arc variables d. It turns out that this is important to obtain a polynomial *path* formulation. More precisely, the existence of a spanning arborescence is important.

Let us restrict our attention to 'LP compatible' norms, i.e. $||\cdot||_1$ and $||\cdot||_\infty$. In this case, standard techniques can be used to linearize the objective. We do not carry out this reformulation explicitly here. We write (3) in its non-linear form but still treat it as an LP, i.e. we refer to the LP obtained when the linearization is carried out. Consequently, (3) can be solved directly as an LP.

Instead of solving (3) directly we consider its LP dual. For each destination t and each arc $(i,j) \in \tilde{A} = (A_t \cup \bar{A}_t) \uplus D_t$, the dual variable associated with the corresponding reduced cost constraint is θ_{ij}^t. The auxiliary (linearizable) constraint, $v_{ij} = |c_{ij} - \bar{c}_{ij}|$, used for the objective as mentioned above has dual variable γ_{ij}. Finally, $||\cdot||^*$ denotes the dual norm of $||\cdot||$. Observe that γ_{ij} is unrestricted in sign for all $(i,j) \in A$ and that θ_{ij}^t is unrestricted in sign for arborescence arcs, i.e. when $(i,j) \in A_t$ or $(i,j) \in D_t$.

$$
\begin{aligned}
\max \quad & \sum_{(i,j)\in A} \gamma_{ij}\bar{c}_{ij} \\
s.t. \quad & \sum_{j:(i,j)\in A \uplus D_t} \theta_{ij}^t - \sum_{j:(j,i)\in \uplus D_t} \theta_{ji}^t = 0, && i \in N, t \in T, \\
& \sum_{t\in T} \theta_{ij}^t \leq -\gamma_{ij}, && (i,j) \in A, \\
& \sum_{t\in T} \theta_{ij}^t \leq 0, && (i,j) \in D, \\
& \theta_{ij}^t \geq 0, && (i,j) \in \bar{A}_t, t \in T, \\
& ||\gamma||^* \leq 1.
\end{aligned}
\tag{4}
$$

Except for the unrestricted variables and the unit ball constraint on γ, the structure in (4) is very similar to a multicommodity flow problem, We exploit the existence of unrestricted variables in a cycle basis reformulation.

2 A Fundamental Cycle Basis Formulation

The advantage of a cycle basis formulation for circulation problems is that the flow conservation constraints are automatically satisfied. Hence, only capacity constraints and flow bounds will have to be modeled. A major drawback is that the flow bounds will no longer be simple variable bounds, but rather generalized bound constraints. For (4) we overcome this flaw. The key to this improvement is to carefully take unrestricted flow variables into account when choosing the cycle bases.

We obtain an equivalent model without flow balance constraints. Our preliminary computational experiments indicate that the compact model can be solved more efficiently. The formulation also has interesting relations to constraint/column generation approaches to solve ISP.

Fundamental Cycle Bases To present our cycle basis model we first introduce fundamental cycle bases. For a comprehensive treatment, we refer to [4, 11].

For each $t \in T$, let $R_t \subseteq A_t \uplus D_t$ be a reverse spanning arborescence rooted at t. For any $s \in N$, there is a unique (s,t)-path in E_t. Denote this path by $\tilde{P}_s^t$. Each arc $(u,v) \notin E_t$ induces an oriented circuit $C_{uv}^t = F_{uv}^t \cup B_{uv}^t$ where $F_{uv}^t = (u,v) \cup \tilde{P}_v^t \setminus \tilde{P}_u^t$ is a 'forward' path and $B_{uv}^t = \tilde{P}_u^t \setminus \tilde{P}_v^t$ is a 'backward' path. This circuit is the fundamental cycle induced by arc (u,v). The incidence vector, $\gamma_{t,uv} \in \{-1, 0, 1\}^{\tilde{A}}$, associated with C_{uv}^t has entry 1 (-1) in component (i,j) if it is a forward (backward) arc of C_{uv}^t.

Let Γ_t be the matrix formed by all incidence vectors induced by an arc $(u,v) \notin E_t$. Any circulation, $\theta \in \mathbb{R}^{\tilde{A}}$ can be expressed as $\theta = \Gamma_t x$ for some $x \in \mathbb{R}^{\tilde{A} \setminus E_t}$, i.e. Γ_t is a basis for the cycle space of $\tilde{G}$. A cycle matrix example is given in Figure 1.

The Fundamental Cycle Basis Model Our aim is to derive a compact circulation based model equivalent to (4). We first illustrate how cycle bases are used to model circulations in $\tilde{G}$.

For any $t \in T$, denote by x_{uv}^t the amount of flow sent in the fundamental cycle induced by $(u,v) \notin E_t$ and let θ_{ij}^t be the circulation induced by $x \in \mathbb{R}^{\tilde{A} \setminus E_t}$. This yields the bijective relation $\theta^t = \Gamma_t x^t$, between x^t and θ^t, or equivalently

$$\theta_{ij}^t = \sum_{(u,v)\in\tilde{A}\setminus E_t} \gamma_{t,uv}^{ij} x_{uv}^t, \qquad (i,j) \in \tilde{A}. \tag{5}$$

In particular, note that $\theta_{ij}^t = x_{ij}^t$ for arcs $(i,j) \in \tilde{A} \setminus E_t$ not in the arborescence E_t.

Using the relation in (5) we can translate the part of (4) that involves θ-variables. By construction, θ^t defined by (5) satisfy the node balance constraints. To model capacity constraints it suffices to consider the total flow on arc $(i,j) \in \tilde{A}$,

$$\sum_{t\in T} \theta_{ij}^t = \sum_{t\in T} \sum_{(u,v)\in\tilde{A}\setminus E_t} \gamma_{t,uv}^{ij} x_{uv}^t, \qquad (i,j) \in \tilde{A}. \tag{6}$$

For lower bound constraints we note that,

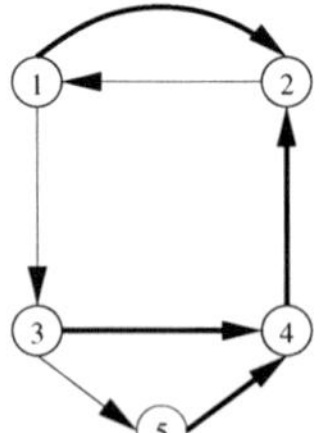

Fig. 1 The fundamental cycle matrix associated with the arborescence rooted at node 2 (bold arcs) is given in the table. Bold entries correspond to arborescence arcs.

	C_{13}^2	C_{21}^2	C_{35}^2
12	-1	1	
13	1		
21		1	
34	1		-1
35			1
42	1		
54			1

$$\theta_{ij}^t = x_{ij}^t, \qquad (i,j) \in \tilde{A} \setminus E_t. \tag{7}$$

Finally, let $\tilde{\gamma}$ be the column vector $(-\gamma', 0')' \in \mathbb{R}^{\tilde{A}}$. Using the above relations, the cycle basis model of (4) becomes

$$\begin{aligned} \max \ & \bar{c}'\gamma \\ s.t. \ & \sum_{t \in T} \Gamma_t x^t \leq \tilde{\gamma}, \\ & x_{uv}^t \geq 0, \qquad (u,v) \in \tilde{A} \setminus (A_t \uplus D_t), \ t \in T, \\ & \|\gamma\|^* \leq 1. \end{aligned} \tag{8}$$

Observe that x_{uv}^t is unrestricted for $(u,v) \in (A_t \uplus D_t) \setminus E_t$.

Theorem 1. *Models* (4) *and* (8) *are equivalent.*

We emphasize that the choice of arborescences is very important. The crucial observation to obtain a compact model is that the flow, θ_{ij}^t is unrestricted for arcs in the arborescence, i.e. when $(i,j) \in E_t$. Utilizing this, we avoid generalized lower bound constraint that would otherwise be required on these arcs, cf. (5). An arc that requires a lower bound constraint is never in the arborescence, therefore $\theta_{ij}^t = x_{ij}^t$, cf. (7). Ordinary nonnegativity constraints suffice to model flow lower bounds for these variables.

Note that the existence of an arborescence relies on the auxiliary arcs D in (3).

A Fundamental Path Basis Model Consider the dual of (8). Recall that a column in Γ_t corresponds to a fundamental cycle $C_{uv}^t = F_{uv}^t \cup B_{uv}^t$ induced by an arc $(u,v) \notin E_t$. Also, $F_{uv}^t = (u,v) \cup \tilde{P}_v^t \setminus \tilde{P}_u^t$ and $B_{uv}^t = \tilde{P}_u^t \setminus \tilde{P}_v^t$ are paths from u to a common apex.

From this description, the following *path based* LP dual of (8) is derived.

$$\begin{aligned} \min \ & \|c^A - \bar{c}\| \\ s.t. \ & c_{uv} + \sum_{(i,j) \in P_v^t \setminus P_u^t} c_{uv} - \sum_{(i,j) \in P_u^t \setminus P_v^t} c_{uv} \geq 0, \qquad (u,v) \in \tilde{A} \setminus (A_t \uplus D_t), \ t \in T, \\ & c_{uv} + \sum_{(i,j) \in P_v^t \setminus P_u^t} c_{uv} - \sum_{(i,j) \in P_u^t \setminus P_v^t} c_{uv} = 0, \qquad (u,v) \in (A_t \uplus D_t) \setminus E_t, \ t \in T, \\ & c_{uv} \geq 0, \qquad (u,v) \in \tilde{A}, \end{aligned} \tag{9}$$

where c^A denotes the restriction of c to arcs in G.

Theorem 2. *Models* (1), (3) *and* (9) *are equivalent.*

For 'LP compatible' norms, Theorem 2 follows directly from basic LP theory. However, it straightforward to obtain (9) from (3) by eliminating the unrestricted variables, i.e. the node potentials. This proves that *the equivalence of models* (1), (3) *and* (9) *holds for any norm.*

3 Further Remarks and Conclusion

The most interesting remark on (9) in relation to (1) may be that it suffices to use a fixed, polynomial size, subset of paths. This also implies that no separation problem has to be solved. Alternatively, the separation problem can be solved by enumerating only the fundamental paths in (9). Equivalent statements hold for the dual problems in terms of cycle pricing and column generation. We have observed in preliminary computational tests that (8) can be solved more efficiently than (3) and (4). We have not compared (8) to the constraint/column generation approach. However, our models to some extent reduce the need for these latter approaches.

In this paper, we used LP duality to derive (9). However, we pointed out that it is a valid formulation of ISP for any norm. In particular, for $|| \cdot ||_2$ considered in [6]. Hence, there is a polynomial size path based formulation for ISP.

Finally, we also mention that our model above uses reverse arborescences. It is of course feasible to use arborescences instead, or a combination thereof. What changes is the set of auxiliary arcs and their 'meaning'.

Conclusion We present a novel formulation of the inverse shortest path problem (and its LP-dual, when applicable) based on fundamental paths (and fundamental cycle bases). This approach leads to polynomial sized models that can be solved more efficiently than previous models from the literature. It also yields the first polynomial size *path* formulation of the problem.

References

1. R.K. Ahuja and J.B. Orlin. Inverse optimization. *Operations Research*, pages 771–783, 2001.
2. A. Bley. Inapproximability results for the inverse shortest paths problem with integer lengths and unique shortest paths. *Netw.*, 50(1):29–36, 2007.
3. A. Bley, B. Fortz, E. Gourdin, K. Holmberg, O. Klopfenstein, M. Pióro, A. Tomaszewski, and H. Ümit. Optimization of OSPF routing in IP networks. *Graphs and Algorithms in Communication Networks*, pages 199–240, 2010.
4. B. Bollobás. *Modern graph theory*. Springer Verlag, 1998.
5. D. Burton. *On the inverse shortest path problem*. PhD thesis, 1993.
6. D. Burton and P.L. Toint. On an instance of the inverse shortest paths problem. *Mathematical Programming*, 53:45–61, 1992.
7. M. Call. Inverse Shortest Path Routing Problems in the Design of IP Networks. Linköping Studies in Science and Technology. Thesis No. 1448, 2010.
8. M. Call and K. Holmberg. Complexity of Inverse Shortest Path Routing. In *Network Optimization: International Network Optimization Conference (INOC 2011), Hamburg, Germany*, volume 6701. Springer, 2011.
9. T. Cui and D.S. Hochbaum. Complexity of some inverse shortest path lengths problems. *Networks*, 56(1):20–29, 2010.
10. C. Heuberger. Inverse combinatorial optimization: A survey on problems, methods, and results. *Journal of Combinatorial Optimization*, 8:329–361, 2004.
11. C. Liebchen and R. Rizzi. Classes of cycle bases. *Discrete Applied Mathematics*, 155(3):337–355, 2007.
12. J. Zhang, Z. Ma, and C. Yang. A column generation method for inverse shortest path problems. *Mathematical Methods of Operations Research*, 41(3):347–358, 1995.

Parallel algorithms for the maximum flow problem with minimum lot sizes

Mujahed Eleyat, Dag Haugland, Magnus Lie Hetland, and Lasse Natvig

Abstract In many transportation systems, the shipment quantities are subject to minimum lot sizes in addition to regular capacity constraints. This means that either the quantity must be zero, or it must be between the two bounds. In this work, we prove that the maximum flow problem with minimum lot-size constraints on the arcs is strongly NP-hard, and we enhance the performance of a previously suggested heuristic. Profiling the serial implementation shows that most of the execution time is spent on solving a series of regular max flow problems. Therefore, we develop a parallel augmenting path algorithm that accelerates the heuristic by an average factor of 1.25.

1 Introduction

Planning models for production, storage and transportation of goods often have to reflect minimum lot size constraints. The essence of such constraints is that if the activity level is not above some given lower bound, the operation in question must be inactive. Frequently, the computational consequence is that an otherwise tractable model becomes NP-hard.

In this work, we study the maximum flow problem subject to minimum lot size constraints. The purpose of the work is to suggest efficient computational meth-

Mujahed Eleyat
Miriam AS, Halden, Norway & IDI, NTNU, Trondheim, Norway, e-mail: mujahed@miriam.as

Dag Haugland
Department of Informatics, University of Bergen, Bergen, Norway, e-mail: Dag.Haugland@ii.uib.no

Magnus Lie Hetland
Department of Computer and Information Science (IDI), Norwegian University of Science and Technology (NTNU), Trondheim, Norway, e-mail: mlh@idi.ntnu.no

Lasse Natvig
Department of Computer and Information Science (IDI), Norwegian University of Science and Technology (NTNU), Trondheim, Norway, e-mail: lasse@idi.ntnu.no

D. Klatte et al. (eds.), *Operations Research Proceedings 2011*, Operations Research Proceedings, 83
DOI 10.1007/978-3-642-29210-1_14, © Springer-Verlag Berlin Heidelberg 2012

ods that, despite the proven intractability of the problem, are able to produce near-optimal solutions with modest computational effort. To that end, we suggest a heuristic method based on a parallel implementation of a max flow algorithm.

It is well known that the maximum flow problem and its variants are challenging to parallelize. In fact, the maximum flow is P-complete, even for acyclic graphs [5, p. 152], which means that finding a highly parallel algorithm is very unlikely. This does not mean, however, that achieving speedups with a concurrent solution is impossible. Indeed, several such solutions exist. These algorithms tend to avoid the augmenting path approach, which at least intuitively seems inherently sequential. The other common approach, of pushing a preflow through the network [9] seems more amenable to parallelism. For example, already in 1982, Shiloach introduced a special-purpose concurrent max flow algorithm (with cubic sequential running time) based on this idea [11]. The classic algorithm of Goldberg [3] was also intended to be parallel even in its original version, and its parallel execution has since been improved [1]. The heuristic method suggested in the current work is based upon a parallel augmenting path algorithm.

1.1 Problem definition

We let $G = (N, A)$ be a directed graph with node set N and arc set A, and we let ℓ and u be non-negative integer vectors of respectively lower and upper flow bounds. We assume that G contains a unique source $s \in N$ and a unique sink $t \in N$, and that A contains a circulation arc (t, s) with $\ell_{ts} = 0$ and $u_{ts} = \infty$ from the sink to the source.

Let $x \in \mathbb{Z}_+^A$ be a flow vector, and consider the problem of maximizing x_{ts} such that for all $(i, j) \in A$, either $x_{ij} = 0$ or $x_{ij} \in [\ell_{ij}, u_{ij}]$.
Define $F(G) = \left\{ x \in \mathbb{Z}_+^A : \sum_{j:(i,j)\in A} x_{ij} - \sum_{j:(j,i)\in A} x_{ji} = 0 \ \forall i \in N \right\}$. For any arc set $X \subseteq A$, let $F_X(G) = \left\{ x \in F(G) : x_{ij} \in [\ell_{ij}, u_{ij}] \ \forall (i,j) \in X, x_{ij} = 0 \ \forall (i,j) \in \bar{X} \right\}$, where $\bar{X} = A \setminus X$. The problem is then expressed as

$$[P] \quad \max_{X,x} \{ x_{ts} : X \subseteq A, x \in F_X(G) \}$$

We say that X is feasible if $F_X(G) \neq \emptyset$. We let G_X denote the subgraph with node set N and arc set X.

2 Computational complexity

That network flow problems with semi-continuous flow variables are NP-hard is folklore. A formal proof was given in [6], and in this section, we improve the result by demonstrating strong NP-hardness.

Proposition 1. *Problem [P] is strongly NP-hard.*

Proof. The proof is by a polynomial reduction from EXACT COVER BY 3-SETS (X3C), which is strongly NP-complete [**?**]. For the reduction to confer NP-hardness in the *strong* sense, we must also make sure that all numbers are polynomially bounded in the problem size and maximal numerical magnitude of the original (trivially satisfied in the following). Given a finite set $Y = \{y_1, \ldots, y_n\}$, where $n/3 = q$ is an integer, and a set $C = \{C_1, \ldots, C_m\}$ of subsets of Y, where $|C_1| = \cdots = |C_m| = 3$, the X3C problem is to decide whether there exists some $C' \subseteq C$ of pairwise disjoint subsets such that C' covers Y. Consider any X3C-instance, and define the directed graph with node set $N = \{s, v_1^c, \ldots, v_m^c, v_1^y, \ldots, v_n^y, t', t\}$. Define the arc sets: $A_1 = \{(s, v_1^c), \ldots, (s, v_m^c)\}$, $A_2^i = \{(v_i^c, v_j^y) : y_j \in C_i\}$ $(i = 1, \ldots, m)$, $A_3 = \{(v_1^y, t'), \ldots, (v_n^y, t')\}$, and $A_4 = \{(t', t)\}$, and $A = A_1 \cup A_2^1 \cup \cdots \cup A_2^m \cup A_3 \cup A_4$. An illustration of $G = (N, A)$ is given by Parmar in his discussion of the subject [10], where it is used to prove NP-hardness of the EQUAL-SPLIT NETWORK FLOW PROBLEM. All arcs in A_1 have lower and upper flow bounds 1, whereas all other arcs, except (t', t), have zero lower flow bound and upper flow bound equal to $\frac{1}{3}$. Arc (t', t) has zero lower flow bound and capacity q. It is easily verified that the X3C-instance is a yes-instance if and only if the optimal solution to problem $[P]$ applied to G is to send q units of flow along arc (t', t). Hence, there exists a polynomial reduction from X3C to the decision version of $[P]$, and the proof is complete. □

3 A heuristic method

Since Proposition 1 discourages an exact algorithm, we will rather apply a heuristic solution method that may fail to find the optimal solution. The method is discussed in detail in [6], and reviewed briefly here. The idea is to start with $X = \{(t, s)\}$, and gradually extend the set. Finding a larger, feasible X is non-trivial, and we will consider two kinds of extensions of X: (1) Extending an infeasible X in order to reduce constraint violations, and (2) extending a feasible X in order to increase x_{ts}.

Checking whether X is feasible, is accomplished by defining the digraph $G_X' = (N', A')$, with the node set $N' = N \cup \{s', t'\}$ consisting of all nodes in G and a new source s' and a new terminal t'. The arc set A' consists of the arcs in X, the circulation arc (t', s'), and the arcs (s', i) and (i, t') for all $i \in N$. Define the arc capacities $c_{ij} = u_{ij} - \ell_{ij}$ for all $(i, j) \in X$, $c_{t's'} = +\infty$, $c_{s'i} = \sum_{j:(j,i) \in X} \ell_{ji}$, and $c_{it'} = \sum_{j:(i,j) \in X} \ell_{ij}$ for all $i \in N$.

Proposition 2. *Assume $x' \in \mathbb{Z}^{A'}$ is a solution to the* MAXIMUM FLOW PROBLEM *in G_X' with source s', sink t', and arc capacities c. Then the arc set X is feasible if and only if $x_{t's'}' = \sum_{(i,j) \in X} \ell_{ij}$.*

Proof. See Theorem 10.2.1 in [8]. □

For feasibility checking, Proposition 2 suggests that a max flow problem is solved in each iteration of our heuristic (Algorithm 1). If X is feasible, we find an $x \in F_X(G)$ maximizing x_{ts}, and compare it to the best solution so far. We also try to improve x by searching for a flow-augmenting path in G. If X is infeasible, then $x'_{t's'} < \sum_{(i,j)\in X} \ell_{ij}$, and we search for a path from s' to t' in G' that can increase $x'_{t's'}$. In conclusion, steps 3, 5, 9 and 11 involve computation of augmenting paths in G or G'. For a fast implementation of the heuristic, we give a parallel augmenting path algorithm in the next section.

Algorithm 1 Heuristic($G,\ell,u,(t,s)$)

1:   $X \leftarrow \{(t,s)\}$, $z^* \leftarrow 0$, $X^* \leftarrow X$
2: **repeat**
3:      Compute x' as given in Proposition 2.
4:      **if** $x'_{t's'} = \sum_{(i,j)\in X} \ell_{ij}$ **then**
5:          Compute $x \in \arg\max\{x_{ts} : x \in F_X(G)\}$
6:          **if** $x_{ts} > z^*$ **then**
7:              $z^* \leftarrow x_{ts}$, $X^* \leftarrow X$
8:          **end if**
9:          Let B be the set of arcs of a flow-augmenting path from s to t in G
10:     **else**
11:          Let B be the set of arcs of a flow-augmenting path from s' to t' in G'
12:     **end if**
13:      $X \leftarrow X \cup B$
14: **until** $B = \emptyset$
15: **return** X^*

---

4 Parallel implementation

We chose to parallelize the augumenting path method and not the push-relabel one because the latter can not be used to perform step 5 of Algorithm 1. Another reason is that recent research [7] shows that parallel push-relabel methods have low speed-up in sparse instances.

Algorithm 2 shows our parallelization technique. In step 1, a set $S \subseteq N$ of start nodes is selected such that they have the same distance from the source. The selection is achieved using a modified version of breadth first search (BFS). In step 3, the paths from the source to $m \leq |S|$ start nodes are found using one BFS operation, and paths to the sink are found using m BFS operations.

The efficiency of the algorithm is greatly dependent on the residual capacities of arcs shared by two or more different paths. Ideally, the residual capacity of each arc a should be no smaller than the sum of the flow capacities of augmenting paths intersecting a.

Algorithm 2 Parallel path augmentation max flow algorithm

---

1. Main thread: Select a set S of start nodes in the residual graph $G(x)$
repeat
    2. Main thread: Randomly choose m nodes from S
    3. In parallel: main thread finds the shortest paths from the source to the m nodes while other threads find the shortest paths from the m nodes to the sink
    4. Main thread: Perform m flow augmentations through the shortest paths
until no more paths

---

5 Performance and conclusion

We implemented the parallel algorithm using C++ and Pthreads and executed it on 2×6 Opteron processor. The instances were obtained using the RMFGEN-generator of Goldfarb [4] and 40% of the arcs were selected randomly to have nonzero lower bounds.

The speedup, calculated using the execution time of the serial algorithm as a baseline, is shown in Table 1. The speedup is not high mainly due to the fact that the paths found in parallel usually share an arc that gets saturated by augmenting flow through one of the paths. Moreover, and because we need to synchronize the threads every time they perform the small task of finding paths in a small sparse subgraph, parallel overhead is another reason for little speedup, especially when solving relatively small problems. In fact, this is the reason we chose not to parallelize BFS, which requires synchronizing the threads many times while finding a single path.

The performance can be enhanced by finding paths that don't share arcs with small residual capacity. Finding such paths may not be easy and might require a lot of synchronization overhead. However, heuristic that let threads select different groups of arcs may also lead to better performance without increasing parallel overhead.

Table 1 Performance Results

Instance	a	b	c_1	c_2	$|N|$	$|A|$	Speedup(6 cores)
$G1$	25	15	2000	10000	9375	44750	1.12
$G2$	30	10	2000	10000	9000	42900	1.28
$G3$	20	18	2000	10000	7200	34160	1.45
$G4$	20	15	2000	10000	6000	28400	1.13

References

1. Anderson, R.J., Setubal, J.C.: On the parallel implementation of Goldberg's maximum flow algorithm. In: Proceedings of the fourth annual ACM symposium on Parallel algorithms and architectures, SPAA'92, pp. 168-177, ACM (1992)
2. Garey, M., Johnson, D.: Computers and Intractability: A Guide to the Theory of NP-Completeness. Freeman (1979)
3. Goldberg, A.V., Tarjan, R.E.: A new approach to the maximum-flow problem. J. ACM. **35**, 921–940 (1988)
4. Goldfarb, D., Grigoriadis, M.D.: A Computational Comparison of the Dinic and Network Simplex Methods for Maximum Flow. Annals of Operations Research. **78**, 83–123 (1988)
5. Greenlaw, R., Hoover, H.J., Ruzzo, W.L.: Limits to Parallel Computation: P-Completeness Theory. Oxford University Press (1995)
6. Haugland, D., Eleyat, M., Hetland, M.L.: The maximum flow problem with minimum lot sizes. In: Proceedings of the Second international conference on Computational logistics, ICCL'11, pp. 170-182. Springer-Verlag (2011)
7. Hong, B., He, Z.: An Asynchronous Multithreaded Algorithm for the Maximum Network Flow Problem with Nonblocking Global Relabeling Heuristic. IEEE Trans. Parallel Distrib. Syst. **22**, 1025–1033 (2011)
8. Jungnickel, D.: Graphs, Networks and Algorithms. Springer (2008)
9. Karzanov, A.V.: Determining the maximal flow in a network by the method of preflows. Soviet Math. Dokl. **15**, 434–437 (1975)
10. Parmar, P.: Integer Programming Approaches for Equal-Split Network Flow Problems. PhD thesis, Georgia Institute of Technology (2007)
11. Shiloach, Y., Vishkin, U.: An $\mathcal{O}(n^2 \log n)$ parallel max-flow algorithm. Journal of Algorithms. **3**, 128–146 (1982)

Estimating trenching costs in FTTx network planning

Sebastian Orlowski, Axel Werner, and Roland Wessäly

Abstract In this paper we assess to which extent trenching costs of an FTTx network are unavoidable, even if technical side constraints are neglected. For that purpose we present an extended Steiner tree model. Using a variety of realistic problem instances we demonstrate that the total trenching cost can only be reduced by about 5 percent in realistic scenarios. This work has been funded by BMBF (German Federal Ministry of Education and Research) within the program "KMU-innovativ".

1 Introduction

Broadband internet access is a key infrastructure since both the demand for high-data services rates and the private and commercial dependency on all-time broad-band connectivity are continuously rising. In this light, several telecommunication carriers are already realizing *fiber-to-the-home* (FTTH) or *fiber-to-the-building* (FTTB) projects, and in addition, many companies are seriously investigating how such a network could be deployed. Given the vast investments that have to be made – an estimated amount of 40 to 60 billion Euro in total for Germany alone – the careful preparation of such a deployment project is indispensable.

The planning of an *FTTx network* is a highly complex task comprising numerous interesting optimization problems. In the German BMBF-project FTTX-PLAN (see [2], cf. [8], [9]), we developed methods to compute technologically feasible and cost-optimized FTTx networks including hardware, installation and trenching cost. Since the latter cost are considered to be the lion's share of the expenses involved in the roll-out of an FTTx network (cf. [11, Section 6.1]), we focus in this paper on

Sebastian Orlowski, Roland Wessäly
atesio GmbH, Bundesallee 89, 12161 Berlin, Germany
e-mail: orlowski,wessaely@atesio.de

Axel Werner
Zuse Institute Berlin, Takustraße 7, 14195 Berlin, Germany
e-mail: werner@zib.de

D. Klatte et al. (eds.), *Operations Research Proceedings 2011*, Operations Research Proceedings, 89
DOI 10.1007/978-3-642-29210-1_15, © Springer-Verlag Berlin Heidelberg 2012

an evaluation of the incurred trenching cost in overall cost-optimized networks as computed by the FTTX-PLAN software.

To this end, we use an extended Steiner tree model to determine lower bounds on the trenching costs for a given FTTx instance. This model is presented in Section 3, after a high-level description of the FTTx network planning problem is given in Section 2. Finally, Section 4 compares trenching costs in solutions obtained from FTTX-PLAN with the lower bounds obtained by the Steiner tree model on various realistic test instances.

2 Problem formulation

The FTTx network planning problem can be described as the task to connect a number of given customers to central offices of the telecommunication carrier, using optical fibers and various active and passive components. Fibers can be laid out in different types of cables which themselves are direct-buried or embedded into different types of ducts, which are eventually buried into the ground along specified trails of the deployment area.

The topological structure of the input is represented by the *trail network*, an undirected graph whose edges designate the trails along which connections (fibers, cables, ducts) can be laid. To each edge is assigned a cost value that determines the trenching costs for the trail in question. The trails can also have existing infrastructure, such as dark fibers or ducts, which can be used for planning.

Some nodes in the trail network can be of a special type – each such node is associated with a setup cost which is incurred if it is included in the final network. Figure 1 shows an example of a trail network with special nodes BTPs and COs.

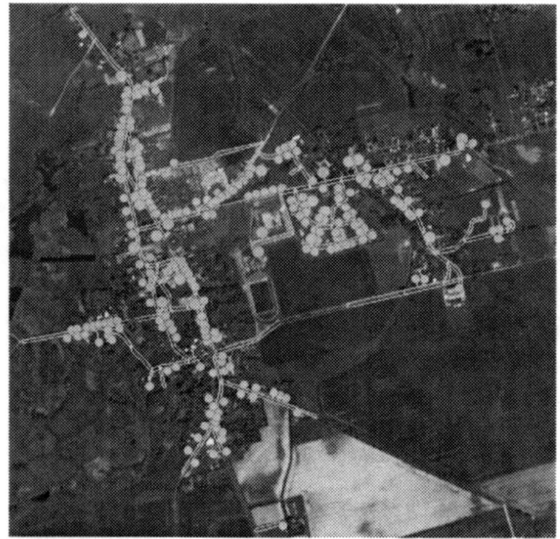

Fig. 1 Trail network of an instance, projected onto a satellite image of the deployment area. Yellow dots indicate BTPs, red diamonds possible CO locations; trails are colored according to their trenching costs – green for low, red for high costs.

- BTPs (*Building Termination Points*) represent potential customer locations. Each such location contains a number of residents of (possibly) different types, with certain demand values. These comprise several parameters, such as the number of required fibers.
- DPs (*Distribution Points*) are intermediate locations to (logically) connect customer locations to central offices. At DPs joint closures and splitters can be installed in order to concentrate fibers from individual customers into cables with a higher number of fibers.
- COs (*Central Offices*) are locations where customer connections are terminated using active equipment.

Global parameters formulate requirements to be fulfilled by any feasible network. These include length restrictions for optical connections, bounds on the number of customers connected to a CO or another concentrator location, or targeted percentage rates of connected customers.

Eventually, a hardware catalogue describes the usable active and passive hardware components – with investment and installation costs. These components impose capacity restrictions, for instance, on the number of cables or fibers that can be attached to a joint closure or the number of downstream ports that are available at a splitter. This further restricts the feasibility of a technical valid solution network.

The software developed within FTTX-PLAN computes cost-optimized solutions to the described planning problem using a variety of methods in different stages, including IP models and heuristics.

3 Estimating trenching costs

Costs for trenching are the most substantial budget item in the deployment of an FTTx network. Given such a network it can naturally be asked to which extent the incurred trenching costs are unavoidable since all BTP locations must be topologically connected to a CO. To answer this question, we compute lower bounds on the trenching costs by solving an extended Steiner tree problem that connects all BTPs to COs, where the CO locations must obey additional capacity constraints and incur setup cost.

Let (V,A) be the directed graph that is constructed from the trail network by replacing each edge $e \in E$ with two arcs (with the end-nodes of e) in opposite directions and augmenting this graph with an artificial root node (representing a virtual CO) connected by artificial arcs to every potential CO location. Each artificial arc a has a capacity k_a (the maximal number of BTPs to be connected to the CO in question) and a cost c_a (the setup cost of the CO). Let $A_0 \subset A$ denote the set of artificial arcs. The problem to solve is to find a cost-minimal tree in (V,A) connecting all BTPs to the artificial root node.

The following mixed-integer-programming flow formulation for this problem can be given, where V_B denotes the set of BTP locations, all of which have to be connected, N_C the maximal number of COs to open, $\delta^+(v)$ and $\delta^-(v)$ the set of outgoing and incoming arcs of node $v \in V$, respectively, and e^+ and e^- the two opposite arcs that originated from an edge $e \in E$.

$$\min \quad \sum_{e \in E} c_e w_e + \sum_{a \in A_0} c_a x_a$$

$$\text{s.t.} \quad \sum_{a \in \delta^-(v)} f_a - \sum_{a \in \delta^+(v)} f_a = \begin{cases} 1 & \text{if } v \in V_B \\ 0 & \text{otherwise} \end{cases} \quad \forall v \in V \tag{1}$$

$$f_a \leq |V_B| x_a \qquad \forall a \in A \tag{2}$$

$$x_{e^+} + x_{e^-} = w_e \qquad \forall e \in E \tag{3}$$

$$\sum_{a \in \delta^-(v)} x_a = 1 \qquad \forall v \in V_B \tag{4}$$

$$\sum_{a \in \delta^-(v)} x_a \leq 1 \qquad \forall v \in V \setminus V_B \tag{5}$$

$$\sum_{a \in \delta^-(v)} x_a \leq \sum_{a \in \delta^+(v)} x_a \qquad \forall v \in V \setminus V_B \tag{6}$$

$$\sum_{a \in \delta^-(v)} x_a \geq x_{a'} \qquad \forall v \in V \setminus V_B,\, a' \in \delta^+(v) \tag{7}$$

$$f_a \leq k_a \qquad \forall a \in A_0 \tag{8}$$

$$\sum_{a \in A_0} x_a \leq N_C \tag{9}$$

$$f_a \geq 0 \qquad \forall a \in A \tag{10}$$

$$x_a \in \{0,1\} \qquad \forall a \in A \tag{11}$$

$$w_e \in \{0,1\} \qquad \forall e \in E \tag{12}$$

Constraints (4) – (7) are the so-called *flow-balance inequalities* from [5]. Note that most planning and capacity restrictions are relaxed in this model, with the sole exception of the number of BTPs connected to a CO location and the total number of COs to open (constraints (8) and (9)). Furthermore, no kind of length restrictions are respected; since connection lengths have to have a strict upper bound in an optical fiber network, this is quite a severe relaxation from a practical point of view. Therefore we cannot expect that lower bounds on trenching costs obtained with the extended Steiner tree model can be realized in practically feasible FTTx networks.

Finally, we note that similar models have been presented for a slightly different problem involving existing copper infrastructure, the (Capacitated) Connected Facility Location Problem; see [6] for a recent contribution and [4] for an extensive treatment of this subject. Steiner tree problems themselves have been studied for quite some time and models similar to the one used here have been given in the literature; see [3, Section 9.6] or [10]. In [7] a multi-commodity flow formulation is mentioned, which can be seen as the basis of the model presented above.

4 Results and conclusion

We assessed trenching costs in 7 instances, given in Table 1. The first three trail networks were artificially generated using GIS information from OpenStreetMap [1]. The last four are modified networks provided by different German city carriers. Figure 2 shows an example of an FTTX-PLAN solution and the trenched trails attaining the lower bound for the instance a2.

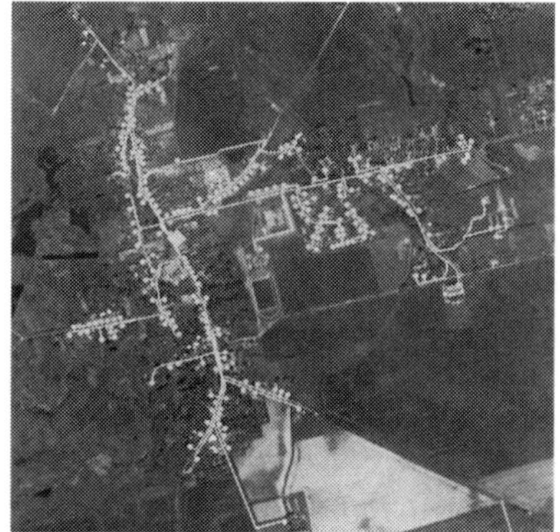

Fig. 2 Solution network (left) to instance a2 and trenched trails attaining the lower bound (right).

Table 1 shows the results from the computations. Besides the size of the instances, we have given the trenching costs incurred by the solution networks, the lower bound obtained with the model in Section 3, and the relative gap between these two values. Additionally, we list the average and maximal connection lengths for the networks and the lower bound.

Instance:	a1	a2	a3	c1	c2	c3	c4
# nodes	637	1229	4110	1051	1151	2264	6532
# edges	826	1356	4350	1079	1199	2380	7350
# customers	39	238	1670	345	315	475	1947
# potential COs	4	5	6	4	5	1	1
network trenching cost	235640	598750	2114690	322252	1073784	2788439	4408460
average connection length	717.1	1186.6	733.4	589.6	1969.2	995.3	2094.1
maximal connection length	1320	2257	2334	1180	4369	2452	4809
lower bound	224750	575110	2066190	312399	1063896	2743952	4323196
relative gap	4.8%	4.1%	2.3%	3.2%	0.9%	1.6%	2.0%
average connection length	299.2	887.5	504.6	363.7	1531.0	590.2	1092.9
maximal connection length	1049	3142	3215	1394	7315	2586	4786

Table 1 Comparison of trenching costs for different instances.

It can be observed that trenching costs in the solution networks are in an acceptable range with respect to the obtained lower bounds (up to 5 percent). Furthermore, length restrictions can be the major problem in the extended Steiner tree model. The maximal length from a BTP to its CO is in extreme cases almost doubled with respect to the solution networks.

Additionally, we compared trenching costs for an instance where existing infrastructure is provided. As can be seen from Table 2, both network trenching costs as well as the lower bound give smaller values, and the solution network uses a fair amount of the provided empty ducts.

	trenching costs	used existing infrastructure		
		total length	# trails used	% trails used
solution network	4063839	7131	352	95%
lower bound	3990065	7531	360	97%
relative gap	1.8%			

Table 2 Trenching costs with existing infrastructure (instance c4 with empty ducts on 371 trails).

Further research. Next we will investigate which and how many practical restrictions are violated by the solution computed with the extended Steiner tree model. Based on this analysis, further extensions of the model will be formulated in order to tighten the lower bound beyond the bounds presented in this paper.

References

1. OpenStreetMap project. www.openstreetmap.org.
2. BMBF project FTTX-PLAN, 2009–2011. www.fttx-plan.de.
3. M.O. Ball, T.L. Magnanti, C.L. Monma, and G.L. Nemhauser, editors. *Network models*, volume 7 of *Handbooks in Operations Research and Management Science*. Elsevier, Amsterdam, 1995.
4. S. Gollowitzer and I. Ljubić. MIP models for connected facility location: A theoretical and computational study. *Computers & Operations Research*, 38(2):435–449, 2011.
5. T. Koch and A. Martin. Solving Steiner tree problems in graphs to optimality. *Networks*, 32(3):207–232, 1998.
6. M. Leitner and G.R. Raidl. Branch-and-cut-and-price for capacitated connected facility location. *Journal of Mathematical Modelling and Algorithms*, 2011. online first, http://dx.doi.org/10.1007/s10852-011-9153-5.
7. N. Maculan. The Steiner problem in graphs. *Ann. Discrete Math.*, 31:185–211, 1987.
8. M. Martens, S. Orlowski, A. Werner, R. Wessäly, and W. Bentz. FTTx-PLAN: Optimierter Aufbau von FTTx-Netzen. In *Breitbandversorgung in Deutschland*, volume 220 of *ITG-Fachbericht*. VDE-Verlag, März 2010.
9. S. Orlowski, A. Werner, R. Wessäly, K. Eckel, J. Seibel, E. Patzak, H. Louchet, and W. Bentz. Schätze heben bei der Planung von FTTx-Netzen: optimierte Nutzung von existierenden Leerrohren – eine Praxisstudie. In *Breitbandversorgung in Deutschland*, ITG-Fachbericht. VDE-Verlag, März 2011.
10. T. Polzin and S.V. Daneshmand. A comparison of Steiner tree relaxations. *Discrete Appl. Math.*, 112(1-3):241–261, 2001.
11. P. Rigby, editor. *FTTH Handbook*. FTTH Council Europe, 4th edition, 2011.

Stream III
Decision Analysis, Decision Support

Eranda Dragoti-Çela, TU Graz, ÖGOR Chair
Jutta Geldermann, University of Göttingen, GOR Chair
Roland Scholz, ETH Zurich, SVOR Chair

Why pairwise comparison methods may fail in MCDM rankings

PL Kunsch

Abstract This article is based on previous author's articles on a correct way for ranking alternatives in Multi-criteria Decision-Making (MCDM). A straightforward but robust Weighed-Sum Procedure (WSP) is presented. It consists in combining additively ranks or scores evaluated against a dimensionless scale, common to all criteria. MCDM methodologies based on pair-wise comparison methods are on the contrary not providing the correct rankings in many practical situations. The roots of the observed failures are made clearer for the well-known Promethee method by means of a simple example.

1 Introduction

Practical Multi-criteria Decision-Making (MCDM) problems are mostly concerned with ranking n alternatives $(a_i), i = 1, 2, ..., n$. Assume: (1) the (nxK) pay-off table (P_{ik}) containing the dimensioned evaluations of n alternatives for K criteria, $(C_k), k = 1, 2, ..., K$, assumed with no loss of generality not to depend on the decision-maker(s) (DMs) preferences. (2) The preference relationship between the K criteria provided by DMs, for example by ranking the criteria with the possibility of indifference. It has been shown by the author [2] [3] that the following Weighed-Sum Procedure (WSP) provides all possible global rankings, and only them, compatible with these preferences:

- To start with, the pay-off table (P_{ik}) with different scales and units is replaced by a dimensionless (nxK) scoring table (S_{ik}) in which the alternatives are evaluated on a common dimensionless rank, or integer-score scale. Commensurability is gained, as required for solving the MCDM problem: the columns of (S_{ik}) may be combined the adequate way;
- In the second step the global ranking of the alternatives is obtained by combining additively the ranks or scores for all criteria, in accordance with the preference

P.L. Kunsch
Vrije Universiteit Brussel, Pleinlaan 2 BE-1050 Brussels, e-mail: pkunsch@vub.ac.be

D. Klatte et al. (eds.), *Operations Research Proceedings 2011*, Operations Research Proceedings,   97
DOI 10.1007/978-3-642-29210-1_16, © Springer-Verlag Berlin Heidelberg 2012

relationship. Rounding off rules are used in order to identify indifference ties between two or more alternatives.

These two steps are elaborated in the sections 2 and 3 respectively. In section 4 some failure mechanisms of pair-wise comparison methods are demonstrated by using the popular *Promethee* method [1].

2 The choice of a common scale for the criteria

A straightforward scoring scale is the ordinal scale in which the alternatives are ranked from the best ones, which receive the rank=score 1, to the least good ones which receive ranks=scores $1 \leq s_{ik} = l_{ik} \leq n$, where l_{ik} is the rank position of a_i in C_k. Some technical rounding rules are used for identifying indifference ties from the pay-off table evaluations, e.g., evaluations differing by less than 1 % in absolute are considered as giving indifference between two alternatives. Commensurability is gained, as required for solving the MCDM problem: the rank vectors S_k, i.e., the columns of (S_{ik}) may be combined. Note that the commensurability properties of the ordinal rank scale are three:

1. The scale is dimensionless;
2. The scale is unique and common to all criteria;
3. The values on the scale are enumerable and finite; they are integer numbers; these numbers are $\leq n$ in the ranking case.

The last property means that the raw data from the pay-off table have been put into n classes: the classes of the first ranked, second ranked, etc. Several classes may be empty in case of ties. Any different scale choice for making the criteria evaluations commensurable should at least share these three properties. While the ordinal scale is unique up to the choice of indifference thresholds, the choice of a cardinal scale is no longer unique. A quite sensible choice for fulfilling the above properties is the (inverse) Likert scale [2] [3] with 5 scores, in which 1 is the best score (rather than 5 in the direct Likert scale):

$$L_5 = [1 = Excellent 2 = Good 3 = Satisfactory 4 = Sufficient 5 = JustAcceptable] \tag{1}$$

L_5 indeed provides five distinct classes in which to put the (P_{ik}) evaluations. Some classes may be empty, also the first one (score 1), which would not be possible with ordinal ranking. To achieve the scoring classification DMs have to provide lower and upper limits to each of the five classes. The choice of only five classes is a sensible one because the human mind has difficulties in making judgements on too many objects at the time. Also because evaluations have many times no perfect accuracy, assuming non-integer scores for gaining a thinner granularity is often unnecessary, but may be considered. Because some alternatives might become arbitrarily bad a new class *Unacceptable* is added to L_5. This addition defines the basic 6-point German Grading Scale G_6:

$$G_6 = L_5 \cup [6 = Unacceptable] \tag{2}$$

With this scale a preliminary screening is possible in order to eliminate unacceptable alternatives from the global ranking.

3 The global WSP rankings

The second step of the WSP requires normalised weights within $(Kx1)$ column vectors Ws compatible with the preferences of the DMs. For instance assume that within a group of DMs with similar preferences regarding five criteria C_k , $k = 1, 2, \cdots, 5$ a complete ranking is provided as follows:

$$C_1 \succ C_4 \succ C_2 = C_5 \succ C_3 \tag{3}$$

where $\succ$ indicate strict preference and $=$ indicates indifference. Sets of normalised random numbers ordered like in (3) are generated. The global $(nx1)$ score vector $G(W)$ is calculated for each set of normalised weights contained in the vector W, and the score marix S as follows:

$$G(W) = SW \tag{4}$$

The global ranking with respect to W is herewith provided with the possibility of indifference by using adequate rounding rules. When exact values for the weights are known (a rare case, except for complete indifference between the criteria) a unique computation is sufficient. In the case the weights are random numbers, e.g., corresponding to the ranking (3), statistics on the ranks occupied by the alternatives are obtained by means of Monte-Carlo calculations. Let us consider for illustration the following case with equal weights. In a company six typists $T_i, i = 1, 2, \cdots, 6$ must be ranked in order to determine promotions and salary increases with respect to Typing Speed C_1 and Accuracy C_2 – both of equal importance – as measured by objective indicators. Using the supposedly known limits of the six classes on two $G_6 = L_5$ scales (none of the candidates gets excluded from the ranking) the following $(2x6)$ transposed score table S' is obtained:

$$S' = \begin{vmatrix} 1 & 2 & 2 & 2 & 3 & 3 \\ 4 & 3 & 1 & 3 & 1 & 2 \end{vmatrix} \qquad W = (0.5, 0.5) \tag{5}$$

From (4) the global scores and ranking are obtained:

$$(2.5, 2.5, 1.5, 2.5, 2, 2.5) \to T_3 \succ T_5 \succ T_1 = T_2 = T_4 = T_6 \tag{6}$$

This ranking is clearly correct according to the very ancient additive formula for evaluating pupils at school. Note that each typist gets a combined score with does not depend on the scores obtained by the other candidates. This thus excludes the possibility of illogical rank reversals (RRs). RRs appear when the ranking of two alternatives are reversed or changed from preference to indifference, when third irrelevant alternatives are introduced, removed, or replaced by less desirable alternatives. RRs affect all pair-wise comparison methods [5]. In the next section the roots of this anomaly are analysed by means of an example, by using the Promethee method [1], which is representative of outranking methods based on pair-wise comparisons.

4 A comparison with Promethee

The Promethee methodology is now briefly introduced [1]. The relative or absolute difference in performance of each couple of alternatives $d_k(a,b) \geq 0$ regarding each criterion C_k is evaluated by means of a preference function $0 \leq P_k(a,b) \leq 1$, always vanishing when $d_k(a,b) < 0$. The most common type of preference function used in Promethee is the linear type represented in Fig. 1.

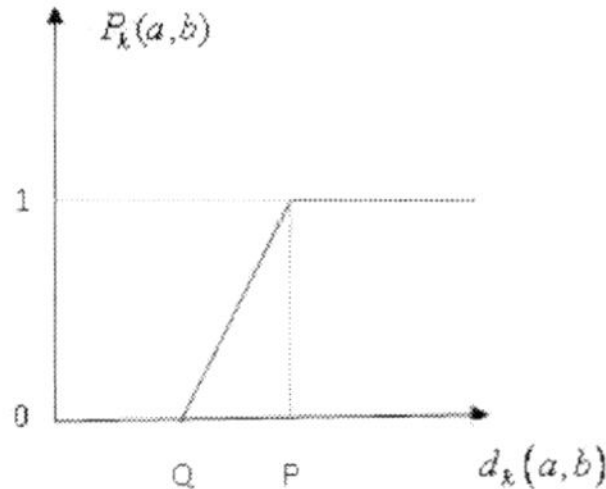

Fig. 1 The linear preference function of Promethee between a pair of alternatives (a,b) in function of $d_k(a,b) \geq 0$. Q is the indifference threshold and P the strict preference threshold.

The global preference index of a over b is given by a weighed sum using the normalised weight vector $(w_k), k = 1,2,\cdots,K$:

$$\pi(a,b) = \sum_{k=1}^{K} w_k P_k(a,b) \leq 1 \tag{7}$$

From (7) the net flow of alternative a is calculated as follows:

$$0 \leq \phi(a) = \frac{1}{(n-1)} \sum_{b \neq a} [\pi(a,b) - \pi(b,a)] \leq 1 \tag{8}$$

The net flow of alternative a represents the degree of preference of this alternative over all other alternatives in the considered set. A global ranking is established by ordering the net flows in decreasing order. The ranking will in general depend on the choice of the technical thresholds Q, P. To illustrate the problems that may arise, let us consider the typist example introduced in section 3. Because the scores are integer numbers we start by considering that $Q = P = 0$, or any value $0 \leq P \leq 1$ for the preference function $P_k(a,b), k = 1, 2, ...K$ in Fig. 1. This is equivalent to working with ordinary criteria with no thresholds. This results in the following Promethee net flow vector for the six typists, up to a sign inversion from maximum to minimum and an affine transformation not affecting the ranking. The latter is established accordingly:

$$(3.50, 3.75, 2.25, 3.75, 3.50, 4.25) \rightarrow T_3 \succ T_1 = T_5 \succ T_2 = T_4 \succ T_6 \qquad (9)$$

Promethee's global ranking is wrong, compare with (6)[1]. Though integer scoring data in the pay-off table have been used right from the beginning, similar conclusions would be obtained with dimensioned absolute data: working with ordinary criteria illicitly changes the pay-off table. In the present case with integer scores and ordinary criteria:

1. The pay-off table is replaced by the table of ranks;
2. In case of a tie of between p alternatives the average rank at position l is used, being equal to:

$$r_l^{avg} = l + \frac{(p-1)}{2} \qquad (10)$$

Promethee thus replaces the scoring table (5) by the following one:

$$S_1^{'(P)} = \begin{vmatrix} 1 & 3 & 3 & 3 & 5.5 & 5.5 \\ 6 & 4.5 & 1.5 & 4.5 & 1.5 & 3 \end{vmatrix} \qquad (11)$$

giving the ranking (9) from (4). This easily induces RRs because the Promethee-adjusted score of any alternative in some criterion depends on the tie multiplicity p. In this example, the global Promethee ranking gives $T_1 = T_5$ in (9). Moving now the third alternative from the first to the second position in criterion C_2 (thus given a worse score to this alternative), the new scoring table becomes:

$$S_2^{'(P)} = \begin{vmatrix} 1 & 3 & 3 & 3 & 5.5 & 5.5 \\ 6 & 4.5 & 2 & 4.5 & 1 & 3 \end{vmatrix} \rightarrow T_3 \succ T_5 \succ T_1 \succ T_2 = T_4 \succ T_6 \qquad (12)$$

and now $T_5 \succ T_1$ due to this change of a third alternative. Thus the global ranking is not only wrong, but it is also sensitive to changes in the multiplicities of the existing ties, creating an important risk of causing undesirable RRs. This result confirms the conclusion of a recent article [5] exploring RRs in outranking methods, particularly *Electre* [4]. The basic root of this anomaly is recognised there as being the artificial interdependency created between alternatives because of the pair-wise

---

[1] For example why is $T_2 \succ T_6$ in (9) while the scores in the two equally important criteria are $(2,3)$ and $(3,2)$ respectively, see (5)!

comparisons. Though Promethee is very different from Electre, this general observation is also verified in our example. Alternatives which are very close together for some criterion C_k – or even are building a tie like in the typist example – mutually influence themselves. The final scores change and with them possibly the global ranking, compare (6) with (9) or (12). The only way for eliminating this artificial interference between alternatives is avoiding pair-wise comparisons altogether. Instead direct comparisons of alternatives must be made with independent anchoring values. This is verified for the typist example by using the preference function in Fig. 1 with $Q = 0$, and a linear preference relationship covering the whole range of evaluations; the maximum range of scores$= 3$, so that $P \geq 3$ must be chosen. Doing so, it is observed that the net flows (8) are now equivalent to the global scores in (6): the pair-wise comparisons are made unnecessary, because they reduce to a direct comparison of the alternatives with the best one with score 1, for each criterion.

5 Conclusion

When solving MCDM problems one should be aware of difficulties arising from pair-wise comparisons. The presented typist example, though simple and tested here only against Promethee has a general validity: it is not just pinpointing an isolated atypical anomaly of some most popular outranking methods. A fundamental flaw is inherent to all pair-wise comparison methods, of which Rank Reversals are only one manifestation [5], [2], [3]. The correct way is making direct comparisons of individual alternatives with independent anchoring points representing the limits of performance classes. The German Grading Scale, which has been successfully used for centuries in schools, is very useful for serving this purpose.

References

1. Brans, J.P., Mareschal, B.: PROMETHEE methods. In Figueira J., Greco S., Ehrgott M. (eds.) Multiple Criteria Decision Analysis: state of the art surveys, pp. 163-195. Springer, New York (2005)
2. Kunsch, P.L.: A Statistical Approach to Complex Multi-Criteria Decisions. In: Ruan D. (ed.) Computational Intelligence in Complex Decision Systems, pp. 147-182. Atlantis Press, Paris (2010)
3. Kunsch, P.L.: A Statistical multi-criteria procedure with stochastic preferences. International Journal of Multicriteria Decision Making **1**, 49-73 (2010)
4. Roy B.: The Outranking Approach and the Foundations of the ELECTRE Methods. Theory and Decision **31**, 49-73 (1991)
5. Wang, X., Triantaphyllou, E.: Ranking irregularities when evaluating alternatives by using some ELECTRE methods, Omega **36**, 45-63 (2008)

Ranking of qualitative decision options using copulas

Biljana Mileva-Boshkoska, Marko Bohanec

Abstract We study the ranking of classified multi-attribute qualitative options. To obtain a full ranking of options within classes, qualitative options are mapped into quantitative ones. Current approaches, such as the Qualitative-Quantitative (QQ) method, use linear functions for ranking, hence performing well for linear and monotone options; however QQ underperforms in cases of non-linear and non-monotone options. To address this problem, we propose a new QQ-based method in which we introduce copulas as an aggregation utility instead of linear functions. In addition, we analyze the behavior of different hierarchical structures of bivariate copulas to model the non-linear dependences among attributes and classes.

1 Introduction

Qualitative decision making, as part of our daily life, represents a difficult task. This is even more pronounced in cases when the decision maker is faced with unfamiliar problems, problems described with many attributes or in cases when there are many choices for attribute values at hand. To support a consistent decision making in this kind of situations, the Qualitative-Quantitative (QQ) method [2, 5], was developed as an extension to DEX [4] method and DEXi computer program [3]. In this paper, we propose a new QQ-based method that overcomes the limitations of the current approach.

We address qualitative decision problems described with a set of options. As a starting point, options are given in tabular format constructing a qualitative mapping from multiple qualitative attributes to one output attribute. The output attribute represents a class which is preferentially ordered. Options that belong to the same class are almost equally appreciated by the decision maker. Input attributes, on the other

Biljana Mileva-Boshkoska
Jožef Stefan Institute, Jamova 39, 1000 Ljubljana, Slovenia e-mail: biljana.mileva@ijs.si

Marko Bohanec
Jožef Stefan Institute, Jamova 39, 1000 Ljubljana, Slovenia e-mail: marko.bohanec@ijs.si
University of Nova Gorica, Vipavska 13, 5000 Nova Gorica, Slovenia

D. Klatte et al. (eds.), *Operations Research Proceedings 2011*, Operations Research Proceedings, 103
DOI 10.1007/978-3-642-29210-1_17, © Springer-Verlag Berlin Heidelberg 2012

hand, may, but need not be preferentially ordered. In this way, decision maker's preferences are defined qualitatively and options are ranked only partially.

The goal of the method is to obtain a full ranking of options, i.e. options should be ordered from best to worst. In order to achieve that, values of ordered qualitative attributes are mapped into quantitative ones using ordered discrete numbers and forming a quantitative tabular function. Current approaches, such as the QQ method, use linear regression function to approximate the values of the quantitative tabular function and then use normalization of the obtained regression values to ensure that the final quantitative model is consistent with the original qualitative one.

The linear regression function allows good approximation when options form linear and monotone tabular functions; however it underperforms in cases of non-linear and non-monotone tabular functions, especially when the linear approximation is inadequate. This is particularly the case when the values of input attributes are not preferentially ordered. In order to respond to the challenge of solving non-liner and/or non-monotone tabular functions, we propose a new QQ-based method that uses copula functions as its aggregation utility.

2 Introduction to Copulas

Copulas are functions that manage to capture non-linear dependences among variables. They were introduced by Sklar [11] who proved that the joint distribution function of two random variables is equal to the copula of their marginal distributions on the unit interval. Hence to use copulas, we will consider the attributes and the class as random variables. Next, we determine the dependences among attributes and the class by using copulas.

From the available copula families, we focus on the Archimedean copula family [9], or more precisely on the Clayton copula which is defined as:

$$C_\theta(u,v) = [\max(u^{-\theta} + v^{-\theta} - 1, 0)]^{-\frac{1}{\theta}} \tag{1}$$

where $u = F(x)$ and $v = G(y)$ are cumulative distribution functions of the variables x and y on the interval $[0, 1]$. The estimation of the dependence parameter θ in (1) is performed using the maximum likelihood approach.

Equation (1) represents a bivariate copula that defines the joint probability of two random variables. The extension to multivariate copulas and the process of making decision goes through several steps. Firstly we construct a hierarchical structure of bivariate copulas. In such manner we are able to model the dependences among all available attributes and classes. Afterwards, we use the dependency obtained from the hierarchical structure to define a copula-based non-linear quantile regression function. Next we rank the options by using the obtained regression function. In what follows we will present details for each of these steps.

2.1 Different hierarchical structures of copulas

Although copulas are mainly used to define the dependency between two random variables, they can be extended to include more marginal distributions, hence determining the multivariate dependency. There are two approaches of their extension.

In the first one, we build one copula that has one dependence parameter θ for all marginal distributions of the n random variables. This construction is known as exchangeable multivariate Archimedean copula [1]. This approach requires many mathematical calculations and frequently it requires numerical solving of the final copula value [13].

The second approach uses at most $n - 1$ dependence parameters θ_i for n random variables. In this approach, bivariate copulas are formed and nested in a hierarchical construction. One such construction represents the fully nested Archimedean construction (FNAC). FNAC allows different groupings of the marginal distributions of the variables. For example, if $n = 3$, the following constructions are possible: $(u_3, [u_1, u_2; \theta_2]; \theta_1)$, $(u_2, [u_1, u_3; \theta_2]; \theta_1)$, $(u_1, [u_2, u_3; \theta_2]; \theta_1)$, as shown in Figure 1. In order to ensure that the obtained construction represents a copula itself, the following condition has to be fulfilled:

$$\theta_i > \theta_{i-1} > ... > \theta_1 \tag{2}$$

where θ_i is the most nested dependence parameter, the θ_{i-1} is the second most nested parameter and so on.

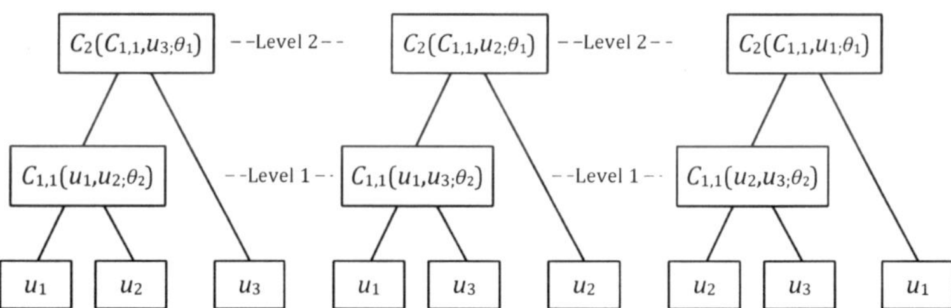

Fig. 1 Three fully nested Archimedean constructions using different orders of input distribution functions.

There are cases when it is not possible to build FNAC due to the θ constraints. In cases when the number of variables is $n > 3$ and the fully nested construction breaches the condition given in (1), we may use the partially nested Archimedean construction (PNAC) [8, 12]. Two examples of PNACs for a decision problem with four attributes and class are presented in Figure 2. In order the PNAC to represent a copula itself, the condition is that parameters θ_i must decrease with the level of nesting, while there are no constraints on their values when two or more copulas are built on the same level. For example, for the two PNACs in Figure 2, there are no conditions regarding the values of θ_{11} and θ_{12}. However, the following has to be fulfilled regarding copula structure given in Figure 2 (left): $\theta_2 < \theta_{11}$ and $\theta_2 < \theta_{12}$.

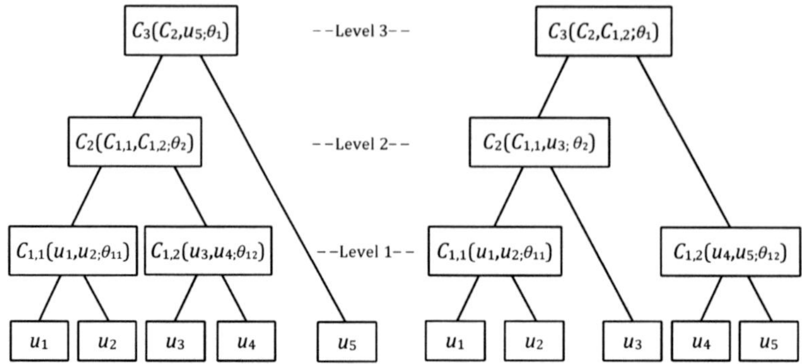

Fig. 2 Partially nested Archimedean copula: construction 1 (left) and construction 2 (right)

2.2 Regression using copula functions

Performing regression using copulas may be found in [7, 14]. In this paper we will use the regression approach for bivariate copulas as described in [6, 10], which leads to the following quantile regression function:

$$y = G^{-1}[1 - (F^{-1}(x))^{-\theta} + (q(F^{-1}(x))^{1+\theta})^{-\frac{\theta}{1+\theta}}]^{-\frac{1}{\theta}} \tag{3}$$

Here, $F^{-1}(x)$ and $G^{-1}(x)$ represent inverse cumulative distribution functions. The quantile q may receive values in the interval $[0,1]$, and for $q = 0.5$ we get median regression, which is used in the example that follows. The parameter θ is the one of the copula obtained on the highest level in the hierarchical structure of copulas, as shown in Figures 1 and 2.

One disadvantage of using PNAC is that it requires more steps in the regression process than in FNAC. Additionally, as the number of attributes increases, the number of possible different PNACs increases, too. Unlike FNAC, each new PNAC requires definition of additional computer procedures for determination of the hierarchical copula and for the regression task.

3 Example

To illustrate this approach, we consider an example of qualitatively classified options, each one described with four qualitative attributes. Attributes of the options and the corresponding class may get values from the preferentially ordered qualitative set, such as {*not acceptable, acceptable, good, very good, excellent*} where the value *not acceptable* is the least preferred and the value *excellent* is the most preferred. In order to obtain a quantitative representation of the example, each qualitative value is mapped into the ordered discrete set of numbers {1, 2, 3, 4, 5}. This mapping ensures consistency with the qualitative model in a way that the higher the number the greater the preference. The quantitative representation of the example is given in Table 1. In Table 1 the first column is the option number, the next four columns are the quantitative values of attributes followed by the corresponding class. The last three rows represent calculations obtained with regression functions with Clayton copula, normalized Clayton copula and QQ method, respectively.

If we examine the calculations obtained with QQ regarding options 8 and 9, we notice that option 9 is better or at least as good as option 8 for all attributes. Consequently we would expect that QQ will rank option 9 better than option 8, which is not the case.

No.	QA_1	QA_2	QA_3	QA_4	Class	Clayton	Clayt$_{norm}$	QQ
1	5	5	5	1	1	0.2467	0.5647	0.6553
2	4	3	3	2	1	0.4534	1.3540	0.7971
3	3	2	4	2	1	0.4393	1.2999	0.7536
4	2	4	1	4	1	0.2298	0.5000	1.3447
5	1	2	5	4	1	0.2382	0.5320	1.0699
6	3	3	5	5	1	0.4917	1.5000	1.0797
7	1	1	1	1	2	0.0590	1.5000	1.7010
8	3	4	1	2	2	0.2271	1.8862	1.9046
9	3	4	5	4	2	0.4943	2.5000	1.8960
10	3	4	1	5	2	0.2382	1.9117	2.2990
11	2	3	5	5	2	0.4754	2.4567	2.0391
12	1	1	1	1	3	0.0590	2.5000	2.7489
13	3	5	4	1	3	0.2383	2.9290	2.7111
14	5	4	2	4	3	0.4769	3.5000	2.9426
15	2	2	1	5	3	0.2233	2.8932	3.2889
16	2	1	3	5	3	0.2346	2.9203	3.0460
17	1	1	4	2	4	0.1560	3.5000	3.8290
18	4	5	4	2	4	0.4769	4.5000	3.8840
19	5	1	2	3	4	0.2284	3.7254	3.6826
20	5	3	3	3	4	0.4751	4.4944	3.7943
21	4	4	2	4	4	0.4708	4.4809	4.1594
22	2	5	4	4	4	0.4769	4.4998	4.3174
23	2	1	2	5	4	0.2284	3.7255	4.2133
24	1	4	4	4	5	0.2405	4.5000	4.8583
25	1	5	5	5	5	0.2467	5.5000	5.1417

Table 1 Quantitatively described problem

It is this kind of failures of the QQ method that we want to address with the regression using copula functions. As there are 4 attributes, we first examine the FNAC. However, none of the obtained FNACs fulfilled equation (1). On the other hand, the PNAC given in Figure 2 (left) fulfills the equation (1) hence we use this structure to obtain the results presented in Table 1, in the row *Clayton*. Calculations show that monotonicity is not breached within classes for any pair of options and hence we consider the ranking correct.

In the next step we examine the overall ranking of options. Examination of options 11 and 12 show that option 11 has a higher value than option 12 according to *Clayton*, although according to the original model, option 12 (classified in class 3) is more preferred than option 11 (classified in class 2). In order to obtain overall consistent ranking with the original qualitative model, the final step of normalization into the classes $c \pm 0.5$ is performed. Calculations of the last step are given in column $Clayt_{norm}$ of Table 1.

4 Conclusion

In this paper we presented a new QQ-based method for full ranking of non-linear and non-monotone qualitative decision options. In the new method, we introduced copulas as aggregation utility of attributes and classes. Additionally, we defined a copula regression function capable of modeling non-linear and non-monotone decision preferences. Furthermore we examined and demonstrated the usage of FNAC and PNAC on an illustrative example which showed that when FNAC does not form correct copula, PNAC is a good candidate to look for solution.

Acknowledgements The research of the first author was supported by Ad futura Programme of the Slovenian Human Resources and Scholarship Fund.

References

1. Berg, D., Aas, K.: Models for construction of multivariate dependance: A comparison study. European Journal of Finance **15**(7-8), 639–659 (2009)
2. Bohanec, M.: Odločanje in modeli. DMFA, Ljubljana (2006)
3. Bohanec, M.: DEXi: Program for Multi-Attribute Decision Making: User's manual : version 3.03. IJS Report DP-10707, Jožef Stefan Institute, Ljubljana (2011)
4. Bohanec, M., Rajkovič, V.: DEX: An expert system shell for decision support. Sistemica **1**, 145–157 (1990)
5. Bohanec, M., Urh, B., Rajkovič, V.: Evaluation of options by combined qualitative and quantitative methods. Acta Psychologica **80**, 67–89 (1992)
6. Bouyé, E., Salmon, M.: Dynamic copula quantile regressions and tail area dynamic dependence in forex markets. European Journal Of Finance **15**, 721–750 (2009)
7. Brown, L., Cai, T., Zhang, R., Zhao, L., Zhou, H.: A root-unroot transform and wavelet block thresholding approach to adaptive density estimation. unpublished (2005)
8. Fischer, M., Kock, C., Schluter, S., Weigert, F.: An empirical analysis of multivariate copula models. Quantitative Finance **9**(7), 839–854 (2009)
9. Joe, H.: Multivariate Models and Dependence Consepts. Chapman and Hall (1997)
10. Kolev, N., Paiva, D.: Copula based regression models: A survey. Journal of Statistical Planning and Inference **139**(11), 3847 – 3856 (2009)
11. Nelsen, R.B.: An Introduction to Copulas, 2nd edn. Springer, New York (2006)
12. Savu, C., Trade, M.: Hierarchical archimedean copulas. In: International Conference on High Frequency Finance. Konstanz, Germany (2006)
13. Trivedi, P., David, Z.: Copula Modeling: An Introduction for Practitioners. World Scientific Publishing (2006)
14. Wasserman, L.: All of Nonparametric Statistics. Springer Texts in Statistics, USA (2006)

Learning utility functions from preference relations on graphs

Géraldine Bous

Abstract The increasing popularity of graphs as fundamental data structures, due to their inherent flexibility in modeling information and its structure, has led to the development of methods to efficiently store, search and query graphs. Graphs are nonetheless complex entities whose analysis is cognitively challenging. This calls for the development of decision support systems that build upon a measure of 'usefulness' of graphs. We address this problem by introducing and defining the concept of 'graph utility'. As the direct specification of utility functions is itself a difficult problem, we explore the problem of learning utility functions for graphs on the basis of user preferences that are extracted from pairwise graph comparisons.

1 Introduction

In Decision Theory, supervised learning has played a central role in estimating utility functions for problems involving choice and rankings of alternatives. The general principle of such methods is to use *ordinal regression* to adjust a *parametric utility model* to the preferences of a decision maker (DM). Preferences are generally obtained by direct questioning of the DM, for example by requiring her to make pairwise comparisons on a small subset of alternatives (the *learning set*).

Research in Multicriteria Decision Theory has mainly addressed the problem of learning *multiattribute utility functions* [6] for alternatives characterized by multiple criteria. However, many recent applications in areas like image recognition, network analysis or bioinformatics use graphs as fundamental data structures. In the same manner, for recent developments in business intelligence, representing alternatives as 'loose' sets of objects is insufficient to model the structure of information. Graphs offer richer possibilities to capture both the structure and relational dimension in knowledge and information models. Nonetheless, using graphs as knowledge structures has a major disadvantage: their complexity for the user, as it is more diffi-

Géraldine Bous
SAP Research Sophia Antipolis, Business Intelligence Practice, Av. Dr. M. Donat 805, 06250 Mougins, France, e-mail: geraldine.bous@sap.com

cult to compare two graphs than two simple alternatives. Hence, there is a real need to develop decision support and recommendation systems for graph-based information models. It is thus necessary to establish frameworks that characterize the utility of a graph, as well as to develop supervised learning methods to compute estimates of graph utility. Once utility functions have been learned, it is simple to use the value functions to evaluate broader sets of alternatives in view of solving ranking, choice or sorting problems.

In this paper we discuss a first approach to establishing such a framework by extending the theory of multiattribute utility learning to sets of graphs. First, we discuss the notion of graph utility and describe the model used. In the second part, we describe an algorithm to learn utility functions from a preference relation provided by the DM. Finally, we conclude and discuss directions for future work.

2 Additive multiattribute utility model for graphs

In this section, we define a model of utility for graphs. The first part addresses the utility model of the graph itself; the second part extends the model by defining multiattribute utility functions for vertices and edges.

2.1 The graph utility model

Informally, the utility u of an item is a numerical value that reflects its desirability: the higher the utility, the better an item is considered to be. Consider for instance a finite set of objects S; a utility function $u : S \mapsto R$ is a function that, for $x, y \in S$, satisfies $u(x) \geq u(y) \Leftrightarrow x \succsim y$, where $\succsim$ denotes that x is 'preferred to or equivalent to' y. Traditional decision theoretic literature has focused on establishing axioms and models for the utility of abstract entities (e.g. item x) or, as a special case, for items that have one or more measurable attributes [2, 6]; however, to the author's best knowledge, there is no previous work addressing the definition and modeling of utility functions for complex objects like graphs, that are composed of several connected entities.

Let $G = \{G_i \mid i = 1, ..., |G|\}$ denote a set containing $|G| < \infty$ graphs $G_i = (V_i, E_i)$ of vertices V_i and edges E_i. In addition, we use the notations $V = \cup_{i=1}^{|G|} V_i$ to denote the set of all vertices and, analogously, $E = \cup_{i=1}^{|G|} E_i$ to denote the set of all edges. In general terms, the notion of utility of a graph is identical to that evoked above: it is a function $u : G \mapsto R$ that for any two $G_i, G_j \in G$, $i \neq j$, satisfies $u(G_i) \geq u(G_j) \Leftrightarrow G_i \succsim G_j$. The open research question is how to define $u(G_i)$ so that it reflects the 'contents' and structure, i.e. vertices and edges, of the graph. In other words, what in the graph makes its value? In order to make this explicit, it is necessary to express the utility of a graph as a function of the utility of its components, i.e. $u(G_i) = f(u(V_i), u(E_i))$ with $u(V_i) = f(u(v), \forall v \in V_i)$ and $u(E_i) = f(u(e), \forall e \in E_i)$. What is ultimately needed is hence a utility function on the sets containing all vertices and edges, i.e. $u : V \mapsto R$ and $u : E \mapsto R$.

The function f is an aggregation function, as it stands for the 'combination' of the utilities of subcomponents (vertices, edges) into a unique utility value for the supercomponent (the graph). The most commonly used aggregation approach is the additive model [2], which is applicable provided that the different components whose utility is aggregated are mutually independent. The use of the additive model in the context of graph utilities yields

$$u(G_i) = \sum_{v \in V_i} u(v) + \sum_{e \in E_i} u(e) \qquad \forall\, G_i \in G. \tag{1}$$

2.2 Vertex and edge utility model

The graph utility model of equation (1) describes the utility of a graph as being the sum of the individual components it is composed of. While this reflects the utility of the graph structure, it is still necessary to define the utility of the actual contents of the graph, i.e. to define a model and the corresponding utility of vertices and edges. To this purpose, we make the following the assumption: every node and edge is an instance of a collection of objects that have multiple attributes. For example, a node could represent cities that are characterized by attributes like attractiveness, cost of the stay, etc., while edges would represent traveling from one city to another with e.g. plane or train and have attributes like cost, travel duration, etc. In brief, the modeling of utilities for nodes and edges falls into the context of multiattribute utility theory (MAUT) [6]. Let us therefore consider that vertices are m-dimensional objects and that $x(v) = (x_1(v), x_2(v), ..., x_m(v))$ is the vector that describes the values of vertex v on the m attributes. If the attributes are independent, we may define a marginal utility for each attribute, i.e. $u_k(v)$ for $k = 1, 2, ..., m$. In MAUT, marginal utilities are frequently modeled as piecewise linear functions in view of allowing enough flexibility to approximate any type of functions by linear segments [1,3,5,8]. To keep things simple, we restrict ourselves to the case of a single linear segment. In order to define this function, we have to distinguish between attributes that are to be maximized (e.g. attractiveness) and minimized (e.g. cost). Let

$$\bar{x}_k = \begin{cases} \max_{v \in V} x_k(v) & \text{if } k \text{ is a maximization criterion} \\ \min_{v \in V} x_k(v) & \text{if } k \text{ is a minimization criterion} \end{cases} \tag{2}$$

and

$$\underline{x}_k = \begin{cases} \min_{v \in V} x_k(v) & \text{if } k \text{ is a maximization criterion} \\ \max_{v \in V} x_k(v) & \text{if } k \text{ is a minimization criterion.} \end{cases} \tag{3}$$

The marginal linear utility function for criterion k, where $\omega_k \geq 0$ is the weight of the criterion, is then given by:

$$u_k(v) = \omega_k \cdot \frac{x_k(v) - \underline{x}_k}{\overline{x}_k - \underline{x}_k}. \tag{4}$$

Using the additive model to aggregate the marginal utilities of a vertex then yields

$$u(v) = \sum_{k=1}^{m} u_k(v). \tag{5}$$

For the edges, with vector $y(e) = (y_1(e), y_2(e), ..., y_n(e))$ describing their n attributes and ϕ_l to denote the weights for the marginal utility functions, an analogous development leads to the utility model

$$u(e) = \sum_{l=1}^{n} u_l(e). \tag{6}$$

Finally, combining the graph utility model (1) with (5) and (6) yields

$$u(G_i) = \sum_{v \in V_i} \sum_{k=1}^{m} u_k(v) + \sum_{e \in E_i} \sum_{l=1}^{n} u_l(e) \qquad \forall\, G_i \in G. \tag{7}$$

3 Learning the utility of a graph from pairwise comparisons

The graph utility model (7) defined above requires determining the weights for each and every marginal utility function, which limits the applicability of our approach in practice, e.g. for the development of a recommendation or decision aiding system that provides rankings of graphs in view of solving choice or subset choice problems. Besides the definition of the utility model, it is therefore important to establish a methodology to learn the weights with a semi-supervised approach. The class of UTA methods [1,5,8] for learning utility functions in the MAUT context provides a solid framework for this purpose, that we here extend to the case of graph problems.

Let $G_L \subset G$ denote the learning set; it contains a series of graphs that the DM is asked to rank in order of decreasing preference. To keep notations simple, we assume that the graphs of G_L are labeled according to their rank, i.e. $G_L = \{G_i \mid i = 1, ..., |G_L|\}$ (the remaining labels are then arbitrarily assigned to the graphs of $\{G \setminus G_L\}$), which leads to

$$G_1 \succsim G_2 \succsim ... \succsim G_{|G_L|}. \tag{8}$$

For every adjacent pair in the ranking, we can write a preference constraint in terms of utilities, that is

$$u(G_i) \geq u(G_{i+1}) \qquad \text{for } i = 1, ..., |G_L| - 1. \tag{9}$$

If the preference constraints define a nonempty polytope, it means that there is more than one feasible solution to define the weights of the vertex and edge utility functions. Previous research in utility learning (on 'simple' multiattribute objects) has shown that, in such cases, the analytic center [4, 9] of the preference polytope is a suitable choice to address the indetermination and provide a unique solution [1]. For a non-empty polyhedron $\mathscr{P} = \{x \in \mathrm{R}^m \mid Ax \leq b\}$, the analytic center x^a is defined as

$$x^a = \arg\max_{x, s > 0} \sum_{i=1}^{m} \ln s_i, \tag{10}$$

where $s_i = b_i - a_i.x$ is the (strictly positive) slack variable of the ith constraint and $a_i.$ is the ith line of matrix A. The logarithm in the maximization function above is known as a *barrier function*, whose effect will be to 'push' the solution away from the boundaries of the polyhedron, yielding a central point (see [7, 10] for details).

Applying this approach to graph utility learning requires rewriting the preference constraints with explicit slack variables, i.e.

$$s_{i,i+1} = u(G_i) - u(G_{i+1}) \qquad \text{for } i = 1, ..., |G_L| - 1. \tag{11}$$

The utility function learning problem can then be expressed with the following non-linear optimization problem:

$$\max \quad \sum_{i=1}^{|G_L|-1} \ln s_{i,i+1} \tag{12}$$

$$\text{s.t.} \quad s_{i,i+1} = u(G_i) - u(G_{i+1}) \quad i = 1, ..., |G_L| - 1 \tag{13}$$

$$\omega_k, \phi_l \geq 0 \quad \forall k, l \tag{14}$$

$$\sum_{v \in V} \sum_{k=1}^{m} \omega_k + \sum_{e \in E} \sum_{l=1}^{n} \phi_l = 1. \tag{15}$$

The last constraint is known as the *normalization constraint*; it ensures that the maximum utility, taking into account all nodes and edges at their maximum value (i.e. $u(\bar{x}_k) = \omega_k$ and $u(\bar{y}_l) = \phi_l$), is equal to one. The optimization problem (12) can be solved without further modifications with Newton's method using projective methods to handle equality constraints (see [1] for details). Note, however, that if the preference polytope is empty, then the method above is inadequate. In such cases, it is necessary to extend the model to include error variables that allow a violation of preference constraints, violation which is then addressed by minimizing the sum of error variables in the objective function (the reader may refer to [1, 8] for further details).

The solution provided by the learning algorithm provides a fully determined utility model that allows to compute the utility of vertices, edges and all graphs of the set G. In this manner, the utility of each graph may be used to, e.g., rank these from best to worst in view of making recommendations that take into account the preferences of the DM.

4 Conclusions and directions for future research

The increasing amount of data and information models that are based on graph structures require the development of theoretical foundations and models that allow developing recommendation systems for graphs. In this paper we have presented an additive utility model for graphs that are composed of vertices and edges characterized multiple attributes. We have extended this proposal by presenting a suitable algorithm to learn the utility model on the basis of a user provided ranking on a subset of graphs.

The novelty of the research topic makes this proposal a first approach to an area that still has many challenges to offer. Among the most important points that must be mentioned as requiring further research is the analysis and development of other models. The additive model used here requires independence; it is hence worthwhile to formalize the conditions under which this assumption can be made in case of graphs. Second, it is important to explore other models, including approaches that would make it possible to take into account structural constraints (paths, subgraphs), instead of fully decomposing the graph into all its fundamental components. Finally, it is worthwhile noting that the graph model used here, which assumes one utility model for nodes and one for edges, may not be descriptive enough in applications that involve different types of objects.

References

1. Bous, G., Fortemps, Ph., Glineur, F., Pirlot, M.: ACUTA: A novel method for eliciting additive value functions on the basis of holistic preference statements. EJOR **206**, 435–444 (2010)
2. Fishburn, P.C.: Utility Theory for Decision Making. Wiley, New York (1970)
3. Greco, S., Mousseau, V., Słowiński, R.: Ordinal regression revisited: multiple criteria ranking using a set of additive value functions. EJOR **191**, 415–435 (2008)
4. Huard, P.: Resolution of mathematical programming with nonlinear constraints by the method of centers. In: Abadie, J. (ed.) Nonlinear Programming, pp. 209–219. Wiley, New York (1967)
5. Jacquet-Lagrèze, E., Siskos, Y.: Assessing a set of additive utility functions to multicriteria decision-making: the UTA method. EJOR **10**, 151–164 (1982)
6. Keeney, R.L., Raiffa, H.: Decisions with multiple objectives: Preferences and value tradeoffs. Wiley, New York (1976)
7. Nesterov, Y.E., Nemirovskii, A.S.: Interior-point Polynomial Algorithms in Convex Programming. SIAM, Philadelphia (1994)
8. Siskos, Y., Grigoroudis, E., Matsatsinis, N.: UTA Methods. In: Figueira, J., Greco, S., Ehrgott, M. (eds.) Multiple Criteria Decision Analysis: State of the Art Surveys, pp. 297–334. Springer, New York (2005)
9. Sonnevend, G.: An analytical centre for polyhedrons and new classes of global algorithms for linear (smooth, convex) programming. In: Prekopa, A., Szelezsan, J., Strazicky, B. (eds.) LNCIS, pp. 866–876. Springer, Heidelberg (1985)
10. Ye, Y.: Interior Point Algorithms: Theory and Analysis. Wiley, New York (1997)

Dealing with conflicting targets by using group decision making within PROMETHEE

Ute Weissfloch and Jutta Geldermann

Abstract This paper presents an approach to applying PROMETHEE for group decision making to support strategic decision making for the implementation of product-service systems (PSS). Such strategic decisions are not only dependent on profit but also on other economic and ecological criteria, i.e. gain or loss of know-how and chances as well as risks concerning cooperation. Therefore, the application of a multi criteria decision support methodology is necessary, in order to include both, the provider's and the customer's perspective. In contrast to the application of PROMETHEE by a single decision maker, the weighting process includes the share of the vote of the different decision makers – as already shown in other approaches – as well as contrary targets of the decision makers for the defined set of criteria.

1 Introduction

There exists a variety of MCDM-methods with different advantages and disadvantages. The outranking-methods have been developed to handle inconsistent information as the decision maker is not totally aware of his preferences [4]. One of the outranking-methods is the decision making method PROMETHEE (Preference Ranking Organisation Method for Enrichment Evaluations). Most applications are strategic decisions as e.g. location problems [2] or the selection of PSS [9]. This method has its advantages in "simplicity, clearness and stability" [2, p. 228]. In general it was developed for single decision makers. However, in those decision making processes quite often, more than one decision maker is involved. Therefore, the methodology has to be extended to more decision makers. There already exist different approaches on group decision making within PROMETHEE e.g. [3, 7].

Ute Weissfloch
Fraunhofer Institute for Systems and Innovation Research ISI, Breslauer Strae 48, 76139 Karlsruhe, Germany;, e-mail: ute.weissfloch@isi.fraunhofer.de

Jutta Geldermann
Georg-August-University of Goettingen, Platz der Goettinger Sieben 3, 37073 Goettingen, Germany e-mail: geldermann@wiwi.uni-goettingen.de

D. Klatte et al. (eds.), *Operations Research Proceedings 2011*, Operations Research Proceedings, 115
DOI 10.1007/978-3-642-29210-1_19, © Springer-Verlag Berlin Heidelberg 2012

Mancharis et al. [7] developed two different approaches, which are shortly introduced. In the beginning a problem structuring process is necessary to define the alternatives and the set of criteria, so that each decision maker has the same evaluation table [7]. For one suggested approach each decision maker first carries out an individual evaluation based on PROMETHEE. In the next step, the criteria are replaced by the decision makers where the characteristics of the criteria are the net flows of the alternatives of the individual evaluations [1, 7]. In the other approach, the decision table is enlarged, so that the evaluation of each decision maker is in one large decision matrix and PROMETHEE is applied to this table [3, 7]. However, no further literature on the application of these approaches exists. In other approaches each decision maker takes his own decision in advance and afterwards discusses the results and negotiates with the other decision makers [5,6]. Though, in these approaches it is not really pointed out for which criteria the decision makers have different aims and how this impinges on the relevance of those criteria in the decision process. This paper suggests an approach to overcome these lacks. It is explained in the following chapter, whereas in chapter 3 the applicability for strategic problems is demonstrated using the example of PSS.

2 Group decision making within PROMETHEE

As the problem structuring process is not target of this paper, it can be assumed, that a certain decision problem already exists, i.e. it is transparent, who are the decision makers and what are the alternatives. How to structure a decision problem is described e.g. [7]. To support a decision making process for a group, a weighting process similar to the approach of Macharis et al. [7] is proposed. First of all, the share of the vote v^r for every decision maker r is defined. Then, each decision maker decides on his set of criteria. These criteria are harmonized for all decision makers whereby all criteria are kept in the final set except those which are doubled or quite similar. In this case, the final criteria have to be defined within the group. When the definition of the set of criteria is completed, every decision maker elicits his own weighting and states, if the criteria have to be maximized or minimized. To determine the weights for the outranking process out of the individual weightings, the following approach is suggested:

- The share of vote multiplied by the weighting is the individual contribution to the total weighting
- Criteria, which should be maximized are taken into account with '+', such, which are to be minimized are considered as '−'. This is especially important for criteria with conflicting targets, which should be maximized according to one or more decision makers, but a minimization is requested by the other(s).

$$w_{t,i}^r = +1 \times w_i^r \times v^r \qquad \text{, if criterion i has to be maximized} \qquad (1)$$

$$w_{t,i}^{r} = -1 \times w_{i}^{r} \times v^{r} \qquad \text{, if criterion i has to be minimized} \qquad (2)$$

with $w_{t,i}^{r}$: individual contribution to weighting of decision maker r to criterion i, w_{i}^{r}: original weighting of the decision maker r for criterion i and v^{r}: share of vote for decision maker r

- To get the group weighting, the individual contributions are summed up, considering the algebraic sign for the maximization (eq. 1) or minimization (eq. 2) of the criterion requested by the decision maker.

$$w_{t,i}^{g} = \sum_{r} w_{t,i}^{r} \qquad (3)$$

with $w_{t,i}^{g}$: weighting of the group for criterion i.

- In the end, the algebraic sign of $w_{t,i}^{g}$ decides, whether this criterion has to be maximized or minimized for the group decision.
- For reaching a better comparability between the group weighting and the individual weightings the sum of the absolute values of the weighting is normalized to 1.

$$w_{n,i} = \frac{|w_{t,i}^{g}|}{\sum_{i} |w_{t,i}^{g}|} \qquad (4)$$

with $w_{n,i}$: normalized weighting of the group for criterion i.

- In case of conflicting targets each decision maker can define a minimum – in case of maximization of the criterion – resp. maximum value – for minimization of this criterion – which has to be fulfilled by this criterion. This should lead to a target range, i.e. the maximal value should lie above the minimal value; otherwise it is not possible to reach a consensus. If the criterion falls below or exceeds this corridor for an alternative, this alternative is blocked by a veto and is not considered in the decision process any more.
- On basis of this weighting, the PROMETHEE method is applied [1]. The decision makers have to agree on a preference function and the indifference/preference thresholds. The proposal would be to choose one standard preference function for quantitative criteria and one for qualitative criteria. Both PROMETHEE I and PROMETHEE II can be applied for this approach.

3 Exemplary case study

To demonstrate the applicability of the suggested weighting process a simple example is calculated in this chapter. The case study presents a company, which considers the introduction of innovative PSS. PROMETHEE already proved to be a good methodology to decide upon the implementation of a certain business concept for one decision maker [9]. But such a decision will also involve the customers or potential cooperation partners in praxis. Therefore, an easy example to demonstrate the approach of group decision making was chosen.

Decision makers

A manufacturer of technical equipment considers the implementation of PSS. The decision problem includes the question which type of PSS should be offered. To learn about the acceptance of different PSS, a customer takes part in this decision. Therefore, two decision makers are assumed, whereof both have the same share of votes, i.e. 50 %.

Criteria

For this example, the decision makers agree on three criteria, which are:

- the profit for the provider – as this value also influences the profit for the customer – measured in Euro,
- the energy consumption of the technical equipment, measured in the percentage of the original value without any new business concept, and
- the advantages of cooperation between provider and customer measured in a qualitative scale from 1 to 6.

Alternatives

The decision makers have a choice of three different types of PSS, whereof the first option is the usual business model, which they already used up to now. The other alternatives could e.g. represent a use oriented and a result oriented business concept. For further information about this classification see e.g. [8]. Table 1 shows the decision matrix.

Table 1 Decision matrix

Criterion \ Alternative	PSS 1	PSS 2	PSS 3
Profit of provider	5000	7000	10000
Energy consumption	100	95	80
Cooperation	1	6	3

Weighting

Table 2 depicts the weightings of the decision makers and of the group. The profit of the provider is the highest weighted criterion for both of the decision makers. However, the provider wants to maximize it, whereas the customer wants to minimize it, as a higher profit for the provider normally means less profit for the customer. However, in the normalized group weighting the profit has only a very low weight of 0.06, as the rule regarding the different optimization directions applies and therefore the difference of the individual contribution of provider and customer is calculated. The other two criteria are quite similarly weighted by both decision makers and also the direction of optimization is the same. In combination with the situation of the profit of the provider, this is why the other two criteria gain a bit of importance in the normalized group weighting. It is assumed that all values of the profit for the provider lie in the target range.

	Provider		individual contribution provider	Customer		individual contribution customer	group weighting	normalized group weighting	
Profit provider	Max	0.65	0.33	Min	0.60	-0.30	0.03	Max	0.06
Energy consumption	Min	0.10	-0.05	Min	0.25	-0.13	-0.18	Min	0.44
Cooperation	Max	0.25	0.13	Max	0.15	0.08	0.20	Max	0.50

Table 2 Weighting

Results

Figure 1 depicts the net flows and the rankings of the single decision makers and the group. There are no differences in the results between PROMETHEE I and PROMETHEE II, therefore, the results of the partial ranking are not shown separately. Considering the results of the single decision makers, the provider prefers PSS 3 most. After this follows PSS 2 and on the last position is PSS 1. For the customer the ranking is the other way round. He prefers PSS 1 most and PSS 3 least. Regarding the results of group decision, PSS 2 is the best compromise for both the decision makers. This is ranked on the second position for both decision makers, so no one has to cope with the worst choice. According to the group weighting PSS 3 is at second position with only little distance to the first one. The last-ranked business concept is PSS 1, where the negative flows predominate, what already can be seen in the decision matrix, as this alternative is worse than PSS 3 in all criteria, as long as the profit for the provider is maximized.

Fig. 1 Results single decision maker and group

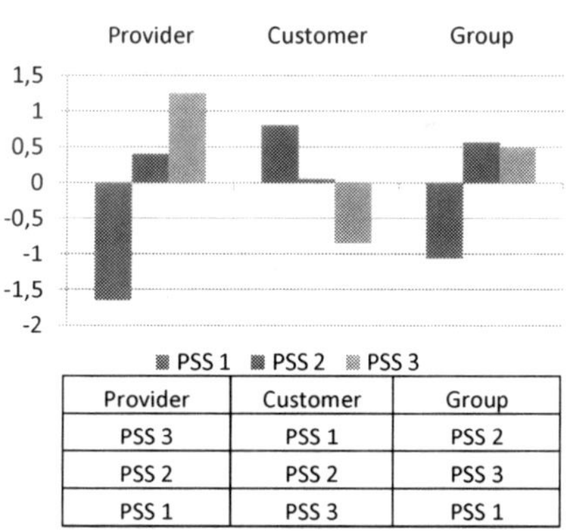

Provider	Customer	Group
PSS 3	PSS 1	PSS 2
PSS 2	PSS 2	PSS 3
PSS 1	PSS 3	PSS 1

Discussion

If the criteria with conflicting targets are highly ranked by all decision makers they are losing importance as shown above. Therefore, the decision is mainly based on criteria with equal targets of the decision makers whereby the veto function helps to find a compromise concerning criteria with conflicting targets. In case that only one decision maker considers the conflicting criteria as important and the other decision maker has rather low weights for these special criteria, the calculated difference is not that small so the importance in the group weighting decreases less strongly. In this case, criteria with conflicting targets are akin to other criteria with equal targets.

4 Conclusion

A new approach has been described, which demonstrates the influence of conflicting targets for certain criteria in a group decision support system. It clearly shows the different aims of the decision makers as well as the relevance of those criteria on the decision. The case study shows, that the influence of those criteria is decreasing, and they can be regulated by upper resp. lower boundaries, which are acceptable for the decision makers. Problematic for this approach is the necessity for the decision makers to agree on the data for the qualitative criteria, even though this normally is a subjective impression. Moreover, they also have to select a global preference function for each criterion, which is not necessarily required (but also suggested) in the existing approaches [7]. However, these limitations help to overcome the large size of the global evaluation approach, so a clear view of the problem is possible, which especially includes the influence of conflicting targets.

Acknowledgements This research is conducted in the context of a research and development project (*www.eneffah.de*) funded by the German Federal Ministry of Economics and Technology (BMWi) and managed by the Project Management Agency Juelich (PTJ).

References

1. Brans, J.-P., Mareschal B.: PROMETHEE Methods. In: Figueira, J., Greco, S., Ehrgott, M. (eds) Multiple Criteria Decision Anlaysis, pp. 163 – 195, Springer, New York (2005)
2. Brans, J.-P., Vincke, P., Mareschal, B.: How to Select and how to Rank Projects: The PROMETHEE Method. Europ. J. of Operational Research **24**, 228 – 238 (1986)
3. Geldermann, J.: Mehrzielentscheidungen in der industriellen Produktion. Universitaetsverlag, Karlsruhe (2006)
4. Geldermann, J., Rentz, O.: Integrated technique assessment with imprecise information as a support for the identification of best available techniques (BAT). OR Spektrum **23** (1), 137 – 157 (2001)
5. Ghafghazi, S., Sowlati, T., Sokhansanj, S., Melin, S.: A multicriteria approach to evaluate district heating system options. Applied Energy, **87** (4), 1134 – 1140 (2009)
6. Haralambopoulos, D.A., Polatidis, H.: Renewable energy projects: Structuring a multi-criteria group decision-making framework. Renewable Energy, **28** (6), 961 – 973 (2003)
7. Macharis, C., Brans, J.-P., Mareschal, B.: The GDSS PROMETHEE Procedure. J. of Decision Systems, **7**, 283 – 307 (1998)
8. Tukker, A.: Eight types of Product-service System: Eight ways to sustainability? Experiences from SUSPRONET. Bus. Strat. Env. **13**, 246 – 260 (2004)
9. Weissfloch, U., Mattes, K., Schroeter, M.: Multi-criteria evaluation of service-based new business concepts to increase energy efficiency in compressed air systems. Innovation for Sustainable Production 2010, Proceedings, 61 – 65, Bruges (2010)

Relation between the Efficiency of Public Forestry Firms and Subsidies: The Swiss Case

Bernur Açıkgöz Ersoy and J. Alexander K. Mack

Abstract This study is an empirical analysis of the productive efficiency of public forestry firms in Switzerland. Because of the available database, the period under examination extends from 1998 to 2006, and is based on a balanced panel of 100 firms. We inquire whether public subsidies exert any significant impact on the efficiency of these firms. In order to determine the productive efficiency, a nonparametric method (DEA) is applied. Panel unit root tests and panel cointegration technique are employed to establish the long-run relationship between efficiency and selected variables. Results show that subsidies had a positive influence on technical efficiency.

1 Introduction

Since the mid 1980s, forestry firms globally record increasing deficits in their core business, i.e., forest management and wood production. A scissors effect appeared between revenues and expenses, i.e., wood prices tend to decline and production costs increase [2]. Wood production and sales have ceased to cover the costs of the other tasks and activities of forest management entrusted to some 2600 forestry firms. The recent recovery of the wood market certainly improved the financial prospects of the sector, but, as shown by the empirical results (*infra*), there remains considerable room to improve the technical efficiency of Swiss forestry firms. So far it seems that subsidies play a negative role in this sense [1, 11]. Despite the precautions taken in the monitoring of projects, it is accepted practice that any subsidy has risks of perverse effects, i.e., information asymmetry between the project initiators and public authorities can lead to errors of choice in logging projects. Indeed,

Bernur Açıkgöz Ersoy
Celal Bayar University, School of Applied Sciences, International Trade Department, 45100 Manisa, Turkey, e-mail: bernur.acikgoz@bayar.edu.tr

J. Alexander K. Mack
Bern University of Applied Sciences, Engineering and Information Technology, Mathematics, Natural Sciences and Humanities, 2501 Biel, Switzerland, e-mail: alexander.mack@bfh.ch

D. Klatte et al. (eds.), *Operations Research Proceedings 2011*, Operations Research Proceedings, 121
DOI 10.1007/978-3-642-29210-1_20, © Springer-Verlag Berlin Heidelberg 2012

the subsidy creates the illusion that the project costs less than in reality and, thus, the risk exists to support rather expensive projects. Moreover, the availability of external financing may lead to prefer projects that receive a subsidy, to the detriment of other projects that meet some intrinsic criteria. Finally, the firms' fuzzy budget constraints will do the rest as deficits will be covered by the public owners. On the other hand, one could expect that subsidies increase the technical efficiency if they provide firms with an incentive to innovate or switch to new technologies [6].

Originally limited to the financial support of investments, the Swiss subsidization regime of forestry activities mainly covered the reparation of extraordinary damages, the care and management of tree populations, and protective measures. Generally, during most of the time period analyzed, subsidies covered only net costs, i.e., those generated by the project, minus possible wood receipts. The definition of the net costs is based on packages that have been determined by taking into account the actual costs in the cantons, and a minimum participation of the forestry firm. This financial instrument, largely used following hurricane Lothar in December 1999, concerns mainly the prevention and reparation of natural disasters as well as secondary damages (diseases or bark beetle attacks). Currently, it is rather usual to separate the commercial production of wood from other forest benefits likely to receive proper compensation, for example through a service contract. These improvements can lead to cost reductions and improved competitiveness of Swiss forests. The aim of this study is to examine the long-term impact of subsidies on the technical efficiency of Swiss public forestry firms in wood production, and help policy makers introduce better targeted forestry policies.

2 Database and methods

The data for this study was provided by the Swiss forest owners association, which maintains a standardized operating account of about 700 forestry firms. These firms are, in most cases, in public hands, i.e., cantons and local authorities. Indeed, in Switzerland, large private forest firms hardly exist and are not included in the database. Because of the available database, the period under examination runs from 1998 to 2006. Besides the general data that characterizes an individual firm, or more specifically, an "accounting unit" which can include several owners and operators, data mining has focused on variables related to the second level of production. The latter includes the logging, hauling, tending, transportation to / from the storage, and other ancillary activities and operations. After careful examination of the data, a balanced panel was formed. It consists of 100 firms of relatively similar size, allowing a comparison of results over time. The variables, which are potentially useful in determining the productive efficiency, can be divided into three groups, i.e., those that describe the production, those used to measure the resources used (inputs), and environmental variables that affect the conditions for wood production, regardless of the volume and quality of resources (labor, machinery, administration, etc.).

The nonparametric method used to determine the efficiency frontier is called "Data Envelopment Analysis" (DEA). Developed by Charnes, Cooper and Rhodes [3], and derived from the work of Farrell [5], it generalizes the concept of efficiency for multiple inputs and outputs to construct a mathematical optimization program, whose solution provides a measure of efficiency, relative to the frontier. The units considered here, i.e., the forestry firms, take, by assumption, autonomous decisions, particularly regarding the production and the factors used. They transform, given the existing technology, inputs (labor, machinery and vehicles, administration) into outputs (wood).

In other words, DEA measures the efficiency of a decision unit by calculating the relative difference between the point representing the value of observed inputs and outputs, and a hypothetical point on the frontier of that same production. The method is then used to identify best practices with respect to all observations, that is to say the production frontier, and thus to measure the efficiency degree (score) of each unit. To account, then, for the effect of the environment likely to influence the performance of the forestry firms, the efficiency scores obtained from the DEA analysis are regressed, in the second stage, on environmental variables, i.e., (exogenous) factors that are not or only indirectly under the control of a firm.

To establish the long-run relationship between efficiency and some selected variables, panel cointegration technique is employed. Before proceeding to the estimation using the cointegration technique, the first step is to investigate the stationarity properties of the variables. For this purpose, the Levin, Lin and Chu [7], ADF - Fisher Chi-square and PP - Fisher Chi-square tests [4, 8], which are the most widely used methods among panel data unit root tests in the literature, are performed.

The second step is to test for the existence of a long-run relationship between selected variables and the firms' technical efficiency scores. To examine the long-run relationship, the Pedroni panel cointegration test [9, 10], which takes into consideration heterogeneity by using specific parameters, is performed. Pedroni proposes seven test statistics for the null hypothesis of no cointegration for panel cointegration. At last, after finding cointegration in the second step, the coefficients are estimated by applying fully modified ordinary least squares (FMOLS) method.

3 Empirical results

Table 1 shows the technical efficiency scores obtained for the 100 forestry firms of the sample during the period 1998–2006 by applying the DEA method. The model is input oriented, and assumes constant returns to scale (CRS). The output is the total annual wood production in m^3; the inputs that have been selected are the hours worked by the wood production personnel, the machine hours performed by all forest vehicles, the administrative costs (in Swiss francs) for wood production, and third-party services (in Swiss francs) which were employed by the firm (fees paid to private forestry companies specialized in wood production).

Table 1 Technical efficiency scores (*N*=900)

Year	Mean	Median	S.D.	Min.	Max.	$\geq 50\%$	Efficient
1998	43.40%	38.40%	18.76%	18.40%	100%	27	3
1999	41.73%	36.50%	18.54%	19.40%	100%	22	4
2000	42.90%	39.45%	16.35%	16.30%	100%	28	1
2001	39.35%	33.05%	20.10%	11.90%	100%	24	3
2002	39.97%	33.60%	18.03%	17.00%	100%	22	2
2003	43.07%	38.45%	17.41%	19.60%	100%	24	1
2004	47.05%	38.55%	20.26%	17.90%	100%	37	4
2005	45.89%	40.45%	17.45%	20.20%	100%	32	1
2006	45.99%	39.30%	19.50%	17.60%	100%	27	5

It appears that over the nine-year period, only a minority of firms has an efficient production (between one and five firms per year), i.e., reach an efficiency score of 100%. For these firms, the nonparametric analysis cannot identify any potential for improvement. Conversely, more than two thirds of the firms have efficiency scores below 50%, reflecting significant gaps. The average efficiency scores are above 40% between 1998 and 2000, before falling below 40% in 2001 and 2002. From 2003 on, these averages reach values higher than 40%, and reach even 47% in 2004. The decline in the technical efficiency scores in 2001 and 2002 can be explained by the occurrence and aftermath of hurricane Lothar in late 1999. From 2003 on, the operating conditions appear to normalize again. Thus, it seems that Lothar had, at least in the aftermath, a weak negative effect, but which became apparent only one year later, perhaps due to an excess of production factors which were put in place to deal with the consequences of the hurricane. To determine the long-run impact of public subsidies (in Swiss francs), but also total wood growth (in estimated standing m^3), other receipts for ancillary operations (in Swiss francs), training expenditures (in Swiss francs), and investments (in Swiss francs), on the technical efficiency of the forestry firms, the first step is to investigate the stationarity properties of the variables. Panel unit root tests for stationarity are performed on both levels and first differences for all six variables of the model as indicated in Table 2.

The second step is to test for the long-run relationship between the technical efficiency of the firms and subsidies, total wood growth, other receipts, training expenditures, and investments. The results of the Pedroni panel cointegration tests and FMOLS estimations in its most general form are presented in Table 3. The panel cointegration tests point to the existence of a long-run relationship between efficiency and explanatory variables. Indeed, the results point out that the null hypothesis of no cointegration is strongly rejected in all cases for panel v, panel ρ, panel PP, group ρ, and group PP statistics. As a result, subsidies, total wood growth, other receipts, and training expenditures, have a significant positive effect on efficiency. On the other hand, investments exert a negligible but significant negative effect on efficiency. One explanation might be that the investments considered here (due to the available data for the period analyzed) still include investments in all three functions of forests, i.e., not only in the economic function but also in the protective and social functions. Hence, investments in the protective and social functions of forests might

Table 2 Summary of panel unit root tests

VARIABLES	Pool unit root tests	Level		1st Difference	
		Statistic	Prob.*	Statistic	Prob.*
ln Efficiency	Levin, Lin and Chu t	-3.46975	0.0003	-42.5208	0.0000
	ADF - Fisher Chi-square	184.458	0.7777	1275.25	0.0000
	PP - Fisher Chi-square	193.167	0.6226	1439.34	0.0000
ln Subsidies	Levin, Lin and Chu t	-0.60100	0.2739	-42.0917	0.0000
	ADF - Fisher Chi-square	125.214	1.0000	1283.82	0.0000
	PP - Fisher Chi-square	149.859	0.9967	1508.14	0.0000
ln Total wood growth	Levin, Lin and Chu t	-1.41891	0.0780	-49.5461	0.0000
	ADF - Fisher Chi-square	125.806	1.0000	1322.12	0.0000
	PP - Fisher Chi-square	128.231	1.0000	1440.80	0.0000
ln Other receipts	Levin, Lin and Chu t	1.97469	0.9758	-77.0780	0.0000
	ADF - Fisher Chi-square	87.2617	1.0000	1269.00	0.0000
	PP - Fisher Chi-square	83.6518	1.0000	1369.83	0.0000
ln Training expenditures	Levin, Lin and Chu t	0.67627	0.7506	-39.6059	0.0000
	ADF - Fisher Chi-square	143.777	0.9914	1108.88	0.0000
	PP - Fisher Chi-square	145.667	0.9871	1247.40	0.0000
ln Investments	Levin, Lin and Chu t	-2.74779	0.0030	-50.2561	0.0000
	ADF - Fisher Chi-square	219.913	0.0450	1156.71	0.0000
	PP - Fisher Chi-square	245.915	0.0021	1295.83	0.0000

Exogenous variable: None
Sample: 1998-2006

* Probalities for Fisher tests are computed using an asymptotic Chi-square distribution. All other tests assume asymptotic normality.

Table 3 Long-run estimation results

Dependent variable	ln Efficiency
Method	FMOLS
Independent variables	$N=100$, $T=9$
ln Subsidies	0.49
	(166.21)
ln Total wood growth	0.42
	(86.9)
ln Other receipts	0.25
	(159.25)
ln Training expenditures	0.26
	(111.33)
ln Investments	-0.00
	(-587.67)
Panel statistics	**Statistic** **Prob.**
Panel v-stat*	-5.342982 0.0000
Panel ρ-stat*	9.370455 0.0000
Panel PP-stat*	-8.215251 0.0000
Panel ADF-stat*	-6.775049 0.0000
Group ρ-stat**	14.97219 0.0000
Group PP-stat**	-12.71712 0.0000
Group ADF-stat**	-6.950149 0.0000

(t-stats in parantheses)
* Alternative hypothesis: common AR coefs. (within-dimension)
** Alternative hypothesis: individual AR coefs. (between-dimension)

well exert a negative effect on the productive efficiency of forestry firms. Another reason could be that the existing system of public support for investments creates incentives to invest in unprofitable operations.

4 Conclusions

The analysis of the performance of Swiss public forestry firms shows a potential for improvement in efficiency for most firms. In particular, the analysis of the technical efficiency shows that the number of efficient firms is extremely low, i.e., more than two thirds of the firms have scores below 50%. Regarding the efficiency over time, it appears that initially the efficiency scores are relatively stable (1998–2000), then they experience a decline (2001–2002), before rising again (2003–2006). The lower efficiency scores in 2001 and 2002 can be explained by the aftermath of hurricane Lothar. By 2003, the operating conditions appear to normalize again.

The long-run analysis investigates the influence of different factors on the efficiency of Swiss public forestry firms. The results show that in the long term, public subsidies appear to have a strong positive effect. Reasons might be linked to learning-by-doing effects, the quality of the inputs, or the particular context of forestry economics. Moreover, total wood growth, revenues received by forestry firms, and training expenditures, also have a positive effect on efficiency. On the other side, investments have a negligible but significant negative effect on efficiency. Reasons for this might be linked to the available data or the existing system of public investment support.

References

1. Aoyagi, S., Managi, S.: The Impact of Subsidies on Efficiency and Production: Empirical Test of Forestry in Japan. International Journal of Agricultural Resources, Governance and Ecology **3**(3/4), 216–230 (2004)
2. BAFU: Jahrbuch Wald und Holz. Federal Office for the Environment (FOEN), Berne (2010)
3. Charnes, A., Cooper, W.W., Rhodes, E.: Measuring the Efficiency of Decision Making Units. European Journal of Operational Research **2**, 429–444 (1978)
4. Choi, I.: Unit Root Tests for Panel Data. Journal of International Money and Finance **20**, 249–272 (2001)
5. Farrell, J.: The Measurement of Productive Efficiency. Journal of the Royal Statistical Society: Series A **120**(3), 253–281 (1957)
6. Kumbhakar, S.C., Lien, G.: Impact of Subsidies on Farm Productivity and Efficiency. In: V.E. Ball et al. (eds.), The Economic Impact of Public Support to Agriculture, Studies in Productivity and Efficiency, Vol. 7, pp. 109-124. Springer, New York (2010)
7. Levin, A., Lin, C.F., Chu, C.: Unit Root Tests in Panel Data: Asymptotic and Finite-Sample Properties. Journal of Econometrics **108**(1), 1–24 (2002)
8. Maddala, G.S., Wu, S.: A Comparative Study of Unit Root Tests with Panel Data and a New Simple Test. Oxford Bulletin of Economics and Statistics **61**, Issue S1, 631–652 (1999)
9. Pedroni, P.: Critical Values for Cointegration Tests in Heterogeneous Panels with Multiple Regressors. Oxford Bulletin of Economics and Statistics **61**, Issue S1, 653–670 (1999)
10. Pedroni, P.: Fully Modified OLS for Heterogeneous Cointegrated Panels. In: B.H. Baltagi (ed.), Advances in Econometrics: Nonstationary Panels, Panel Cointegration, and Dynamic Panels, Vol. 15, pp. 93-130. JAI Press (Elsevier), Amsterdam (2000)
11. Schoenenberger, A., Mack, A., von Gunten, F.: Efficacité technique des exploitations forestières publiques en Suisse. Strukturberichterstattung, No. 42, State Secretariat for Economic Affairs (SECO), Berne (2009)

Stream IV
Energy, Environment and Climate

Gerold Petritsch, e&t Energie Handelsgesellschaft m.b.H., Vienna, ÖGOR Chair
Stefan Pickl, Universität der Bundeswehr München, GOR Chair
Thomas Rutherford, ETH Zurich, SVOR Chair

Modeling impacts of the European Emission Trade System on the future energy sector

Carola Hammer

Abstract This study focuses on regulatory instruments and especially on the European Emission Trade System (ETS). It is in our interest to forecast the effects of the ETS on different economic indicators and its costs. Therefore we use an optimization model, in which we consider the regulation framework, the market parameters and technical constraints for the German energy market as well as an endogenous price for emission allowances, running times of plants and capacity enlargements.

1 Introduction

The Emission Trade System (ETS) was implemented in the year 2005 by the EU to control CO_2 emissions in setting annual restrictions and in demanding emission allowances (EA) for every emitted ton of CO_2.[1] Thereby the ETS internalizes environmental costs and adds this new cost type to the production function of companies in the energy sector and energy intensive industry.[2] In contrast to legal requirements (Command-and-Control instruments) the CAP-and-Trade system ETS sets only incentives and remains relatively untested in its economic effects because of its novelty. Therefore we are interested in the reaction of the energy sector on these new costs and the economic consequences of this reaction, what raises the questions:

- Is the ETS an effective regulation instrument for reaching environmental and economic policy goals?
- What will be the costs of the ETS and the EA price subject to the given CAP?
- What impact has the ETS on the future plant portfolio as well as on the power and heat generation?
- What further effects does the ETS have on the German economy
  (e.g. power price, import dependency, supply security, efficiency increase)?

Carola Hammer
Chair of Management Accounting and Control, Technical University Munich e-mail: Carola.Hammer@tum.de

[1] Cf. Geiger et al. (2004).

[2] Cf. Fichtner (2005).

D. Klatte et al. (eds.), *Operations Research Proceedings 2011*, Operations Research Proceedings, 129
DOI 10.1007/978-3-642-29210-1_21, © Springer-Verlag Berlin Heidelberg 2012

2 Approach

For answering these questions we are using a linear programming model to optimize production[3] and investment decisions[4] of the energy sector. Thereby the objective is to maximize the profit, calculated with a contribution margin accounting, and the decision variables are running times of plants and capacity enlargements. For reliable results, we first have to consider the political, economic and technical framework conditions, what gives us the input parameter, constraints and appropriate model structure. Accordingly, we do not simplify but respect the difference between the ETS and taxes in taking the EA price as uncertain and endogenous variable. This results in a calculation of the EA price as intersection of the supply and demand function. The first is given by the CAP. However, the second is deduced by the optimal quantity of carbon emissions depending on the adaption of the merit order to variations in the EA price. Since the switch load effect leads to a discontinuous slope, we calculate the demand function for EA with a sensitivity analysis.

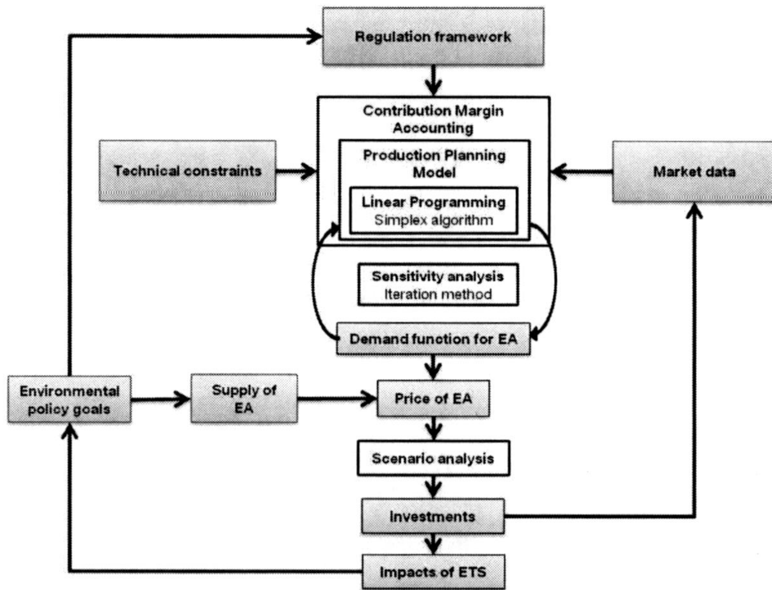

Fig. 1 Approach of the production and investment planning model

In the long run not only running times of plants are variable, but also changes in the plant portfolio are possible due to construction and wear. Since in the long-term perspective the framework conditions may change additionally, we consider the endogenous modifications of the input parameter in a feedback loop on the one hand and use a scenario analysis for the exogenous changes on the other hand. In

---

[3] Cf. Fichtner (2005).

[4] Cf. Roth (2008).

our forecast with time horizon 2010-2020 we look at five different scenarios: referent scenario without ETS (0), basis scenario with ETS (1), enlargement of grid capacity at the boarders (2), without phasing out nuclear energy program (3), with market incentive program (4). After knowing the running times of plants, the EA price and the capacity enlargements, the question after the achievement of environmental and economic policy objectives can be answered by deriving the abatement of CO_2-emissions, the share of renewables and combined heat and power plants (CHP) in power generation, the power import balance, the primary energy demand, the efficiency increase, the import quota of primary energy, the supply security and the power price.

3 Model

3.1 Objective function

The objective function is to maximize the annual profits of the the German energy sector, which consist of the profits from electricity and heat generation in power plants, CHP ($G_{i,a}$) and heat only plants ($G_{h,a}$) as well as profits from power exchange ($G_{b,a}$) within the Union for the Coordination of the Transmission of Electricity (UCTE).

$$\max G_a = \sum_{i=1}^{I} G_{i,a} + \sum_{h=1}^{H} G_{h,a} + \sum_{n=1}^{N} G_{n,a} \tag{1}$$

The profit per power plant or per CHP plant (i) is the contribution margin ($DB_{i,a}$) multiplied by the amount of generated power ($P_{el,i,a} \cdot t_{i,a}$) minus the fixed costs ($K_{F,i,a}$) of the plant. To exclude the decision variables the installed capacities are disaggregated in the already existing installations ($P_{el,i,a}$) and the new capacity enlargement or wear ($Z_{el,i,a}$), in which the existing installations are the capacity enlargements or wear of the past ($P_{el,i,a} = \sum_{\alpha=0}^{a-1} Z_{el,i,\alpha}$). The amount of generated power consists of the installed capacities ($P_{el,i,a} + Z_{el,i,a}$) multiplied with the running time of a plant ($t_{i,a}$). The fixed costs are calculated by the multiplication of the fixed costs per unit ($K_{spezF,i,a}$) and the installed capacities.

$$G_{i,a} = (DB_{i,a} \cdot t_{i,a} - K_{spezF,i,a}) \cdot (P_{el,i,a} + Z_{el,i,a}) \tag{2}$$

Same counts for heat only plants (h) but sales, costs and capacities are referring to the amount of heat. Since there is no heat exchange with neighboring countries, the annual demand constraint for heat ($D_{th,a}$) has to be fulfilled by domestic generation. Therefore heat, which is not produced by CHP [$D_{th,a} - \sum_{i=1}^{I} \frac{1-\sigma_i}{\sigma_i} \cdot (P_{el,i,a} + Z_{el,i,a}) \cdot t_{i,a}$], has to be generated by heat only plants. Thereby the amount of generated heat in CHP is based on the power production and calculated by the power ratio of a plant (σ_i).

$$G_{h,a} = DB_{h,a} \cdot \left[D_{th,a} - \sum_{i=1}^{I} \frac{1-\sigma_i}{\sigma_i} \cdot (P_{el,i,a} + Z_{el,i,a}) \cdot t_{i,a} \right]$$
$$- K_{spezF,h,a} \cdot (P_{th,h,a} + Z_{th,h,a}) \tag{3}$$

In contrast to heat distribution, the losses in power transmission are lower, which offers the opportunity to import or export power. This leads to the possibility of producing more or less power than the annual domestic demand for power ($D_{el,a}$). Whether it is advantageous to import or to export depends on the price difference of the domestic ($p_{el,a}$) and the foreign electricity price ($p_{el,f,a}$).

$$G_{b,a} = (p_{el,a} - p_{el,f,a}) \cdot \left[D_{el,a} - \sum_{i=1}^{I} (P_{el,i,a} + Z_{el,i,a}) \cdot t_{i,a} \right] \tag{4}$$

3.2 Constraints

Demand constraint: The produced power plus the balance of import and export (net import) must meet the domestic power demand. The range of importable or exportable power amount depends on the available grid capacity at the borders ($P_{el,b,a} + Z_{el,b,a}$). Also here the existing grid capacities are the capacity enlargements of the past ($P_{el,b,a} = \sum_{\alpha=0}^{a-1} Z_{el,b,\alpha}$).

$$\left| D_{el,a} - \sum_{i=1}^{I} (P_{el,b,a} + Z_{el,b,a}) \cdot t_{i,a} \right| \leq 8,760 \cdot \sum_{b=1}^{B} v_{b,a} \cdot (P_{el,b,a} + Z_{el,b,a}) \tag{5}$$

Capacity constraint for heat only plants: The sum of generated heat in CHPs as well as in heat only plants has to fulfill the demand function for heat, but must not require a heat production above the available capacities.

$$0 \leq D_{th,a} - \sum_{i=1}^{I} \frac{1-\sigma_i}{\sigma_i} \cdot (P_{el,i,a} + Z_{el,i,a}) \cdot t_{i,a} \leq v_{h,a} \cdot (P_{th,h,a} + Z_{th,h,a}) \cdot 8,760 \tag{6}$$

Running time constraint: In one year the maximal running time of a plant is 8,760 h multiplied by its availability ($v_{i,a}$). The availability of a plant depends on down time, revision time and, in the case of renewable energy, on the availability of the primary energy. Switch off of most of the plants involves revision and start-up costs. Hence most of the plants normally run at least at a certain minimum of full load hours ($\underline{t}_i$).

$$\underline{t}_i \leq t_{i,a} \leq 8,760 \cdot v_{i,a} \quad \forall i \tag{7}$$

Expansion constraint: Especially for renewables, restrictions ($\overline{P}_{el,i}$ respectively $\overline{P}_{th,h}$) are existing in the expansion potential (mainly for geographic reasons). But also difficult or expensive access to fossil primary energies, cooling water for power plants, permits or legal requirements can set an upper limit for constructions.

$$0 \leq \sum_{i=1}^{I} \left(P_{el,i,a} + Z_{el,i,a} \right) \leq \overline{P}_{el,i} \tag{8}$$

$$0 \leq \sum_{h=1}^{H} \left(P_{th,h,a} + Z_{th,h,a} \right) \leq \overline{P}_{th,h} \tag{9}$$

$$0 \leq \left(P_{el,i,a} + Z_{el,i,a} \right) \quad \forall i \tag{10}$$

$$0 \leq \left(P_{th,h,a} + Z_{th,h,a} \right) \quad \forall h \tag{11}$$

For cross-border grid capacities, the EU commission is planning network investments, which are allocated according to demand and international trading. Dismantling is excluded in order to build an European wide electricity market.

$$0 \leq \sum_{b=1}^{B} Z_{el,b,a} \leq \overline{Z}_{el,b,a} \tag{12}$$

$$0 \leq Z_{el,b,a} \quad \forall b \tag{13}$$

4 Results

After solving the optimization model with a simplex algorithm, we receive similar but too low investment levels in scenario 0 to 3. This underinvestment problem leads to cutbacks in total capacity. However scenario 4 with lower capital costs for renewables due to subsidized interest rates shows that countermeasures are possible.

In scenarios 1 to 4 (all with ETS) is the enlargement of power production by biomass and the decline of coal remarkable. Additionally, in scenario 1 and 2 the increasing use of gas is significant both in power and heat production. Investments in wind, water and geothermy plants play only in scenario 4 a special role, which is the only scenario which reaches the EU and national goals of 22 % respectively 30 % of renewables in the energy mix.

Furthermore the total amount of power production decreases in all scenarios, but especially in 3 with the expansion of the EU internal electricity market, since in balance there will be an electricity import instead of an export as before.

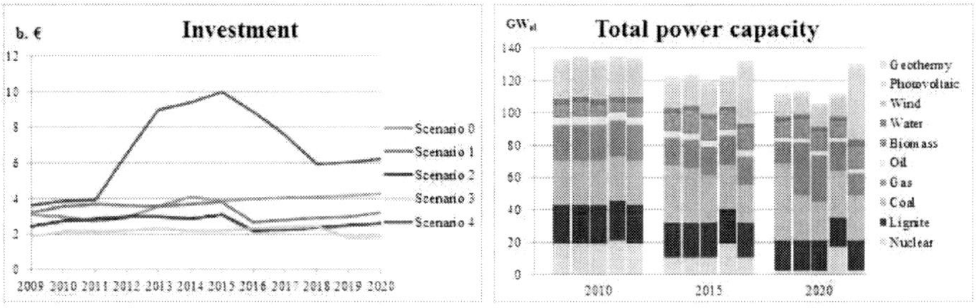

Fig. 2 Investments in capacities and development of the plant portfolio

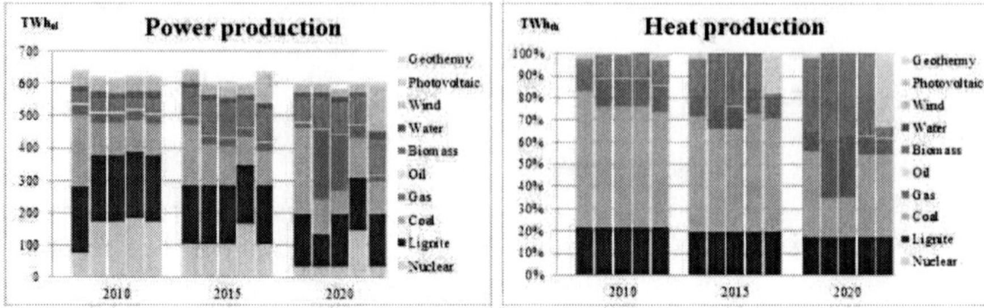

Fig. 3 Power and heat production per technology

But what are the results for the ecological effectiveness and economic efficiency of the ETS? The emission allowance price shows a big range between 0 and 85 Euro in 2020 and so the emission costs with 0 to 24 b. Euro. This means specific emission costs of 0 to 4.01 ct/kWh$_{el}$. Since emission costs are reflected in the power price and the ETS triggers a fuel switch from lignite and coal to more expensive but low emission gas, in 2020 the power price can be 6.16 ct/kWh$_{el}$ higher in comparison to the scenario 0 without ETS. Renewables, nuclear plants and an EU internal electricity market can reduce this price increase effect, but not the emission amount, since the CAP is always used. Nevertheless is the emission abatement ascribed to the ETS 96 to 119 m. t CO_2 or 0.146 to 0.201 t CO_2/MWh$_{el}$. Other reasons for emission abatement can be a technological efficiency increase (about 2 to 15 % points on average over the total plant portfolio) and so a decline in fuel consumption. However, an increase of the primary energy import quota of up to 7 % points in scenario 1 and a decrease of the supply security of up to 8 % points in scenario 4 can also be observed.

5 Discussion and conclusion

Since the ETS has strong impacts on production decisions, it is expedient for emission control. But low interest rates (scenario 4) influence more effective long-term decisions like plant investments. So effective the ETS is in achieving emission goals, so costly it can be on the other hand. Emission costs can go up to 24 b. Euro in 2020 and can noticeably drive up power prices. But choosing the right regulatory framework can lower this costs.

References

1. Fichtner, W.: Emissionsrechte, Energie und Produktion Verknappung der Umweltnutzung und produktionswirtschaftliche Planung. Berlin (2005).
2. Geiger, B., Hardi, M., Brückl, O., Roth, H., Tzscheutschler, P.: CO_2-Vermeidungskosten im Kraftwerksbereich, bei den erneuerbaren Energien sowie bei nachfrageseitigen Energieeffizienzmanahmen. IfE Schriftenreihe Heft 47, München (2004).
3. Roth, H.: Modellentwicklung zur Kraftwerksparkoptimierung mit Hilfe von Evolutionsstrategien. München (2008).

Feasibility on using carbon credits:
A multiobjective model

Bruna de Barros Correia, Natália Addas Porto and Paulo de Barros Correia

Abstract This paper aims to examine the economic feasibility on trading Certified Emission Reductions (CERs) from Clean Development Mechanisms (CDM) projects that are related to electricity generation from renewable energy sources in Brazil. Its purpose is to identify favorable conditions for combining CERs trade obtained by generating electricity from wind power, biomass cogeneration and small hydro-power plants, in replace of fossil fuel plants. As those are all seasonal sources, which means that the energy offers swing along the months of the years, some risks arise associated with the CER's net benefit. Instead of being examined alone, given that some sources can hedge others, the projects are analyzed in a portfolio framework.

1 Introduction

Certified Emission Reductions (CERs) from Clean Development Mechanisms (CDM) projects, that are related to the production and use of renewable energy sources, can be a good opportunity for market investments, as they represent an alternative solution for countries in the ANNEX I, that might not fulfill their emissions targets under the Kyoto Protocol. The activities related to renewable energy can also contribute to reducing Greenhouse Gases (GHG) emissions, as well as helping developing countries to construct or maintain a clean energy mix.

Developing an optimization module for carbon credits contracts can be a useful tool to mitigate the risks for investors and, at the same time, to maximize the benefits, since marketing decisions are directed by the price of negotiated credit.

Bruna de Barros Correia
University of Campinas – UNICAMP – Brazil, e-mail: brunabc@fem.unicamp.br

Natália Addas Porto
University of Campinas – UNICAMP – Brazil, e-mail: natalia@fem.unicamp.br

Paulo de Barros Correia
University of Campinas – UNICAMP – Brazil, e-mail: pcorreia@fem.unicamp.br

D. Klatte et al. (eds.), *Operations Research Proceedings 2011*, Operations Research Proceedings, 135
DOI 10.1007/978-3-642-29210-1_22, © Springer-Verlag Berlin Heidelberg 2012

The main purpose of the present paper is to present a multiobjective model that indicates favorable conditions for combining CERs trade that are obtained by generating electricity from wind power, biomass cogeneration and small hydro-power plants.

2 Quantifying carbon credits from CDM projects

CDM projects are provided by ARTICLE 12 of the Kyoto Protocol and bring about the possibility of partnerships between countries in ANNEX I and NON-ANNEX I countries. It is important to note that activities related to CDM projects shall demonstrate environmental benefits, which have to be compensatory, measurable, focused long-term and, also be related to the reduction or removal of GHG emissions in the atmosphere.

Once a CDM project design document (PDD) is approved and its activity has started, it is possible to identify the energy generated each month by each project. Multiplying the energy generated each month by the emission factor gives the environmental benefit (reduction or removal of GHG emissions) for each project activity. From the moment these benefits are calculated and confirmed, they are used as a financial asset, which are tradeable and are called Certified Emissions Reduction (CERs), as shown in the Data Module (Fig.1). Each unit of CER is equal to one tCO2e and represents a carbon credit. Each carbon credit can be traded under the international carbon market.

Given the CERs emission for each project activity, it is possible to establish the mean benefit reached by each activity using the net present value (NPV) criterion, as is shown in the Optimization Module (Fig.1).

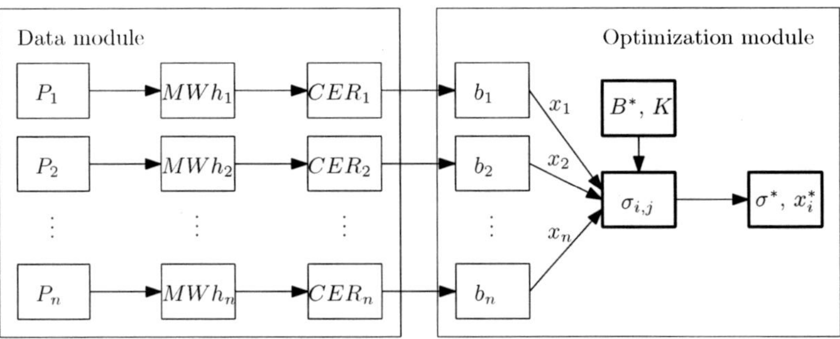

Fig. 1 Model structure: data and optimization modules

3 Portfolio model for selecting projects

Beyond the NPV analyses, the portfolio approach also attempts to focus on the financial risk involved in selecting alternatives for investment, since this risk can be reduced as the assets are diversified. Markowitz [3] proposed a mean-variance model, seeking to increase the mean return and to reduce its variance. The model assumes that investors can choose among n risk assets to seek the investment weight in each asset, which results in a portfolio with an efficient mean-variance compromise.

Despite its multiobjective programming (MOP) conception [7], the portfolio selection problem is often dealt with a nonlinear programming (NLP) approach [4], which minimizes the variance parametrized for a mean return target (Eq.1).

$$\sigma^{*2} = \min\left\{\sum_{i=1}^{n}\sum_{j=1}^{n}\omega_i\sigma_{ij}\omega_j : \sum_{i=1}^{n}b_i\omega_i \geq B^*, \sum_{i=1}^{n}\omega_i = 1, \omega_i \geq 0\right\} \tag{1}$$

where: σ is the portfolio standard deviation; B^* the portfolio benefit target; b_i the asset expected benefit; ω_i the asset weight in the portfolio; σ_{ij} the asset covariance.

This paper works with an integer variant of the Markowitz model [1] where projects are selected for investment. Again the decision maker is looking for an efficient compromise between the mean return and its variance. As additional constraints to Eq.1 the budget is limited and the projects must be integrally accepted or rejected. Thus, the problem of selecting a portfolio of projects is dealt with solving the nonlinear integer programming (NLIP) [9] parametrized in the benefit target B^* (Eq.2).

$$\rho^{*2} = \min\left\{\sum_{i=1}^{n}\sum_{j=1}^{n}x_i\rho_{ij}x_j : \sum_{i=1}^{n}b_ix_i \geq B^*, \sum_{i=1}^{n}c_ix_i \leq K, x_i \in \{0,1\}\right\} \tag{2}$$

where: B^* is the benefit target of the CDM project portfolio; K is the budget target of the CDM project portfolio; c_i is the cost of the CDM project i; x_i is the boolean variable of project i. If $x_i = 1$, then the CMD project i is included in the portfolio, else the project i is not included if $x_i = 0$. To improve numerical stability, the objective function is defined in terms the correlation ρ_{ij} instead of the covariance σ_{ij}, with $\rho_{ij} = \frac{\sigma_{ij}}{\sqrt{\sigma_i\sigma_j}}$.

4 Case study

Table 1 presents the six CDM projects for renewable energy generation that are considered for the present case study: *Água Doce* Wind Power Generation Project (WIN-1), *Horizonte* Wind Power Generation Project (WIN-2), *Bioenergy S.A.* Cogeneration (BIO-1), *Irani Celulose* Cogeneration (BIO-2), *Santana I* SHP Project (SHP-1) and *Cachoeirão* SHP Project (SHP-2).

Table 1 The main data of the CDM projects

	WIN-1	WIN-2	BIO-1	BIO-2	SHP-1	SHP-2
Power (MW)	9	4.8	31	9.43	28.05	14.76
Life (years)	20	20	20	30	30	28
Emission factor (tCO2/MWh)	0.5258	0.5258	0.2677	0.526	0.1635	0.1842
CER Cost (10^6 US$)	0.166	0.166	0.326	0.166	0.326	0.166
Transaction Cost (annual)	7%	7%	7%	7%	7%	7%

It is important to note that the proceedings used to quantify the CERs for each project activity are based on the scheme presented in (Fig.1). The database takes into account the installed power, the plants lifetime, as well as the emission factors obtained from each PDD [5, 8]. The initial costs are based on World Bank estimates over small and large-scale CDM projects, which comprehends the project registration at CDM Executive Board and also periodic verification costs before the CERs issuances. The transaction costs are incurred to complete the CERs issuance and represent 7% on the amount of the emissions emitted annually by each project activity [10]. And, finally, the CERs'price has been estimated in accordance with prices determineted Point Carbon [6], wich was recorded on 15.38 US$/tCO2.

Table 2 Generation (MWh) and CER (tCO2) from the CDM projects

Month	WIN-1 MWh	WIN-2 MWh	BIO-1 MWh	BIO-2 MWh	SHP-1 MWh	SHP-2 MWh	WIN-1 tCO2e	WIN-2 tCO2e	BIO-1 tCO2e	BIO-2 tCO2e	SHP-1 tCO2e	SHP-2 tCO2e
1	1709	575	0	3002	10456	5570	899	302	0	10101	1709	1026
2	1563	498	0	2671	10456	5570	822	262	0	9124	1709	1026
3	1499	487	0	3147	10456	5570	788	256	0	11820	1709	1026
4	1499	481	789	3034	10456	5570	788	253	211	11723	1709	1026
5	1791	640	7901	3159	10456	5570	942	337	2115	12703	1709	1026
6	1840	703	9392	2936	10456	5570	967	370	2514	11906	1709	1026
7	1861	857	8464	3120	10456	5570	979	451	2266	12637	1709	1026
8	1931	875	9928	3265	10456	5570	1016	460	2658	12758	1709	1026
9	1927	909	9876	3103	10456	5570	1013	478	2644	11531	1709	1026
10	1824	805	9553	3154	10456	5570	959	423	2557	11633	1709	1026
11	1776	721	7277	3067	10456	5570	934	379	1948	11805	1709	1026
12	1628	654	2825	3229	10456	5570	856	344	756	10144	1709	1026
TOTAL	20848	8205	66006	36886	125475	66843	10962	4314	17669	137884	20513	12311

Table 2 displays the monthly energy generation values as well as the CERs obtained from the six projects activities. Exceptionally, BIO-2 also includes avoided methane emissions, as cellulose biomass production is not disposed of in landfills.

Sequentially, it is established the mean net benefit reached by each activity using the NPV criterion. The investment rate of return on the financial market is weighed

Table 3 Monthly mean benefit from the CDM projects

Month	WIN-1 US$	WIN-2 US$	BIO-1 US$	BIO-2 US$	SHP-1 US$	SHP-2 US$
1	8235	1168	-4834	115006	15883	10101
2	7104	733	-4834	108549	15710	9998
3	6524	772	-4834	148246	15540	9895
4	6484	767	-2624	147505	15371	9794
5	8515	1844	19469	161502	17215	9693
6	8789	2116	23964	149920	17033	9593
7	8854	3328	22527	156559	16851	9494
8	9285	3194	28206	157231	16671	9396
9	9211	3461	28182	139561	16493	9299
19	9347	2681	27024	137697	16316	9202
11	8903	2213	19291	139587	16140	10256
12	7702	1711	5315	115796	15966	10151

at 10% per year. The monthly mean benefit data for the six projects are described in Table 3.

Table 4 Annual correlation ρ_{ij} and mean benefit b_i

ρ_{ij}	WIN-1	WIN-2	BIO-1	BIO-2	SHP-1	SHP-2
WIN-1	1,00000	0,88046	0,89767	0,28322	0,76430	-0,53686
WIN-2	0,88046	1,00000	0,91854	0,43685	0,69677	-0,68046
BIO-1	0,89767	0,91854	1,00000	0,56237	0,81081	-0,70244
BIO-2	0,28322	0,43685	0,56237	1,00000	0,56644	-0,53690
SHP-1	0,76430	0,69677	0,81081	0,56644	1,00000	-0,52023
SHP-2	-0,53686	-0,68046	-0,70244	-0,53690	-0,52023	1,00000
b_i	98951	23990	156852	1677159	195191	116873

Table 4 shows the annual correlation coefficients (ρ_{ij}) and mean benefit (b_i) for the six projects, which are the input data of the optimization module (Fig. 1).

The nonlinear integer programming is dealt with by a commercial solver, the Lingo 11.0, which contemplates an integrated suite of optimization modules [2]. The main results are the Pareto efficient solutions identified for several values of the benefit target B^*, as is shown in Table 5. The first column indicates the selected projects by the $\mathbf{x}^*$ vector components, and columns two and three show the associated risk σ^* and mean net benefit B^*.

As expected, the risk σ^* increases with the expected benefit target B^*. The minimum risk is achieved when the projects SHP-1 and SHP-2 are selected, with σ^* equal to US\$ 491 and B^* to US\$ 25,000. At the other extreme, the maximum net benefit is attained when all projects, but WIN-2 are selected. It results in σ^* equal to US\$ 27,901 and B^* to US\$ 185,000.

Table 5 Pareto efficient frontier

Selected Projects x^*	Standard Deviation σ^*	Mean Benefit B^*
000011	491	25000
100011	1328	30000
110011	2179	35000
011011	14694	40000
000101	16805	140000
100101	17109	150000
000111	17134	160000
100111	17457	170000
110111	17947	175000
001111	27256	178000
101111	27901	185000

5 Conclusions

Investing in the carbon credits market through the implementation of CDM projects might be a good opportunity for investors. The present paper case study proposes a new model for investment analysis attached to projects design documents, which identifies favorable conditions for combining CERs trade obtained with generating electricity from wind power, biomass cogeneration and small hydro-power plants. Considering that those are all seasonal sources, the energy offer swings along the month of the years. As a consequence, there is a loss of ability to manage risk as the expected benefit rises. Given that some sources can hedge others, the projects are analyzed in a portfolio framework. The entire formulation presented is included in a commercial solver. And the results are demonstrated through a Pareto frontier, which indicates the non-dominated set of solutions.

References

1. Castro, F., Gago, J., Hartillo, I., Puerto, J., Ucha, J.M.: An algebraic approach to integer portfolio problems. European Journal of Operational Research **210**, 647–659 (2011).
2. Lindo Systems Inc. http://www.lindo.com.
3. Markowitz, H.M.: Portfolio Selection. Journal of Finance **7**, 77–91 (1952).
4. Markowitz, H.M.: Mean-Variance Analysis in Portfolio Choice and Capital Markets. Wiley, New York (2000)
5. MCT, Ministério de Cincia e Tecnologia. http://www.mct.gov.br
6. Point Carbon. http://www.pointcarbon.com
7. Steuer, Ralph E.: Multiple Criteria Optimization: Theory, Computation, and Application. Wiley, New York (1986)
8. UNFCCC, United Nations Framework Convention on Climate Change. http://unfccc.int
9. Wolsey, L. A.; Nemhauser, G. L., *Integer and combinatorial optimization*, Wiley-Interscience. (1999).
10. World Bank: State and Trends of the Carbon Marker. The World Bank, Washington DC (2008)

Integration of Fluctuating Renewable Energy in Europe

Stephan Spiecker and Christoph Weber

Abstract With increasing amounts of power generation from intermittent sources like wind and solar, capacity planning has not only to account for the expected load variations but also for the stochastics of volatile power feed-in. Moreover investments in power generation are no longer centrally planned in deregulated power markets but rather decided on competitive grounds by individual power companies. This poses particular challenges when it comes to dispatch. Within this article an approach is presented which allows assessing electricity market development in the presence of stochastic power feed-in and endogenous investments of power plants and renewable energies.

1 Introduction

In recent years, climate protection has played a growing if not even dominating role in the European energy policy and also in the energy policy of many member states. Beside the introduction of the EU certificate trading scheme far-reaching policies and measures have been developed for expansion of renewable energies right up to a scenario with 100% electricity production from renewable energy. Many of these concepts give the impression that target scenarios are developed neglecting any potential risk and imponderables along the appreciated way. This applies especially for the fluctuating feed-in of energy sources like wind and solar energy.

In Europe large amounts of intermittent generation capacities, most notably wind turbines and solar modules have been installed in recent years. Technologies with intermittent production makes system planning more difficult, because their capacity credit is limited compared to conventional generation and their stochastic behav-

Stephan Spiecker
Chair for Management Sciences and Energy Economics, University of Duisburg-Essen , 45117 Essen, Germany, e-mail: stephan.spiecker@uni-due.de

Christoph Weber
Chair for Management Sciences and Energy Economics, University of Duisburg-Essen , 45117 Essen, Germany, e-mail: christoph.weber@uni-due.de

D. Klatte et al. (eds.), *Operations Research Proceedings 2011*, Operations Research Proceedings, 141
DOI 10.1007/978-3-642-29210-1_23, © Springer-Verlag Berlin Heidelberg 2012

ior impacts both the dispatch of plants and also the long run investment planning in power systems [3], [6]. Therefore a stochastic power system market model is presented in the following that takes the intermittent characteristics of wind and solar radiation into account and is capable of modeling the whole European power market in order to evaluate future power system developments. We use the model to assess the influence of intermittent production of renewable energies on future power market dispatch based on several scenarios.

2 Methodology

For a quantification of the effects of fluctuating renewable energies concerning power plant operation as well as investments the stochastic European electricity market model (E2M2s) is used. Assuming functioning competitive markets, the E2M2s model determines market results through optimization. As a result cost efficient power plants are used to cover (price-inelastic) demand. This approach is formulated as a linear, stochastic programming model encompassing different regions and different time steps (typical days and typical hours). It is implemented in the General Algebraic Modeling System (GAMS). A detailed description of the model equations can be found in [4] and [5].

The model determines endogenously the optimal power plant operation and transmission line loading, but also investments in new generation capacities. Therefore annualized investment costs for newly built capacities are considered in the system objective function. This leads to investment decisions in line with the Peak load Pricing approach as developed e.g. by [1], with the yearly full load hours being a key driver for technology selection. Investments in RES are further restricted by cost potential curves, their stochasticity is depicted by recombining trees [2].

3 Scenarios

Besides uncertainty due to stochastic renewable energy infeed, other uncertain factors are taken into account by considering different scenarios (cf. Tab. 3). These scenarios focus on different aspects of the triangle of energy policy targets, consisting of security of supply, sustainability and economic efficiency. For the scenario calculations, different parameter sets for demand, fuel price, CO_2 reduction targets, acceptance of nuclear power and RES electricity import (from Africa), are defined (cf. Table I). For RES, development paths are exogenously set until the year of policy change. After that renewable energy technologies are only endogenously invested under market conditions.

In a first scenario, environmental protection and renewable energies are given high priority (climate - market). In a sensitivity analysis, the impact of government driven expansion of wind, solar and biomass on the integration of renewable energies is investigated (climate - policy). In an alternative scenario, the focus is on preserving competitiveness without compromising environmental protection (efficiency). A fourth scenario focuses on security of supply in connection with strong economic growth (security). Finally, a scenario is investigated where conflicting interests and competing objectives lead to an energy policy without any clear priorities (conflict).

	Climate -Policy	Climate -Market	Efficiency	Security	Conflict
Demand	low	low	low	high	mid
Politically driven RES development	high	high	mid	low	mid
RES imports from North Africa	high	high	(-)	(-)	low
Fuel price coal and natural gas	high	high	high	coal: low gas: high	mid
CO_2-reduction compared to 1990	95%	95%	80%	30%	60%
Acceptance of nuclear power	low	low	high	mid	low
RES policy change [year]	(-)	2020	2015	2025	2020

Table 1 scenario overview

4 Results

4.1 Common development

Development of power production capacity is mainly driven by future electricity demand. With increasing demand new power plant capacities are needed and old power plants have to be replaced. Beside these market driven investments some investments are politically motivated - especially the extension of renewable energies. According to the development of electricity demand the highest capacity occurs in the security-scenario (cf. Fig. 1).

The flexibility of conventional power plants hardly represents a barrier for the expansion of renewable energies. However, fluctuation and limited forecasting possibilities of renewable energy feed-in cause additional costs in cases of high penetration. Full load hours of conventional power plants are dramatically reduced and especially gas fired power plants are used as back up technologies that are held available for security reasons. But also for renewable energies a decrease in full-load hours occurs, if low demand and high availability of renewable energies coincide. Advantages of gas fired power plants are their flexibility, their low CO_2 emissions and their low fix costs which make them an adequate substitute to fluctuating energies.

The main challenge for the conventional generation system are high shares of fluctuating renewable energy. Therefore, the climate-market scenario as a scenario with high capacities of fluctuating renewable energy is in the focus of the following analysis.

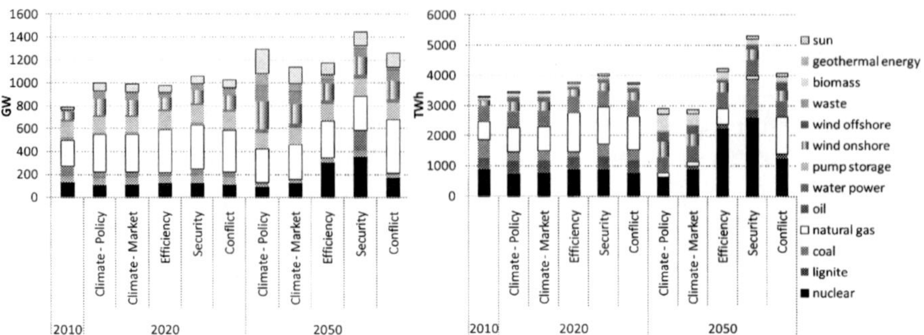

Fig. 1 Development of European power plant capacities and power production

4.2 Production

In Fig. 2 typical production patterns for European power production in the year 2050 (climate-market scenario) are presented. In the upper figures a typical weekend day in summer is depicted. Here, load is low compared to other days of the year and especially solar infeeds are high. On the right side infeed from renewables on a summer day are low while they are high on the left side. The adjustment of the

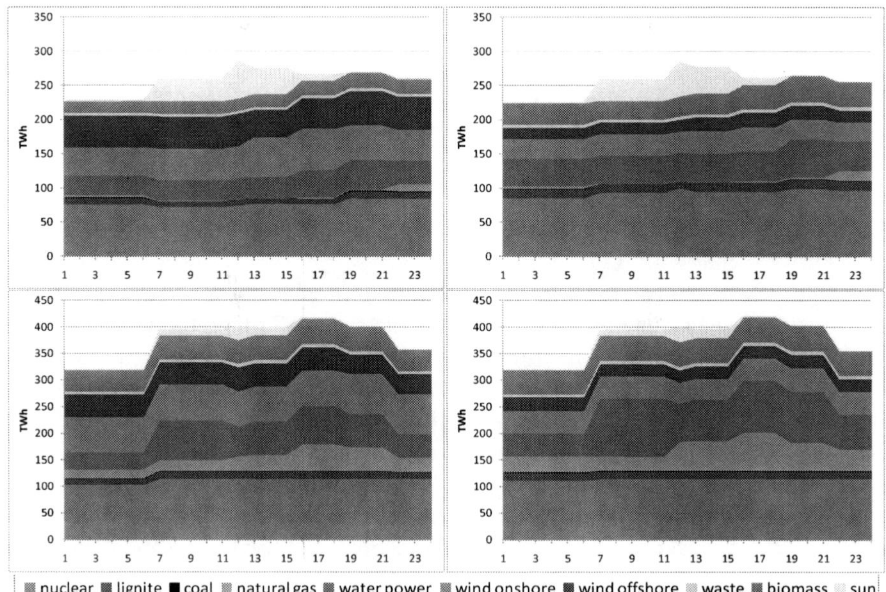

Fig. 2 European power production at the twelfth hour of a typical summer weekend day with a) high and b) low RES infeeds and an evening of a typical winter workingday with c) high and d) low RES infeeds

residual power plant is done by biomass fired power plants on the one side but also by nuclear and lignite fired power plant. To offer this flexibility they often have to work in part load. Remaining adjustments are done by natural gas and hydro power plants. Hydro power plants are partly restricted because of their missing storage capacity in the case of run-of-river power plants.

The lower pictures show a working day in winter. There, load is high and infeeds especially from solar production are low. Again, a situation with typical high and a situation with typical low infeeds are presented. Adjustments are mainly achieved by hydro and thermal power plants namely natural gas and nuclear. Especially for reduction of fluctuations between single hours natural gas fired power plants are used.

Comparing summer and winter day the main differences in the conventional power system occur for seasonal hydro storage and natural gas fired power plants. They are predestinated for such employment arising from changes between high and low residual load situations due to their flexibility and cost structure.

4.3 Transmission

Fig. 2 shows different load flow situations in Europe for the climate-market scenario in 2050. Again the upper pictures depict a situation at a weekend in summer at the twelfth hour. On the left side RES infeed is low while it is high on the left side. Still in 2050 Germany is a main producer of fluctuating renewable energy. But other countries like France and Spain also raise their potentials. But due to better weather conditions especially for solar production but also for wind production they are less effected by fluctuations and uncertainty. In the case of high infeeds the Scandinavian especially hydro based system is used to adjust the fluctuating production on the European mainland. When renewable infeeds are low electricity from Scandinavian hydro power plants is exported to continental Europe. But also the flows on Germany's western border are affected where imports are inverted to exports of several GWh in high wind situations. In general high infeeds lead to increasing power flows to the European peripheral regions. Independent from the infeed situation a North-South-flow can be observed with constant flows to Italy.

The lower pictures present a similar situation on an evening in winter with high and low wind infeed. In comparison with the situation in summer changes in RES infeed have a lower impact on power system flows. But again a reversal of flows on the western German border can be observed. Main changes occur within Scandinavia where Norwegian hydro power reduces supply shortfalls in the other countries when infeeds from RES are low. It can also be observed that Austrian hydro storages - seasonal storages but also pump storages - are used to compensate the fluctuations from RES infeed. Comparing all four situations it is shown that Austrian hydro power is especially used for fluctuations within a day while Scandinavian hydro power is used for longer term fluctuations.

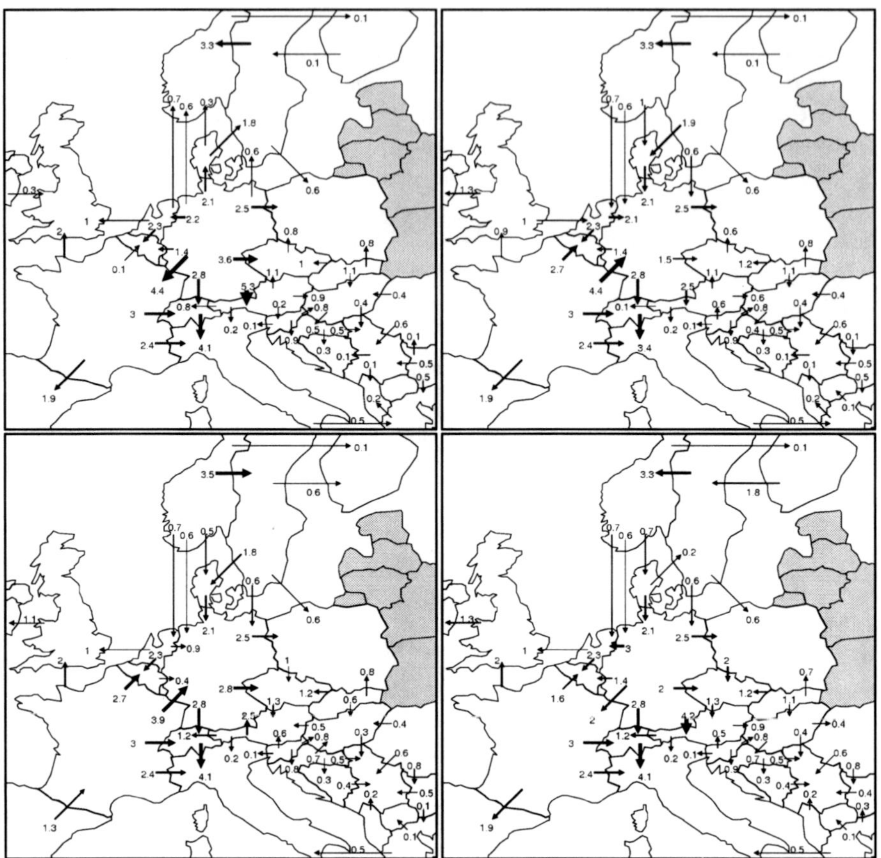

Fig. 3 Transmission on a typical summer weekend day with a) high and b) low RES infeeds and a typical winter workingday with c) low and d) high RES infeeds

References

1. Boiteux, M.: Peak load-pricing. Journal of Business. **33**, 157–179 (1960)
2. Kchler, C., Vigerske, S.: Decomposition of Multistage Stochastic Programs with Recombining Scenario Trees. Stochastic Programming E-Print Series 9, (2007)
3. Mst, D., Fichtner, W.: Renewable energy sources in European energy supply and interactions with emission trading. Energy Policy. **38**, 2898–2910 (2010)
4. Spiecker, S., Weber, C.: Integration of Fluctuating Renewable Energy - a German Case Study. EWL Working Paper (2010)
5. Swider, D., Weber, C.: The costs of wind's intermittency in Germany: application of a stochastic electricity market model. Euro. Trans. Electr. Power. **17**, 151–172 (2007)
6. Tuohy, A., Meibom, P., Denny, E., O'Malley, M.: Unit commitment for systems with significant wind penetration. IEEE Trans. Power Syst. **24**, 592–601 (2009)

Price-Induced Load-Balancing at Consumer Households for Smart Devices

Cornelius Köpp, Hans-Jörg von Metthenheim and Michael H. Breitner

Abstract The rising share of renewable energies in today's power grids poses challenges to electricity providers and distributors. Renewable energies, like, e.g., solar power and wind, are not as reliable as conventional energy sources. The literature introduces several concepts of how renewable energy sources can be load-balanced on the producer side. However, the consumer side also offers load-balancing potential. Smart devices are able to react to changing price signals. A rational behavior for a smart device is to run when electricity rates are low. Possible devices include washing machines, dryers, refrigerators, warm water boilers, and heat pumps. Prototypes of these devices are just starting to appear. For a field experiment with 500 households we simulate adequate device behavior. The simulation leads to a mapping from price signal to load change. We then train a neural network to output an appropriate price signal for a desired load change. Our results show that even with strong consumer-friendly constraints on acceptable price changes the resulting load change is significant. We currently implement the results with a leading energy services provider.

1 Introduction

The demand of electrical energy fluctuates permanently. Over the years typical patterns of aggregate energy consumption were observed and serve as the base for the planning of energy production, so the energy supply is regulated according to fluc-

Cornelius Köpp
Institut für Wirtschaftsinformatik, Leibniz Universität Hannover, e-mail: koepp@iwi.uni-hannover.de

Hans-Jörg von Mettenheim
Institut für Wirtschaftsinformatik, Leibniz Universität Hannover, e-mail: mettenheim@iwi.uni-hannover.de

Michael H. Breitner
Institut für Wirtschaftsinformatik, Leibniz Universität Hannover, e-mail: breitner@iwi.uni-hannover.de

D. Klatte et al. (eds.), *Operations Research Proceedings 2011*, Operations Research Proceedings, 147
DOI 10.1007/978-3-642-29210-1_24, © Springer-Verlag Berlin Heidelberg 2012

tuating demand. Due to the development of renewable energy sources like wind turbines and photovoltaic systems, there is a steadily increasing proportion of uncontrollable variations [4] on the supply side. In contrast to conventional power plants, these energy sources do not produce precisely defined quantities. Electrical energy can now be saved only with great losses, unlike, for example gas. Known memory options such as pumped-storage power plants have been built in the past at appropriate places. A further expansion is hardly possible. As the unpredictable flows [10] will continue to increase in future, alternative solutions are required.

To deal with this problem there are several concepts on the producer side, like virtual power plants; see for example [6, 9]. However, the consumer side also offers load-balancing potential, so both sides should be part of solutions for the future of energy grids [6]. Variable electricity rates can provide customers with an incentive to shift their electricity consumption to times in which, for example, there is a surplus of wind energy.

2 Dynamic Electricity Rates

In the past – and in general still now – electricity rates are constant all over the day. Well known exceptions are night storage heaters with a two-rate electricity meter. This will change in the future: In Germany, for example, since 2011 the energy suppliers have to offer special electricity rates to encourage people to save energy or to shift their power consumption to times of lower consumption. This can be realized by high prices for peak-load times (typically during the day) and low prices for other times (typically during the night). That concept (similar to night storage heaters in the past) is still very static – unlike the electricity production of renewable energies.

Short term, variable, dynamic electricity prices allow the reduction of the variance between demand and supply [5]. Real time pricing is considered the most direct and efficient price based approach for demand side management [1]. Basic idea of dynamic rates is to reflect the actual costs of electricity generation [1]. Following [7] the electricity rates will be higher when there is high demand, and lower when there is low demand. [8] assume that the load is proportional to the costs. This is suitable for a constant generation, but not for an integration of fluctuating renewable energy sources. For proper load shifting, [8] propose a rate model independent of actual energy costs. The energy provider announces the price for each time period up to one day in advance.

There is as potential for controlling more than 10% of electricity, as about 50% of load in households is manageable with only little discomfort [9] and as households for example in Germany consumed nearly 29% of electrical energy in 2009 [3]. In literature typical devices listed as suitable for load management are dishwashers, washing machines, dryers, refrigerators, freezers and heaters with storage. Other devices like night storage heaters, air conditioners, heat pumps and batteries, especially in plug-in hybrid electrical vehicles are also mentioned; see for example [1, 5–9]. Many concepts include still a manual operation of the devices by the user. This leads to a significant loss of comfort and requires the presence of the user. In addition, the user must actively inform about the current and future prices.

Dynamically priced electricity rates are much more complex than constantly priced electricity rates: A dynamically priced electricity rate model must define possible prices, steps as well as upper and lower limits, and must also determine how far in advance consumers will receive the price information. The model must not be too complex, so that it remains understandable to consumers and doesnt violate regulatory restrictions [2]. The price model planned for the field experiment and used in the simulation prototype is kept very simple: The base price is set to 20Ct/kWh and there is a range between a defined minimum (lower than acutal price) of 15Ct/kWh and maximum (higher than acutal price)of 25Ct/kWh (11 1Ct/kWh steps). The price is defined for each hour and will be constant for at least two hours. The price is announced and fixed for six hours (preview time) in advance. This enables a fast response to changes in wind and solar power and still offers the consumer a reasonable certainty about the costs incurred for its electricity consumption.

3 Simulation of Smart Devices

As smart devices for consumer households are not widely available yet, we simulate their behavior as preparation for the field experiment with 500 households. Smart devices are able to react to changing price signals. A rational behavior for a smart

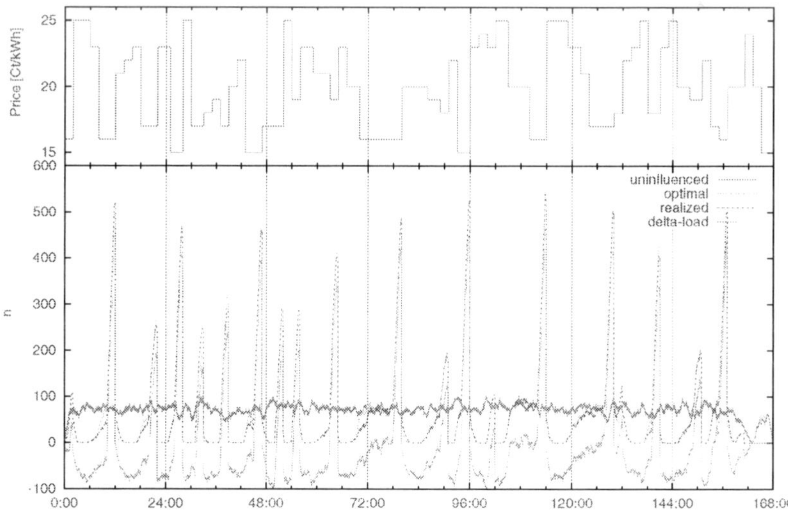

Fig. 1 Simulation result of devices running once (like dishwashers, washing machines, dryers)

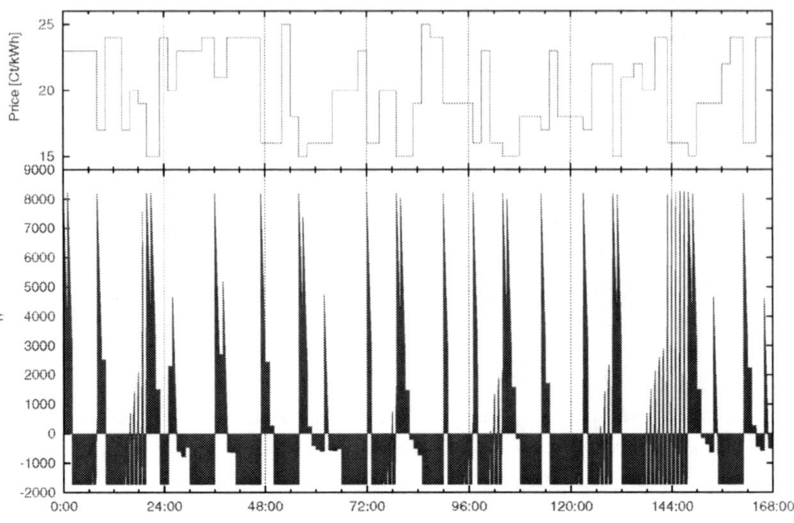

Fig. 2 Simulation reasult of heating devices (like water boilers with storage)

device is to run when electricity rates are low. The exact optimization stragegies used by real smart devices are still unknown, but simulation allows to test several strategies and parameters. The realized load is heavyly dependent on the optimization stragegy used by the devices. We implement two classes of smart devices:

- Devices running once (for a few minutes to several hours) in a defined limited time slice (of several hours, typically much longer than running time), as simplified model of real devices like dish washer, washing machine, or dryer. These devices start at once as soon as the current time is the unique cheapest time within the preview time. Otherwise the devices wait for are cheaper time. It is ensured that the device execution is complete at the end of the limited time slice. Figure 1 show examplary results from simulation of this device class.
- Devices without time restriction and a given daily runtime, as a simplified model of water heaters with large storage or heat pumps. The daily running time is based on running times of real water heaters. These devices have a storage capacity of several hours. The storage is filled completely while the current time is the cheapest time within the preview time. Otherwise the device will only run (for minimal required time) when the storage runs empty in the beginning hour.

The results (examplarily see figure 1 and 2) of simulation show that the price significantly influences the load. The two implemented device classes result in different delta-loads (difference between uninfluenced and influenced load). The load change is asymmetric: There are high positive peaks and smaller negative troughs. The very high peaks are the result of cheapest solution for the device owners, but are not in the interest of the energy provider.

4 Neural Network Training

The result from the simulation is a mapping from price signals to resulting load changes. In the final application we are, however, interested in getting this information the other way round. That is, we need a mapping that associates a desired load change (delta load) with a corresponding price signal. With a limited number of devices with known behavior it would have been possible to explicitly formulate and solve an optimization problem. In practice, we will at least have several thousands devices with unknown behavior. This makes the explicit approach unrealizable. Instead, we want to be able to learn from aggregated observed device behavior on the fly.

The underlying problem is complex and nonlinear. The nonlinearity results from a supposed switch-like behavior of devices to price signals. That is, there will be price signals which cause several devices to switch on or off almost at once. To solve this task we choose a neural network due to its universal approximation ability which includes the ability to model swith-like behavior well via the transfer function tanh.

Our network type is the classic feedforward three layer perceptron. A recurrent topology is another possibility as we are dealing with time series. However, the three layer perceptron already gives us very satisfactory results. The final implementation will be in the control room of an energy provider and a side goal is therefore, to get a solution which integrates well with other software. The three layer perceptron does not need to maintain state information and was considered an acceptable choice. A deeper investigation of recurrent topologies is left to further research.

The inputs to our network consist of

- price signals of the last 12 hours (history)
- fixed price signals of the next 1 up to 6 hours (future, preview time)
- realized delta loads of the past 12 hours (history)
- the desired delta load for the following 1 hour (future)

while the output is the

- price signal for 1 hour, following the 1st up to 6th hour (future, depends on preview time)

To reduce complexity our dataset is prepared to reflect archetypical price patterns. This means, that we use only few price levels, e.g., a mean price with a several hours peak (varied from two to 24 hours), and then again a flat price. For each tupel of device class, preview time and peak duration the simulation is used to to generate the training data consists of about 3500 datas set while another about 1000 data sets are used for cross-validation. We use early-stopping, that is, training stops when the error on the validation data sets increases twice in a row. The optimization algorithm is BFGS, a quasi-Newton method. We experimented with different input data sets and different numbers of hidden neurons. The above input set and one hidden neurons gave us the best results. Overall, results are dependen on the shape of the price curve and the preview time.

5 Conclusions and Outlook

Our results indicate that the approach produces usable price signals. The resulting load changes are significant. This holds true, even with strong consumer-friendly constraints like a look-ahead time window for prices. An important advantage is that we can incorporate new data by simply retraining our network. We can therefore cope with an increasing pool of smart devices and with shifting device behavior. Further, only unidirectional communication is needed. I.e., the price signal is broadcast to all devices but there is no need for an individual response by the device. This alleviates privacy concerns.

A topic for further investigation is the limitation of too strong a reaction. There may be cases where a price signal produces too large load swings, because of all devices switching on or off simultaneously. This challenge will not be our chief concern in the field experiment with a still limited number of households. However, it has to be taken into account for future deployment of our approach. Grouping of consumers offers a natural solution: different groups can get different price signals. This can transform a sudden impact on load in a more gradual one.

References

1. Albadi, M.H., El-Saadany, E.F.: Demand response in electricity markets: An Overview. In: Proc Power Engineering Society General Meeting (2007) doi: 10.1109/PES.2007.385728
2. Benz, S. Energieeffizienz durch intelligente Stromzähler – Rechtliche Rahmenbedingungen. Zeitschrift für Umweltrecht. **10**, 457–463 (2008)
3. BMWi: Zahlen und Fakten - Energiedaten - Nationale und Internationale Entwicklung (2011) http://www.bmwi.de/BMWi/Redaktion/Binaer/energie-daten-gesamt,property= blob,bereich=bmwi,sprache=de,rwb=true.xls Cited 28 Feb 2011
4. Cardell, J.B., Anderson, C.L.: Analysis of the System Costs of Wind Variability Through Monte Carlo Simulation. In: Proceedings of the 43rd Hawaii International Conference on System Sciences (2010)
5. Chen, C., Kishore, S., Snyder, L.V.: An innovative RTP-based residential power scheduling scheme for smart grids. In: Proceedings of the IEEE Conference on Acoustics, Speech, and Signal Processing (2011)
6. Köpp, C., von Mettenheim, H.-J., Klages, M., Breitner, M.H.: Analysis of Electrical Load Balancing by Simulation and Neural Network Forecast. In: Operations Research Proceedings 2010, pp. 519–524 (2010) doi: 10.1007/978-3-642-20009-0_82
7. Lee, J., Jung, D.K., Kim, Y., Lee, Y.W., Kim, Y.M.: Smart grid solutions, services, and business models focused on telco. In: Proceedings of Network Operations and Management Symposium Workshops (2010) doi: 10.1109/NOMSW.2010.5486554
8. Mohsenian-Rad, A.H., Wong, V.W.S., Jatskevich, J., Schober, R.: Optimal and autonomous incentive-based energy consumption scheduling algorithm for smart grid. In: Proceedings of Innovative Smart Grid Technologies (2010) doi: 10.1109/ISGT.2010.5434752
9. Molderink, A., Bakker, V., Bosman, M. G.C., Hurink, J., Smit, G. J.M.: A three-step methodology to improve domestic energy efficiency. In: Proceedings of the Innovative Smart Grid Technologies (2010) doi: 10.1109/ISGT.2010.5434731
10. Sioshansi, R.: Evaluating the Impacts of Real-Time Pricing on the Cost and Value of Wind Generation. In: IEEE Transactions on Power Systems (2010) doi: 10.1109/TP-WRS.2009.2032552

Integration of renewable energy sources into the electricity system with new storage and flexibility options

Dr. Matthias Koch and Dierk Bauknecht

Abstract The integration of renewable energy sources is one of the key challenges for the transition of the energy supply chain towards a low carbon society. Due to the intermittency of wind and solar power additional storage and flexibility is required for their integration into the electricity system. Based on a scenario analysis for Germany, the economic and environmental effects of flexibility from electrical cooling as an example for demand side management in a smart grid as well as traditional flexibility from pumped storage power plants is evaluated with the PowerFlex model developed by Öko-Institut. In the 2030 scenario with new flexibility, about 500 GWh of renewable electricity could be integrated in addition which leads to cost reduction effects as well as a decrease of carbon dioxide emissions.

1 Introduction

The integration of renewable energy sources into the electricity system is one of the major tasks for the transition of the economy towards a low carbon production. In Germany for example, electricity generation from fossil fuels was responsible for nearly 40 % of the overall national climate gas emissions in 2009 [5]. Beside of energy efficiency measures to reduce electricity consumption, electricity generation from renewable energy sources like hydro, wind, solar and geothermal power as well as solid, liquid and gaseous biomass is a key element of a sustainable electricity system. Due to the opportunity to store biomass, the integration of electricity generated from biomass is relatively easy. Compared to that, intermittent electricity supply from wind and solar power plants requires additional flexibility and storage capacities in the electricity system. Until now this is mainly provided by thermal

Dr. Matthias Koch
Öko-Institut e.V., Merzhauser Str. 173, D-79100 Freiburg, e-mail: m.koch@oeko.de

Dierk Bauknecht
Öko-Institut e.V., Merzhauser Str. 173, D-79100 Freiburg, e-mail: d.bauknecht@oeko.de

D. Klatte et al. (eds.), *Operations Research Proceedings 2011*, Operations Research Proceedings,   153
DOI 10.1007/978-3-642-29210-1_25, © Springer-Verlag Berlin Heidelberg 2012

power plants and pump storage power plants. But with an expected share of up to 35 % wind power and up to 10 % solar power in 2030 [7], additional flexibility and storage options are needed.

2 Methodology and model description

The PowerFlex model is a bottom-up partial model of the power sector developed by Öko-Institut. It calculates the dispatch of thermal power plants, feed-in from renewable energy sources, pumped storage hydro power plants and flexible power consumption at minimal costs to meet the electricity demand and the necessary reserve capacity. Grid restrictions are not taken into account. The PowerFlex model is designed as a linear mixed-integer optimisation problem, which considers three different operating conditions for thermal power plants with an installed electrical capacity exceeding 100 MW (start-up and shutdown, partial load and full load). Alongside technology-specific ramp rates, electrical efficiency ratios are also distinguished in the different operating conditions based on linear approximation.

Smaller thermal power plants are grouped together according to technology and construction year and ascribed characteristics with the help of type-specific parameters. For these power plants ramp rates are not taken into account. The same applies to pumped storage hydro power plants, which are grouped together according to comparable relations of storage capacity to installed electrical capacity. Biomass power plants using biogas, wood or plant oil are part of the thermal power plant fleet as well as one import power plant. This virtual power plant represents the electricity exchange with the European transmission grid. The marginal costs of the import power plant are set slightly above the most expensive thermal power plant of the national fleet.

The available electricity produced from run-of-river, offshore wind, onshore wind and photovoltaic is predefined using generic feed-in patterns in hourly resolution. The actual quantity of feed-in from hydro, wind and photovoltaic power is determined endogenously, with the result that the available yield of fluctuating electricity can also be curtailed (e.g. in the case of negative residual load and insufficient storage capacity).

The production pattern for electricity from combined heat and power is based on a typical pattern for district heating and an assumed uniform distribution in the case of industrial CHP plants. This produces a specific CHP pattern for each major energy source. For must-run power plants like blast furnace gas or waste incineration plants, an uniformly distributed feed-in of electricity is assumed.

A crucial aspect is the generic representation of demand side flexibility in the PowerFlex model as storage. Flexibility is thereby described with storage capacity, installed electrical capacity and the time series of demand as technical parameters.

Electricity demand is predefined in hourly resolution analogously to fluctuating feed-in from renewable energy sources. The demand pattern is composed of the system load for the considered year plus the industry share of electricity. In order to meet the demand for primary reserve capacity, taking into account the minimum partial load and maximum ramp rates of the power plants, a year-round minimum capacity of thermal power plants is predefined, derived from pre-qualification conditions for primary regulation and technology specific minimum capacity of typical plants. Therefore unit commitment and plant dispatch are implemented as one step in the model.

Based on perfect foresight, the minimal cost dispatch of thermal power plants, feed-in from renewable energy sources and pumped storage hydro power plants is then defined within the scope of mixed-integer linear optimisation, taking into account technical and power industry constraints. The optimisation problem is implemented in GAMS and solved using the Cplex solver. The optimisation horizon is 24 hours (day-ahead dispatch). At the beginning and the end of an optimisation period storages are set to 50 % of the available volume. The power of the thermal plants within the first time step of a new optimisation period depends on the power of the last time step of the previous period. This ensures a consistent plant dispatch. The relative optimality criterion for the termination of the MIP problem is set to 2 %.

Based on a scenario analysis, which compares a baseline scenario without additional flexibility options with new flexibility scenarios, the total costs and carbon dioxide emissions of electricity generation as well as the generation mix, electricity prices, the curtailed amount of wind and solar power and profiles of storage and flexibility utilisation are determined.

3 Case study and scenario definition

This case study considers the German electricity system and takes a time horizon from 2010 to 2020 and 2030 into account. In this section, cost parameters as well energy-economic parameters concerning the thermal power plant fleet, electricity demand and supply from renewable energy sources and additional flexibility options like smart appliances with electrical cooling and a new pumped storage hydro power plant considered in this case study are described.

3.1 Cost parameters

The development of carbon dioxide costs is assumed to increase from 15 Euro/t in 2010 to 20 Euro/t in 2020 and 25 Euro/t in 2030 [2]. The growth rate for fuel cost is assumed to be within a range of 0.7 %/a to 1.4 %/a, starting in 2010 with 4.1 Euro/MWh for lignite, 13.6 Euro/MWh for hard coal, 35.2 Euro/MWh for natural gas and 52 Euro/MWh for light oil [6].

3.2 Thermal power plant fleet

In 2010, the German thermal power plant fleet is dominated by coal fired (45 GW), natural gas fired (23 GW) and nuclear (20 GW) power plants followed by oil and biomass fired plants (5 GW and 4 GW). Due to the recent nuclear face out decision of the German parliament, the capacity of nuclear power plants is going to decrease to 8 GW in 2020 and zero in 2030. According to [7] the capacity of coal fired power plants is going to decrease to 43 GW in 2020 and 28 GW in 2030 while for biomass fired power plants an increase to 8 GW in 2020 and 9 GW in 2030 is assumed. The capacity of natural gas fired power plants is considered to be roughly constant (29 GW in 2020 and 27 GW in 2030) [7]. The overall thermal power plant fleet in the PowerFlex model is composed of approx. 250 individual power plants and 150 technology aggregates.

3.3 Electricity demand and supply from renewable energy sources

The electricity demand is assumed to decrease from 594 TWh in 2010 to 567 TWh in 2020 and 558 TWh in 2030. The electricity supply from hydro power is stable at about 22 TWh/a, while the electricity generation from biomass doubles between 2010 (32 TWh) and 2030 (56 TWh) and wind and solar power increase fourfold from 2010 (43 TWh wind, 12 TWh solar) to 2030 (182 TWh wind, 57 TWh solar) [7].

3.4 Additional flexibility options

As additional flexibility options smart appliances with electrical cooling as an example of the smart grid concept are taken into account as well as the currently in Germany planned pumped storage hydro power plant "Atdorf" as an example of conventional storage technology in the electricity sector.

3.4.1 Smart appliances with electrical cooling

Processes with electrical cooling are mainly included in cold storage warehouses and in cooling and refrigeration appliances in households and food retailing. Electrical cooling in food and other industry is not taken into account in this case study. The assumed total electricity consumption for electrical cooling from the considered processes is about 30,000 GWh/a, which represents 5 % of the total German electricity consumption in 2010 [1] [10] [9].

The installed electrical power of refrigerating machines is about 8 GW, mainly located in cooling appliances in households. The assumed electrical storage capacity is set to 3 GWh. The average electricity demand derived from different technology specific profiles is about 3.3 GW [8]. Sensitive input data of these flexibility parameters are the variable temperature interval of cooling and refrigeration, the mass and specific heat capacity of refrigerated goods, the coefficient of performance of the refrigerating machine, power scaling factors and full load hours. As a result of a sensitivity analysis, an uncertainty of up to +/- 50 % for the installed electrical power and electrical storage capacity has to be taken into account.

The overall efficiency of this kind of demand side management is set to 95 %.

3.4.2 Pumped storage hydro power plant "Atdorf"

The pumped storage hydro power plant "Atdorf", which is actually planned in the south-western part of Germany, consists of 1,400 MW pump and generation capacity each, a storage capacity of approx. 10 GWh and an overall efficiency of about 80 %. This plant would increase the German capacity of pumped storage hydro power plants by up to 20 % [3].

4 Results and conclusions

Beside of the economic and environmental effects some key figures concerning model statistics are discussed in order to give an impression of the complexity of the optimisation problem.

4.1 Model statistics

Each optimisation problem consists of about 15,000 binary variables, depending on the amount of thermal power plants with an installed capacity exceeding 100 MW. The overall solution time for all 365 optimisation periods in one scenario year is about 10,000 to 30,000 CPU seconds. The relative optimality criterion is in the range of 0.01 % to 2.6 % with 1,3 % in average. In up to 5 optimisation periods per year the solver terminates the branch and bound process due to memory restrictions.

4.2 Economic and environmental effects

In the 2010 scenario, all renewable electricity available can be included into the electricity system. Additional flexibility leads in general to a small increase of electricity generation from lignite and nuclear power plants and a corresponding decrease of the overall generation costs. The efficiency of storage is thereby a sensitive input parameter. With an increasing share of intermittent and renewable electricity generation, the amount of wasted electricity through downwards regulation increases as well. In the 2030 baseline scenario about 4 % of the available renewable electricity could not be included into the electricity system. With new demand side flexibility from electrical cooling and the planned pumped storage hydro power plant "Atdorf", about 500 GWh of renewable electricity could be integrated in addition. This leads to cost reduction effects as well as a decrease of carbon dioxide emissions.

Acknowledgements This paper is a result of the ongoing E-Energy research project eTelligence (model region Cuxhaven), funded by the German Federal Ministry of Economics and Technology [4].

References

1. Bürger,V.: Identifikation, Quantifizierung und Systematisierung technischer und verhaltensbedingter Stromeinsparpotenziale privater Haushalte (2009)

2. Cames, M., Matthes, F., Healy, S.: Functioning of the ETS and the Flexible Mechanism. European Parliament's Committee on Environment, Public Health and Food Safety (2011)

3. Deutsche Energie-Agentur GmbH: Analyse der Notwendigkeit des Ausbaus von Pumpspeicherwerken und anderen Stromspeichern zur Integration der erneuerbaren Energien. (2010)

4. Krause, W., Bauknecht, D., Bischofs, L., Erge, T., Klose, T., Rüttinger, H., Stadler, M.: Das Leuchtturmprojekt eTelligence. Energy 2.0. (2009)

5. Federal Environment Agency, Germany: National Inventory Report for the German Greenhouse Gas Inventory 1990 - 2009 (2011)

6. Matthes, F.: Energiepreise fr aktuelle Modellierungsarbeiten. Regressionsanalytisch basierte Projektionen. Teil 1: Preise fr Importenergien und Kraftwerksbrennstoffe (2010)

7. Nitsch, J., Pregger, T., Scholz, Y., Naegler, T., Sterner, M., Gerhardt, N., von Oehsen, A., Pape, C., Saint-Drenan, Y., Wenzel, B.: Langfristszenarien und Strategien für den Ausbau der erneuerbaren Energien in Deutschland bei Berücksichtigung der Entwicklung in Europa und global (2010)

8. Stamminger, R. (ed.): Synergy Potential of Smart Domestic Appliances in Renewable Energy Systems. Herzogenrath (2009)

9. Umweltbundesamt (ed.): Vergleichende Bewertung der Klimarelevanz von Kälteanlagen und -geräten für den Supermarkt. Forschungsbericht 206 44 300 (2008)

10. Verband Deutscher Kühlhäuser und Kühllogistikunternehmen e. V.: Erhebung über die Höhe und die Strukturen der gesamten Kühl- und Tiefkühllagerkapazitäten in Deutschland. Bonn (2009)

Stream V
Financial Modeling, Risk Management, Banking

Erich-Walter Farkas, University of Zurich, SVOR Chair
Roland Mestel, University of Graz, ÖGOR Chair
Daniel Rösch, Leibniz Universität Hannover, GOR Chair

Performance Evaluation of European Banks Using Multicriteria Analysis Techniques

Michael Doumpos and Kyriaki Kosmidou

Abstract The recent crisis has reaffirmed the need to strengthen the procedures for monitoring the performance and risk taking behavior of banks. In this paper multicriteria analysis methods are used to evaluate the financial performance of a sample of European banks over the period 2006–2009. Different evaluation schemes are used, including a value function model, an outranking approach, and a cross-efficiency technique. The obtained results provide useful insights on the performance of the banks, the effects of the recent crisis, the relationship between bank performance and their size and country of origin, the stability of the evaluations over time and the factors that describe the dynamics of bank performance.

1 Introduction

The stability and effectiveness of the banking sector promote economic stability, and contribute to the increase of the standard of living. The recent financial crisis, however, has reaffirmed the need for further reinforcing the global financial system, as well as the need to strengthen the procedures for monitoring bank performance and the risks to which banks are exposed to.

Methodologies for bank performance evaluation are of major importance for a widely diversified group of stakeholders. Multicriteria techniques are particularly useful in this context. Statistical and data mining methods have also been used, mainly for predicting bank failures and rating analysis. Considerable work has also been devoted to bank efficiency measurement using data envelopment analysis (DEA) and other relevant techniques. Reviews on past relevant research can be

Michael Doumpos
Technical University of Crete, Dept. of Production Engineering and Management, Financial Engineering Laboratory, University Campus, 73100 Chania, Greece, e-mail: mdoumpos@dpem.tuc.gr

Kyriaki Kosmidou
Aristotle University of Thessaloniki, Dept. of Economics Division of Business Administration, 54124 Thessaloniki, Greece, e-mail: kosmid@econ.auth.gr

D. Klatte et al. (eds.), *Operations Research Proceedings 2011*, Operations Research Proceedings, 161
DOI 10.1007/978-3-642-29210-1_26, © Springer-Verlag Berlin Heidelberg 2012

found in [1,4], whereas an overview of bank rating systems used by central banks for supervisory risk assessment can be found in [6].

Though the recent crisis has global effects, US and European banks seem to the ones most affected. The European banking sector, in particular, still continues to be in turmoil. Thus, an up to date analysis of the performance of European banks is of major importance. Within this context, the aim of this study is to perform a "re-analysis" of the past performance of a sample of European banks, using up to date financial data spanning the recent crisis. The evaluation is based on multicriteria techniques, which are well-suited for aggregating multiple performance appraisal criteria. In particular, an additive value function model, an outranking relation approach, as well as a DEA-based cross-efficiency approach are used. The consideration of multiple methods enables the verification of the results under different evaluation schemes.

The rest of the paper is structured as follows. Section 2 describes the multicriteria evaluation methods that are employed in the analysis. Section 3 presents the sample data and the obtained results. Finally, section 5 concludes the paper and discusses some future research directions.

2 Multicriteria Evaluation Methods

The methods used to perform the evaluation process are based on different paradigms of multicriteria analysis, including the value function approach and outranking relations. These approaches are implemented within the simulation framework of SMAA-2 (Stochastic Multicriteria Acceptability Analysis). A DEA-based evaluation technique is also employed.

SMAA-2 [5] is based on a simulation framework for multicriteria evaluation problems. The simulation approach implemented in SMAA-2 is particularly useful in the context of this study, given the lack of a decision maker or an expert analyst who would provide preferential information on the parameters of the evaluation process. SMAA-2 is implemented in this study in combination with two evaluation models. The first is an additive value function (AVF):

$$V(\mathbf{x}_i) = w_1 v_1(x_{i1}) + w_2 v_2(x_{i2}) + \cdots + w_n v_n(x_{in}) \tag{1}$$

where $\mathbf{x}_i = (x_{i1}, \ldots, x_{in})$ is the data vector for bank i over n evaluation criteria, $w_k > 0$ ($k = 1, \ldots, n$) is the trade-off constant for criterion k (with $w_1 + \cdots + w_n = 1$) and $v_k(\cdot)$ is the marginal value function for criterion k (normalized between 0 and 1).

The second evaluation approach is the PROMETHEE II outranking model [2], which can be expressed as follows:

$$\Phi(\mathbf{x}_i) = \frac{1}{m-1} \sum_{j=1}^{m} \sum_{k=1}^{n} w_k \left[\pi_k(x_{ik}, x_{jk}) - \pi_k(x_{jk}, x_{ik}) \right] \tag{2}$$

where m is the number of banks in the sample and $\pi_k(x_{ik}, x_{jk})$ is the partial preference index indicating (in a [0, 1] scale) the strength of the preference for bank i over bank j on criterion k. In this study the Gaussian function is used to define all partial preference indices:

$$\pi_k(x_{ik}, x_{jk}) = \begin{cases} 1 - e^{-(x_{ik}-x_{jk})^2/2\gamma_k^2} & \text{if } x_{ik} > x_{jk} \\ 0 & \text{otherwise} \end{cases} \tag{3}$$

where $\gamma_k > 0$ is a user-defined constant.

With these evaluation models, the simulation framework of SMAA-2 is used to evaluate the banks under different scenarios of the evaluation parameters. The simulated parameters include the criteria trade-off constants (simulated over the unit simplex), the marginal value functions in model (1), and the parameters $\gamma_1, \ldots, \gamma_n$ in the PROMETHEE model. In the current study, we implemented 10,000 simulation runs for these parameters. The rankings of the banks in each simulation are aggregated using the holistic acceptability index proposed in [5].

Except for the above multicriteria models, a DEA-based cross-efficiency (CE) model is also employed. DEA is a popular approach for bank efficiency analysis and CE models are particularly useful in obtaining a ranking of all decision making units (banks) under consideration. In this study we use a variant of the CE model proposed by Doyle [3]. Assuming a weighted average evaluation model, Doyle's procedure requires the solution of a linear program for each bank i in the sample, in order to find a set of criteria trade-offs that maximize the performance of bank i relative to the others. The results are then averaged to obtain a global evaluation. Compared to Doyle's model, the approach used in this study has the following differences: (a) instead of the weighted average, we use a piece-wise linear additive model of the form (1), (b) the averaging of the results is only based on a sub-sample of "efficient" banks (a bank i is assumed efficient iff there exists a value function according to which bank i is compared favorable to all other banks).

3 Evaluation of European Banks' Financial Performance

3.1 Data and Variables

The data set used in this study includes commercial European banks over the period 2006–2009. We collected data on bank-specific variables from financial statements available in the Bankscope database. The sample includes commercial banks operating in the 27 countries members of the European Union (EU) for which complete data were available for all years in the period of the analysis. Overall, the sample consists of 146 banks from 21 EU countries.

The banks are evaluated over six financial ratios (Table 1), that take into account the profitability/efficiency of their banks and their capital adequacy, as suggested other studies in the literature. The results of a preliminary analysis has indicated that (on average) during 2008–2009 the loan loss reserves/impaired loans ratio has decreased considerably accompanied with a major decline in the profitability of the banks (ROAA ratio). Furthermore, the impaired loans/gross loans ratio has increased considerably, thus indicating a decline in the asset quality of banks.

Table 1 The selected financial ratios

LLR / IL: Loan loss reserves / Impaired loans	NIE / AA: Non interest expenses / Average assets
IL / GL: Impaired loans / Gross loans	ROAA: Return on average assets
EQ / L: Equity / Liabilities	NL / TA: Net loans / Total assets

3.2 Results

The overall results are summarized in Table 2 averaged for each year as well as for different regions and bank size (i.e., total assets). For AVF and PROMETHEE, the results refer to the holistic acceptability (HA) indices whereas for the CE model the average scores of the banks are presented. It should be noted that the HA figures are not directly comparable to the CE's global scores. Nevertheless, the results of the methods were found to be highly correlated (Kendall's τ rank correlation coefficient between 0.71 for the pair PROMETHEE-CE up to 0.86 for the pair AVF-CE). The evaluations obtained from all models clearly indicate that 2006 and 2007 were the two years where banks (overall) performed best. The effect of the recent crisis is evident in the results of 2008 and 2009, where the overall evaluation results show a considerable deterioration.

Table 2 Summary of evaluation results

		AVF				PROMETHEE				CE			
		2006	2007	2008	2009	2006	2007	2008	2009	2006	2007	2008	2009
Region	Eurozone	0.14	0.14	0.12	0.11	0.14	0.14	0.13	0.11	0.47	0.47	0.46	0.44
	West Europe	0.26	0.27	0.21	0.18	0.27	0.26	0.20	0.19	0.58	0.58	0.54	0.51
	East Europe	0.20	0.20	0.16	0.10	0.19	0.18	0.15	0.08	0.52	0.52	0.49	0.42
Size	Large	0.20	0.16	0.17	0.16	0.21	0.18	0.19	0.19	0.53	0.51	0.52	0.50
	Small	0.16	0.16	0.13	0.11	0.16	0.15	0.13	0.11	0.48	0.49	0.47	0.44
	Overall	0.16	0.16	0.14	0.11	0.16	0.16	0.14	0.12	0.49	0.49	0.47	0.45

The results of Table 2 also show important regional differences. In particular banks from west Europe (e.g., United Kingdom, Sweden, Denmark) performed consistently better in all years. The average performance of the banks from this area decreased considerably in 2008 (by about 8–22%, depending on the evaluation method), with an additional (smaller) decrease in 2009 by about 5–16%. Banks from east Europe, performed well in 2006 and 2007, but their performance dropped with increasing rates in the following years. It is also interesting to note that the effect of the crisis on banks from the Eurozone area picked during 2009 (in contrast to other west European banks who were affected most in 2008).

Furthermore, the results indicate that large banks (i.e., banks with above average assets) performed consistently better than small ones, with the differences being significant (according with the Mann-Whitney test), at the 5% level. Furthermore, the differences between small and large banks increased considerably in 2008 and 2009, thus confirming the vulnerability of smaller banks to adverse economic conditions.

Bank evaluations such as the one obtained in this study are often used to define bank ratings [6]. Thus, we defined a three-class rating of the banks on the basis of the obtained evaluations. Rating class A includes the top performing banks, whose evaluation falls into the top 25% percentile. Rating class B includes banks from the 25-75% interquartile range, whereas rating class C consists of banks at the bottom 25% percentile. In to analyze the stability and mobility of the ratings, an average migration matrix is constructed. The average migration rates presented in Table 3 are obtained by averaging all year-by-year migration matrices for each method. The rows correspond to the rating of the banks in any year t and the columns to the ratings of year $t+1$. Thus, each entry (R_t, R_{t+1}) represents the expected probability that a bank from the rating class R_t in year t (row) migrates to rating class R_{t+1} in the next year (column). The results of Table 3 show that the constructed ratings have a satisfactory stability, as a significant percentage of the banks retain their rating over two consecutive years. The upgrades (lower part of the migration matrices) are less frequent than the downgrades, with the frequency of downgrades increasing in 2008 and 2009 (about 15–21% in 2008 and 21–31% in 2009). Finally, it is worth noting that there was no case of a C rated bank migrating to the A rating class, whereas the opposite was observed in only very few cases.

Table 3 Average migration rates (in %)

	AVF			PROMETHEE			CE		
	A	B	C	A	B	C	A	B	C
A	67.8	30.5	1.7	67.5	28.1	4.4	70.5	28.6	0.9
B	7.2	75.6	17.2	9.6	71.3	19.1	11.4	72.3	16.4
C	0.0	18.2	81.8	0.0	18.1	81.9	0.0	21.1	78.9

In order to further analyze the explanatory power of the selected evaluation criteria on describing the changes in the performance of the banks, we classified the banks as "successful" and "unsuccessful". In particular, for each year t the banks whose performance improved compared to year $t-1$ are classified as "successful", whereas all other banks are considered "unsuccessful". The annual changes of the criteria were then tested against this classification using stepwise logistic regression. The results of Table 4 involve the coefficients of the lagged variables in each model together with the step in which they entered the model (all variables are important at the 5% level). From this analysis, it is evident that the changes in ROAA, non-interest expenses/average assets, and net loans/total assets ratio, are the strongest explanatory variables for the changes in the performance of the banks.

Table 4 Logistic regression results on the relationship between the evaluation criteria and the changes in the estimated performance of the banks

	AVF	PROMETHEE	CE
Δ(LLR / IL)	–	−0.053 (5)	−0.004 (4)
Δ(IL / GL)	–	−0.571 (6)	–
Δ(EQ / L)	0.220 (4)	0.841 (3)	–
Δ(NIE / AA)	−0.976 (2)	−5.227 (2)	−0.647 (3)
Δ(ROAA)	2.002 (1)	9.837 (1)	1.804 (1)
Δ(NL / TA)	−0.061 (3)	−0.346 (4)	−0.065 (2)
Constant	−0.149 (3)	0.258 (3)	−0.010 (3)

4 Conclusions

This study implemented different multicriteria techniques for building sound bank performance evaluation models. The analysis has clearly demonstrated the negative effects of the crisis on the performance of European banks. Banks from East Europe were severely affected, whereas the negative trend intensified during 2009 for banks in the Eurozone. Smaller banks were also found be more vulnerable. The obtained ratings have shown good stability properties over time and their analysis provided indications on the factors describing the dynamics of bank performance.

Future research could focus on extending the analysis to cover the entire European banking sector. Also, the use of such multicriteria techniques within the stress testing framework implemented for European banks would lead to a powerful integrated tool for analyzing and monitoring bank performance. The combination with efficiency and productivity appraisal techniques should also be further explored.

References

1. Berger, A. and Humphrey, D.B.: Efficiency of financial institutions: International survey and directions for future research. European Journal of Operational Research. **98**, 175–212 (1997).
2. Brans, J. and Vincke, P.: A preference ranking organization method. Management Science. **31**, 647–656 (1985).
3. Doyle, J.R.: Multiattribute choice for the lazy decision maker: Let the alternatives decide. Organizational Behavior and Human Decision Processes. **62**, 87–100 (1995).
4. Fethi, D. and Pasiouras, F.: Assessing bank efficiency and performance with operational research and artificial intelligent techniques: A survey. European Journal of Operational Research. **204**, 189–198 (2010).
5. Lahdelma, R. and Salminen, P.: SMAA-2: Stochastic multicriteria acceptability analysis for group decision making. Operations Research. **49**, 444–454 (2001).
6. Sahajwala, R. and Van den Bergh, P.: Supervisory risk assessment and early warning systems. Bank of International Settlements, Basel, Switzerland (2000) http://www.bis.org/publ/bcbs_wp4.htm.

Growth Optimal Portfolio Insurance in Continuous and Discrete Time

Sergey Sosnovskiy

Abstract We propose simple solution for the problem of insurance of the Growth Optimal Portfolio (GOP). Our approach comprise two layers, where essentially we combine OBPI and CPPI for portfolio protection. Firstly, using martingale convex duality methods, we obtain terminal payoffs of non-protected and insured portfolios. We represent terminal value of the insured GOP as a payoff of a call option written on non-protected portfolio. Using the relationship between options pricing and portfolio protection we apply Black-Scholes formula to derive simple closed form solution of optimal investment strategy. Secondly, we show that classic CPPI method provides an upper bound of investment fraction in discrete time and in a sense is an extreme investment rule. However, combined with the developed continuous time protection method it allows to handle gap risk and dynamic risk constraints (VaR, CVaR). Moreover, it maintains strategy performance against volatility misspecification. Numerical simulations show robustness of the proposed algorithm and that the developed method allows to capture market recovery, whereas CPPI method often fails to achieve that.

1 Introduction

In a multiperiod portfolio allocation it is an old and natural idea to choose as an optimization criterion the average growth rate (i.e. the geometric mean of return). This idea already dates back to the famous St. Petersburg paradox and later was developed by Kelly [8], Thorp [11], Hakansson and Ziemba [7] and many others. The resulting portfolios are known as the growth optimal portfolios (GOP).

Growth optimal portfolio strategies have many attractive features, however they face a significant shortcoming - high riskiness, especially in short terms. As potential remedy, the so called fractional Kelly strategy has been suggested, where only

Sergey Sosnovskiy
Frankfurt School of Finance and Management,
Sonnemannstrasse 9-11, Frankfurt am Main, 60314, Germany,
e-mail: s.sosnovskiy@fs.de

D. Klatte et al. (eds.), *Operations Research Proceedings 2011*, Operations Research Proceedings,   167
DOI 10.1007/978-3-642-29210-1_27, © Springer-Verlag Berlin Heidelberg 2012

fraction of log-optimal investment proportion is used, however it is unclear how to choose the optimal fraction.

Grossman and Vila [4] were among the first to address the issue of portfolio insurance in combination with utility maximization. The approach was later developed in Grossman, Zhou [5] and Cvitanic, Karatzas [3] who considered drawdown hedging. Basak and Shapiro [1] analyzed optimal portfolios restricted by VaR and CVaR constraints on terminal wealth and compared them to portfolio insurance with static constraint. In their paper risk is modeled only at the last stage, and it is not reevaluated after starting point. Moreover, their results are obtained in a pure diffusion continuous time setting and under this assumption there is no jump risk.

Versions of growth optimal portfolio optimization with dynamic risk constraints were considered in Cuoco, He and Isaenko [2] and in Pirvu, Žitković [9]. Their approach is in continuous time, where they consider local intervals $[t, t + \tau]$, under the assumption that portfolio composition and market parameters do not change within those intervals they control portfolio risk.

This paper contributions are (proofs can be found in [10]):

1) By exploring the link between options pricing and portfolio insurance we derive portfolio protection strategies in continuous and discrete time in a simplified way. The trade off between protection level and the potential losses (price of insurance) is illustrated.

2) We propose a new handling of dynamic risk constraints which differs from Cuoco, He and Isaenko [2] and Pirvu, Žitković [9]. Our starting point is the observation that the risk we have to reduce essentially arises from the discreteness of trading. To cope with that, we adjust the continuous time solution of portfolio insurance problem in discrete time such as to obey risk constraints at every step of a discretization. A simple hedging formula allows to control jumps not exceeding some prespecified level.

Optimization criterion Our optimization criterion throughout the paper is maximization of the average growth rate. Assuming that rebalancing is performed k times per annum and T denotes the number of years in the investment period, the geometric mean of investment return is $\left(\frac{V_T}{V_0} \right)^{1/kT}$

Assuming that $V_0 = 1$ and by letting $\gamma = 1 - \frac{1}{kT}$ maximization of the average growth rate is equivalent to maximization of the power utility:

$$U(V) = \frac{V^{1-\gamma} - 1}{1 - \gamma}$$

When the number of rebalancing periods k increases, and therefore $\gamma \to 1$ this utility function converges to $\ln(V_T)$.

2 Growth optimal portfolio insurance in continuous time

We solve the growth rate maximizing portfolio insurance problem in 3 stages. At first, we find value of the GOP for all times $t \leq T$. Then, using dual martingale approach we find the insured portfolio payoff at the terminal stage. Finally, using standard BS options theory we derive the optimal strategy which replicates terminal payoff. This also gives hedging strategy.

Our market consists of a risk-free bond with dynamics $dB_t = rB_t dt$ and a risky stock with dynamics $dS_t = \mu S_t dt + \sigma S_t dZ_t$, with constant drift and volatility coefficients. Market price of risk is $\theta = \frac{\mu - r}{\sigma}$. If a trading strategy x_t denotes the proportion of wealth invested to the risky asset, then the dynamics of the portfolio value is given by:

$$\frac{dV_t}{V_t} = x_t \frac{dS_t}{S_t} + (1 - x_t)\frac{dB_t}{B_t}$$

2.1 Value of the Growth Optimal Portfolio

First lemma reveals the value of the GOP and its properties.

Lemma 1. *i) With the power utility* $U(V) = \frac{V^{1-\gamma}-1}{1-\gamma}$ *value of a growth optimal portfolio* $\overline{V}_t$ *solving the terminal optimization problem*

$$\max_{V_T} \mathbb{E}U(V_T(\cdot)) \tag{1}$$

$$s.t. \quad e^{-rT}\mathbb{E}_Q V_T = V_0 \tag{2}$$

at time $t \leq T$ is given by

$$\overline{V}_t(Z_t) = V_0 e^{rt} e^{\frac{\theta^2}{2}\frac{2\gamma-1}{\gamma^2}t} e^{\frac{\theta}{\gamma}Z_t} \tag{3}$$

ii) The grwoth optimal investment strategy is to keep constant proportion of wealth invested into the risky asset

$$\overline{x} = \frac{1}{\gamma}\frac{\mu - r}{\sigma^2}$$

iii) The growth optimal portfolio (GOP) has constant volatility $\sigma_{gop} = \frac{\theta}{\gamma} = \overline{x}\sigma$.

2.2 Final payoff with insurance constraint

An insurance constraint ensures that the portfolio value at the terminal stage does not fall below some floor level F_T. We express terminal payoff of protected growth optimal portfolio in terms of adjusted non-protected GOP. We derive a solution using convex duality arguments.

Lemma 2. *Optimal final stage payoff of protected GOP, which a is solution to*

$$\max_{V_T(\cdot) \geq F_T} \mathbb{E}U(V_T(.)) \tag{4}$$

$$s.t. \quad e^{-rT}\mathbb{E}_Q V_T = V_0 \tag{5}$$

is given by

$$V_T^*(z) = \max(F_T, \eta \overline{V}_T(z)), \tag{6}$$

where the multiplier η is chosen such as to satisfy the budget constraint (5) and $\overline{V}_T$ is terminal value of GOP given by (3).

The coefficient η has an important interpretation as the *price of insurance*. This is the relative loss to the original unprotected portfolio $\overline{V}_T$ in case when the GOP ends up "in the money". From the other side if the unprotected GOP falls below F_T the protected one delivers the promised floor value. There is a tradeoff between protection level and potential relative loss, and η gives the price for it.

2.3 Protection strategy and its link with options pricing

Standard approach for finding strategy which replicates terminal payoff requires calculation of conditional expectation and matching diffusion coefficients by application of Ito's lemma.

We would like to show that there's a shortcut for finding the optimal trading strategy which gives the result in a much less analytically involved way. It is easy to see that (6) can be represented as

$$V_T^*(z_T) = \max(F, \eta \overline{V}_T(z_T)) = [\eta \overline{V}_T(z_T) - F_T]^+ + F_T = C_T + F_T,$$

which is equivalent to the payoff of a call option with strike F_T written on $\eta \overline{V}_T$ plus the strike value F_T. By Lemma 1.(iii) the volatility of the growth optimal portfolio is constant so that we can apply the standard Black-Scholes formula to price the insured portfolio at time t:

$$V_t^* = e^{-r(T-t)}\mathbb{E}_Q(V_T|\mathscr{F}_t) = C_t(F_T, \overline{x}\sigma) + F_t = \eta \overline{V}_t \Phi(d_1) - F_t \Phi(d_2) + F_t \tag{7}$$

where $d_{1,2}$ are defined as:

$$d_{1,2}(t) = \frac{\ln(\eta V_t/F_t) \pm \frac{\bar{x}^2 \sigma^2}{2}(T-t)}{\bar{x}\sigma\sqrt{T-t}}$$

Simplification yields

$$V_t^* = \eta \overline{V}_t \Phi(d_1(t)) + F_t \Phi(-d_2(t)) \tag{8}$$

Applying this formula for $t = 0$ gives closed form expression for the budget constraint (5). Since $d_{1,2}(0)$ both depend on η, this equation must be solved numerically, which yields price of insurance η.

Optimal proportion - hedging strategy. From the BS theory it is known that number of shares (delta) necessary to replicate the optional part of the insured strategy V_t^* is $\Phi(d_1)$. Since the option is written on fictitious asset $\eta \overline{V}_t$ this means that investment to this asset is $\eta \overline{V}_t \Phi(d_1)$. In the non-protected GOP $\overline{V}_t$ fixed fraction is invested to the risky asset/fund , thus exposure (amount of money in the risky fund S_t) is

$$e_t^* = \eta \bar{x} \overline{V}_t \Phi(d_1)$$

Expressed in relative terms the proportion of the protected portfolio V_t^* invested into the risky asset is

$$x_t^* = \frac{e_t^*}{V_t^*} = \bar{x}\frac{\eta \overline{V}_t \Phi(d_1)}{V_t^*} \tag{9}$$

Substitution of $\eta \overline{V}_T \Phi(d_1) = V_t^* - F_t \Phi(-d_2)$ from (8) finally gives

$$x_t^* = \bar{x}\left(1 - \frac{F_t}{V_t^*}\Phi(-d_2)\right) \tag{10}$$

3 Portfolio Protection in Discrete Time

In discrete time there is no longer continuity of hedging decisions and here naturally appears gap risk. To avoid cases when portfolio value might sink below the floor at next stage, we find what conditions must be satisfied by the fraction of wealth invested to the risky fund such as to avoid risk (or minimize it). In discrete time investment proportion x_t should be chosen such as to guarantee that given some worst-case asset return r_w and constant (overnight) bond return r_0 next period portfolio value V_{t+1} stays above the floor $F_{t+1} = F_t r_0$:

$$V_{t+1} = V_t \left(x_t r_w + (1-x_t)r_0\right) \geq F_t r_0 \tag{11}$$

Simplification yields maximum value of proportion to be kept in the risky asset:

$$x_t \leq m\left(1 - \frac{F_t}{V_t}\right) =: x_t^{cppi}, \quad \text{where } m = \frac{r_0}{r_0 - r_w} \tag{12}$$

For all investment proportions which exceed value prescribed by (12) one cannot guarantee that hedged portfolio stays above the next stage floor with assumed worst-case return r_w. The developed approach can be easily extended for handling dynamic VaR and CVaR constraints. Finally, combining formulas (10) and (12) we obtain rule for discrete portfolio hedging:

$$x_t^{adj} = \min[x_t^*, x_t^{cppi}] \tag{13}$$

4 Computational results

Numerical simulations have shown robustness of proposed method not only to the presence of jumps in portfolio returns, but also to volatility misspecification.

5 Conclusion

Developed growth optimal portfolio insurance method is robust and simple. Also it can be easily extended for protection of portfolios with constant rebalanced weights. An application to drawdown hedging is in a paper to follow.

References

1. S. Basak and A. Shapiro. Value-at-risk-based risk management: optimal policies and asset prices. *Review of Financial Studies*, 14(2):371, 2001. ISSN 0893-9454.
2. D. Cuoco, H. He, and S. Isaenko. Optimal Dynamic Trading Strategies with Risk Limits. *Operations Research*, 56(2):358–368, 2008. ISSN 0030-364X.
3. J. Cvitanic and I. Karatzas. On portfolio optimization under" drawdown" constraints. *IMA volumes in mathematics and its applications*, 65:35–35, 1995. ISSN 0940-6573.
4. S.J. Grossman and J.L. Vila. Portfolio insurance in complete markets: A note. *The Journal of Business*, 62(4):473–476, 1989. ISSN 0021-9398.
5. S.J. Grossman and Z. Zhou. Optimal investment strategies for controlling drawdowns. *Mathematical Finance*, 3(3):241–276, 1993. ISSN 1467-9965.
6. S.J. Grossman and Z. Zhou. Equilibrium analysis of portfolio insurance. *Journal of Finance*, 51(4):1379–1403, 1996. ISSN 0022-1082.
7. N.H. Hakansson and W.T. Ziemba. Capital growth theory. *Handbooks in Operations Research and Management Science*, 9:65–86, 1995. ISSN 0927-0507.
8. J.L. Kelly. A new interpretation of information rate. *Information Theory, IRE Transactions on*, 2(3):185–189, 1956. ISSN 0096-1000.
9. T.A. Pirvu and G. Žitković. Maximizing the growth rate under risk constraints. *Mathematical Finance*, 19(3):423–455, 2009. ISSN 1467-9965.
10. S. Sosnovskiy, U. Walther. Insurance of the Growth Optimal Portfolio. In *Frankfurt School - Working Papers Series*.
11. E.O. Thorp. Portfolio choice and the Kelly criterion. *Stochastic models in finance*, pages 599–619, 1971.

The Regularization Aspect of Optimal-Robust Conditional Value-at-Risk Portfolios

Apostolos Fertis and Michel Baes and Hans-Jakob Lüthi

Abstract In portfolio management, Robust Conditional Value - at - Risk (Robust CVaR) has been proposed to deal with structured uncertainty in the estimation of the assets probability distribution. Meanwhile, regularization in portfolio optimization has been investigated as a way to construct portfolios that show satisfactory out-of-sample performance under estimation error. In this paper, we prove that optimal-Robust CVaR portfolios possess the regularization property. Based on expected utility theory concepts, we explicitly derive the regularization scheme that these portfolios follow and its connection with the scenario set properties.

1 Introduction

The portfolio management problem refers to the decision on the proportion of wealth invested in a set of available assets such that the uncertain return is optimized according to some metric taking into account both its magnitude and the risk in its potential values. Markowitz was the first one to formulate a quantitative approach to this problem [17]. The mean-variance approach that he followed set the path for the development of the fields of financial engineering and asset pricing, drastically affected by the milestones of the Capital Asset Pricing Model [16, 19, 24], the Arbitrage Pricing Theory [22], the Black-Scholes model [3] and Merton's theory of option pricing [18].

Much concern has been expressed about the effects of errors in the estimation of the parameters that characterize the probability distribution of the asset returns. Constraining portfolio norms has been proposed as a technique to tackle this uncertainty in the estimated parameters with the resulting regularized portfolios showing improved out-of-sample performance [8, 9, 13]. Bayesian techniques have been also

Apostolos Fertis
Institute for Operations Research (IFOR), Eidgenössische Technische Hochschule Zürich, Rämistrasse 101, HG G 21.3, 8092, Zürich, Switzerland, tel.: +41 44 632 4031, e-mail: afertis@ifor.math.ethz.ch

D. Klatte et al. (eds.), *Operations Research Proceedings 2011*, Operations Research Proceedings, 173
DOI 10.1007/978-3-642-29210-1_28, © Springer-Verlag Berlin Heidelberg 2012

suggested [2, 12]. Estimation error insensitive portfolios derived through Huber's robust statistics methods [15] have been investigated [10, 12].

Meanwhile, risk management has been connected with robust optimization. Robust optimization deals with uncertainty in the data of an optimization problem. It seeks the solution that optimizes the worst-case value of the objective function when the data vary in some uncertainty set. Such a solution can be efficiently computed for many cases of uncertainty sets in linear or general convex optimization problems [3–5]. Given that in portfolio management the parameters of the assets probability distribution are estimated with limited accuracy, their true value can be assumed to reside in an uncertainty set surrounding the estimated value. Based on problems that optimize a trade-off between the mean and the variance of the portfolio, corresponding robust optimization problems that optimize the worst-case trade-off when the probability distribution parameters vary in uncertainty sets have been proposed and solved [14]. The robust problems have been solved by transforming them to equivalent problems that impose regularization constraints on the portfolio weights [14].

Furthermore, risk measures have been considered to be mappings of random financial positions to real numbers imposing a preferential order among financial positions that encapsulates the trade-off between their mean and risk. Coherent risk measures have been defined as a particular class of risk measures satisfying certain desirable properties [1]. It has been proved that any coherent risk measure evaluated for some random position can be expressed as the worst-case expectation of this position when the probability measure used in the expectation varies in some uncertainty set, called scenario set [1]. In this way, portfolio optimization using a coherent risk measure is a robust optimization problem. The dual representation of coherent risk measures has been connected to the expected utility theory of von Neumann and Morgenstern [20] through the concept of the Optimized Certainty Equivalent (OCE) [6].

The most well-known coherent risk measure is Conditional Value-at-Risk (CVaR) [21]. CVaR concentrates on the worst-case tail of the position's probability distribution. Portfolio optimization using CVaR is implemented by dualizing its representation via the scenario set [23]. When the probability distribution is unknown but can be restricted to a scenario set containing all the potential distributions, Robust Conditional Value-at-Risk (Robust CVaR) can be defined as the worst-case CVaR when the distribution varies in this set [11, 25]. The scenario set is constructed by structuring randomness in two stages and concentrating uncertainty at the first stage. Again, dualizing the representation enables portfolio optimization using Robust CVaR [11, 25].

Although Robust CVaR has been shown to be resistent against probability distribution uncertainty, the properties of optimal-Robust CVaR portfolios have not been investigated. Given that portfolio regularization has been proposed to deal with parameter estimation uncertainty, the question of whether optimal-Robust CVaR portfolios possess some kind of regularization property arises.

In this paper, we prove that optimal-Robust CVaR portfolios possess the regularization property. The regularization scheme that they follow is connected with the expected utility theory, where the expectation used for the utility is taken with respect to the probability measures that determine the structure of the scenario set. The representation that we prove offers a crisp explanation of the optimal-Robust CVaR portfolio properties with respect to the format and the size of the scenario set.

2 Robust CVaR

Consider a set of m financial assets which have random discounted return rates in a one-period horizon. The probability distribution of the returns is uncertain. We describe the stochastic framework. Let P_k, $k = 1, 2, \ldots, r$, and $\mathbb{P}$ be probability measures defined in σ-field $(\Omega, \mathscr{F})$, such that P_k, $k = 1, 2, \ldots, r$, is absolutely continuous with respect to $\mathbb{P}$. The driving probability measure P is assumed to be a mixture of P_k with unknown coefficients. More specifically, the uncertainty in the probability distribution will be determined using a norm $\|\cdot\|$ in $\mathbb{R}^r$, $\xi_0 = (\xi_{0,1}, \xi_{0,2}, \ldots, \xi_{0,r})^T \in \mathbb{R}^r_+$ with $\sum_{k=1}^r \xi_{0,k} = 1$ and $\phi \in \mathbb{R}_+$. If we define the compact set

$$\Xi := \left\{ \xi = (\xi_1, \xi_2, \ldots, \xi_r)^T \in \mathbb{R}^r_+ \;\middle|\; \|\xi - \xi_0\| \leq \phi, \; \sum_{k=1}^r \xi_k = 1 \right\}, \qquad (1)$$

we have that

$$P \in S := \left\{ \sum_{k=1}^r \xi_k P_k \;\middle|\; \xi = (\xi_1, \xi_2, \ldots, \xi_r)^T \in \Xi \right\}. \qquad (2)$$

Set S contains all the potential probability distributions, which reside in an uncertainty set centered at $P_0 = \sum_{k=1}^r \xi_{0,k} P_k$. For some level β, $0 \leq \beta < 1$, we denote with $\mathrm{CVaR}_P(X)$, $\mathrm{RCVaR}(X)$ the CVaR, Robust CVaR, respectively, of some random position $X \in L^1(\Omega, \mathscr{F}, \mathbb{P})$ under some probability measure $P \in S$. Random variables $X_1, X_2, \ldots, X_m \in L^1(\Omega, \mathscr{F}, \mathbb{P})$ are equal to the discounted return rates of the assets. An investment portfolio over these assets can be represented by $\theta = (\theta_1, \theta_2, \ldots, \theta_m)^T \in \mathbb{R}^m_+$, where $\sum_{j=1}^m \theta_j = 1$. The discounted return rate of the portfolio is $\sum_{j=1}^m \theta_j X_j$.

The portfolio problem using the Robust CVaR criterion is

$$\begin{array}{ll} \min_{\theta \in \mathbb{R}^m_+} \mathrm{RCVaR}\left(\sum_{j=1}^m \theta_j X_j\right) = & \min_{\theta \in \mathbb{R}^m_+} \max_{P \in S} \mathrm{CVaR}_P\left(\sum_{j=1}^m \theta_j X_j\right) \\ \text{s.t.} \quad \sum_{j=1}^m \theta_j = 1 & \text{s.t.} \quad \sum_{j=1}^m \theta_j = 1. \end{array} \qquad (3)$$

We use $\|\cdot\|_*$ to denote the dual norm to $\|\cdot\|$. Problem (3) is equivalent [11] to

$$
\begin{aligned}
\min \ &\eta + t + \phi \|v\|_* + \xi_0^T v \\
\text{s.t.} \ \ &t + v_k \geq \frac{1}{1-\beta} E_{P_k}[Z], \ k = 1, 2, \ldots, r \\
&Z \geq -\eta - \sum_{j=1}^m \theta_j X_j, \ \mathbb{P} - a.s. \\
&\sum_{j=1}^m \theta_j = 1 \\
&\eta, t \in \mathbb{R}, \ v \in \mathbb{R}^r, \ Z \in L_+^1(\Omega, \mathscr{F}, \mathbb{P}), \ \theta \in \mathbb{R}_+^r.
\end{aligned}
\tag{4}
$$

Problem (4) can be used to compute optimal-Robust CVaR portfolios. If Ω is finite, then Problem (4) is a finite-dimensional linear optimization problem and the result follows from linear optimization duality. The derivation of the result is much more difficult in the case that Ω is infinite. In this case, to solve Problem (4) the Sample Average Approximation or the Stochastic Approximation methods can be used [23]. The resulting portfolio is resistent against varying the probability distribution in the scenario set S. Despite that portfolio regularization has been successfully used to protect against parameter estimation uncertainty, it is not known whether optimal-Robust CVaR portfolios are regularized.

3 Regularization of optimal-Robust CVaR portfolios

The regularization aspect of optimal-Robust CVaR portfolios will be depicted using the expected utility theory of von Neumann and Morgenstern [20] and, in particular, utility function

$$
u(x) = (1-\beta)^{-1} \min(0,x),
\tag{5}
$$

where $x \in \mathbb{R}$ is some discounted return rate. Function u imposes a prefential order among financial positions. If $\theta_1 = (\theta_{1,1}, \theta_{1,2}, \ldots, \theta_{1,m})^T$, $\theta_2 = (\theta_{2,1}, \theta_{2,2}, \ldots, \theta_{2,m}^T$ are two asset portfolios, $P \in S$ is the probability measure that drives the asset evolution and $E_P[u(\sum_{j=1}^m \theta_{1,j} X_j)] \geq E_P[u(\sum_{j=1}^m \theta_{2,j} X_j)]$, portfolio θ_1 is preferred to portfolio θ_2.

The next theorem expresses Problem (4) as a trade-off between two functions involving the expected utilities of the investor when present consumption is possible under probability measures P_0, P_1, ..., P_r. The vector in $\mathbb{R}^r$ containing 1 at all positions is denoted by $\mathbf{e}$.

Theorem 1. *Problem (4) is equivalent to*

$$
\begin{aligned}
\min_{\eta, \theta, v} \ &-E_{P_0}[\eta + u(\textstyle\sum_{j=1}^m \theta_j X_j - \eta)] + \phi \min_t \|v - t\mathbf{e}\|_* \\
\text{s.t.} \ \ &v_k = -E_{P_k}[\eta + u(\textstyle\sum_{j=1}^m \theta_j X_j - \eta)], \ k = 1, 2, \ldots, r \\
&\textstyle\sum_{j=1}^m \theta_j = 1 \\
&\eta \in \mathbb{R}, \ \theta \in \mathbb{R}_+^m, v \in \mathbb{R}^r.
\end{aligned}
\tag{6}
$$

Proof. Due to lack of space, we present the basic points of the proof. By applying the transformation $\eta := -\eta$, $t := t + \eta$, Problem (4) becomes

$$\min \ \xi_0^T(v + te) + \phi \|v\|_*$$
$$\text{s.t.} \ \ v_k + t \geq -\eta + \frac{1}{1-\beta} E_{P_k}[Z], \ k = 1, 2, \dots, r \tag{7}$$
$$Z \geq \eta - \sum_{j=1}^m \theta_j X_j, \ \mathbb{P} - a.s.$$
$$\sum_{j=1}^m \theta_j = 1, \ \eta, t \in \mathbb{R}, \ v \in \mathbb{R}^r, \ Z \in L_+^1(\Omega, \mathscr{F}, \mathbb{P}), \ \theta \in \mathbb{R}_+^r.$$

Consider an optimal solution $(\eta^o, t^o, v^o, Z^o, \theta^o)$ to Problem (7). We claim that it should satisfy

$$Z^o = \max(0, \eta^o - \sum_{j=1}^m \theta_j^o X_j) \ and \ v_k^o + t^o = -\eta^o + (1-\beta)^{-1} E_{P_k}[Z^o] \ for \ all \ k.$$

Suppose that this is not true. Then, since $(\eta^o, t^o, v^o, Z^o, \theta^o)$ is a feasible solution as well, it holds that

$$Z^o > \max(0, \eta^o - \sum_{j=1}^m \theta_j^o X_j) \ or \ v_k^o + t^o > -\eta^o + (1-\beta)^{-1} E_{P_k}[Z^o] \ for \ some \ k.$$

If we set $Z' = \max(0, \ \eta^o - \sum_{j=1}^m \theta_j^o X_j)$, $t' = \min_k \{-v_k^o - \eta^o + \frac{1}{1-\beta} E_{P_k}[Z^o]\} < t^o$, then, $(\eta^o, t', v^o, Z', \theta^o)$ is feasible with a smaller objective value, which is a contradiction. Thus, the claim is true. Consequently, the inequalities in Problem (7) can be substituted with equalities. The result follows after a few steps.

If the investor decides to consume a fraction η of their wealth at present, then, their future return is $\sum_{j=1}^m \theta_j X_j - \eta$ and thus, their total utility is $\eta + u(\sum_{j=1}^m \theta_j X_j - \eta)$ [6]. If $\phi = 0$, the problem becomes the portfolio optimization problem using the CVaR criterion, since $E_{P_0}[\eta + u(\sum_{j=1}^m \theta_j X_j - \eta)]$ is the expected utility with present consumption η under probability measure P_0. If $\phi > 0$, the regularization term $\phi \min_t \|v - te\|_*$ is also affecting the portfolio. The k-th coordinate v_k of the regularization vector v is equal to the opposite of the expected utility with present consumption η under probability measure P_k. Thus, the regularization term is equal to the distance of v from the line $te, t \in \mathbb{R}$, according to norm $\|\cdot\|_*$. This implies that the optimal-Robust CVaR portfolios are regularized in such a way that the expected utilities under the probability measures $P_1, P_2, \dots, P_r$ are close to each other. The metric used to measure their proximity is the dual norm $\|\cdot\|_*$ to the norm $\|\cdot\|$ used to define the uncertainty set S. This is a different kind of regularization than the ones already considered in the literature [8, 9, 13]. The strength of the regularization is controlled by the size ϕ of the uncertainty set S.

4 Conclusions

We have shown that optimal-Robust CVaR portfolios that deal with a kind of structured uncertainty in the probability distribution that the assets follow possess the regularization property. Since regularized portfolios have already been shown to deal with uncertainty in the parameter estimation, such an observation allows a broader view of the importance of regularization in facing probability distribution uncertainty. The regularization properties were explicitly connected with the properties of the uncertainty set, enabling the choice of the regularization scheme according to the format and the size of the uncertainty set.

References

1. Artzner, P., Delbaen, F., Eber, J.-M., Heath, D.: Coherent risk measures. Mathematical Finance **9**(3), 203-228 (1999).
2. Black, F., Litterman, R.: Asset allocation: Combining investor views with market equilibrium. Technical Report, Fixed Income Research, Goldman Sachs & Co., New York (1990).
3. Black, F., Scholes, M.: The Pricing of Options and Corporate Liabilities. Journal of Political Economy **81**(3), 637654 (1973).
4. Ben-Tal, A., Nemirovski, A.: Robust convex optimization. Mathematics of Operations Research **23**(4), 769-805 (1998).
5. Ben-Tal, A., Nemirovski, A.: Robust solutions of uncertain linear programs. Operations Research Letters **25**(1), 1-13 (1999).
6. Ben-Tal, A. Teboulle, M.: An old-new concept of convex risk measures: the optimized certainty equivalent. Mathematical Finance **17**(3), 449-476 (2007).
7. Bertsimas, D., Sim, M.: The price of robustness. Operations Research **52**(1), 35-53 (2004).
8. Chopra, V. K.: Improving optimization. Journal of Investing **8**, 51-59 (1993).
9. DeMiguel, V., Garlappi, L., Nogales, F. J., Uppal, R.: A Generalized Approach to Portfolio Optimization: Improving Performance by Constraining Portfolio Norms. Management Science **55**(5), 798-812 (2009).
10. DeMiguel, V., Nogales, F. J.: Portfolio Selection with Robust Estimation. Operations Research **57**(3), 560-577 (2009).
11. Fertis, A., Baes, M., Lüthi, H.-J.: Robust Risk Management. Submitted for publication (2011).
12. Frost, P. A., Savarino, J. E.: An empirical Bayes approach to efficient portfolio selection. Journal of Financial and Quantitative Analysis **21**, 293-305 (1986).
13. Frost, P. A., Savarino, J. E.: For better performance: Constrain portfolio weights. Journal of Portfolio Management **15**, 29-34 (1988).
14. Goldfarb, D., Iyengar, G.: Robust Portfolio Selection Problems. Mathematics of Operations Research **20**(1), 1-38 (2003).
15. Huber, P. J.: Robust Statistics. John Wiley & Sons, New York (1981).
16. Lintner, J.: Valuation of risky assets and the selection of risky investments in stock portfolios and capital budgets. Review of Economics and Statistics **47**, 13-37 (1965).
17. Markowitz, H. M.: Portfolio selection. Journal of Finance **7**, 77-91 (1952).
18. Merton, R. C.: Theory of Rational Option Pricing. Bell Journal of Economics and Management Science **4**(1), 141-183 (1973).
19. Mossin, J.: Equilibrium in capital asset markets. Econometrica **34**(4), 768-783 (1966).
20. von Neumann, J., Morgenstern, O.: Theory of Games and Economic Behavior. Princeton University Press, Princeton (1947).
21. Rockafellar, R. T., Uryasev, S.: Optimization of conditional value-at-risk. The Journal of Risk **2**(3), 21-41 (2000).
22. Ross, S.: The Arbitrage Theory of Capital Asset Pricing. Journal of Economic Theory **13**(3), 341-360 (1976).
23. Shapiro, A., Dentcheva, D., Ruszczynski, A.: Lectures on Stochastic Programming: Modeling and Theory. SIAM, Philadelphia (2009).
24. Sharpe, W.: Capital asset prices: A theory of market equilibrium under conditions of risk. Journal of Finance **19**(3), 425-442 (1964).
25. Zhu, S., Fukushima, M.: Worst-case conditional value-at-risk with application to robust portfolio management. Operations Research **57**(5), 1155-1168 (2009).

The effect of different market opening structures on market quality – experimental evidence

Gernot Hinterleitner, Philipp Hornung, Ulrike Leopold-Wildburger, Roland Mestel, and Stefan Palan

Abstract This experimental study focuses on the effect of opening call auctions on market quality. Our examination extends the existing market microstructure literature in two respects. First, in addition of analyzing the market opening, we examine the effect of an opening call auction on the subsequent continuous trading phase. Second, we analyze two different kinds of call auctions with different pre-opening transparency levels. We find that an opening call auction significantly increases the informational efficiency of opening prices and leads to higher market quality of the subsequent continuous double auction market as compared to the continuous double auction alone.

1 Introduction

The opening is a crucial time of a trading day because it marks the start of the price discovery process and the incorporation of overnight information into prices. For

Gernot Hinterleitner
Institute of Statistics and Operations Research, Karl-Franzens-University Graz, Austria
e-mail: gernot.hinterleitner@edu.uni-graz.at

Philipp Hornung
Institute of Statistics and Operations Research, Karl-Franzens-University Graz, Austria
e-mail: philipp.hornung@edu.uni-graz.at

Ulrike Leopold-Wildburger
Institute of Statistics and Operations Research, Karl-Franzens-University Graz, Austria
e-mail: ulrike.leopold@uni-graz.at

Roland Mestel
Institute of Banking and Finance, Karl-Franzens-University Graz, Austria
e-mail: roland.mestel@uni-graz.at

Stefan Palan
Institute of Banking and Finance, Karl-Franzens-University Graz, Austria
e-mail: stefan.palan@uni-graz.at

D. Klatte et al. (eds.), *Operations Research Proceedings 2011*, Operations Research Proceedings,   179
DOI 10.1007/978-3-642-29210-1_29, © Springer-Verlag Berlin Heidelberg 2012

this reason, uncertainty about the fundamental value is high at the opening compared to intraday trading [2], leading to lower market quality (i.e. lower informational efficiency and liquidity).

Several stock exchanges around the world use a call auction procedure as their opening mechanism. This is due to the attributes of call auction trading (collecting orders from different traders, generating liquidity and determination of one price), which can be expected to improve market quality at the market opening, as it is mentioned in several studies [7, 15].

This study analyzes the impact of different opening call auction institutions in terms of market quality at the time of the market opening and during the subsequent continuous trading (spillover effect) using laboratory experiments. To our knowledge this is the first experimental study analyzing the spillover effect of an opening call auction.

2 Pre-trade-transparency and spillover effect

The designs of the call auctions employed by stock markets outside the laboratory is relatively heterogeneous. One main design question of call auctions is the pre-trade-transparency [8]. Higher levels of pre-trade-information, like indicative prices and an open order book, are promoted by some theoretical analyses ([3, 17]) because they seem to generate a better price discovery process. In contrast, other studies [10, 16] find that more pre-trade-information can motivate strategic behavior of informed investors in the form of price manipulation or a reduced activity of order entry until close to the end of the pre-opening, both of which can lead to lower market quality. Some of the experimental literature with no informed investors [1, 12] similarly suggests that higher pre-trading transparency leads to a lower market quality of the call auction because of price manipulation by participants.

Beside the discussion of the ideal level of call auction transparency for the opening prices, its effect on the subsequent continuous trading is also of interest (spillover effect). The introduction of an opening call auction at the Singapore Exchange [5] and at NASDAQ [18] has been found to lead to higher price efficiency in the subsequent continuous double auction, indicating a positive spillover effect from the opening call auction. Comparing the Australian and the Jakarta Stock Exchange leads to a similar result [6].

A negative spillover effect of the opening call auction was found at the Hongkong Exchange [9] and was conjectured to have been caused by the unattractive design of the opening call market. An investigation of the National Stock Exchange in India [4] also reports a negative spillover effect.

In terms of transparency, [14] higher market quality at the start of continuous trading was found at the Euronext Paris compared to the Deutsche Brse. This result is dedicated to higher pre-opening transparency at the Euronext Paris [14]. Analyzing the introduction of indicative prices during the pre-opening at the Shanghai Stock Exchange leads to similar results [13].

However, the empirical literature is subject to some shortcomings: Comparing stock exchanges with different call auction designs is difficult because of different trading rules and a different trading environment which can influence the results. In addition, informational efficiency can only be measured using proxies because the fundamental values of stocks are not known. In our study, we use the experimental method to overcome these problems.

3 Experimental design

The design of the experiment conducted in this study is based on [19], who compares single trading mechanisms. In contrast to Theissen we examine informational efficiency and liquidity by comparing a Continuous Double Auction alone (CDA alone) with complement markets. The complement markets are made up of an opening call auction and the subsequent continuous double auction. We refer to the complement market with a non-transparent opening call auction as CM and to the one with a transparent opening call auction as CMT.[1]

The non-transparent call auction does not display any information about order flow or indicative prices whereas all orders and indicative prices are displayed to the participants in the transparent version. In both call auctions, the opening price is calculated such that trading volume is being maximized.[2] The market size is fixed at twelve traders, each of which starts with the same endowment of shares and virtual currency and each of which receives a private price signal at the beginning of each period. The price signal lies within 10 % of the fundamental value of the asset. Every experimental session has six trading periods.

At no time during the experiment is the fundamental value made known to the participants. Furthermore, for every period a new fundamental value is randomly determined. The same settings of fundamental values and price signals are used once in every treatment. During each period traders can conduct long or short transactions by creating orders or accepting orders created by others. After each trading period the amount of shares and virtual currency was re-initialized for the next trading period. The call auctions last two minutes; all continuous double auctions (stand-alone and in complement markets) are limited to five minutes.

After the experiment, participants received a payment in accordance to their ranking. This rank-dependent payment lies between 0 and 47 Euros for the complement markets and between 0 and 40 Euros for the CDA alone markets; the show-up-fee for the CDA alone markets was 8 Euros and for the complement markets 10 Euros. This design feature is due to the higher duration of the complement markets. [3]

---

[1] The experiment was programmed and conducted with the software z-Tree (Zurich Toolbox for Readymade Economic Experiments), the state-of-the-art program for developing economic laboratory experiments [11].

[2] If trading volume is maximized for two or more prices, the average of these prices is determined as opening price. Price-Time-Priority exists concerning rationing of orders.

[3] The average payment was about 9 Euros per hour and participant for each treatment.

4 Results

We first analyze the efficiency of the opening price for all treatments by calculating the absolute relative deviation of the opening price from the fundamental value for each period. This measure is also called "Market-Relative-Error" (MRE) [19]. Table 1 shows the average MRE values for each experimental session. A Kruskal-Wallis-ANOVA rejects the hypothesis of equal medians of the MRE values using period data (p-value: 0.038).

Table 1 Average MRE of the opening prices of each series

Series	Call – CM	Call – CMT	CDA alone
1	1.09 %	5.49 %	3.88 %
2	1.51 %	0.82 %	4.67 %
3	2.17 %	2.70 %	2.03 %
4	1.49 %	2.71 %	3.35 %
5	1.82 %	1.70 %	2.58 %
6	3.92 %	2.01 %	8.25 %
7	1.43 %	2.84 %	2.74 %
Average	1.92 %	2.61 %	3.93 %

A pairwise Wilcoxon test confirms that the call auctions generate a significantly more efficient opening price than the CDA alone market (p-values-value<0.05). Comparing the MRE values of the opening prices of the two call auctions, we find no significant difference between the transparent and non-transparent call auction markets (p-value: 0.192).

This result is in such a manner in line with literature, as no clear answer about the advantage of transparent opening call auctions can be dedicated from different experimental and theoretical studies [e.g. [1], [3] or [17]).

Detailed analysis about the order flow shows evidence of price manipulation by participants in some periods, leading to higher MRE values for the opening price. We also analyzed the spillover effect of an opening call auction on CDA market quality. The MRE measure used for the CDA markets was calculated as the absolute relative deviation of the transaction price from the fundamental value of the period. In contrast to the opening price, the MRE of each transaction price is weighted by the ratio of the trading volume of this transaction over the total trading volume in the period. The MRE values of each transaction in a period are summed up to a period MRE value for the CDA market [19]. Table 2 shows descriptive statistics and results of the Wilcoxon tests of the CDA market MRE values.

Table 2 documents a positive spillover effect from an opening call auction. The MRE values in the CDA markets after an opening call auction are significantly lower than the MRE values in the CDA alone market. However a CDA after a transparent call auction does not lead to more efficient prices than a CDA after a

Table 2 Descriptive statistics and results of the pairwise Wilcoxon test on MRE values

Panel A: Descriptive Statistics	CDA alone	CDA–CM	CDA–CMT
Mean	3.60 %	2.68 %	3.09 %
Median	3.22 %	2.65 %	3.15 %
Standard Deviation	1.95 %	1.15 %	1.15 %
Panel B: Wilcoxon test	CDA alone vs. CDA–CM	CDA alone vs. CDA–CMT	CDA–CM vs. CDA–CMT
Z	−2.682	−1.963	−0.869
p-value	0.006	0.05	0.384

non-transparent call auction. These results are clearly supported by analyzing the bid-ask-spreads during continuous trading. The CDA markets after an opening call auction have significantly lower spreads (Wilcoxon test, p-value-value<0.10) than the CDA alone markets. However, the CDA market after our transparent call auction has higher spreads than the CDA after the non-transparent call auction, as confirmed by a Wilcoxon test (p-value: 0.024).

5 Conclusion

According to our experimental study, the practice of stock exchanges to determine opening prices with call auctions is supported by analyzing the opening price itself as well as examining the subsequent continuous trading. Therefore, the main results can be summarized as follows:

Firstly, call auctions generate a significantly more efficient opening price than using a continuous double auction alone. Secondly, the usage of call auctions at the opening is additionally supported by analyzing the effects of opening call markets on subsequent continuous trading. An opening call auction leads to higher informational efficiency and lower spreads compared with the CDA alone market. Thirdly, higher pre-opening transparency does not lead to higher market quality at the opening, nor in the following continuous double auction, as compared to the treatment with a non-transparent opening auction.

Acknowledgements Financial support from the Faculty of Social and Economic Sciences and the Research Management and Service at the Karl-Franzens-University Graz is gratefully acknowledged.

References

1. Arifovic, Jasmina / Ledyard, John (2007): Call market book information and efficiency, Journal of Economic Dynamics and Control 31, S. 1971-2000.
2. Barclay, M. / T. Hendershott (2004): Liquidity Externalities and Adverse Selection: Evidence from Trading after Hours, Journal of Finance, 59 (2004), 681710.
3. Baruch, Shmuel (2005): Who benefits from an open limit order book?, Journal of Business, Vol. 78, No. 4, S. 1267-1306.
4. Camilleri, Silvio John / Green, Christopher J. (2009): The impact of the suspension of opening and closing call auctions: Evidence from the national stock exchange of India, International Journal of Banking, Accounting and Finance 1, No. 3, S. 257-284.
5. Chang, Rosita P. / Rhee, S. Ghon / Stone, Gregory R. / Tang, Ning (2008): How does the call market method affect price efficiency? Evidence from the Singapore Stock Market, Journal of Banking and Finance 32, S. 2205-2219.
6. Comerton-Forde, Carole (1999): Do trading rules impact on market efficiency? A comparison of opening procedures on the Australian and Jakarta Stock Exchange, Pacific-Basin Finance Journal 7, S. 495-521.
7. Comerton-Forde, Carole / Lau, Sie Ting / McInish, Thomas (2007): Opening and Closing behavior following the introduction of call auctions in Singapore, Pacific-Basin Finance Journal 15, S. 18-35.
8. Comerton-Forde, Carole / Rydge, James (2004): A Review of Stock Market Microstructure, SIRCA, April 2004.
9. Comerton-Forde, Carole / Rydge, James / Burridge, Hayley (2007): Not all call auctions are created equal: evidence from Hongkong, Review of Quantitative Financial Accounting 29, S. 395-413.
10. Domowitz, Ian / Madhavan, Ananth (2001): Open Sesame: Alternative Opening Algorithms in Securities Markets, Schwartz, Robert A. (Hrsg.): The Electronic Call Auction: Market Mechanism and Trading, Building a Better Stock Market, Kluwer Academic Publishers, Boston, S. 375394.
11. Fischbacher, Urs (2007): z-Tree Zurich Toolbox for Ready-made Economic Experiments, Experimental Economics, Vol. 10, No. 2, 171-178.
12. Friedman, Daniel (1993): How trading institutions affect financial market performance: Some laboratory evidence, Economic Inquiry 31, S. 410-435.
13. Gerace, Dionigi / Tian, Gary Gang / Zheng, Willa (2009): Call Auction Transparency and Market Liquidity: The Shanghai Experience, Working Paper.
14. Hoffmann, Peter / van Bommel, Jos (2010): Pre-Trade Transparency in Call Auctions: Quantity Discovery vs. Price Efficiency, Working Paper.
15. Madhavan, Ananth (1992): Trading Mechanisms in Securities Markets, Journal of Finance, 37, 2, S. 607-641.
16. Medrano, L. / Vives, X. (2001): Strategic behavior and price discovery, RAND Journal of Economics 32, S. 221248.
17. Pagano, Marco / Rell, Ailsa (1996): Transparency and Liquidity: A Comparison of Auction and Dealer Markets with informed trading, The Journal of Finance, Vol. 51, No. 2, S. 579-611.
18. Pagano, Michael S. / Peng, Lin / Schwartz, Robert A. (2010): Market Structure and Price Formation at Market Openings and Closings: Evidence from Nasdaqs Calls, Working Paper, Villanova University / Baruch College.
19. Theissen, Erik (2000): Market Structure, informational efficiency and liquidity: An experimental comparison of auction and dealer markets, Journal of Financial Markets 3, S. 333-363.

Reassessing recovery rates – floating recoveries

Chris Kenyon and Ralf Werner

Abstract An important but rather neglected input to the pricing of credit derivatives is the recovery rate. It is a priori not clear how an (expected) recovery rate should be chosen for standard credit derivatives pricing models. The problem is especially acute when quoted prices contradict previous recovery rate assumptions. Therefore, in this paper we consider a dynamic calibration of traditional static CDS models. We start from the usual recalibration and propose heuristics for a market-consistent calibration of the recovery rate. Thus the market-implied recovery rate that is now an output can be used as an input for stochastic recovery rate models.

1 Introduction

The elephant in the room around CDS, as well as CMBS, CMBX, and CDO pricing is the recovery rate. Practically, whilst recovery swaps exist they are not liquid. The problem how to choose the recovery rate is especially acute when upfront payments, or quoted prices, contradict previous recovery rate assumption, see Figure 1. In the same figure, it can also be observed that although recoveries usually decrease as ratings drop, ratings are not always in sync with market movements.

Theoretical developments have run ahead of market reality for calibration. Whilst CDS and CDO reduced-form models exist for stochastic hazard rates and stochastic recovery rates [2,5,6], actual calibration is problematic, as they require the expected recovery rate *as input*. Although these models are still important in that they move the theory forward and make it more expressive of potential situations, market risk practitioners may not always be able to find sufficient evidence to calibrate and support such more expressive models.

In this paper we take a step backwards and consider a dynamic calibration of traditional static CDS models, where we focus on the most popular *deterministic*

Chris Kenyon
Credit Suisse AG, e-mail: chris.kenyon@credit-suisse.com

Ralf Werner
Hochschule München, Fakultät für Informatik und Mathematik, Lothstr. 64, 80335 München
e-mail: ralf.werner@hm.edu

D. Klatte et al. (eds.), *Operations Research Proceedings 2011*, Operations Research Proceedings, 185
DOI 10.1007/978-3-642-29210-1_30, © Springer-Verlag Berlin Heidelberg 2012

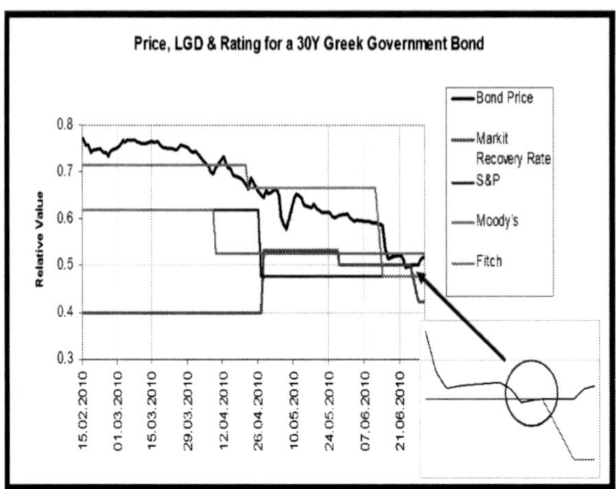

Fig. 1 Inconsistency between price and recovery rate for a Greek government bond.

hazard rate model[1]. This suggested dynamic calibration is in contrast to usual usage of static CDS models, where a deterministic hazard rate function is derived from market prices for a given recovery rate. Instead, the market-implied recovery rate now becomes an output of the model rather than an input. In this way we aim to make practical the recent more expressive models [2, 5, 6], however, we limit the scope of this exposition to market-consistent calibration of static recovery rates for brevity of presentation.

2 Modeling

Let us start by recalling the most common definitions of the *recovery rate* δ of a bond, see for example [4], Section 5.7 or [1]:

- *Recovery of face value*[2]: A fraction $\delta \in [0; 1]$ of the outstanding notional N is paid at the *time of default* τ.
- *Recovery of treasury*: A fraction $\delta \in [0; 1]$ of the outstanding notional N is paid at the *bond maturity T*.
- *Recovery of market value*: A fraction $\delta \in [0; 1]$ of the value of the claim just prior to default (denoted by $PV_{Bond}(\tau^-)$) is paid at time of default τ.

---

[1] See for example [4], Section 8.4 or `http://www.cdsmodel.com/cdsmodel/`

[2] Recovery of face value is the standard assumption used in most CDS pricing models, for example in the already mentioned hazard rate model which is also the standard ISDA model, see `http://www.cdsmodel.com/cdsmodel/`

At time of default τ, the present values of the recovery payments are given by

$$R_{fv} = \delta \cdot N, \qquad R_{tr} = P(\tau,T) \cdot \delta \cdot N, \quad \text{and} \quad R_{mv} = \delta \cdot PV_{Bond}(\tau^-),$$

where $P(\tau,T)$ denotes the discount factor at time τ for time T.

For better illustration, let us consider a simple toy model[3] where a bond with future coupon payment dates $t_1 < \cdots < t_n = T$ (t_0 = today) can default at any point in time. The following cash flow is received by the bond holder of a floating rate bond with $N = 1$ at time t_i if we assume recovery of face value[4]:

$$CF_{Bond}(t_i) = \Delta_i \cdot (f_i + s_{Bond}) \cdot \mathbf{1}_{\{\tau > t_i\}} + \delta \cdot \mathbf{1}_{\{\tau \in (t_{i-1}, t_i]\}} + \mathbf{1}_{\{\tau > t_i\}} \cdot \mathbf{1}_{\{t_i = T\}}. \quad (1)$$

In this context Δ_i denotes the time in years between two coupon dates t_{i-1} and t_i, f_i is the corresponding forward rate and s_{Bond} is the spread over the risk free rate. In case a fixed coupon bond is considered the coupon C replaces $(f_i + s_{Bond})$.

2.1 Theoretical limits on recovery rates

From Equation (1) the present value of the bond can be obtained via

$$PV_{Bond}(t_0) = \mathbf{E_Q}\left[\sum_{i=1}^{n} P(t_0, t_i) \cdot CF_{Bond}(t_i) \mid \mathscr{F}_{t_0} \right].$$

From this, it immediately follows that

$$PV_{Bond}(t_0) \geq P(t_0, t_1) \cdot \delta$$

as the worst-case for the bond holder can only occur by an immediate default. This yields an upper bound on the recovery rate δ depending on the bond price, which shows that not all recovery rates are feasible for pricing corresponding credit derivatives like CDS, etc. In contrast, if the alternative approach of recovery of market value is used all recovery rates between 0 and 1 are feasible.

Let us now consider a CDS in the toy model. In this context, a CDS shall be characterized by the following cash flows:

- *Protection leg*: The protection leg pays the loss-given-default $1 - \delta$ at the coupon payment date following default plus the corresponding coupon payment (forward rate + spread) which has been lost due to default:

---

[3] In this model, a few simplifying modeling assumptions were taken, for instance, accrued interest is ignored; however, this is only made for brevity of presentation and can be easily generalized to more realistic settings.

[4] A similar formula holds if recovery of treasury is considered. For recovery of market value, both the bond price and the individual recovery payments have to be calculated by backward induction through a binomial tree.

$$CF_{CDS}^{Prot}(t_i) = \Delta_i \cdot (f_i + s_{Bond}) \cdot \mathbf{1}_{\{\tau \in (t_{i-1},t_i]\}} + (1-\delta) \cdot \mathbf{1}_{\{\tau \in (t_{i-1},t_i]\}}.$$

- *Premium leg*: In return for the protection, a fixed premium s_{CDS} has to be paid by the protection buyer until a default has been observed:

$$CF_{CDS}^{Prem}(t_i) = \Delta_i \cdot s_{CDS} \cdot \mathbf{1}_{\{\tau \geq t_i\}}.$$

Then the present value of the CDS can be obtained similarly to the price of the bond

$$PV_{CDS}(t_0) = \mathbf{E_Q}\left[\sum_{i=1}^{n} P(t_0,t_i) \cdot CF_{CDS}^{Prot}(t_i)\right] - \mathbf{E_Q}\left[\sum_{i=1}^{n} P(t_0,t_i) \cdot CF_{CDS}^{Prem}(t_i) \mid \mathscr{F}_{t_0}\right],$$

and it can be shown (by comparison of the cash flows to a default-free floating rate bond) that it holds

$$\forall t : PV_{Bond}(t) + PV_{CDS}(t) = 1 \quad \Longleftrightarrow \quad s_{CDS} = s_{Bond},$$

independent of the specific recovery value. More generally, for a given recovery rate δ, the bond price can be deduced from the CDS price and vice versa, as both prices are only depending on the hazard rate (i.e. default probabilities). Thus it is obvious that if the bond trades below the recovery rate, no reasonable CDS price or par spread can be obtained any more for the corresponding CDS, see Figure 2.

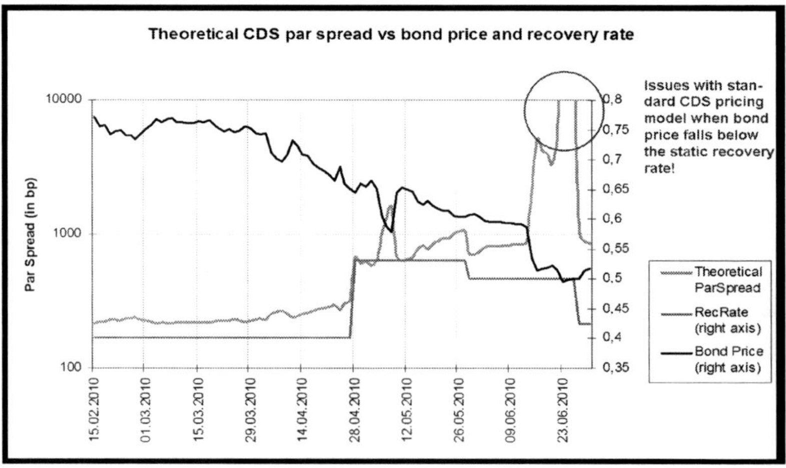

Fig. 2 Inconsistency between price and recovery rate leads to non-existing CDS par spread.

In practice, if the bond price drops towards the current recovery assumption, the assumption is reviewed and the recovery rate is adjusted downwards. If the bond price rebounds, the recovery rate may be reviewed and re-adjusted. Such a behaviour is illustrated in Figure 1, where it can be observed how Markit updates their estimate for the recovery rate δ.

2.2 *Market-consistent heuristics*

Although such a manual adjustment of recovery rates is still best practice and usually adequate for front office purposes, it has some important implications for risk management processes. Firstly, significant market moves in bond prices might lead to infeasible recovery estimates, which in turn would lead to a break-down of the corresponding CDS pricing model. Secondly, as a change of the recovery rate has a price impact on all credit derivatives which are not at par, recovery risk needs to be considered within the daily market risk calculations. As recovery rates exhibit a jump process behaviour this usually leads to cumbersome calibration and computation issues.

A feasible alternative to avoid the issues of infeasible recovery rates would be a switch of the pricing methodology to recovery of market value. Alas, there are two difficulties in this approach: switching pricing models might not be an easy task from a technical or IT perspective, especially, if a large variety of product types has to be considered; further, recovery of market value does not seem to describe reality in a more exact manner than the ordinary recovery of face value, see [1].

Therefore, we propose another alternative, which allows banks to keep in place pricing models based on recovery of face value, while at the same time incorporating some of the features of recovery of market value into the models. Instead of the manual adjustment of the recovery rate, we propose a mechanism which adjusts the recovery rate automatically with the price of the bond. After numerical experiments with B-splines and other parametric curves, cf. [3], two very simple formulas for the mechanism are suggested as starting point:

$$\delta(PV_{Bond}) = \delta_{init} \cdot PV_{Bond}, \quad \text{or} \quad \delta(PV_{Bond}) = PV_{Bond} \cdot (1 - (1 - \delta_{init}) \cdot PV_{Bond}),$$

where δ_{init} denotes the initial estimate of the recovery rate when the bond was issued. In general, such a mechanism should satisfy a few key properties:

- the recovery rate should depend monotonously and smoothly on the bond price,
- the hazard rate should increase as the bond price decreases,
- the hazard rate should increase proportionately as the price decreases, at least for prices around par,
- the implied hazard rate should be convex in the bond price for a large range of bond prices,
- the price deviation between the new and the traditional choice of the recovery rate should be small as long as the bond price is far away from the original recovery estimate, and

- only a very low number of parameters should be necessary to describe the mechanism.

Both choices satisfy all of the above requirements quite well with the exception that the first choice is not close to the traditional model for low bond prices. The second model fits all of the requirements, but it has to be noted that it is only valid if $\delta_{init} \geq 50\%$. Using slightly more complicated B-splines, the performance of the models can be further improved, see [3].

In these models, given the initial recovery rate, bond prices and deterministic hazard rates and thus CDS prices are in a one-to-one relationship. For risk management purposes, the hazard rate is the only necessary risk factor, as recovery is now linked to the hazard rate by the above relationship.

3 Conclusions

Market price movements can be far more dramatic than can be accommodated by just about any choice of constant recovery rates. Practically this leads to a necessity of recovery rates that can be daily re-calibrated so as to be consistent with market observations — even for static models. This has led various authors to propose stochastic recovery rate models [2, 5, 6]. However, a missing starting point has been the actual determination of the expected recovery rate to start these models off.

We have proposed simple heuristics to permit market-consistent recovery rates which satisfy the above list of key properties. This approach leads to dynamic recovery rates that are stochastic in time but deterministically linked to the equivalent bond prices and is thus widely applicable in risk management.

Acknowledgements The authors would like to thank Donal Gallagher for useful discussions and commentary. Further, we are indebted to Manuela Spangler her support with numerical calculations.

References

1. Guo, X., Jarrow, R., Lin, H.: Distressed debt prices and recovery rate estimation. Review of Derivatives Research **11**(3), 171–204 (2008)
2. Höcht, S., Zagst, R.: Pricing credit derivatives under stochastic recovery in a hybrid model. Appl. Stoch. Model. Bus. Ind. **26**(3), 254–276 (2010)
3. Kenyon, C., Werner, R.: A Short Note on Market-Consistent Calibration of Static Recovery Rates. (2010) Available at SSRN: http://ssrn.com/abstract=1551387
4. Lando, D.: Credit risk modeling: theory and applications. Princeton Series in Finance, New Jersey (2004)
5. Li, Y.: A Dynamic Correlation Modelling Framework with Consistent Stochastic Recovery. (2010) Available at SSRN: http://ssrn.com/abstract=1351358
6. Werpachowski, R.: Random Recovery for Both CDOs and First-to-Default Baskets. (2010) Available at SSRN: http://ssrn.com/abstract=1468376

Downside Risk Approach for Multi-Objective Portfolio Optimization

Bartosz Sawik

Abstract This paper presents a multi-objective portfolio model with the expected return as a performance measure and the expected worst-case return as a risk measure. The problems are formulated as a triple-objective mixed integer program. One of the problem objectives is to allocate the wealth on different securities to optimize the portfolio return. The portfolio approach has allowed the two popular in financial engineering percentile measures of risk, value-at-risk (VaR) and conditional value-at-risk (CVaR) to be applied. The decision maker can assess the value of portfolio return and the risk level, and can decide how to invest in a real life situation comparing with ideal (optimal) portfolio solutions. The concave efficient frontiers illustrate the trade-off between the conditional value-at-risk and the expected return of the portfolio. Numerical examples based on historical daily input data from the Warsaw Stock Exchange are presented and selected computational results are provided. The computational experiments show that the proposed solution approach provides the decision maker with a simple tool for evaluating the relationship between the expected and the worst-case portfolio return.

1 Introduction

The portfolio problem involves computing the proportion of the initial budget that should be allocated in the available securities, is at the core of the field of financial management. A fundamental answer to this problem was given by Markowitz ([4]) who proposed the mean-variance model which laid the basis of modern portfolio theory. Since the mean-variance theory of Markowitz, an enormous amount of papers have been published extending or modifying the basic model in three directions (e.g. [1], [2], [3], [5], [9], [10], [11], [12], [13], [14], [15], [16]). The proposed multi-criteria portfolio approach allows two percentile measures of risk in financial engineering: value-at-risk (VaR) and conditional value-at-risk (CVaR) to be applied for managing the risk of portfolio loss. The proposed mixed integer and linear pro-

AGH University of Science and Technology, Faculty of Management, Department of Applied Computer Science, Al. Mickiewicza 30, PL-30-059 Krakow, Poland. e-mail: b_sawik@yahoo.com

D. Klatte et al. (eds.), *Operations Research Proceedings 2011*, Operations Research Proceedings, 191
DOI 10.1007/978-3-642-29210-1_31, © Springer-Verlag Berlin Heidelberg 2012

gramming models provide the decision maker with a simple tool for evaluating the relationship between expected and worst-case loss of portfolio return.

2 Definitions of Percentile Measures of Risk

VaR and CVaR have been widely used in financial engineering in the field of portfolio management (e.g. [8]). CVaR is used in conjunction with VaR and is applied for estimating the risk with non-symmetric cost distributions. Uryasev ([17]) and Rockafellar and Uryasev ([6], [7]) introduced a new approach to select a portfolio with the reduced risk of high losses. The portfolio is optimized by calculating VaR and minimizing CVaR simultaneously.

Let $\alpha \in (0, 1)$ be the confidence level. The percentile measures of risk, *VaR* and *CVaR* can be defined as below:

- Value-at-Risk (*VaR*) at a $100\alpha\%$ confidence level is the targeted return of the portfolio such that for $100\alpha\%$ of outcomes, the return will not be lower than *VaR*. In other words, *VaR* is a decision variable based on the α-percentile of return, i.e., in $100(1 - \alpha)\%$ of outcomes, the return may not attain *VaR*.
- Conditional Value-at-Risk (*CVaR*) at a $100\alpha\%$ confidence level is the expected return of the portfolio in the worst $100(1 - \alpha)\%$ of the cases. Allowing $100(1 - \alpha)\%$ of the outcomes not exceed *VaR*, and the mean value of these outcomes is represented by *CVaR*.

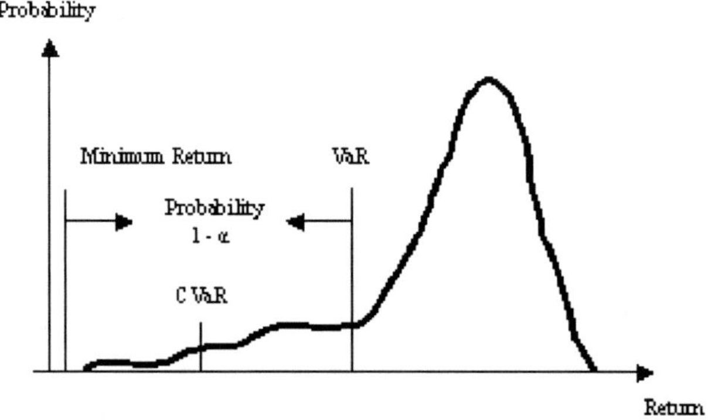

Fig. 1 Value-at-Risk and Conditional Value-at-Risk

Fig. 1 illustrates value-at-risk and conditional value-at-risk for a given portfolio and the confidence level α.

3 Problem Formulation

This section presents mathematical programming formulation for the multi-objective selection of optimal portfolio. The presented portfolio model provides flexibility in how a decision maker wants to balance its risk tolerance with the expected portfolio returns and number of stocks in optimal portfolio. In the risk averse objective, VaR is a decision variable denoting the Value-at-risk. The nondominated solution set of the multi-objective portfolio can be found by the parametrization on β. Constraint (2) ensures that all capital is invested in the portfolio (the selected securities). Risk constraint (3) defines the tail return for scenario i. Constraints (4-6) defines relation between constraints x_j, z_j. Constraint (7) eliminates from selection the assets with non-positive expected return over all scenarios. Eq. (8) and (9) are non-negativity conditions. Constraint (10) defines binary variable z_j.

Table 1 Notation

	Indices
i	= historical time period i, $i \in M = \{1,\ldots,m\}$ (day)
j	= security j, $j \in N = \{1,\ldots,n\}$
	Input Parameters
r_{ij}	= observed return of security j in historical time period i
p_i	= probability of historical portfolio realization i
α	= confidence level
$\beta_1, \beta_1, \beta_1$	= weights in the objective functions
	Decision Variables
R_i	= tail return, i.e., the amount by which VaR exceeds returns in scenario i
VaR	= Value-at-risk of portfolio return based on the α-percentile of return, i.e., in $100\alpha\%$ of historical portfolio realization, the outcome must be greater than VaR
x_j	= amount of capital invested in security j
z_j	= 1 if capital is invested in jth security, 0 otherwise

Maximize

$$\beta_1\left(VaR - (1-\alpha)^{-1}\sum_{i=1}^{m} p_i R_i\right) + \beta_2\left(\sum_{i=1}^{m} p_i\left(\sum_{j=1}^{n} r_{ij}x_j\right)\right) + \beta_3\left(\sum_{j=1}^{n} z_j\right) \tag{1}$$

subject to

$$\sum_{j=1}^{n} x_j = 1 \tag{2}$$

$$R_i \geq VaR - \sum_{j=1}^{n} r_{ij}x_j; i \in M \tag{3}$$

$$x_j \geq 0.005 z_j; j \in N \tag{4}$$

$$x_j \leq z_j; j \in N \tag{5}$$

$$\sum_{j=1}^{n} z_j \geq 1; \tag{6}$$

$$x_j = 0; j \in N : \sum_{i=1}^{m} p_i r_{ij} \leq 0 \tag{7}$$

$$x_j \geq 0; j \in N \tag{8}$$

$$R_i \geq 0; i \in M \tag{9}$$

$$z_j \in \{0,1\}; j \in N \tag{10}$$

4 Computational Examples

In this section the strength of presented portfolio approach is demonstrated on computational examples (Fig. 2– 2).

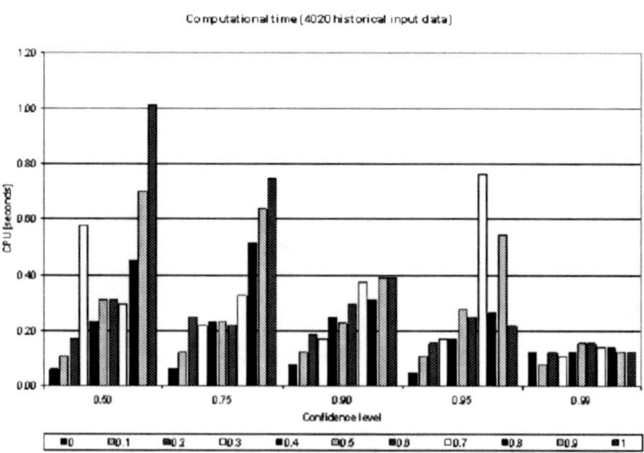

Fig. 2 Computational time

The data sets for the example problems were based on historic daily portfolios of the Warsaw Stock Exchange 4020 days (from 30th April 1991 to 30th of January 2009) and 240 securities. In the computational experiments the five levels of the confidence level was applied $\alpha \in \{0.99, 0.95, 0.90, 0.75, 0.50\}$.

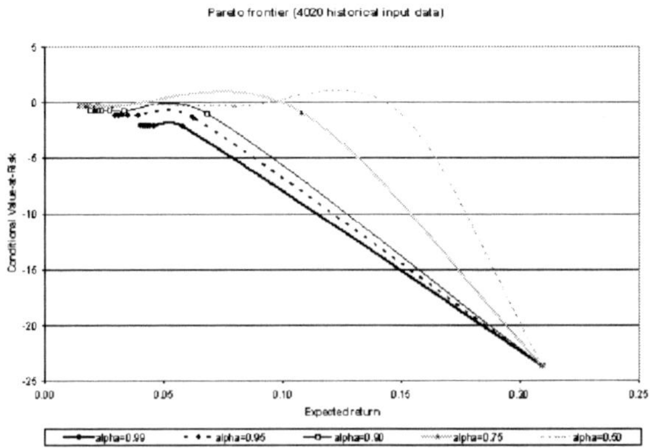

Fig. 3 Pareto frontier

All computational experiments were conducted on a laptop with IntelCore 2 Duo T9300 processor running at 2.5GHz and with 4GB RAM. For the implementation of the multi-objective portfolio models, the AMPL programming language and the CPLEX v.11 solver (with the default settings) were applied.

5 Conclusion

The portfolio approach presented in this paper has allowed the two popular in financial engineering percentile measures of risk, value-at-risk (VaR) and conditional value-at-risk (CVaR) to be applied. The computational experiments show that the proposed solution approach provides the decision maker with a simple tool for evaluating the relationship between the expected and the worst-case portfolio return. The decision maker can assess the value of portfolio return and the risk level, and can decide how to invest in a real life situation comparing with ideal (optimal) portfolio solutions. A risk-aversive decision maker wants to maximize the conditional value-at-risk. Since the amount by which losses in each scenario exceed VaR has been constrained of being positive, the presented models try to increase VaR and hence positively impact the objective functions. However, large increases in VaR may result in more historic portfolios (scenarios) with tail loss, counterbalancing this effect. The concave efficient frontiers illustrate the trade-off between the conditional value-at-risk and the expected return of the portfolio. In all cases the CPU time increases when the confidence level decreases. The number of securities selected for the optimal portfolio varies between 1 and more than 50 assets. Those numbers show very little dependence on the confidence level and the size of historical portfolio used as an input data.

Acknowledgements This work has been partially supported by Polish Ministry of Science & Higher Education (MNISW) grant # 6459/B/T02/2011/40 and by AGH.

References

1. Alexander G.J., Baptista A.M.: A Comparison of VaR and CVaR Constraints on Portfolio Selection with the Mean-Variance Model. Management Science, **50** (9), 1261–1273 (2004)
2. Benati S., Rizzi R.: A mixed integer linear programming formulation of the optimal mean/Value-at-Risk portfolio problem. European Journal of Operational Research, **176**, 423–434 (2007)
3. Mansini R., Ogryczak W., Speranza M. G.: Conditional value at risk and related linear programming models for portfolio optimization. Annals of Operations Research, **152**, 227–256 (2007)
4. Markowitz H.M.: Portfolio selection. Journal of Finance, **7**, 77–91 (1952)
5. Ogryczak W.: Multiple criteria linear programming model for portfolio selection. Annals of Operations Research, **97**, 143–162 (2000)
6. Rockafellar R.T., Uryasev S.: Optimization of conditional value-at-risk. The Journal of Risk, **2**(3), 21–41 (2000)
7. Rockafellar R.T., Uryasev S.: Conditional value-at-risk for general loss distributions. The Journal of Banking and Finance, **26**, 1443–1471 (2002)
8. Sarykalin S., Serraino G., Uryasev S.: Value-at-Risk vs. Conditional Value-at-Risk in Risk Management and Optimization. In: Z-L. Chen, S. Raghavan, P. Gray (Eds.) Tutorials in Operations Research, INFORMS Annual Meeting, Washington D.C., USA, October 12–15 (2008)
9. Sawik B.: Conditional value-at-risk and value-at-risk for portfolio optimization model with weighting approach. Automatyka, **15**(2), 429–434 (2011)
10. Sawik B.: A Bi-Objective Portfolio Optimization with Conditional Value-at-Risk. Decision Making in Manufacturing and Services, **4**(1-2), 47–69 (2010)
11. Sawik B.: Selected Multi-Objective Methods for Multi-Period Portfolio Optimization by Mixed Integer Programming. In: Lawrence K.D., Kleinman G. (Eds.) Applications of Management Science, Volume 14, Applications in Multi-Criteria Decision Making, Data Envelopment Analysis and Finance, Emerald Group Publishing Limited, UK, USA, pp. 3–34 (2010)
12. Sawik B.: A Reference Point Approach to Bi-Objective Dynamic Portfolio Optimization. Decision Making in Manufacturing and Services, **3**(1-2), 73–85 (2009)
13. Sawik B.: Lexicographic and Weighting Approach to Multi-Criteria Portfolio Optimization by Mixed Integer Programming. In: Lawrence K.D., Kleinman G. (Eds.) Applications of Management Science, Vol. 13, Financial Modeling Applications and Data Envelopment Applications, Emerald Group Publishing Limited, UK, USA, pp. 3–18 (2009)
14. Sawik B.: A weighted-sum mixed integer program for bi-objective dynamic portfolio optimization. Automatyka, **13**(2), 563–571 (2009)
15. Sawik B.: A Three Stage Lexicographic Approach for Multi-Criteria Portfolio Optimization by Mixed Integer Programming. Przeglad Elektrotechniczny, **84**(9), 108–112 (2008)
16. Speranza M.G.: Linear programming models for portfolio optimization. Finance, **14**, 107–123 (1993)
17. Uryasev S.: Conditional value-at-risk: optimization algorithms and applications. Financial Engineering News, Issue **14**, February (2000)

Stream VI
Game Theory, Computational and Experimental Economics

Thomas Burkhardt, University of Koblenz, GOR Chair
Felix Kübler, University of Zurich, SVOR Chair
Ulrike Leopold-Wildburger, University of Graz, ÖGOR Chair

Social Values and Cooperation. Results from an Iterated Prisoner's Dilemma Experiment.

Jürgen Fleiß and Ulrike Leopold-Wildburger

Abstract The following article deals with the question of cooperation in dilemma situations. We ran an iterated Prisoner's Dilemma experiment and measured the players social value orientation using the Ring Measure of Social Value. We then analyze the players behavior in the Prisoner's Dilemma in relation to their social value orientation to test the hypotheses that prosocial players are more likely to cooperate. We find evidence that this is indeed the case. We do not find evidence that if two prosocial players interact with each other they achieve higher cooperation rates than two proself players or one prosocial and one proself player.

1 Introduction

In dilemma situations people are faced with a decision between their selfinterest and the interest of a group. Cooperation would lead to mutual benefits but there are incentives for the individual to defect [12]. In such dilemma situations standard economic theory predicts defective action by all players leading to a Pareto inferiour outcome [4]. One of the most prominent example for such a dilemma situation is of course the Prisoner's Dilemma (PD).

But even if the theoretical look at such problems is rather grim, experimental economists have shown that cooperation appears much more frequently than the theory predicts [8].

One solution to solve the puzzle of cooperation is the assumption that some people have 'social preferences' and thus also care about the outcome of others [1]. A number of techniques have been developed to measure such preferences [11].

Jürgen Fleiß
Department of Statistics and Operations Resarch, University of Graz, Universitätsstraße 15, 8010 Graz, e-mail: juergen.fleiss@uni-graz.at

Ulrike Leopold-Wildburger
Department of Statistics and Operations Resarch, University of Graz, Universitätsstraße 15, 8010 Graz, e-mail: ulrike.leopold@uni-graz.at

D. Klatte et al. (eds.), *Operations Research Proceedings 2011*, Operations Research Proceedings, 199
DOI 10.1007/978-3-642-29210-1_32, © Springer-Verlag Berlin Heidelberg 2012

In our paper we will report on a series of experiments which consist of 2 parts: a questionaire which aims to measure the social value orientation of the players and an infinitely repeated PD game by which the cooperative and defective behavior of the players can be measured.

We will analyze the behavior of the players in the PD in relation to their measured social value orientation and show that socially oriented players cooperate more often.

The rest of the paper is organized as follows. Section 2 gives a short overview of the social value concept used in this paper and the according measure developed by Liebrand and McClintock [6]. In section 3 we describe the experimental design. In section 4 we will present the results of our experiments. The paper concludes with a short summary in section 5.

2 Social Value Orientation

Standard economic theory assumes players to be selfish in a sense that they are only concerned about their own outcome. This assumption is challenged by studies which have shown that individual differences in social value orientations exist. These social values can be understood as preferences for certain patterns of outcomes to oneself and others [8]. Social values are assumed to be a personality trait that plays an important role for an individual's behavior in settings of outcome interdependence [1], [5].

One major strand of research on social values is associated with the work of Liebrand and McClintock [6]. They define an individual's utility function as follows: $U_i = w_1 * (\text{Outcomes for self}) + w_2 * (\text{Outcomes for other})$.

This definition allows the construction of an infinite number of utility functions by varying w_1 and w_2. Four utility functions are especially emphasised because they have been identified as significant for the behavior of individuals in decision tasks [6]. If $w_1 = 0$ and $w_2 = 1$ a person is considered to be altruistic. Cooperative persons take into account the own payoff and the others equally ($w_1 = 1$, $w_2 = 1$). Individualists maximize their own outcome ($w_1 = 1$, $w_2 = 0$) while competitors maximize the relative advantage ($w_1 = 1$, $w_2 = -1$). The Ring Measure of Social Value [6] allows to classify our players accordingly.

The distribution of the angles (Fig. 1) implies that we are dealing with two different groups. Thus we classify our player in two groups, split at an angle of 22.5 degree. We call the group containing players with an angle below 22.5 degree proself and above 22.5 degree prosocial. In doing so we follow Van Lange and Liebrand [10]. Table 1 shows the classification of the players.[1]

---

[1] A detailed describtion of the Ring Measure of Social Values can be found in [6] and [7].

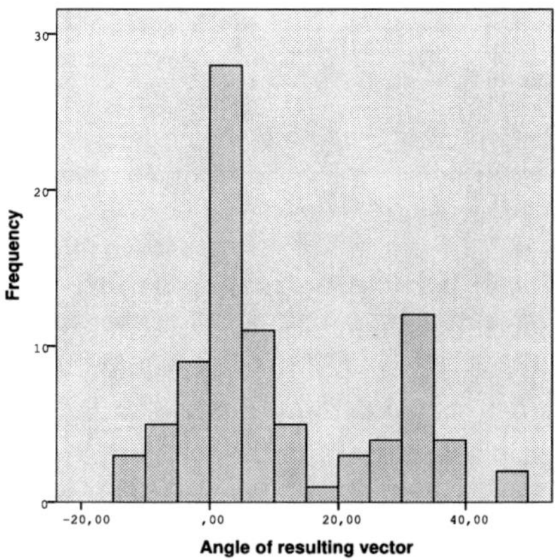

Fig. 1 Distribution of the angles of the social value orientation as measured by the Ring Measure of Social Values

Table 1 Classification of the players into prosocial and proself according to their social value orientation

	Number of cases	%
Proself	62	71.3
Prosocial	25	28.7

3 Experimental Design

We conducted a computerized PD experiment using the payoffs in Table 2. Players interacted repeatedly with the same partner and were rematched with a new partner two or three times. In 9 cases players are rematched once, in 14 cases the players are rematched twice. Thus we had two or three matchings per session. A matching we call a number of successive rounds in which one player interacts with the same partner. A player is matched with a different partner after a number of rounds not known to the participants in advance. One matching lasts between 10 and 35 rounds.

The experiment is conducted in the Experimental Economics Laboratory of the Department of Statistics and Operations Research at the University of Graz. The participants are recruited in classes and by campus advertisementes. Potential participants are informed that they can take part in a paid experiment consisting of an decision making task. We arranged the experiments in a way that 8 or 12 participants take part at the same time. 92 subjects took part in the experiments.[2]

These participants were then randomly assigned in groups of four that were led to computer workstations. The workstations are seperated from each other so participants cannot directly interact with each other. Subjects do not know which one of

---

[2] 5 Players were excluded from the analysis because either they either were outliers in the social preferences distribution or their comments suggested that they did not understand the game properly.

Table 2 Payoffs Treatment 1 and 2

Treatment 1			Treatment 2		
Row-player/column-player	A	B	Row-player/column-player	A	B
A	4/4	0/5	A	3/3	0/4
B	5/0	1/1	B	4/0	1/1

the other participants is their partner. Subjects are then given introductory material on the experiment which they read. After that the experiment started.

In the first part of the experiment the subjects fill out the Ring Measure of Social Value described above. They decided between 24 alternative pairs of outcomes for themselves and a hypothetical partner. These outcomes were imaginary amounts of money (both positive and negative). After completing the first part they are matched with a partner and begin to play an iterated Prisoner's Dilemma with unknown lenght. Subjects chose between the two alternatives A and B and could take as much time as they needed. Decisions are made simultaneously and after all four players in a group have taken their decision a player is informed about the decision of his or her partner. When subjects are matched with a new partner they are informed about the change.

After all players interacted with their partners the experiment ends and the participants are paid a substantial monetary reward according to their own decisions and the decisions of their partners. On average a session lasted about 80 minutes.

4 Results

First we take a look at the players behavior in relation to his or her social value orientation as measured by the Ring Measure of Social Values described above. To do this, we will use the angle of the players resulting vector and not the classification. A higher angle is interpreted to correspond with a stronger social value orientation.

Some key characteristics of the players behavior in the PD will be studied: the action a player realizes in the first interaction with his or her first partner, the relative number of cooperative actions in the first matching and the number of defective actions in the first matching. In addition we will look at the success of player in form of the points he or she earned.

Table 3 shows the correlations of a players social value angle with the described aspects of his or her behavior.

Table 3 Correlations between a players social value orientation and his or her behavior

	A in first round (1=yes)	Share of A (matching 1)	Share of B (matching 1)	Points achieved
Spearman-Rho	.250*	.240*	-.241*	.218*

*... $p < .05$

Table 4 Mean and standard deviation for the differently composed matchings

Composition of matching	Share of AA	Share of BB	#matchings
2 Prosocial	.482 (.344)	.212 (.234)	8
2 Proself	.487 (.384)	.287 (324)	57
Mixed	.507 (.400)	.328 (.328)	38

We see a positive correlation between the first action a player realizes and his or her social value orientation. Thus a more socially oriented player is more likely to cooperate in the first interaction with his or her first partner. Also, the share of cooperative (A) and defective (B) actions in the first matching correlates as expected with the social value orientation. Players with a stronger social value orientation (a higher angle) play A more and B less frequently. The effect sizes vary between .24 and .25. This corresponds with the results of a recent meta-analysis that found an average effect size of .23 for paid experiments [1].

We also find a significant positive effect between social value orientation and the points a player achieved. This suggests that players with a stronger social value orientation are more succesfull in the IPD.

4.1 Analysis of matchings

In the last step we will test the hypotheses that the composition of matchings influences the the share of mutually cooperative actions in a matching. To do so, we coded for each player in each pair their social value orientation. We use the dichotomous classification into prosocial and proself described in section 2.

Table 4 shows the mean share of outcomes (AA, BB, BA and AB) in the different matchings (standard deviation in brackets) and the number of matchings with different compositions. 57 of the matchings consisted of 2 proself players, 38 matchings were composed of 1 proself and 1 prosocial player. 8 matchings were composed of 2 prosocial players.

We see that the means of AA and BB interactions are quite similar for all three groups. The highest average of mutual cooperation can be found in the mixed groups, which also yield the highest average of mutually defective outcomes. A Kruskal-Wallis-Test shows no statistically significant results.[3] Thus with our sample we cannot proof differences in cooperation rates for matchings with different group composition.

---

[3] A group size of 8 for matchings with 2 prosocial players is rather small. In the literature a minimum group size of 5 is demanded [3].

5 Summary

Our contribution deals with the question of cooperation in social dilemma situations. We conducted a experiment that consisted of two parts: a questionare to measure the players social value orientations and an iterated Prisoner's Dilemma to measure cooperative and defective behavior. Our analysis shows that the more socially oriented a player is, the more likely he is to show cooperative behavior in the Prisoner's Dilemma. The observations and effect strength are consistent with similar studies [1].

In a next step we looked into the effect of the social value orientation of two players in a matching. In this case our hypotheses that if two prosocial players are matched higher cooperation rates can be observed could not be proven.

References

1. Balliet, D., Parks, C., Joireman, J.: Social Value Orientation and Cooperation in Soical Dilemmas: A Meta-Analysis. Group Processes & Ingroup Relations (2009) doi: 10.1177/1368430209105040
2. De Cremer, D., Van Lange, P.A.M.: Why Prosocials Exhibit Greater Cooperation than Proselfs: The Roles of Social Responsibility and Reciprocity. Eur. J. Pers. **15**, 5-18 (2001).
3. Doli, D.: Statistik mit R. Einfhrung fr Wirtschafts- und Sozialwissenschaftler. Oldenbourg, Munich/Vienna (2004).
4. Grling, T.: Value Priorities, social value orientations and cooperation in social dilemmas. British Journal of Social Psychology **38**, 397-408 (1999).
5. Hulbert, L.G., Corrła da Silva, M.L., Adegboyegy, G.: Cooperation in social dilemmas and allocentrism: a scial value approach. Eur. J. Soc. Psychol. **31**, 641-657 (2001).
6. Liebrand, W.B.G, McClintock, C.G.: The ring measure of social values: a computerized procedure for assessing individual differences in information processing and social value orientation. European Journal of Personality **2**, 217-230 (1988).
7. Offerman, T., Sonnemans, J., Schram A.: Value Orientations, Expectations and Voluntary Contributions in Public Goods. The Economic Journal **106**, 817-845 (1996).
8. Simpson, B.: Social Values, Subjective Transformations, and Cooperation in Social Dilemmas. Social Psychology Quarterly **67**, 385-395 (2004).
9. Van Lange, P.A.M.: On perceiving morality and potency: Social values and the effects of person perception in a give-some dilemma. European Journal of Personality **3**, 209-225 (1989).
10. Van Lange, P.A.M., Liebrand, W.B.G., The Influence of Other's Morality and Own Social Value Orientation on Cooperation in The Netherlands and the U.S.A. International Journal of Psychology **26**, 429-449 (1991).
11. Van Lange, P.A.M.: The Persuit of Joint Outcomes and Equality in Outcomes: An Integrative Model of Social Value Orientation. Journal of Personality and Social Psychology **2**, 337-349 (1999).
12. Weber, J.M.: Suckers or Saviors? Consistent Contributors in Social Dilemmas. Journal of Personality and Social Psychology **95**, 1340-1353 (2008).

Honesty in Budgetary Reporting - an Experimental Study

Arleta Mietek and Ulrike Leopold-Wildburger

Abstract In the experiment we investigate how an agent's participation in the budgetary process influences his willingness to truthful communication of private information, if there are financial incentives to misrepresentation. The paper focuses on the trade-off between the agent's preferences for wealth and honesty. The objective is to find and examine the relation between social preferences and honesty in managerial reporting.

1 Introduction

Budgetary reporting and the budgeting process are important issues in management accounting. Reporting facilitates the flow of information within an organization. Due to the budgeting the information asymmetry between upper and lower management can be offset. The (generally) better informed local management communicates its information to the upper management, which can be further used for planning, resource allocation and coordination in the organization and also for performance measurement. And exactly that described use of budgets caused and still causes many discussions. Agency theory assumes a rational agent who reports dishonestly. Some authors (e.g. [8]) criticize the missing reasonability of using budgets for performance evaluation because of its incentives to misrepresentation of information.

The presented experiment is one part of a greater study in which we examine the effects of communication and the degree of the principal's control on the agent's reporting behavior. However in this paper the focus is on the trade-off between the agent's preferences for wealth on the one hand and his preference for honesty on

Arleta Mietek
Department of Statistics and Operations Research, Universitätsstraße 15, A-8010 Graz, e-mail:
arleta.mietek@uni-graz.at

Ulrike Leopold-Wildburger
Department of Statistics and Operations Research, Universitätsstraße 15, A-8010 Graz, e-mail:
ulrike.leopold@uni-graz.at

D. Klatte et al. (eds.), *Operations Research Proceedings 2011*, Operations Research Proceedings, 205
DOI 10.1007/978-3-642-29210-1_33, © Springer-Verlag Berlin Heidelberg 2012

the other hand. The objective is to find and examine the relation between social preferences and honesty in managerial reporting.

In particular, there are two parties in the experiment: the supervisor or principal (upper manager) and the better informed subordinate or agent (lower manager). The agent has private, perfect information regarding the division's production and should report it to the supervisor. Whereupon, the subordinate's compensation depends both on the actual production and the reported production. The parameters are set such that the subordinate gets the highest payoff by reporting the possibly lowest production, which means the highest level of misrepresentation. We want to investigate how the agent's participation in the budgetary process influences his willingness to truthful communication of private information if there are financial incentives to misrepresentation.

2 Background

Budgetary reporting and the budgeting process are important issues in management accounting and they have recently become a prevalent issue also in experimental accounting (compare for a review e.g. [3], [15], [4], [2]). The traditional principal-agent framework assumes both parties to be rational and to maximize their payoffs. The agency theory predicts that in cases of hidden information, the agent's reporting behavior depends only on his preference for wealth maximization, (compare e.g. [1]). Motivated by experimental research Mittendorf [13] extends the model of Antle and Eppen [1]. He incorporates agent's preferences for honesty in their model and he demonstrates that manager's cost of misrepresenting exceeds the marginal benefit from doing so.

Experimental literature has consistently demonstrated that agents are willing to sacrifice wealth to report honestly or partially honestly (compare e.g. [5], [7], [14]). In the experiment we use the following slack-inducing pay scheme for the agent A:

$$P(X,B)^A = \begin{cases} F + \alpha_2 \cdot B + \alpha_1 \cdot (X - B) \text{ for } X \geq B \\ F + \alpha_2 \cdot B + \alpha_3 \cdot (X - B) \text{ for } X < B \end{cases} \text{ with } 0 < \alpha_2 < \alpha_1 \leq \alpha_3$$

where X and B are the actual production and the agent's self-set report, respectively and F is the fixed salary. The report B is the information regarding production, which is communicated by the agent in the process of budgetary reporting. The parameters $\alpha_2, \alpha_1, \alpha_3$ are set such that the contract gives the agent a financial incentive to misrepresent his private information. Thus it imposes a direct trade-off between the agent's preferences for wealth and honesty on the agent. The last part of the compensation refers to the deviation of actual production X from the report B: a positive deviation – overfulfillment – leads to a bonus, while a negative deviation – underfulfillment – is associated with a penalty. Because of $\alpha_2 < \alpha_3$ the possible penalty is weighted higher than the bonus associated with the reported production.

Arnold [2] states that the experimental investigation of slack-inducing contracts is meaningful for at least two reasons: (1) Despite the common, practical use of budgets in incentive contracts, firms do not regularly use truth-inducing contracts. (2) The supervisor can not credibly commit herself not to use the budgeting information for performance evaluation or for control of the division, even if this information is not explicitly included in the pay scheme. Thus, the slack-inducing contract reproduces the conflict of interest, which arises because of the lacking commitment.

The principal P appears in the study as a residual claimant and receives the following period's return:

$$P(X,B)^P = D \cdot X - P(X,B)^A$$

where D is the profitability of the project and $P(X,B)^A$ the agent's compensation. Both payoff functions point the principal-agent conflict up: while the agent goes for a minimum budget in order to maximize his wealth, the principal is interested in maximizing the budget.

3 Experiment

Procedure The experiment was programmed with z-Tree [6], a software that enables to program and to conduct economic experiments. For the baseline treatment presented in this paper we recruited 38 participants. They all were drawn from the University of Graz campus and included undergraduate, graduate and post-graduate students. The study consists of two parts: (1) the budgetary reporting experiment including a short post-experimental questionnaire and (2) a questionnaire examining the participants' social value orientation. After the experiment the subjects were rewarded for their participation according to their performance.

Experimental Task The multi-period experiment reproduces the relationship between two parties: the supervisor or principal P and the better informed subordinate or agent A. All participants play the role of the agent, while the computer[1] takes the role of the principal. It is common knowledge that the actual period's production X ranges between 200 and 400 pieces, in steps of 20. Agents, however, are told to be able, due to their experience, to forecast the actual period's production for sure. Thus, they receive at the beginning of each period private, perfect information regarding the production and shall report it to the supervisor. Participants are aware that the supervisor expects them to report their best estimate of the actual production, but she will never get to know the actual production. The supervisor has no control - the proposed budget is incorporated in the pay scheme without any possibility to reject it.

---

[1] In the baseline treatment the principal has no control and even no possibility to reject the report/budget, and thus it is not necessary to employ real participants for this role. In this treatment we concentrate on agent's reporting behavior. Compare e.g [5].

Agents are compensated due to the before mentioned slack-inducing contract. In particular in the experiment $\alpha_1 = \alpha_3 = 0.3$ and $F = 10$ such that the pay scheme is adopted to:

$$P(X,B)^A = 10 + 0.2 \cdot B + 0.3 \cdot (X - B)$$

Under this pay scheme the agent has a financial incentive to report the lowest possible production in order to make the bonus easily attainable. The principal receives the following payoff:

$$P(X,B)^P = 0.55 \cdot X - P(X,B)^A$$

Some examples in the instructions illustrate the agent's payoffs according to different combinations of the actual (X) and the reported (B) production. The instructions also provide the payoff-matrix (such as Table 1), including payoffs of both parties in all possible X-B-combinations.

		\multicolumn{11}{c}{**Actual production X**}									
		200	**220**	**240**	**260**	**280**	**300**	**320**	**340**	**360**	**380**	**400**
	200	60/50	65/56	70/62	75/68	80/74	85/80	90/86	95/92	100/98	105/104	110/110
	220	62/48	67/54	72/60	77/66	82/72	87/78	92/84	97/90	102/96	107/102	112/108
	240	64/46	69/52	74/58	79/64	84/70	89/76	94/82	99/88	104/94	109/100	114/106
	260	66/44	71/50	76/56	81/62	86/68	91/74	96/80	101/86	106/92	111/98	116/104
	280	68/42	73/48	78/54	83/60	88/66	93/72	98/78	103/84	108/90	113/96	118/102
Report B	**300**	70/40	75/46	80/52	85/58	90/64	95/70	100/76	105/82	110/88	115/94	120/100
	320	72/38	77/44	82/50	87/56	92/62	97/68	102/74	107/80	112/86	117/92	122/98
	340	74/36	79/42	84/48	89/54	94/60	99/66	104/72	109/78	114/84	119/90	124/96
	360	76/34	81/40	86/46	91/52	96/58	101/64	106/70	111/76	116/82	121/88	126/94
	380	78/32	83/38	88/44	93/50	98/56	103/62	108/68	113/74	118/80	123/86	128/92
	400	80/30	85/36	90/42	95/48	100/54	105/60	110/66	115/72	120/78	125/84	130/90

Table 1 Payoff-matrix: Principal/**Agent**

After reading the instructions, the subjects had to solve two examples examining whether they had understood the task, their compensation and the consequences of their decisions for the principal.

Finally the participants filled in a questionnaire including questions regarding the information asymmetry between agent and principal and the general understanding of the pay scheme. The questionnaire contained also questions testing whether the participants understood that all responses during the experiment were anonymous.

Social value orientation We use the ring measure of social values (e.g. [11], [10]) to analyze whether there is any connection between the agent's reporting behavior and his social value orientation. This method, due to a series of decisions regarding the allocation of a resource to oneself and to others, classifies subjects in four social types: altruistic (maximizing other's outcomes), cooperative (maximizing joint outcomes), individualistic (maximizing own outcomes) and competitive (maximizing relative advantage), (compare [11], p. 219). We predict that cooperative agents report more honestly than individualistic and competitive types.

4 Results and Conclusions

All participants understood that their information regarding the production was a perfect information, while the principal knew only the range of 200-400 units. Participants were aware that the principal would never find out whether they reported honestly or not.

Related to Evans et. al. [5] we measure honesty as $h = 1 - \dfrac{(X - B)}{(X - B_{\min})}$, where $X - B$ is the actual misrepresentation and $X - B_{\min}$ is the highest possible misrepresentation. The average percentage honesty in the baseline experiment is 38.05% which is obviously inconstistent with the theoretical prediction of 0%. Subjects in the experiment do not report only to maximize their wealth, because they do not always report the minimum possible production of 200 units. Only 12 of 38 agents report a minimum production of 200 in each experimental period. In 5 cases (13.16%) we observe exclusively honest reports ($X = B$ in each period). The remaining sample of 21 subjects (55.26%) reports partially honest, which means that their reporting behavior is not consistent during the experiment. Our observations confirm findings of Evans et.al [5] and Matuszewski [12] regarding the inconsistency with the types model by Koford and Penno [9]: there are not only ethical (always honest) and rational (never honest) agents but additionally also partially honest agents, whereupon the average percentage honesty of the "partially honest" type is relative high and amounts to 46.96%. Additionally, the regression of the reported production B on the actual production X indicates that reports are not random but they are rather positive related to the actual production X ($\beta = .624$, $p < .001$), which confirms that subjects do have other preferences than only wealth maximization.

Using the ring measure of social Vales [11] we can categorize 37 participants into two social groups: cooperative (14 or 36.84%) and individualistic (23 or 60.53%), one participant remains unclassified. The crosstabulation of agent types and their

		Social Value Orientation		
		Cooperative	Individualistic	Total
Agent Types	Honest	5	0	5
		13.5%	0.0%	13.5%
	Partially honest	7	12	19
		18.9%	32.4%	51.4%
	Rational	2	11	13
		5.4%	29.4%	35.1%
		14	23	37
		37.8%	62.2%	
Honesty (h)	Mean	71.4%	19.4%	39.1%
	Median	83.6%	4.0%	30.0%
	Std. Deviation	35.3%	24.8%	38.5%

Table 2 Crosstabulation of Agent Types and SVO

SVO (Table 2) suggests that there is a relation between social preferences and the honesty in reporting. Cooperative agents indicate a higher mean (median) of 71.4% (83.6%) than individualistic agents. Kruskal-Wallis test supports those findings, indicating that honesty is significantly higher for cooperative than for individualistic agents ($p < .001$). Our prediction that cooperative agents report more honestly than individualistic agents is also verified by the regression of h on SVO. With $\beta = -52\%$ and $p < .001$ the model validates a significant negative association between honesty and social preferences: percentage average honesty of individualistic agents significantly lower than the honesty of cooperative subjects.

References

1. Antle, R., Eppen, G. D.: Capital rationing and organizational slack in capital budgeting. Management Science. 31, 415–444 (1985)
2. Arnold, M. C. Experimentelle Forschung in der Budgetierung - Luegen, nichts als Luegen? Journal fuer Betriebswirtschaft. 57, 69–99 (2007)
3. Brown, J. L., Evans III, J. H., Moser, D. V.: Agency Theory and Participative Budgeting Experiments. Journal of Management Accounting Research. 21, 317–345 (2009)
4. Covaleski, M. A., Evans III, Jo. H., Luft, J. L., Shields, M. D.: Budgeting Research: Three Theoretical Perspectives and Criteria for Selective Integration. Handbook of Management Accounting Research. 2, 587–624 (2007)
5. Evans III, J. H., Hannan, R. L., Krishnan, R., Moser, D. V.: Honesty in Managerial Reporting. Accounting Review. 76, 537-559 (2001)
6. Fischbacher, U.: z-Tree: Zurich Toolbox for Ready-made Economic Experiments. Experimental Economics. 10, 171–178 (2007)
7. Hannan, R. L.,Rankin, F. W., Towry, K. L.: The Effect of Information Systems on Honesty in Managerial Reporting: A Behavioral Perspective. Contemporary Accounting Research. 23, 885–918 (2006)
8. Jensen, M. C.: Paying People to Lie: the Truth about the Budgeting Process. European Financial Management. 9, 379–406 (2003)
9. Koford, K., Penno, M.: Accounting, principal-agent theory, and self-interested behavior. In: Bowie, N. E., Freeman, R. E. (eds.) Ethics and Agency Theory, pp. 127142. Oxford University Press, New York (1992)
10. Liebrand, W. B. G.: The effect of social motives, communication and group size on behaviour in an N-person multi-stage mixed-motive game. European Journal of Social Psychology. 14, 239–264 (1984)
11. Liebrand, W. B. G., McClintock, C. G.: The ring measure of social values: a computerized procedure for assessing individual differences in information processing and social value orientation. European Journal of Personality. 2, 217–230 (1988)
12. Matuszewski, L. J.: Honesty in Managerial Reporting: Is It Affected by Perceptions of Horizontal Equity? Journal of Management Accounting Research. 22, 233–250 (2010)
13. Mittendorf, B.: Capital Budgeting when Managers Value both Honesty and Perquisites. Journal of Management Accounting Research. 18, 77–95 (2006)
14. Rankin, F. W., Schwartz, S. T., Young, R. A.: The Effect of Honesty and Superior Authority on Budget Proposals. Accounting Review. 83, 1083–1099 (2008)
15. Sprinkle, G. B.: Perspectives on experimental research in managerial accounting. Handbook of Management Accounting Research. 1, 415–444 (2007)

Points Scoring Systems in Formula 1: A Game Theoretic Perspective

Krystsina Bakhrankova

Abstract An efficient points scoring system is often vital to encouraging competition in a sport. It is also related to the viewership and spectator numbers crucial to lasting popularity and financial success. This paper analyzes the three points scoring systems of the last twenty seasons in Formula 1 (1991-2010) from a game theoretic perspective to examine their competitive impact.

1 Introduction

With an estimated annual turnover of 4 billion USD and some 50 000 employees in over 30 countries, Formula 1 is a truly global sport and business. Started as a primarily European competition, it is "the longest-established motorsport championship series in the world" with races presently spanning five continents. Surpassed only by the Olympic Games and the World Cup, it is the third most watched sport broadcast in 185 countries for some 850 million fans [8].

In recent years, challenged by legislative changes in the advertising regulation and diminishing public interest, Formula 1 was forced to look for new global sponsors as well as circuits in the emerging markets of Asia and the Middle East. Another attempt to reinvent the competition is related to two transitions in the scoring system in the last twenty years. From its inception in 1950, various rules have existed with respect to the number of races that counted towards the World Championship, team (constructor) points, and points for the fastest lap. Eventually, the system settled on awarding points to the top six finishers (10-6-4-3-2-1) in 1991, only to be altered to eight scoring places (10-8-6-5-4-3-2-1) in 2003 and to ten places (25-18-15-12-10-8-6-4-2-1) in 2010. This paper uses game theory to analyze these two shifts and verify their competitive effects.

Krystsina Bakhrankova
SINTEF Technology and Society, Applied Economics, Postboks 4760 Sluppen, 7465 Trondheim, Norway, e-mail: krystsina.bakhrankova@sintef.no

D. Klatte et al. (eds.), *Operations Research Proceedings 2011*, Operations Research Proceedings, 211
DOI 10.1007/978-3-642-29210-1_34, © Springer-Verlag Berlin Heidelberg 2012

2 Relevant Literature

Due to its highly competitive and constantly changing environment, Formula 1 is often considered an appropriate arena for lessons in optimal business performance and strategy (e.g., [8], [4], and [7]). However, none of the existing works analyze the competitive impact of the recent changes in the points scoring system. The relevant literature applying game theory in this context comes from soccer (e.g., [6] and [2]).

3 The Game Theoretical Model

Building on the game theoretic framework developed in [5] and extended in [6], the following assumptions are made:
1. Two drivers referred to as $D1$ and $D2$ are analyzed;
2. Their performance levels – driving skills and car capabilities – are known;
3. The Grand Prix outcome is uncertain;
4. The reciprocal uncertainty assessments are agreed on (complete information);
5. Two choices of strategy referred to as A (aggressive) and D (defensive);
6. Both drivers make their strategic choices simultaneously before the race start;
7. The objective of each driver is to maximize his expected point score.

3.1 Competition During a Race

First, an in-race rivalry between two drivers for each place from the 1st to the 10th is analyzed as a separate game. The following Table 1 demonstrates such a situation for two significantly unequal drivers ($D1$ is better than $D2$). Here, p_{12} is a probability that $D1$ finishes 1st (ahead) and $D2 - 2^{nd}$ (behind), p_{21} is vice versa, and p_{out} is a collusion probability, when both are ousted and score 0 points. In addition, the pay-offs are calculated for each driver and each system: $S1$(10-6-4-3-2-1), $S2$ (10-8-6-5-4-3-2-1), and $S3$ (25-18-15-12-10-8-6-4-2-1).

Table 1 Two significantly unequal drivers – contest for the 1st and the 2nd place

D1	D2	p_{12}	p_{21}	p_{out}	$P(D1)_{S1}$	$P(D2)_{S1}$	$P(D1)_{S2}$	$P(D2)_{S2}$	$P(D1)_{S3}$	$P(D2)_{S3}$
A	A	0.70	0.10	0.20	7.6	5.2	7.8	6.6	19.30	15.10
D	A	0.60	0.30	0.10	7.8	6.6	8.4	7.8	20.40	18.30
A	D	0.85	0.05	0.10	8.8	5.6	8.9	7.3	22.15	16.55
D	D	0.80	0.15	0.05	8.9	6.3	9.2	7.9	22.70	18.15

When normal form games are constructed for each system, it is observed that the unique pure-strategy Nash equilibrium shifts from (D, A) in $S1$ to (D, D) in $S2$

and comes back to (D, A) in $S3$. Further, differences in expected point scores are compared for each pair of Nash equilibria, where a are points for a place ahead (1st) and b – for a place behind (2nd):

1. From $S1$ to $S2$: $E(D1^{DD}) - E(D2^{DD}) > E(D1^{DA}) - E(D2^{DA})$
$0.80a + 0.15b - (0.15a + 0.80b) > 0.60a + 0.30b - (0.30a + 0.60b) =>$
$0.35a > 0.35b$ – this expression is always true as $a > b$ in any system.

Hence, the competitive imbalance has increased from $S1$ to $S2$. It is reasonable as this shift increases the 1st-place chances of a better driver $D1$ and reduces those of a poorer one $D2$ (cf. Table 1).

2. From $S2$ to $S3$: $E(D1^{DA}) - E(D2^{DA}) > E(D1^{DD}) - E(D2^{DD})$
$0.60a + 0.30b - (0.30a + 0.60b) > 0.80a + 0.15b - (0.15a + 0.80b) =>$
$-0.35a > -0.35b$ – this expression is always false as $a > b$ in any system.

Hence, the competitive imbalance has decreased from $S2$ to $S3$. The result is sensible as this shift improves the 1st-place chances of a poorer driver $D2$ and diminishes those of a better one $D1$ (cf. Table 1).

It is notable that in the absence of these calculations, the conclusions would have been the opposite for each of the shifts similarly to the soccer analysis (cf. [6]). Yet, contrary to soccer, both drivers are able to score in Formula 1 – if one of them is a winner of a place duel, it does not mean that the other one scores nothing (unless he is ousted from the race).

Using the same probabilities, similar analysis has been carried out for each place contest, where $Imbalance_1$ assesses the shift from $S1$ to $S2$ and $Imbalance_2$ – from $S2$ to $S3$ (cf. Table 2).

Table 2 Two significantly unequal drivers – contest for each scoring place from the 1st to the 10th

Contest/Places	$(NashEq)_{S1}$	$(NashEq)_{S2}$	$(NashEq)_{S3}$	$Imbalance_1$	$Imbalance_2$
1st vs. 2nd	(D, A)	(D, D)	(D, A)	**increase**	*decrease*
2nd vs. 3rd	(D, A)	(D, D)	(D, D)	**increase**	no change
3rd vs. 4th	(D, D)	(D, D)	(D, D)	no change	no change
4th vs. 5th	(D, A)	(D, D)	(D, D)	**increase**	no change
5th vs. 6th	(D, A) (A, D)	(D, D)	(D, D)	**increase** *decrease*	no change
6th vs. 7th	(A, A)	(D, A)	(D, D)	*decrease*	**increase**
7th vs. 8th	–	(D, A) (A, D)	(D, A)	–	no change *decrease*
8th vs. 9th	–	(A, A)	(D, A) (A, D)	–	*decrease* **increase**
9th vs. 10th	–	–	(D, A) (A, D)	–	–
10th vs. 11th	–	–	(A, A)	–	–

Based on the above calculations, it is concluded that the shift from $S1$ to $S2$ has mostly increased competitive imbalance. Concurrently, the following shift from $S2$ to $S3$ has not significantly improved the situation apart from the contest for the 1st place (less feasible for $D2$). Yet, the imbalance has worsened for the 6th place (more feasible for $D2$), while others remained unchanged on the $S2$ levels. The expansion of scoring places has had an overall positive effect, while more aggressive strategies prevail as one moves down the grid.

Another set of probabilities for two marginally unequal drivers was generated similarly to Table 1. The corresponding analysis and calculations of competitive imbalance are presented in the following Table 3.

Table 3 Two marginally unequal drivers – contest for each scoring place from the 1st to the 10th

Contest / Places	$(NashEq)_{S1}$	$(NashEq)_{S2}$	$(NashEq)_{S3}$	$Imbalance_1$	$Imbalance_2$
1st vs. 2nd	(A, D)	(D, D)	(D, A)	*decrease*	*decrease*
2nd vs. 3rd	(D, A)	(D, A)	(D, D)	no change	**increase**
3rd vs. 4th	(D, A)	(D, D)	(D, D)	**increase**	no change
4th vs. 5th	(D, A)	(D, D)	(D, D)	**increase**	no change
5th vs. 6th	(A, D)	(D, A)	(D, D)	*decrease*	**increase**
6th vs. 7th	(A, A)	(D, A)	(D, A)	*decrease*	no change
7th vs. 8th	–	(A, D)	(D, A)	–	*decrease*
8th vs. 9th	–	(A, A)	(A, D)	–	**increase**
9th vs. 10th	–	–	(A, D)	–	–
10th vs. 11th	–	–	(A, A)	–	–

In this case, the shift from $S1$ to $S2$ has mostly decreased competitive imbalance. However, the following shift from $S2$ to $S3$ has worsened the situation apart from the contest for the 1st place which has improved. Yet, the imbalance has worsened for the 2nd and the 5th places, while others remained unchanged on the $S2$ levels. Here, the expansion of scoring places has had both a positive and a negative effect, while more aggressive strategies also prevail as one moves down the grid.

If an example with two theoretically equal drivers[1] is constructed in line with Table 1, the only strategy used regardless of the system is (D, D).

Summing up, the shift from $S1$ to $S2$ has mostly increased competitive imbalance for two significantly unequal drivers and has mostly decreased it for two marginally unequal participants. Yet, the following shift from $S2$ to $S3$ has not significantly improved the situation of the former, while made it worse for the latter. Strategies and competitive balance have not changed for two theoretically equal drivers.

3.2 Overall Race Competition

In addition to the in-race rivalry, it is interesting to examine how the overall competition changes depending on the points scoring system. Here, each driver is assigned a certain probability distribution based on performance as well as his and his rival's choice of strategy before the race start. The following Table 4 exemplifies two significantly unequal drivers together with their calculated pay-offs for each system.[2]

---

[1] Two drivers of the same team arguably closest resemble this situation.

[2] For each pair of strategies and each system, the sums of the distributions for all scoring places follow the same pattern as in Table 1. The 11th place is the lowest considered as it always yields

Table 4 Two significantly unequal drivers – overall race competition

Place	p_{D1}^{AA}	p_{D2}^{AA}	p_{D1}^{DA}	p_{D2}^{DA}	p_{D1}^{AD}	p_{D2}^{AD}	p_{D1}^{DD}	p_{D2}^{DD}
1st	0.25	0.005	0.2	0.015	0.325	0.004	0.3	0.01
2nd	0.175	0.0075	0.15	0.02	0.225	0.006	0.2	0.015
3rd	0.14	0.01	0.125	0.025	0.15	0.008	0.12	0.02
4th	0.085	0.015	0.1	0.035	0.12	0.012	0.1	0.025
5th	0.08	0.02	0.09	0.045	0.07	0.0175	0.075	0.03
6th	0.07	0.05	0.08	0.06	0.03	0.0325	0.055	0.05
7th	0.05	0.06	0.075	0.1	0.0225	0.045	0.035	0.075
8th	0.025	0.1325	0.03	0.135	0.02	0.075	0.0275	0.15
9th	0.015	0.15	0.015	0.18	0.015	0.17	0.025	0.175
10th	0.01	0.25	0.01	0.21	0.0125	0.275	0.02	0.2
11th	0.1	0.3	0.125	0.175	0.01	0.355	0.0425	0.25
P_{S1}	4.595	0.27	3.96	0.625	5.73	0.2115	5.185	0.455
P_{S2}	5.82	0.7275	5.23	1.33	6.985	0.5285	6.3825	1.035
P_{S3}	14.32	2.63	12.925	4.17	17.0625	2.092	15.68	3.42

When normal form games are constructed for each system, it is observed that the unique pure-strategy Nash equilibrium is always (A, A) regardless of the system. The same is true if two marginally unequal drivers are considered. Yet, if two equal drivers are analyzed, then the unique Nash equilibrium for all the systems is either (A, D) or (D, A). Thus, as confirmed by the experimental results, either shift in the points scoring system does not alter the corresponding strategic choices.[3]

4 Data Analysis

Subsequent to the preceding discussions, it is essential to analyze the actual points scored during the last twenty seasons in Formula 1 (1991-2010). The following Table 5 shows some comparative statistics (ΔPoints is the winner's margin to the runner-up and ΔRaces is the margin of the title's award to the season's end).[4]

Here, the winner's margin has varied between 1 and 67 points, the race margin – from 0 to 6 races, Wins (%) – from 31 to 65 and Podiums (%) – within 47-100 (1991-2002, S_1). Most of these measures have decreased in 2003-2009 with S_2. It is in line with the modeling predictions, where the competitive imbalance has mostly decreased for two marginally unequal drivers, while has mostly increased for two significantly unequal participants. Besides, in 2005-2010, the Wins (%) has dropped drastically. Perhaps indicative of greater competition, it contradicts the pop-

---

0 points; it accounts for all other non-scoring places and, in effect, for probability of collusion or withdrawal. The probabilities can be also represented as compound lotteries – cf. [1].

[3] Similarly to [6], a generalized analysis can be carried out for both in-race rivalry and overall competition.

[4] Data were sourced from www.formula1.com.

Table 5 Statistics of the Formula 1 World Drivers' Champions (1991-2010)

Season	'91	'92	'93	'94	'95	'96	'97	'98	'99	'00	'01	'02	'03	'04	'05	'06	'07	'08	'09	'10
ΔPoints	**24**	52	26	1	33	19	3	14	2	19	58	67	**2**	34	21	13	1	1	11	**4**
ΔRaces	**1**	5	2	0	2	0	0	0	0	1	4	6	**0**	4	2	0	0	0	1	**0**
Wins (%)	**44**	56	44	50	53	50	41	50	31	53	53	65	**38**	72	37	39	35	28	35	**26**
Podiums (%)	**75**	75	75	63	65	63	47	69	63	71	82	100	**50**	83	79	78	71	56	53	**53**

ular claim that a champion should be the winner of most races. It remains to be seen what actual effects will result from the latest shift.

5 Conclusion

This paper analyzed two shifts in the points scoring system of the last twenty seasons in Formula 1 from a game theoretic perspective to verify their competitive effects. Moreover, the relevant data were examined to confirm some of the modeling predictions. The results should be carefully interpreted as these shifts were not the only changes introduced (cf. [3]). As such, they do not imply that every driver will change his strategy accordingly. Besides, the technological component is essential in Formula 1 dominance of teams and engine manufacturers can be often used as a proxy for shifting competitive balance. Further, pit stop strategies and weather conditions also impact race outcomes. Yet, the approach presented provides insight into the workings of the points scoring system and can be a good start for designing an optimal system. Given multiple players and numerous complexities, this task is far from trivial and requires a thorough analysis of the intertwined real-world interactions.

References

1. Binmore, K.: Fun and Games. D.C. Heath, Lexington, MA, and Toronto, Canada (1992)
2. Brocas, I., Carrillo, J.D.: Do the "three-point victory" and "golden goal" rules make soccer more exciting? J. Spor. Econ. **5(2)**, 169–185 (2004) doi: 10.1177/1527002503257207
3. *Fédération Internationale de l'Automobile* (FIA): 2011 Formula One Sporting Regulations. http://www.fia.com/sport/Regulations/f1regs.html. Cited 19 Jun 2011
4. Gilson, C., Pratt, M., Roberts, K., Weymes E.: Peak Performance: Inspirational Business Lessons from the World's Top Sports Organizations. Texere, New York (2000)
5. Haugen, K.K.: An economic model for player trade in professional sports. J. Spor. Econ. **7(3)**, 309–318 (2006) doi: 10.1177/1527002504272941
6. Haugen, K.K.: Point score systems and competitive imbalance in professional soccer. J. Spor. Econ. **9(2)**, 191–210 (2008) doi: 10.1177/1527002507301116
7. Hotten, R.: Winning: The Business of Formula One. Texere, New York (2000)
8. Jenkins, M., Pasternak, K., West, R.: Performance at the Limit: Business Lessons from Formula 1 Motor Racing. Cambridge University Press, Cambridge (2009)

An experiment examining insider trading with respect to different market structures

Philipp Hornung, Gernot Hinterleitner, Ulrike Leopold-Wildburger, Roland Mestel, and Stefan Palan

Abstract As theoretical and empirical approaches suffer some shortcomings if it comes to analyzing insiders' behavior, we conducted an adequate experiment. The experimental design incorporates traders with perfect information of the fundamental value of the tradeable asset. These insiders compete with regular uninformed participants on different market structures, in particular on a transparent and on an intransparent call market as well as on a continuous double auction. The results shed first light on a couple of interesting issues, including whether and how insiders try to stay undetected, how their profits are accumulated and what market structure is more advantageous for insiders.

1 Introduction

There is an extensive academic literature on the topic of insider trading. Theoretical, empirical and experimental studies have focused on a number of different aspects

Philipp Hornung
Institute of Statistics and Operations Research, Karl-Franzens-University Graz, Austria
e-mail: philipp.hornung@edu.uni-graz.at

Gernot Hinterleitner
Institute of Statistics and Operations Research, Karl-Franzens-University Graz, Austria
e-mail: gernot.hinterleitner@edu.uni-graz.at

Ulrike Leopold-Wildburger
Institute of Statistics and Operations Research, Karl-Franzens-University Graz, Austria
e-mail: ulrike.leopold@uni-graz.at

Roland Mestel
Institute of Banking and Finance, Karl-Franzens-University Graz, Austria
e-mail: roland.mestel@uni-graz.at

Stefan Palan
Institute of Banking and Finance, Karl-Franzens-University Graz, Austria
e-mail: stefan.palan@uni-graz.at

D. Klatte et al. (eds.), *Operations Research Proceedings 2011*, Operations Research Proceedings, 217
DOI 10.1007/978-3-642-29210-1_35, © Springer-Verlag Berlin Heidelberg 2012

within this field of research. Nevertheless, the existing body of evidence does not yet contain reliable findings on the impact of different market microstructures in the presence of insiders. This study analyzes key questions regarding the influence of a complimentary call auction on insiders' behavior and profits. The next secion will give an introduction into the research questions. Section 3 describes the experimental design used for this study. The results of the experiment are presented in section 4. The final section concludes.

2 Research Questions

The few previous studies which combine the two areas of insider trading and market microstructure either focus on a special market mechanism or compare different single market mechanisms side-by-side. But on stock exchanges throughout the world a combination, i. e. a sequence of single mechanisms, is commonly being used. Therefore our study scrutinizes insider trading under the most basic, yet widely implemented form of these combinations: A call auction is carried out as a complimentary market before continuous trading begins. In addition to this we vary the level of transparency of the opening call auction. This possibility allows us to analyze the first central aspect, namely the discouraging impact of transparency on traders with inside information [5]. A higher level of transparency could get regular traders to know that they are compcting with better informed investors [3]. This in turn may lead them to act in a way to reduce their losses to insiders and therefore diminish insiders' gains. Given that insiders still earn profits in the call auction, to what extent do they keep up against the profits in the continuous trading? Previous experimental work has shown results either way [7, 8]. Irrespective of the extent of the profits, we proceed by asking how insiders go about trying to maximize their profits in the continuous trading in general. The literature mentions two basic strategies insiders can pursue. On the one hand it may be favorable for an informed trader to make use of the informational advantage as soon as possible, especially under competition with other insiders [9]. Aggressive trading at the beginning may lead to considerable profits, because prices are far away from the fundamental value [1]. But even with increasing efficiency, possibilities for financial gains may not be reduced completely [4]. This leads to the second possibility which states that it might be better for insiders to reap profits gradually, given that an information-withholding strategy does not reveal the special status of being an insider [2]. As these research questions deal with insiders' decisions, an experimental study seems to be the appropriate means to approach them. Despite their advantages, theoretical models are subject to some limitations when it comes to human behavior, and the empirical approach can not supply us with data on insider trading free from specific and unknown influences.

3 Experimental Design

The experimental design is a comprehensive extension of Theissen's study [10]. In contrast to his experiment, which compared a single continuous double auction with an intransparent call auction, our experiment includes a transparent call market as well. Furthermore, the call auctions (CA) act as complimentary markets at the market opening and are followed by a phase of continuous trading. In each of these three treatments, 12 participants can trade multiple units of a single asset. In the continuous double auction (CDA), all traders can create buy and sell orders or accept submitted orders by other traders. It lasts 5 minutes during which only the best order of each trading side is displayed. In the intransparent complimentary market (CM), all submitted buy and sell orders are collected for 2 minutes after which they are executed at one price. In contrast to this version of the CA, the complimentary market with transparency (CMT) features an order book which is open at all times. Additionally the indicative price and indicative volume are displayed. Analogous to previous experiments, insiders in the experiment know the fundamental value of the asset. Regular traders receive noisy price signals only, each of which does not differ by more than 10 % from the fundamental value. Insiders can get ahold of inside information if they successfully pass through a two stage process at the beginning of each trading period. First they have to show general interest in illegal insider trading. Afterwards they are confronted with the probability of detection of the inside status. This probability is displayed in steps of 10 % uniformly distributed between 10 % and 90 %. The penalty is set to a very high level – 10 % of the portfolio value – and participants were made aware of this. Each of the three treatments comprehends six periods. The only values which change between the periods are the fundamental value and the price signals. The endowments are reset at the beginning of a new period. All in all we conduct seven sessions of each treatment. Therefore our sample comprises 252 participants. All of them are students of the social and economic sciences at the Karl-Franzens-University Graz. The experiment was programmed and conducted using the software z-Tree [6]. All participants are paid in cash dependent on their overall trading results. The experiment lasted about 3 hours which resulted in an average payment of nearly 9 Euros per hour and participant. This innovative arrangement allowed for the testing of certain assumptions addressing the questions of when traders show interest in the special information and the influence of the different trading mechanisms.

4 Results

Result 1 *A transparent call auction prevents insider trading more effectively than an intransparent version.*

Insiders know that they have an informational advantage over regular traders because of their more accurate information. Since traders are paid according to their trading results, it has to be assumed that they try to maximize their experimental profits. Insiders pursuit the capitalization on their special information. In order to maximize their gains, they try to earn profits as much and as long as possible to the detriment of the non-insiders. In an intransparent call auction they can submit their orders in a hidden fashion. But when the order book is visible to every trader, insiders' actions may pose some aid for the orientation of the other investors. Therefore it can be conjectured that insiders are attracted by intransparency – given that the use of their special information is subject to some kind of cost, be it a financial fee or a penalty as in our experiment.

Table 1 Numbers of insiders in the different markets

	No. of Ins.	No. of Non-Ins.	Total	% of Ins.
CDA alone	114	390	504	22,6 %
CM	142	362	504	28,2 %
CMT	90	414	504	17,9 %

If the distributions of the numbers of insiders per period are compared between the three treatments, we notice a highly significant overall difference (Kruskal-Wallis test, p-value: 0.011). Further comparisons using Mann-Whitney tests prove that there is no significant difference between the CDA alone and the intransparent CA (p-value: 0.190), whereas the differences between the CDA alone and the transparent CA (p-value: 0.059) and between the intransparent and the transparent CAs (p-value: 0.004) are significant.

If we take the decision process into account by which the participants can obtain the inside information, we are not able to discover any bias. For example it could be conceivable that participants in the transparent treatment were shown lower probabilities of detection. If that were the case, the number of insiders would be lower in this treatment, but the influence of the transparency itself would lessen. As a couple of statistical tests show, there is not only no significant difference between the probabilities which were shown to the participants before the decision. Even the distributions of the probabilities at which the insiders decide in favor of inside information is not different between the three markets. We can therefore conclude that traders with an informational advantage try to avoid a transparent CA.

Result 2 *While submitting orders in the continuous double auction insiders try to stay undetected.*

A continuous analysis of the trading process is not appropriate, because more than one order can be submitted each second. Likewise, there are seconds in which not even a single order is being entered. We therefore divide the duration of the trading period into quartiles.

The data indicates that the numbers of orders created by insiders do not differ much between the quartiles. Friedman tests are carried out for each treatment separately. They state that there is no significant difference in any market (p-value: 0.321;

Table 2 Submitted orders in the different markets

		total	Q1	Q2	Q3	Q4
CDA alone	Orders submitted	2941	797	655	624	865
	submitted by Ins.	744	210	165	145	224
	% of Ins. Orders	25 %	26 %	25 %	23 %	26 %
CDA of CM	Orders submitted	3641	853	936	831	1021
	submitted by Ins.	1050	243	252	244	311
	% of Ins. Orders	29 %	28 %	27 %	29 %	30 %
CDA of CMT	Orders submitted	3624	977	890	896	861
	submitted by Ins.	733	184	192	184	173
	% of Ins. Orders	20 %	19 %	22 %	21 %	20 %

0.820; 0.492 respectively). In order to obtain a more detailed picture, the analysis is repeated using deciles. Again, we do not find any significant differences.[1]

In addition we analyze insiders' orders with respect to different percentages of insiders in the individual markets that are listed in table 1 on a periodic basis. A Wilcoxon test does not result in a significant difference between the proportions of insiders and the proportion of insiders' orders (p-values: 0.242; 0.802; 0.232 respectively). In none of the three CDAs do insiders submit disproportionately more orders than uninformed traders.

Even though we do not recognize aggressive trading at the beginning of the continuous trading as suspected by the literature, we can look for further support by determining what kind of trader submits the very first order in a period.

Using a chi-squared test for goodness of fit we compare the sample of first orders against the expected frequency given by the proportions of insiders in table 1. The results show that there is no significant difference (p-value: 0.856; 0.460; 0.549 respectively). The insiders do not try to use their informational advantage in an aggressive manner when the submission of the first order is taken as the measure for aggressiveness.

Result 3 *Insiders accumulate their profits primarily in the continuous double auction.*

We analyze the profits of all insiders who trade in both the CA and the CDA. More precisely we compare to what extend each market structure contributes to the overall profits, examining the two treatments with complimentary markets separately. Both Wilcoxon tests show a significant difference in favor of the CDA (p-value: 0.049 (CM); 0.001 (CMT)).

Not all insiders can be included: Although almost all traders submit orders in the CA, a substantial number of these orders are never executed. The fact that the percentage of unsuccessful insiders is higher in the intransparent market suggests that insiders have a higher probability of generating profits in the transparent than

---

[1] The corresponding table is not displayed due to space limitations.

in the intransparent CA. The uncertainty inherent in the intransparent CA reduces the probability of order execution. The ratio of profits is 41 % (CA) to 59 % (CDA) in the CM and 35 % (CA) to 65 % (CDA) in the CMT. This finding suggests that the intransparent CA is advantageous for insiders. This is the case although the probability of order execution is lower – but whenever insiders' orders are executed, they promise high profits.

5 Conclusion

Our experiment analyzes insider trading in different market structures and shows that insiders try to avoid a transparent market – although the probability of profit-making is higher than in an intransparent market. In order to stay undetected, insiders even place orders gradually instead of making full use of their information early on.

Acknowledgements Financial support from the Faculty of Social and Economic Sciences and the Research Management and Service at the Karl-Franzens-University Graz is gratefully acknowledged.

References

1. Bloomfield, R. , O'Hara, M. , Saar, G.: The "make or take" decision in an electronic market: Evidence on the evolution of liquidity. Journal of Financial Economics 75(1): 165–199, 2005.
2. Buffa, A. M.: Insider Trade Disclosure, Market Efficiency, and Liquidity. Working Paper, London Business School, 2010.
3. Brünner, T. , Levínský, R.: Do prices in the unmediated call auction reflect insider information? – An experimental analysis. Jena Economic Research Papers No. 2008 – 090, Friedrich Schiller University, Max Planck Institute of Economics, Jena, 2008.
4. Chau, M. , Vayanos, D.: Strong-Form Efficiency with Monopolistic Insiders. Review of Financial Studies 21(5): 2275–2306, 2008.
5. Domowitz, I. , Madhavan, A.: Open Sesame: Alternative Opening Algorithms in Securities Markets. Schwartz, R. A. (Ed): The Electronic Call Auction: Market Mechanism and Trading: Building a Better Stock Market. Kluwer Academic Publishers, Boston, 375–394, 2001.
6. Fischbacher, U.: z-Tree: Zurich Toolbox for Ready-made Economic Experiments. Experimental Economics 10(2): 171–178, 2007.
7. Krahnen, J. P. , Rieck, C. , Theissen, E.: Insider Trading and portfolio structure in experimental asset markets with a long-lived asset. European Journal of Finance 5(1): 29–50, 1999.
8. Schnitzlein, C. R.: Call and Continuous Trading Mechanisms Under Asymmetric Information: An Experimental Investigation. Journal of Finance 51(2): 613–636, 1996.
9. Schnitzlein, C. R.: Price Formation and Market Quality When the Number and Presence of Insiders Is Unknown. Review of Financial Studies 15(4): 1077–1109, 2002.
10. Theissen, E.: Market Structure, informational efficiency and liquidity: An experimental comparison of auction and dealer markets. Journal of Financial Markets 3(4): 333–363, 2000.

Markov Simulation of an Iterated Prisoners' Dilemma Experiment

Thomas Burkhardt, Armin Leopold and Ulrike Leopold-Wildburger

Abstract We study cooperation and its development over the time in a repeated prisoner's dilemma experiment with unknown length and unknown continuation probability. Subjects are re matched with new partners several times.
The study examines the influence of past decisions to future behavior.
In the simulation we raise the question whether the experimentally observed time pattern of cooperation can be reconstructed with a simple Markov model of individual behavior. The model parameters are inferred from the experimental data. The resulting Markov model is used to simulate experimental outcomes. Our model indicates that a simple Markov model can be used as a reasonable description of transition probabilities for successive states of cooperation.

1 Introduction

In social cognitive science we have to deal with the following unsolved problem: reliability of human behavior in cooperative situations. In human societies, however, cooperation is in principal based on social norms and social values and a considerable amount of cooperation follows from the legal basis.

In animal societies cooperation is mainly based on affinity and genetic relationship. Job sharing within animal societies as ants, termites, bees and other animals are called polyethism. Passing forward liquid and food as a sort of cooperation is called trophallaxis as well for human as for animal societies.

Thomas Burkhardt
Department for Management, Chair of Finance, University of Koblenz e-mail: tburkha@uni-koblenz.de

Armin Leopold
Department for Informatics, Mathematics and Operations Research, University BW Munich e-mail: armin.leopold@unibw.de

Ulrike Leopold-Wildburger
Department of Statistics and Operations Research, University of Graz e-mail: ulrike.leopold@uni-graz.at

D. Klatte et al. (eds.), *Operations Research Proceedings 2011*, Operations Research Proceedings, 223
DOI 10.1007/978-3-642-29210-1_36, © Springer-Verlag Berlin Heidelberg 2012

Even there is lot of research on cooperation there is still little empirical evidence why and under which conditions people cooperate at all in situations such as the prisoner's dilemma game. In situations in which there is always a future as it can be simulated by infinitely repeated games the credible threat of future retaliation can cause opportunistic behavior and it can be a reason for supporting cooperation. Dal Bo (2005) calls it 'the shadow of the future' which affects a specific behavior, thereby solving the tension between private incentives and the common good.

Experimental evidence on infinitely repeated games is still rare. In our paper we report on a series of IPD experiments with an unknown number of repetitions, by which the cooperative and defective behavior can be measured, respectively. It will be shown that according to Dal Bo the possibility of future action modifies the players behavior.

Section 2 describes the experimental design and elaborates some reasonable hypotheses. Section 3 presents the results of the simulation study using parameters from the experiments. Our observations suggest to use simple but stochastic behavioral rules. Therefore we utilize a Markov model and simulation which uses the transition probabilities for future cooperation and we find stochastic patterns of cooperation over time.

2 Experimental design and hypotheses

We conducted a prisoner's dilemma experiment with the payoffs (3/3) in case of cooperation and (1/1) in case of defection; (0/5) and (5/0) in case of no accordance. Players interacted repeatedly with the same partner. The continuation rule, however, is unknown to the participants. After a certain number of rounds not known to the participants in advance, each player was matched with a new partner (s)he did not play against before. Four players participated in a session. Therefore each player was matched with a different opponent three times, i.e. we had three matchings per session. Each matching consisted of 6 repeated interactions. We thus run a partner design within a matching, and a stranger design between matchings.

The experimental design we used in all sessions was as follows: Each subject has to make a decision. (S)he repeatedly plays against another subject. The decision of each subject is made independently of each other, inter-subjects contacts are anonymous, the choices of the opponent are shown on the screen. Each subject is paid the amount retained by his/her decision. Each point (payoff) is converted into Euro at an exchange rate of 35 Eurocents/point.

A well-known characteristic of the prisoner's dilemma game is the ambiguous relation between cooperation and defection. On the one hand, each player can increase her/his payoff by defecting instead of cooperating. On the other hand, the payoffs of all interacting players increase if a higher share of them cooperates. The theoretical basis for these contradicting tendencies is clearly established. In this study, we will analyze the empirical implications and characteristics of these ambiguous motives.

H1   There is a positive correlation between share cooperation and payoff. Result:
      The Pearson correlation yields a value of .807 (***) for the 96 matchings.
H2   Whenever players start with CC and arrive at CC after a certain number of
      rounds they do not defect in between, moving along a stable path of mutual co-
      operation. Result: The probability of mutual cooperation in at least five rounds
      is significantly higher when starting with CC than with any other combination of
      initial moves (Fisher exact test p = 0.01).

3 A stochastic approach to describe behavior: a Markov model

Usual normative models of the behavior of players in IPD games have difficulties
to describe the experimentally observed behavior. Other approaches, which try to
capture more or less deterministic behavioral patterns based on experimental data
are capable to reflect aspects of the recorded considerations of players within their
decision making process, but lack the flexibility to capture the stochastic aspects of
the decisions finally made.

Our approach focuses on those stochastic aspects. We try to devise a stochastic
description of the experimentally observed behavior of our players. This does not
mean that we claim that players do not act rationally, or according to well defined
considerations. We just assume that the joint outcome of these considerations will
generate results in the iterated game which show stochastic patterns, which may
reasonably be described by an appropriate stochastic model, that provides us with
an improved predictive power compared to other possible descriptions. Furthermore,
we hope that the revealed stochastic structure will contribute to our understanding
of individual behavior. Following this idea, we try to describe the observed behavior
with a particularly simple stochastic model, the Markov model.

A Markov model explains the state of a system s_{t+1} stochastically only based
on its current state s_t. We consider the joint decision of the players in the PD game
in a given round as the state of the system IPD. The possible states are therefore
{CC, CD, DC, DD} (coded by {3,2,1,0}) in each decision round. The development
of cooperation over time is modeled by a state transition matrix $M = \{M_{i,j}\}$, whose
entries give the probability to be in state j in the next round, given that the current
state is i. This means that the decision behavior in a given round is modeled by a
probability distribution, and the players take only into account the decision behavior
of the previous round. In this sense, the model is memoryless. We will see that
even such a simple (and obviously "'unrealistic'") model is capable to capture some
essential features of observed behavior.

Obviously, this basic Markov approach allows for several specifications of the
model. First of all, the transition matrix could be time dependent. Given enough
data, one could estimate a transition matrix for each individual round to get a detailed picture of a possible time dependence and the learning mechanisms which
determine it. Within our limited data set, it seems more appropriate to estimate transition matrices for the three matchings individually, and an overall transition matrix

based on the first 6 rounds of all three matchings. The last 6 rounds of matching 3 have not been used, as the data reflects an obvious structural break. Players did assume that matching 3 would end after 6 rounds, based on previous experience. This approach allows us to capture possible structural changes which are to be expected by re-matching players, and to get a first impression if a Markov model could be suitable to capture features of the decisions in a way that may be generalized.

Second, there are several estimation methods which could be used to infer the transition matrix from the data. As our primary interest here is to find generalizable features of player behavior, we used a symmetric estimation approach which assumes that all players behave stochastically equivalent. Then, we can estimate the probabilities $\{p_i\}$ that a player will cooperate in the next round, given the state of the previous round. The overall result is $(p_{CC}, p_{CD}, p_{DC}, p_{DD}) = (.93, .27, .24, .11)$. This vector can be used to calculate the overall transition matrix that corresponds to it, given our assumption that both players behave stochastically the same way. The decision probabilities for the second symmetric player are the same by assumption, but the vector components p_{CD} and p_{DC} have to be interchanged. For this reason, the transition matrix is given by

$$\begin{pmatrix} p_{CC}^2 & (1-p_{CC})\,p_{CC} & (1-p_{CC})\,p_{CC} & (1-p_{CC})^2 \\ p_{CD}\,p_{DC} & p_{CD}(1-p_{DC}) & (1-p_{CD})\,p_{DC} & (1-p_{CD})(1-p_{DC}) \\ p_{CD}\,p_{DC} & (1-p_{CD})\,p_{DC} & p_{CD}(1-p_{DC}) & (1-p_{CD})(1-p_{DC}) \\ p_{DD}^2 & (1-p_{DD})\,p_{DD} & (1-p_{DD})\,p_{DD} & (1-p_{DD})^2 \end{pmatrix}$$

Please observe the symmetries $M_{CC,CD} = M_{CC,DC}$, $M_{CD,CC} = M_{DC,CC}$, $M_{CD,DD} = M_{DC,DD}$, $M_{DD,CD} = M_{DD,DC}$, $M_{CD,CD} = M_{DC,DC}$, $M_{CD,DC} = M_{DC,CD}$, which result by construction. The transition matrix has therefore only four free parameters. Our experimental data gives (rounded to two digits):

$$M = \begin{pmatrix} .86 & .07 & .07 & .01 \\ .06 & .21 & .17 & .56 \\ .06 & .17 & .21 & .56 \\ .01 & .10 & .10 & .80 \end{pmatrix}$$

This overall matrix shows essentially the structure that we find for the corresponding matrices M_1, M_2 and M_3 estimated for the matchings 1 to 3 individually:

$$M_1 = \begin{pmatrix} .89 & .05 & .05 & .003 \\ .11 & .17 & .29 & .43 \\ .11 & .29 & .17 & .43 \\ .01 & .08 & .08 & .84 \end{pmatrix}, M_2 = \begin{pmatrix} .90 & .05 & .05 & .002 \\ .01 & .21 & .04 & .73 \\ .01 & .04 & .21 & .73 \\ .01 & .10 & .10 & .78 \end{pmatrix}, M_3 = \begin{pmatrix} .82 & .09 & .09 & .008 \\ .06 & .25 & .13 & .56 \\ .06 & .13 & .25 & .56 \\ .02 & .13 & .13 & .72 \end{pmatrix}$$

The estimates use the data of the first six rounds of all three matchings. That is $20 \times 5 \times 2$ observations for each matchwise symmetric players C-response probabilities to calculate M_1 to M_3, and all this data to estimate M. Furthermore, the data from the first round of each matching (20 observations) has been used to estimate the initial distribution d_j for each matching $j = 1, 2, 3$:

$$d_1 = (.40, .20, .30, .10)$$
$$d_2 = (.60, .10, .25, .05)$$
$$d_3 = (.80, .00, .15, .05)$$

We clearly see an increase in the initial probability of joint cooperation from matching to matching. We like to denote this as a 'restart-effect', because it seems likely that a new partner is credited with an increasing level of trust, once the individual learns that mutual cooperation has a tendency to increase returns. On the other hand, within one matching we find a decreasing probability of joint cooperation over time. This tendency can be exemplified by the steady state distribution generated by M: $(.16, .14, .14, .56)$. So in the long run there seems to develop a dominating probability for joint defection.

This simple model seems to contribute to a better understanding of experimental observations in several ways. Firstly, it mimics the high probabilities of rather long runs of either CC or DD states, which are, e.g., about 60% for a run length of 4, given an initial state of either CC or DD. Secondly, the model can be used to simulate hypothetical experimental results which can be compared with the actual ones. This allows for a whole range of advanced analysis.

The following figure shows the probability of joint cooperation (CC) found in 20 simulated experiments, using the matching-specific transition matrices in connection with the initial distribution for each matching, together with the actual experimental result (thick black line). The simulations seem to capture the experimental observations reasonably well, although we have a rather short length of the matching, and the first two matchings had equal length of 6 rounds. It is quite obvious that players assumed that the third matching would also have 6 rounds, which led them to defect more often because of expected end effects, which could not be captured with transition matrices estimated for whole matches. For the first two matchings, the experimental observations are well covered by the range of the simulated observations. The restart-effect is also prominent.

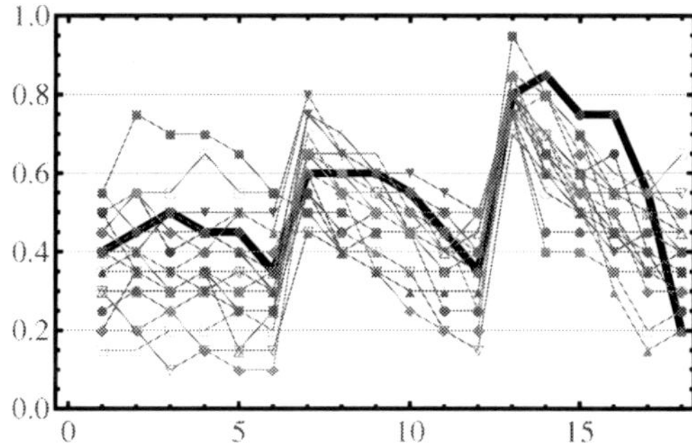

The simulated experiments show rather wide fluctuations. This indicates that we should expect similar fluctuations in the results of actual experiments if we repeat them. This should make us careful in interpreting certain patterns in experimental data, as they could prove as the mere result of stochastic fluctuations.

The next figure shows these probabilities as inferred directly from the estimated transitions matrices, again together with the actual experimental result (thick black line). The apparent deviation is likely to have two reasons. The fitted model is not flexible enough to reflect learning effects properly. We have to live with that because of data limitations. More important is the insight, that it could well be possible that the typical behavior that we might find based on large data sets might well be represented by Markov matrices, but that just the realization of the actually conducted experiment deviates from this average, although its stochastic behavior is well represented.

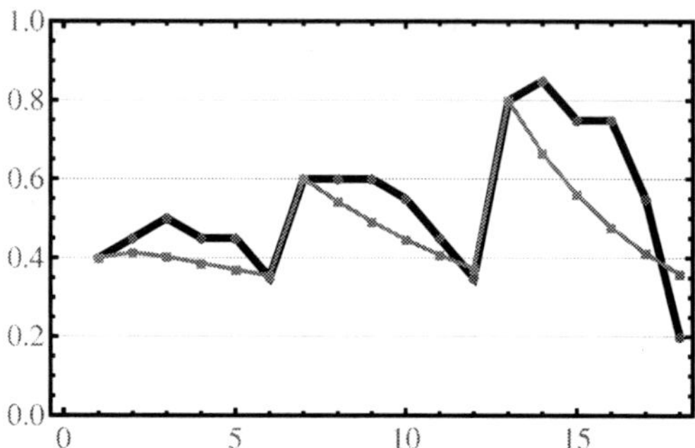

To clarify these issues we need to develop test criteria for the accuracy of Markov models, and it is likely advantageous to use time dependent models if enough data is available. Tests need then to be performed out-of-sample.

4 Conclusion

We conducted a series of IPD experiments with unknown length and re-matching of players and observed the patterns of cooperation over time. This pattern shows apparent stochastic behavior, which calls for an attempt of stochastic modeling. We present descriptive as well as simulation results for a very simple Markov model derived under the assumption of stochastically equivalent behavior of all players. They indicate that various patterns in the data may well be the results of an in fact simple memoryless stochastic behavior. With more data, likely learning behavior could also be investigated in more detail.

References

1. Dal Bo P.: Cooperation under the shadow of the future. Experimental evidence from infinitely repeated games, Am. Econ. Rev. **95**, 1591-1604 (2005)
2. Hennig-Schmid, H., Leopold-Wildburger U.: How the shadow of the past affects the future - An iterated prisoners dilemma experiment, working paper, University of Graz, (2011)

Determining the Optimal Strategies for Antagonistic Positional Games in Markov Decision Processes

Dmitrii Lozovanu and Stefan Pickl

Abstract A class of stochastic antagonistic positional games for Markov decision processes with average and expected total discounted costs' optimization criteria are formulated and studied. Saddle point conditions in the considered class of games that extend saddle point conditions for deterministic parity games are derived. Furthermore, algorithms for determining the optimal stationary strategies of the players are proposed and grounded.

1 Introduction and Problem Formulation

In this paper we study a class of stochastic antagonistic positional games for Markov decision processes with average and expected total discounted costs' optimization criteria [5]. The basic game models we formulate are using the framework of Markov decision processes (X, A, p, c) with a finite set of states X, a finite set of actions A, a transition probability function $p : X \times X \times A \rightarrow [0, 1]$ that satisfies the condition

$$\sum_{y \in X} p^a_{x,y} = 1 \qquad \forall x \in X, \ \forall a \in A$$

and a transition cost function $c : X \times X \rightarrow R$ which represents the costs $c_{x,y}$ of states' transitions for the dynamical system if it makes a transition from the state $x \in X$ to another state $y \in X$. We assume that the Markov decision processes are controlled by two players as follows: The set of states X is divided into two disjoint subsets X_1 and X_2, $X = X_1 \cup X_2$ $(X_1 \cap X_2 = \emptyset)$, where X_1 represents the position set of the first player and X_2 represents the position set of the second player.

Dmitrii Lozovanu
Institute of Mathematics and Computer Science, Academy of Sciences of Moldova, Academy str. 5, Chisinau, MD–2028, Moldova, e-mail: lozovanu@math.md

Stefan Pickl
Institute for Theoretical Computer Science, Mathematics and Operations Research, Universität der Bundeswehr München, 85577 Neubiberg-München, Germany, e-mail: stefan.pickl@unibw.de

D. Klatte et al. (eds.), *Operations Research Proceedings 2011*, Operations Research Proceedings, DOI 10.1007/978-3-642-29210-1_37, © Springer-Verlag Berlin Heidelberg 2012

Each player fixes actions in his positions, i.e., if the dynamic system at the given moment of time is in the state x (which belongs to the position set of the first player) then the action $a \in A$ is fixed by the first player; otherwise the action is fixed by the second player. Each player in the decision process fixes actions in the states from his position set using stationary strategies. We define the stationary strategies of the players as two maps:

$$s_1 : x \to a \in A^1(x) \quad \text{for} \quad x \in X_1,$$
$$s_2 : x \to a \in A^2(x) \quad \text{for} \quad x \in X_2,$$

where $A^1(x)$ is the set of actions of the first player in the state $x \in X_1$ and $A^2(x)$ is the set of actions of the second player in the state $x \in X_2$. Without loss of generality we may consider $|A^i(x)| = |A^i| = |A|$, $\forall x \in X_i$, $i = 1, 2$. In order to simplify the notation we denote the set of possible actions in a state $x \in X$ for an arbitrary player by $A(x)$.

If the players fix their stationary strategies s_1 and s_2, respectively, then we obtain a situation $s = (s_1, s_2)$. This situation corresponds to a simple Markov process determined by the probability distributions $p_{x,y}^{s_i(x)}$ in the states $x \in X_i$ for $i = 1, 2$. We denote by $P^s = (p_{x,y}^s)$ the matrix of probability transitions of this Markov process. If the starting state x_0 is given, then for the Markov process with the matrix of probability transitions P^s and the matrix of transition costs $C = (c_{x,y})$ we can determine the average cost per transition $M_{x_0}(s_1, s_2)$ that corresponds to the situation $s = (s_1, s_2)$. Therefore, on the set of situations $S^1 \times S^2$ we can define the payoff function

$$F_{x_0}(s_1, s_2) = M_{x_0}(s_1, s_2).$$

In such a way we obtain an antagonistic positional game which is determined by the corresponding finite sets of strategies S_1, S_2 of the players and the payoff function $F_{x_0}(s_1, s_2)$ defined on $S = S_1 \times S_2$. This game is determined uniquely by the set of states X, the position sets X_1, X_2, the set of actions A, the cost function $c : X \times X \to R$, the probability function $p : X \times X \times A \to [0, 1]$ and the starting position x_0. Therefore, we denote it by $(X, A, X_1, X_2, c, p, x_0)$ and call this game *stochastic antagonistic positional game with average payoff function*. We show that for the players in the considered game there exist the optimal stationary strategies, i.e., we show that there exist the strategies $s_1^* \in S_1, s_2^* \in S_2$ that satisfy the condition

$$F_x(s_1^*, s_2^*) = \max_{s_1 \in S_1} \min_{s_2 \in S_2} F_x(s_1, s_2) = \min_{s_2 \in S_2} \max_{s_1 \in S_1} F_x(s_1, s_2) \qquad \forall x \in X. \qquad (1)$$

In the case $p_{x,y}^a = 0 \vee 1$ $\forall x, y \in X$, $\forall a \in A$ the stochastic positional game is transformed into the parity game studied in [1, 3]. In [2, 4] the stochastic positional game of m players have been formulated where a Nash equilibria condition is proven. However, for a stochastic positional game with m players in the general case Nash equilibria may not exist. The main result we present in this paper shows that a saddle point for stochastic positional games with average payoff functions always exists.

In our paper we consider also the *stochastic antagonistic positional game with discounted payoff function*. We define this game in a similar way as the game above.

We consider Markov decision processes (X,A,c,p) that may be controlled by two players with the corresponding position sets X_1 and X_2, where the players fix actions in their position sets using stationary strategies. Here, we assume that for the Markov decision process the discount factor λ, $0 < \lambda < 1$ is given, where the cost of system's transition from a state $x \in X$ to another state y at the moment of time t is discounted with the rate λ^t, i.e., the cost of system's transition from the state x at the moment of time t to the state y at the moment of time $t+1$ is equal to $\lambda^t c_{x,y}$. For fixed stationary strategies s_1, s_2 of the players we obtain a situation $s = (s_1, s_2)$ that determines a simple Markov process with the transition probability matrix $P^s = (p^s_{x,y})$ and the matrix of transition costs $C = (c_{x,y})$. Therefore, if the starting state x_0 is known then we can determine the expected total discounted cost $\sigma^\lambda_{x_0}(s^1, s^2)$ that corresponds to the situation $s = (s_1, s_2)$. So, on the set of situations $S_1 \times S_2$ we can define the payoff function

$$\overline{F}_{x_0}(s_1, s_2) = \sigma^\lambda_{x_0}(s_1, s_2).$$

We denote the stochastic antagonistic positional game with discounted payoff function by $(X, A, X_1, X_2, c, p, \lambda, x_0)$. We show that for the players in this game there exist the optimal stationary strategies, i.e., there exist the strategies $s_1^* \in S_1, s_2^* \in S_2$ that satisfy the condition

$$\overline{F}_{x_0}(s_1^*, s_2^*) = \max_{s_1 \in S_1} \min_{s_2 \in S_2} \overline{F}_{x_0}(s_1, s_2) = \min_{s_2 \in S_2} \max_{s_1 \in S_1} \overline{F}_{x_0}(s_1, s_2).$$

2 Determining the Saddle Points for Antagonistic Games with Average Payoff Function

In this section we show for an arbitrary stochastic antagonistic positional game $(X, A, X_1, X_2, c, p, x)$ with average payoff function $F_x(s^1, s^2)$ that there exists a saddle point. We prove this result by using the results from [2, 4, 5]. For an arbitrary $x \in X$ and a fixed $a \in A(x)$ we denote $\mu_{x,a} = \sum_{y \in X(x)} p^a_{x,y} c_{x,y}$.

Theorem 1. *Let* $(X, A, X_1, X_2, c, p, \overline{x})$ *be an arbitrary stochastic positional game with average payoff function* $F_{\overline{x}}(s_1, s_2)$. *Then the system of equations*

$$\begin{cases} \varepsilon_x + \omega_x = \max_{a \in A(x)} \left\{ \mu_{x,a} + \sum_{y \in X} p^a_{x,y} \varepsilon_y \right\} & \forall x \in X_1, \\ \varepsilon_x + \omega_x = \min_{a \in A(x)} \left\{ \mu_{x,a} + \sum_{y \in X} p^a_{x,y} \varepsilon_y \right\} & \forall x \in X_2, \end{cases} \quad (2)$$

has a solution under the set of solutions of the system of equations

$$\begin{cases} \omega_x = \max_{a \in A(x)} \left\{ \sum_{y \in X} p^a_{x,y} \omega_x \right\} & \forall x \in X_1, \\ \omega_x = \min_{a \in A(x)} \left\{ \sum_{y \in X} p^a_{x,y} \omega_x \right\} & \forall x \in X_2, \end{cases} \quad (3)$$

i.e., the system of equations (3) *has such a solution* ω_x^*, $x \in X$ *for which there exists a solution* ε_x^*, $x \in X$ *of the system of equations*

$$\begin{cases} \varepsilon_x + \omega_x^* = \max_{a \in A(x)} \left\{ \mu_{x,a} + \sum_{y \in X} p_{x,y}^a \varepsilon_y \right\} & \forall x \in X_1, \\[2em] \varepsilon_x + \omega_x^* = \min_{a \in A(x)} \left\{ \mu_{x,a} + \sum_{y \in X} p_{x,y}^a \varepsilon_y \right\} & \forall x \in X_2. \end{cases} \tag{4}$$

The optimal stationary strategies of the players

$$s_1^* : x \;\to\; a^1 \in A(x) \quad for \quad x \in X_1,$$
$$s_2^* : x \;\to\; a^2 \in A(x) \quad for \quad x \in X_2$$

in the stochastic positional game can be found by fixing arbitrary maps $s_1^*(x) \in A(x)$ *for* $x \in X_1$ *and* $s_2^*(x) \in A(x)$ *for* $x \in X_2$ *such that*

$$s_1^*(x) \in \left(\arg\max_{a \in A(x)} \left\{ \sum_{y \in X} p_{x,y}^a \omega_x^* \right\} \right) \cap \left(\arg\max_{a \in A(x)} \left\{ \mu_{x,a} + \sum_{y \in X} p_{x,y}^a \varepsilon_y^* \right\} \right) \tag{5}$$
$$\forall x \in X_1$$

and

$$s_2^*(x) \in \left(\arg\min_{a \in A(x)} \left\{ \sum_{y \in X} p_{x,y}^a \omega_x^* \right\} \right) \cap \left(\arg\min_{a \in A(x)} \left\{ \mu_{x,a} + \sum_{y \in X} p_{x,y}^a \varepsilon_y^* \right\} \right) \tag{6}$$
$$\forall x \in X_2.$$

For the strategies s^{1*}, s^{2*} *the corresponding values of the payoff function* $F_{\bar{x}}(s^{1*}, s^{2*})$ *coincides with the values* $\omega_{\bar{x}}^*$ *for* $\bar{x} \in X$ *and* (1) *holds.*

Proof. Let $x \in X$ be an arbitrary state and consider the stationary strategies $\bar{s}_1 \in S_1$, $\bar{s}_2 \in S_2$ for which

$$F_x(\bar{s}_1, \bar{s}_2) = \min_{s_2 \in S_2} \max_{s_1 \in S_1} F_x(s_1, s_2).$$

We show that

$$F_x(\bar{s}_1, \bar{s}_2) = \max_{s_1 \in S_1} \min_{s_2 \in S_2} F_x(s_1, s_2),$$

i.e., we show that (1) holds and $\bar{s}_1 = s_1^*$, $\bar{s}_2 = s_2^*$.

According to the properties of the bias equations from [5] for the situation $\bar{s} = (\bar{s}_1, \bar{s}_2)$ the system of linear equations

$$\begin{cases} \varepsilon_x + \omega_x = \mu_{x,a} + \sum_{y \in X} p_{x,y}^a \varepsilon_y & \forall x \in X_1,\ a = \bar{s}^1(x), \\[1.5em] \varepsilon_x + \omega_x = \mu_{x,a} + \sum_{y \in X} p_{x,y}^a \varepsilon_y & \forall x \in X_2,\ a = \bar{s}^2(x), \\[1.5em] \omega_x = \sum_{y \in X} p_{x,y}^a \omega_x & \forall x \in X_1,\ a = \bar{s}^1(x), \\[1.5em] \omega_x = \sum_{y \in X} p_{x,y}^a \omega_x & \forall x \in X_2,\ a = \bar{s}^2(x) \end{cases} \tag{7}$$

has the solution ε_x^*, ω_x^* $(x \in X)$ which for a fixed strategy $\bar{s}_2 \in S^2$ satisfies the condition

$$
\begin{cases}
\varepsilon_x^* + \omega_x^* \geq \mu_{x,a} + \sum_{y \in X} p_{x,y}^a \varepsilon_y^* & \forall x \in X_1,\ a \in A(x), \\[2mm]
\varepsilon_x^* + \omega_x^* = \mu_{x,a} + \sum_{y \in X} p_{x,y}^a \varepsilon_y^* & \forall x \in X_2,\ a = \bar{s}_2(x), \\[2mm]
\omega_x^* \geq \sum_{y \in X} p_{x,y}^a \omega_x^* & \forall x \in X_1,\ a \in A(x), \\[2mm]
\omega_x^* = \sum_{y \in X} p_{x,y}^a \omega_x^* & \forall x \in X_2,\ a = \bar{s}_2(x)
\end{cases}
$$

and $F_x(\bar{s}_1, \bar{s}_2) = \omega_x^* \ \forall\, x \in X$.

Taking into account that $F_x(\bar{s}_1, \bar{s}_2) = \min_{s^2 \in S^2} F_x(\bar{s}_1, s_2)$ then for a fixed strategy $\bar{s}_1 \in S_1$ the solution ε_x^*, ω_x^* $(x \in X)$ satisfies the condition

$$
\begin{cases}
\varepsilon_x^* + \omega_x^* = \mu_{x,a} + \sum_{y \in X} p_{x,y}^a \varepsilon_y^* & \forall x \in X_1,\ a = \bar{s}^1(x), \\[2mm]
\varepsilon_x^* + \omega_x^* \leq \mu_{x,a} + \sum_{y \in X} p_{x,y}^a \varepsilon_y^* & \forall x \in X_2,\ a \in A(x), \\[2mm]
\omega_x^* = \sum_{y \in X} p_{x,y}^a \omega_x^* & \forall x \in X_1,\ a = \bar{s}^1(x), \\[2mm]
\omega_x^* \leq \sum_{y \in X} p_{x,y}^a \omega_x^* & \forall x \in X_2,\ a \in A(x).
\end{cases}
$$

So, the following system

$$
\begin{cases}
\varepsilon_x + \omega_x \geq \mu_{x,a} + \sum_{y \in X} p_{x,y}^a \varepsilon_y & \forall x \in X_1,\ a \in A(x), \\[2mm]
\varepsilon_x + \omega_x \leq \mu_{x,a} + \sum_{y \in X} p_{x,y}^a \varepsilon_y & \forall x \in X_2,\ a \in A(x), \\[2mm]
\omega_x \geq \sum_{y \in X} p_{x,y}^a \omega_x & \forall x \in X_1,\ a \in A(x)(x), \\[2mm]
\omega_x \leq \sum_{y \in X} p_{x,y}^a \omega_x & \forall x \in X_2,\ a \in A(x)
\end{cases}
$$

has a solution, which satisfies the condition (7). This means that $s_1^* = \bar{s}_1$, $s_1^* = \bar{s}_1$ and

$$
\max_{s_1 \in S_1} \min_{s_2 \in S^2} F_x(s_1, s_2) = \min_{s_2 \in S^2} \max_{s_1 \in S_1} F_x(s_1, s_2) \qquad \forall x \in X,
$$

i.e., the theorem holds. $\qquad\qquad\qquad\qquad\qquad\qquad\qquad\qquad\qquad\qquad\quad \square$

Thus, the optimal strategies s^{1*}, s^{2*} for an average antagonistic positional game can be found using the solution of the systems (2)–(4) and conditions (5), (6).

3 Determining the Saddle Points for Antagonistic Games with Discounted Payoff Function

The saddle point conditions for stochastic antagonistic positional games $(X, A, X_1, X_2, c, p, \lambda, x_0)$ can be derived from [2, 4] in the case $m = 2$, $c = c^1 = -c^2$, i.e., we obtain the following

Theorem 2. *Let* $(X, A, X_1, X_2, c\ p, \gamma, \bar{x})$ *be an arbitrary stochastic antagonistic positional game with discounted payoff function* $\widehat{F}_{\bar{x}}(s^1, s^2)$. *Then there exist values* σ_x *for* $x \in X$ *that satisfy the conditions:*

$$1)\ \max_{a \in A(x)} \left\{ \mu_{x,a} + \gamma \sum_{y \in X} p^a_{x,y} \sigma_y - \sigma_x \right\} = 0 \qquad \forall x \in X_1,$$

$$2)\ \min_{a \in A(x)} \left\{ \mu_{x,a} + \gamma \sum_{y \in X} p^a_{x,y} \sigma_y - \sigma_x \right\} = 0 \qquad \forall x \in X_2.$$

The optimal stationary strategies s^{1*}, s^{2*} *of the players in the game can be found by fixing the maps*

$$s^{1*}(x) = a^* \in \arg \max_{a \in A(x)} \left\{ \mu_{x,a} + \gamma \sum_{y \in X} p^a_{x,y} \sigma_y - \sigma_x \right\} \qquad \forall x \in X_1,$$

$$s^{2*}(x) = a^* \in \arg \min_{a \in A(x)} \left\{ \mu_{x,a} + \gamma \sum_{y \in X} p^a_{x,y} \sigma_y - \sigma_x \right\} \qquad \forall x \in X_2,$$

where
$$\widehat{F}_{\bar{x}}(s^{1*}, s^{2*}) = \sigma_{\bar{x}} \qquad \forall \bar{x} \in X.$$

Based on this theorem we can determine the optimal strategies of the players.

References

1. Ehrenfeucht, A., Mycielski, J.: Positional strategies for mean payoff games. International Journal of Game Theory **8**, 109–113 (1979)
2. Lozovanu, D.: The game-theoretical approach to Markov decision problems and determining Nash equilibria for stochastic positional games. Int. J. Mathematical Modelling and Numerical Optimization **2(2)**, 162–164 (2011)
3. Lozovanu, D., Pickl, S.: Discrete control and algorithms for solving antagonistic dynamic games on networks. Optimization **58(6)**, 665–683 (2009)
4. Lozovanu, D., Pickl, S., Kropat, E.: Markov decision processes and determining Nash equilibria for stochastic positional games. Proceedings of 18th World Cogress IFAC-2011, 320–327 (2011)
5. Puterman, M.: Markov Decision Processes: Stochastic Dynamic Programming. John Wiley, New Jersey (2005)

Stream VII
Health, Life Sciences, Bioinformatics

Rolf Krause, University of Lugano, SVOR Chair
Teresa Melo, HTW Saarland, GOR Chair
Marion Rauner, University of Vienna, ÖGOR Chair

Analyses of Two Different Regression Models and Bootstrapping

Fulya Gokalp

Abstract Regression methods are used to explain the relationship between a single response variable and one or more explanatory variables. Graphical methods are generally the first step and are used to identify models that can be explored to describe the relationship. Although linear models are frequently used and they are user friendly, many important associations are not linear and require considerably more analytical effort. This study is focused on such nonlinear models. To perform statistical inference in this context, we need to account for the error structure of the data. The experimental design for the nutrition data that we use called for each subject to be studied under two different values of the explanatory variable. However, some participants did not complete the entire study and, as a result, the data are available for only one value of the explanatory variable for them. In the analysis section, the bootstrapping method will be used to re-sample the data points with replacement, which will then be used with a nonlinear parametric model. The confidence intervals for the parameters of the nonlinear model will be calculated with the reflection method for the nutrition data set. In addition, the break point of the spline regression will be determined for the same data set. Although the nutrition data set will be used for this study, the basic ideas can be used in many other fields such as education, engineering and biology.

1 Introduction

The most common approach to explain the nonlinear relationships between a response (dependent) variable and one or more explanatory (independent) variables is the nonlinear regression method. This approach is more appropriate for particular data sets. The growth from birth to maturity in human subjects, for example, typically is nonlinear in nature, characterized by rapid growth after birth, pro-

Fulya Gokalp
Yildiz Technical University, Department of Statistics, Davutpasa Cad., 34210, Esenler, Istanbul, Turkey, e-mail: fulyagokalp@gmail.com

D. Klatte et al. (eds.), *Operations Research Proceedings 2011*, Operations Research Proceedings,   237
DOI 10.1007/978-3-642-29210-1_38, © Springer-Verlag Berlin Heidelberg 2012

nounced growth during puberty, and a leveling off sometime before adulthood (Kutner, 2005)[6].

While setting a nonlinear regression model, the number of parameters does not need to be the same as the number of variables. According to Seber and Wild (1989)[7], since parameter terms often have physical interpretations, a major aim of the investigation is to estimate the parameters as precisely as possible. For a nonlinear regression model, statistical softwares use iterative methods to estimate the regression coefficients by the least squares method which corresponds to maximum likelihood estimation when the errors are independent and normally distributed.

We use bootstrapping in a nonlinear regression model to infer the parameters of the model. "This technique may be useful for analyzing smallish expensive-to-collect data sets where prior information is sparse, distributional assumptions are unclear, and where further data may be difficult to acquire" (Henderson, 2005)[4]. In addition, bootstrapping is a computer based iteration method to resample the data with the Monte Carlo Sampling Method. It requires re-sampling the data with replacement, in which each new sample has the same number of subjects as the population data. This method can provide standard errors, confidence intervals, or an estimate of the bias of an estimator.

Spline Regression is another regression method used to fit the data with regression lines separated by a break point (a.k.a. changing point, knot). Each line has different slopes for different ranges of the independent variable. Eubank (1999)[2] famously used a smoothing spline to estimate the mean function of the response variable. Spline modeling is generally appropriate when any nonlinearity is observed in the data. Greenland (1995)[3] suggests spline regression as it is based on more realistic category-specific models that are especially worthwhile when nonlinearity is expected.

In this study, we used nutrition data for the implementation of the spline and the nonlinear regression with bootstrapping. Detailed information about the data set and data collection procedures can be found in L.A. Jackman et al. (1997)[5]. The study protocols were approved by the Use of Human Subjects Research Committees of Purdue University and Indiana University School of Medicine for this data. Detailed explanation for the data and analysis can be found in following sections.

2 Analyses

This part of the study is divided into two sections: the first section consists of the explanation of the data, and the second section consists of data manipulation and data analysis.

2.1 Data

The L.A. Jackman et al. (1997)[5] study was based on the idea that the achievement of maximal calcium retention during adolescence may influence the magnitude of peak bone mass and subsequently lower risk of osteoporosis. Calcium retention is generally considered to reach a plateau at a certain calcium intake. To test this idea, subjects were given two controlled diets with a different amount of calcium in each. However, some individuals completed the study and provided two observations while others provided only one observation due to drop outs.

In our data, the total number of observations is 202 which includes 44 subjects with a single observation, and 79 subjects with two observations. Besides, there are no significant differences between these subjects in the following areas: medical problems, use of medications, pregnancy, abortion, eating disorder, or current use of tobacco. The distribution of the response variable (calcium retention) has clear center, and is slightly symmetric. On the other hand, the distribution of the independent variable (calcium intake) is not normal. Since some subjects have two observations for calcium intake levels (low/high), the distribution for this variable has two peaks. Moreover, there is a nonlinear relationship between the dependent and independent variable.

Figure 1 shows that there are positive correlations between high/low calcium intake levels and high/low retention levels for most of the subjects. These correlations demonstrate the dependence between these two levels of variables for the same subject. This dependence affected the re-sampling part of the study.

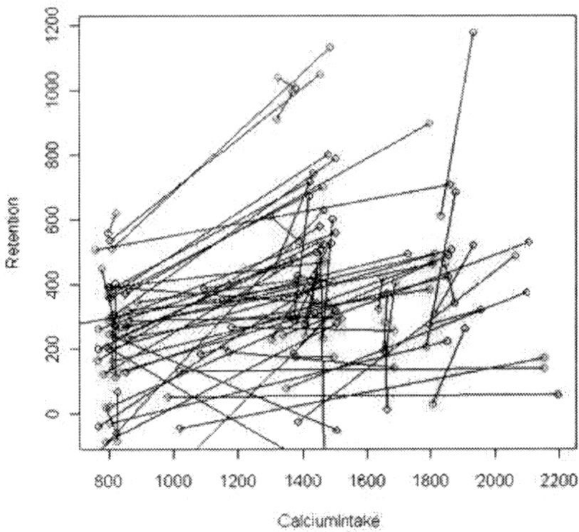

Fig. 1 Two levels of calcium intake versus calcium retention for subjects with double observations

The nonlinear-regression model was found by the study of Jackman et al. (1997)[5] after evaluating many models. The nonlinear regression model which explains the relationship between calcium intake and calcium retention is as follows:

$$\text{Retention} = ae^{b+c(intake)} / (1 + e^{b+c(intake)})$$

In this model, a, b, and c are parameters. The error term of the model is assumed to be normally distributed with constant variance. Calcium retention is the dependent (response) variable and, calcium intake is the independent (explanatory) variable. Parameter a is the limiting value of retention for arbitrarily large intakes and represents the mean maximal calcium retention. The expression $e^{b+c(intake)} / (1 + e^{b+c(intake)})$ represents the proportion of maximal retention for any given value of intake (Jackman et al., 1997)[5]. The correlation between high/low calcium intake and high/low calcium retention makes it necessary to use the iteration methods when analyzing this data. We used the bootstrapping as a re-sampling method in this study. The crucial point of the bootstrapping part is to re-sample paired observations at once for double observed subjects.

2.2 Data Manipulation and Data Analysis

Data manipulation was carried out in both R and SAS prior to data analysis. The implementation of the analysis are similar in both statistical packages, even though the re-sampling of the data points differs.

2.2.1 Nonlinear Regression

Nonlinear regression with bootstrapping, and spline regression (segmented regression) were both carried out in R and SAS following the manipulation process. Bootstrapping samples were used with 'nlm' and 'nls' functions in R to find the least-squares estimates of the parameters. The main difference between analyzing linear models and the nonlinear models in R is that the exact nature of the equation as part of the model has to be specified when non-linear modeling is used (Crawley, 2009)[1]. Therefore, the nonlinear regression model from Jackman et al.'s (1997)[5], was used with initial estimates of parameters as follows:

a = 473, b = -2.77, c = 0.0025

Additionally, the 'nlin' procedure was used to find out the nonlinear least-squares estimate of the parameters with given initial values in SAS. The Gauss-Newton method is used for the iteration part in 'nlin' procedure. After performing this procedure in SAS, the parameter estimations were obtained by bootstrap sample size times. After applying nonlinear regression method to bootstrapping samples, SAS gives the output for each of the samples. We used these outputs to determine a bootstrap confidence interval for each parameter. One of the procedures for setting up a $1 - \alpha$ confidence interval is the reflection method (Kutner, 2005)[6]. To apply this method the $(\alpha/2)100$ and $(1 - \alpha/2)100$ percentiles of the bootstrapping distribution of the parameters are calculated. According to Kutner (2005)[6], the approximate $1 - \alpha$ confidence interval for parameter (β) is as follows:

$b - d_2 \leq \beta \leq b + d_1$

$d_1 = b - ((\alpha/2)100 percentile) \; d_2 = ((1 - \alpha/2)100 percentile) - b$

The reflection method was illustrated to find out the confidence intervals of the parameters. The confidence intervals for each parameter are below (alpha=0.05):

The confidence intervals of the first parameter:

$d_1 = 473 - 370.8 = 102.2 \; d_2 = 639.52 - 473 = 166.52$

$306.47 \leq \beta_1 \leq 575.2$

The confidence intervals of the second parameter:

$d_1 = -2.77 - (-6.40) = 3.63 \; d_2 = -1.57 - (-2.77) = 1.19$

$-3.97 \leq \beta_2 \leq 0.86$

The confidence intervals of the third parameter:

$d_1 = 0.0025 - 0.001493 = 0.001 \; d_2 = 0.0077 - 0.0025 = 0.0052$

$-0.0027 \leq \beta_3 \leq 0.0035$

Since the confidence intervals of the first parameter do not include zero, this parameter is statistically significant for the nonparametric model.

2.2.2 Spline Regression

The second model we covered in data analysis is the spline regression. The plot of the dependent and the independent variables indicates a point where change occurs in the data. This break point is one of the parameters of the spline regression. Up to this point, calcium retention tends to increase along with the increase in calcium intake. After this point of calcium intake, calcium retention remains stable.

The spline regression is as follows:

Retention $= \beta_0 + \beta_1(CalciumIntake)$ for calcium intake $<$ break point

Retention $= \beta_0 + \beta_1(BreakPoint)$ for calcium intake $\geq$ break point

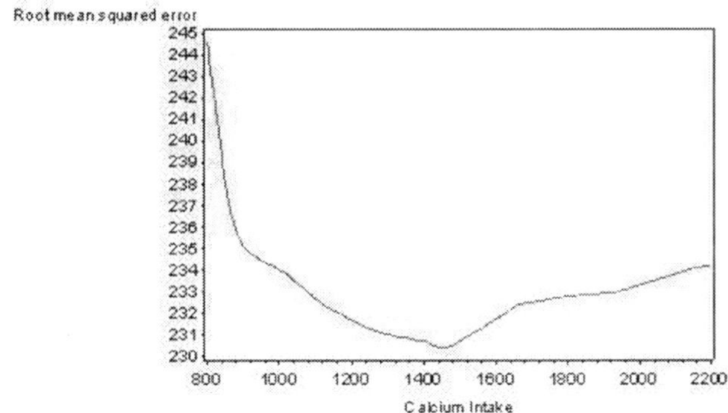

Fig. 2 The root mean squared error (RMSE) versus calcium intake

The test values of the calcium intake were taken from 800 to 2200 by 1 to compute the spline regression models with these break points. Then, the root mean squared errors were calculated and plotted for each of these break points. See the break point's relationship with calcium intake in Figure 2. The graph shows that the root mean squared error reaches to the minimum point for a particular value of calcium intake. This point was determined as '1452' and used as a break point for the spline regression. The spline regression was implemented once more with the break point. Results indicated that the value of the dependent variable remained stable after the break point value of the independent variable (calcium intake).

3 Conclusions

This study focused on the data sampling method for a special case which has two observations for some subjects in a particular data set. Preliminary analysis strongly suggested that the residuals from the same individual were correlated. The bootstrap method provided a convenient way to account for this characteristic of the data. This study also focused on applying the bootstrap in a non-linear setting to obtain confidence intervals for the estimated parameters. In addition to non-linear regression, the data were used to select the break point of the spline regression. This break point was used for the spline regression, and the predictions of this model were used to show them on the same graph with the data points. Results indicated that calcium intake above the breakpoint was not associated with any additional calcium. In other words, calcium retention reached a plateau at 1452mg calcium intake.

References

1. Crawley, M.J.: The R Book. John Wiley and Sons, England (2009)
2. Eubank,R.L.: Nonparametric Regression and Spline Smoothing. Marcel Dekker, New York (1999)
3. Greenland, S.: Dose-response and trens analysis in epidemiology: alternatives to categorical analysis. Epidemiology. **6(4)**, 356–365 (1995)
4. Henderson, A.R.: The bootstrap: A technique for data-driven statistics. Using computer-intensive analyses to explore experimental data. Clinica Chimica Acta. **359**, 1–26 (2005)
5. Jackman, L.A., Millane, S.S., Martin, B.R., Wood, O.B., McCabe, G.P., Peacock, M., and Weaver, C.: Calcium Retention in relation to calcium intake and postmenarcheal in adolescent females. American Society for Clinical Nutrition. **66**, 327–333 (1997)
6. Kutner, M.H.: Remedial Measures for Evaluating Precision in Nonstandard Situations – Bootstrapping. In: Kutner, M.H., Applied Linear Statistical Models, pp. 458-460. McGraw-Hill Irwin, New York, (2004)
7. Seber, G.A.F. and Wild, C.J.: Nonlinear Regression. Wiley Series in Probability and Statistics (Paperback), New Jersey (1989)

Modifications of BIC for data mining under sparsity

Florian Frommlet

Abstract In many research areas today the number of features for which data is collected is much larger than the sample size based on which inference is made. This is especially true for applications in bioinformatics, but the theory presented here is of general interest in any data mining context, where the number of "interesting" features is expected to be small. In particular mBIC, mBIC1 and mBIC2 are discussed, three modifications of the Bayesian information criterion BIC which in case of an orthogonal designs control the family wise error (mBIC) and the false discovery rate (mBIC1, mBIC2), respectively. In a brief simulation study the performance of these criteria is illustrated for orthogonal and non-orthogonal regression matrices.

1 Introduction

In many research areas today one is confronted with high dimensional data sets. Genetic association studies might serve as an example. Let X_j denote the state of one out of p genetic markers, and let Y be the value of some trait, either quantitative (for example height) or dichotomous (like in case control studies). The aim is to find those genetic markers which are associated with the trait. This is a difficult task because usually the number of markers vastly exceeds the sample size.

In [8] a model selection approach based on modifications of the Bayesian information criterion (mBIC) was introduced to analyze genome-wide association studies (GWAS), and in [7] certain asymptotic optimality results under sparsity (ABOS) of these selection criteria were derived in case of orthogonality. The simulation study of this article illustrates the behavior of these selection criteria for orthogonal and non-orthogonal regression matrices. Results are only presented for quantitative traits. Simulation studies for logistic regression can be found in [10], where the theory of mBIC is developed for generalized linear models.

Florian Frommlet
Medical University Vienna, Spitalgasse 23, 1090 Vienna,
e-mail: Florian.Frommlet@meduniwien.ac.at

D. Klatte et al. (eds.), *Operations Research Proceedings 2011*, Operations Research Proceedings, 243
DOI 10.1007/978-3-642-29210-1_39, © Springer-Verlag Berlin Heidelberg 2012

2 Modifications of BIC

Consider the problem of variable selection in a regression setting, with p regressors $X_1, \ldots, X_p$, and n measurements of some regressand $Y = (y_1, \ldots, y_n)^T$. Let each potential model be denoted by an index set $M \subset \{1, \ldots, p\}$, and X_M be the corresponding design matrix. We only consider models including an intercept term, thus the first column of the $n \times (k_M + 1)$ matrix X_M always consists of ones, where k_M is the size of M. In case of real valued y_i the corresponding linear regression model in vector form is given by

$$Y = X_M \beta + \varepsilon \,, \tag{1}$$

where $\beta = (\beta_0, \beta_1, \ldots, \beta_{k_M})^T$ and the error terms $\varepsilon_i \sim \mathcal{N}(0,1)$ are usually assumed to be iid normal. For dichotomous traits one might consider logistic regression models.

Classical approaches to model selection are often based on penalized likelihoods, where a penalty term which increases with model complexity is added to the negative log-likelihood of the data. The two most prominent examples are Akaike's information criterion AIC and Schwarz's Bayesian information criterion BIC. It is well known that in a classical setting (constant p and $n \to \infty$) AIC has advantages in terms of prediction, whereas BIC is consistent if the correct model is among the set of all considered models. In view of our applications mentioned above selection based on BIC should therefore be more appealing.

Given the maximum likelihood L_M of a certain model, the BIC criterion can be formulated as

$$\mathrm{BIC}(M) = -2 \log L_M + k_M \log n \,, \tag{2}$$

where model selection is performed by finding the model which minimizes $\mathrm{BIC}(M)$. Contrary to the classical results it turns out that BIC is no longer consistent when $p \geq n$ and when the true model is rather small (see [6]). In this situation, which is commonly referred to as sparsity, BIC has a tendency to select too complex models, which lead Bogdan et al. [5] to introduce the following modified version of BIC:

$$\mathrm{mBIC}(M) = -2 \log L_M + k_M (\log n + 2 \log p - 2 \log 4) \,. \tag{3}$$

The additional penalty term $2 \log p$ of mBIC corrects for the large number of potential models when $p \geq n$. In case of orthogonal design matrices it has been shown that mBIC is closely related to the Bonferroni multiple testing rule. In that sense mBIC can be thought of as controlling the family wise error rate (FWER). For further details and the motivation of the constant $-2 \log 4$ see [6].

Recent theoretical results [1,3] suggest that under sparsity multiple testing procedures controlling the false discovery rate (FDR) are preferable to FWER controlling procedures. This was the motivation in [7] to introduce the following modifications of BIC:

$$\text{mBIC1}(M) = -2\log L_M + k_M(\log n + 2\log p) - 2\log(k_M!) - \sum_{i=1}^{k_M} \log\log(np^2/i^2),$$

$$(4)$$

and

$$\text{mBIC2}(M) = -2\log L_M + k_M(\log n + 2\log p - 2\log 4) - 2\log(k_M!).\qquad(5)$$

Both penalty terms were derived taking inspiration from [1]. In an orthogonal setting they are closely related to the popular Benjamini-Hochberg multiple testing procedure [2]. In [7] conditions for asymptotic Bayes optimality under sparsity (ABOS) are given for the selection criteria (3), (4) and (5). It turns out that the FDR controlling criteria mBIC1 and mBIC2 are ABOS under much less restrictions on the sparsity levels than mBIC.

3 Simulation Study

The asymptotic optimality results for modifications of BIC were proven in [7] only for orthogonal designs. The first part of the simulation study will illustrate these findings. The orthogonality assumption simplifies both the analytical approach as well as simulation considerably, but it has certain obvious limitations. In most applications orthogonality is out of question, in particular it does not allow for $p > n$.

Concerning non-orthogonal cases the performance of mBIC has been studied thoroughly in the context of QTL mapping, where one is dealing with a fairly particular correlation structure of markers (see for example [4]). Recently in [8] a comprehensive simulation study was evaluating the performance of mBIC and mBIC2 for analyzing GWAS data, where correlation is typically observed in form of a block structure. This kind of correlation structure motivates the second simulation scenario, which systematically addresses the impact of correlated regressors on the performance of different selection criteria.

Orthogonal design: The first part of the simulation study with orthogonal design uses Hadamard matrices with elements equal to 1 or -1. Sample size and number of markers ranges within $n = p + 1 = \{128, 256, 512, 1024, 2048, 4096\}$. Two scenarios were considered, first keeping the number of nonzero regressors k constant, and second keeping the sparsity parameter k/p constant. Given k, the values of nonzero coefficients $\beta_1, \ldots, \beta_k$ were simulated from a normal distribution $N(0, \tau^2)$, with $\tau = 0.3$. The response variable was then simulated according to the multiple regression model (1), with 2000 replicates for each scenario.

In the first scenario the number of regressors was kept constant at $k = 12$ for all values of p, thus k/p being essentially proportional to p^{-1}. This type of asymptotic behavior for $p \to \infty$ was referred to as 'extreme sparsity' in [3] and [7]. Theory shows that mBIC is doing well under extreme sparsity, which is confirmed in Figure 1a). FDR of mBIC is close to zero, which reflects the fact that mBIC is controlling FWER. In comparison mBIC1 and mBIC2 have slightly larger FDR, which however is getting smaller with increasing p. For very small sample size mBIC1 is controlling FDR better than mBIC2, whereas for increasing p the criteria become more or less indistinguishable. The slightly larger type 1 error of mBIC1 and mBIC2 compared with mBIC leads to a significant increase in power. While mBIC1 has the smallest misclassification rate, in case of extreme sparsity mBIC performs almost equally well. In comparison the results of BIC are really poor. Its larger power does not compensate for the hugely increased false discovery rate, leading to a misclassification rate which is significantly larger than for the modified criteria.

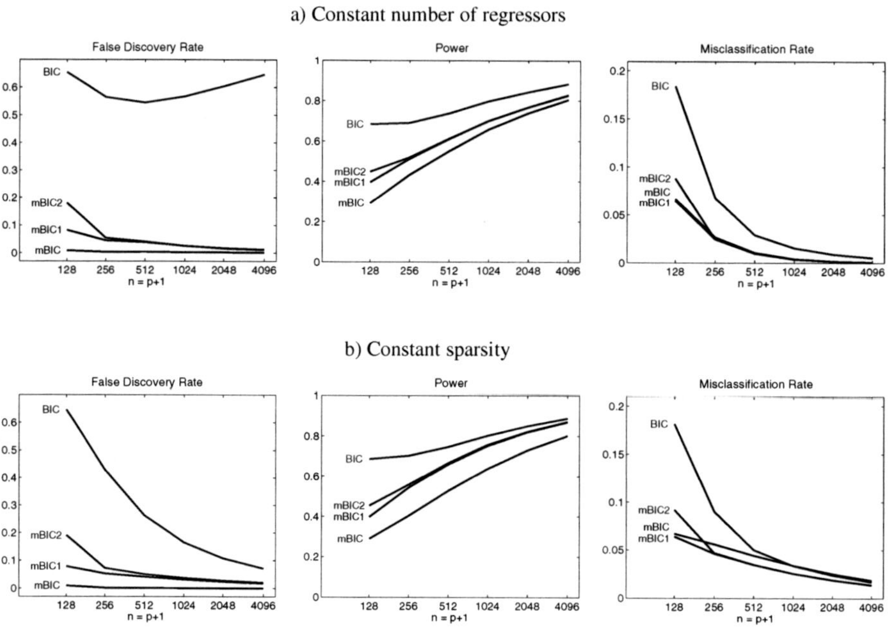

Fig. 1 Simulations results for orthogonal design matrix. Part a) shows estimated FDR, power and misclassification rate for different model selection criteria in case of extreme sparsity. Part b) shows the same statistics when the sparsity parameter k/p is kept fixed.

In the second scenario we have chosen $k \in \{12, 24, 48, 96, 192, 384\}$, which means that the sparsity parameter is kept constant roughly at $k/p = 0.094$. This situation is no longer included in the asymptotic analysis of [3] and [7], where sparsity is assumed to converge to 0 with increasing p. However, it is interesting to see the performance of the different criteria at the dense end of the spectrum of sparsity.

Figure 1b) illustrates that the general behavior of mBIC, mBIC1 and mBIC2 in terms of FDR is quite similar to the case of extreme sparsity, though for large p the differences remain a little bit larger. On the other hand the gain in power for mBIC1 and mBIC2 compared with mBIC is larger than in case of extreme sparsity, leading to misclassification rates of the FDR controlling rules which are now visibly smaller than those of mBIC. It is noticeable that BIC is doing slightly better in the denser case. Its FDR is now converging towards 0 for large p, and its misclassification rate approaches that of mBIC.

Effect of correlations: For the second part of the simulation study a design matrix with $n = p = 256$ was obtained by simulating from a multivariate normal distribution $X \sim \mathcal{N}_p(0, \Sigma)$. Figure 2 shows the block structure of Σ. Correlation within blocks was set constant, varying between $\rho = 0$ and $\rho = 0.8$ for different matrices. Trait values of 2000 replicates were simulated similarly to the previous section, where effect sizes were chosen slightly larger with $\tau = 0.5$. The positions of the $k = 24$ non-zero regressors are shown in Figure 2.

For orthogonal designs the minimization of selection criteria is particularly simple, because the estimate of a regression coefficient for a given explanatory variable does not depend on other regressors. It is enough to find the minimum among nested models according to the order obtained from p-values of one-factor regression models. In the non-orthogonal case one has to consider instead more sophisticated search procedures, leading to more time consuming procedures. Here we applied backward elimination starting from the top 40 regressors of one-factor models, followed by forward selection. This heuristic procedure is not guaranteed to find the model which minimizes a given selection criterion, but more sophisticated search procedures would become prohibitively time consuming for a simulation study. The focus of this article is on comparing different criteria, and to this end the relatively simple search strategy serves well.

Figure 3 illustrates the effects of correlation on the performance of different selection procedures. The larger the correlation, the more often regression models select a strongly correlated regressor instead of the correct one, resulting in a loss of power. However, this effect only becomes recognizable for $\rho \geq 0.3$, and really severe for $\rho \geq 0.5$. In the literature on GWAS thresholds on correlation have been used to distinguish between correct detections and wrong detections [8,9]. The presented results indicate the usefulness of a threshold somewhere between $\rho = \sqrt{0.05}$ of [9] and $\rho = 0.7$ of [8].

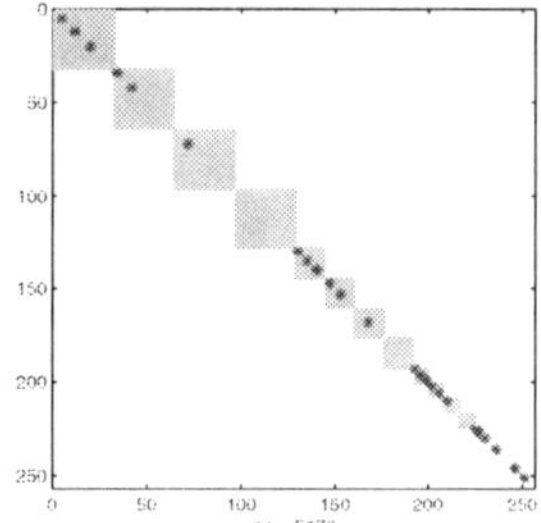

Fig. 2 The correlation structure of regressors consists of several blocks of different sizes. Within each block correlation was ρ, otherwise 0. Stars indicate the position of non-zero regressors.

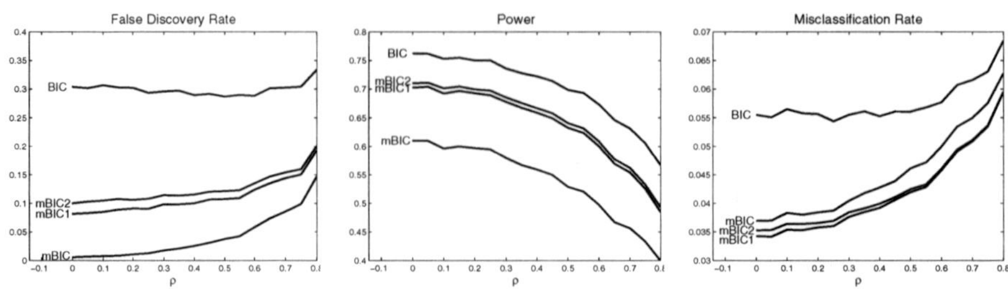

Fig. 3 Dependence of FDR, power and misclassification rate for different model selection criteria on the correlation of regressors.

4 Conclusion

The simulation study of this article verifies the theoretical results of [7] for the orthogonal case, and indicates to which extent these results might be extended to the more general situation of non-orthogonal regressors. Developing the theory of ABOS for correlated regressors remains an interesting topic for further research.

Acknowledgements This research was funded by WWTF project MA09-007a.

References

1. Abramovich, F., Benjamini, Y., Donoho, D.L., Johnstone, I.M.: Adapting to unknown sparsity by controlling the false discovery rate. Ann. Statist. **34**, 584–653 (2006)
2. Benjamini, Y., Hochberg, Y.: Controlling the false discovery rate: a practical and powerful approach to multiple testing. J. Roy. Statist. Soc. Ser. B. **57**, 289–300 (1995)
3. Bogdan, M., Chakrabarti, A., Frommlet, F., Ghosh, J. K.: Asymptotic Bayes-Optimality under sparsity of some multiple testing procedures. Ann. Statist. **39**, 1551-1579 (2011)
4. Bogdan, M., Frommlet, F., Biecek, P., Cheng, R., Ghosh, J.K. and Doerge, R.W.: Extending the Modified Bayesian Information Criterion (mBIC) to dense markers and multiple interval mapping. Biometrics **64**, 1162 – 1169 (2008)
5. Bogdan, M., Ghosh, J.K., Doerge, R.W.: Modifying the Schwarz Bayesian Information Criterion to locate multiple interacting quantitive trait loci. Genetics **167**, 989–999 (2004)
6. Bogdan, M., Żak-Szatkowska, M., Ghosh, J.K.: Selecting explanatory variables with the modified version of Bayesian Information Criterion. Quality and Reliability Engineering International **24**, 627–641, (2008)
7. Frommlet, F., Chakrabarti, A., Murawska, M., Bogdan, M.,: Asymptotic Bayes optimality under sparsity for generally distributed effect sizes under the alternative. Technical report, arXiv:1005.4753 (2011)
8. Frommlet, F., Ruhaltinger, F., Twaróg, P., Bogdan, M.,: Modified versions of Bayesian Information Criterion for genome-wide association studies. CSDA, in print, doi:10.1016/j.csda.2011.05.005 (2011)
9. Hoggart, C.J., Whittaker, J.C., De Iorio, M., Balding, D.J.: Simultaneous Analysis of All SNPs in Genome-Wide and Re-Sequencing Association Studies. PLOS Genetics **4**(7), e1000130. doi:10.1371/journal.pgen.1000130, (2008)
10. Żak-Szatkowska M., Bogdan, M.: Modified versions of Bayesian Information Criterion for sparse Generalized Linear Models. CSDA, in press, doi:10.1016/j.csda.2011.04.016 (2011).

Stream VIII
Location, Logistics, Transportation and Traffic

Knut Haase, University of Hamburg, GOR Chair
Richard Hartl, University of Vienna, ÖGOR Chair
Ulrich Weidmann, ETH Zurich, SVOR Chair

Scheduling of outbound luggage handling at airports

Torben Barth and David Pisinger

Abstract This article considers the outbound luggage handling problem at airports. The problem is to assign handling facilities to outbound flights and decide about the handling start time. This dynamic, near real-time assignment problem is part of the daily airport operations. Quality, efficiency and robustness issues are combined in a multi-criteria objective function. We present important requirements for the real world usage of the model and compare different solution techniques. One solution method is a heuristic approach oriented on the logic of GRASP (Greedy randomized adaptive search procedure). Another solution method is a decomposition approach. The problem is divided into different subproblems and solved in iterative steps. The different solution approaches are tested on real world data from Frankfurt Airport.

1 Introduction

The problem investigated is to schedule the outbound luggage handling at airports. For every outbound flight the handling facility and the handling start time has to be chosen.

We consider the following setup: The baggage infrastructure department of the airport is responsible for the selection of handling facilities and the handling at the facilities. All handling facilities are split up in different regions. Today one dispatcher is responsible for each region. The flights are assigned to the different regions according to the aircraft parking positions. In peak times the dispatchers try to help each other by reassigning flights to another dispatcher when necessary. The presented solution approaches are designed to assign the flights to regions and handling facilities and to find the best start time.

Torben Barth
DTU Management, Produktionstorvet 24, DK-2800 Kgs.Lyngby, e-mail: tocb@man.dtu.dk
Fraport AG, D-60547 Frankfurt am Main, e-mail: t.barth@fraport.de

David Pisinger
DTU Management, Produktionstorvet 24, DK-2800 Kgs.Lyngby, e-mail: pisinger@man.dtu.dk

D. Klatte et al. (eds.), *Operations Research Proceedings 2011*, Operations Research Proceedings,   251
DOI 10.1007/978-3-642-29210-1_40, © Springer-Verlag Berlin Heidelberg 2012

The operational process can be described as follows: Bags arrive from check-in counters and from feeding flights for different outbound flights. If the handling of an outbound flight has not started, the bags will be stored in storage spaces. After the start of the handling process, the bags are transported by a transport system e.g. conveyor belts, to the handling facilities. There, the bags will be loaded into containers, pallets or carts for the final transport to the aircraft. The existing handling resources are scarce and have to be used in an efficient way respecting the locations of the facilities and the parking positions of the aircraft.

In general, the dispatching process starts at the beginning of a shift in the morning. At first, the dispatchers generate a schedule for all flights in their region. The final decision for each flight is made when the handling of a flight should start. The handling start time is calculated backwards from the scheduled time of departure. It depends in general on the aircraft type, the number of expected bags and a number of different sorting criteria. Examples of sorting criteria are bags from first class or connecting passengers.

The handling facilities can be organized centrally or decentrally. If the handling facilities are organized decentralized, one handling facility is connected with every parking positions and the handling facilities are located in close proximity to the parking positions at the terminal building. In the centralized situation there exist handling facilities in large halls where the bags are handled from different flights at the same time. In this case, it is necessary to transport the bags or containers from the handling facilities to the aircraft over longer distances. These transports can be in the responsibility of a different department or company. At decentral facilities parallel handling of flights is usually not possible. In contrast, up to five flights can be handled at the same facility at the same time at central facilities.

Except for the different locations of the handling facilities, additional important properties of the planning process are the handling capacity (number of flights and bags), the handling speed and space for containers or carts (depending on the number of sorting criteria).

In this article we present two different solution approaches to solve the scheduling problem of outbound luggage. The approaches are applicable at the day of operation to generate a schedule for a whole shift and for the final decision about the handling place and handling start time of each single flight.

Related Work Abdelghany et al. investigated the problem of scheduling outbound luggage for a major US-carrier at one of its hubs [1]. Frey and Kolisch presented a MIP-model for the scheduling of outbound luggage at airports [7]. Frey et al. presented a heuristic decomposition approach for this problem [6].

Both approaches have in common that the focus is on central handling facilities and not on geographical aspects. Abdelghany et al. assumed fixed handling start times and considered a simple model of the handling facilities corresponding to the situation at airports in the US. The objective in the work of Frey et al. is the balanced usage of the handling facilities. The schedule is generated for one day in advance. In contrast, we look on a multi-criteria problem with decentral and central handling facilities in a real-time environment.

Barth and Franz [2] presented a solution approach for the related transfer luggage handling problem in a real-time environment.

2 Problem Description

We consider the problem of assigning handling facilities and handling start times for airport layouts with centralized and decentralized handling facilities. The objective function consists of several criteria:

1. Minimize transportation time (distance between handling facility and aircraft parking position)
2. Maximize stability of solution (number of unchanged suggestions between two optimization runs)
3. Maximize robustness of schedule (buffer in handling time)
4. Balanced distribution of flights to handling facilities
5. Balanced distribution of flights to regions
6. Minimize excess usage of sorting space at a facility
7. Minimize excess flights at facilities

The main constraints of the model are the sorting capacity at the handling facilities, and the demand that all bags of every flight are handled in time. Furthermore, it is necessary to introduce side constraints for modeling objective criteria (4) - (7). The different objective criteria are aggregated into one objective by the use of appropriate weights. The problem is to find an assignment from flights to facilities and a starting time of the handling that optimizes the solution respecting the given constraints.

Requirements for a Real-world Implementation In the studied setup, the system should act as a decision support tool. Hence the algorithm proposes a solution and a dispatcher is responsible for the final decision. Therefore, it is critical to reach a high acceptance of the suggestions generated. Our experiences at Frankfurt Airport show that the following points are the key for a successful implementation.

It is necessary to combine different objective terms to satisfy the different needs at the airport (*Multi-criteria objective*). Furthermore, the generated schedules should be robust against changes during the day of operation (*Robustness*) and should be calculated quick to react as fast as possible to the occurring changes (*Currentness*). Another important point is that the solution method is transparent and enables the examination of the generated suggestions (*Transparency*). Finally, the management demands to design flexible (to changes in business logic) / extensible models (*Flexibility*) and to develop systems which are easy to maintain (*Maintenance*).

Case study We tested the presented approaches with historic data sets from Frankfurt Airport. A detailed description of baggage handling at an infrastructure level in Frankfurt is given in [5]. In our tests, we solved the problem on static data for a whole eight hour shift. In a dynamic real time environment the problem could be solved regularly on the most recent data for the next hours. This approach showed promising results for the control of transfer luggage [2].

A typical day at Frankfurt Airport has up to 650 outbound flights. The handling resources consist of about 50 halls with 80 handling facilities. The numbers show that there is a large amount of decentral handling. In the future, over 60,000 bags per day are expected. The expected arrival times of the individual bags have impact on the planning process.

We assume that the number of expected bags is given at the start of the shift. If the exact number is not available the number is estimated with the help of past experience, bags already in the baggage handling system at the airport and the number of passengers.

3 Solution Approaches

In this section we present two different solution approaches on the presented outbound luggage handling problem. GRASP was chosen since it showed promising results on related problems (see [4], [8]). The experiences at Frankfurt Airport showed that a MIP formulation is well suited to represent luggage handling problems [2]. Based on these experiences we expected a high fulfilment of the real world requirements with a MIP approach. Since there seems no feasible formulation for the whole model we decided to test a decomposition approach based on different MIPs.

GRASP Approach This approach is based on a Greedy Randomized Adaptive Search Procedures (*GRASP*) [3]. In each iteration of the approach the flights are assigned in a randomized greedy way to the handling facilities. After the insertion, the solution is optimized with a local search. The algorithm is described below in detail.

- *step 0*: Initialization: The job size is estimated for each flight (the number of different sorting criteria is selected as a first and the number of bags as a second criterion). The minimum handling time for each flight and handling facility is calculated. The regions are ranked for each flight according to distance. For each region an empty flight list is created. Each flight is inserted in each region's list. The lists are ordered by the rank (asc) as first and the job size (desc) as second criterion.
- Repeat steps 1–5 until stopping criterion is met.
- *step 1*: Do until all flights are inserted.
    step 1.1: Select region with lowest utilization.
    step 1.2: Select one of the first n flights from the region's list randomly.
    step 1.3: Insert flight at the best feasible handling facility of the selected region.
    step 1.4: Delete flight from every region's lists.
- *step 2*: Try to increase handling time to introduce robustness in the plan without violating feasibility.
- *step 3*: For each pair of flights check if swapping of the handling facilities improves the solution. Repeat step 2.
- *step 4*: Try for each flight if a better handling facility is available. Repeat step 2.
- *step 5*: Compare current solution with best solution. If current solution has better objective values the current solution is saved.

Decomposition Approach The decomposition approach (*DECOMP*) is based on splitting the problem in different subproblems which can be modelled as MIP and which are easy to solve. The subproblems are oriented at the today existing business structure and reflect the way of the decions of the human decision makers. The advantage of the approach is that – in contrast to the dispatchers decisions – it is possible to calculate the optimal solution of every subproblem and enhance solution quality by iterating over the different subproblems.

Subproblem 1 deals with the distribution of the flights to geographical regions. The objective is to find a balanced distribution over all regions that minimizes the transportation time. The transportation time can be calculated with minimum or average transportation time per region. Subproblem 2 is stated for each region. For each region a balanced distribution of the flights to the handling facilities is calculated. Subproblem 3 is stated for every handling facility. The objective of subproblem 3 is to find the best handling start time for each flight of one handling facility optimizing the robustness of the schedule at the handling facility. Subproblem 4 searches the flight which spoils the objective function most for each region (the flight with most impact over all regions is called flight f^-). Subproblem 5 reassigns flight f^- to the region where it causes the least cost. Subproblems 3–5 are iterated until a stopping criterion is met.

4 Solution Approach Comparison on Real-world Data

We compared the two approaches with respect to solution quality (objective values), solution speed and the criteria presented in Section 2 on static real world data. Results produced with *GRASP* were about 1% worse than results produced with *DECOMP*. We observed that *GRASP* had very good objective values for transportation time and robustness which are the main focus of the algorithm. *DECOMP* dominated *GRASP* since the different criteria were better balanced to an overall solution.

The runtime for one iteration of the *GRASP* approach is under one second. The solution quality improves with a high number of iterations, but the solution quality is already good after a few iterations. The worst observed value after one iteration was only 2 % worse than the best objective value after 1000 iterations. The decomposition approach (*DECOMP*) with five iterations of subproblems 3–5 took about a quarter of an hour run time. At least one iteration is necessary for satisfying results – one iteration of subproblems 1–5 takes about ten minutes.

The assessment of the different requirements for real-world implementation (see Section 2) is summarized in the following. For the issue of *Transparency* both approaches have easily accessible objective values and allow to rerun old optimization runs. Furthermore, it is possible to follow the solution process more closely in *GRASP*, since the problem is not divided in different steps. *GRASP* is much quicker and can therefore react quicker to changes of the data (*Currentness*). *DECOMP* can be more standardized with the help of modelling languages (*Maintenance*). *GRASP* has to be designed specific for each objective criterion. The overall objective is only easy to include in the local search (*step 2 and 4*). In general, *DECOMP* showed better results if the objective combines different criteria (*Multi-criteria objective*).

For the issue of *Flexibility GRASP* was more robust for general changes. *DECOMP* showed advantages when objectives were changed or inserted. The *Robustness* depends on the weights of the different objective criteria and not on the approach.

5 Conclusion and Future Work

We presented two heuristic approaches to support the scheduling of outbound luggage handling at airports. One approach is based on the logic of GRASP and the other on a decomposition approach. In our opinion both approaches can help to improve the process and the management has to weigh their goals for the system to choose one of them. The assessment in Section 4 showed that both approaches have advantages and disadvantages. In general, it is necessary to find a trade-off between the different requirements for the real-world usage. It showed that the approach based on GRASP is much faster, but it is better suited for cases with only one or two criteria in the objective.

In a next step the behavior of the approaches in a dynamic environment should be examined. In the year 2012 an implementation of the approach based on decomposition is planned at Frankfurt Airport. The approach will be adapted to handle the specific requirements in Frankfurt. Only subproblem 2 will be run regularly. All other subproblems will only be run if it is necessary or there is enough time for example before the start of the shifts.

There are two interesting extensions and their impact on solution quality should be examined in the future. One idea is that a change of handling facilities is possible during the handling and the other is that a interruption of the handling is possible after the handling start.

References

1. Abdelghany, A., Abdelghany, K., Narasimhan, R.: Scheduling baggage-handling facilities in congested airports. Journal of Air Transport Management **12,2**, 76–81 (2006)
2. Barth, T., Franz, M.: Gemischt-ganzzahlige Optimierung in Echtzeit am Beispiel der Steuerung des Transfergepäckumschlags am Frankfurter Flughafen (In German). Gemeinsame Fachtagung der GOR und der Deutschen Gesellschaft für System Dynamics. (2008)
3. Feo, T., Venkatraman, K., Bard, J.: A GRASP for a difficult single machine scheduling problem. Computers Ops Res. **18**, 635–643 (1991)
4. Festa, P., Resende, M.: GRASP: basic components and enhancements. Telecommun Syst. **46**, 253–271 (2011)
5. FRAPORT AG: Baggage Management and Infrastructure. Available at http://www.fraportgroundservices.com (2007)
6. Frey, M., Artigues, C., Kolisch, R., Lopez, P.: Scheduling and Planning the Outbound Baggage Process at International Airports. IEEM 2010.
7. Frey, M., Kolisch, R.: Scheduling of outbound baggage at airports. PMS 2010.
8. Norin, A., Andersson, T., Värbrand, P., Yuan, D.: A GRASP Heuristic for Scheduling De-icing trucks at Stockholm Arlanda Airport. INO 2007.

A tree search heuristic for the container retrieval problem

Florian Forster, Andreas Bortfeldt

Abstract The problem of finding a sequence of crane moves in order to retrieve all containers from a block of container stacks in a predefined order with a minimum of crane move time is called the container retrieval problem. In this article, we describe a tree search heuristic based on a natural classification scheme of crane moves that is able to solve the container retrieval problem for typical problem dimensions in very short computation times. A comparison with the approach recently published by Lee and Lee (2010) shows that the heuristic is a very competitive method for solving the container retrieval problem. The mean crane move time for the tested problem instances could be reduced by 15.7 %, the average number of crane moves was reduced by 8.7 %, while the computation speed could be drastically improved from several hours to less than 10 seconds on average, the latter being especially important for the application in container terminals where the retrieval decisions have to be made in short periods of time.

1 Introduction

Container shipping is a key driver of the globalization. The number of containers globally shipped each year is increasing at impressive rates and has reached nearly 140 million TEU in 2010, the equivalent of 140 million standard ISO containers [9]. Meanwhile, the global container ship fleet exceeds 4200 container ships with a total capacity of more than 10 million TEU [8]. An important success factor of the container transport lies in the fact that it allows for a uniform transport chain from the producer to the consumer, with no need for repackaging the goods. Container terminals play a crucial role in this chain, as they must provide for an efficient

Florian Forster
University of Hagen, Profilstraße 8, D-58084 Hagen, Germany, e-mail: florianforster@mac.com

Andreas Bortfeldt
University of Hagen, Profilstraße 8, D-58084 Hagen, Germany, e-mail: andreas.bortfeldt@fernuni-hagen.de

container flow between ships and landside means of transport, e.g., trucks and trains. A container terminal basically consists of three elements:

- The seaside, where containers are unloaded from and loaded to container ships. The unloaded containers are called import containers, while the loaded containers are called export containers.
- The yard, where all containers are temporarily stored for future transportation.
- The landside, where export containers are unloaded from and import containers are loaded to landside means of transport.

Depending on various factors (e.g., container weight and destination port), export containers need to be loaded to a ship in a certain order. Within the yard, containers are usually stored in stacks to improve space utilization. The future loading order of the containers can not be considered when the containers are stored in the yard, since it is generally unknown or at least uncertain at this point of time. As a consequence, situations occur in which a given container that needs to be taken out of the yard is blocked by one or more containers at higher positions in the stack. These blocking containers need to be relocated to other stacks before the given container can be unloaded from the yard. Considering the fact that the retrieval process in the yard is often the bottleneck of the entire container flow at container terminals, time-consuming and unproductive relocations have to be avoided in order to ensure high container flow rates.

2 The container retrieval problem

Within a container yard, containers are stored in stacks, multiple stacks in a row are called a bay, and multiple adjacent bays form a block. At most container terminals, blocks are served by gantry cranes. The frame height of a gantry crane imposes a natural limit for the maximum number of containers per stack (maximum stack height). A gantry crane can pick up the topmost container from any stack in the block and either reposition it to another stack that has not yet reached the maximum stack height or move it to a dedicated spot where a vehicle is waiting to transport the container to the seaside or to the landside. The time necessary to execute a crane move is proportional to the number of bays b and stacks s the crane has to cross between his last position, the starting stack and the target stack, with time factors t_b and t_s respectively. The acceleration and deceleration time t_{ad} of the crane as well as the pickup and place-down time t_{pp} can be considered constants. Hence, the total time for a crane move is calculated as $b \cdot t_b + s \cdot t_s + ad \cdot t_{ad} + t_{pp}$, with $ad = 1$ if $b \geq 1$ and $ad = 0$ otherwise.

Every container in the block belongs to a container group $g \in N$. The container group determines the order in which the containers have to be retrieved from the block. A container c with container group g_c can only be removed if all containers with container groups less than g_c have already been removed from the block.

The container retrieval problem is to find a sequence of crane moves for a given container block that:

- removes every container from the block,
- observes the order implied by the container groups, and
- minimizes the time necessary to execute the moves.

The container retrieval problem was first formulated by Lee and Lee [7]. They propose a heuristic with three phases. In the first phase, an initial solution is constructed using basic greedy decision rules. In the second phase, it is tried to reduce the number of crane moves in the initial solution using a binary integer program. With the resulting solution as input parameter, the third phase attempts to reduce the crane move time using a mixed integer program.

The container relocation problem is an optimization problem similar to the container retrieval problem, differing mainly in two aspects. First, the container relocation problem deals with emptying a single bay, while the container retrieval problem aims at emptying an entire block. Second, the optimization goal of the container relocation problem is to minimize the number of crane moves, while the container retrieval problem is about finding a solution with minimum crane move time. Several heuristic algorithms for solving the container relocation problem were proposed so far [6] [1] [2] [3] [4], also recently from the authors of this paper [5].

3 A tree search procedure for the container retrieval problem

In this section, we present a tree search procedure for the container retrieval problem. Noting the proximity between the container retrieval problem and the container relocation problem, we evolve the tree search procedure we developed for the container relocation problem (single bay) to make it applicable to the container retrieval problem (multiple bays). Please refer to [5] for a detailed description of the algorithm. In the sequel, we first point out the basic elements of the tree search procedure for the container relocation problem, followed by the description on how to use the described tree search procedure to solve the container retrieval problem.

3.1 A tree search procedure for the container relocation problem

As the starting point for our solution approach, we define two predicates for containers. A container c is *badly placed* if there is at least one container with a lower container group in the stack below c. Otherwise, the container is *well placed*. All badly placed containers need to be relocated at least once, as they are blocking a container with a lower container group, hindering this container to be removed from the stack. Hence, the predicates are well suited to calculate a lower bound for the number of relocations necessary to empty a bay.

Given the exponential growth of nodes in the enumeration tree of a full tree search and computational experiments with a complete branch-and-bound algorithm for the container relocation problem, we conclude that for real-world problem dimensions, only a partial tree search seems practical. Therefore, we develop a classification scheme differentiating between productive and unproductive crane moves and restrict our tree search to productive moves only. Moreover, instead of branching after each single crane move, we suggest to build sequences of crane moves (compound moves) and only further investigate a fixed number of promising compound moves in each branching step. This way, the tree search's enumeration tree is limited both in width and height. Finally, we developed a greedy solution algorithm that is executed before the actual tree search is performed, so the greedy solution can act as an initial upper bound.

Benchmarking our tree search with other approaches for solving the container relocation problem, we found that the tree search yields solutions at a competitive solution quality in short computation times.

3.2 Using the tree search procedure for the container relocation problem to solve the container retrieval problem

In this subsection, we show how the tree search procedure developed for the container relocation problem can be used to solve the container retrieval problem. A container yard's block consists of a certain number of bays. For each of the bays, we can apply the tree search algorithm for the container relocation problem in order to get a sequence of crane moves emptying the respective bay. It is worth mentioning that these solutions could be computed in parallel, as they are completely independent of each other.

The remaining task is to merge the crane moves of individual bay solutions into a single solution valid for the entire block. Therefore, we start with an empty solution for the container retrieval problem. Then, we repetitively identify a bay in the block currently holding a container with lowest container group in the entire block. From the corresponding solution of the container relocation problem for the bay, we execute all moves up to the remove of the container with lowest container group, delete them from the bay solution, and append them to the solution for the container retrieval problem. This procedure is repeated until the block is empty. A formal description of this idea is given in algorithm 1.

4 Evaluation

In order to evaluate the competitiveness of our approach, we performed a benchmark with the test cases published by Lee and Lee [7]. There are two different sets of test cases. The test case set R consists of 10 testcases, each containing 5 test instances. The test case set U consists of 10 testcases, each containing 2 test instances. Throughout all test instances, the bays have 16 stacks, while the test cases within a test case set differ in the number of bays, the maximum stacking height and the

---

Algorithm 1 Merging individual bay solutions to a single block solution

---

    baySolutions ← solutions from tree search for the individual bays of the block
    blockSolution ← ∅
    while block is not empty **do**
        lowestGroupBlock ← lowest container group in block
        for all bay *b* in block **do**
            lowestGroupBay ← lowest container group in bay *b*
            if *b* is not empty and *lowestGroupBay* = *lowestGroupBlock* **then**
                baySolution ← *baySolutions*(*b*)
                removed ← *false*
                while not *removed* **do**
                    move ← first move in *baySolution*
                    execute *move*
                    delete first move from *baySolution*
                    blockSolution ← *blockSolution* ∘ *move*
                    if *move* is remove **then**
                        removed ← *true*
                    end if
                end while
            end if
        end for
    end while

---

number of containers in the block. Moreover, for the test case set U, all containers residing in the second slot or higher within a stack are initially badly placed, while for the test case set R, the container groups are randomly assigned. We calculated solutions for the individual test instances using the tree search heuristic proposed in this paper, and calculated crane move times based on the crane time parameters given by Lee and Lee[1] [7]. As the results in table 1 show, the tree search algorithm was able to find solutions with considerable less moves and a considerably reduced crane move time for each of the test cases. In average over all test instances, the mean number of crane moves was reduced by 8.7 % and the mean crane move time was reduced by 15.7 %. Moreover, the calculation time was tremendously lowered. On average, the tree search algorithm needed 8.8 seconds to find a solution for a test instance, compared to 5.4 hours for the algorithm of Lee and Lee - mainly because the tree search algorithm does not rely on computation-intensive integer optimiza-tion programs.

5 Conclusion

In this paper,we described how a tree search approach for one stacking problem in the yard of container terminals (container relocation problem, single bay) can be adapted to solve another related stacking problem (container retrieval problem, multiple bays) and showed that this adapted tree search algorithm is able to find bet-ter solutions in shorter computation times than the currently best-known approach for the problem. To even further improve the results, we consider calculating the

---

[1] t_b : 3.5s, t_s : 1.2s, t_{ad} : 40s, t_{pp} : 30s.

Table 1 Benchmark between the tree search algorithm and the algorithm from [7]. CMT: avg. crane move time (s), M: avg. number of moves, CT: avg. computation time (s).

Test case	Tree search			Lee and Lee [7]			Difference (%)		
	CMT	M	CT	CMT	M	CT	CMT	M	CT
R011606-0070	9534.1	108.0	8.2	11445.5	125.4	8204.3	-16.7	-13.9	-99.9
R011608-0090	12501.6	148.4	10.1	16943.1	191.4	9597.2	-26.2	-22.5	-99.9
R021606-0140	19412.7	215.6	9.1	21949.6	230.2	9607.9	-11.6	-6.3	-99.9
R021608-0190	27366.5	324.6	10.1	34146.0	367.8	12824.2	-19.9	-11.7	-99.9
R041606-0280	41103.3	439.8	9.5	45663.6	454.2	12427.4	-10.0	-3.2	-99.9
R041608-0380	58853.0	657.8	10.8	73978.6	769.2	13353.4	-20.4	-14.5	-99.9
R061606-0430	67417.1	684.2	10.2	74885.0	709.8	15015.6	-10.0	-3.6	-99.9
R061608-0570	92363.0	988.2	11.9	122612.8	1242.0	17105.1	-24.7	-20.4	-99.9
R081606-0570	93556.5	910.2	10.7	104954.7	945.4	17104.8	-10.9	-3.7	-99.9
R101606-0720	122450.9	1134.0	11.9	136716.5	1169.2	20009.5	-10.4	-3.0	-99.9
U011606-0070	10077.4	125.0	1.2	11539.2	127.5	21579.6	-12.7	-2.0	-100.0
U011608-0090	13244.6	167.0	10.0	15975.6	177.5	21574.1	-17.1	-5.9	-100.0
U021606-0140	20902.1	251.5	2.9	23982.2	257.0	21563.3	-12.8	-2.1	-100.0
U021608-0190	28676.5	353.5	4.5	36452.6	400.5	21555.0	-21.3	-11.7	-100.0
U041606-0280	44406.2	501.0	6.1	50041.9	507.5	21541.5	-11.3	-1.3	-100.0
U041608-0380	60499.3	706.0	5.3	76112.4	800.0	21527.7	-20.5	-11.8	-100.0
U061606-0430	70906.4	772.5	5.6	81893.9	793.0	21515.9	-13.4	-2.6	-100.0
U061608-0570	95397.4	1059.5	4.6	122541.5	1234.0	21515.4	-22.2	-14.1	-100.0
U081606-0570	97740.1	1019.5	3.8	113705.4	1044.0	21510.5	-14.0	-2.3	-100.0
U101606-0720	128218.5	1291.0	6.9	153100.5	1345.0	21519.7	-16.3	-4.0	-100.0
Weighted avg.	55184.7	579.2	8.8	65531.0	634.2	19276.6	-15.7	-8.7	-100.0

single bay results in parallel, giving each individual tree search instance more time to explore the solution space. A second option to advance the algorithm is to take relocations of containers between bays into account, which in certain situations may lead to even shorter move sequences and crane move times.

References

1. Caserta M., Voß S., Sniedovich M.: Applying the corridor method to a blocks relocation problem. OR Spectrum (2009) doi: 10.1007/s00291-009-0176-5
2. Caserta M., Schwarze S., Voß S.: A new binary description of the blocks relocation problem and benefits in a look ahead heuristic. In: Proc. of the EvoCOP 2009, pp. 37-48. Springer, Heidelberg (2009)
3. Caserta M., Voß S.: A cooperative strategy for guiding the corridor method. Studies in Computational Intelligence **236**, 273–286 (2009)
4. Caserta M., Voß S.: Corridor selection and fine tuning for the corridor method. In: Proc. of the LION 2009, pp. 163-175. Springer, Heidelberg (2009)
5. Forster F., Bortfeldt A.: A heuristic for retrieving containers from a yard. Computers & Operations Research **39**, 299–309 (2012, already available online)
6. Kim K., Hong G.: A heuristic rule for relocating blocks. Computers & Operations Research **33**, 940–954 (2006)
7. Lee Y., Lee Y-J.: A heuristic for retrieving containers from a yard. Computers & Operations Research **37**, 1139–1147 (2010)
8. Stopford M.: Maritime Economics. Routledge (2009)
9. UNCTAD Review of Maritime Transport (2010)

Spatial and temporal synchronization of vehicles in logistics networks

Dorota Slawa Mankowska and Christian Bierwirth and Frank Meisel

Abstract In logistics networks, service providers offer on-site services to the customers. Often, there is a need to synchronize the routes of vehicles that provide such services. Synchronization is required for example in case of a heterogeneous fleet of serving vehicles, where large vehicles carry goods and supply small vehicles, which distribute the goods to customers. We model such a routing problem for a network of customers that are served through a fleet of heterogeneous vehicles. We present a new model for synchronizing cargo exchanges on vehicle routes and provide first computational results for this problem.

1 Introduction

Synchronization of vehicle routes means coupling the routes of two or more transport means in time and space. A practical example of synchronization requirements occurs, when vehicles meet each other for exchanging load, so called truck-meets-truck traffic, see [4]. In this paper, we provide a model for vehicle routing under such a synchronization requirement. More precisely, we consider a routing problem for two heterogeneous vehicles. The vehicles differ in load capacities and their capability to access customer locations. A large vehicle that is incapable of moving to some of the customers is used to carry the goods and supplies the small vehicle that can access these customers. Although vehicle routing problems are one of the most considered optimization problems, see [3], synchronization requirements among vehicles have hardly been considered so far, see [1,2].

The goal of this paper is to present a mixed-integer programming model for synchronizing the routes of two vehicles together with an illustrative example (Section 2) and to provide first numerical results when tackling the model with the CPLEX MIP Solver (Section 5). The tests address the solvability of the model and the sensitivity of solutions with respect to three different objectives. Potential for future research is outlined in Section 4.

Martin-Luther-University Halle-Wittenberg, Große Steinstraße 73, 06108 Halle, Germany
e-mail: {dorota.mankowska;christian.bierwirth;frank.meisel}@wiwi.uni-halle.de

D. Klatte et al. (eds.), *Operations Research Proceedings 2011*, Operations Research Proceedings, 263
DOI 10.1007/978-3-642-29210-1_42, © Springer-Verlag Berlin Heidelberg 2012

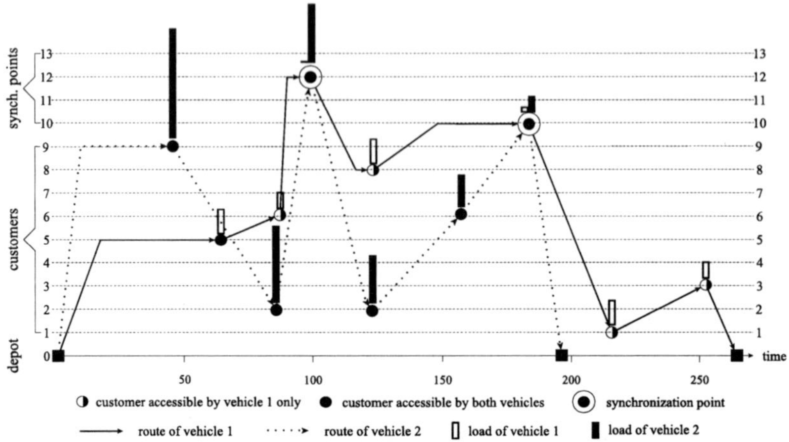

Fig. 1 Time-space diagram of vehicle routes with synchronization for load exchange.

2 Problem Description and Mathematical Formulation

The following notation is used to model the routing of vehicles under synchronization constraints. The problem is formulated on a complete undirected network $G = (N, E)$ with N being the set of nodes and E being the set of edges. The set of nodes is $N = \{0\} \cup C \cup S$, where node 0 refers to a depot, C is a set of customers to be served, and S is a set of potential synchronization points, where vehicles can meet to exchange load. Each customer $i \in C$ demands q_i units of a good that is distributed by the vehicles. The delivery must take place within a service time window $[e_i, l_i]$. At the depot, a heterogeneous fleet V of two vehicles is available. The capacity of the first vehicle is Cap_1 and the capacity of the second vehicle is Cap_2. We assume that $q_i \leq max\{Cap_1, Cap_2\}$ for all $i \in C$. Furthermore, we denote by p_{iv} the service duration that is needed, if customer i is served by vehicle $v \in V$.

To model the accessibility of customers i by vehicles v, we have given a binary matrix $[a_{iv}]$ with $a_{iv} = 1$, if vehicle v can move to customer i. If vehicle v cannot access customer i, e.g. due to deficient infrastructure, we set $a_{iv} = 0$. Clearly, for the depot and the potential synchronization points in S we have $a_{iv} = 1$, for all $i \in \{0\} \cup S$ and all $v \in V$. The distances between nodes $i, j \in N$, i.e. the weights of edges $(i, j) \in E$, are denoted d_{ij}. We assume that traveling times are equal to these distances.

The following decision variables are used. Binary variable x_{ijv} is set to 1 if vehicle $v \in V$ moves directly from node $i \in N$ to node $j \in N$, 0 otherwise. The time variable t_{iv} denotes the service start of vehicle v, if i refers to a customer, the start of a load exchange operation, if i refers to a synchronization point, and the return time, if i is the depot. The binary variables u_i take value 1, if the vehicles use the potential synchronization point $i \in S$ for exchanging load on their routes. Q_{iv} denotes the amount of cargo loaded in vehicle v, when v arrives at node $i \in N$. We consider the following three objectives for the routing problem:

- Minimizing the total distance traveled by all vehicles (Z_1),

$$\min \rightarrow Z_1 = \sum_{v \in V} \sum_{i \in N} \sum_{j \in N} d_{ij} \cdot x_{ijv} \tag{1}$$

- Minimizing the maximal tour duration $T = \max_{v \in V}\{t_{0v}\}$ (Z_2),

$$\min \rightarrow Z_2 = T \tag{2}$$

- Minimizing total vehicle waiting times due to time windows and synchronization (Z_3), as is of relevance for companies that aim at minimizing vehicle idle time.

$$\min \rightarrow Z_3 = \sum_{v \in V} W_v \tag{3}$$

The problem is considered under following constraints:

$$\sum_{j \in N} x_{0jv} = \sum_{j \in N} x_{j0v} = 1 \qquad\qquad \forall v \in V \tag{4}$$

$$\sum_{v \in V} \sum_{j \in N} x_{ijv} = 1 \qquad\qquad \forall i \in C \tag{5}$$

$$\sum_{j \in N} x_{jiv} = \sum_{j \in N} x_{ijv} \qquad\qquad \forall i \in N, v \in V \tag{6}$$

$$t_{iv} + (p_{iv} + d_{ij}) \cdot x_{ijv} \le t_{jv} + M_1 \cdot (1 - x_{ijv}) \qquad \forall i, j \in N, v \in V \tag{7}$$

$$e_i \cdot \sum_{j \in N} x_{ijv} \le t_{iv} \le l_i \cdot \sum_{j \in N} x_{ijv} \qquad \forall i \in C, v \in V \tag{8}$$

$$Q_{iv} - q_i \ge Q_{jv} - M_2 \cdot (1 - x_{ijv}) \qquad \forall i \in C, j \in N, v \in V \tag{9}$$

$$\sum_{j \in C} x_{jiv} = u_i \qquad\qquad \forall i \in S, v \in V \tag{10}$$

$$t_{i1} = t_{i2} \qquad\qquad \forall i \in S \tag{11}$$

$$Q_{i1} + Q_{i2} \ge Q_{j1} + Q_{k2} - M_3 \cdot (3 - u_i - x_{ij1} - x_{ik2}) \forall i \in S, j, k \in N \tag{12}$$

$$T \ge t_{0v} \qquad\qquad \forall v \in V \tag{13}$$

$$W_v = t_{0v} - \sum_{i \in N} \sum_{j \in N} (p_{iv} + d_{ij}) \cdot x_{ijv} \qquad \forall v \in V \tag{14}$$

$$x_{ijv} \in \{0, a_{iv} \cdot a_{jv}\} \qquad\qquad \forall i, j \in N, v \in V \tag{15}$$

$$u_i \in \{0, 1\} \qquad\qquad \forall i \in S \tag{16}$$

$$t_{iv} \ge 0 \qquad\qquad \forall i \in N, v \in V \tag{17}$$

$$Cap_v \ge Q_{iv} \ge 0 \qquad\qquad \forall i \in N, v \in V \tag{18}$$

Constraints (4) ensure that the route of each vehicle starts and ends in the depot. Through (5), every customer is visited exactly once by one of the vehicles. Restrictions (6) balance the vehicle flow. Constraints (7) and (8) determine the service start times at customers. (7) calculates the arrival of vehicle v at a node j by taking into account the service start time at a node i plus the service duration of this node and the travel time from i to j, if v visits i directly before j. Here, M_1 must be set sufficiently large and is set to $M_1 = l_i$ to derive a tight formulation. (8) ensures that the services start within the customers' time window. Constraints (9) control the reduction of vehicle cargo when serving customers, where $M_2 = 2 \cdot Cap_v$. Synchronization of vehicles is handled by (10) to (12). More precisely, (10) ensures that each vehicle moves to synchronization point $i \in S$, if and only if this point is used for synchronization in the solution (i.e. if $u_i = 1$). Here, we set $M_3 = Cap_1 + Cap_2$. From (11), starting times for load exchange operations at synchronization points must be the same for both vehicles. (12) balance the cargo volume of vehicles at synchronization point i, such that the total cargo of vehicles when arriving at the synchronization point equals the total cargo of both vehicles when arriving at their subsequent route points j and k. To measure the value of objective functions (2) and (3), we add Constraints (13) and (14). (13) determines the maximum tour duration T. (14) determines the waiting times of the vehicles. The Constraints (15) to (18) define the domains of the decision variables. Note that, in the definition of x_{ijv}, $a_{iv} \cdot a_{jv}$ takes value 1, if and only if vehicle v can access both nodes i and j, which avoids trips to or from nodes that the vehicle cannot access.

Example: We consider a network with one depot, nine customers $C = \{1, 2, \ldots, 9\}$, and four potential synchronization points $S = \{10, 11, 12, 13\}$ for exchanging load among vehicles. All nodes are randomly placed in a square area 50×50. The distances and travel times between the nodes are Euclidean. Time windows are of length 150. The begin of each customer time window is randomly drawn from $[0, \ldots, 130]$. Customer demands are given by vector $q = [6, 9, 10, 4, 5, 5, 5, 7, 9]$. The capacities of the vehicles are $Cap_1 = 10$ and $Cap_2 = 50$ units. Figure 1 shows the optimal solution for this instance when pursuing objective Z_1. The bars at each node show the current cargo of the vehicles when arriving at this node. It can be seen that the cargo of the small vehicle (vehicle 1) and the large vehicle (vehicle 2) decreases from serving the customers. For supplying the small vehicle, synchronization takes place at nodes ten and twelve. Here, load is exchanged, such that vehicle 1 receives five cargo units from vehicle 2 at node 12 and seven units at node 10. After the second replenishment, vehicle 2 returns to the depot and vehicle 1 serves the remaining customers 1 and 3. Waiting times at customer nodes are a consequence of time windows. Furthermore, vehicle 1 has to wait at both synchronization points until vehicle 2 arrives. The example illustrates how synchronization of vehicles is planed in time and space to enable serving the customers. The opportunity to exchange load directly on their routes allows the vehicles to continue their tours without having to return to the depot for replenishment.

3 Numerical experiments

We execute computational tests using randomly generated test instances to investigate the model's solvability under different objectives and varied problem size. Test set A contains ten instances, each with 12 nodes (1 depot, 9 customers, and 2 potential synchronization points). Test set B contains ten instances, each with 19 nodes (1 depot, 14 customers, and 4 potential synchronization points). The nodes are randomly located in a square area 50×50 for all instances in sets A and B. The distances and travel times between all nodes are Euclidean. We chose the accessibility a_{iv} randomly with $\sum_{v \in V} a_{iv} \geq 1$ for customer nodes $i \in C$. The processing times p_{iv} for nodes $i \in C$ and vehicles $v \in V$ are random integers from the interval $[1, \ldots, 7]$.

No.	Z_1 pursued				Z_2 pursued				Z_3 pursued			
	Z_1	Z_2	Z_3	CPU	Z_1	Z_2	Z_3	CPU	Z_1	Z_2	Z_3	CPU
1	227.0	227.0	160.9	00:01	261.3	199.4	101.5	00:26	392.1	253.0	67.4	01:47
2	256.9	165.0	32.2	00:01	259.9	164.5	33.2	00:26	381.9	235.0	2.5	02:08
3	203.6	215.0	119.8	00:01	220.8	168.0	84.2	02:27	372.0	227.0	12.7	18:58
4	207.8	205.0	100.5	00:01	235.0	161.9	55.9	02:54	334.4	205.0	2.8	40:46
5	274.5	214.0	69.3	00:03	285.5	191.0	56.4	00:21	403.7	236.0	16.7	02:59
6	215.5	208.0	156.6	00:01	271.9	160.4	16.8	02:23	321.7	187.0	5.7	23:37
7	232.4	187.0	24.4	00:01	235.4	180.2	100.1	01:06	354.5	289.0	4.3	11:00
8	228.9	217.0	142.2	00:01	288.4	185.2	67.6	02:18	324.0	208.0	17.8	05:42
9	186.5	244.0	159.9	00:01	197.9	194.5	154.2	01:20	377.9	236.0	2.2	03:13
10	246.0	191.0	93.9	00:01	248.1	177.6	68.2	00:07	446.4	269.0	16.2	01:41
∅	227.9	207.3	106.0	00:01	249.2	175.9	73.8	01:29	368.5	232.4	14.8	12:14

Table 1 Numerical results for instance set A and three different objective functions.

The demand q_i of customer i is a random integer from the interval $[1, \ldots, 10]$. The capacity of vehicle 1 is set to $Cap_1=15$ for set A and to $Cap_1=25$ for set B. The capacity of the vehicle 2 is $Cap_2 = \sum_{i \in C} q_i - Cap_1$ for all instances. The customer time windows are of length 140 for set A and of length 160 for set B. We use ILOG CPLEX 12.2 on a 2.80 GHz Intel Core Duo.

In the first test, we solve set A once for each objective Z_1, Z_2, and Z_3. The computational results of final solutions are shown in Table 1. The table gives the observed values of the three performance measures and the computation time of CPLEX [mm:ss] for each instance and each pursued objective. It can be seen that optimal solutions under objective Z_1 are achieved within at most three seconds. When turning to Z_2, the average computation time increases to 1.5 minutes per instance and even to 12 minutes when pursuing Z_3. Hence, we find that minimizing the total distance traveled is comparably easy for CPLEX, whereas minimizing the maximal tour duration or the total waiting times appears to be more difficult. The explanation is that the route duration t_{0v} is highly underestimated in the LP-relaxation of the problem. Strengthening the model with respect to a better estimation of t_{0v}, e.g. by developing suitable valid inequalities, appears as one promising direction for future research. Regarding the objective function values, the results reveal that the three objectives are conflicting. When pursuing Z_1, this performance measure takes a minimum (as expected), but maximal tour duration and total waiting time are higher compared with solutions obtained under objectives Z_2 and Z_3, respectively. Interestingly, also Z_2 and Z_3 appear to be conflicting as is revealed by the strong increase of the average Z_2 value that is observed when Z_3 is pursued. The changes in the performance measures are drastic. E.g. when pursuing Z_3 instead of Z_1, the average Z_1-value increases by 62% whereas the average Z_3-value decreases by 86%. This result shows that the particular objective pursued in the vehicle routing problem with synchronization constraints must be selected carefully to obtain useful results in a particular practical application.

For the second test, we solve instance sets A and B under objective Z_1. Furthermore, we vary the number of synchronization points in these sets. Table 2 shows the results for set A with $|S| = 2$ and $|S| = 4$ potential synchronization points and for set B with $|S| = 4$ and $|S| = 6$. It can be seen that the large instances in set B are also solved to optimality within minutes. For both sets, the average objective value

	set A				set B			
	$\|S\|=2$		$\|S\|=4$		$\|S\|=4$		$\|S\|=6$	
No.	Z_1	CPU	Z_1	CPU	Z_1	CPU	Z_1	CPU
1	227.0	00:01	227.0	00:02	287.7	00:10	271.6	01:22
2	256.9	00:01	237.5	00:12	246.4	00:29	246.4	03:36
3	203.6	00:01	203.6	00:01	248.0	00:50	248.0	02:11
4	207.8	00:01	207.8	00:01	226.2	17:02	214.6	08:09
5	274.5	00:03	273.9	00:07	280.6	04:58	280.6	14:49
6	215.5	00:01	215.5	00:05	241.7	00:34	241.7	00:49
7	232.4	00:01	232.4	00:01	268.4	00:29	268.4	00:49
8	228.9	00:01	209.6	00:01	283.0	00:26	277.5	01:01
9	186.5	00:01	186.5	00:01	232.0	09:02	232.0	11:18
10	246.0	00:01	213.7	00:01	266.5	00:02	266.5	00:05
∅	227.9	00:01	220.1	00:03	254.8	03:46	252.9	04:45

Table 2 Numerical results for instance sets A and B and different synchronization opportunities.

decreases when the set of synchronization points is enlarged. This result shows that the additional potential for synchronization is exploited successfully in several instances. However, the average runtime grows when increasing the number of synchronization points for the instances. Hence, we see that the set of synchronization points that are considered in a problem must be selected carefully such that good solutions can be achieved without deteriorating the solvability of the problem.

4 Concluding remarks

We have proposed a model for the vehicle routing problem with synchronization constraints under various objectives together with first computational results. Future research will address developing valid inequalities for this model. Furthermore, we will generalize the model towards an arbitrary number of vehicles, which is a challenging task because it raises questions like which of the vehicles actually perform a particular service and which of them require synchronization. Finally, heuristics are to be developed for solving large problem instances.

Acknowledgements This research is funded by the Deutsche Forschungsgemeinschaft (DFG) under project reference B02110263.

References

1. Bredström, D., Rönnqvist, M.: Combined vehicle routing and scheduling with temporal precedence and synchronization constraints. European Journal of Operational Research **191**,19-31 (2008)
2. Del Pia, A., Filippi, C.: A variable neighborhood descent algorithm for a real waste collection problem with mobile depots. International Transactions in Operational Research **13**,125-141 (2006)
3. Golden, B., Raghavan, S., Wasil, E. A.: The Vehicle Routing Problem: Latest Advances and New Challenges. Operations research/Computer science interfaces series **43**, Springer, New York (2008)
4. Weise, T., Podlich, A., Reinhard, K., Gorldt, Chr., Geihs, K.: Evolutionary Freight Transportation Planning. In: Giacobini, M. et al.: Applications of Evolutionary Computing – Proceedings of EvoWorkshops 2009 **2484/2009**,768-777 (2009)

Sequential Zone Adjustment for Approximate Solving of Large p-Median Problems

Jaroslav Janacek and Marek Kvet

Abstract The p-median problems are used very often for designing optimal structure of most public service systems. Since particular models are characterized by big number of possible facility locations, current exact approaches often fail. This paper deals with an approximate approach based on specific model reformulation. It uses an approximation of the network distance between a service center and a customer by some of pre-determined distances. The pre-determined distances are given by dividing points separating the range of possible distances. Deployment of these points influences the accuracy of the approximation. To improve this approach, we have developed a sequential method of the dividing point deployment. We study here the effectiveness of this method and compare the results to the former static method.

1 Introduction

The family of public service systems [6], including medical emergency system [5], public administration system [3] and many others, can be designed making use of the p-median problem. Particular mathematical models are characterized by considerably big number of possible service center locations. The number of served customers takes the value of several thousands and the number of possible locations can take this value as well. It was found that the number of possible service center locations impacts the computational time necessary for finding the optimal solution [4]. Concerning the problem size, it is obvious that the location-allocation model of the p-median problem constitutes such mathematical programming problem which resists to any attempt for fast solution.
On the other side, large covering problems are easily solvable by common opti-

Jaroslav Janacek
University of Zilina, Faculty of Management Science and Informatics, Department of Transportation Networks, Univerzitna 8215/1, 010 26 Zilina, Slovakia, e-mail: Jaroslav.Janacek@fri.uniza.sk

Marek Kvet
University of Zilina, Faculty of Management Science and Informatics, Department of Transportation Networks, Univerzitna 8215/1, 010 26 Zilina, Slovakia, e-mail: Marek.Kvet@fri.uniza.sk

D. Klatte et al. (eds.), *Operations Research Proceedings 2011*, Operations Research Proceedings, 269
DOI 10.1007/978-3-642-29210-1_43, © Springer-Verlag Berlin Heidelberg 2012

mization software. The necessity of solving large p-median problems leads to the approximate approach which enables to solve real-sized instances in admissible time [2]. This covering approach pays for short computational time by a loss of accuracy. Nevertheless, the accuracy can be improved by suitable determination of so called dividing points which are used in the problem reformulation. In this paper, we present a sequential approach to the dividing point determination, study the impact of dividing point selection to the solution accuracy and compare suggested method to the former static method.

2 Covering Formulation of the p-Median Problem

To formulate the p-median problem on a discrete network, we denote a set of served customers by symbol J and a set of possible facility locations by symbol I. In this paper, we deal only with the following formulation of the p-median problem: Determine at most p facility locations from I so that the sum of network distances from each element of J to the nearest located facility is minimal. The distance between a possible location i and a customer j is denoted by d_{ij}. In the location-allocation model, there must be decided which served object is assigned to which located facility and also where these service centres should be located [2], [3].

According to [2], the approximate approach is based on a relaxation of the assignment of a service center to a customer. In this method, the distance between a customer and the nearest facility is approximated unless the assigned facility must be determined. The range $< 0, max\{d_{ij} : i \in I, j \in J\} >$ of all possible distances is partitioned into $r + 1$ zones. The zones are separated by finite ascending sequence of dividing points $D_1, D_2, \ldots, D_r$ where $0 = D_0 < D_1$ and $D_r < D_m = max\{d_{ij} : i \in I, j \in J\}$. A zone k corresponds with the interval $(D_k, D_{k+1} >$, the zone one corresponds with the interval $(D_1, D_2 >$ and so on, till the r-th zone which corresponds with the interval $(D_r, D_m >$. The width of the k-th interval is denoted by e_k for $k = 0, \ldots, r$. We introduce a variable $y_i \in \{0, 1\}$ for each $i \in I$, which takes the value of 1 if a facility should be located at the location i and which takes the value of 0 otherwise. In addition, an auxiliary zero-one variable x_{jk} for $k = 0, \ldots, r$ is introduced. This variable takes the value of 1 if the distance of the customer $j \in J$ from the nearest located center is greater than D_k and this variable takes the value of 0 otherwise. Then the expression $e_0 x_{j0} + e_1 x_{j1} + e_2 x_{j2} + e_3 x_{j3} + \cdots + e_r x_{jr}$ is an upper approximation of d_{ij}. If the distance d_{ij} belongs to the interval $(D_k, D_{k+1} >$, it is estimated by the upper bound D_{k+1} with a possible deviation e_k.

Similarly to the classical covering model, we introduce a zero-one constant a_{ij}^k for each triple $< i, j, k >\in I \times J \times \{0, \ldots, r\}$. The constant $a_{ij}^k = 1$ if the distance between the customer j and the possible facility location i is less or equal to D_k, otherwise $a_{ij}^k = 0$. Then the covering model can be formulated as follows:

$$Minimize \quad \sum_{j \in J} \sum_{k=0}^{r} e_k x_{jk} \tag{1}$$

$$Subject\ to: \qquad x_{jk} + \sum_{i \in I} a_{ij}^k y_i \geq 1 \qquad for\ j \in J\ and\ k = 0, \ldots, r \qquad (2)$$

$$\sum_{i \in I} y_i \leq p \tag{3}$$

$$x_{jk} \geq 0 \qquad for\ j \in J\ and\ k = 0, \ldots, r \tag{4}$$

$$y_i \in \{0, 1\} \qquad for\ i \in I \tag{5}$$

In this model, the objective function (1) gives the upper bound of the sum of original distances. The constraints (2) ensure that each variable x_{jk} is allowed to take the value of 0 if there is at least one service center located in radius D_k from the customer j. The constraint (3) limits the number of located facilities by p.

3 Partitioning of Actual Distance Range

It is obvious that only limited number r of dividing points $D_1, D_2, \ldots, D_r$ keeps the model size in a mediate extent. This restriction impacts the accuracy of the solution. Let us confine to the networks with integer network distances in this study. The elements of the distance matrix $\{d_{ij}\}$ form a finite ordered set of values $d_0 < d_1 < \cdots < d_m$ where $D_0 = d_0$ and $D_m = d_m$. Let the value d_h have a frequency N_h of its occurrence in the matrix $\{d_{ij}\}$. If there are only r different values between d_0 and d_m, then the dividing points $D_1, D_2, \ldots, D_r$ can be set equal to these values. Then the exact solution can be obtained by mere solving the problem $(1) - (5)$.
Otherwise, the distance d between a customer and the nearest located center can be only estimated knowing that it belongs to the interval $(D_k, D_{k+1} >$ given by a pair of dividing points. Let us denote $D_k^1, D_k^2, \ldots, D_k^{r(k)}$ the values of the sequence $d_0, d_1, \ldots, d_m$ which are greater than D_k and less then D_{k+1}. Maximum deviation of the upper estimation D_{k+1} from the exact value d is $D_{k+1} - D_k^1$.
If we were able to anticipate the frequency n_h of each d_h in the optimal solution, we could minimize the deviation using dividing points obtained by solving the problem:

$$Minimize \qquad \sum_{t=1}^{m} \sum_{h=1}^{t} (d_t - d_h) n_h x_{ht} \tag{6}$$

$$Subject\ to: \qquad x_{(h-1)t} \leq x_{ht} \qquad for\ t = 2, \ldots, m\ and\ h = 2, \ldots, t \tag{7}$$

$$\sum_{t=h}^{m} x_{ht} = 1 \tag{8}$$

$$\sum_{t=1}^{m-1} x_{tt} = r \tag{9}$$

$$x_{ht} \geq 0 \qquad for \ \ t = 1,\ldots,m \ \ and \ \ h = 1,\ldots,t \tag{10}$$

If the distance d_h belongs to the interval ending by a dividing point d_t, then the decision variable x_{ht} takes the value of 1. Link-up constraints (7) ensure that the distance d_{h-1} can belong to the interval ending with d_t only if all distances between d_{h-1} and d_t belong to this interval. Constraints (8) assure that each distance d_h belongs to some interval and constraint (9) enables to set only r dividing points. After the problem (6) – (10) is solved, the nonzero values of x_{tt} indicate the distances d_t which correspond with the dividing points. The static approach to the dividing points determination comes out from the hypothesis that the frequency n_h of d_h may be proportional to N_h and to some weight which decreases with increasing value of d_h.

$$n_h = N_h e^{-d_h/T} \tag{11}$$

In the expression (11), T is a positive parameter and N_h is the mentioned occurrence frequency, where only $|I| - p + 1$ smallest distances of each matrix column are included.

4 The Sequential Zone Adjustment Method

The sequential zone adjustment method is based on the idea of making the estimation of individual distance d_h relevancy more accurate. The distance relevancy means here a measure of our expectation that this distance value is the distance between a customer and the nearest service center in some optimal solution.

The initial relevancy estimation described in the previous section comes out from the hypothesis (11). It assumes that the relevancy n_h of the distance d_h is proportional to the frequency of the distance d_h in the matrix of distances. This value of relevance is reduced according to the hypothesis that the shorter distance value appears in the optimal solution more often than the bigger one. This approach comprehends only our general expectation, but it does not reflect real property of the solved instance. To make the relevance estimation more realistic, we make use of the optimal solution of the approximate problem (1) – (5), where the dividing points are initialized by arbitrary ad-hoc method. Assuming that the optimal solution is known, especially knowing which y_i is equal to one, only the associated rows of the matrix $\{d_{ij}\}$ are taken into account. These rows are denoted as active rows. Then each column of this matrix is inspected and the minimal value over the active rows is included into the set of relevant distances and their occurrence frequencies. By this way a new sequence of distances d_k and their frequencies n_k is obtained. The minimal and maximal distances d_0 and d_m are taken from the former sequence. This new sequence is used in the problem (6) – (10) and the optimal solution of this problem brings a new set of dividing points. Based on the new dividing point set, the sequence of zones is readjust and the approximate problem (1) – (5) is resolved for the new zone sequence. This process can be repeated while better solution of the original problem is obtained.

5 Numerical Experiments

We performed a sequence of numerical experiments to test the effectiveness of suggested sequential method. Solved instances of the p-median problem were obtained from the $OR - Lib$ set of benchmarks [1]. To make the results more comparable, we preserve the original notation of the instances, e.g. $pmed21$, and if we modified the instance in the number of facilities p, then we denote the modified instance name by an asterisk, e.g. $pm * d21$. An individual experiment is organized so that the initial sequence of dividing points is obtained according to the model (6) – (10), where the relevancy n_h of a distance d_h is computed according to (11) for $T = 1000$. The sequential zone adjustment method starts from the initial sequence of dividing points, defines associated constants a_{ij}^k and e_k and solves the problem (1) – (5). Resulting values of y_i are used to obtain reduced sequence of the relevant distances d_h with their frequencies n_h as described in the previous section. Since the optimal value of the objective function (1) is only upper bound of the original objective function value, we use the following formula (12) to obtain the real objective function value.

$$\sum_{j \in J} min\{d_{ij} : i \in I, \ y_i = 1\} \tag{12}$$

This step is repeated for new sequences d_h and n_h until either no improvement of (12) is obtained or the number of relevant distances drops below 20. The results are reported in table 1 divided into the following row-blocks.

Block *Desc* comprises a concise description of an instance which includes name of the instance (*Name*), the number of customers (*Cust*), which is equal to the number of possible locations, and the number p denoting the maximal number of located centres. The denotation *Exact* indicates the row of the objective function values (12) for the individual instances obtained by the optimization environment XPRESS-IVE. The row-block *Seque* refers on the sequence of solutions of the problem (1) – (5) obtained from step-by-step improved set of the dividing points. In the rows denoted by *Run1*, *Run2*, etc., there are given the percentage deviations of the solution values (12) from the exact solution. The section *CompT* contains computational time (*ETime*) in seconds necessary for obtaining the exact solution and the sum of computational times (*STime*) for the sequence of problems (1) – (5).

The experiments were performed using the optimization software XPRESS-IVE. The associated code was run on a personal computer equipped with the Intel Core 2 6700 processor with parameters: 2.66 GHz and 3 GB RAM.

Table 1 Numerical experiments for medium instances of the p-median problem

	Name	pmed5	pmed8	pm*d12	pm*d16	pm*d21	pm*d29	pm*d31	pm*d37	pm*d40
Desc	*Cust*	100	200	300	400	500	600	700	800	900
	p	33	20	100	128	200	193	292	240	400
	Exact	1434	4459	1706	1750	1426	1882	1289	2278	1397
Seque [%]	*Run1*	6.7	2.0	21.5	31.3	51.4	31.9	64.1	28.3	48.2
	Run2	0.5	0.4	1.1	0.2	0.0	0.1	0.0	0.4	0.0
	Run3	0.5	0.4	1.1	0.2		0.1		0.4	
CompT [s]	*ETime*	0	4	12	28	56	75	119	185	264
	STime	0	5	7	12	8	10	12	16	15

6 Conclusions

We have presented the sequential zone adjustment method which improves the accuracy of the approximate approach to the p-median problem. The results show that our suggested method constitutes a promising way of solving large instances of the p-median problem. As shown in table 1, only two steps of the sequential method are enough to obtain a result which differs from the optimal solution by less than one percent. The computational time of the sequential approach is comparable to the computational time of location-allocation approach for small instances, but it is considerably lower if the size of instances grows. This phenomenon deserves further research especially for larger instances of the p-median problem.

Acknowledgements This work was supported by the research grant VEGA 1/0361/10 - Optimal design of public service systems under uncertainty.

References

1. Beasley J. E.: OR Library: Distributing Test Problems by Electronic Mail, Journal of the Operational Research Society, 41(11), pp 1069-1072 (1990)
2. Janacek, J.: Approximate Covering Models of Location Problems. In: Proceedings of the 1st International Conference on Applied Operational Research ICAOR 08, Vol. 1, September 2008, Yerevan, Armenia, pp 53-61 (2008)
3. Janacek, J., Linda, B., Ritschelov, I.: Optimization of Municipalities with Extended Competence Selection. Prager Economic Papers-Quarterly Journal of Economic Theory and Policy, Vol. 19, No 1, pp 21-34 (2010)
4. Janackova, M., Szendreyova, A.: An impact of set of centres of distribution system design problem on computational time of solving algorithm. In: Mikulski J (ed). Advances in Transport Systems Telematics, Katowice, pp 109-112 (2006)
5. Janosikova, L.: Emergency Medical Service Planning. In: Communications Scientific Letters of the University of Zilina, Vol. 9, No 2, pp 64-68 (2007)
6. Marianov, V., Serra, D.: Location problems in the public sector. In: Drezner Z (ed.) et al. Facility location. Applications and theory. Berlin: Springer, pp 119-150 (2002)

A new alternating heuristic for the $(r \mid p)$–centroid problem on the plane

Emilio Carrizosa, Ivan Davydov, Yury Kochetov

Abstract In the $(r \mid p)$-centroid problem, two players, called leader and follower, open facilities to service clients. We assume that clients are identified with their location on the Euclidian plane, and facilities can be opened anywhere in the plane. The leader opens p facilities. Later on, the follower opens r facilities. Each client patronizes the closest facility. Our goal is to find p facilities for the leader to maximize his market share. For this Stackelberg game we develop a new alternating heuristic, based on the exact approach for the follower problem. At each iteration of the heuristic, we consider the solution of one player and calculate the best answer for the other player. At the final stage, the clients are clustered, and an exact polynomial-time algorithm for the $(1 \mid 1)$-centroid problem is applied. Computational experiments show that this heuristic dominates the previous alternating heuristic of Bhadury, Eiselt, and Jaramillo.

1 Introduction

This paper addresses a Stackelberg facility location game on a two–dimensional Euclidian plane. It is assumed that the clients demands are concentrated at a finite number of points in the plane. In the first stage of the game, a player, called the leader, opens his own p facilities. At the second stage, another player, called here the follower, opens his own r facilities. At the third final stage, each client chooses the closest opened facility as a supplier. In case of ties, the leader's facility is preferred. Each player tries to maximize his own market share. The goal of the game is to find p points for the leader facilities to maximize his market share.

Emilio Carrizosa
Facultad de Matemáticas, Universidad de Sevilla, Sevilla, Spain, e-mail: ecarrizosa@us.es

Ivan Davydov
Sobolev Institute of Mathematics, Novosibirsk, Russia, e-mail: vann.davydov@gmail.com

Yury Kochetov
Sobolev Institute of Mathematics, Novosibirsk, Russia, e-mail: jkochet@math.nsc.ru

D. Klatte et al. (eds.), *Operations Research Proceedings 2011*, Operations Research Proceedings, 275
DOI 10.1007/978-3-642-29210-1_44, © Springer-Verlag Berlin Heidelberg 2012

This Stackelberg game was studied by Hakimi in 1981 [8, 9] for location on a network. Following Hakimi, the leader problem is called a *centroid problem* and the follower problem is called a *medianoid problem*. In [1] the centroid problem with another behavior of clients was considered. In [7] an exact polynomial time algorithm is presented for these problems in case $p = r = 1$. Similar models with locational constraints are studied in [3,4]. For arbitrary p and r, an alternating heuristic is presented in [2]. A greedy and a minimum-differentiation algorithm are used for approximation of the follower market share. A comprehensive review of complexity results and properties of the problems can be found in [10, 11, 13].

In this paper we improve the alternating heuristic from [2] using an exact approach for the follower problem. We reduce it to the discrete maximum capture problem and apply the branch and bound method. At the end of the alternating process, the final solution for the leader is improved by using an exact algorithm for the $(1 \mid 1)$-centroid problem. All clients are clustered into p subsets. For each subset we relocate the leader facility using the optimal solution for the $(1 \mid 1)$-centroid problem. Computational results for randomly generated instances [6] show that the new approach dominates the benchmark procedures.

2 Mathematical model

Let us consider a two–dimensional Euclidian plane in which n clients are located. We assume that each client j has a positive demand w_j. Let X be the set of p points where the leader opens his own facilities and let Y be the set of r points where the follower opens his own facilities. The distances from client j to the closest facility of the leader and the closest facility of the follower are denoted as $d(j,X)$ and $d(j,Y)$ respectively. The client j prefers Y over X if $d(j,Y) < d(j,X)$ and prefers X over Y otherwise. By

$$U(Y \prec X) := \{j \mid d(j,Y) < d(j,X)\}$$

we denote the set of clients preferring Y over X. The total demand captured by the follower by locating his facilities at Y while the leader locates his facilities at X is given by

$$W(Y \prec X) := \sum (w_j \mid j \in U(Y \prec X)).$$

For X given, the follower tries to maximize his own market share. The maximal value $W^*(X)$ is defined to be

$$W^*(X) := \max_{Y, |Y|=r} W(Y \prec X).$$

This maximization problem will be called the *follower problem*. The leader tries to minimize the market share of the follower. This minimal value $W^*(X^*)$ is defined to be

$$W^*(X^*) := \min_{X, |X|=p} W^*(X).$$

For the best solution X^* of the leader, his market share is $\sum_{j=1}^{n} w_j - W^*(X^*)$. In the $(r \mid p)$–*centroid problem,* the goal is to find X^* and $W^*(X^*)$.

3 The follower problem

Let us first describe an exact approach for the follower problem. Such problem will be rewritten as an integer linear programming problem, and solved using a branch and bound method.

For each client j, we introduce a disk D_j with radius $d(j,X)$ and center in the point where this client is located. Let us consider the resulting intersection of each set of two or more such disks. These disks and their intersections will be called *regions.* The total number of regions is large, but we can eliminate some, and consider the maximal regions as those defined by intersections only. In any case, we have at most $n^2 + n$ regions. Now we define a binary matrix (a_{kj}) to indicate the clients which will patronize a facility of the follower if it is opened inside a region. Formally, define $a_{kj} := 1$ if a facility of the follower in region k captures the client j and $a_{kj} := 0$ otherwise. In order to present the follower problem as an integer linear program we introduce two sets of the decision variables:

$$y_k = \begin{cases} 1 & \text{if the follower opens his own facility inside of region } k, \\ 0 & \text{otherwise,} \end{cases}$$

$$z_j = \begin{cases} 1 & \text{if the follower captures client } j, \\ 0 & \text{otherwise.} \end{cases}$$

Now the follower problem can be written as the maximum capture problem:

$$\max \sum_{j=1}^{n} w_j z_j$$

$$\text{subject to} \qquad z_j \le \sum_{k=1}^{n^2+n} a_{kj} y_k, \quad j = 1,\ldots,n,$$

$$\sum_{k=1}^{n^2+n} y_k = r,$$

$$y_k, z_j \in \{0,1\}.$$

The objective function gives the market share of the follower, to be minimized. The first constraint guarantees that client j will patronize a facility of the leader only if the follower has no facility at the distance less than $d(j,X)$. The second constraint allows the follower to open exactly r facilities.

In [2] it is claimed that the problem is NP-hard, and two heuristics are developed. We note that the integrality gap is small for this problem in the case of the two-dimensional Euclidian plane. The branch and bound method [5] easily finds the optimal solution. For this reason, the exact value $W^*(X)$ is used in our heuristic for the centroid problem.

4 Alternating heuristic

In this section we present an improved alternating heuristic. The idea of alternating methods is well-known [2, 12]. Given a solution X for the leader, the best-possible solution Y for the follower is computed. Once that is done, the leader may tentatively assume the role of the follower and reoptimize his set of facilities by solving the corresponding problem for the given solution Y. This process is then repeated until a termination condition is satisfied. In other words, the players alternately solve a follower problem. Convergence results of similar alternating algorithms for equilibrium problems can be found in [14].

In our case, a key issue is that an exact-polynomial time method by Drezner [7] for the $(1 \mid 1)$-centroid problem is applied. The method is described as follows.

Improved alternating heuristic

1. Create a starting solution X for the leader.

2. **While** not termination condition **do**:

    2.1. Find the best solution Y for the follower against the solution X.

    2.2. Find the best solution X for the leader against the solution Y.

  end while

3. Improve the final solution X by solving exactly the $(1 \mid 1)$–centroid problem.

The starting solution is generated at random. Calculations are terminated after a sufficiently large number of iterations. Note that the optimal solution of the follower problem shows us a subset of regions only. However we need the exact coordinates for the facilities. For the follower, all points inside each region are equivalent. But it is not the case for the leader at Step 2.2. of alternating process.

In order to minimize the market share of the follower, we should compute the coordinates of the leader facilities inside of the regions very carefully. To reduce the running time of the iterative process, we take the center points of the regions. At the final iteration the current solution X is modified as follows. All clients are clustered in p subsets according to X. Clients are allocated to the same subset if their closest leader facility is the same. For each subset, the $(1 \mid 1)$–centroid problem is solved, assuming that the follower will attack each subset by opening one facility. Optimal solutions for these subsets generate a new solution for the leader. As a result, we may get a new clustering of the clients and the procedure is repeated. The best found solution is the result of the method.

5 Computational experiments

We have coded the improved alternating algorithm in Delphi 7.0 environment and tested it on benchmark instances from the electronic library *Discrete Location Problems* [6]. For all instances we have $n = 100$, and demand points are randomly distributed among the square 7000×7000 uniformly. Two types of weights are considered: $w_j = 1$ and $w_j \in [1, 200]$. For all instances the behavior of the algorithm with $p = r$ is studied.

Two types of experiments were performed. In the first experiment we wanted to measure the influence of the starting solution of the leader at Step 1 of the algorithm. Different random solutions and *corner* solutions when all facilities are concentrated near a corner of the square were created. For all cases we observe that the algorithm produces the same final facilities locations for the leader for both types of weights. We guess that for higher dimensions, $n > 100$ and $p \neq r$, we may get another behavior of the algorithm, but now we are observing a fast convergence to the same equilibrium.

Table 1 Comparison of alternating heuristics

Instance number	Heuristic of Eiselt et al.	Improved heuristic	Procedure of clustering
111	1404 (31%)	1581 (35%)	1671 (37%)
211	1591 (28%)	1820 (32%)	1992 (35%)
311	1379 (29%)	1662 (35%)	1756 (37%)
411	1541 (29%)	1749 (33%)	1917 (36%)
511	1418 (31%)	1574 (35%)	1668 (37%)

In the second experiment our algorithm and the alternating heuristic from [2] are compared. Our goal is to understand the influence of the exact approach for the follower problem at Step 2 and idea of clustering at Step 3. Table 1 presents the computational results for the case $w_j \in [1, 200]$, $p = r = 10$. The second column of the Table 1 presents the market share of the leader according to [2]. Actually, we apply this algorithm to create a solution for the leader and then the exact value of the leader market share is computed. We show in brackets such values as percentages. The third column shows the same values for our algorithm without Step 3. The last column presents the leader market share for the final solution. As we can see, the exact method for the follower problem and the clustering procedure are important and they increase the leader market share. The same conclusions were obtained in the case $w_j = 1$. These values seem to be optimal, though we do not have a proof. Constructing sharp upper bounds for the global optimum is a very interesting and important direction for further research.

6 Conclusions

We have considered the well-known Stackelberg facility location game on the two-dimensional Euclidian plane. An improved alternating heuristic is presented. In this new heuristic, we have used the exact method for the follower problem, and a clustering procedure with an exact polynomial-time method for the $(1 \mid 1)$-centroid problem is used. Computational results for random generated instances show advantages of the proposed approach against benchmark procedures.

Acknowledgements

The first author is partially supported by grants MTM2009-14039 and FQM329, Spain. The second and third authors are partially supported by RFBR grants 09-01-00059 and 11-07-00474.

References

1. Bauer, A., Domshke, W., Pesch, E.: Competitive location on a network. European Journal of Operational Research. **66**, 372-391 (1993)
2. Bhadury, J., Eiselt, H.A., Jaramillo, J.H.: An alternating heuristic for medianoid and centroid problems in the plane. Computers and Operations Research. **30**, 553-565 (2003)
3. Carrizosa, E., Conde, E., Munoz-Marquez, M., Puerto, J.: Simpson points in planar problems with locational constraints. The round-norm case. Mathematics of Operations Research. **22**, 276–290 (1997)
4. Carrizosa, E., Conde, E., Munoz-Marquez, M., Puerto, J.: Simpson points in planar problems with locational constraints. The polyhedral–gauge case. Mathematics of Operations Research. **22**, 291–300 (1997)
5. CPLEX: http://www-142.ibm.com/software/products/ru/ru/ilogcple/
6. *Discrete Location Problems*. Benchmark library. http:// math.nsc.ru/AP/benchmarks/index.html
7. Drezner, Z.: Competitive location strategies for two facilities. Regional Science and Urban Economics. **12**, 485–493 (1982)
8. Hakimi S.L.: On locating new facilities in a competitive environment. ISOLDE Conference, Skodsborg, Denmark, (1981)
9. Hakimi S.L.: On locating new facilities in a competitive environment. European Journal of Operational Research. **12**, 29–35 (1983)
10. Hakimi, S.L.: Locations with spatial interactions: competitive locations and games, in: P. Mirchandani, R. Francis (Eds.) *Discrete Location Theory*, Wiley, 439–478 (1990)
11. Hansen, P., Thisse, J.F., Wendell, R.E.: Equilibrium analysis for voting and competitive location problems, in: P. Mirchandani, R. Francis (Eds.) *Discrete Location Theory*, Wiley, 479–502 (1990)
12. Hotelling, H.: Stability in competition. Economic J. **39**, 41–57 (1929)
13. Kress, D., Pesch, E.: Sequential competitive location on networks. European Journal of Operational Research. (in print)
14. Sherali, H.D., Soyster, A.L.: Convergence analysis and algorithmic implications of twodynamic processes toward an oligopoly — competitive fringe equilibrium set. Computers and Operations Research. **15**, 69-81 (1988)

Column Generation for Multi-Matrix Blocking Problems

Robert Voll and Uwe Clausen

Abstract The blocking problem in railroad freight traffic is a tactical routing problem. The problem was modeled as an arc-based MIP in former publications. We reformulate the problem and introduce a new model. The modified objective function can be handled easier without loss of structural information. Nevertheless, real instances cannot be solved by standard algorithms. Therefore, we present a column generation approach. The model is reformulated in a path-based way. We exploit the new model structure in the pricing step. Sets of new paths for each relation are generated in polynomial time.

1 Introduction

Railway freight traffic can be divided into three modes of transport: block train, single wagon load, and intermodal traffic. We will provide a mathematical approach for a special subproblem within the single wagon load traffic. We briefly describe this mode of transport:

A customer orders the transport of a single wagon or a small group of wagons from a particular origin to a destination. Hence, the number of wagons is not large enough to justify the assignment of a locomotive and its personal for the whole route. Therefore, it is necessary to consolidate wagons of different relations for parts of their ways. The bundling and separation of wagons is possible in so called shunting yards all over the network. We consider and optimize so called *blocking plans*. These plans contain information about routings and intermediate stops in shunting yards for every relation, i.e., pair of nodes in the network. The blocking plan shall be constant

Robert Voll
Institut fuer Transportlogistik, TU Dortmund, Leonhard-Euler-Str. 2, 44227 Dortmund
e-mail: voll@itl.tu-dortmund.de

Uwe Clausen
Institut fuer Transportlogistik, TU Dortmund, Leonhard-Euler-Str. 2, 44227 Dortmund and
Fraunhofer-Institut fuer Materialfluss und Logistik, Joseph-v.-Fraunhofer-Str. 2, 44227 Dortmund
e-mail: Uwe.Clausen@iml.fraunhofer.de

D. Klatte et al. (eds.), *Operations Research Proceedings 2011*, Operations Research Proceedings, 281
DOI 10.1007/978-3-642-29210-1_45, © Springer-Verlag Berlin Heidelberg 2012

for the whole planning horizon in spite of daily changing requests. Timetabling is much easier if railways are used periodically.

The following chapters are organised as follows: The scientific state of research about the optimization of blocking plans is summarized in section 2. In chapter 3 we modify an arc-based formulation for the blocking problem developed earlier. In section 4 a corresponding path-based formulation is described. It is used to derive a column generation approach with a polynomial-time pricing step. Section 5 summarizes the new results and gives an outlook on further research needs.

2 Literature Review

The blocking problem has been formulated in several papers in the last three decades ([1], [2], [3]). They all concentrate on North American railroad infrastructure and neglect special European business rules. The German railroad market is considered in [4] and [7]. Special hierarchy rules, which are used nowadays, are included in [4]. Important properties of a blocking model were identified and treated in [7] . The usage of multiple request matrices was presented for the first time.
The interesting aspect of robustness in blocking plan optimization was investigated in [5].
Column generation was introduced in several books and articles. In most cases we follow the notation of [6].

3 Optimization Model

The model presented in this section is an improved version of the optimization model presented in [7]. A new restriction on the number of reclassifications per relation simplifies the problem structure. The following notation is introduced:

Sets

$\mathcal{N}$	nodes
$\mathcal{M} = \{(i,j) \in \mathcal{N} \times \mathcal{N}\}$	arcs
$\mathcal{D}$	time periods
$\mathcal{K}$	possible relations
$\mathcal{K}(d) \subset \mathcal{K}$	active relations in time period d

Decision Variables

$$x_{ij}^k = \begin{cases} 1, & \text{if relation } k \text{ uses arc } (i,j) \\ 0, & \text{otherwise} \end{cases}$$

n_{ij} number of trains using arc (i,j)

Parameters

$(c_{ij}) \in \mathbb{R}^{|\mathcal{N}| \times |\mathcal{N}|}$	costs per train on arc (i,j)
$(l_k^d) \in \mathbb{R}^{|\mathcal{K}|}$	aggregated length of the wagons on relation k in time period d
$(w_k^d) \in \mathbb{R}^{|\mathcal{K}|}$	aggregated weight of the wagons on relation k in time period d
$(R(i)) \in \mathbb{N}^{|\mathcal{N}|}$	number of trains which can be classified at i in one time period
$(S(k)) \in \mathbb{N}_0^{|\mathcal{K}|}$	upper limit on reclassifications for relation k

$$B(i,k) \in \{-1,0,1\} = \begin{cases} 1, & \text{if } i \text{ is origin of relation } k \\ -1, & \text{if } i \text{ is destination of relation } k \\ 0, & \text{otherwise} \end{cases}$$

(OPT_{arc})

$$\min \quad \sum_{(i,j) \in \mathcal{M}} c_{ij} n_{ij}$$

$$\text{s.t.} \quad \sum_{j \in \mathcal{N}} x_{ij}^k - \sum_{j \in \mathcal{N}} x_{ji}^k = B(i,k) \qquad \forall i \in \mathcal{N}, \forall k \in \mathcal{K} \qquad (A1)$$

$$\sum_{k \in \mathcal{K}(d)} l_k^d x_{ij}^k \leq 700 n_{ij} \qquad \forall (i,j) \in \mathcal{M}, \forall d \in \mathcal{D} \qquad (A2)$$

$$\sum_{k \in \mathcal{K}(d)} w_k^d x_{ij}^k \leq 1600 n_{ij} \qquad \forall (i,j) \in \mathcal{M}, \forall d \in \mathcal{D} \qquad (A3)$$

$$\sum_{j \in \mathcal{N}} n_{ij} \leq R(i) \qquad \forall i \in \mathcal{N} \qquad (A4)$$

$$\sum_{(i,j) \in \mathcal{M}} x_{ij}^k \leq S(k) \qquad \forall k \in \mathcal{K} \qquad (A5)$$

$$x_{ij}^k \in \{0,1\} \qquad \forall (i,j) \in \mathcal{M}, \forall k \in \mathcal{K} \qquad (A6)$$

$$n_{ij} \in \mathbb{N} \qquad \forall (i,j) \in \mathcal{M} \qquad (A7)$$

There are two sets of decision variables. The x_{ij}^k-variables decide about the routings. The n_{ij}-variables determine the number of trains on arc (i,j). The objective function measures up train costs. This does not destroy the consolidation effect described in [7].

Equations (A1) provide flow conservation. Constraints (A2) and (A3) ensure train length and weight conditions. The number of trains classified at each yard is restricted by (A4). Constraints (A5) manage the number of reclassifications per wagon. The integrality constraints are necessary, because sendings with same origins and destinations shall not be divided into several ones.

4 Column Generation

We develop a column generation algorithm for the presented model. At first, we divide the constraints into two groups. The first one contains information about the routing of a single relation. Constraints (A1) and (A5) correspond to a particular relation. If other relations are omitted, these two *local contraints* force the corresponding relation on feasible paths. All other constraints have *global character*. They link more relations. We will ignore them for now and concentrate on (A1) and (A5).

Definition 1. A vector $x_{ij}^{k'} \in \{0,1\}^{|\mathcal{N}|^2}$ is called a *path* for the relation k' if $x_{ij}^{k'}$ achieves constraints (A1) and (A5). $\mathscr{P}_k$ is the set of all paths for k.

The set of feasible x_{ij}^k in (OPT_{arc}) is a subset of the convex hull of all path-arcs $x_{pij}^{k'}$ for k. Hence, every x_{ij}^k is a convex combination of the flow variables $x_{pij}^{k'}$:

$$x_{ij}^k = \sum_{p \in P_k} x_{pij}^k \lambda_p^k \qquad \forall (i,j) \in \mathcal{M}, \forall k \in \mathcal{K}$$

$$\sum_{p \in P_k} \lambda_p^k = 1 \qquad \forall k \in \mathcal{K}$$

$$\lambda_p^k \geq 0 \qquad \forall p \in \mathscr{P}_k, \forall k \in \mathcal{K}$$

We use this knowledge to reformulate (OPT_{arc}) in a path-based way. This reformulation is called *master problem* in the context of column generation.

(OPT_{path})

$$\min \quad \sum_{(i,j) \in \mathcal{M}} c_{ij} n_{ij}$$

$$s.t. \quad \sum_{k \in \mathcal{K}(d)} \sum_{p \in \mathscr{P}(k)} l_k^d x_{pij}^k \lambda_p^k \leq 700 n_{ij} \quad \forall (i,j) \in \mathcal{M}, d \in \mathcal{D} \quad (B1)$$

$$\sum_{k \in \mathcal{K}(d)} \sum_{p \in \mathscr{P}(k)} w_k^d x_{pij}^k \lambda_p^k \leq 1600 n_{ij} \quad \forall (i,j) \in \mathcal{M}, d \in \mathcal{D} \quad (B2)$$

$$\sum_{j \in \mathcal{N}} n_{ij} \leq R(i) \qquad \forall i \in \mathcal{N} \qquad (B3)$$

$$\sum_{p \in \mathscr{P}_k} \lambda_p^k = 1 \qquad \forall k \in \mathcal{K} \qquad (B4)$$

$$\lambda_p^k \geq 0 \qquad \forall k \in \mathcal{K}, \forall p \in \mathscr{P}_k \quad (B5)$$

$$\lambda_p^k \in \{0,1\} \qquad \forall p \in \mathscr{P}_k, \forall k \in \mathcal{K} \quad (B6)$$

Constraints (B1) and (B2) reflect the length and weight restrictions, respectively. In (B3) the number of classified trains at each yard is considered. Constraints (B4) and (B5) are convexity constraints for the paths. The last constraint enforces the choice of exactly one path for each relation. The constraints concerning flow conservation and the bound on the number of reclassifications are not necessary anymore. Their structural information is coded in the path definition.

The column generation process is given as follows. The first step is the solution of the LP relaxation of the master problem OPT_{path} with an initial set of paths. This program is called the *restricted master problem* (RMP). Let μ, η, τ denote the dual variables to the equations (B1), (B2) and (B4), respectively. In order to find variables λ_p^k, which can improve the current solution, we calculate the *reduced costs* c_p^k of the variable λ_p^k:

$$c_p^k = \sum_{ij} [\Omega_{ij}^k x_{pij}^k] - \tau(k)$$

$$\text{with} \qquad \Omega_{ij}^k = - \sum_{d \in \mathscr{D}} [l_k^d \mu_{ij}^d + w_k^d \eta_{ij}^d] \geq 0 \qquad \forall (i,j) \in \mathscr{M}, \forall k \in \mathscr{K}$$

If $\tau(k) \leq 0$ no improving paths can be found for k. Otherwise the search for a the path with minimal negative reduced costs can be reformulated as a shortest path problem on the graph with the arc weights Ω_{ij}^k. This process is called the *pricing subproblem*. Conditions (A1) and (A5) from the path definition must be included:

$$c_p^* = \min_{(A1),(A5)} \sum_{(i,j) \in \mathscr{M}} [\Omega_{ij}^k x_{ij}^k] - \tau(k)$$

Most shortest path algorithms like Dijkstra cannot handle the limitation (A5) on the number of allowed arcs. The Bellman-Ford algorithm finds the shortest path using at most L arcs in cubic time. Moreover, the use of this algorithm has a very convenient side effect. Bellman-Ford even calculates shortest paths with at most $1, ..., L$ arcs, respectively. Therefore, it is possible to create a set of new feasible paths with varying lengths for each relation in polynomial time.
The new paths are added to the path sets and the master problem (MP) is reoptimized. This process is repeated until no new paths can be added. The relaxed MP is solved to proven optimality.

First numerical tests are quite promising. We tested a scenario provided by our research partners at Deutsche Bahn. It contains 21 nodes and about 250 relations on one day. The column generation approach was implemented in MATLAB 7 using GAMS and the CPLEX LP-solver for the RMPs. Our algorithm solved the path-based root LP within 113 seconds, while CPLEX needed 51 seconds for the solution of the relaxed arc-based model. A more sophisticated implementation will accelerate our algoirthm.

Although our algorithm cannot beat CPLEX so far, the approach might become successful. The column generation algorithm provides more information than the solution of the relaxed arc-based model. In addition to the optimal solution a set of feasible paths is generated by the shortest path algorithm during the pricing steps. Those paths can be used in the master problem to compute integer solutions. The structure of the paths carries polyhedral information about the set of feasible arc-flows, which can provide tighter lower bounds in the nodes of the branching tree.

5 Summary and Future Work

In this paper we presented a new formulation for the blocking problem. It simplifies models developed earlier without loss of structural details. We reformulated the problem again as a path-based model. This model can be handled by column generation. The pricing step has a simple structure and creates a set of new paths for each relation in polynomial time.
In first numerical experiments the column generation algorithm was not faster than modern commercial software. Hence, it will be necessary to improve our solution techniques. We achieve the extension of the here shown results to a Branch-and-Price algorithm in the future. The research focus lies on effective branching rules, which do not destroy the pricing structure.

Acknowledgements The presented results arised from the project "A User-Guided Planning-Tool For Car Grouping in Railroad Freight Traffic" within the "Effienzcluster Ruhr" funded by the German Federal Ministry for Science and Education (BMBF) in cooperation with the DB Mobility Logistics AG, Grant-ID: 01IC10L26A.

References

1. Ahuja, R. K., Jha, K. C. and Liu, J.:Solving Real-Life Railroad Blocking Problems.", Interfaces, Vol. **37**, Nr. 5, pp. 404-419 (2007)
2. Barnhart, C., Jin, H., and Vance, P. H. :Railroad Blocking: A Network Design Application. Operations Research, Vol. **48**, Nr. 4, pp. 603-614.(2000)
3. Bodin, L. D., Golden, B. L., Schuster, A. D. and Romig, W.: A model for the blocking of trains. Transportation Research Part B: Methodological. Elsevier, **14(1-2)**, 115-120 (1980).
4. Fuegenschuh, A., Homfeld H. and Schuelldorf, H.: Single Car Routing in Rail Freight Transport. Dagstuhl Seminar Proceedings 09261, Leibniz-Zentrum Informatik, Germany (2009)
5. Jin, H.: Designing robust railroad blocking plans. Massachusetts Institute of Technology, Massachusetts Institute of Technology, Dept. of Civil and Environmental Engineering (1998)
6. Luebbecke, M., Desrosiers, J.: A Primer in Column Generation. Column Generation, GERAD, New York (2005)
7. Voll, R., Clausen, U.: A Blocking Model with Bundling Effects Respecting Multiple OD-Matrices. Proceedings of the 4th International Conference on Experiments/Process/System Modeling/Simulation/Optimization, Athens (2011)

Reduction of Empty Container Repositioning Costs by Container Sharing

Herbert Kopfer and Sebastian Sterzik

1 Problem Description

Empty container repositioning is a major cost driver for international container operators. While the share of empty containers of total sea-based container flows is around 20%, consultants estimate the rate of empty container transport for the hinterland even twice as high [1]. As [4] stated, most of the routes in the hinterland consist of pendulum tours comprising one full and one empty container move between the port, a trucking company's depot and the customers (receiver/shipper). Solutions to shrink the amount of empty container movements focus on moving empty containers which are available at some customer locations directly to places where they will be needed next. Based on this approach, this contribution investigates the rising possibilities to reposition empty containers in the seaport hinterland if trucking companies cooperate with each other. In the following we present two scenarios and solve several data sets to quantify the emerging cost savings for companies who exchange their empty containers. Both scenarios lead to an integrated model considering empty container repositioning and vehicle routing simultaneously.

Imagine a hinterland transportation scenario (see Figure 1) in which full and empty containers have to be moved between different locations by at least two trucking companies. Each trucking company serves its own customer base with a homogeneous fleet of vehicles. We assume 40-feet-containers, i.e. a vehicle can only move one container at a time. The fleet is parked at a depot in which, moreover, empty containers can be stacked. Each vehicle starts and ends its tour at the depot of its trucking company. Additionally, it is assumed that there is a seaport terminal to which the trucks carry full and empty containers from customers' places and vice versa. Time windows at the terminal as well as time windows at the customer nodes need to be taken into account. We distinguish two types of customers. On the one hand shippers offer freight which is to be transported to a foreign region via the seaport. The flow of a full container from the shipper to the terminal is defined as outbound full (OF) container. On the other hand receivers require the transport of

Chair of Logistics, Department of Economics and Business Studies, University of Bremen, Wilhelm-Herbst-Str. 5, 28359 Bremen, Germany, e-mail: {kopfer, sterzik}@uni-bremen.de

D. Klatte et al. (eds.), *Operations Research Proceedings 2011*, Operations Research Proceedings, 287
DOI 10.1007/978-3-642-29210-1_46, © Springer-Verlag Berlin Heidelberg 2012

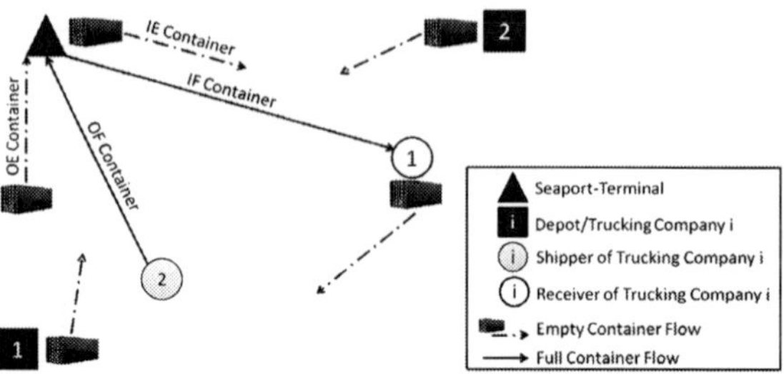

Fig. 1 Basic Scenario

their goods from an outside region via the terminal. The flow of a full container from the terminal to the receiver is called inbound full (IF) container. It is obvious that a shipper requires an empty container before he can fill the freight in it. Additionally, an empty container remains at the receiver's location after the IF container is emptied. We define two time windows at each customer location: During the first time window the full/empty container has to be delivered to the customer location. After the container is emptied/filled, it can be picked up by a vehicle during the second time window. Moreover, trucking companies have to consider two additional transportation requests. Due to the imbalance between import- and export-dominated areas, they need to take care of outbound empty (OE) or inbound empty (IE) containers which either have to be moved to the terminal or derive from it. For these requests only the terminal as the destination or as the origin is given in advance. Hence, the locations that can provide empty containers for the OE requests and the destinations for the imported empty containers need to be determined during the solution process. This scenario was defined by Zhang et al. (2009) as the multi-depot container truck transportation problem (CTTP) [5]. The objective is to minimize the carriers' total fulfillment costs consisting of fixed and variable costs. Hence, in a first step the number of used vehicles should be minimized while in the second step the optimization of the vehicles' total operating time symbolizing the transportation costs should be pursued [3]. In this paper the first objective is formulated as a constraint and the minimization of the operating time is chosen as the objective function of our models. Thereby, the number of used vehicles within the employed models is raised iteratively until a feasible solution is found. The resulting solution approach leads to an integrated model which does not only consider vehicle routing but also empty container repositioning. Thus, solving the problem determines: a) where to deliver the empty containers released after inbound full/empty loads, b) where to pick up the empty containers for outbound full/empty loads, and c) in which order and by which truck the loads should be carried out.

In this paper, two scenarios for the hinterland transportation are analyzed. In the first scenario empty containers are uniquely assigned to trucking companies, i.e. they can solely be used by the company they are assigned to. For instance, an empty container obtained at a receiver location served by a certain trucking company can exclusively be used for transportation requests of this company. This scenario is given by the above CTTP. In the second scenario companies can use empty containers of cooper-

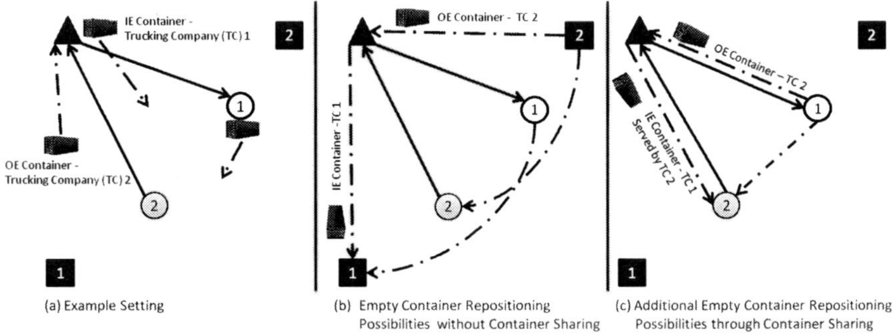

Fig. 2 Possible Benefits through Container Sharing

ating trucking companies. I.e. companies share their information at which locations empty containers are currently stacked and they agree with the mutual exchange of these containers. Thus, the cooperating companies can improve their routes and increase their profit by decreasing transportation costs in return. The permission of sharing empty containers among trucking companies leads to the multi-depot container truck transportation problem with container sharing (CTTP-CS).

Figure 2 demonstrates the rising possibilities to reposition empty containers of two trucking companies cooperating with each other. Trucking company 1 has to serve a receiver and needs to move an IE container to the hinterland. Trucking company 2 has to serve a shipper and needs to move an OE container to the terminal (see Figure 2 (a)). In the non-cooperative case, as illustrated in Figure 2 (b), the only opportunity for both companies to reposition empty containers is moving them either to (trucking company 1) or from (trucking company 2) their own depot. If the exchange of containers is permitted, both companies could benefit through the emerging additional flexibility to allocate empty containers to a vehicle's tour (see Figure 2 (c)). The amount of the benefit of a cooperation highly depends on time and place conditions given by the time windows for pickup and delivery and by the locations of the terminal and customers.

2 Simultaneous Solution Approach

Let V denote the nodes of a directed graph, consisting of customer node set V_C, terminal node set V_T and depot node set V_D. There are two types of customers ($V_C = V_S \cup V_R$), the node sets $V_S = V_{Si} \cup V_{So}$ and $V_R = V_{Ri} \cup V_{Ro}$ describe the shipper and the receiver node sets. V_{Si} and V_{Ri} refer to the first time window of the shipper/receiver, in which an empty/full container has to be made available. After the container has been completely filled or emptied, respectively, container $c \in C$ can be picked up by a vehicle $k \in K$ during the second time window (V_{So} and V_{Ro}). The terminal node set V_T consists of V_{Ti} and V_{To} which correspond to the amount of ingoing and outgoing containers. The number of all customer and terminal nodes is defined by n. In the trucking companies' depots a large number a of empty containers can be stacked. Hereby, the depot set V_D is subdivided into the start and end depot node sets V_{D^s} and V_{D^e}. Node $i \in V_C \cup V_T$ has to be reached during its time window, determined by the interval $[b_i / e_i]$. We consider d trucking companies. The trucking company of vehicle k and customer i, respectively, is determined by d_k and d_i. For each two distinct stop locations, t_{ij} represents the travel time from location i to location j. At node $i \in V_C \cup V_T$ a service time s_i for the loading/unloading operation of a container is considered. While the binary decision variables y_{ijc} and x_{ijk} define whether container c/vehicle k traverses the arc from location i to j, L_{ic} and T_{ik} specify the arrival time of a container/vehicle at a location.

$$\min z = \sum_{k \in K} \left(T_{(n+d+d_k)k} - T_{(n+d_k)k} \right) \tag{1}$$

$$\sum_{j \in V} \sum_{c \in C} y_{ijc} = 1 \ \ \forall i \in V_C \cup V_{T^i} \tag{2}$$

$$\sum_{i \in V} \sum_{c \in C} y_{ijc} = 1 \ \ \forall j \in V_{T^o} \tag{3}$$

$$\sum_{j \in V_{R^i} \cup V_{S^i} \cup V_{D^e}} \sum_{c \in C} y_{ijc} = 1 \ \ \forall i \in V_{T^i} \tag{4}$$

$$\sum_{i \in V_{D^s}} \sum_{j \in V} \sum_{c \in C} y_{ijc} = a \tag{5}$$

$$\sum_{i \in V} \sum_{c \in C} y_{ijc} = 1 \ \ \forall j \in V_{T^o} \cup V_{D^e} \tag{6}$$

$$\sum_{j \in V} y_{ijc} - \sum_{j \in V} y_{ijc} = 0 \ \ \forall i \in V_C, c \in C \tag{7}$$

$$L_{jc} \geq L_{ic} + t_{ij} + s_i - M(1 - y_{ijc}) \ \ \forall i, j \in V, c \in C \tag{8}$$

$$\sum_{j \in V} \sum_{k \in K} x_{ijk} = 1 \ \ \forall i \in V_C \cup V_T \tag{9}$$

$$\sum_{j \in V} x_{(n+d_k)jk} = 1 \ \ \forall k \in K \tag{10}$$

$$\sum_{i \in V} x_{i(n+d+d_k)k} = 1 \ \ \forall k \in K \tag{11}$$

$$\sum_{j \in V} x_{jik} - \sum_{j \in V} x_{ijk} = 0 \ \ \forall i \in V_C \cup V_T, k \in K \tag{12}$$

$$T_{jk} \geq T_{ik} + t_{ij} + s_i - M(1 - x_{ijk}) \ \ \forall i, j \in V, k \in K \tag{13}$$

$$b_i \leq T_{ik} \leq e_i \ \ \forall i \in V_C \cup V_T, k \in K \tag{14}$$

$$x_{ijk} d_k = x_{ijk} d_i \ \ \forall i \in V_C \cup V_T, j \in V, k \in K \tag{15}$$

$$\sum_{k \in K} x_{ijk} \geq y_{ijc} \ \ \forall i \in V_{S^o} \cup V_{R^o} \cup V_T, j \in V, c \in C \tag{16}$$

$$\sum_{k \in K} x_{ijk} \geq y_{ijc} \quad \forall i \in V_C \cup V_T, j \in V_{S^i} \cup V_{R^i} \cup V_T \cup V_{D^e}, c \in C \tag{17}$$

$$T_{ik} = L_{ic} \quad \forall i \in V_C \cup V_T, k \in K, c \in C \tag{18}$$

$$x_{ijk}, y_{ijc} \in 0, 1 \quad \forall i, j \in V, k \in K, c \in C \tag{19}$$

$$T_{ik}, L_{ic} : real\ variables \quad \forall i \in V, k \in K, c \in C \tag{20}$$

By considering empty container repositioning on the one hand (2-8) and vehicle routing and scheduling (9-15) on the other hand, the presented model pursues the minimization of the vehicles' total travel time. Since a container cannot drive on its own, it has to be assured that every container is transported by a vehicle. Equations (16)-(18) require that the vehicles are interlinked with the containers and that both pass every location at the same time. Thereby, the vehicles cover the containers' routes but can skip the filling and emptying process of the container at a customer location. Restrictions (2)-(3) and (9) assure that every customer node is visited once and that a container enters/leaves the terminal nodes according to the number of outbound and inbound containers. While a vehicle starts and ends its route at the corresponding depot of the trucking company (10-11), the start and end node of a container can either be the terminal or the depot (4-6). Furthermore, the route continuity (7 and 12) and the time restrictions (8 and 13-14) have to be held. Restriction (15) defines the CTTP in which container sharing is not permitted. If the exchange of empty containers is allowed (CTTP-CS), the equation changes to:

$$x_{ijk}d_k = x_{ijk}d_i \quad \forall i \in V_S \cup V_{R^i} \cup V_{T^o}, j \in V, k \in K \tag{21}$$

3 Computational Results

We generated ten test instances based on Solomon's benchmark vehicle routing problem with time windows data sets [2]. Each data set consists of two trucking companies, five customer clusters (comprising one shipper and one receiver) and, additionally, one IE container and one OE container, whereas each company has to serve six requests. To fit the factor that a cooperation is more profitable if the customer and terminal nodes are located close to each other, we randomly chose a coordinate for each customer type from one cluster of the c1-Solomon data sets. According to Figure 2 the receiver and the shipper of a cluster belong to different trucking companies. Moreover, we adapted the time windows of these customers so that the receiver's second time window is consistent with the shipper's first time window. The obtained results in Table 1 indicate that the rising possibilities to reposition empty containers within a cooperation can lead to a huge reduction of fulfillment costs. Regarding the fixed costs, remarkable 37% less vehicles are used in the CTTP-CS compared to the CTTP on average. Besides, a reduction of 38% according to the number of used containers can be determined. The variable costs illustrated by the total travel time are decreased by 23%.

Table 1 Benefit of Container Sharing

Data Set	CTTP (Objective value/ Number of used vehicles/ Number of used containers)	CTTP - CS (Objective value/ Number of used vehicles/ Number of used containers)	Benefit of container sharing according to the objective values
1	1581.85/ 12/ 12	1159.77/ 8/ 8	26.88%
2	1671.90/ 12/ 12	1275.37/ 6/ 7	23.72%
3	1654.77/ 12/ 12	1373.42/ 8/ 8	17.00%
4	1620.80/ 12/ 12	1248.06/ 8/ 7	23.00%
5	1851.47/ 12/ 12	1287.73/ 8/ 7	30.45%
6	1714.19/ 12/ 12	1398.87/ 10/ 8	18.39%
7	1556.53/ 12/ 12	1247.86/ 8/ 8	19.83%
8	1583.17/ 12/ 12	1225.57/ 6/ 8	22.59%
9	1484.20/ 12/ 12	1181.74/ 8/ 7	20.38%
10	1827.75/ 12/ 12	1296.84/ 6/ 7	29.05%

4 Conclusions

In this contribution, we illustrated the potential of container sharing in the hinterland of seaports. By implementing a holistic solution approach which considers empty container repositioning and vehicle routing and scheduling simultaneously, remarkable savings regarding fixed and variable costs could be obtained. However, the data sets were modified to highlight the advantages of the CTTP-CS compared to the CTTP. In a further step, we are going to develop heuristic approaches in order to analyze the benefit of a cooperation in bigger test instances and to investigate the precondidtions for successful container sharing scenarios.

Acknowledgements This research was supported by the German Research Foundation (DFG) as part of the Collaborative Research Centre 637 "Autonomous Cooperating Logistic Processes - A Paradigm Shift and its Limitations" (Subproject B7).

References

1. Konings, R.: Foldable containers to reduce the costs of empty transport? A cost-benefit analysis from a chain and multi-actor perspective. Maritime Economics & Logistics **7**, 223–249 (2005)
2. Solomon, M.M.: Algorithms for the vehicle routing and scheduling problems with time window constraints. Operations Research **35**(2), 254–265 (1987)
3. Vahrenkamp, R.: Logistik: Management und Strategien, 6th edn. Oldenbourg, Munich (2007)
4. Veenstra, A.W.: Empty container repositioning: the port of Rotterdam case, chap. 6, pp. 65–76. Springer, Heidelberg (2005)
5. Zhang, R., Yun, W., Moon, I.: A reactive tabu search algorithm for the multi-depot container truck transportation problem. Transportation Research Part E: Logistics and Transportation Review **45**(6), 904–914 (2009)

Optimizing strategic planning in median systems subject to uncertain disruption and gradual recovery

Chaya Losada and Atsuo Suzuki

Abstract This paper addresses the capacity planning of facilities in median systems at an early stage so that, if potential disruptions arise and forecast demand is misestimated, overall costs are minimized. We consider some realistic features such as uncertainty in the magnitude, time and location of disruptions as well as gradual recovery of disrupted facilities over time. The proposed two-stage stochastic linear program (2-SLP) is solved in a computationally efficient way via an enhanced implementation of the stochastic decomposition method. To ascertain the quality of the solutions obtained some deterministic bounds are calculated.

1 Introduction

Reliability models for facility location problems often focus on warranting a highly risk averse decision making criterion based on protecting some of the existing facilities in order to minimize the maximum overall potential cost/damage [5, 7]. However, precautionary strategies designed to minimize the impact of worst-case disruptive events may be fruitless if they focus on preventing the occurrence of disruption scenarios that are as highly disruptive as they are improbable.

Here, we assume that some probabilities on disruption occurrence are available as well as probability information on possible misestimations of the forecast demands. This allows us to consider a rather risk neutral decision criterion based on minimizing both precautionary investments and expected costs. We formulate the problem as a two-stage stochastic linear program (2-SLP) where in the first stage

Chaya Losada
Nanzan University, Department of Mathematical Sciences and Information Engineering, Nanzan University, Seto, Aichi, 489-0863, Japan, e-mail: chayalosada@gmail.com

Atsuo Suzuki
Nanzan University, Department of Mathematical Sciences and Information Engineering, Nanzan University, Seto, Aichi, 489-0863, Japan, e-mail: atsuo@nanzan-u.ac.jp

D. Klatte et al. (eds.), *Operations Research Proceedings 2011*, Operations Research Proceedings, 293
DOI 10.1007/978-3-642-29210-1_47, © Springer-Verlag Berlin Heidelberg 2012

the amount of capacity redundancy for each facility is decided and implemented. The recourse subproblem (second stage) minimizes the overall travel distance in serving customers over a time horizon given a particular realization of the uncertain parameters and the precautionary measures implemented in the first stage.

We assume that the consequences of disruption on a given facility do not propagate to unaffected facilities and that disruptive events are independent. The magnitude of the disruption is measured as a reduction rate in the nominal capacity of the affected facility. Also we assume that the additional capacity due to first stage decisions is not subject to disruption (although, this assumption could be easily lifted without affecting the solution approach). We consider that from the first time period that follows the disruption, a disrupted facility recovers towards its nominal capacity level gradually over time. The speed at which a disrupted facility recovers depends on the magnitude of the disruption and on whether or not additional disruptive events strike that facility within the time horizon. Thus, the total recovery time of a disrupted facility is uncertain. We also view system disruption as an unexpected increase in the request for service to some facilities if this increased demand cannot be satisfied with the nominal capacity of the involved facilities.

The number of random variables for small to medium size instances of the problem renders computationally impractical well-known exact based solution approaches for 2-SLP, such as the L-Shaped method [11] or column generation. To solve the problem, we apply Stochastic Decomposition (SD) [3], a statistically based decomposition method. We provide computational examples of the problem for some variations of the classical method and give some estimates on the quality of the solutions obtained.

The structure of this document is as follows: in Section 2 we present the formulation of the model. Section 3 summarizes the solution approach. In Section 4 we report some preliminary results and evaluate the quality of the solutions via upper and lower bounds to the optimal objective function value. Finally, in Section 5 we point out directions for future research.

2 A Two-Stage Stochastic Model: Uncertain Disruption and Demand

Let N be the set of customers, F the set of facilities, T the number of time periods in the planning horizon, h_{it}^o the forecast demand of customer $i \in N$ in time unit t, d_{ij} the distance between customer $i \in N$ and facility $j \in F$, c_{jt}^o the nominal capacity of facility $j \in F$ in time unit t, p^o the penalty paid per unit of unmet demand and r the rate of facility recovery. Also we let $i_k \in F$ denote the k-th closest facility from customer i, where in case of equidistant facilities, and without loss of generality, ties can be broken arbitrarily. In addition, let $\xi = (\widetilde{\mathbf{w}}, \widetilde{\mathbf{s}}, \widetilde{\mathbf{e}})$ be a vector of random variables defined on a probability space $(\Omega, \mathscr{F}, \mathscr{P})$, where $\widetilde{\mathbf{w}}$ and $\widetilde{\mathbf{s}}$ are independent random vectors. Then, $\widetilde{h}_{it} = h_{it}^o(1 + \widetilde{w}_{it})$ defines the random demand of customer $i \in N$ in time t. Similarly, $\widetilde{c}_{jt} = c_{jt}^o(1 - \widetilde{e}_{jt})$ stands for the random capacity of facility j in time t, where

$\widetilde{e}_{j1} = \widetilde{s}_{j1}, \forall j \in F$ and if $\widetilde{e}_{jt-1} \neq 0$ then $\widetilde{e}_{jt} = \min(1, \max(0, \widetilde{e}_{jt-1} - (r/\widetilde{e}_{jt-1})) + \widetilde{s}_{jt})$ and else $\widetilde{e}_{jt} = \widetilde{s}_{jt}, \forall j \in F, 2 \leq t \leq T$. That is, $\widetilde{e}_{jt}$ is a dependent random variable representing the rate of capacity loss for facility j in time period t due to the cumulative impact of multiple disruptive events occurring during and before time period t and due to a recovery process where $(r/\widetilde{e}_{jt-1})$ is the recovery rate from time period $t-1$ to time period t and $r \in [0,1]$ is a fixed parameter. For computational convenience, the median system is formulated via a network structure [6] (instead of the conventional median formulation [2]) where related decision variables are: $x_{ii_kt} \in \mathbb{R}_+, \forall i \in N, i_k \in F, 1 \leq t \leq T$, $\delta_{ii_kt} \in \mathbb{R}_+, \forall i \in N, i_k \in F, 1 \leq t \leq T$, and $\hat{x}_{it} \in \mathbb{R}_+, \forall i \in N, 1 \leq t \leq T$. Decision variables x_{ii_kt} state the amount of demand served by the k-th closest facility from customer i in time unit t, δ_{ii_kt} indicate the amount of demand of customer i that cannot be served by its k-th closest facility and $\hat{x}_{it}$ is the amount of demand of customer i that cannot be met with the available facilities' capacity levels.

On the other hand, we let decision variables g_j, ($\mathbf{g}$ in vector form), be the amount of additional capacity assigned to facility j before the planning horizon begins and q be a decision variable for the total expenditure in strategic investments. Then we define $\Gamma \equiv \{(\mathbf{g},q) | \sum_{j \in F} f_j g_j \leq q, g_j \in \mathbb{R}_+ \forall j \in F, q \in \mathbb{R}_+\}$, where f_j is the cost per unit of redundant capacity implemented in facility j. The 2-SLP is described by the following formulation:

$$\text{Min}_{(\mathbf{g},q) \in \Gamma} \quad q + E_{\widetilde{\xi}}[h(\mathbf{g},q,\widetilde{\xi})] \tag{1}$$

where

$$h(\mathbf{g},q,\widetilde{\xi}) = \text{Min}_{\mathbf{x},\delta,\hat{\mathbf{x}}} \quad \sum_{i \in N} \sum_{k=1}^{|F|} \sum_{t=1}^{T} d_{ij} x_{ii_kt} + \sum_{i \in N} \sum_{t=1}^{T} p^0 \hat{x}_{it} \tag{2}$$

$$x_{ii_1t} + \delta_{ii_1t} = \widetilde{h}_{it}, \forall i \in N, 1 \leq t \leq T \tag{3}$$

$$\delta_{ii_{k-1}t} = x_{ii_kt} + \delta_{ii_kt}, \forall i \in N, 1 < k \leq |F|, 1 \leq t \leq T \tag{4}$$

$$\delta_{ii_{|F|}t} = \hat{x}_{it}, \forall i \in N, 1 \leq t \leq T \tag{5}$$

$$\sum_{i \in N} x_{ijt} \leq \widetilde{c}_{jt} + g_j, \forall j \in F, 1 \leq t \leq T \tag{6}$$

$$(x_{ii_kt}, \delta_{ii_kt}, \hat{x}_{it}) \in \mathbb{R}_+, \forall i \in N, 1 \leq k \leq |F|, 1 \leq t \leq T \tag{7}$$

It is important to note that $h(\mathbf{g},q,\widetilde{\xi})$ is finite for any $(\mathbf{g},q) \in \mathbb{R}_+^{|F|+1}$ and hence the problem has complete recourse. The objective function (6) minimizes the total investments and the expected operational costs defined by (2). Constraints (3), (4) and (5) are demand based flow balance constraints for the closest facility, all but the closest facility and the dummy facility used to calculate the unmet demand, respectively. Constraints (6) specify the capacity limitations of a given facility based on the occurrence of a disruption during time t, the remanent capacity if disruptions occurred in previous time periods and the redundant capacity $\mathbf{g}$ due to strategic investments.

3 A Solution Approach: Stochastic Decomposition

Stochastic decomposition (SD) combines the L-Shaped method [11] and Monte Carlo sampling to obtain a piece-wise linear approximation of the supports of the objective function by producing a cut per iteration [3]. For a minimization problem, SD recursively approximates the expected recourse function from below providing a statistical lower bound to the sample mean value. SD may be specially suitable when the probability law of the random vector cannot be defined precisely but one can obtain observations or realizations of the random vector [4]. Although SD comprises two type of errors, (one is made in the approximation of the sample average function while the second is statistical in nature and based on the sample used), the algorithm is guaranteed to converge asymptotically to an optimal value for a subsequence of iterates. For the most part, computational enhancements of the basic SD method rely on counterpart enhancements for the L-Shaped method (e.g., these enhancements consider the inclusion of a regularized term in the objective function of the master problem [3, 9] or a multi-cut version of the L-Shaped method [1, 4]). These techniques are usually targeted at improving the approximation of the support function which results in better solutions that lead to a reduction in the total number of iterations at the expense of incrementing the computational effort related to the master problem.

We apply a version of the SD algorithm to the proposed 2-SLP, where some of the aforementioned enhancements and some different ones are considered, such as a regularized term and valid inequalities.

4 Computational Results

We tested the computational performance of the proposed approach on an Intel Core2 Duo T6400 2.0 GHz processor with 4 GB of RAM. The algorithm was implemented in C++ using CPLEX 12.1 (IBM) callable libraries. The computational tests were conducted on data obtained from the Swain dataset [10] taking the first 20 demand points only. In our preliminary results, we solved each problem for different number of facilities ($|F| = 2, 4, 6$) and a time horizon of 7 time units ($T = 7$). The forecast demands for the first time unit ($h_{i1}^o, \forall i \in N$) were those given in the Swain dataset. For all other time units, we assumed that demand increased in 1% with respect to the demand of the previous time period. On the other hand, the nominal capacity of facilities at every time unit was established to be the needed quantity to have all forecast demands served by the closest facility from each demand point. This affords to view increments in capacity of the facilities in the first stage of the 2-SLP as only due to uncertainty associated to demand and disruption levels. The penalty paid for every unit of unmet demand was set to the maximum traveling distance for a customer to reach its farthest facility times a factor of 11. The cost of increasing one unit of the nominal capacity was set to half the price of the penalty and r was set to 0.1. Finally, we assumed a discrete support of the random variables such that $w_{it} \in \{0.03, 0.18.0.29\}, \forall i \in N, 1 \leq t \leq T$ and $s_{jt} \in \{0, 0.2, 0.5\}, \forall j \in F, 1 \leq t \leq T$. Then, the probability mass function for both types of random variables was $\{0.7, 0.2, 0.1\}$.

Table 1 shows, for the three instances tested with the number of facilities ranging in $\{2, 4, 6\}$, the number of independent random variables, dependent random variables and possible scenarios. The number of scenarios if there were only independent random vectors $\widetilde{s}$ and $\widetilde{w}$ is indicated in parenthesis in the fourth column.

Table 1 Comparison between different problem instances of the number of random variables and possible scenarios

$\|F\|$	No. Indep. Rnd. Vars	No. Dep. Rnd. Vars	No. Scenarios
2	154	14	$3^2 5^2 8^2 13^2 23^2 46^2 90^2 3^{140} > (3^{154})$
4	168	28	$3^4 5^4 8^4 13^4 23^4 46^4 90^4 3^{140} > (3^{168})$
6	182	42	$3^6 5^6 8^6 13^6 23^6 46^6 90^6 3^{140} > (3^{182})$

Swain dataset, $\|N\| = 20$, and $T = 7$

It is important to replicate the results obtained from any solution approach that is based on sampled data. Here, we run 30 replications of any instance and report the average and standard deviation in parenthesis for every field. Fields are defined from left to right in Table 2 as the objective function value, computational time in seconds, total number of iterations required, the relative error in estimating the mean sample value function and deterministic upper and lower bounds to the optimal solution. We set a minimum number of iterations required prior existing the algorithm equal to 1000 and a maximum number of iterations or limit on the total iterations elapsed equal to 4000. All tests conducted (for every instances and run) attained convergence according to the stopping rule used to establish algorithmically asymptotic convergence. The termination criteria was based on asymptotic properties and an estimation of the objective value stability [3].

Table 2 Preliminary computational results and evaluation of errors

$\|F\|$	Obj. Val.	Time(s)	No. Iterations	Error in Obj. Estimate	Upper/Lower Bounds
2	65,219.27 (611.47)	232.32 (154.99)	2,225.03 (519.42)	0.0080 (0.0057)	92,156.83/38,807.32
4	76,399.84 (839.14)	146.49 (29.33)	1,038.59 (25.54)	0.0104 (0.0078)	130,695.35/46,305.30
6	65,987.71 (680.68)	185.54 (29.19)	1,031.1 (29.99)	0.0107 (0.0083)	124,224.57/40,589.83

Swain dataset, $\|N\| = 20$, and $T = 7$

From Table 2 one can observe that the solution is very stable as shown by a small value in the standard deviation of the objective function value, (e.g., for the instance with $\|F\| = 2$, the standard deviation is 611.47, implying that due to the law of large numbers the objective function value will have a maximum error of 1198.46

with a 95% of confidence, which results in an absolute deviation rate of 0.018). In addition, the averaged relative error in the estimate is at most slightly over 1% with small standard deviation values. Finally, we note that obtaining the optimal solution to these instances is virtually impossible, and thus, we cannot compare our solutions against the optimal ones. Thus, we evaluate the quality of the solutions obtained via some deterministic upper and lower bounds. We use the Jensen inequality and a combination of it with penalty functions to generate, respectively, a lower and upper bound to the optimal solution [8]. It can be seen that the objective function value of our solution approach is actually within the bounds.

5 Conclusions and Future Research

The proposed model stands as a further step towards the realistic modelling of critical infrastructure under disruption and the improvement of managerial strategies thereof. We have considered uncertainty associated with the magnitude, time and location of disruptions as well as gradual recovery of disrupted facilities over time. Also, we considered uncertainty in the forecast demand. Ongoing research is extending this model to include different protective measures and to have probabilistic protection where protective measures may only be successful with some probability.

References

1. Birge, J.R., Louveaux, F.: A multi-cut algorithm for two stage linear programs. European Journal of Operational Research. **34**, 384–392 (1988)
2. Hakimi, S.L.: Optimum location of switching centers and the absolute centers and medians of a graph. Operations Research, **113**, 544–559 (1964)
3. Higle, J.L., Sen, S.: Stochastic Decomposition: A Statistical Method for Large Scale Stochastic Linear Programming. Kluwer Academic Publishers (1996)
4. Higle, J.L., Lowe, W.W., Odio, R.: Conditional stochastic decomposition: an algorithm interface for optimization and simulation. Operations Research **42(2)**, 311-322 (1994)
5. Liberatore, F., Scaparra, M.P., Daskin, M.S.: Optimization methods for hedging against disruptions with ripple effects in location analysis. Omega **40**, 21–30 (2012)
6. Losada, C., Scaparra, M.P., Church, R.L., Daskin, M.S.: The stochastic interdiction median problem with disruption intensity levels. Working paper no. 187, University of Kent, Kent Business School, Canterbury, Kent, UK (2010)
7. Losada, C., Scaparra, M.P., O'Hanley, J.R. Optimizing system resilience: a facility protection model with recovery time. Forthcoming in the European Journal of Operational Research, DOI information: 10.1016/j.ejor.2011.09.044
8. Morton, D.P., Wood, R.K.: Restricted-recourse bounds for stochastic linear programming. Operations Research **47(6)**, 943–956 (1999)
9. Ruszczyński A. A regularized decomposition method for minimizing a sum of polyhedral functions. Mathematical Programming **35**, 309–333 (1986)
10. Swain, R.: A decomposition algorithm for a class of facility location problems. Ph.D. Diss., Cornell University (1971)
11. Slyke, R.M. Van, Wets, R.J.B.: L-Shaped programs with applications to optimal control and stochastic linear programming. SIAM Journal of Applied Mathematics **17**, 638–663 (1969)

A new simulated annealing algorithm for the freight consolidation problem

Erk Struwe, Xin Wang, Christopher Sander, and Herbert Kopfer

Abstract Logistics services are one of the most outsourced activities today. Freight charges for the outsourced transportation requests are mostly calculated based on tariff agreements depending on their quantities and distances. Due to economies of scale, it is beneficial for the outsourcing company to consolidate their requests, which is called the Freight Consolidation Problem (FCP). In this paper, the FCP is formally defined and modeled. To solve large-scale instances, a simulated annealing algorithm is proposed and tested on newly generated instances. Results show that the algorithm can provide good solutions within acceptable computational times.

1 Introduction

Nowadays, many companies outsource transportation activities to logistics service providers (LSPs). In order to reduce transactional costs, freight tariff agreements between the outsourcing company and LSPs are signed, based on which freights are charged. Due to economies of scale, it is beneficial for the outsourcing company to bundle their requests before releasing them to LSPs. The problem of bundling requests to minimize the total charges is called the Freight Consolidation Problem (FCP). Early research on the FCP has been done by Kopfer [1, 2]. Recently, FCP has been studied as part of the Integrated Transportation Planning Problem [3, 4].

Due to the complex bundling structures, FCP for pickup-and-delivery (PDP) requests can hardly be well solved efficiently yet. This paper aims to fill this research gap and is organized as follows. The problem is formally defined and modeled in Sect. 2. An efficient Simulated Annealing (SA) algorithm is proposed in Sect. 3 and tested in Sect. 4. At last, some conclusions are drawn in Sect. 5.

Erk Struwe
University of Bremen, Chair of Logistics, Wilhelm-Herbst-Str. 5, D-28359 Bremen
e-mail: estruwe@estruwe.de

Xin Wang
University of Bremen, Chair of Logistics, Wilhelm-Herbst-Str. 5, D-28359 Bremen
e-mail: xin.wang@uni-bremen.de

D. Klatte et al. (eds.), *Operations Research Proceedings 2011*, Operations Research Proceedings,   299
DOI 10.1007/978-3-642-29210-1_48, © Springer-Verlag Berlin Heidelberg 2012

2 The Freight Consolidation Problem

The FCP can be formally defined on a graph $G = (N, A)$. Assume a set $R = \{1, \ldots, n\}$ of n PDP requests, let $P = \{1, \ldots, n\}$ be the set of pickup locations and $D = \{n+1, \ldots, 2n\}$ be the set of delivery locations. Each request $r \in R$ represents the transportation of the amount of goods q_r from the pickup node $r^+ = r \in P$ to the according delivery node $r^- = r + n \in D$. Define $N = P \cup D$, which is the set of all customer nodes. For each arc $(i, j) \in A = N \times N$, a distance $d_{ij} \geq 0$ is assigned. If an arc with distance d is used for the freight calculation of load q, the freight charge c will be determined according to a cost function $c(q, d) = f(q, d)$, with $c(q_1, d) + c(q_2, d) \geq c(q_1 + q_2, d)$ and $c(q, d_1) + c(q, d_2) \geq c(q, d_1 + d_2)$. Suppose a path $\pi_r = (r, \ldots, r+n)$ for a request r, which begins at its pickup location r, over some nodes belonging to other requests, and ends at its delivery location $r+n$. If request r is not consolidated with another request for the freight calculation, its path is $(r, r+n)$. Let A_r be the set of arcs in path π_r. Freight charges c_{ij} will be determined for each arc $(i, j) \in \{(i, j) \mid (i, j) \in \cup_{r \in R} A_r\}$. For a given arc $(i, j) \in A$, let $R_{ij} \subseteq R$ be the set of requests with arc (i, j) in their paths. c_{ij} will be calculated as $c(\sum_{r \in R_{ij}} q_r, d)$ instead of $\sum_{r \in R_{ij}} c(q_r, d)$, while $\sum_{r \in R_{ij}} q_r$ must not exceed the flow capacity Q on the arc. For those arcs with $R_{ij} = \varnothing$, $c_{ij} = c(0, d) = 0$. The objective of the FCP is to minimize the total freight charges $C = \sum_{i \in N} \sum_{j \in N} c_{ij}$. Fig. 1 shows how cost savings can be achieved by taking advantage of the degressive objective function of the FCP. Note that $c_{2^+ 2^-} = 200$ in Fig. 1b is less expensive than $2 \cdot c_{2^+ 2^-} = 240$ in Fig. 1a.

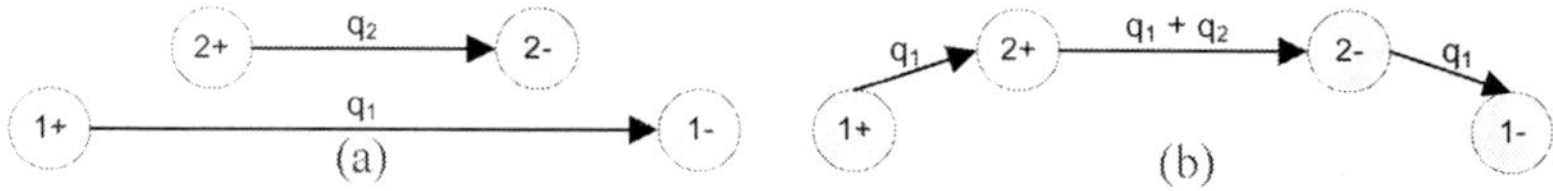

Fig. 1 Example of freight consolidation, $q_1 = q_2$. (a) Requests 1 and 2 are carried out independently: $C = 280 + 120 = 400$. (b) Requests 1 and 2 are consolidated on arc $(2^+, 2^-)$: $C = 80 + 200 + 80 = 360$.

The FCP consists of two dependent subproblems. The first one is to divide R into m request bundles with $\cup_{i=1}^{m} B_i = R$ and $B_i \cap B_j = \varnothing, \forall i, j = 1, \ldots, m, i \neq j$. A bundle with only one request is a single request bundle. A subgraph $G_b = (N_b, A_b)$ of G can be defined for each bundle $b = B_1, \ldots, B_m$. N_b is the customer nodes set of requests in b, $A_b = N_b \times N_b$. The second subproblem is to find a spanning tree $T_b = \{[i, j] \mid (i, j) \in \cup_{r \in b} A_r\}$ of undirected edges $[i, j]$ ignoring the directions of the arcs for each bundle b on G_b, such that the bundle freight charges $C_b = \sum_{i \in T_b} \sum_{j \in T_b} c_{ij}$ are minimized. Denote T'_b as the set of arcs used in bundle b according to T_b. It is further to be ensured that none of the nodes in N_b is merely used as a (de)consolidation point for other requests, without being (de)consolidated with those requests (see Fig. 2).

In order to model the FCP, we extend G by introducing a fictive depot 0 to $G' = (V, A')$, with $V = N \cup \{0\}$, $A' = V \times V$ and $d_{i0} = d_{0i} = 0 \; \forall i \in N$. The path π_r can be extended to a fictive round tour $\theta_r = (0, r, \ldots, r+n, 0)$, which starts from and ends at the fictive depot 0. Let $y_{ij} \; \forall i, j \in N$ be the loads over arc (i, j). The binary variable

Fig. 2 Example of forbidden consolidation. Bundle 2 (dashed) consolidates requests 3 and 4 at the pickup location of request 2 though it belongs to bundle 1 (solid).

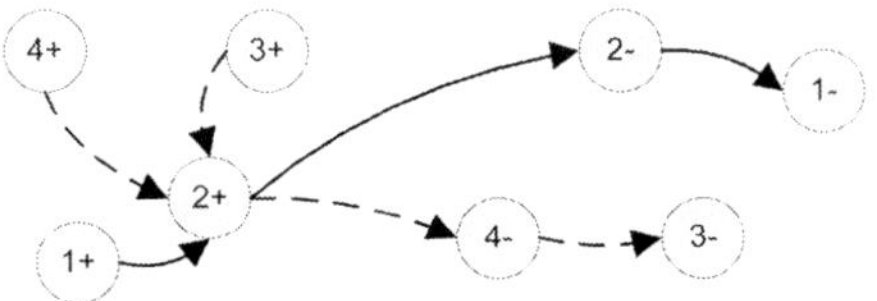

$x_{ijr} \; \forall i,j \in V, r \in R$ will be one iff arc (i,j) is part of θ_r. Variable $z_{ij} \in \{0,1\} \; \forall i,j \in N$ will be one iff arc (i,j) is used in any path. The FCP aims to create n fictive round tours which yield minimal freight charges and can be modeled as follows:

$$\min C = \sum_{i \in N} \sum_{j \in N} c_{ij} = \sum_{i \in N} \sum_{j \in N} c\,(y_{ij}, d_{ij}) \tag{1}$$

subject to:

$$x_{0rr} = x_{(r+n)0r} = 1 \quad \forall r \in R \tag{2}$$

$$\sum_{i \in V} x_{0ir} = \sum_{i \in V} x_{i0r} = 1 \quad \forall r \in R \tag{3}$$

$$\sum_{h \in V} x_{hir} = \sum_{j \in V} x_{ijr} \quad \forall r \in R, i \in N \tag{4}$$

$$u_{ir} - u_{jr} + M \cdot x_{ijr} \leq M - 1 \quad \forall r \in R, i, j \in N, i \neq j \tag{5}$$

$$M \cdot z_{ij} \geq \sum_{r \in R} x_{ijr} \quad \forall i, j \in N \tag{6}$$

$$\sum_{h \in N} z_{hi} \leq 1 \quad \forall i \in D \tag{7}$$

$$\sum_{j \in N} z_{ij} \leq 1 \quad \forall i \in P \tag{8}$$

$$y_{ij} = \sum_{r \in R} x_{ijr} \cdot q_{(r+n)} \quad \forall i, j \in N \tag{9}$$

$$y_{ij} \leq Q \cdot z_{ij} \quad \forall i, j \in N \tag{10}$$

$$x_{iir} = 0 \quad \forall i \in V, r \in R \tag{11}$$

$$z_{ii} = 0 \quad \forall i \in N \tag{12}$$

$$x_{ijr} \in \{0,1\} \quad \forall i, j \in V, r \in R \tag{13}$$

$$y_{ij} \geq 0 \quad \forall i, j \in V \tag{14}$$

$$z_{ij} \in \{0,1\} \quad \forall i, j \in V \tag{15}$$

$$u_{ir} \geq 0 \quad \forall i \in N, r \in R \tag{16}$$

Constraints (2) to (4) imply that θ_r is a round tour which starts and ends with arcs $(0,r)$ and $(r+n,0)$ respectively. Constraint (5) eliminates subtours. Constraints (6) to (8) limit the number of incoming or outgoing arcs connected with delivery or pickup nodes to exclude pure (de)consolidation points. Constraints (9) and (10) ensure the capacity restriction is hold. Constraints (11) and (12) forbid self-cycles.

3 Solution methodology

A Simulated Annealing algorithm is proposed to solve the FCP even for large-scale instances. Solutions to the FCP are represented by m bundles. Each bundle b is an array. Each element in the array consists of an arc $(i, j) \in T_b'$ and its corresponding R_{ij}. An initial solution s_0 is obtained by inserting requests successively in random order into the solution. While inserting a request, the algorithm tries to integrate it into existing bundles at all possible positions and to create a single request bundle for its own. Let ΔC_b^* be the insertion costs of inserting request $r \in U$ into an existing bundle b at the position that increases the objective value the least. Let $C_{\{r\}}$ be the insertion costs of creating a single request bundle for r. U is the set of unplanned requests. The algorithm evaluates ΔC_b^* for all existing bundles and $C_{\{r\}}$ in a parallel fashion and afterwards inserts r at the position with the lowest insertion costs.

In order to insert the new request's pickup location r into an existing bundle b, the routine creates candidates by adding arc (r, i) for every $i \in b \cap P$ (Fig. 3a). Additionally, it tries to replace every $(i, j) \in T_b'$ by (i, r, j) (Fig. 3b). The delivery location $r + n$ is inserted by adding $(j, r + n)$ for (i, j) (Fig. 3c) or by replacing it with $(i, r + n, j)$ (Fig. 3d), $\forall (i, j) \in T_b'$. Feasible candidate bundles are created by recursively connecting every combination of links to r and $r + n$ by every possible path on the arcs in T_b' in between. Thus, the amount of candidate bundles grows exponentially with the amount of arcs in T_b'. Finally, also the replacement of every $(i, j) \in T_b'$ by $(i, r, r + n, j)$ is tried (Fig. 3e).

The removal of a request r from a bundle b may lead to an infeasible solution (Fig. 3f). To avoid this, the bundle b will be reconstructed after the removal of r by reinserting the remaining requests into it in the order as it was constructed before. To diversify the search, the order is shuffled randomly at a small probability of α.

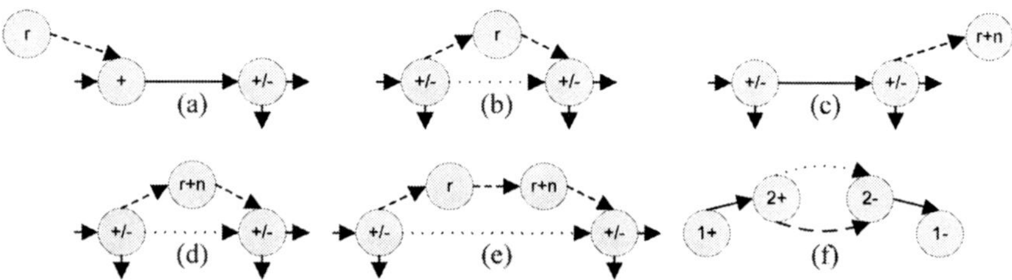

Fig. 3 Request insertion and removal. Pickup nodes are marked with +, delivery nodes with −. +/− nodes may be both. In (f), dotted arcs are replaced by dashed arcs, solid arcs remain.

In each iteration, one of two operators is used to generate neighborhood solutions. The *Single-Move* operator is used at a probability of $1 - \beta$. It randomly selects a request, removes it from its current bundle and relocates it at its best position except in its previous bundle. The *Swap* operator, which is used at a probability of β, randomly selects two requests from different bundles and swaps them. If at least one request cannot be inserted into the new bundle, the current iteration is skipped.

4 Computational results

For computational analysis, the algorithm has been tested on three sets of instances with 10 requests each. All requests have their pickup locations in one area and delivery locations in another. Set 2 features substantially lower average quantities than the other sets. In Set 3, the distance between the two areas is increased. Optimal solutions have been obtained by using the commercial solver IBM CPLEX.

As the objective function, $\sum_{i \in N} \sum_{j \in N} c_{ij} = \sum_{i \in N} \sum_{j \in N} cf \cdot p_{ij} \cdot d_{ij}$ is used. cf is a basic freight rate. p_{ij} is a price factor that considers the utilization of the vehicle and is typically a piecewise linear function of y_{ij}. For our tests, $cf = 1$, $Q = 40$, and the price factor function $p_{ij} = f(y_{ij})$ is depicted in Fig. 4.

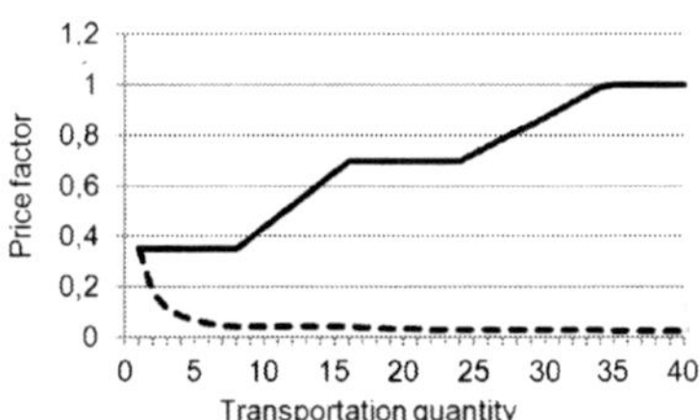

Fig. 4 Price factor p_{ij} as a piecewise linear function of transportation quantity y_{ij} (solid line) with $Q = 40$. The dashed line shows the average value per transportation quantity unit.

The SA algorithm uses a geometric cooling schedule with an initial temperature t_0. After n_{rep} iterations in each temperature step, t is decreased by $\gamma \cdot t$, while γ is the cooling factor. The start temperature t_0 is set such that a solution that is 15% worse than the initial solution is accepted with probability 0.15. n_{rep} and γ are set to 50 and 0.98 respectively. Both the *Swap* operator and the random shuffle of reinsertion order for bundle reconstruction are used at a probability of $\beta = \alpha = 0.2$. The algorithm stops when a worsening of 1% is accepted at a probability lower than 0.01. This leads to an average of about 9,100 iterations.

Computational results obtained on an Intel Core 2 Quad Q6600 CPU are shown in Table 1. Total freight charges with and without freight consolidation obtained by CPLEX solver are given in columns three and four respectively. Column five shows the cost-saving potentials that could be achieved by performing freight consolidation. Especially for small loads (Set 2) and long transportation distances between the pickup and delivery regions (Set 3), freight charges can be reduced considerably up to 42%. This result indicates the practical importance of the FCP.

For the performance evaluation, the proposed SA algorithm has been tested five times on each instance. The obtained FCP solutions and the realized cost-savings are shown in columns six and seven respectively. Our algorithm takes advantage of almost the entire cost-saving potentials in Sets 1 and 3 with average deviations from the optimum of only 0.76% and 0.27%. In contrast, instances in Set 2 are much more difficult to solve due to more complex spanning trees for the low-volume requests.

Additionally, the algorithm has been tested on a very large problem instance with 5,000 requests in a 400 × 400 coordinate system. Even such an instance can be solved with our heuristic in less than 10 minutes.

Table 1 Computational results (instances marked with * are not included in averages)

Set	Instance	Solution without consolidation	Optimal solution CPLEX	Improvement potential [%]	Best solution SA	Improvement SA [%]	Distance to optimum SA [%]	Time [s]
1	1	97.81	77.71	20.55	77.71	20.55	0.00	2.1
1	2	94.84	77.19	18.61	77.75	18.01	0.72	2.6
1	3	82.45	66.07	19.87	66.86	18.91	1.18	2.7
1	4	80.34	68.48	14.76	68.84	14.31	0.52	2.5
1	5	87.25	73.84	15.37	74.10	15.07	0.35	2.1
1	6	103.46	90.80	12.24	90.80	12.24	0.00	1.5
1	7	67.21	57.41	14.57	57.66	14.21	0.43	1.9
1	8	81.42	62.74	22.94	64.53	20.75	2.77	2.4
1	9	90.12	77.27	14.26	77.29	14.24	0.03	1.8
1	10	96.53	83.22	13.79	84.55	12.41	1.58	1.8
2	1	56.00	37.78	32.54	39.96	28.65	5.45	7.0
2	2	55.88	32.15	42.46	34.37	38.49	6.45	10.2
2	3*	54.27	n/a	n/a	35.02	35.48	n/a	8.4
2	4	62.77	36.55	41.78	40.94	34.77	10.74	9.1
2	5	54.07	37.26	31.09	39.66	26.66	6.04	6.9
3	1	266.62	201.55	24.41	201.55	24.41	0.00	2.0
3	2	249.27	185.71	25.50	187.37	24.83	0.88	2.7
3	3*	228.87	n/a	n/a	169.16	26.09	n/a	2.7
3	4	241.19	179.49	25.58	179.85	25.43	0.20	2.3
3	5	265.20	199.19	24.89	199.19	24.89	0.00	2.1
1	average	—	—	16.70	—	16.07	0.76	2.1
2	average	—	—	36.97	—	32.14	7.17	8.3
3	average	—	—	25.09	—	24.89	0.27	2.3

5 Conclusions

In this paper, the FCP for PDP requests has been formally defined and modeled. In order to solve this extremely complex combinatorial optimization problem for large-scale instances, we present an SA heuristic approach. Computational results show that through request consolidation, freight charges can be reduced to considerable extent, depending on different characteristics of the instances. Compared with the optimal solutions for small-scale instances, the SA algorithm is very efficient and effective. Nevertheless, even very large-scale instances can be solved in reasonable time with this heuristic. In the future, more computational tests will be conducted, especially for the solution of very large-scale instances using our heuristic.

References

1. Kopfer, H.: Der Entwurf und die Realisierung eines A*-Verfahrens zur Lösung des Frachtoptimierungsproblems. OR Spektrum 12/4, 207–218 (1990)
2. Kopfer, H.: Konzepte genetischer Algorithmen und ihre Anwendung auf das Frachtoptimierungsproblem im gewerblichen Güterverkehr. OR Spektrum 14/3, 137–147 (1992)
3. Kopfer, H., Wang, X.: Combining Vehicle Routing with Forwarding — Extension of the Vehicle Routing Problem by Different Types of Sub-contraction. Journal of the Korean Institute of Industrial Engineers 35(1), 1–14 (2009)
4. Krajewska, M.A., Kopfer, H.: Transportation planning in freight forwarding companies — Tabu search algorithm for the integrated operational transportation planning problem. European Journal of Operational Research 197, 741–751 (2009)

Models and Algorithms for Intermodal Transportation and Equipment Selection

Christina Burt and Jakob Puchinger

Abstract We study a cement transportation problem with asset-management (equipment selection) for strategic planning scenarios involving multiple forms of transportation and network expansion. A deterministic mixed integer programming approach is difficult due to the underlying capacitated multi-commodity network flow model and the need to consider a time-space network to ensure operational feasibility. Considering uncertain aspects is also difficult but important. In this paper we propose three approaches: solve a deterministic mixed integer program optimally; solve stochastic programs to obtain robust bounds on the solution; and, study alternative solutions to obtain their optimal cost vectors using inverse programming.

1 Introduction

We consider a strategic network design problem with asset management, inspired by the requirements of a leading European construction group currently delivering cement in a truck transportation system. The company is considering opportunities to expand the network to incorporate intermodal transportation by adding rail and/or barge links. Additional factories are also considered and as such, the production schedule is not known.

Given a set of cement orders, we wish to select transportation equipment and related arcs; select services and associated schedules; and build flow paths through the network such that a production schedule at each factory can also be generated. We aim to achieve the optimum strategic network design at minimum cost given that certain aspects of costing are uncertain. Since we permit multi-commodity flows to satisfy orders, and the selection of services and equipment requires a fixed charge in the objective function, this problem is strongly related to and is likely to be at least as difficult as the *fixed charge capacitated multi-commodity flow network problem*—known to be $\mathcal{NP}$-hard [7]. Adding to this difficulty is the scale of the time-space network.

AIT Austrian Institute of Technology, Mobility Department, Vienna
e-mail: {christina.burt.fl|jakob.puchinger}@ait.ac.at

D. Klatte et al. (eds.), *Operations Research Proceedings 2011*, Operations Research Proceedings,   305
DOI 10.1007/978-3-642-29210-1_49, © Springer-Verlag Berlin Heidelberg 2012

Similar problems have been modelled as multi-commodity flow mixed integer program on time-space networks [2]; however, it has been long known that the bound provided by the *LP* relaxation is weak [4]. Worse, the time-space network can lead to an inhibiting number of variables, such as in [11]. Since there are uncertain aspects to our problem—notably costs and demands—a stochastic programming approach is desirable. In the closely related context of service network design, a two-stage stochastic programming model [10] has been succesfully used. Preprocessing (see [5]), decomposition (see [3, 9]), and cutting planes (see [4–6]) have helped improve the tractability of the underlying network model. However, in genuine applications, such as [2], this has been insufficient to produce a thorough stability analysis.

We propose a portfolio of solution approaches that generate a complementary set of future options. In this paper we propose three approaches: solve a deterministic mixed integer program optimally (Section 2); generate solutions to a simplified stochastic program (Section 4); and study the expected solutions to obtain their optimal cost vectors (Section 5). We briefly outline future challenges in Section 6.

2 Model

We define the network on a graph $G = (\mathcal{N}, \mathcal{A})$ where $\mathcal{N}$ is the set of nodes and $\mathcal{A}$ is the set of paths conncecting each pair. Each arc is indexed by the nodes, i and j, and equipment, e, that moves between, giving the tuple $a = (i, j, e)$. We model the problem as a multi-commodity network flow mixed integer program, where the commodities represent the capability to satisfy orders from more than one factory.

The flow variables, $\mathbf{x}^t_{a,k}$, capture the flow of cement on arc a at the start of period t for commodity k. The equipment index is important in ensuring that the commodity is not split once it has left the factory in one package (since it is not practical to split cement packaging). The service variables, $\mathbf{y}^t_a$, indicate whether a service by equipment e ($\in a$) runs on arc a at the start of period t. These variables can be used to also indicate the time of purchase of a particular machine by comparing the existence of services provided in period t with the existence of services provided in possible previous time periods, $t - t(a')$, where $t(a')$ is the service time along a preceding arc a' (see Figure 1). Salvage time is indicated for each equipment by binary variable $\mathbf{s}^{t,e}$.

The objective function minimises the cost of owning (f^e) and operating (v^a) the equipment. All costs are discounted back to the present using a discount factor, $d^t = \frac{1}{(1+I)^t}$, for some interest rate I. Transshipment nodes may incur a fixed fee when the flow changes mode of transport, as indicated by binary variable $\mathbf{z}^t_k$.

The movement of cement through the network is modelled with flow constraints: *demand* constraints for each order, κ, with associated due date, τ, implied by the order (Constraint 1); *node balancing* for interior nodes (Constraint 2) and the demand nodes (Constraint 3)—this captures the ability of equipment to drop off partial orders; and, *capacity* constraints on the production nodes, $\mathcal{N}^O$ (Constraint 4).

Since the production schedule is not known *a priori*, the latter is an inequality. Decision variables are linked by Constraint (5), where c_e is the maximum capacity of a provided service. The disaggregated right hand side prevents commodity splitting across equipment within an arc. Special-ordered set constraints (type 1) prevent flow splitting across arcs (Constraint 6) and across equipment (Constraint 7).

$$\min \sum_e f^e \sum_a \mathbf{y}_a^0 + \sum_{i,e,t>0} d^t f_\pi^e \left[\sum_{a=(i,j,e)} \mathbf{y}_a^t - \sum_{a'=(i',i,e)} \mathbf{y}_{a'}^{t-t(a')} + \mathbf{s}^{t,e} \right] + \sum_{a,t} d^t v_y^a \mathbf{y}_a^t$$
$$+ \sum_{a,k,t} d^t v_x^a \mathbf{x}_{a,k}^t + \sum_{k,t} d^t f^\tau \mathbf{z}_k^t - \sum_{e,t} d^t d_2^t f^e \mathbf{s}^{t,e}$$

subject to

$$\sum_{\substack{\{a=(i,j,e)|j\in\kappa\},k, \\ \tau^\kappa-\tau^0<t<\tau^\kappa}} \mathbf{x}_{a,k}^{t-t(a)} \geq \sum_{\substack{\{a=(j,j',e)|j\in\kappa\},k, \\ \tau^\kappa-\tau^0<t<\tau^\kappa}} \mathbf{x}_{a,k}^t + D^\kappa \qquad \forall \kappa, \tag{1}$$

$$\sum_{a=(i,j,e)} \mathbf{x}_{a,k}^{t-t(a)} = \sum_{a'=(j,j',e)} \mathbf{x}_{a',k}^t \qquad \forall j \in \mathcal{N}\setminus\{\mathcal{N}^O\cup\mathcal{N}^D\},k,t, \tag{2}$$

$$\sum_{a=(i,j,e),k,t} \mathbf{x}_{a,k}^t = \sum_{a'=(j,j',e),k,t} \mathbf{x}_{a',k}^t + \sum_{\kappa|j\in\kappa} D^\kappa \qquad \forall j \in \mathcal{N}^D, \tag{3}$$

$$\sum_{a=(i,j,e),k} \mathbf{x}_{a,k}^t \leq P_i^t \qquad \forall i \in \mathcal{N}^O,t, \tag{4}$$

$$\sum_k \mathbf{x}_{a,k}^t \leq c_e \mathbf{y}_a^t \qquad \forall a,t, \tag{5}$$

$$SOS1(\mathbf{x}_{a,k}^t|(i,j)\in\mathscr{A}) \qquad \forall e,k,t, \tag{6}$$

$$SOS1(\mathbf{x}_{a,k}^t|e\in\mathscr{E}) \qquad \forall (i,j),k,t, \tag{7}$$

$$\sum_a \mathbf{y}_a^t \leq 1 \qquad \forall e,t, \tag{8}$$

$$\mathbf{x}_{a,k}^{t-t(a)} + \sum_{a'=(j,j',e')|e'\in\mathscr{E}\setminus e} \mathbf{x}_{a',k}^t - \sum_{a'=(j,j',e')|e'\in\mathscr{E}} \mathbf{x}_{a',k}^t \leq M\mathbf{z}_k^t \qquad \forall a,k,t \in [t(a),T], \tag{9}$$

$$\mathbf{y}_a^{t-t(a)} \leq \sum_{a'\in(j,j')} \mathbf{y}_{a'}^t + s^{t-1,e} \qquad \forall a,t \in [t(a),T], \tag{10}$$

$$\sum_t \left(\sum_{a=(i,j,e)} \mathbf{y}_a^t - \sum_{a'=(i',i,e)} \mathbf{y}_{a'}^{t-t(a')} \right) \leq 1 \qquad \forall e, \tag{11}$$

$$\mathbf{s}^{t,e} \leq \sum_{a,h\leq t} \mathbf{y}_a^h \qquad \forall e,t, \tag{12}$$

$$\sum_t \mathbf{s}^{t,e} \leq 1 \qquad \forall e, \tag{13}$$

$$\sum_{a,h>t} \mathbf{y}_a^h \leq (1-\mathbf{s}^{t,e})M \qquad \forall e,t, \tag{14}$$

$$\mathbf{y}_a^t \in \{0,1\} \qquad \forall a,t, \tag{15}$$

$$\mathbf{z}_k^t \in \{0,1\} \qquad \forall t,k, \tag{16}$$

$$\mathbf{s}^{t,e} \in \{0,1\} \qquad \forall t,e, \tag{17}$$

$$\mathbf{x}_{a,k}^t \in \mathscr{R}^+ \qquad \forall a,k,t. \tag{18}$$

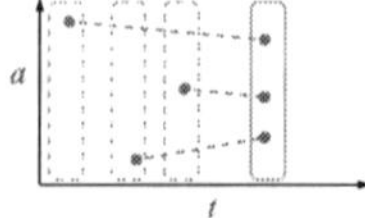

Fig. 1 Purchases are determined by looking at all arc service variables in one time period and comparing it with service variables in the immediately preceding possible time period.

SOS1 constraints allow at most one variable in the set to be non-zero. Since the service variables are binary, we use a knapsack constraint to prevent equipment splitting across arcs (Constraint 8). The time of transshipment is captured in Constraint (9). Consistency in the time-space network is provided for service by the restriction (10). We restrict purchasing in Constraint (11), using a similar argument as illustrated in Figure 1. Finally, we ensure consistency in the salvage variables (Constraints (12)–(14)). We reduce the domains of the constraints by considering edge conditions, and exclude variables never appearing in the optimal solution.

The literature suggests that solving the full, deterministic mixed-binary model without simplification will be difficult. Both preprocessing and decomposition approaches are required to achieve tractability. Obvious preprocessing includes disjoint sets, precedence, cycles and covering. However, we need to be mindful that the preprocessing doesn't break the strong structure provided for decomposition by the time-space network. Therefore much analysis is needed for this approach—especially if the final implementation is to avoid the use of commercial solvers.

3 Stochastic Programming Approach

We propose a two-stage stochastic programming approach. The equipment and salvage variables are fixed in the first stage. In the second stage, some random events $\omega \in \Omega$ are realised and determine the random vector χ consisting of: the operating cost coefficients $v^a(\omega)$; and, the demand D^κ. The recourse action then consists of decisions about the service and flow variables. Assuming a fixed number of realisations/scenarios for χ the stochastic program can be defined in its extensive form. Furthermore, all binary variables are fixed in the first stage. Therefore the proposed stochastic program can be solved using a standard algorithm such as the L-shaped method. The model defined in Section 2 includes the possibility of outsourcing transport at a high cost, ensuring the feasibility of the second stage.

Since the linear program of the second stage can be solved efficiently, it remains to be shown how to fix the first stage variables. On the one hand this could be done by exatly solving the first stage at every iteration of the L-shaped method using a mixed integer programming solver. On the other hand for problem instances of realistic size the use of heuristics might become necessary [8]. In our case, first stage decisions could be taken by first solving the flow problem and then allowing the flow to imply the services required for each scenario. Then the decision about the equipment and factories can be taken based on the required services.

4 Scenario-scaping Heuristic

An alternative approach is to develop an understanding of when alternate solutions would become optimal. To achieve this, we propose a heuristic to approximate the cost vector that would make an initial feasible solution optimal. The proposed heuristic itself is outlined in Algorithm 1. We will begin with a good feasible solution proposed by the industry partner. We will then use inverse programming (as proposed in [1]) under the L_1 norm to find the nearest cost vector that makes the feasible solution optimal. This is a type of reverse stochastic programming: instead of determining the best decision under uncertainty, we provide the conditions under which each decision will be best. This is particularly useful for decision-making if the industry partner has a way of influencing the cost vector or has undisclosed cost information.

Algorithm 1 Scenario-scaping with inverse programming.

1: Define the capacitated network design problem as a mixed integer program, $\mathcal{P}$, with objective function z.

2: **for** each scenario $s \in \mathcal{S}$ **do**

3:     Heuristically determine the equipment selection solution for s and denote this h_s.

4:     Fix all equipment variables according to h_s—this fixes all integer variables in the model. Fix some of the continuous variables (flow) as implied by h_s and the location solution in s.

5:     Solve the remaining reduced LP problem $\mathcal{P}^s$ (exactly or approximately) with the fixed variables to obtain the full scenario solution, x^0. Note that x^0 is now feasible for $\mathcal{P}$.

6:     Consider the set of binding constraints, $\mathcal{B}$, for x^0 in $\mathcal{P}$. Remove all non-binding constraints to obtain the reduced problem, $\mathcal{P}_B^s$—keep all integer variables fixed according to h_s but restore the bounds on all continuous variables.

7:     Solve the reduced problem $\mathcal{P}_B^s$ exactly. Note that, different to $\mathcal{P}^s$, $\mathcal{P}_B^s$ does not contain any fixed continuous variables.

8:     Let the vector π be the optimal dual variables associated with the constraints in $\mathcal{B}$. The elements of the new cost vector for scenario s are given by:

$$d_j^* = \begin{cases} c_j - |c_j^\pi| \ \forall \ \{j : c_j^\pi > 0 \text{ and } x_j^* > 0\} \\ c_j + |c_j^\pi| \ \forall \ \{j : c_j^\pi < 0 \text{ and } x_j^* < u_{i,j}\} \\ c_j \qquad \text{otherwise} \end{cases}$$

where $c_j^\pi = c_j - \sum_{i \in \mathcal{B}} a_{i,j}\pi_i$; $u_{i,j}$ is the upper bound on variable j in $\mathcal{P}$; and i corresponds to constraints in $\mathcal{P}$.

9: **end for**

We obtain the initial feasible solution x^0 by first fixing many (integer) variables when we define the scenario and solving the remaining reduced mixed integer program approximately. The SOS constraints create a non-convex feasible region. To combat this we propagate the SOS constraints to obtain the finite set of convex LP's that comprise the feasible region—since these *LP*'s are already significantly reduced it is trivial to enumerate through the set to obtain the optimal solution.

5 Conclusions and Discussion

In this paper we have studied a difficult network design problem and proposed several approaches, which combine to provide supporting information in a holistic manner. This is important given the cost of making these decisions and the uncertainty involved in the cost vector. In this way, we meet our goal of obtaining quality solutions while taking into account that the genuine optimal solution is influenced by events we cannot foresee. The complementary solutions we provide will empower our industry partner to make good and near-optimal decisions even if they need to withhold (or cannot quantify) confidential data.

Acknowledgements This work is part of the project I2Bau, partially funded by the Austrian Federal Ministry for Transport, Innovation and Technology (BMVIT) within the strategic programme I2VSplus under grant 826188.

References

1. R. K. AHUJA AND J. B. ORLIN, *Inverse optimization*, Operations Research, 49 (2001), pp. 771–783.
2. J. ANDERSON, T. G. CRAINIC, AND M. CHRISTIANSEN, *Service network design with asset management: Formulations and comparative analyses*, Transportation Research, 17 (2009), pp. 197–207.
3. C. BARNHART, N. KRISHNAN, D. KIM, AND K. WARE, *Network design for express shipment delivery*, Computational Optimization and Applications, 21 (2002), pp. 239–262.
4. D. BIENSTOCK AND G. MURATORE, *Strong inequalities for capacitated survivable network design problems*, Mathematical Programming, 89 (2000), pp. 127–147.
5. N. BOLAND, M. KRISHNAMOORTHY, A. ERNST, AND J. EBERY, *Preprocessing and cutting for multiple allocation hub location problems*, European Journal of Operational Research, 155 (2004), pp. 638–653.
6. J.-F. CORDEAU, M. IORI, G. LAPORTE, AND J. J. S. GONZALEZ, *A branch-and-cut algorithm for the pickup and delivery traveling salesman problem with lifo loading*, tech. report, CIRRELT, 2007.
7. B. GENDRON, T. CRAINIC, AND A. FRANGIONI, *Telecommunications Network Planning*, Kluwer Academic, 1998, ch. Multicommodity capacitated network design.
8. A. HOFF, A.-G. LIUM, A. LØKKETANGEN, AND T. CRAINIC, *A metaheuristic for stochastic service network design*, Journal of Heuristics, 16 (2010), pp. 653–679.
9. S. IRNICH, *Netzwerk-Design für zweistufige Transportsysteme und ein Branch-and-Price-Verfahren für das gemischte Direkt- und Hubflugproblem*, PhD thesis, Lehrstuhl für Unternehmensforschung, Rheinisch-Westfälische Technische Hochschule, 2002.
10. A.-G. LIUM, T. G. CRAINIC, AND S. W. WALLACE, *A study of demand stochasticity in service network design*, TRANSPORTATION SCIENCE, 43 (2009), pp. 144–157.
11. Y. ZHANG, *Advances in LTL load plan design*, PhD thesis, Georgia Institute of Technology, 2010.

Contraction Hierarchies with A* for digital road maps

Curt Nowak, Felix Hahne, and Klaus Ambrosi

Abstract One of the most successful recent approaches for solving single source - single destination problems in graphs is the Contraction Hierarchies (CH) algorithm, originally published by [3]. The general algorithm consists of two phases: Firstly, a total order on the nodes in the graph is calculated. Secondly for queries, a modified bi-directional Dijkstra-search is performed on the hierarchy implied hereby. Relying on Dijkstra's algorithm, CH makes no use of the geometric information contained within digital road maps. We propose A*-like modifications of the original query algorithm that double query speed. Results are presented in a benchmark using a map from the OpenStreetMap project.

1 Introduction

Contraction Hierarchies (CH) are a means for single source - single destination optimal path searches in directed graphs. Every node is given a distinct rank inducing a total order on all nodes. Then nodes are *contracted* in increasing order of rank: They are (temporarily) deleted from the graph. At this point, optimal paths between two different nodes in the graph are likely to be affected. In these cases artifical *shortcut edges* are added to the graph 'bridging the gap'. When every node (but the last) has been contracted this way, the CH is finished and contains all (now ordered) nodes and edges of the original graph plus all shortcut edges.

Curt Nowak (M.Sc.)
Universität Hildesheim, Institut für Betriebswirtschaft und Wirtschaftsinformatik, Marienburger Platz 22, 31141 Hildesheim e-mail: nowak@bwl.uni-hildesheim.de

Dr. Felix Hahne
Universität Hildesheim, Institut für Betriebswirtschaft und Wirtschaftsinformatik, Marienburger Platz 22, 31141 Hildesheim e-mail: hahne@bwl.uni-hildesheim.de

Prof. Dr. Klaus Ambrosi
Universität Hildesheim, Institut für Betriebswirtschaft und Wirtschaftsinformatik, Marienburger Platz 22, 31141 Hildesheim e-mail: ambrosi@bwl.uni-hildesheim.de

D. Klatte et al. (eds.), *Operations Research Proceedings 2011*, Operations Research Proceedings, 311
DOI 10.1007/978-3-642-29210-1_50, © Springer-Verlag Berlin Heidelberg 2012

Queries on a CH are executed using a modified bi-directional Dijkstra search. The modification is twofold:

- When a node is expanded, only those edges are relaxed that lead to nodes with a higher rank than the expanded node. This greatly reduces the search space.
- The bi-directional search works iteratively switching between forward search and backward search after each Dijkstra iteration.

In a CH, forward and backward search will always meet at the highest ranked node of all nodes in the optimal path, as shown in [3].

The performance of CH is determined by the pre-processing and the query. Different node ordering during pre-processing will lead to very different query times. [3] present a couple of heuristics that we will rely on for our benchmarks in section 4. While there is most likely room for improvement to be found for node sorting criteria, this paper puts the focus on improving the query algorithm.

2 Related Work

The task of solving the single source (s) - single destination (t) optimal path problems arises in many real world applications like vehicle routing. When it comes to solve huge instances, e.g. such based upon large digital roadmaps, the basic Dijkstra algorithm in [2] is unsuitable in terms of runtime and memory usage. Therefore many speed-up techniques have been developed, which can be classified into three groups, that can also be combined (see [1]):

- *Bi-directional Search:* Starting the search simultaneously both from s and t until the intersection of the search-spaces is nonempty.
- *Hierarchical Approach:* In a pre-processing step, a hierarchical structure of the given network is extracted and made use of in the following queries.
- *Goal-Directed Approach:* The search is directed towards t and those parts of the network, that cannot possibly contain the shortest path, are excluded. In most cases, there is also some pre-processing necessary.

One of the most recent approaches is CH, which uses the hierarchical and the bi-directional search approach. It has a remarkable tradeoff between pre-processing effort and query speed-up. By finally including a goal-direction as the third technique, we show how to further improve query speed.

3 Applying A* to Contraction Hierarchies

In this section we will show how to achieve A*-like behavior for CH queries. W.l.o.g. we will discuss shortest path searches.

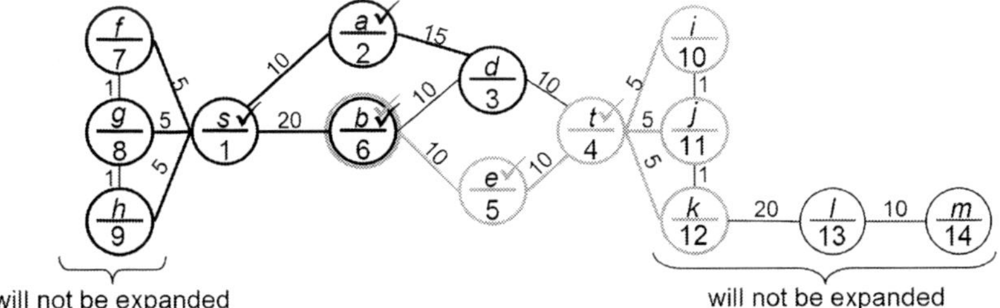

Fig. 1 Example for search from s to t using the combined approach. The state after the sixth iteration is shown. Numbers denote node ranks; bold black (blue) color denotes the forward (backward) search; highlighted nodes are *reached*, checks denote *settled* nodes. Air distances are omitted for brevity. *Bridge node* (b) yields an upper bound of 40 units. Nodes l and m are never reached.

Whenever the search trees of the forward and the backward search meet at what we will call a *bridge node*, a possible path is found. Its length l_{upper} provides an upper bound for the optimal path: the search trees of the forward and backward search can both be clipped with the help of l_{upper}. Also nodes that are *reached* at a longer distance in a future Dijkstra iteration in either direction will not be included any more. This creates a goal-directed behavior for the bi-directional Dijkstra search.

3.1 Query speed-up 1: A* clipping

Consider the forward search during a query for a shortest path between nodes s and t. Let v be the node *settled* at distance $d_{forw}(v)$ during the latest Dijkstra iteration. In the spirit of the A* algorithm, the air distance $d_{air}(v,t)$ between v and t can be used at this point to calculate a simple lower bound $d_{low}(s,v,t)$ for the length of the shortest path from s to t via v:

$$d_{low}(s,v,t) = d_{forw}(v) + d_{air}(v,t).$$

If a bridge node has been found beforehand, an upper bound for the length of the shortest path is also known: l_{upper}. Thus, if $d_{low}(s,v,t) \geq l_{upper}$, then v does not need to be expanded as it cannot lead to a shorter path. This approach works for the backward search accordingly and strongly reduces the search space. We will refer to it in the following as *A* clipping*.

3.2 Query speed-up 2: A* search for bridge nodes

Another query improvement can be achieved by employing the A* algorithm for the search for bridge nodes. We have implemented a combination of both a modified bi-directional A* and Dijkstra search by splitting the query into two chronological stages:

1. the search for a bridge node and
2. the remaining search.

During the first stage, our query algorithm performs the bi-directional search as described in [3]. However, instead of the modified Dijkstra search, it performs an equally modified A* search. In the majority of cases – depending on the layout of the underlying road network – this will find a bridge node much faster. With this comes the advantage of being able to bound the forward and backward search trees earlier as described above.

Once a bridge node is found, the algorithm moves to the second stage, namely the bi-directional Dijkstra search of the original CH algorithm. For this, the priority queues containing the *reached* nodes of the forward and backward search, respectively, must be re-sorted according to Dijkstra's algorithm. Note that the *A* clipping* approach can be applied during the rest of the search nevertheless (see fig.1). The benchmarks in section 4 will prove the combined approach superior to the classic CH algorithm.

The pseudocode for the approach is given in algorithm 1. (We assume that the A* algorithm performs a *NOP* when asked to iterate while having no *reached nodes* left to expand. The same holds for Dijkstra's algorithm.)

Algorithm 1 Combined bi-directional A* and Dijkstra search in CH

Require: Graph $G = (V, E)$, with full Contraction Hierarchy present (i.e. ranks and shortcut edges are established), starting node s and target node t and $s, t \in V$

1: **while** Either forward or backward A* search has nodes left to expand **do**   ▷ Stage 1
2:      perform one forward A* iteration
3:      **if** *bridge node* was found **then**
4:          GOTO line 12
5:      **end if**
6:      perform one backward A* iteration
7:      **if** *bridge node* was found **then**
8:          GOTO line 12
9:      **end if**
10: **end while**

11: Error: No path exists between s and t

12: re-sort search trees according to Dijkstra's algorithm   ▷ Stage 2
13: calculate l_{upper} using the *bridge node*
14: **while** Either forward or backward Dijkstra search has nodes left to expand **do**
15:      perform one forward Dijkstra iteration with *A* clipping* using l_{upper} for bounding
16:      **if** better *bridge node* was found **then**
17:          re-calculate l_{upper}
18:      **end if**
19:      perform one backward Dijkstra iteration with *A* clipping* using l_{upper} for bounding
20:      **if** better *bridge node* was found **then**
21:          re-calculate l_{upper}
22:      **end if**
23: **end while**
24: **return** optimal path from s to t via the best known *bridge node* and length l_{upper}

In order to find a bridge node even quicker, it is possible to use the parameterized version of the A* algorithm as described in [4]. As this approach is a heuristic, however, the path found in stage 1 is not necessarily optimal. Therefore, in stage 2, all previously *settled nodes* must lose this status and become *reached nodes* again. Only then, optimality can be guaranteed for the delivered paths.

4 Benchmarks and Results

For our benchmark we use a map of Lower Saxony, Germany, created from the OpenStreetMap (OSM) project[1] containing 251.889 nodes and 323.170 edges.[2] As mentioned above, the node sorting criterion has a strong impact on CH performance, both for its creation as for query time. We use three different criteria in order to rule out misleading result interpretation that may be caused by node sorting effects. Our three criteria are

- *edge difference (E)*,
- *edge difference + deleted neighbors (ED)*, and
- 190* *edge difference* +120* *deleted neighbors+cost of query (EDS)*.[3]

Note that these CH contain the same nodes; they only differ in edges.

We took a random sample of 100 of these nodes and ran shortest path searches for every pair (9900 queries). Results are measured in aggregated number of Dijkstra iterations for these queries.[4]

For the queries we utilize an implementation of

1. an uni-directional standard Dijkstra algorithm [5]
2. the original bi-directional Dijkstra algorithm described in [3],
3. the original bi-directional Dijkstra algorithm with *A* clipping*,
4. our combined approach as described in section 3.2,
5. our combined approach as described with *A* clipping*, and
6. our combined approach as described with *A* clipping* using the parameterized version of the A* algorithm with parameter $\alpha = 2$.

Table 1 summarizes the results with (2.) as reference for each node sorting criterion. While *A* clipping* alone (3.) already increases query speed by about 24%, the *A* search for bridge nodes* alone (4.) only leads to a 1% improvement. However, the combination of the two approaches (5.) yields the largest speed-up with **less than 50%** of the iteration required by the original bi-directional Dijkstra (2.). For less tuned CHs, algorithm 5. outperforms the original approach even more.

---

[1] www.openstreetmap.org

[2] We merge the original map's edges so that it only contains nodes with a degree of 3 or more.

[3] A more detailed description of these can be found in [3].

[4] Query times are not provided, as they are too dependant on hardware and implementation. However, to provide an order of magnitude: the best algorithm took on average less than 2ms per query.

[5] Note: As the underlying map is a CH, this Dijkstra implementation may use shortcut edges.

The parameterized version of the A* algorithm, however, did not bring further improvement. Apparently, the faster finding of a bridge node cannot make up for the additional overhead induced by having to settle many nodes twice.[6]

5 Conclusion and Outlook

We have presented two ways to incorporate the advantages of the A* algorithm for queries in Contraction Hierarchies for graphs in which an appropriate estimator for distances is available, such as the air distance is for real-world street maps. Combined, they increase the query speed by a factor of two without sacrificing optimality.

Taken that *EDS* is the best of the tested CH[7], our results suggest a negative correlation between the quality of the CH (determined by node sorting) and the improvement of algorithm 5. versus the original approach.

	E		ED		EDS	
Impl.	Iterations	%Iter.	Iterations	%Iter.	Iterations	%Iter.
1.	1.267.599.917	11,800.15%	1.267.599.917	43,977.41%	1.267.599.917	44,091.08%
2.	10.742.238	100.00%	2.882.389	100.00%	2.874.958	100.00%
3.	7.268.666	67.66%	2.197.199	76.23%	2.210.792	76.90%
4.	10.598.592	98.66%	2.854.785	99.04%	2.849.282	99.11%
5.	**4.439.970**	**41.33%**	**1.362.598**	**47.27%**	**1.410.839**	**49.07%**
6.	5.823.862	54.21%	1.809.129	62.76%	1.869.239	65.02%

Table 1 Benchmark results for 9900 shortest path queries on the map of Lower Saxony for node sortings *E*, *ED*, and *EDS*

References

1. BAUER, R., DELLING, D., SANDERS, P., SCHIEFERDECKER, D., SCHULTES, D., AND WAGNER, D. Combining hierarchical and goal-directed speed-up techniques for dijkstra's algorithm. *ACM Journal of Experimental Algorithmics 2010*, 15 (2010).
2. DIJKSTRA, E. W. A Note on Two Problems in Connexion with Graphs. *Numerische Mathematik 1* (1959), 269–271.
3. GEISBERGER, R., SANDERS, P., SCHULTES, D., AND DELLING, D. Contraction hierarchies: Faster and simpler hierarchical routing in road networks. *Proceedings of the 7th Workshop on Experimental Algorithms (WEA'08) 5038 of Lecture Notes in Computer Science* (2008), 319–333.
4. HAHNE, F. *Kürzeste und schnellste Wege in digitalen Straßenkarten (Dissertation)*. Universität Hildesheim, Hildesheim, 2000.

---

[6] This evalutation also holds for other values for α.

[7] [3] supports this assumption.

An IP approach to toll enforcement optimization on German motorways

Ralf Borndörfer, Guillaume Sagnol, and Elmar Swarat

Abstract This paper proposes the first model for toll enforcement optimization on German motorways. The enforcement is done by mobile control teams and our goal is to produce a schedule achieving network-wide control, proportional to spatial and time-dependent traffic distributions. Our model consists of two parts. The first plans control tours using a vehicle routing approach with profits and some side constraints. The second plans feasible rosters for the control teams. Both problems can be modeled as Multi-Commodity Flow Problems. Adding additional coupling constraints produces a large-scale integrated integer programming formulation. We show that this model can be solved to optimality for real world instances associated with a control area in East Germany.

1 Introduction

In 2005 Germany introduced a distance-based toll for commercial trucks weighing twelve tonnes or more in order to fund growing investments for maintenance and extensions of motorways. The enforcement of the toll is the responsibility of the German Federal Office for Goods Transport (BAG). It is implemented by a combination of 300 automatic stationary gantry bridges and by tours of 300 control vehicles on the entire highway network. The control tours are operated by teams, composed of one or two inspectors. The vehicles and crews are based at a number of depots and are responsible for certain control areas. The goal of our approach is to construct a set of tours that guarantee a network-wide control whose intensity is proportional to spatial and time dependent traffic distributions. The tours must fit

Ralf Borndörfer
Zuse Institute Berlin, Optimization, Takustr. 7, 14195 Berlin, Germany, e-mail: borndoerfer@zib.de

Guillaume Sagnol
Zuse Institute Berlin, Optimization, Takustr. 7, 14195 Berlin, Germany, e-mail: sagnol@zib.de

Elmar Swarat
Zuse Institute Berlin, Optimization, Takustr. 7, 14195 Berlin, Germany, e-mail: swarat@zib.de

D. Klatte et al. (eds.), *Operations Research Proceedings 2011*, Operations Research Proceedings, 317
DOI 10.1007/978-3-642-29210-1_51, © Springer-Verlag Berlin Heidelberg 2012

within a feasible crew roster, respecting all legal rules, over a time horizon of several weeks. An important restriction is that each team and vehicle can only control highway sections in their associated control area, close to their base depots. Because of this restriction we can not use sequential and partly anonymous planning approaches for duty scheduling and rostering like in public transport (see [8] ch. 1), since those would lead to infeasible staff rosters. Hence, personalized duty roster planning must be used in our model. These two components are combined into an integrated model using suitable constraints.

To the best knowledge of the authors no optimization approach for toll enforcement has appeared in the literature. Related publications deal with problems such as tax evasion or ticket evasion in public transport; they mainly discuss the expected behaviour of evaders or payers from a theoretical point of view, e.g. [1], or optimal levels of inspection, see [2]. A recent approach to inspector scheduling in public transport was proposed DSB S-tog in Denmark [5], but in contrast to our problem they focus on temporal scheduling of the inspectors and not on their routes through the network.

The paper is structured as follows: In Section 2 we present models for tour and roster planning, and build from these components an integrated toll enforcement optimization model. This model is used in Section 3 for a computational study. We present results for two instances of our *Toll Enforcement (Optimization) Problem (TEP)* which are part of the MIPLIB 2010 [4]. Our computations show that real-world instances of the TEP can be solved to proven optimality for an entire control region over a time horizon of four weeks.

2 Optimal Toll Enforcement

Our toll enforcement optimization model consists of two parts. Section 2.1 presents a model for tour planning, and Section 2.2 a model for the assignment of feasible staff rosters to all inspectors. These two components are combined into an integrated model using suitable constraints.

2.1 A Graph and IP-Model for the Planning of Inspector Tours

The TEP can be described in terms of a *section graph* $G = (S, N)$, in which the nodes $s \in S$ represent *control sections*, which are sub-parts of the motorway network with a length of approximate 25-50 km. An edge $n \in N$ connects two sections, if they have at least one motorway junction or motorway exit in common. Furthermore, there is a given planning horizon T, e.g., four weeks, and some given time discretization Δ. Since it is required to define both the spatial routing and the temporal sequence of the tours, we extend G to a space-time digraph $D = (V, A)$, the *tour planning graph*. Its nodes $v \in V$ are either defined as a pair of a section and a point in time,

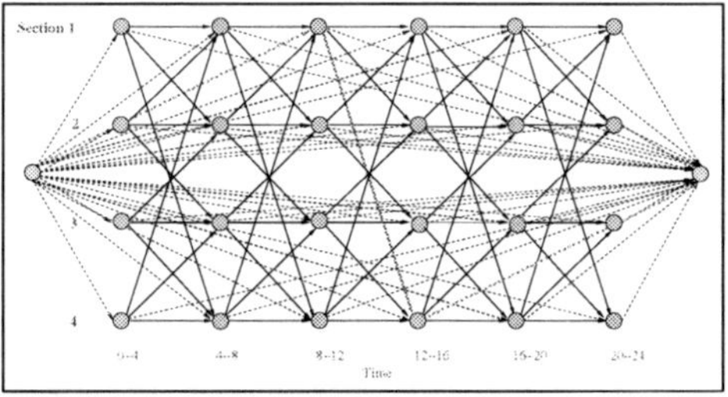

Fig. 1 Construction of the tour planning graph.

i.e., $v = (s,t) \in S \times Z := \{0,\Delta,\dots,T-\Delta,T\} \subset [0,T]$, or they represent artifical start and end nodes d_s and d_t for the vehicle paths (depot nodes). Directed arcs connect either adjacent time intervals of the same section, or they connect adjacent sections, i.e., $\forall s \in S$ there is $((s,t_1),(s,t_2)) \in A$ with $t_2 = t_1 + \Delta$, starting at $t_1 = 0$ until $t_2 = T$, and if $(s_1,s_2) \in N$ it holds that $((s_1,t),(s_2,t+\Delta)) \in A \,\forall t \in Z \setminus \{T\}$. Figure 1 illustrates this construction. In addition, arcs are inserted from the start depot to all other non-depot nodes and from all non-depot nodes to the end depot node. Finally, a profit value is associated with each node $v = (s,t)$. We consider the problem of finding a feasible (d_s,d_t)-path in D for each vehicle f on each day, that respects a tour length restriction of 8h30. Each control tour corresponds to such a *control path*. The profit w_p of a control path p is the sum of the profit of its visited nodes. This approach could be seen as a vehicle routing problem with profits under some additional constraints. Vehicle routing is an well-established research area, see [6] for an overview. For the case of dealing with profits Feillet et al. [3] give a literature survey.

There are restrictions on the feasible starting and ending times of the control tours; the feasible times are defined by the *(Working) Time Windows*. Let P be set of all control paths in D and $P_{f,j} \subset P$ the set of all paths that are feasible for vehicle $f \in F$ and start at day $j \in J$. Furthermore for a given section $s \in S$, the set of all paths $p \in P$ that visit a node $v = (s,t_i) \in V$ is denoted by P_s and the minimum control quota is named by κ_s. The *Tour Planning Problem (TPP)* is then formulated as a 0/1 multi-commodity flow problem in D, where vehicles f represent the commodities. We introduce 0/1-variables z_p, $p \in P$, that indicate if tour p is chosen or not. Then the TPP can be modeled by the following integer program:

$$\max \sum_{p \in P} w_p z_p \tag{1}$$

$$\sum_{p \in P_{f,j}} z_p \leq 1, \qquad \forall (f,j) \in F \times J \tag{2}$$

$$\sum_{p \in P_s} z_p \geq \kappa_s, \qquad \forall s \in S \tag{3}$$

$$z_p \in \{0,1\}, \quad \forall p \in P. \tag{4}$$

In this formulation the objective function (1) maximizes the profit of all selected tours. The requirement that each vehicle f can do at most one tour per day is assured by Constraints (2). Constraints (3) guarantee that at least κ_s paths, that traverse section s, are chosen in any feasible control schedule. The last contraints (4) are the integrality constraints.

2.2 Integration of Duty Roster Planning

The second task in the TEP is the planning of the rosters, called the *Inspector Rostering Problem (IRP)*. There, the objective is to minimize the costs, which can be real costs of a duty or artifical costs that penalize some feasible but inappropriate sequence of duties. An example for this is if a duty on the next day respects the minimum rest times, but starts earlier than the day before, e.g., Mo 8-17, Tu 6-15.

We formulate the IRP as a multi-commodity flow problem in a directed graph $\tilde{D} = (\tilde{V} := (\hat{V} \cup \{s,t\}), \tilde{A})$ with two artificial start and end nodes s,t. The nodes $v \in \hat{V}$ represent duties as a pair of day and time-window. The arcs $(u,v) \in \tilde{A} \subseteq \tilde{V} \times \tilde{V}$ model a feasible sequence of two duties according to legal rules. Again the depot nodes s and t are connected with all non-depot nodes. Furthermore, let M be the set of all inspectors and W the set of all weeks. Then $\hat{V}_w \subseteq \hat{V}$ is defined as the set of all duties in week $w \in W$. In addition, let t_v be the duration of duty $v \in \hat{V}$, while a^m_{min} and u^m_{max} indicate the mimimum or maximum weekly labor time of inspector $m \in M$. Introducing flow variables $x^m_{u,v}$ for each arc (u,v) and inspector m, we propose the following integer programming formulation for the IRP:

$$\min \sum_{m \in M} \sum_{(u,v) \in \tilde{A}} c_{u,v} x^m_{u,v} \tag{5}$$

$$\sum_v x^m_{s,v} = 1, \qquad \forall m \in M \tag{6}$$

$$\sum_k x^m_{v,k} - \sum_u x^m_{u,v} = 0, \qquad \forall v \in \hat{V}, m \in M \tag{7}$$

$$\sum_{u \in \hat{V}_w} \sum_v t_u x^m_{u,v} \leq a^m_{max}, \qquad \forall w \in W, m \in M \tag{8}$$

$$\sum_{u \in \hat{V}_w} \sum_v t_u x^m_{u,v} \geq a^m_{min}, \qquad \forall w \in W, m \in M \tag{9}$$

$$x^m_{u,v} \in \{0,1\}, \quad \forall (u,v) \in \tilde{A}, m \in M. \tag{10}$$

The objective function (5) minimizes the cost. Constraints (6) guarantee that exactly one roster path is chosen for each inspector. Such a path is called *Inspector Roster Path*. By (7) we model the flow conservation in each non-depot node. Maximum and minimum weekly working times are enforced by inequalities (8,11). Finally, in (9) we have the integrality constraints for the flow variables.

Table 1 IP-Solution analysis of some instances of the control region Brandenburg. The parameter "dc" indicates the use of duty costs, "dm" a pre-given duty mix according to the time windows and $\Delta(h)$ the time discretization. The value of the root LP is denoted by v(lp) and the best integer solution by v^*. The solution time limit equals $6h = 21600sec$.

instance	$\Delta(h)$	dc	dm	columns	rows	v(lp)	v^*	gap(%)	time(sec.)
T1	4	x	x	136264	17245	368997.56	350288.76	0.06	12283.0[1]
T2	4	x	-	136264	17237	449883.49	435343.00	opt.	181.62
T3	4	-	x	128808	17257	533597.42	499561.80	1.84	21600.00
T4	2	x	x	376328	22877	677133.58	644458.66	0.05	21600.00
T5	2	x	-	376328	22869	732316.22	709276.99	0.04	3554.2[1]
T6	2	-	x	368872	22889	846722.87	796495.47	1.52	21600.00
bab5	4	x	x	21600	4964	117062.16	106411.84	opt.	640.34
bab3	2	x	x	393800	23069	686288.87	655559.30	0.45	21600.00

The TPP and the IRP are connected by coupling constraints into an integrated formulation for the TEP. To this purpose, we denote by $P_{f,u}$ the set of all control paths feasible for vehicle f and duty $u \in \hat{V}$. The parameter n_f gives the number of inspectors in vehicle f and by $m \in f$ it is meant, that the inspector m uses vehicle f (which is a fixed assignment). This leads to the following coupling constraint.

$$\sum_{p \in P_{f,u}} n_f z_p - \sum_{m \in f} \sum_v x_{u,v}^m = 0 \ \forall f \in F, u \in \hat{V} \tag{11}$$

The constraints (10) ensure that for each control path p in D all inspectors in the corresponding team have a feasible inspector roster path with a duty in the time horizon of the planned tour. The objective function is a combination of collecting the profit (1) and minimizing the cost (5).

3 Computational Results and Conclusion

We tested our model on instances associated with the control region Brandenburg that are based on real-world data. Six instances (T1,...,T6) model four week planning periods and basic legal rules like minimum rest times, but they differ in some parameter settings, see Table 1. The others belong to the MIPLIB 2010, see below. In the table, columns "dc" and "dm" characterize the instances as follows: "dc" stands for using direct duty costs while "dm" demands a duty mix regarding to the time windows in the roster, e.g., approx. 40% of all duties must begin at 6am and end at 15pm. All computations were done on a PC with an Intel i7 Quad-Core processor with 2.97 GHz and 16 GB RAM. CPLEX 12.2 [7] was applied as an IP solver using four threads. Furthermore, we used a time limit of 6h and a limit of 10 GB for the Branch&Bound tree.

---

[1] Memory-Limit reached

Table 1 shows the results of our computations. The term v(lp) denotes the optimal value of the root LP and v^* the best integer value. An important result is that all duty-cost instances could be solved to near-optimality within the given time limit of six hours. An interesting observation is that a larger number of rows and columns does not always lead to an increase of solution times or higher integrality gaps. Especially in the 4h-case (T1,T2), a major part of the complexity originates from the duty mix constraints. In fact, the solution time is reduced drastically if we omit them. Another important observation is that the problem becomes more difficult if we replace duty costs by coefficients that penalize inappropriate rosters (T3, T6).

The MIPLIB 2010 problem library [4] contains two instances from the TEP. Both model the TPP in a slightly different way then described in Section 2.1, which, however, produces equivalent results. The first instance is bab5[2]: It is similar to T1, but models a plan for 8 days, see Table 1. It can be solved optimally within 15 minutes by CPLEX 12.2. The second instance bab3 has the same parameters as T4 (see Table 1), except for a small difference with respect to the length of a duty. It is more difficult to solve compared to T4 since it has a gap of 0.45% after 6 hours of running CPLEX. Even after several days of computation this instance could not be solved to optimality.

We conclude that we are able to model the TEP by two graph models and then to formulate it by an IP that integrates all important legal rules. Solving this IP results in high quality solutions for real-world instances with time discretizations of two or four hours. In the future we want to integrate additional rules and implement advanced algorithms to solve more complex instances.

References

1. M.G. ALLINGHAM AND A. SANDMO (2006) *Income Tax Evasion: A Theoretical Analysis.* Intl. Library of Crit. Writings in Economics Vol 195 No. 3 2006. Edward Elgar Publ. Ltd.
2. C. BOYD AND C. MARTINI AND J. RICKARD AND A. RUSSELL (1989). *Fare Evasion and Non-Compliance: A Simple Model.* Journal of Transport Economics and Policy. London School of Economics and Political Science and the University of Bath.
3. D. FEILLET AND P. DEJAX AND M. GENDREAU (2005) *Traveling Salesman Problems with Profits.* Transportation Science Vol. 39 No. 2 2005. Informs.
4. T. KOCH AND T. ACHTERBERG AND E. ANDERSEN AND O. BASTERT AND T. BERTHOLD AND R. E. BIXBY AND E. DANNA AND G. GAMRATH AND A. M. GLEIXNER AND S. HEINZ AND A. LODI AND H. MITTELMANN AND T. RALPHS AND D. SALVAGNIN AND D. E. STEFFY AND K. WOLTER (2011) *MIPLIB 2010.* Mathematical Programming Computation Vol. 3 No. 2 2011. Springer.
5. P. THORLACIUS AND J. CLAUSEN AND K. BRYGGE (2009). *Scheduling of Inspectors for Ticket Spot Checking in Urban Rail Transportation.* DSB S-tog. Copenhagen.
6. P. TOTH AND D. VIGO (2002) *The Vehicle Routing Problem.* Society for Industrial and Applied Mathematics. Philadelphia.
7. *User-Manual CPLEX 12.2* (2010) IBM ILOG CPLEX. IBM Software Group.
8. S. WEIDER (2007) *Integration of Vehicle and Duty Scheduling in Public Transport.* Technische Universität Berlin. Dissertation.

---

[2] "BAB" is the offical abbreviation of the German motorways

Driver scheduling based on "driver-friendly" vehicle schedules

Viktor Árgilán, János Balogh, József Békési, Balázs Dávid,
Miklós Krész, Attila Tóth

Abstract In the area of optimization of public transportation there are several methods for modeling and solving vehicle and driver scheduling problems. We designed a sequential heuristic method for solving the combined (vehicle and driver scheduling) problem. Our model is based on a modification of the vehicle schedules to satisfy driver requirements. We introduced a driver friendly approach of the optimization of the scheduling, which is closer to the practice used by public transportation companies. We give test results for this, which are shown as illustrative example for the method in the scheduling of the bus trips of Szeged city, Hungary.

1 Introduction

In this paper we consider the combined vehicle and driver scheduling problem, one of the main daily scheduling tasks of transportation companies. Public transportation usually consists of bus or other vehicle lines, which connect some stations. The lines and the daily trips of the lines are usually given by a timetable. This is given in advance in practice, based on the travel demands and logistic decisions. The timetable explicitly gives the departure and arrival time of the trips for each line. It is also well-known, that we can form several series of trips, which can be executed continuously by a vehicle. If we can cover all the trips with such chains, then we get a set of feasible schedules. Another task is to assign drivers to these vehicle schedules. This should be done in such a way, that all the natural requirements given by transportation companies are satisfied. In addition to this, our aim is to find the best possible schedule, supposing that we can measure the goodness of them.

There are several methods for modeling and solving vehicle and driver scheduling problems. The classical way is the so called sequential method, when first we

Viktor Árgilán, János Balogh, József Békési, Balázs Dávid,
Miklós Krész, Attila Tóth
University of Szeged, Juhász Gyula Faculty of Education, Department of Applied Informatics,
e-mail: {gilan,balogh,bekesi,davidb,kresz,attila}@jgypk.u-szeged.hu

D. Klatte et al. (eds.), *Operations Research Proceedings 2011*, Operations Research Proceedings, 323
DOI 10.1007/978-3-642-29210-1_52, © Springer-Verlag Berlin Heidelberg 2012

solve the vehicle scheduling phase, then based on the given vehicle schedules we calculate the driver schedules. Other approaches combine the two steps and handle them as sub-problems of the combined problem. Powerful solution methods have been given for this integrated problem, see Borndörfer et al. [3] for an example. However, the solution of these combined models is complicated and requires remarkable computation time, especially for large problems. In addition to this, the solutions given by the above methods do not consider the phisycal capacity of the drivers (e.g. drivers need to change buses several times a day, which is very demanding for them). This aspect can arise an actual need in the schedule of transportation companies, as their drivers can usually be assigned to a single bus during their whole shift.

The aim of our paper is to present a solution algorithm for the combined problem which can be used as the part of an interactive decision support system of a company. Such a system has to provide a solution for the problem in adequate time, as executing several solutions with different parameter settings can aid the planning process of the company. We developed a sequential heuristics method for solving the combined problem, with regards to the before mentioned capacity problem of the drivers. The method is based on a modification of the vehicle schedules to satisfy the driver requirements. We refer to these transformed schedules as "driver-friendly" schedules. Using these modified schedules, our algorithm can create the final driver schedules easily.

2 Models of vehicle and driver scheduling

The most frequently used model for solving the vehicle scheduling is the so called MDVSP model. It was clearly defined by Bodin et al. [2] and was shown to be NP-hard by Bertossi et al. [1]. The model represents the most important components of the real-life cases. The driver scheduling task can be handled as a set partition problem. Fischetti et. al. [5, 6] proved that this problem is also NP-hard.

The aim of the integrated vehicle and driver scheduling is to find a minimum cost vehicle schedules and driver duties such that they are *compatible*. Vehicle schedules and driver duties are compatible if each schedule event is executed by exactly one vehicle and driver. The cost function of the model should represent the whole vehicle and driver costs. Combined models are powerful, because they give the exact optimal solutions of the problem. However their solution may take long running times, especially for large real-life problems. A mathematical model for the integrated vehicle and driver scheduling problem is given by Huisman et al. [7, 8], which is based on the combination of a network and set partition models.

3 The "driver-friendly" algorithm

The methods presented in Section 2 usually give a solution where the drivers have to switch buses several times a day. This procedure is a demanding task for the drivers, and also requires the introduction of certain driver events (e.g. administration). This is the reason why we introduced an approach which is closer to the practice used by public transportation companies: drivers are assigned exactly one bus that they drive continuously during their shift. This approach is an iterated method, which uses a sequential solution of the subproblems (see Figure 1).

The first step creates the vehicle schedules from the trips of the input data. The resulting schedules are sorted into different classes according to their length. Depending on their classes, some of the schedules are divided into two parts, if their length is greater then the maximal length of the shifts. These parts are then joined together into new pairs (if possible), to form feasible shifts with regards to their length. Finally, a transformation step is executed, where the intervals of the driver breaks are determined in the shifts. If trips have to be deleted because of the intervals of the breaks, they become our new input, and the steps of the algorithm are called iteratively, until all schedules satisfy the driver rules.

Fig. 1 Main steps of the algorithm

Step 1: Solving the vehicle scheduling problem The first step of our sequential heuristic is the solution of the vehicle scheduling problem. The applied method should give a good quality solution, with a fast running time, as this will be called iteratively. Using the MILP solver for a model of the problem would require a long running time, so heuristics were introduced to get a result in a significantly shorter time. We examined several different solution algorithms, all based on the time-space network introduced in [9]. An overview of these algorithms can be found in [4]. An important aspect studied at all of these methods is the applicability on real-life problems. For effective use in public transportation, solution should be obtained fast, and the cost should also be close to the optimal cost of the problem.

Step 2: Classification, cut and join The basic idea behind the heuristic is the classification of vehicle schedules considering human limits, since there are EU regulations which maximize the working time of a driver on a day. For this, we examine the length of the different schedules, and determine the theoretical number of drivers needed to execute them. The definition of the classes is based on two types of shifts usually used by the companies. The first one is the continuous full-time shift which contains only short breaks. The other one is the split shift which contains a long break, when the driver can go home and has to come back later to continue his shift. According to this, we introduce the three following classes:

$$A = \{s \in S \,|\, l(s) \leq M\},$$
$$B = \{s \in S \,|\, M < l(s) < 1.5M\},$$
$$C = \{s \in S \,|\, 1.5M < l(s)\},$$

where s is a schedule, S is the set of the schedules, M is the maximal length of the shifts and $l(s)$ gives the length of the shift s. The value of $l(s)$ is calculated by adding the driving time of the vehicle schedule and the worktime of the associated driver events (passengers getting on and off the vehicle, etc.). The value of M is the maximal driving time minus the working time of driver events that are needed to be performed by all drivers regardless of their schedule (check-in, check-out, technical maintenance, etc.). Class A contains schedules which can be executed by one driver. Schedules in class B can be executed by two drivers (one in full-time, and one in part-time), so they have to be divided into two parts. The dividing points are alternating between around $1/3$ and $2/3$ of the length of the schedule. Alternating the place of the dividing point is needed so that the part-time shifts could be matched together later on to form so called split shifts. Class C contains schedules which also need two drivers to execute, but as opposed to schedules in class B, the two drivers both have to be in full-time. These part-shifts are addressed as work-pieces in the future. Note that each schedule in class C can be assigned two drivers as its length does not exceed the maximal shift length of two drivers.

There are regulations not only for the length of the shifts but also for the minimal resting time in a day. In our case, we also have a minimal time for total resting. If the shift is split shift, then there is a minimal length for the long break and for the minimal length of the resting time after the shift as well. The sum of these two values must be at least so long as the required total resting time. Accordingly, during the joining (and producing split shifts), we introduced further rules for finding possible matching work-pieces.

1. There must be enough time for long break between the two work-pieces.
2. There must be enough time for resting after the shift on that day.
3. The sum of the resting time and long break must be at least the minimal daily resting time.

Using the above rules, a graph can be built where the nodes are the work-pieces, and there is an edge between two work-pieces if they are compatible. Performing a maximal matching on this graph, we join those work-pieces as a split shift which are in the solution, and leave those work-pieces as a single, full-time shift which are not.

Step 3: Transformation The aim of this step is providing a free time interval for short breaks inside the shifts. In the regulations there is a minimal and maximal length for the short break, a time interval according to the working time when it must be performed and exact geographical places where it can be performed. First if there is enough time between two events of a shift within the required time interval of the break, then it can be assigned in that gap. Otherwise, tasks have to be removed from the corresponding interval. The algorithm examines the tasks in that interval, and determines the best one or two subsequent trips that can be removed, producing the required time gap. The duration of the possible arising deadhead trips is also taken into consideration, when the algorithm chooses the trips to be removed. The best case if these trips can be moved to another shift without any conflict. If such moves are possible, then these tasks are removed from the shift and inserted to the other shifts considering the possible deadhead trips. Otherwise, such trips are deleted from the shift, and put to a list. This list contains trips that are needed to be rescheduled, and will be refered to as free-list in the future. Trips put to the list are chosen to have minimal overlaps in time with the trips already on the free-list.

Step 4: Rearranging the free-list The free-list is basically a set of trips, which have to be re-scheduled after the transformation step executed for all shifts. This is executed using an iteration method. We can easily see, that the number of trips on the free-list is strictly decreasing in every iteration, so the algorithm stops in finite steps (all results on our test database show that we only need one iteration step). If there are no more trips on the free-list, the algorithm stops, and all of the resulting schedules are "driver-friendly" vehicle schedules.

4 Test results

We tested our algorithm on real-life instances taken from the public transportation of Szeged, Hungary (a middle-sized city with ~ 170.000 inhabitants and ~ 2.000 trips on a usual workday). We evaluated our results using two different cost functions. The vehicle cost is the linear combination of the number of vehicles, and the distance covered by the trips and deadhead trips. The driver cost is the sum of working time required to satisfy all events in the resulting shifts.

Our results are compared to the corresponding costs given by the schedules of the company, while we also used statistical and algorithmic methods to give theoretical lower bounds for both types of costs. The lower bound for the vehicle schedules was calculated by solving the vehicle scheduling problem with an MILP solver. The lower bound for the working time was given by a statistical method, which used the distribution of the trips to determine a theoretical lower bound on the number of drivers, and an algorithm that computed the minimal working time of the vehicle tasks (independent of the number of vehichles). The above method will be described in a forthcoming paper.

Our experience shows that both vehicle and driver costs are about 3% better than the corresponding costs of the company. This is a satisfactory result, as it gives a significant decrease in the operative cost of the company. Also, our costs are only about 8% away from the calculated theoretical lower bounds. These results were achieved on a vehicle schedule produced by a greedy variable fixing heuristic proposed in [4]. The running time of our algorithm varied between 5 to 10 minutes, depending on the test case. The computer configuration used for solving the test instances was a PC of two cores, with each processor running at 2.0 GHz, and supplied with 2 GB of memory.

5 Conclusions

Our paper proposed a combined vehicle and driver scheduling algorithm. We introduced the concept of "driver-friendliness", and solved our problems using this as our main goal. The test instances for the method were real-life data taken from the public transportation of Szeged. The quality of the solutions and the short running time both show that the algorithm can be applied efficiently for real-life cases.

References

1. Bertossi, A.A., Carraresi, P., and Gallo, G.: On Some Matching Problems Arising in Vehicle Scheduling Models. Networks **17**(3), 271–281 (1987)
2. Bodin, L., Golden, B., Assad, A., and Ball, M.: Routing and Scheduling of Vehicles and Crews: The State of the Art. Computers and Operations Research **10**, 63–211 (1983)
3. Borndörfer, L., Löbel, A., Weider, S.: A Bundle Method for Multi-Depot Integrated Vehicle and Crew Scheduling in Public Transport. Lecture Notes in Economics and Mathematical Systems **600**, 3–24 (2008)
4. Dávid, B.: Heuristics for the Multiple-Depot Vehicle Scheduling Problem. In: Proceedings of the MATCOS-10 Conference, accepted for publication
5. Fischetti, M., Lodi, A., Toth, P.: A branch-and-cut algorithm for the multi depot vehicle scheduling problem. Technical report (1999)
6. Fischetti, M., Toth, P.: An additive bounding procedure for combinatorial optimization problems. Operations Research **37**(2), 319–328 (1989)
7. Huisman, D.: Integrated and Dynamic Vehicle and Crew Scheduling. PhD thesis, Erasmus University of Rotterdam (2004)
8. Huisman, D., Freling, R., Wagelmans, A.P.: Multiple-depot integrated vehicle and crew scheduling. Transportation Science **39**(4), 491–502 (2005)
9. Kliewer, N., Mellouli, T., Suhl, L.: A timespace network based exact optimization model for multi-depot bus scheduling. European Journal of Operational Research **175**(3), 1616–1627 (2007)

Optimizing ordered median functions with applications to single facility location

Victor Blanco, Justo Puerto, and Safae El Haj Ben Ali

Abstract This paper considers the problem of minimizing the ordered median function of finitely many rational functions over compact semi-algebraic sets. Ordered median of rational functions are not, in general, neither rational functions nor the supremum of rational functions. We prove that the problem can be transformed into a new problem embedded in a higher dimension space where it admits a convenient representation. This reformulation admits a hierarchy of SDP relaxations that approximates, up to any degree of accuracy, the optimal value of those problems. We apply this general framework to a broad family of continuous location problems solving some difficult problems.

1 Introduction

Ordered Median Function (OMF) operators provide a parameterized class of mean type aggregation operators (see [5, 11] and the references therein for further details). Many notable mean operators such as the max, arithmetic average, median, k-centrum, range and min, are members of this class. They have been widely used in location theory and computational intelligence because of their ability to represent flexible models of modern logistics and linguistically expressed aggregation instructions in artificial intelligence. In spite of its intrinsic interest, as far as we know, a common approach for solving this family of problems is not available. Nevertheless, one can find in the literature different methods for solving particular instances of problems within this family, see e.g. [2, 5–9]. The first goal of this paper is to develop a unified tool for solving this class of optimization problems. In this line, we prove that the general problem can be transformed into a new problem embedded

Victor Blanco
Universidad de Granada, Dep. of Algebra, e-mail: vblanco@ugr.es. Thanks to Grant JCI-2009-03896

Justo Puerto · Safae El Haj Ben Ali
Universidad de Sevilla, IMUS e-mail: puerto@us.es. Thanks to FQM-5849 (Junta de Andalucía\FEDER) and MTM2010-19576-C02-01 (MICINN, Spain)

in a higher dimension space where it admits a convenient representation that allows to arbitrarily approximate or to solve it as a minimization problem over an adequate closed semi-algebraic set.

We apply this general tool to design a common approach also to solve the ordered median family of location problems, for different distances and in any finite dimension. This is essentially the second goal of this paper. Of course, we know that the problem in its full generality is $NP - hard$ since it includes general instances of concave minimization. Therefore, we cannot expect to obtain polynomial algorithms for this class of problems. Rather, we will apply a new methodology first proposed by Lasserre [3], that provides a hierarchy of semidefinite problems that converge to the optimal solution of the original problem, with the property that each auxiliary problem in the process can be solved in polynomial time.

2 Minimizing the ordered weighted average of finitely many rational functions

Let $\mathbf{K} \subset \mathbb{R}^d$ be a basic semi-algebraic set defined as $\mathbf{K} := \{x \in \mathbb{R}^d : g_j(x) \geq 0, \quad j = 1,\ldots,\ell\}$, for $g_1,\ldots,g_\ell \in \mathbb{R}[x]$. Let us introduce the function $\mathrm{OM}(x) = \sum_{k=1}^m \lambda_k(x) f_{(k)}(x)$, for some rational functions $(f_j) \subset \mathbb{R}[x]$, being $f_k = p_k/q_k$ rational functions with $p_k, q_k \in \mathbb{R}[x]$, $\lambda_k(x) \in \mathbb{R}[x]$, and $f_{(k)}(x) \in \{f_1(x),\ldots,f_m(x)\}$ such that $f_{(1)}(x) \geq f_{(2)}(x) \geq \cdots \geq f_{(m)}(x)$ for $x \in \mathbb{R}^n$. We assume that $\mathbf{K}$ satisfies Putinar's property (see [10] for the definition of this property and its importance in this theory) and that $q_k > 0$ on $\mathbf{K}$, for every $k = 1,\ldots,m$.

Consider the following problem:

$$\rho_\lambda := \min_x \{\mathrm{OM}(x) : x \in \mathbf{K}\}, \tag{OMRP$_\lambda^0$}$$

Associated with the above problem we introduce an auxiliary problem. For each $i = 1,\ldots,m$, $j = 1,\ldots,m$ consider the decision variables w_{ij} that model for each $x \in \mathbf{K}$

$$w_{ij} = \begin{cases} 1 & \text{if } f_i(x) = f_{(j)}(x), \\ 0 & \text{otherwise.} \end{cases}$$

Now, we consider the problem:

$$\overline{\rho}_\lambda = \min_{x,w} \sum_{j=1}^m \lambda_j(x) \sum_{i=1}^m f_i(x) w_{ij} \tag{OMRP$_\lambda$}$$

$$\text{s.t. } \sum_{j=1}^m w_{ij} = \sum_{j=1}^m w_{ji} = 1, \text{ for } i = 1,\ldots,m, \tag{1}$$

$$w_{ij}^2 - w_{ij} = 0, \text{ for } i, j = 1, \ldots, m,$$

$$\sum_{i=1}^{m} w_{ij} f_i(x) \geq \sum_{i=1}^{m} w_{ij+1} f_i(x), \ j = 1, \ldots, m, \tag{2}$$

$$\sum_{i=1}^{m} \sum_{j=1}^{m} w_{ij}^2 \leq m, \tag{3}$$

$$w_{ij} \in \mathbb{R}, \text{ for } i, j = 1, \ldots, m, \quad x \in \mathbf{K}. \tag{4}$$

The first set of constraints ensures that for each x, $f_i(x)$ is sorted in a unique position. The second set ensures that the jth position is only assigned to one rational function. The next constraints are added to assure that $w_{ij} \in \{0, 1\}$. The fourth one states that $f_{(1)}(x) \geq \cdots \geq f_{(m)}(x)$. The last set of constraints ensures the satisfaction of Putinar's property of the new feasible region. (Note that this last set of constraints are redundant but it is convenient to add them for a better description of the feasible set.)

These two problems, $(\mathrm{OMRP}_\lambda^0)$ and (OMRP_λ) satisfy the following relationship that is proved in [1].

Theorem 1. *Let x be a feasible solution of $(\mathrm{OMRP}_\lambda^0)$ then there exists a solution (x, w) for (OMRP_λ) such that their objective values are equal. Conversely, if (x, w) is a feasible solution for (OMRP_λ) then there exists a solution x for $(\mathrm{OMRP}_\lambda^0)$ having the same objective value. In particular $\rho_\lambda = \hat{\rho}_\lambda$.*

Then, we observe that $f_i = p_i/q_i$ for each $i = 1, \ldots, m$. Therefore, the constraint $\sum_{i=1}^{m} w_{ij} f_i(x) \geq \sum_{j=1}^{m} w_{ij+1} f_i(x)$ can be written as a polynomial constraint as:

$$\sum_{i=1}^{m} w_{ij}\, p_i(x) \prod_{k \neq i}^{m} q_k(x) \leq \sum_{i=1}^{m} w_{ij+1}\, p_i(x) \prod_{k \neq i}^{m} q_k(x) \text{ for } j = 1, \ldots, m.$$

Let us denote by $\overline{\mathbf{K}}$ the basic closed semi-algebraic set that defines the feasible region of (OMRP_λ).

Lemma 1 ([1]). *If $\mathbf{K} \subset \mathbb{R}^m$ satisfies Putinar's property ([10]) then $\overline{\mathbf{K}} \subset \mathbb{R}^{n+m^2}$ satisfies Putinar's property.*

In addition, the objective function of (OMRP_λ) can be written as a quotient of polynomials in $\mathbb{R}[x, w]$. Indeed, take

$$p_\lambda(x, w) = \sum_{j=1}^{m} \lambda_j(x) \sum_{i=1}^{m} w_{ij}\, p_i(x) \prod_{k \neq i}^{m} q_k(x) \text{ and } q_\lambda(x, w) = \prod_{k=1}^{m} q_k(x). \tag{5}$$

Then, $\sum_{j=1}^{m} \lambda_j(x) \sum_{i=1}^{m} f_i(x) w_{ij} = \dfrac{p_\lambda(x, w)}{q_\lambda(x, w)}$. Hence, we can transform Problem (OMRP_λ) in an infinite dimension linear program on the space of Borel measures defined on $\overline{\mathbf{K}}$.

Proposition 1 ([1]). *Let $\overline{\mathbf{K}} \subset \mathbb{R}^{n+m^2}$ be the closed basic semi-algebraic set defined by the constraints (1-4). Consider the infinite-dimensional linear optimization problem*

$$\mathscr{P}_\lambda: \quad \widehat{\rho}_\lambda = \min_{x,w} \left\{ \int_{\overline{\mathbf{K}}} p_\lambda d\mu : \int_{\overline{\mathbf{K}}} q_\lambda d\mu = 1, \mu \in M(\overline{\mathbf{K}}) \right\},$$

being p_λ, $q_\lambda \in \mathbb{R}[x,w]$ as defined above. Then $\rho_\lambda = \widehat{\rho}_\lambda$.

The reader may note the great generality of this class of problems. Depending on the choice of the polynomial weights λ we get different classes of problems. For instance, $\lambda = (1,0,\ldots,0,0)$ corresponds to minimize the maximum of a finite number of rational functions; $\lambda = (1,\overset{(k)}{\ldots},1,0,\ldots,0)$ corresponds to minimize the sum of the k-largest rational functions; $\lambda = (0,\overset{(k_1)}{\ldots},0,1,\ldots,1,0,\overset{(k_2)}{\ldots},0)$ models the minimization of the (k_1,k_2)-trimmed mean of m rational functions; $\lambda = (1,\alpha,\ldots,\alpha)$ corresponds to the α-centdian; $\lambda = (1,\ldots,-1)$ corresponds to minimize the range of a set of rational functions.

3 A convergence result of semidefinite relaxations 'à la Lasserre'

We are now in position to define the hierarchy of semidefinite relaxations for solving the **MOMRF** problem. Let $\mathbf{y} = (y_\alpha)$ be a real sequence indexed in the monomial basis $(x^\beta w^\gamma)$ of $\mathbb{R}[x,w]$ (with $\alpha = (\beta,\gamma) \in \mathbb{N}^n \times \mathbb{N}^{m^2}$). Let $p_\lambda(x,w)$ and $q_\lambda(x,w)$ be defined as in (5).

Let $h_0(x,w) := p_\lambda(x,w)$, and denote $\xi_j := \lceil (\deg g_j)/2 \rceil$, $v_j := \lceil (\deg h_j)/2 \rceil$ and $v'_j := \lceil (\deg h'_j)/2 \rceil$ where $\{g_1,\ldots,g_\ell\}$ are the polynomial constraints that define $\mathbf{K}$ and $\{h_1,\ldots,h_m\}$ and $\{h'_1,\ldots,h'_m\}$ are, respectively, the polynomial constraints (2) and (3) in $\overline{\mathbf{K}} \setminus \mathbf{K}$, respectively.

Let us denote by $I(0) = \{1,\ldots,n\}$ and $I(j) = \{(j,k)\}_{k=1,\ldots,m}$, for all $j = 1,\ldots,m$. With $x(I(0))$, $w(I(j))$ we refer, respectively, to the monomials x, w indexed only by subsets of elements in the sets $I(0)$ and $I(j)$, respectively. Then, for g_k, with $k = 1,\ldots,\ell$, let $\mathbf{M}_r(y,I(0))$ (respectively $\mathbf{M}_r(g_k y,I(0))$) be the moment (resp. localizing) submatrix obtained from $\mathbf{M}_r(y)$ (resp. $\mathbf{M}_r(g_k y)$) retaining only those rows and columns indexed in the canonical basis of $\mathbb{R}[x(I(0))]$ (resp. $\mathbb{R}[x(I(0))]$). Analogously, for h_j and h'_j, $j = 1,\ldots,m$, as defined in (2) and (3), respectively, let $\mathbf{M}_r(y,I(0) \cup I(j) \cup I(j+1))$ (respectively $\mathbf{M}_r(h_j y,I(0) \cup I(j) \cup I(j+1))$, $\mathbf{M}_r(h'_j y,I(0) \cup I(j) \cup I(j+1))$) be the moment (resp. localizing) submatrix obtained from $\mathbf{M}_r(y)$ (resp. $\mathbf{M}_r(h_j y)$, $\mathbf{M}_r(h'_j y)$) retaining only those rows and columns indexed in the canonical basis of $\mathbb{R}[x(I(0)) \cup w(I(j)) \cup w(I(j+1))]$ (resp. $\mathbb{R}[x(I(0)) \cup w(I(j)) \cup w(I(j+1))]$). (See [4] for convergence results of sparse SDP relaxations.)

For $r \geq \max\{r_0,v_0\}$ where $r_0 := \max_{k=1,\ldots,\ell} \xi_k$, $v_0 := \max\{ \max_{j=1,\ldots,m} v_j, \max_{j=1,\ldots,m} v'_j \}$, we introduce the following hierarchy of semidefinite programs:

$$\min_{\mathbf{y}} \; L_{\mathbf{y}}(p_\lambda)$$

$$
\begin{aligned}
\text{s.t.} \quad & M_r(\mathbf{y},I(0)) & \succeq \; & 0, \\
& M_{r-\xi_k}(g_k\mathbf{y},I(0)) & \succeq \; & 0, \quad k=1,\dots,\ell, \\
& M_r(\mathbf{y},I(0)\cup I(j)\cup I(j+1)) & \succeq \; & 0, \quad j=1,\dots,m, \\
& M_{r-v_j}(h_j\mathbf{y},I(0)\cup I(j)\cup I(j+1)) & \succeq \; & 0, \quad j=1,\dots,m, \qquad (\mathbf{Q}_r) \\
& M_{r-v'_j}(h'_j\mathbf{y},I(0)\cup I(j)\cup I(j+1)) & \succeq \; & 0, \quad j=1,\dots,m, \\
& L_y(\textstyle\sum_{j=1}^{m} w_{ji}-1) = L_y(\sum_{j=1}^{m} w_{ij}-1) = \; & & 0, \quad i=1,\dots,m, \\
& L_y(w_{ij}^2 - w_{ij}) = \; & & 0, \quad i,j=1,\dots,m, \\
& L_y(q_\lambda) = \; & & 1,
\end{aligned}
$$

with optimal value denoted $\inf \mathbf{Q}_r$ (and $\min \mathbf{Q}_r$ if the infimum is attained).

Theorem 2 (See [1]). *Let* $\overline{\mathbf{K}} \subset \mathbb{R}^{n+m^2}$ *be the feasible domain of* (OMRP$_\lambda$). *Then,*

(a) $\inf \mathbf{Q}_r \uparrow \rho_\lambda$ *as* $r \to \infty$.
(b) *Let* $\mathbf{y}^r$, *be an optimal solution of the SDP relaxation* $(\mathbf{Q}_r)$. *If* $\operatorname{rank} M_r(\mathbf{y}^r,I(0)) = \operatorname{rank} M_{r-r_0}(\mathbf{y}^r,I(0))$ *and* $\operatorname{rank} M_r(\mathbf{y}^r,I(0)\cup I(j)\cup I(j+1)) = \operatorname{rank} M_{r-v_0}(\mathbf{y}^r,I(0)\cup I(j)\cup I(j+1))$ $j=1,\dots,m$, *and if* $\operatorname{rank}(M_r(y^*,I(0)\cup(I(k)\cup I(k+1))\cap(I(j)\cup I(j+1)))) = 1$ *for all* $j \neq k$ *then* $\min \mathbf{Q}_r = \rho_\lambda$.

Moreover, let $\Delta_j := \{(x^*(j),w^*(j))\}$ *be the set of solutions obtained by the ranks conditions in* (b). *Then, every* (x^*,w^*) *such that* $(x_i^*,w_i^*)_{i\in I(j)} = (x^*(j),w^*(j))$ *for some* Δ_j *is an optimal solution of Problem* **MOMRF**.

4 Computational results

Intensive computational results has been done with a series of 5 location problems on the plane and in the 3-dimension space with ℓ_p norms. Namely, Weber, minimax, k-centrum, range and trimmed mean problems. In all cases, we have solved problems with at least 500 demand points with first and second relaxation orders and with an accuracy higher than 10^{-4}. Details of the problems, tables with computational results and further details on the implementations done in SDPT3 can be found in [1]. We tested problems with up to 500 demands points randomly generated in the unit square and the unit cube. We move the number of points, n, between 100 and 500 and ten instances were generated for each value of n. A summary of these computational experiments is shown in Table 1. There, column n stands for the number of points considered in the problem, CPU time is the average running time needed to solve each of the instances, ε_{obj} is the standard error measure with respect to the implementation in [2] and r is the relaxation order.

problem	n	CPUtime	ε_{obj}	r	problem	n	CPUtime	ε_{obj}	r
	100	31.57	0.00000174	3		100	164.37	0.00000008	3
	200	65.75	0.00000190	3		200	275.02	0.00000000	3
Weber $\mathbb{R}^2$	300	102.95	0.00001241	3	Weber $\mathbb{R}^3$	300	501.09	0.00000006	3
	400	145.62	0.00000333	3		400	746.70	0.00000011	3
	500	187.02	0.00000754	3		500	1063.50	0.00000000	3

problem	n	CPUtime	ε_{obj}	r	problem	n	CPUtime	ε_{obj}	r
Center $\mathbb{R}^2$	100	9.87	0.00005552	2	Center $\mathbb{R}^3$	100	164.37	0.00000008	2
	200	38.76	0.00013507	2		200	82.01	0.00011298	2
	300	58.10	0.00033715	2		300	124.47	0.00030316	2
	400	90.22	0.00048347	2		400	172.05	0.00052552	2
	500	151.64	0.00066416	2		500	226.73	0.00059268	2
Range $\mathbb{R}^2$	100	58.36	0.00052427	2	Range $\mathbb{R}^3$	100	97.26	0.00015535	2
	200	98.23	0.00041499	2		200	197.66	0.00040517	2
	300	159.87	0.00014556	2		300	322.21	0.00036845	2
	400	167.74	0.00123896	2		400	361.12	0.00022448	2
	500	228.68	0.00438388	2		500	554.94	0.00028013	2
Tr.mean $\mathbb{R}^2$	100	91.26	0.00218668	2	Tr.mean $\mathbb{R}^2$	100	220.68	0.00011032	2
	200	257.23	0.00032380	2		200	552.19	0.00056138	2
	300	326.52	0.00225994	2		300	884.40	0.00038481	2
	400	582.36	0.00047130	2		400	1165.79	0.00058261	2
	500	685.79	0.00079679	2		500	9151.90	0.00063861	2

	n	$k = \lceil 0.1n \rceil$ CPUtime	ε_{obj}	$k = \lceil 0.5n \rceil$ CPUtime	ε_{obj}	$k = \lceil 0.9n \rceil$ CPUtime	ε_{obj}	r
k-centrum $\mathbb{R}^2$	100	53.68	0.00000084	41.42	0.00000065	39.49	0.00000188	2
	200	123.02	0.00000056	96.40	0.00000075	88.10	0.00000275	2
	300	180.38	0.00000408	161.84	0.00000081	146.22	0.00000349	2
	400	260.27	0.00000079	225.07	0.00003689	201.01	0.00000376	2
	500	345.93	0.00000224	310.19	0.00000119	269.99	0.00000200	2
k-centrum $\mathbb{R}^3$	100	122.01	0.00000088	109.17	0.00000224	118.03	0.00000067	2
	200	305.51	0.00002407	255.54	0.00007106	284.80	0.00000157	2
	300	492.04	0.00046130	433.69	0.00007630	471.78	0.00000174	2
	400	619.97	0.00000091	585.93	0.00000055	523.81	0.00000829	2
	500	817.75	0.00012138	789.77	0.00000087	664.94	0.00000318	2

Table 1: A summary of the computational experiments.

References

1. Blanco V., El-Haj Ben-Ali S. and Puerto J. (2011). *A Semidefinite Programming approach for minimizing ordered weighted averages of rational functions.* arXiv:1106.2407v2.
2. Drezner Z. and Nickel S. (2009). *Solving the ordered one-median problem in the plane.* European J. Oper. Res., 195 no. 1, 46–61.
3. Lasserre J. B. (2001). *Global Optimization with Polynomials and the Problem of Moments.* SIAM J. Optim., 11, 796–817.
4. Lasserre J.B. (2006). *Convergent SDP-relaxations in polynomial optimization with sparsity.* SIAM J. Optim., 17, 822–843.
5. Nickel S. and Puerto J. (2005). *Facility Location - A Unified Approach.* Springer Verlag.
6. Ogryczak W. and Śliwiński T. (2010). *On solving optimization problems with ordered average criteria and constraints.* Stud. Fuzziness Soft Comput. Springer, Berlin, 254, 209–230.
7. Ogryczak W. (2010). *Conditional median as a robust solution concept for uncapacitated location problems.* TOP, 18 (1), 271–285.
8. Ogryczak W. and Zawadzki M. (2002). *Conditional median: a parametric solution concept for location problems.* Ann. Oper. Res., 110, 167–181.
9. Puerto J. and Tamir A. (2005). *Locating tree-shaped facilities using the ordered median objective.* Mathematical Programming, 102 (2), 313–338.
10. Putinar M. (1993). *Positive Polynomials on Compact Semi-Algebraic Sets.* Ind. Univ. Math. J., 42, 969–984.
11. Yager R. and Kacprzyk J. (1997). *The Ordered Weighted Averaging Operators: Theory and Applications,* Kluwer: Norwell, MA.

A Column-and-Row Generation Algorithm for a Crew Planning Problem in Railways

A.Ç. Suyabatmaz and G. Şahin

Abstract We develop a set-covering type formulation for a crew planning problem that determines the minimum sufficient crew size for a region over a finite planning horizon where the periodic repeatability of crew schedules is considered as well. The resulting problem formulation cannot be solved with a traditional column generation algorithm. We propose a column-and-row generation algorithm and present preliminary computational results.

1 Introduction

The crew planning process in railways is composed of three subsequent decision making tasks. The first task is to construct feasible crew schedules that honor regulations and rules imposed by the company and legal liabilities. The next task is to select a set of crew schedules based on a set of criteria such that each duty in the planning horizon is covered in at least one crew schedule. The final task is to assign those selected crew schedules to individual crew members. The problem is solved for a single-region over a finite planning horizon. Within the general approach there are two particular deficient suppositions with respect to the constructed crew schedules: (i) available crew resource in the region is sufficient, (ii) schedules can be repeated over the periodic recurrence of the planning horizon (i.e. the end of a feasible crew schedule can be feasibly connected to the beginning of another one).

With respect to the problem environment we consider, our study is close to Ernst et al. [1] and Şahin and Yüceoğlu [3]. Ernst et al. [1] consider the problem in two phases: the planing stage where the number of crew members is determined and the operational (rostering) stage where the connectivity of the crew schedules is main-

Ali Çetin Suyabatmaz
Manufacturing Systems/Industrial Engineering, Faculty of Engineering and Natural Sciences, Sabanci University, Orhanli, Tuzla, 34956 Istanbul, Turkey e-mail: csuyabatmaz@sabanciuniv.edu

Güvenç Şahin
Manufacturing Systems/Industrial Engineering, Faculty of Engineering and Natural Sciences, Sabanci University, Orhanli, Tuzla, 34956 Istanbul, Turkey e-mail: guvencs@sabanciuniv.edu

D. Klatte et al. (eds.), *Operations Research Proceedings 2011*, Operations Research Proceedings, 335
DOI 10.1007/978-3-642-29210-1_54, © Springer-Verlag Berlin Heidelberg 2012

tained. Ernst et al. [1] consider both deficiencies and develop a two-stage solution methodology that fails to guarantee optimal solutions. Their heuristic two-stage approach minimizes the number of crew members in the first stage and tries to satisfy the connectivity of rosters in the second stage. Şahin and Yüceoğlu [3] focus on the first deficiency and represent the selection of crew schedules as the tactical level counterpart of the construction of crew schedules. They formulate an integer programming network flow model to minimize the number of crew members required in the region to cover the duties.

In this study, we deal with the tactical crew capacity planning problem which determines the minimum required number of crew members. In our setting, the feasibility of crew schedules and the connectivity of rosters are integrated to find a repeatable set of schedules that satisfy the operational rules and regulations.

2 Problem Definition and Mathematical Modeling

The crew capacity planning problem determines the minimum number of crew members required to cover all duties in the region. Şahin and Yüceoğlu [3] develops a network representation of the problem and solves a minimum flow problem to determine the crew capacity. As an alternative approach, one can formulate this problem as a pure set covering problem and solve it with a column generation algorithm, where a feasible crew schedule corresponds to a column in the formulation.

In this study, we consider finding a set of feasible crew schedules that can be connected to other schedules from one period to the other. To formulate this version of the problem, the set covering problem is enriched with additional constraints that represent the connectivity relationship among the schedules.

Let the set of duties in the finite planning horizon be denoted by I and the set of feasible schedules covering the horizon be denoted by J. A binary parameter a_{ij} indicates that duty $i \in I$ is included in schedule $j \in J$ when $a_{ij} = 1$. Parameter $l_{jj'}$ indicates the connectivity relationship between schedules as follows:

$$l_{jj'} = \begin{cases} 1, & \text{if schedule } j \text{ can be connected to schedule } j'; \\ 0, & \text{otherwise.} \end{cases}$$

The decision variables x_j and y_{ij} are defined as

$$x_j = \begin{cases} 1, & \text{if schedule } j \text{ is selected/included in solution;} \\ 0, & \text{otherwise.} \end{cases}$$

$$y_{jj'} = \begin{cases} 1, & \text{if schedule } j \text{ is connected to schedule } j'; \\ 0, & \text{otherwise.} \end{cases}$$

Then, the mathematical programming formulation of the tactical crew capacity planning problem with a set-covering type formulation approach is as follows:

$$\text{minimize} \quad \sum_{j \in J} x_j \tag{1}$$

$$\text{subject to} \quad \sum_{j \in J} a_{ij} x_j \geq 1, \qquad\qquad i \in I, \tag{2}$$

$$\sum_{j' \in J} l_{jj'} \, y_{jj'} - x_j = 0, \qquad\qquad j \in J, \tag{3}$$

$$\sum_{j' \in J} l_{j'j} \, y_{j'j} - x_j = 0, \qquad\qquad j \in J, \tag{4}$$

$$x_j \in \{0,1\}, \qquad\qquad j \in J, \tag{5}$$

$$y_{jj'} \in \{0,1\}, \qquad\qquad j, j' \in J. \tag{6}$$

Objective function (1) minimizes the total number of schedules (i.e. number of crew members). Constraints (2) ensure that each duty is covered by at least one schedule. Constraints (3) guarantee that each selected schedule follows (is connected to) another selected schedule in the solution. Likewise, constraints (4) guarantee that each selected schedule is being followed by (connects to) another selected schedule in the solution. Constraints (5) and (6) represent the domain of the decision variables.

Solving (1) - (6) directly is not practical due to the size of set J, which contains all possible schedules. However, a simultaneous column-and-row generation algorithm may be employed to solve the linear programming relaxation of (1) - (6) since the mathematical formulation stated above belongs to the set of column-dependent-rows (CDR) problems [2].

3 Solution Method

In a classical column generation algorithm, the set of constraints in the restricted master problem (RMP) is fixed, and complete dual information is supplied from the RMP to the pricing sub-problem (PSP). In each iteration, PSP generates a new column (i.e. decision variable) to be added to RMP, by computing the reduced cost of the column using the retrieved dual information. In problem (1) - (6), existence of connectivity constraints (3) - (4) (i.e. rows of the formulation) depends on the existence of columns (represented by x_j) in the formulation; a new column induces new linking constraints to be added to RMP. In order to correctly compute the reduced cost of a column in PSP, we need to use the associated dual variables of these linking constraints (rows) a priori. In this respect, a traditional column generation algorithm would not suffice to solve this problem due to this missing dual information associated with the missing rows that do not yet exist. To overcome this challenge, we develop a column-and-row generation algorithm that simultaneously generates feasible crew schedules and associated connectivity constraints.

The simultaneous column-and-row generation algorithm starts with the RMP that includes only a selected subset of schedules (columns), and then iteratively adds new columns to RMP to improve its objective function value as in the column generation algorithm. At each iteration of the simultaneous column-and-row generation algorithm, RMP is solved to optimality and the optimal dual values from RMP are used to solve PSP which allows generating several variables and their associated linking constraints simultaneously by estimating the dual values of the missing linking constraints.

Let J_c be the set of existing columns in RMP, and the set of complementary columns (i.e. the remaining feasible columns) be $\overline{J_c}$. Let u_i, v_j and w_j be the dual variables corresponding to, respectively, constraints $i \in I$ in (2) and constraints (3) and (4) for only $j \in J_c$. The dual of RMP can then be formulated as

$$\text{maximize} \quad \sum_{i \in I} u_i \tag{7}$$

$$\text{subject to} \quad \sum_{i \in I} a_{ij} u_i - v_j - w_j \leq 1, \qquad j \in J, \tag{8}$$

$$l_{jj'}(v_j + w_{j'}) \leq 0, \qquad j, j' \in J, \tag{9}$$

$$u_i \geq 0, \qquad i \in I, \tag{10}$$

$$v_j, w_j \text{ u.i.s} \qquad j \in J. \tag{11}$$

PSP is to find a column $j \in \overline{J_c}$ that has a negative reduced cost (i.e. violates the corresponding dual constraint in (8)). If we search for the column that makes the most improvement in the objective function value of RMP, the pricing problem becomes a two-stage problem to find

$$j* = \arg\min_{j \in \overline{J_c}} \left\{ 1 - \sum_{i \in I} a_{ij} u_i + v_j + w_j \right\} \tag{12}$$

$$\text{minimize} \quad v_j \tag{13}$$

$$\text{subject to} \quad l_{jj'}(v_j + w_{j'}) \leq 0, \qquad j \in \overline{J_c}, j' \in J_c, \tag{14}$$

$$\text{at least one constraint in (14) is tight;} \tag{15}$$

$$\text{minimize} \quad w_j \tag{16}$$

$$\text{subject to} \quad l_{j'j}(v_{j'} + w_j) \leq 0, \qquad j' \in J_c, j \in \overline{J_c}, \tag{17}$$

$$\text{at least one constraint in (17) is tight.} \tag{18}$$

where problems (13)-(15) and (16)-(18) impose the connectivity relations between the new column $j \in \overline{J_c}$ and the existing columns in J_c.

Constraints (14) and (17) impose that the dual constraints (9) are not violated as we try to find the column with the largest violation in dual constraint (8). The new column, represented by x_{j*} is expected to enter the basis of RMP at the next iteration of the column-and-row generation algorithm. We know that when the x_{j*} enters the basis, the connectivity constraints in (3) and (4) is to be honored with two new basic variables $y_{j*j'} = 1, j' \in J_c$ and $y_{j'j*} = 1, j' \in J_c$. Optimal solution of (13)-(15) indicates which $y_{jj'}$ variable is to enter the basis (take the value of 1) in (3), corresponding to a tight constraint in (14). This is indeed imposed by the complementary slackness condition associated with constraint (15). Similarly, the optimal solution of (16)-(18) indicates which $y_{j'j}$ variable is to enter the basis (take the value of 1) in (4), corresponding to a tight constraint in (17) with respect to the complementary slackness condition associated with constraint (18). Consequently, if $1 - \sum_{i \in I} a_{ij} u_i + v_j + w_j$ is less than zero, then x_{j*} is added to RMP and RMP is augmented along with one new linking constraint of type (3) and one new linking constraint of type (4).

For each column $j \in \overline{J_c}$ (new candidate column), if j can be connected right before the existing column $j' \in J_c$, (i.e. $l_{jj'} = 1, j \in \overline{J_c}, j' \in J_c$) the corresponding constraint (14) appears as $v_j \leq -w_{j'}$. Therefore, for each $j \in \overline{J_c}$ (new candidate column), the solution of (13)-(15) becomes

$$v_j = \min_{j' \in J_c, l_{jj'}=1} (-w_{j'}) = \max_{j' \in J_c, l_{jj'}=1} (w_{j'}). \tag{19}$$

Likewise, for all the existing columns j' in J_c that can be linked right before j (i.e. $l_{j'j} = 1, j' \in J_c, j \in \overline{J_c}$,) the corresponding constraint (17) appears as $w_j \leq -v_{j'}$ and for each $j \in \overline{J_c}$ (new candidate column), the solution of (16)-(18) becomes

$$w_j = \min_{j' \in J_c, l_{j'j}=1} (-v_{j'}) = \max_{j' \in J_c, l_{j'j}=1} (v_{j'}). \tag{20}$$

Then, the two stage problem (12)-(18) can be reformulated as

$$j* = \arg\min_{j \in \overline{J_c}} \left\{ 1 - \sum_{i \in I} a_{ij} u_i + \max_{j' \in J_c, l_{jj'}=1} (w_{j'}) + \max_{j' \in J_c, l_{j'j}=1} (v_{j'}) \right\} \tag{21}$$

As in the column generation algorithm, the column-and-row generation algorithm achieves optimality when the optimal solution of (23) is non-negative (i.e. no column exists with negative reduced cost).

Consequently, the correct termination condition for the column-row-generation algorithm that solves the LP relaxation of (1)-(6) optimally is formalized with the following theorem.

Theorem 1 *Let $u_i, i \in I, v_j, j \in J_c$, and $w_j, j \in J_c$ be the optimal dual solution corresponding to the optimal basis $\mathscr{B}$ of the current RMP. The primal solution associated with $\mathscr{B}$ is optimal for the LP relaxation of (1)-(6) if*

$$1 - \sum_{i \in I} a_{ij} u_i + \max_{j' \in J_c, l_{jj'}=1} (w_{j'}) + \max_{j' \in J_c, l_{j'j}=1} (v_{j'}) \geq 0 \tag{22}$$

for every $j \in \overline{J_c}$.

The proof of this theorem follows from the analysis of the column-and-row generation algorithm developed in Muter et al. [2].

4 Computational Results and Concluding Remarks

To validate and verify our solution method, we implemented the solution method with C++ using ILOG Optimization Studio 12.2 for optimization purposes on a PC with Intel Xenon @3.20 GHz CPU and 32GB RAM. The preliminary computational study is performed on a data set that is representative of three different crew regions in the Turkish State Railways system. In Table 1, we summarize these results for three crew regions, namely Istanbul, Ankara and Eskisehir. In a one-week planning horizon, there are 329, 266 and 329 duties to be covered in Istanbul, Ankara and Eskisehir crew regions respectively. In Table 1, LP-LB corresponds to the objective function value of the optimal solution of the linear programming relaxation of (1)-(6) obtained with our column-and-row generation algorithm. # of Iterations and LP Time shows the number of columns generated and the time spent by the algorithm. Then, with the existing columns in the terminating RMP, we solve the problem (1)-(6) to obtain an integer feasible solution to the problem. We solve the integer programming (IP) problem to optimality with a time limit of two hours. IP-FS shows the objective function value of IP problem where IP Time shows the time required by the CPLEX solver. Problems for Istanbul and Ankara are solved to optimality while the problem for Eskisehir reaches the time limit before the optimality of the integer feasible solution is proven. We note that these regions, although the largest instances

Crew Region	LP-LB	# Iterations	LP Time	IP-FS	IP Time
Istanbul	38	348	8 min	42	44 min
Ankara	39	293	6 min	44	28 min
Eskisehir	49	501	17 min	55	120 min

Table 1 Results for TCDD data

in Turkish State Railways system, may correspond to medium-to-large problems in general. Our preliminary experiments show that, good feasible solutions can be attained without the need for complicated solution method for the IP problem.

References

1. Ernst, A.T., Jiang, H., Krishnamoorthy, M., Nott, H., Sier, D.: Staff scheduling and rostering: A review of applications, methods and models. European Journal of Operations Research **153**, 3–27 (2004).
2. Muter, I., Birbil, S.I., Bulbul, K.: Simultaneous Column-and-Row Generation for Large-Scale Linear Programs with Column-Dependent-Rows. [Working Paper / Technical Report] Sabanci University ID: 0004(2001) SU-FENS-2010 http://research.sabanciuniv.edu/14197/
3. Şahin, G., Yüceoğlu , B.: Tactical crew planning in railways. Transportation Research Part E: Logistics and Transportation Review (2011) doi: 10.1016/j.tre.2011.05.013

Stream IX
Metaheuristics and Biologically Inspired Approaches

Luca Maria Gambardella, IDSIA, Lugano, SVOR Chair
Walter Gutjahr, University of Vienna, ÖGOR Chair
Franz Rothlauf, Johannes Gutenberg University Mainz, GOR Chair

Harmony Search Algorithms for binary optimization problems

Miriam Padberg

Abstract In many theoretical and empirical investigations heuristic methods are the subject of investigation for optimization problems in order to get good and valid solutions. In 2001 Harmony Search, a new music-inspired meta-heuristic, was introduced [1]. It has been applied to various types of optimization problems, for example structural design problems, showing significant improvements over other heuristics. Motivated by the question whether these heuristics represent a new class of meta-heuristics or whether they are only another representation of a well-known technique, a couple of investigations were made.
In this paper we will show that for the class of binary optimization problems this new nature-inspired heuristic is "equivalent" to an evolutionary algorithm.

1 Introduction

Many investigations were made to solve binary optimization (BOPT) problems of the form

$$\text{BOPT:} \qquad \max f(x_1, \ldots, x_n)$$

$$\text{s.t.} \quad x_i \in \{0, 1\} \qquad \forall i = 1, \ldots, n,$$

which is usually NP-hard. Inspired by finding the perfect harmony in music-improvisation-process, Geem et. al published a heuristic called Harmony Search in 2001 [1]. The operators used by this new class of meta-heuristics [2] show a couple of similaries to operators of well-known techniques like evolutionary algorithm [4], [5]. This motivates us to compare these music-inspired heuristics for BOPT to a class of evolutionary algorithm in order to identify differences and similarities as known from the analysis of continuous optimization problems [3]. We define two heuristics to be equivalent if they produce the same output with the same probability and we formulate under which conditions a given Harmony

Miriam Padberg
TU Dortmund – Germany, e-mail: miriam.padberg@tu-dortmund.de

D. Klatte et al. (eds.), *Operations Research Proceedings 2011*, Operations Research Proceedings,   343
DOI 10.1007/978-3-642-29210-1_55, © Springer-Verlag Berlin Heidelberg 2012

Search Algorithm is equivalent to an Evolutionary Algorithm. Therefore, first an introduction in Harmony Search Algorithms is given in section 1. In section 2, a class of Evolutionary Algorithms with a population-wide n-point crossover-operator and mutation-operator with probability p is presented. Finally it will be shown under which conditions a given Harmony Search Algorithm is equivalent to an Evolutionary Algorithmn in section 3.

2 Harmony Search Algorithms

Harmony Search Algorithm imitates the music-improvisation process applied by musicans with the goal to find the perfect harmony. Therefore each musician improves different pitches of his / her instrument in order to accomplish a better state of harmony. Inspired by the music-improvisation process, the Harmony Search Algorithm proceeds in different steps in order to find a "perfect" solution of the given optimization problem BOPT.

step 1: Initializing / Set up
First the parameters of the Harmony Search Algorithm are specified. The algorithm maintains a couple of solution-vectors of size m, called Harmony Memory. The algorithm consists of three different operators in order to generate a new solution-vector in each iteration. These operators are *memory consideration (MC)*, *pitch adjustment (PA)* and *random selection (RS)*. With certain probability at least one of the operators is applied: probability $P(PA)$ for applying PA and $P(MC)$ for applying MC. With $P(RC) = 1 - P(PA) - P(MC)$ the probability for applying RS is given. Furthermore a parameter $\delta \in \{0,1\}$ which is necessary for PA is set. As stopping criteria the total number of iterations NI is set. After assigning values to the parameters, the set of solution-vectors of size m is initialized. Therefore a $m \times n$ matrix $S = (s_{i,j})$ is filled with uniformly choosen binary numbers:

$$s_{i,j} = \begin{cases} 0 \text{ with prob. } 0.5, \\ 1 \text{ with prob. } 0.5, \end{cases} \quad \forall i = 1,\ldots,m, \quad j = 1,\ldots,n.$$

Each row of the matrix S represents a solution vector for BOPT.

step 2: Generating a new solution $(x_1,\ldots,x_n)$
After initializing the set of stored solutions, a new solution vector $(x_1,\ldots,x_n)$ is generated. For each position $j \in \{1,\ldots,n\}$ one of the three operators PA, MC or RS is applied. These operators work as follows:

memory consideration (MC)

If MC is applied to generate the assignment at position j one of the m assignments of column j is choosen uniformly:

$$x_j = \begin{cases} s_{1,j} \text{ with prob. } 1/m, \\ \vdots \qquad \vdots \\ s_{m,j} \text{ with prob. } 1/m. \end{cases}$$

pitch adjustment (PA)

If *PA* is applied to generate the assignment at position j, two phases are performed sequentially. First *MC* is applied to generate x'_j. After this an adjustment process is applied:

$$x_j = \begin{cases} x'_j + \delta & \text{with prob. } 0.5 \\ x'_j - \delta & \text{with prob. } 0.5 \end{cases}$$

To be more precise, the adjustment process will be reformulated: if $x'_j = 0$ after *MC* than $x_j = 1$ with probability $P_{0 \mapsto 1}$ and $x_j = 0$ with probability $P_{0 \to 0}$. Similarly, if $x'_j = 1$, than the new assignment x_j is 1 with probabiltiy $P_{1 \mapsto 1}$ and $x_j = 0$ with probability $P_{1 \mapsto 0}$.

random selection (RS)

The new solution vector at position j is chosen randomly from $\{0, 1\}$, i.e.

$$x_j = \begin{cases} 0 & \text{with prob. } 0.5, \\ 1 & \text{with prob. } 0.5. \end{cases}$$

step 3: Updating Harmony Memory S

If the objective value $f(x_1, \dots, x_n)$ is better than the objective value of the worst solution in the memory, the new solution $(x_1, \dots, x_n)$ replaces uniformly one of the worst solutions vector in memory.

step 4: Output $\mathcal{O}_{HS}$

If the stopping criteria (i.e. the given number of iterations NI) is satisfied, the computation is terminated with output $\mathcal{O}_{HS}$ of the run: the solution $\mathbf{x}^* = (x_1^*, \dots, x_n^*)$ in memory S with maximal value of the objective function $f(\mathbf{x}^*)$. Otherwise steps 2 to 4 are repeated.

3 Evolutionary Algorithms

Evolutionary Algorithms are nature-analogue optimization techniques that imitate biological evolution. In the following four steps an Evolutionary Algorithm is presented. This Evolutionary Algorithm performs a special crossover-operator, also known as global discrete recombination [6], and a conventional mutation-operator in order to generate a new solution for BOPT in each step. Typically two individuals, called parents, are chosen to generate a new individual by the use of a crossover-operator. The following class of Evolutionary Algorithm generates the new individual by recombining the assignments over the whole population.

step 1: Initializing / Set up

The algorithm maintains a copule of solution-vectors of size m, called population. In each iteration a new solution vector is generated. Therefore, two different operators *crossover* and *mutation* are given. Crossover is applied in every iteration with probability 1 and after that mutation is applied at every position $j \in \{1, \dots, n\}$ with probability p. As stopping criteria the total number of iterations NI is set. After assigning the parameters m, NI and p with values, the set of solution-vectors of size m is initialized. Therefore, a $m \times n$-matrix $S = (s_{i,j})$, $1 \le i \le m$, $1 \le j \le n$, where every row presents a suitable solution, is initialized with $s_{i,j} \in \{0,1\}$ uniformly $\forall i = 1, \ldots, m$ and $\forall j = 1, \ldots, n$.

step 2: Generating a new solution $(x_1, \ldots, x_n)$

In order to generate a new solution vector for BOPT two phases are performed sequentially. First a population-wide n-point crossover is applied in order to generate a new solution $(x'_1, \ldots, x'_n)$:

$$crossover : \{0,1\}^{m \cdot n} \mapsto \{0,1\}^n : S \mapsto (x'_1, \ldots, x'_n)$$

with

$$x'_j = \begin{cases} s_{1,j} & \text{with prob. } 1/m \\ \vdots & \quad\quad \vdots \\ s_{m,j} & \text{with prob. } 1/m \end{cases} \quad \forall j \in \{1, \ldots, n\}.$$

After crossover, the solution $(x_1, \ldots, x_n)$ is generated by mutation on $(x'_1, \ldots, x'_n)$. Here, with a probability of p, the assignment at position j is turned into the converse assignment.

$$x_j = \begin{cases} 1 - x'_j & \text{with prob. } p, \\ x'_j & \text{with prob. } 1 - p. \end{cases} \quad \forall j = 1, \ldots, n.$$

step 3: Updating population S

After generating a new solution $(x_1, \ldots, x_n)$ the value of the objective function $f(x_1, \ldots, x_n)$ is computed. The new solution replaces uniformly one of the worst solutions in S, if $f(x_1, \ldots, x_n)$ is better than the objective value of the solution with minimum value in memory S.

step 4: Output $\mathcal{O}_{EA}$

If the stopping criteria (i.e. the given number of iterations NI) is satisfied, the computation is terminated with output $\mathcal{O}_{EA}$ of the run: the solution $\mathbf{x}^* = (x_1^*, \ldots, x_n^*)$ in memory S with maximal value of the objective function $f(\mathbf{x}^*)$. Otherwise steps 2 to 4 are repeated.

4 Equivalence of HS and EA

First, the equivalence of two heuristics is defined.

Definition 1. Given two heuristic procedures A and B for an optimization problem with search space X. Let $\mathcal{O}_A$ be the output of the run of procedure A and $\mathcal{O}_B$ be the output of the run of procedure B.

A and B are called equivalent, if the probability that A and B yield the same solution is equal, i.e. if $\forall \mathbf{x}^* = (x_1^*, \ldots, x_n^*) \in X$

$$P_A(\mathcal{O}_A = (x_1^*, \ldots, x_n^*)) = P_B(\mathcal{O}_B = (x_1^*, \ldots, x_n^*)).$$

Now we can prove the following theorem:

Theorem 1. *Let HS and EA be given for the optimization problem BOPT with memory size m and the number of iterations NI in both heuristics. Furthermore the probabilities $P(RS)$, $P(PA)$, $P_{1\mapsto0}$, $P_{1\mapsto1}$, $P_{0\mapsto0}$ and $P_{0\mapsto1}$ with $P_{1\mapsto0} = P_{0\mapsto1}$ for HS are given. For EA a n-point crossover on m elements and a mutation-operator with probability p are given.*
Then HS and EA are equivalent if $p := P(PA) \cdot P_{1\mapsto0} + 0.5 \cdot P(RS)$.

Proof. To show the equivalence of these randomized heuristics, we first elaborate on their similarities. Both algorithms start by initializing the memory. Because of the uniformly chosen $s_{i,j} \in \{0,1\}$ in both initial matrizes S_0 (step 1), the probability to start with the same matrix $S \in \{0,1\}^{m \times n}$ is equal for both heuristics, i.e. $P_{HS}(S_0 = S) = P_{EA}(S_0 = S)$. Given a $m \times n$-matrix $S \in \{0,1\}^{m \times n}$ and a solution-vector $(x_1, \ldots, x_n) \in \{0,1\}^n$ the probabilities that the updated matrix is $S' \in \{0,1\}^{m \times n}$ are equal, because of the fact that the phase of updating memory is equal in both heuristics (step 3). The operators in step 2 for generating a new solution and in step 3 for updating memory are independent of the observed iteration. Hence it is sufficient to show $P_{HS}(S_1 = S') = P_{EA}(S_1 = S')$. By an induction argument, then $P_{HS}(S_{NI} = S'') = P_{ES}(S_{NI} = S'')$ for all $S'' \in \{0,1\}^{m \times n}$ and with step 4 $P_{HS}(\mathcal{O}_{HS} = \mathbf{x}^*) = P_{EA}(\mathcal{O}_{EA} = \mathbf{x}^*) \; \forall \mathbf{x}^* \in \{0,1\}^n$ follows.
In each heuristic the operations for generating a new solution vector are the same for every position $j \in \{1, \ldots, n\}$ and independent of the observed iteration. So it is sufficient to show that for a given matrix $S = (s_{i,j}) \in \{0,1\}^{m \times n}$ the probability to generate $x_j = 1$, $j \in \{1, \ldots, n\}$ in one iteration is equal in both heuristics, i.e. the equation

$$P_{HS}(x_j = 1 | S_0 = S) = P_{ES}(x_j = 1 | S_0 = S)$$

holds, because of $P_{HS}(x_j = 0 | S_0 = S) = 1 - P_{HS}(x_j = 0 | S_0 = S)$ and $P_{EA}(x_j = 0 | S_0 = S) = 1 - P_{EA}(x_j = 0 | S_0 = S)$. Define for the given $m \times n$ matrix S with $s_{i,j} \in \{0,1\} \; \forall 1 \le i \le n, 1 \le j \le m$

$$P_1(j) := P\left(x_j \in \{s_{1,j}, \ldots, s_{m,j}\} | x_j = 1\right) = \frac{1}{m} \sum_{i=1}^{m} s_{i,j}$$

and $P_0(j) := 1 - P_1(j)$.
On the one hand for *HS* and for every $j \in \{1, \ldots, n\}$ it is

$$P_{HS}(x_j = 1 | S_0 = S) = P(x_j = 1 | \text{ by } PA) + P(x_j = 1 | \text{ by } RS) + P(x_j = 1 | \text{ by } MC)$$
$$= P_0(j) \cdot (P(PA) \cdot P_{0\mapsto1} + P(RS) \cdot 0.5)$$
$$+ P_1(j) \cdot (P(PA) \cdot P_{1\mapsto1} + P(MC) + P(RS) \cdot 0.5).$$

On the other hand for the Evolutionary Algorithm it is

$$P_{EA}(x_j = 1|S_0 = S) = P(x'_j = 1| \text{ after crossover}) \cdot P(x_j = 1| \text{ after mutation})$$
$$+ P(x'_j = 0| \text{ after crossover}) \cdot P(x_j = 1| \text{ after mutation})$$
$$= P_0(j) \cdot p + P_1(j) \cdot (1 - p).$$

With $p := P(PA) \cdot P_{1 \mapsto 0} + 0.5 \cdot P(RS)$ and $P(RS) + P(MC) + P(PA) = 1$ it follows:

$$P_{EA}(x_j = 1|S_0 = S) = P_0(j) \cdot (P(PA) \cdot P_{1 \mapsto 0} + 0.5 \cdot P(RS))$$
$$+ P_1(j) \cdot (P(MC) + P(PA) \cdot (1 - P_{1 \mapsto 0}) + 0.5 \cdot P(RS))$$
$$= P_0(j) [P(PA) \cdot P_{1 \mapsto 0} + 0.5 \cdot P(RS)]$$
$$+ P_1(j) [P(MC) + P(PA) \cdot P_{1 \mapsto 1} + 0.5 \cdot P(RS)].$$

Here we have made use of the fact that $P_{1 \mapsto 0} = P_{0 \mapsto 1}$, and the conclusion directly follows.

5 Conclusions

In this paper a scheme was presented in order to compare two heuristics for the class of binary optimization problems: Harmony Search Algorithms and Evolutionary Algorithms with a special crossover-operator. By looking for similarities between Harmony Search Algorithms and Evolutionary Algorithms, a new way to analyze Harmony Search Algorithm is offered. The main result of this investigation is that the class of Harmony Search Algorithms for binary optimization problems is equivalent to a class of Evolutinary Algorithms with mutation and a special crossover-operator: for binary optimization problems every Harmony Search Algorithm can be written in terms of an Evolutionary Algorithm.

References

1. Geem, Kim and Loganathan: A New Heuristic Optimization Algorithm: Harmony Search, Simulation 76 (2), pages 60–68 (2001)
2. Geem and Lee: A new meta-heuristic algorithm for continuous engineering optimization: harmony search in theory and practice, Computer Methods in applied mechanics and engineering 195, pages 3902–3933 (2004)
3. Weyland: A Rigourous Analysis of the Harmony Search Algorithm: How the Research Community can be Misled by a 'Novel' Methodology,International Journal of Applied Metaheuristic Computing (2010)
4. Spears: Evolutionary Algorithms, Natural Computating Series, Springer (2000)
5. Beyer and Schwefel: Evolution strategies - A comprehensive introduction, Natural Computing 1(1), pages 3–53 (2002)
6. Baeck and Schwefel: An overview of evolutionary algorithms for parameter optimization, Evolutionary computation 1(1), pages 1–23 (1993)

Modeling Complex Crop Management-Plant Interactions in Potato Production under Climate Change

Niklaus Lehmann, Robert Finger, and Tommy Klein

Abstract High water withdrawals for agricultural purposes during summer months led during the last decades repeatedly to intolerable ecological conditions in surface water bodies in the Broye region located in Western Switzerland. In this study, we assess different irrigation water withdrawal policies with respect to a farmer's income and to the optimal irrigation and nitrogen fertilization strategies in potato production under current and future expected climate conditions in the Broye region. To this end, we optimize nitrogen and irrigation management decisions in potato production by the use of a bio-economic model and genetic algorithms. Our results show that limiting the water withdrawal amount leads to a substantial decrease in irrigation use, whereas the reduction in a farmer's income in potato production is supportable.

1 Introduction

Combinatorial optimization problems are characterized by a finite number of feasible solutions. In principle, the optimal solution can be found by simple enumeration. In practice, however, this is often not possible since the number of feasible

Niklaus Lehmann
Institute for Environmental Decisions, ETH Zurich
Sonneggstrasse 33, 8092 Zurich
e-mail: nlehmann@ethz.ch

Robert Finger
Institute for Environmental Decisions, ETH Zurich
Sonneggstrasse 33, 8092 Zurich
e-mail: rofinger@ethz.ch

Tommy Klein
Agroscope Reckenholz-Taenikon ART Research Station
Reckenholzstrasse 191, 8046 Zurich
e-mail: tommy.klein@art.admin.ch

D. Klatte et al. (eds.), *Operations Research Proceedings 2011*, Operations Research Proceedings, 349
DOI 10.1007/978-3-642-29210-1_56, © Springer-Verlag Berlin Heidelberg 2012

solution can be extremely large and the generation of each solution would be too time-consuming. In the last decades, genetic algorithms (GAs) have been increasingly used as optimization technique for combinatorial optimization applications and showed great promise in experimentation and industrial applications of such problems [1].

Based on this background, we apply GAs and optimize nitrogen fertilization and irrigation management decisions in potato production in the Broye region (Switzerland) by the use of a bio-economic modelling approach. Besides profits also production risks are taken into account by the used optimization routine. The application of this optimization technique that is relatively new in the agricultural economic field [5] is necessary due to the large amount of possible combinations of the considered decision variables[1] and the overall model complexity.

The Broye watershed located in Western Switzerland is an important agricultural region in Switzerland. Due to low precipitation levels and increasing market requirements irrigation is indispensable in potato production, which is the main crop in this region [4]. However, in dry and hot years the resulting high water withdrawals for agricultural purposes lead to intolerable ecological conditions for a river's fauna in this region[2]. Thus, in the last decades prohibitions of water withdrawals from the Broye river have been repeatedly imposed during the summer months if the river's flow rates fell below environmental thresholds[3]. Obviously, water withdrawals bans are most likely to occur under exceptionally hot and dry climate conditions in the summer months when crops' water requirements are highest. In the future, climate change (CC) is expected to further intensify this conflict in water utilization.

The goal of this article is to assess different irrigation water withdrawal policy scenarios and their impacts on profitability as well as on nitrogen fertilization and irrigation management decisions in potato cultivation under current and future expected climate in the Broye region.

2 Material and Methods

We use an integrated modeling approach in order to assess the impact of different water withdrawal policy scenarios on the agricultural water consumption and income in potato production in the Broye region. To this end, the crop growth model CropSyst (Version 4.13.09) is linked with an economic decision model representing farmers' decisions with respect to nitrogen and irrigation management. These agricultural inputs are optimized from a risk-averse farmer's point of view under different climate and policy scenarios. CropSyst is a process-based, multi-year, multi-crop cropping simulation model, which simulates biological and environmental above- and belowground processes of a single land block fragment at a daily scale [7]. We

---

[1] Theoretically, more than 10^{10} different combinations are possible.

[2] The water temperature in the Broye river reached in some of the past years at the beginning of July values of $26° - 27°$C, whereas the optimal temperature for trouts lies in a range of $7° - 16°$C [4].

[3] A river's minimum flow rate is defined in the federal water protection act of Switzerland (see [2]).

use a version of CropSyst calibrated against observed regional potato yields in the Broye region obtained from the Swiss Farm Accountancy Data Network (FADN) in the period 1990–2009 (e.g. [8]).

We consider a set of decision variables in potato production with regard to fertilizer use and irrigation strategy from a risk-averse farmer's perspective (see Table 1). Both nitrogen and water availability are important potato yield determining factors. However, both inputs also can be harmful for the environment and thus are of special interest for agri-environmental policy makers [3].

Table 1 Considered Management Options

Management Decision	Options
Total applied nitrogen amount	0–250 $kg \cdot ha^{-1}$ (in 10 $kg \cdot ha^{-1}$ steps)
Date of first fertilization application	1–70 days after sowing (in daily steps)
Date of second fertilization application	1–70 days after sowing (in daily steps)
Date of third fertilization application	1–70 days after sowing (in daily steps)
Allocation of total nitrogen amount on different applications	10 different allocation strategies
Irrigation trigger value	0,0.05,0.1,...,0.95,1[a]
Irrigation refill point	0,0.05,0.1,...,0.95,1[b]

[a] We chose the automatic irrigation option in CropSyst during the whole vegetation period of potato. As soon as the soil moisture is lower than a user-defined value this option triggers irrigation (0=maximum irrigation, 1=no irrigation).

[b] The refill point defines to what extent the soil water content is refilled by each irrigation application (0=minimal refillment, 1=maximum refillment).

The potato yield outputs of CropSyst are integrated into an economic decision model, which represents the decision making process of a risk averse farmer who maximizes his utility by selecting optimal management decisions. In this study, the farmer's utility is defined as the certainty equivalent (CE) in potato production that accounts for profits and production risks and is defined as shown in Equation 1:

$$\max_{N,W} CE = E(\pi) - 0.5 \cdot \gamma \cdot \underbrace{\frac{\sigma_\pi^2}{E(\pi)}}_{RP} \tag{1}$$

Where N is the nitrogen fertilization and W is the irrigation strategy. CE is the certainty equivalent ($CHF \cdot ha^{-1}$), π is the profit margin in potato production ($CHF \cdot ha^{-1}$), $E(\pi)$ is the expected profit margin ($CHF \cdot ha^{-1}$), RP is the risk premium ($CHF \cdot ha^{-1}$), σ_π is the standard deviation of the profit margin ($CHF \cdot ha^{-1}$) and $\gamma = 2$ is the relative risk aversion factor.

The stochastic weather generator LARSWG5 [6] is used to generate 25 years of daily weather data for the climate station Payerne under a Baseline scenario (1990–2009) and under a CC scenario around the year 2050 (A1B2050) assuming a climate change signal consistent with the the IPCC-AR4 scenario HADCM3-SRA1B. The applied scenario A1B2050 assumes an increase in temperature during all months and a decrease in precipitation sum during the summer months.

In order to compare the effects of different governmental irrigation water policies on the agricultural water consumption and income in potato production, we apply for both climate scenarios four different irrigation water policies scenarios (see Table 2).

Table 2 Irrigation Policy Scenarios

Policy	Variable Irrigation Costs	Maximal Yearly Irrigation Amount	Temporal Restrictions
1	$0.1\ CHF \cdot \mathrm{m}^{-3}$	Unlimited	None
2	$0.1\ CHF \cdot \mathrm{m}^{-3}$	Unlimited	Irrigation is banned for a random period in summer if $\sum_{i=1}^{n} WB_i < -200mm$ [a]
3	$0.1\ CHF \cdot \mathrm{m}^{-3}$	$2500\ m^3 \cdot ha^{-1}$	None
4	$2.2\ CHF \cdot \mathrm{m}^{-3}$	Unlimited	None

[a] WB_i is the daily waterbalance (mm) in the months of May, June and July. The random period of water withdrawal ban in summer is defined by a normal distribution of the starting point and of the length of the water withdrawal ban. The mean value and standard deviation of the normal distribution of the starting point and of the length of the water withdrawal bans are derived from the observed water withdrawal bans of the canton Vaud in the period 1998–2010.

We use the C++ based GA library package GAlib [9] in order to maximize the CE in potato production under the different climate and policy scenarios. We apply the following control parameters to the GAs: genome size = 8, population size = 30; proportion of replacement = 0.2; selection routine = roulette wheel; mutation probability = 0.05. The optimization run stops if there is no improvement in the objective function for a sequence of 1000 consecutive generations. For this study the optimizations have been conducted on a normal PC (Intel Pentium Core(TM) i5 at 3.33GHz) and took up to more than one week.

3 Results and Discussion

Figure 1 shows the profitability, the expected physical potato yield levels, the production risks (see plot "Coefficient of Variation"), the optimal applied nitrogen fertilizer and the optimal irrigation amount for each water withdrawal policy (see Table 2) under current and under future expected climate conditions. CC has a general negative effect on the profitability as well as on the physical average potato yield levels for all considered water withdrawal policies. The two most effective irrigation water reducing policies are "Policy 3" where the yearly irrigation water quantity is limited to a maximal amount and "Policy 4" where higher water prices are imposed. Applying "Policy 4", though, leads under current climate conditions to a decrease in income of more than 28%, while "Policy 3" reduces the income only by 6%.

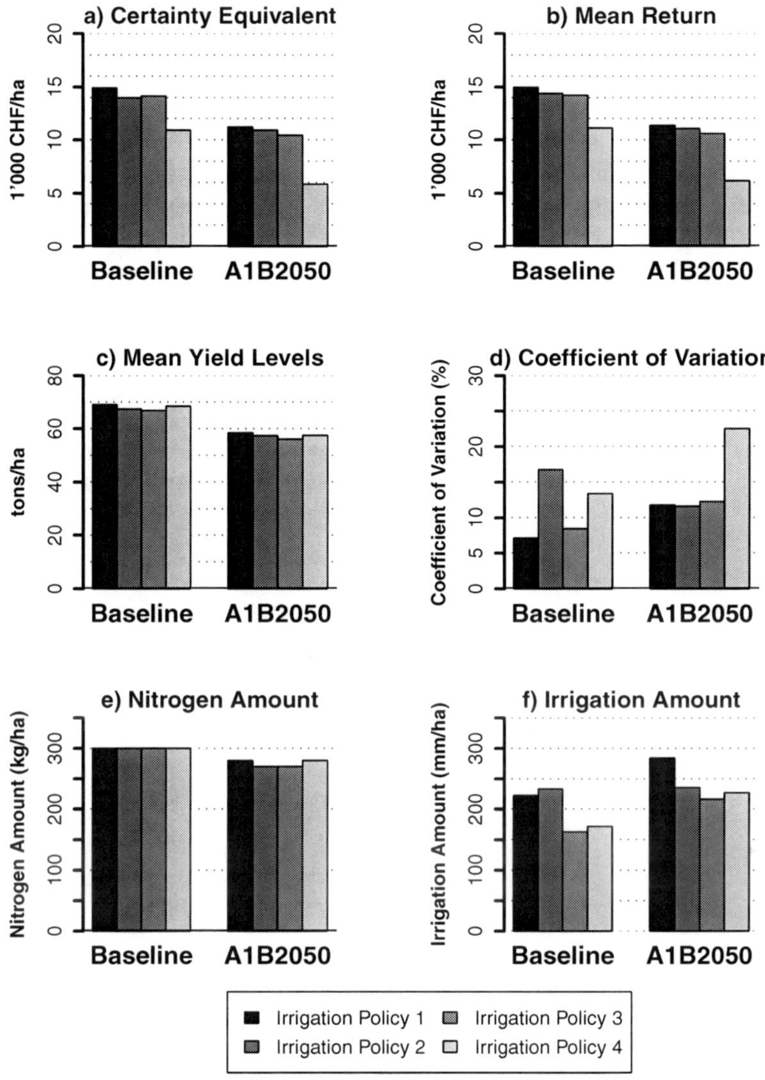

Fig. 1 Figure 1 shows the effects of the four considered irrigation water policies (see Table 2) on the profitability (plot a and b), physical yield levels (plot c), production risks (plot d) and on the optimal nitrogen and water use (plot e and f) under current and future expected climate conditions in potato production in the Broye region.

Under the assumed future climate "Policy 4" even decreases the income by more than 47%, while the income reduction due to "Policy 3" amounts only to 8%. The prohibition of water withdrawal during dry periods ("Policy 2") has the smallest negative effect on the income in potato production (-6% under current climate and

-3% under future expected climate). In contrast to limiting the total irrigation water amount and increasing the water prices, the prohibition of water withdrawal during dry periods ("Policy 2") decreases the average irrigation water amount only under future expected climate conditions substantially.

In conclusion, this study showed that potato production in the Broye region will require more irrigation water under future expected climate. The most effective irrigation water withdrawal policy is a limitation of the maximum yearly applicable irrigation water amount. This policy reduces not only the average irrigation water consumption but also minimizes the reduction in income of potato production assuming an utility maximizing farmer.

Acknowledgements We thank for support of the Swiss National Science Foundation in the framework of the National Research Program 61 and in the framework of the National Centre of Competence in Research on Climate (NCCR Climate). We would like to thank MeteoSwiss and the Research Station Agroscope Reckenholz-Taenikon ART for providing climate and yield data, and Annelie Holzkaemper for support relating to the implementation of the GAs into the optimization model.

References

1. Gen, M., Cheng, R.: Genetic Algorithms and Engineering Optimization. Wiley-Interscience, (2000)
2. Bundesgesetz vom 24. Januar 1991 ber den Schutz der Gewsser (Gewsserschutzgesetz, GSchG)
3. Finger, R., Schmid, S.: Modeling Agricultural Production Risk and the Adaptation to Climate Change. Agricultural Finance Review. **68**, 25–41. 2008
4. Muehlberger de Preux, C.: Broye: Fish or Chips?. UMWELT. **02/08**, 26–27 (2008)
5. Musshoff, O., Hirschauer, N.: Optimizing Production Decisions Using a Hybrid Simulation–Genetic Algorithm Approach. Canadian Journal of Agricultural Economics/Revue canadienne d'agroeconomie. **57**, 35–54 (2009)
6. Semenov, M.A. and Barrow, E.M.: Use of a Stochastic Weather Generator in the Development of Climate Change Scenarios. Climatic Change. **35**, 397–414 (1997)
7. Stoeckle, C.O. and Donatelli, M. and Nelson, R.: CropSyst, a Cropping Systems Simulation Model. European Journal of Agronomy. **18**, 289–307 (2003)
8. Torriani, D.S. and Calanca, P. and Schmid, S. and Beniston, M. and Fuhrer, J.: Potential effects of changes in mean climate and climate variability on the yield of winter and spring crops in Switzerland. Climate Research. **34**, 59–369 (2007)
9. Wall, M.: GAlib: A C++ Library of Genetic Algorithm Components. Technical report, Massachusetts Institute of Technology, (1996)
10. Weber, M., Schild, A.: Stand der Bewsserung in der Schweiz - Bericht zur Umfrage 2006, (2007)

An Enhanced Ant Colony System for
the Sequential Ordering Problem

L.M. Gambardella, R. Montemanni, and D. Weyland

Abstract A well-known Ant Colony System algorithm for the Sequential Ordering Problem is studied to identify its drawbacks. Some criticalities are identified, and an Enhanced Ant Colony System method that tries to overcome them, is proposed. Experimental results show that the enhanced method clearly outperforms the original algorithm and becomes a reference method for the problem under investigation.

1 Introduction

The *Sequential Ordering Problem* (SOP) can model real-world problems such as production planning, single vehicle routing problems with pick-up and delivery constraints and transportation problems in flexible manufacturing systems (Anghinolfi et al. [1]). Among the many exact and heuristic methods presented over the years for the problem, we here mention only those of interest for the present paper (see Anghinolfi et al. [1] for a complete literature review). Gambardella and Dorigo [3] presented an approach based on Ant Colony System (ACS) enriched with a sophisticated Local Search (LS) procedure. Montemanni et al. [4] built on top of this method, adding a Heuristic Manipulation Technique (HMT). A Discrete Particle Swarm Optimization (DPSO) method has been finally discussed in Anghinolfi et al. [1].

In this paper we will critically analyze the ACS method discussed in Gambardella and Dorigo [3] , trying to identify its characteristics and weaknesses, and to propose some improvements to the original schema. The result is what we will refer to as the Enhanced Ant Colony System (EACS) algorithm.

Luca Maria Gambardella, Roberto Montemanni, Dennis Weyland
Istituto Dalle Molle di Studi sull'Intelligenza Artificiale, Galleria 2, CH-6928 Manno, Switzerland,
e-mail: {luca, roberto, dennis}@idsia.ch

D. Klatte et al. (eds.), *Operations Research Proceedings 2011*, Operations Research Proceedings, 355
DOI 10.1007/978-3-642-29210-1_57, © Springer-Verlag Berlin Heidelberg 2012

2 Problem description

The Sequential Ordering Problem, also referred to as the *Asymmetric Travelling Salesman Problem with Precedence Constraints*, can be modeled in graph theoretical terms as follows. A complete directed graph $D = (V,A)$ is given, where V is the set of nodes and $A = \{(i,j) \mid i,j \in V, i \neq j\}$ is the set of arcs. A cost $c_{ij} \in \mathbb{N}$ is associated with each arc $(i,j) \in A$. Without loss of generality it can be assumed that a fixed starting node $1 \in V$ is given. It has to precede all the other nodes. The tour is also closed at node 1, after all the other nodes have been visited ($c_{i1} = 0 \ \forall i \in V$ by definition). Furthermore an additional precedence digraph $P = (V,R)$ is given, defined on the same node set V as D. An arc $(i,j) \in R$, represents a precedence relationship, i.e. i has to precede j in every feasible tour. Such a relation will be denoted as $i \prec j$ in the remainder of the paper. The precedence digraph P must be acyclic in order for a feasible solution to exist. It is also assumed to be transitively closed, since $i \prec k$ can be inferred from $i \prec j$ and $j \prec k$. Note that for the last arc traversed by a tour (entering node 1), precedence constraints do not apply. A tour that satisfies precedence relationships is called *feasible* and the cost of such a tour is the sum of the costs of the edges used in that tour. The objective of the *SOP* is to find a feasible tour with the minimal total cost.

3 The original ACS algorithm

The ACS algorithm is an element of the Ant Colony Optimization (*ACO*) algorithms family. The first *ACO* algorithm, Ant System (*AS*), has been initially proposed in Dorigo et al. [2] and is based on a computational paradigm inspired by the way real ant colonies function. The main element of this metaheuristic algorithm are *ants*, simple computational agents that individually and iteratively construct solutions for the problem. Intermediate partial problem solutions are seen as *states*; each ant moves from a state a to another state b, corresponding to a more complete partial solution. At each step, every ant computes a set of feasible expansions to its current state and moves to one of these probabilistically, according to a probability distribution specified as follows. For ant k the probability p_{ab}^k of moving from state a to state b depends on the combination of two values: the attractiveness μ_{ab} of the move, as computed by some heuristic, indicating the *a priori* desirability of that move and the trail level τ_{ab} of the move, indicating how proficient it has been in the past to make that particular move; it represents therefore an *a posteriori* indication of the desirability of that move. In the reference implementation of the ACS paradigm for the SOP (Gambardella and Dorigo [3]) a state is identified by the last node visited in the solution under construction. With probability q_0 the next node to visit after node a for ant k, is chosen as the feasible node b for which the result of a linear combination of τ_{ab} and μ_{ab} is highest (deterministic rule), while with probability $1 - q_0$ the node b is chosen with a probability given by a combination of τ_{ab} and μ_{ab} (we refer the interested reader to Gambardella and Dorigo [3] for a more detailed explanation). Trails are updated at each iteration, increasing the level of those that correspond to moves included in good solutions, while decreasing all the others. The phase in which ants build up their solutions and update the trails is referred to as the *constructive phase*.

Once an ant has finished constructing its solution, and prior to the evaluation of the solution, a *local search* is generally applied to improve this solution.

The adaptation of the ACS paradigm to the SOP is straightforward: the constructive phase of each ant starts at the origin node and chooses the next nodes probabilistically. The only complication is represented by precedence constraints: each time the next node has to be chosen, nodes that if selected would violate some precedence constraint have to be inserted in the list of forbidden nodes.

The local search adopted by the ACS for the SOP, the *SOP-3-exchange* local search, is extremely efficient. It was originally introduced in Gambardella and Dorigo [3]. The local search makes use of a data structure called *don't push stack* introduced in Gambardella and Dorigo ([3]). The *don't push stack* is a data structure based on a stack, which contains the set of nodes h to be selected. At the beginning of the search the stack is initialized with all the nodes in the sequence (that is, it contains $n + 1$ elements). During the search, node h is popped off the stack and feasible 3-exchanges starting from h are investigated. In case a profitable exchange is executed the six nodes involved in this exchange are pushed onto the stack (if they do not already belong to it). Using this heuristic, once a profitable exchange is executed starting from node h, the top node in the *don't push stack* remains node h. In this way the search is focused on the neighborhood of the most recent exchange.

3.1 Analysis of the ACS algorithm

With reference to the ACS algorithm for the SOP described in Section 3, we can observe that the constructive phase carries out both diversification (exploring new regions of the search space) and intensification (searching very deeply a given region of the search space). On the other hand, the local search procedure is intrinsically a strong intensification phase. In presence of a strong local search procedure, the role of the construction phase can be less prominent and should be revised. In particular the constructive phase could only take care of diversification. Moreover, in the ACS the local search is *added* to the constructive phase, without any effort to have a clever integration of the two components. Both the enhancement are in the direction of speeding up the method, without compromising its search capabilities. We remind the reader that one known drawback of the ACS approach is the large total running time required to build new solutions by each artificial ant. Let n be the number of steps necessary to build a solution. Usually the constructive process takes time $O(n)$ for each of the n steps required.

4 The Enhanced ACS algorithm

Two changes to the original ACS paradigm are proposed with the aim of making the resulting EACS method faster than ACS, without loosing any exploration capability.

4.1 An improved constructive phase

The constructive phase can be made faster by directly considering the best solution computed so far already during the constructive phase. At each phase of the constructive phase, the next node to visit is selected with probability q_0 as the node reached after the current node i in the best solution computed so far (in case this node is not feasible, the classic mechanism adopted by the ACS method is applied). Since the probability q_0 is usually greater than or equal to 0.9, the new approach drastically reduces the running time required to select the next edge to visit (typically from $O(n)$ to something approximable by a constant). With probability $1 - q_0$, the probabilistic criterion used by the original ACS method is adopted to select the next node to visit.

4.2 A better integration between the constructive phase and the local search procedure

A better integration between the constructive phase and the local search routine is achieved as follows. The local search is run only if the cost of the solution produced in the constructive phase is within 20% of the best solution retrieved so far (this value was found experimentally). This prevents the local search routine to waste time on unpromising solutions. Moreover, the *don't push stack* (see Section 3) is initialized in such a way that only the elements that in the current solution are out of sequence with respect to the best solution retrieved so far, are in the stack. This pushes the exploration of the search space towards areas that were potentially unexplored in the previous iterations.

5 Experimental results

The benchmark instances considered are those from the SOPLIB2006 library, commonly adopted in the recent literature[1]. Each instance is identified as n-r-p, where the following naming convention is adopted: n is the number of nodes of the problem, i.e. $V = \{1, 2, \ldots, n\}$; r is the cost range, i.e. $0 \leq c_{ij} \leq r \ \ \forall i, j \in V, i \neq j$; p

---

[1] The library is available at http://www.idsia.ch/~roberto/SOPLIB06.zip.

is the approximate percentage of precedence constraints, i.e. the number of precedence constraints of the problem will be about $\frac{p}{100} \cdot \frac{n(n-1)}{2}$. Instances were created from the following values for the parameters: $n \in \{200, 300, 400, 500, 600, 700\}$; $r \in \{100, 1000\}$; $p \in \{1, 15, 30, 60\}$.

The EACS algorithm has been coded in ANSI C, and the experiments presented have been run on a Dual AMD Opteron 250 2.4 GHz/4GB computer. The parameter settings adopted for both ACS and EACS are those already used in Gambardella and Dorigo [3] (we refer the reader to this paper for more details).

A comparison with ACS and with the best known results are presented in Table 1. For the best known results we indicate the method used to obtain the result, and the cost of the solution (cost). For ACS and EACS we present the average and the best result obtained over 10 runs, where a maximum computation time of 600 seconds for each run is imposed.

The results of Table 1 give clear indications: over the 48 instances considered, EACS was able to improve 45 average results and 43 best results of ACS, never being worse than ACS. When compared with the best-known results, EACS was able to improve 29 of them, while matching the best known results in 15 of the remaining 16 cases.

6 Conclusions

An Enhanced Ant Colony System algorithm for the Sequential Ordering Problem has been presented. The new algorithm was obtained by modifying a well-known Ant Colony System approach. The changes were done according to a preliminary study aiming at highlighting the drawbacks of the adaptation of the original Ant Colony System to the Sequential Ordering Problem. Experimental results have shown the effectiveness of the enhanced method.

The framework proposed is general and can be applied to every combinatorial optimization problem with the proper characteristics.

References

1. D. Anghinolfi, R. Montemanni, M. Paolucci, and L.M. Gambardella. A hybrid particle swarm optimization approach for the sequential ordering problem. *Computers and Operations Research*, 38(7):1076–1085, 2011.
2. M. Dorigo, V. Maniezzo, and A. Colorni. The Ant System: Optimization by a colony of co-operating agents. *IEEE Transactions on Systems, Man, and Cybernetics - Part B: Cybernetics*, 26(1):29–41, 1996.
3. L.M. Gambardella and M. Dorigo. An ant colony system hybridized with a new local search for the sequential ordering problem. *INFORMS Journal on Computing*, 12(3):237–255, 2000.
4. R. Montemanni, D.H. Smith, and L.M. Gambardella. A heuristic manipulation technique for the sequential ordering problem. *Computers and Operations Research*, 35(12):3931–3944, 2008.

Table 1 Computational results.

Instance	Best known		ACS [3]		EACS	
	method	cost	avg	best	avg	best
R.200.100.1	DPSO [1]	64	90.3	88	67.2	63
R.200.100.15	DPSO [1]	1799	2066.0	2002	1818.3	1792
R.200.100.30	HMT [4]	4216	4254.6	4247	4216.0	4216
R.200.100.60	ACS [3]	71749	71749.0	71749	71749.0	71749
R.200.1000.1	DPSO [1]	1414	1549.5	1532	1432.5	1411
R.200.1000.15	DPSO [1]	20481	22602.9	21775	20717.0	20481
R.200.1000.30	HMT [4]	41196	41371.6	41278	41196.0	41196
R.200.1000.60	ACS [3]	71556	71556.0	71556	71556.0	71556
R.300.100.1	DPSO [1]	31	76.4	74	34.6	31
R.300.100.15	DPSO [1]	3167	3738.6	3520	3207.1	3162
R.300.100.30	DPSO [1]	6120	6228.2	6151	6120.0	6120
R.300.100.60	ACS [3]	9726	9726.0	9726	9726.0	9726
R.300.1000.1	DPSO [1]	1338	1586.7	1536	1369.6	1331
R.300.1000.15	DPSO [1]	29475	34447.9	33533	29784.4	29248
R.300.1000.30	DPSO [1]	54147	55013.4	54367	54172.6	54147
R.300.1000.60	ACS [3]	109471	109530.5	109471	109471.0	109471
R.400.100.1	DPSO [1]	21	64.1	59	22.6	21
R.400.100.15	DPSO [1]	3946	5087.1	4838	3986.2	3925
R.400.100.30	DPSO [1]	8165	8476.5	8289	8165.9	8165
R.400.100.60	ACS [3]	15228	15232.4	15228	15228.0	15228
R.400.1000.1	DPSO [1]	1484	1811.1	1783	1475.5	1456
R.400.1000.15	DPSO [1]	40054	46638.6	45055	40122.9	39612
R.400.1000.30	DPSO [1]	85221	85979.6	85579	85203.3	85192
R.400.1000.60	HMT [4]	140816	140994.9	140862	140816.0	140816
R.500.100.1	DPSO [1]	14	55.0	51	16.1	11
R.500.100.15	DPSO [1]	5525	6931.5	6584	5507.8	5431
R.500.100.30	DPSO [1]	9683	10333.2	10047	9668.1	9665
R.500.100.60	HMT [4]	18240	18260.4	18246	18247.4	18240
R.500.1000.1	DPSO [1]	1514	1877.4	1840	1522.5	1501
R.500.1000.15	DPSO [1]	51624	62693.7	60175	51763.0	51091
R.500.1000.30	DPSO [1]	99181	101751.8	100453	99112.0	99018
R.500.1000.60	HMT [4]	178212	178478.1	178323	178212.0	178212
R.600.100.1	DPSO [1]	11	49.8	44	9.4	6
R.600.100.15	DPSO [1]	5923	7806.6	7610	5881.7	5798
R.600.100.30	DPSO [1]	12542	13001.8	12810	12475.5	12465
R.600.100.60	DPSO [1]	23293	23357.4	23342	23293.0	23293
R.600.1000.1	DPSO [1]	1628	1986.7	1936	1598.9	1534
R.600.1000.15	DPSO [1]	59177	72701.1	70454	58281.6	57812
R.600.1000.30	DPSO [1]	127631	132314.3	130244	126961.7	126789
R.600.1000.60	HMT [4]	214608	214970.2	214724	214608.0	214608
R.700.100.1	DPSO [1]	9	42.6	41	7.9	5
R.700.100.15	DPSO [1]	7719	9573	9383	7444.0	7380
R.700.100.30	DPSO [1]	14706	15905.8	15733	14520.0	14513
R.700.100.60	DPSO [1]	24106	24192.3	24151	24172.0	24102
R.700.1000.1	DPSO [1]	1606	1969.2	1912	1614.8	1579
R.700.1000.15	DPSO [1]	72618	85177.7	81439	68630.0	67510
R.700.1000.30	DPSO [1]	136031	141557.9	139769	134651.2	134474
R.700.1000.60	DPSO [1]	245589	246489.6	246128	245684.0	245632

Evolutionary Exploration of E/E-Architectures in Automotive Design

Ralph Moritz, Tamara Ulrich, and Lothar Thiele

Abstract In this work we design an evolutionary algorithm for the exploration of the electric and electronic (E/E-)architectures in a car. We describe a novel way to simultaneously optimize the assignments of function components to electronic control units (ECU) and of ECUs to busses. To this end we present a suitable representation as well as corresponding variation operators. We also provide heuristics for the optimization of the communication infrastructure, i.e. how busses are connected to each other. Preliminary results show that the approach is able to produce architectures similar to those which are used nowadays, as well as promising alternatives.

1 Introduction

In the last decade, the electric and electronic systems (E/E-architectures) of a car have become increasingly complex. Additional functions such as brake assistance, skidding control, or parking aid have been integrated, which in turn lead to an explosion of the number of electronic control units (ECUs) and their communication. As a result, the design of E/E-architectures gets more and more involved and time consuming. Many communication links and dependencies between single functions prohibit a simple decomposition of the architecture design task into independent subtasks. Instead, even minor individual decisions may have a global influence on the whole system evaluation in terms of objective functions and constraints.

In this work, we present a high-level optimization framework which considers all major tasks, from setting up the ECUs and placing them in the car up to connecting them to busses and define the cable routing. It is not the purpose of the proposed approach to fully automate the design of future E/E-architectures. Rather, it should

Ralph Moritz
Robert Bosch GmbH, Schwieberdingen, Germany, e-mail: Ralph.Moritz@de.bosch.com

Tamara Ulrich and Lothar Thiele
Computer Engineering and Networks Laboratory, ETH Zurich, 8092 Zurich, Switzerland e-mail: firstname.lastname@tik.ee.ethz.ch

D. Klatte et al. (eds.), *Operations Research Proceedings 2011*, Operations Research Proceedings, 361
DOI 10.1007/978-3-642-29210-1_58, © Springer-Verlag Berlin Heidelberg 2012

provide decision support for design engineers by presenting alternatives to current designs.

2 The Design of an E/E-Architecture

Figure 1 shows an example of an abstract E/E-architecture design. We assume that the E/E-architecture will have to implement a number of composite functions, which in turn consist of a number of individual components that communicate via signals. Such components can either be sensors, processing components, or actuators. In order to incorporate a function into the car, its components have to be assigned to assembly units, in general ECUs, which in turn have to be placed in an appropriate location in the car and assigned with additional hardware (intelligent semiconductors), depending on the components requirements. Next, the ECUs need to be assigned to busses of a specific type (LIN, CAN, MOST, FlexRay,...), which need to be connected via gateways, such that all digital signals can be routed using the constructed inter-bus topology, i.e. over a sequence of busses connected by gateways. Beside the inter-bus topology, for each bus an intra-bus topology (e.g. linear or star) needs to be defined, that states which ECUs are physically connected.

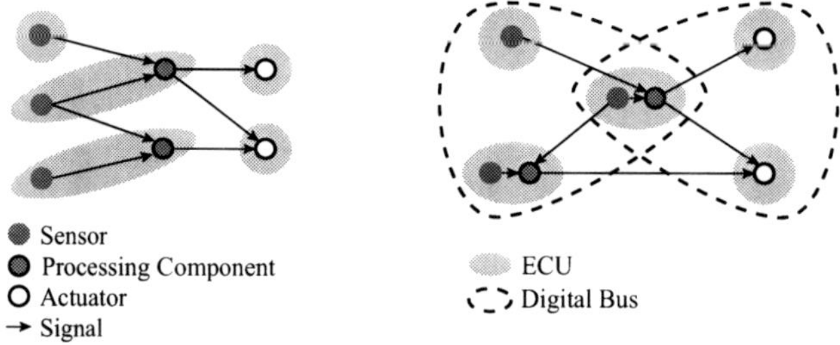

Fig. 1 E/E-Architecture Design Example: A given function containing 7 components and 7 signals is assigned to ECUs (left) and the corresponding ECUs are assigned to digital busses (right), where the middle ECU acts as a gateway between the two busses. Intra-bus topology is not shown.

The resulting architectures are evaluated according to two objectives which are to be minimized. The first one is cost, which is governed by the cable and ECU cost. The cables in turn depend on the signals that have to be routed between the ECUs, on the used communication structure as well as on the placement of the ECUs. The second objective is ECU complexity defined as the average number of different functions assigned to an ECU. A low complexity increases the reliability of a car as the number of functions which are affected by a single ECU failure is minimized.

3 Optimizing E/E-Architectures

Due to the size of the search space and the presence of several objective functions, we propose to design a multiobjective evolutionary algorithm to optimize the E/E-architecture design problem. Evolutionary algorithms (EAs) are considered to be black box optimization algorithms and therefore, they are used mostly for problem classes where classical standard methods fail or can hardly be applied [9]. To make the search most efficient, the EA needs to be adapted to the problem. Problem-specific knowledge enters EAs through (a) embedded local heuristics, (b) an appropriate representation of solutions and (c) corresponding variation operators. The use of local heuristics reduces the search space size of the EA but the available diversity of solutions in the population might be lost [3]. We decided to do the following split, based on the observation that many decisions only affect one objective, namely the cost, and the cost-optimal solution can be found in a straight-forward manner.

EA-optimized decisions :

- Task 1: Assignment of components to ECUs
- Task 2: Physical placement of ECUs
- Task 3: Assignment of ECUs to digital busses

Decisions made by local heuristics :

- Intelligent semiconductor selection for each ECU (simply take the cheapest that satisfies the memory requirements of the processing components of that ECU)
- Inter-bus topology(described in Section 3.3)
- Bus type selection (choose the cheapest type that satisfies the data-rate requirements of the signals that have to be routed over the bus)
- Intra-bus topology (choose such that the cable cost is minimized)

3.1 Representation

Now that we have reduced the number of tasks that have to be optimized by the EA, we need to design a suitable representation to optimize these tasks simultaneously. This is in contrast to the two-stage approach of Limam [6], who first optimizes Task 1 and then takes the best assignment and optimizes Task 2. It also differs from the work of Hardung [4], where physical placement and cable routing is fixed and only the assignment of components to the given ECUs is optimized. Furthermore, the representation must be able to express all possible solutions and allow the design of operators that have a direct and controllable effect [8].

The assignment of components to ECUs corresponds to a partitioning of the components into k clusters, where k, i.e. the number of clusters, is a parameter to be optimized. The assignment of ECUs to busses starts with partitioning the ECUs to busses and later on defining gateways to connect the busses (see section 3.3). This partitioning differs from the first because the number of ECUs as well as the number of busses is not fixed. Therefore the second partitioning depends on the first. One could handle the tasks separately and require additional repair strategies to make both partitioning fit together [1]. In our approach this dependency is handled by the data structure of the representation to reduce the number of infeasible assignments and avoid the bias that may be induced by a repair strategy. This data structure is in form of a hierarchical partitioning, where components are first partitioned into ECUs, and the ECUs are then partitioned to busses, see Figure 2. This data structure is implemented as a tree which allows a direct manipulation by the operators. Representing the placement of ECUs in the car is also straight-forward. We assume that we are given a set of possible mounting spaces and simply add a mounting space field for each ECU node in the representation.

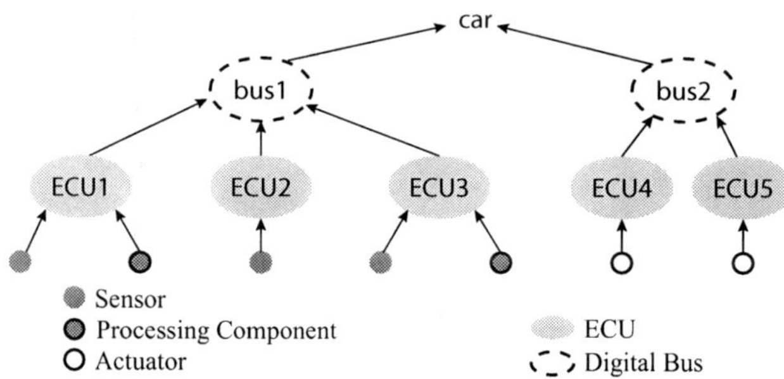

Fig. 2 EA Representation: Hierarchical Partitioning (Component to ECU and ECU to Bus)

3.2 Variation Operators

In general, a mutation operator should perform a slight and random change. Different operators can be characterized according to their exhaustiveness, locality and bias, for which we gave definition and measures in [7]. Focusing on partitioning problems, Falkenauer [2] did extensive research. He found that a mutation operator must be able to merge and split clusters, and optionally to move single elements from one cluster to another. We extended his ideas for the hierarchical partitioning mutation operator as follows: First, a partitioning level (i.e. either the component to ECU or the ECU to bus level) is selected at random. Then, either merge, split or move is selected at random. If a merge operation is selected, two random clusters are merged. In the case of a split operation, a random cluster is split into two random parts. Finally in case of a move operation, a random element is moved to a random cluster. The mutation of the mounting space assignment is straight forward, a random ECU is selected and its mounting space field is randomly changed.

The role of recombination is to improve a solution by adding a certain property of an already good one. Therefore, these properties must be reachable and exchangeable by the recombination operator. If not, a recombination may just perform large changes, without a driving force to improve a solution. For the recombination operator we also make use of Falkenauers work. We again select a partitioning level at random. On this level we randomly select a cluster of the first solution and copy it to the partitioning of the second solution. As the elements of the new cluster now occur twice in this partitioning, the occurrences originating from the second solution are deleted. In the same way, we also add a cluster of the second to the first solution. The proposed recombination therefore creates two new solutions by directly exchanging ECUs or whole busses.

3.3 Gateway Selection Heuristic

The busses resulting from the hierarchical partitioning are not yet connected via gateways and therefore do not ensure that ECUs which implement components that need to communicate to each other and which reside on different busses actually can transmit signals via communication paths on the intra-bus topology. We decided that the busses should be connected in a tree-like manner, such that the routing of signals is unambiguous. This is in accordance with architecture examples from the industry, in which most busses are also connected in a tree shape.

To generate a tree, we first construct a graph where the nodes are the busses and the edges represent the signals that have to be transmitted between the two corresponding busses. Each edge is weighted with the total data rate of all signals that have to be transmitted between sink and source bus of that edge. We then apply a standard algorithm to find a maximum spanning tree of that graph [5]. This spanning tree is unique only if no two edges have the same weight. To remove ties, a given ordering of the signals is used to identify the edge with the lower rank. After calculating the spanning tree, a gateway is created for each edge in the spanning tree by adding an ECU from one bus to the other. This is done in such a way that the cost for additional cables is minimized.

The use of this heuristic has several advantages. First, the resulting topologies assure that for each signal a unique path from sender to receiver exists. This is given because a spanning tree must preserve the connectivity of the initial graph, which is defined by the signals. It is further worth mentioning, that if a bus does not exchange signals with the other busses, the heuristic will not create a gateway. Therefore the resulting topologies are connected as much as needed and as less as possible. Second, the distance of two busses in the topology is related to the data rate of the exchanged signals. Therefore large communication needs are handled as direct as possible. All in all this heuristic reduces the size of the search space by omitting infeasible topologies and further is biased on preferred topologies.

3.4 Handling Infeasibility

There are several situations in which a generated E/E-architecture is infeasible. These architectures could be repaired using heuristics. These repairs may introduce an unintended bias and require extensive problem-specific knowledge. Therefore, we decided to use a generic repeat-strategy. If a mutated architecture is infeasible, the parent is mutated again until a feasible one is found. If no feasible architecture can be found after a fixed number of trials, the original architecture is returned.

4 Conclusions

We presented an evolutionary algorithm that optimizes E/E-architectures in a holistic manner. We first decided which optimization tasks should be optimized by the EA and which by local heuristics. Then, we propose to use a hierarchical partitioning in order to represent the assignment of components to ECUs on the first level and of ECUs to digital busses on the second level. We also define suitable variation operators, as well as heuristics to optimize the communication between digital busses. Preliminary results on a real-world problem with 20 functions and a total of 150 components show that our algorithm is able find both existing architectures which are used in practice, as well as new designs that can be used to improve existing designs. Future work contains the evaluation of operators and selection strategies as well as a more problem specific constraint handling strategy.

References

1. Blickle, T.: Theory of Evolutionary Algorithms and Application to System Synthesis. PhD Thesis, ETH Zurich (1997)
2. Falkenauer, E.: Genetic Algorithms and Grouping Problems. Wiley, New York (1998)
3. Grosan, C. and Abraham, A.: Hybrid Evolutionary Algorithms: Methodologies, Architectures, and Reviews. In: Hybrid Evolutionary Algorithms, Studies in Computational Intelligence Springer 2007
4. Hardung, B.: Optimisation of the Allocation of Functions in Vehicle Networks. PhD Thesis, University of Erlangen-Nürnberg (2006)
5. Kruskal, J.B.: On the Shortest Spanning Subtree of a Graph and the Traveling Salesman Problem. In: Proceedings of the American Mathematical Society, vol. 7, number 1, pp. 48-50, (1956)
6. Limam, M: A New Approach for Gateway-Based Automotive Network Architectures. PhD Thesis, Dresden University of Technology (2009)
7. Moritz, R., Ulrich, T., Thiele, L., Buerklen, S.: Mutation Operator Characterization: Exhaustiveness, Locality, and Bias. In: Proceedings of the 2011 IEEE Congress on Evolutionary Computation, IEEE Press 2011
8. Rothlauf, F.: Representations for Evolutionary Algorithms. In: Proceedings of the 12th Annual Conference Companion on Genetic and Evolutionary Computation, ACM 2010
9. Yu, X. ; Gen, M.: Introduction to Evolutionary Algorithms. Springer (2010)

Stream X
Network Industries and Regulation

Marco Laumanns, IBM Research – Zurich, SVOR Chair
Mikulas Luptacik, Vienna University of Economics and Business, ÖGOR Chair
Grit Walther, University of Wuppertal, GOR Chair

Optimal dimensioning of pipe networks: the new situation when the distribution and the transportation functions are disconnected

Daniel de Wolf and Bouchra Bakhouya

Abstract De Wolf and Smeers [9] consider the problem of the optimal dimensioning of a gas transmission network when the topology of the network is known. The pipe diameters must be chosen to minimize the sum of the investment and operating costs. This model does not reflect any more the current situation on the gas industry. Today, the transportation function and gas buying function are separated. This work considers the new situation for the transportation company. We present here first results obtained on the belgian gas network and on a part of the french network.

1 Introduction

De Wolf and Smeers [5] consider the problem of the *optimal dimensioning of a gas transmission network* when the *topology of the network is known* as a two stages problem: *investment in pipe diameters* in the first stage, *network optimal operations* in the second stage. The second stage problem considered by De Wolf and Smeers [6] was formulated as a cost minimization subject to nonlinear flow-pressure relations, material balances and pressure bounds. This model does not reflect any more the current situation on the gas market. Today, the transportation and gas buying functions are separated.

In Bakhouya et De Wolf [2], the new situation for the exploitation model of the transportation company was presented. The objective for the transportation company is to determine the flows in the network that minimize the energy used for the gas transport. This corresponds to the minimization of the power used in the com-

Daniel de Wolf
TVES/ULCO, 21 quai de la Citadelle, 59.140 Dunkerque, France e-mail: daniel.dewolf@univ-littoral.fr

Bouchra Bakhouya
IESEG, Universite Catholique de Lille, 3 rue de la digue, 59.800 Lille, France e-mail: b.bakhouya@ieseg.fr

D. Klatte et al. (eds.), *Operations Research Proceedings 2011*, Operations Research Proceedings, 369
DOI 10.1007/978-3-642-29210-1_59, © Springer-Verlag Berlin Heidelberg 2012

pressors. In order to reflect this new situation, a modelisation of the compressors was introduced in the exploitation model of De Wolf and Smeers [6]. We present here first results concerning the *dimensioning model*.

2 Problem formulation

The **network is represented** through a graph $G = (N, A)$ where N is the *set of nodes*, and A is the *set of arcs* (pipelines, compressors or valves). We consider *three types of arcs*. The set of arcs A is thus divided in three subsets, namely:

- A_p, the subset of *passive arcs* corresponding to pipelines,
- A_c, the subset of *active arcs* corresponding to compressors,
- A_v, the subset of arcs corresponding to *valves*.

2.1 The investment problem

The transportation company must decide the *pipe diameter* for each pipe, and the *maximal power* for each compression station, in order to minimize the sum of the *investment cost* (in diameters and compression stations) and the *network operations costs*, namely the compressors used power.

We use the following notation for the **investment variables:**

$$D_{ij} \quad \text{notes the pipe diameter, } \forall (i, j) \in A_p,$$
$$P_{ij} \quad \text{notes the maximal power of compressor, } \forall (i, j) \in A_c.$$

For the *investment costs for compressors stations*, we suppose a *fixed installation cost*, noted k_C, and a *marginal cost proportional to the installed power*, $k'_C P_{ij}$. The *total investment cost in compression stations* is thus considered as a linear function of the installed power:

$$C_{inv}(P) = \sum_{(i,j) \in A_c} (k_C + k'_C P_{ij}) \tag{1}$$

For the *investment costs for pipes,* we consider, following De Wolf and Smeers [5], the sum of three terms:

$$C_{inv}(D) = \sum_{(i,j) \in A_p} [k_P + k'_P D_{ij} + k''_P D_{ij}^2] L_{ij} \tag{2}$$

We relax the constraint that the pipe diameters must be chosen in a set of discrete values to avoid the additional difficulty of solving a non linear non convex problem with integer variables. The same assumption is made on the maximal power of compression stations. We only impose a *maximal pipe diameter* and an *upper bound on the maximal power of the compressors :*

$$0 \leq D_{ij} \leq \overline{D}, \ \ 0 \leq P_{ij} \leq \overline{P}$$

2.2 Formulation of second stage problem

The *second stage problem* was already formulated in Bakhouya and De Wolf [2]. The **operation variables** are the following :

f_{ij} notes the gas flow in each arc $(i,j) \in A$,

W_{ij} notes the power dissipated in the compressor $(i,j) \in A_c$,

p_i notes the gas pressure at each node $i \in N$,

s_i notes the net gas supply in each node $i \in N$.

The **objective function** corresponds to minimization of the *energy used in the compressors:*

$$\min Q_{op} = \alpha \sum_{(i,j)\in A_a} \frac{1}{0,9\eta_{therm}} W_{ij} \tag{3}$$

where α is the unitary energy price (in Euro/kW) and η_{therm}, the thermic efficacity.

The **constraints of the second stage problem** are the following. At a **supply node** i, the gas inflow s_i must remain within take limitations specified in the contracts:

$$\underline{s_i} \leq s_i \leq \overline{s_i}$$

The gas transmission company cannot receive gas at a pressure higher than the one insured by the supplier at the entry point. Conversely, at each exit point, the demand must be satisfied at a minimal guaranteed pressure:

$$\underline{p_i} \leq p_i \leq \overline{p_i}$$

The flow conservation equation at node i insures the gas balance at node i:

$$\sum_{j|(i,j)\in A} f_{ij} = \sum_{j|(j,i)\in A} f_{ji} + s_i$$

Now, consider the *constraints on the arcs*. For an arc corresponding to a *pipe*, the relation between the flow f_{ij} in the arc and the pressures at the two ends of the pipe p_i and p_j is of the following form (see O'Neill and al. [8]):

$$sign(f_{ij})f_{ij}^2 = C_{ij}^2(p_i^2 - p_j^2), \ \forall (i,j) \in A_p \tag{4}$$

where C_{ij} is a parameter depending on the *pipe length* L_{ij} and on the *pipe diameter*, D_{ij}:

$$C_{ij} = \pi \frac{D_{ij}^2}{4} \sqrt{\frac{D_{ij}}{\lambda R T L_{ij}}} = K_{ij} D_{ij}^{\frac{5}{2}} \tag{5}$$

For an *active arc* corresponding to a *compressor*, the following expression of the power used by the compressor can be found in André et al [1] :

$$W_{ij} = \gamma_1 f_{ij} \left(\left(\frac{p_j}{p_i} \right)^{\gamma_2} - 1 \right) \; \forall (i,j) \in A_a \tag{6}$$

The power used in the compressor must be lower than the maximal power, P_{ij}:

$$W_{ij} \leq P_{ij}, \; \forall (i,j) \in A_c$$

There is also an upper bound on the maximal pressure increase ratio:

$$\frac{p_j}{p_i} \leq \gamma_3 \; \forall (i,j) \in A_c$$

For active arcs, the direction of the flow is fixed:

$$f_{ij} \geq 0, \; \forall (i,j) \in A_c$$

3 Solution procedure

To solve the investment problem, we propose the following **solution procedure:**

1. We start from a **feasible solution** in terms of the *investment variables D_{ij}* and P_{ij}. For the two practical study cases (namely the belgian gas network and a part of the french network), we start from the actual diameters increased by 20 % and from the actual maximal powers of compression stations increased by 20 %.
2. As explained in De Wolf et Bakhouya [2], we solve an *auxiliary convex problem* to achieve a **good starting point for the second stage problem**. This problem is inspired from Maugis [7]:

$$\min h(f,s) = \sum_{(i,j) \in A} \frac{|f_{ij}| f_{ij}^2}{3 C_{ij}^2}$$

$$\text{s.t.} \quad \sum_{j|(i,j) \in A} f_{ij} - \sum_{j|(j,i) \in A} f_{ji} = s_i \; \forall i \in N \tag{7}$$

$$\underline{s_i} \leq s_i \leq \overline{s_i} \qquad\qquad \forall i \in N$$

It can be shown (See De Wolf and Smeers [4]) that problem (7) has a physical interpretation. Namely, its objective function is the mechanical energy dissipated per unit of time in the pipes. This implies that the point obtained by minimizing the mechanical energy dissipated in the pipes should constitute a good starting point for the complete problem.

3. We solve the *second stage problem* starting from the solution of the problem (7). We obtains thus a feasible solution for the complete problem.

4. Then, we solve the *complete problem* **all in one problem** from this feasible point. Namely, we replace in the two stages formulation of the investment problem, the operations function by its expression and we include all the constraints of the second stage problem in the first stage problem.

4 Numerical results

Our **first study case** is the *belgium gas transport network*. The main characteristics of the belgian gas network are the following:

- there are *24 passive arcs*, and *2 compressors*,
- there are *9 demand nodes*, *6 supply nodes*, and *20 single nodes*,
- there are *no cycle* on this network.

As previously said, we use as *starting point,* the actual network and for the *passive arcs,* we increase the *current diameter* by 20 %, since for *active arcs,* we increase the *maximal power* by 20 %,

The resolution of problem for the *belgian gas network* by GAMS/CONOPT gives the following conclusions:

- The solver *increases some diameters* in order to reduce the use of the compressors.
- The solver *keeps the maximal power of compression stations unchanged.*

We can conclude that the **model prefers to increase the capacity of the passive arcs**, namely to increase the pipe diameters, **in place of increasing the capacity of the active arcs,** namely in place of increasing of the maximal power of compression stations. This is the **first important result** of this work.

Our *second study case* concerns a *cycled network which corresponds to a part of the french network.* The main characteristics are the following:

- there are *41 passive arcs, 7 compressors,* and *10 valves,*
- there are *19 demand nodes, 6 supply nodes,* and *56 single nodes,*
- there are *3 cycles* on this network.

For this second case study, we have received the data from Gaz de France for a period from 2006 to 2021. Two cases must be distinguished:

- *Case of years 2006 to 2011:* for these years, the actual capacities are enough to transport all the demand. The same conclusions can be established as for the belgian gas network. Namely, the solver *increases some diameters* in order to reduce the use of the compressors but it *keeps the maximal power of compression stations unchanged.*

- *Case of years 2012 to 2021:* for these years, a reinforcement of the capacities of the network is needed. In this case also, the model prefers to increase the capacity of the passive arcs (the pipe diameters) in place of increasing the maximal power of compression stations. But the simulations for these years have emphasized the *role of another important design variable*: the **maximal flow in the compression stations** : we have to increase the maximal flow in the compression station to take into account the increase of demand. This is the **second main result** of this work.

5 Conclusions

We have formulated the *optimal dimensioning problem* for a gas transport company as a **two stage problems**: *investment in pipe diameters* and in *maximal power of compression stations* in the first stage, *operations of the network* in the second stage. We have solved this problem for two practical study cases: the *belgian gas network* and *a part of the french gas transmission network*. The mains results of theses simulations are the following. First, the **model prefers to increase the capacity of the passive arcs** (the pipe diameters) **in place of increasing the capacity of the active arcs** (the maximal power of compression stations). Secondly, *another important design variable* is the **maximal flow in the compression stations** which must be increased in accordance with the increase of the demand.

References

1. Jean ANDRE, Laurent CORNIBERT, A tool to optimize the Reinforcement Costs of a Gas Transmission Network, Pipeline Simulation Interest Group, 38th Annual Meeting, Williamsburg (VA), October 2006.
2. B. BAKHOUYA and D. DE WOLF, The gas transmission problem when the merchant and the transport functions are disconnected, HEC Ecole de Gestion de l'ULG WP 2007 01/01, January 2007.
3. BAKHOUYA Bouchra, Optimisation de rseaux de transport de gaz avec considration des compresseurs, thèse de doctorat, 27 juin 2008, Université du Littoral, Dunkerque, France.
4. D. DE WOLF and Y. SMEERS, Mathematical Properties of Formulations of the Gas Transmission Problem, SMG Preprint 94/12, Universit libre de Bruxelles, Juin 1994.
5. D. DE WOLF and Y. SMEERS, Optimal Dimensioning of Pipe Networks with Application to Gas Transmission Networks, Operations Research,Vol 44, No 4, July-August 1996, pp 596-608.
6. D. DE WOLF and Y. SMEERS, The Gas Transmission Problem Solved by an Extension of the Simplex Algorithm, Management Sciences, Vol. 46, No 11, November 2000, pp 1454-1465
7. MAUGIS, M.J.J. 1977, Etude de Réseaux de Transport et de Distribution de Fluides, *R.A.I.R.O.*, Vol.11, No 2, pp 243-248.
8. O'NEILL, R.P., WILLIARD, M., WILKINS, B. and R. PIKE 1979, A Mathematical Programming Model for Allocation Of Natural Gas, *Operations Research 27*, No 5, pp 857-873.

Analysis of Investment Incentives and the Influence of Growth Options in Regulated Industries

Sabine Pallas

Abstract In this contribution, a regulated company's investment incentives are analyzed in a setting with cost-based regulation, a two-stage investment and growth options. First, it is shown that in the first period, at the first stage of the investment, the company will clearly underinvest from the regulator's perspective. Second, a specific formula for the calculation of the regulated price in the second period is derived, which will set incentives for the company to choose the capacity level which the regulator considers efficient.

1 Introduction

The purpose of this paper is to investigate a regulated company's investment choices if it faces a cost-based regulation and, in addition to that, when growth opportunities exist.

In network industries like telecommunications or electricity, a regulated company's investment decisions are affected by the actual regulatory instruments in place. It has been shown in previous research[1] that the widely used cost-based remuneration in combination with straight-line depreciation causes underinvestment from the regulator's perspective.

However, the existing literature does not cover all important aspects of such an investment decision. Especially in widely publicly discussed areas like investments in broadband technology or in smart grids, investment decisions are not static one-time decisions, but the companies will rather invest sequentially: Having gained

Sabine Pallas
Technische Universitaet Muenchen, Arcisstr. 21, D-80333 Muenchen e-mail: sabine.pallas@tum.de

[1] Cf. Friedl (2007), Rogerson (2008), Nezlobin et al. (2008).

D. Klatte et al. (eds.), *Operations Research Proceedings 2011*, Operations Research Proceedings, DOI 10.1007/978-3-642-29210-1_60, © Springer-Verlag Berlin Heidelberg 2012

experience with pilot projects, they will make subsequent investments in later periods. On the one hand, there are papers dealing with the influence of the regulation on investment incentives, like the above mentioned. On the other hand, there are papers analyzing the influence of multi-stage behavior, some considering real options[2]. However, there is no paper combining both streams of literature: It has not been investigated yet how the existing regulation affects investment incentives, if companies structure their investments as described. This paper will contribute to the existing literature by filling this gap.

2 Model description

2.1 Basics

Friedl (2007) shows that underinvestment occurs when straight-line depreciation is used to determine the regulated price. He shows this for a single investment opportunity. However, it is unclear if the same results hold for a two-stage investment. Hence, I use his model as a base-line and extend it to a two-stage investment.

Consider a discrete-time model in which a regulated monopolist faces two subsequent investment opportunities. At time $t = 0$ the company has to decide how many units x_0^C of a product or a service it wants to provide, at $t = 1$ it chooses the capacity x_1^C. The necessary investment outlay is $C_t \cdot x_t^C, t \in \{0, 1\}$ where C_t denotes the outlay for one unit of capacity. Both resulting assets will be fully depreciated after two periods. For simplicity reasons, variable costs are normalized to be zero. This assumption does not change the results and avoids fuzziness about the model dynamics. Furthermore, both investments are assumed to be irreversible. This is a given factor in the considered industries: The expansion of electricity or telecommunication grids can hardly be undone.

At each point in time, there exists a specific demand function on the product market. This function indicates the demanded quantity of the product q_t for a given price p_t (monotonically decreasing in p_t). As in Friedl (2007), linear inverse demand functions are assumed for simplicity. They are given by $p_t(q_t) = a_t - b \cdot q_t$, where a_t is the maximal willingness to pay and b is the slope of the demand function, which is assumed to remain constant over time. If the provided maximal capacities are considered, then the demanded quantities in each period are $q_t(p_t, x_t^C) = \min\{a_t/b - p_t/b, x_t^C\}$. Hence, for low prices, demand is limited by the available capacity and for high prices demand linearly decreases according to the demand function.

---

[2] Cf. McDonald and Siegel (1986), Kulatikula and Perotti (1998), Abel and Eberly (2005).

Clearly, the company will only make a second investment in $t = 1$ if demand has increased. Therefore I assume a weakly increasing demand, which means $a_0 \geq a_1 \geq a_2$.

It is assumed that the company is subject to a cost based regulation: The regulator sets a price cap $\overline{p}_t$ exactly equal to the incurred costs. In this setting, since variable operating costs are normalized to be zero, the only incurred costs are depreciation charges and interest on invested capital.

2.2 First investment decision

It is important to be aware of the fact that the company and the regulator pursue different objectives. The company's goal is to maximize the net present value of an investment project, whereas the regulator's goal is to maximize social welfare. In this setting, Friedl (2007) determines the capacity x_0^C that the company considers efficient, and secondly he determines the regulator's choice of capacity x_0^R using constant demand over time.

The same analysis with increasing demand yields the following capacity choices:

Company	Regulator
$x_0^C = \frac{a_0}{b} - \frac{1}{b} \cdot \left(\frac{C_0}{2} + rC_0 \right)$	$x_0^R = \frac{-C_0(1+r)^2 + a_0(1+r) + a_1}{b(2+r)}$

with r being the interest rate and $\frac{C_0}{2} + rC_0 = \overline{p}_0$. Note that the demand in the second period, i.e. a_1 only appears in the regulator's capacity choice. This is because the regulator maximizes the social welfare over all periods, whereas for the company only the first period is relevant.

Let Δ_0 be the difference between the capacity level the regulator wants to be installed and the company's capacity choice, $\Delta_0 := x_0^R - x_0^C$.

$$\Delta_0 = \frac{2(a_1 - a_0) + C_0 r}{2b(1+r)} > 0. \tag{1}$$

Since $\Delta_0 > 0 \Leftrightarrow x_0^R > x_0^C$ there is underinvestment from the regulator's perspective. Moreover, from this equation, it is straightforward to see that

1. The stronger the demand increases from $t = 0$ to $t = 1$ ($a_1 \gg a_0$), the more severe is the underinvestment.
2. The steeper the demand function is (high value of b), the less severe is the underinvestment.

3. There is underinvestment even for the interest rate r going to zero. In this case $\Delta_0 = (a_1 - a_0)/b$.

To conclude the analysis for the first investment, if straight-line depreciation is used, the regulator always wants to invest more than the company. This result even holds for a vanishing interest rate.

2.3 Second investment decision

Company's capacity choice: At $t = 1$, the company learns about C_1 and $p_1(q_1)$. It then has to choose the capacity x_1^C. If the company makes the second investment, it will face additional costs consisting of depreciation and interest caused by C_1. Consequently, the regulator will allow for a different regulated price. This is done by weighting two parts: A C_0-part[3] $\overline{p}_{0,1} = C_0(1+r)/2$ caused by the first investment and a C_1-part $\overline{p}_{1,1} = C_1(1+2r)/2$ caused by the second investment. An intuitive way to weight the two parts is to use the respective capacities. The resulting regulated price $\overline{p}_{1,w}$ for the second period is[4]

$$\overline{p}_{1,w} = \frac{x_0^C}{x_0^C + x_1^C} \cdot \overline{p}_{0,1} + \frac{x_1^C}{x_0^C + x_1^C} \cdot \overline{p}_{1,1}. \tag{2}$$

It is important to be aware of the fact that at point $t = 1$ the decision on x_0^C has already been made, the company has installed x_0^C according to the previous section. Consequently, x_0^C needs to be incorporated into the investment decision as a fixed constant.

Following the same rationale as before (see Friedl (2007)), the company will want to make sure that the provided capacity will not exceed the demand. Furthermore, the relevant regulated price for its capacity choice will be $\overline{p}_{1,w}$ (not a regulated price in any later period), since it is the highest price and therefore it will cause the lowest demand. Hence, it will choose x_1^C such that the total provided capacity $x_0^C + x_1^C$ will be less or equal than the demanded capacity at price $\overline{p}_{1,w}$:

$$x_0^C + x_1^C \leq \frac{a_1}{b} - \frac{1}{b} \cdot \overline{p}_{1,w} \tag{3}$$

Setting the inequality binding, using $\overline{p}_{1,w}$ as in (2) and solving for x_1^C results in a quadratic equation. Simplification yields:

$$x_1^C = -x_0^C + \frac{2a_1 - C_1(1+2r)}{4b} \pm \sqrt{\frac{\overline{p}_{1,1} - \overline{p}_{0,1}}{b} \cdot x_0^C + \frac{(2a_1 - C_1(1+2r))^2}{16b_1^2}}. \tag{4}$$

---

[3] Note that at $t = 1$ the asset C_0 has not been fully depreciated yet.

[4] The index w stands for "weighted"

Regulator's capacity choice: The rationale for the regulator is again the same as for the first investment decision: The regulator's objective is to choose x_1^R such that it maximizes social welfare given that the company has already decided on x_0^C. Therefore it has to incorporate x_0^C into its decision making[5]. The regulator's objective function for x_1^R hence is:

$$\max_{x_1^R} \left\{ -C_1 x_1^R + \left(a_1(x_0^C + x_1^R) - \frac{b}{2}(x_0^C + x_1^R)^2 \right) \cdot \gamma + \left(a_2 x_1^R - \frac{b}{2}(x_1^R)^2 \right) \cdot \gamma^2 \right\}. \quad (5)$$

Again note that a_2, the demand in the third period, appears in the regulator's objective function, but not in the company's. $\gamma = \frac{1}{1+r}$ is the discount factor

As for the first investment decision, this is a straightforward maximization problem. The optimal x_1^R from the regulator's perspective is

$$x_1^R = \frac{-C_1(1+r)^2 + a_1(1+r) + a_2 - bx_0^C(1+r)}{b(2+r)} \quad (6)$$

Calculation of an optimal $\overline{p}^*$: In section 2.2 the difference between the both capacity choices, Δ_0, was calculated in order to see that there is underinvestment from the regulators perspective. For the second investment this calculation is not straightforward due to the quadratic equation in (4). However, it is obvious that x_1^C and x_1^R do not coincide. This indicates that the choosen weights for $\overline{p}_{1,w}$ were not able to induce the "right" capacity choice from the regulator's perspective.

Still, one can approach the situation differently and calculate the optimal $\overline{p}_1$: Set the company's capacity choice (3) equal to the regulator's (6) and solve for $\overline{p}_1$:

$$x_1^C \overset{!}{=} x_1^R$$

$$\Leftrightarrow \frac{a_1 - \overline{p}_1 - bx_0^C}{b} = \frac{-C_1(1+r)^2 + a_1(1+r) + a_2 - bx_0^C(1+r)}{b(2+r)}$$

$$\Leftrightarrow \overline{p}_1^* = \frac{-a_0 + a_1 - a_2 + C_1(1+r)^2 + \overline{p}_0}{2+r}.$$

This $\overline{p}_1^*$ can also be expressed as a weighted sum of $\overline{p}_{0,1}$ and $\overline{p}_{1,1}$ as in (2), plus two correction terms κ_1 and κ_2:

$$\overline{p}_1^* = \frac{1}{2+r} \cdot \overline{p}_{0,1} + \frac{2}{2+r} \cdot \overline{p}_{1,1} + \underbrace{\frac{-a_0 + a_1 - a_2}{2+r}}_{\kappa_1} + \underbrace{\frac{C_0 r + 2C_1 r^2}{2(2+r)}}_{\kappa_2} \quad (7)$$

---

[5] Note that the total capacity appears in the regulator's objective function, and the total capacity at $t = 1$ is x_0^C (already determined) *plus* x_1^R (to be determined).

This formulation of the regulated price $\overline{p}_1^*$ shows that more weight needs to be put on $\overline{p}_{1,1}$, which is the C_1-part of $\overline{p}_1^*$. The correction term κ_1 is larger, the larger is $|a_1 - a_0|$ compared to $|a_2 - a_1|$. This means, if the regulator expects a big increase in demand from period 2 to period 3, he should allow only a smaller regulated price in period 2. κ_2 only has a minor significance and is even zero for $r = 0$.

Summarizing, for the second investment decision the calculation of the regulated price $\overline{p}_1$ is complex. It consists of two parts: $\overline{p}_{0,1}$, caused by the first investment, and $\overline{p}_{1,1}$, caused by the second investment. Simply building the average of both parts or weighting them with the respective capacities does not induce the company to choose the regulator's investment level. $\overline{p}_1$ needs to be determined like in (7) in order to set incentives for the company to choose the capacity level that is optimal for social welfare.

3 Conclusion

A regulated company's investment incentives for a two-stage investment project are analyzed. Similar to earlier research, it is found that with a cost-based regulation and the use of straight-line depreciation, at the first stage the regulator would want the company to provide a higher capacity than the company considers efficient. For an increasing demand, this is also true for a vanishing interest rate. Furthermore, the steeper the demand function, the less severe is the underinvestment.

For the second stage, the simple calculation of a weighted average to determine the regulated price $\overline{p}_1$ will not induce coinciding capacity levels for the company and the regulator. In order to set the right investment incentives to the company, the regulator needs to determine the regulated price $\overline{p}_1$ following a specific formula.

References

1. Abel, A., Eberly, J. (2005): Investment, Valuation, and Growth Options. Working Paper, Wharton School, University of Pennsylvania.
2. Friedl, G. (2007) : Ursachen und Lsung des Unterinvestitionsproblems bei einer kosten-basierten Preisregulierung. Die Betriebswirtschaft, 67, p. 335-348.
3. Kulatikula, N., Perotti, E. (1998): Strategic Growth Options. In: Management Science, Vol. 44, No. 8, August 1998, p. 1021-1031.
4. McDonald, R., Siegel, D. (1986): The Value of Waiting to Invest. In: The Quarterly Journal of Economics, Vol. 101, p. 707-727.
5. Nezlobin, A., Rajan, M., and Reichelstein, S. (2008). "Dynamics of Rate of Return Regulation", Working paper, Stanford University.
6. Rogerson, W. (2008): Intertemporal Cost Allocation and Investment Decisions. Journal of Political Economy, Vol. 116 (5), p. 931-950.

Robust Optimization in Non-Linear Regression for Speech and Video Quality Prediction in Mobile Multimedia Networks

Charalampos N. Pitas, Apostolos G. Fertis, Athanasios D. Panagopoulos, and Philip Constantinou

Abstract Quality of service (QoS) and quality of experience (QoE) of contemporary mobile communication networks are crucial, complex and correlated. QoS describes network performance while QoE depicts perceptual quality at the user side. A set of key performance indicators (KPIs) describes in details QoS and QoE. Our research is focused specially on mobile speech and video telephony services that are widely provided by commercial UMTS mobile networks. A key point of cellular network planning and optimization is building voice and video quality prediction models. Prediction models have been developed using measurements data collected from live-world UMTS multimedia networks via drive-test measurement campaign. In this paper, we predict quality of mobile services using regression estimates inspired by the paradigm of robust optimization. The robust estimates suggest a weaker dependence than the one suggested by linear regression estimates between the QoS and QoE parameters and connect the strength of the dependence with the accuracy of the data used to compute the estimates.

1 Introduction

Contemporary mobile networks have been based on WCDMA (Wideband Code Division Multiple Access) radio access technology. UMTS (Universal Mobile Telecom-

C. N. Pitas · A. D. Panagopoulos · P. Constantinou
Mobile Radiocommunications Laboratory, School of Electrical and Computer Engineering,
National Technical University of Athens.
9, Iroon Polytechniou Str., Zografou, GR 15773, Athens, Greece.
e-mail: chpitas,thpanag,fkonst@mobile.ntua.gr

A. G. Fertis
Institute for Operations Research, Department of Mathematics (D-MATH),
Eidgenössische Technische Hochschule Zürich, (ETH Zürich).
Rämistrasse 101, HG G 21.3, 8092, Zürich, Switzerland.
e-mail: afertis@ifor.math.ethz.ch

D. Klatte et al. (eds.), *Operations Research Proceedings 2011*, Operations Research Proceedings, 381
DOI 10.1007/978-3-642-29210-1_61, © Springer-Verlag Berlin Heidelberg 2012

munication System) networks have been widely deployed to deliver circuit switched (CS) speech and video telephony services [8]. Both network monitoring systems as well as extended drive-test measurement campaigns are useful tools in radio network planning, performance evaluation and optimization [6,7,14]. While radio coverage prediction models have been developed and widely used in network planning processes, quality prediction is assumed of the recent research areas. Advanced statistical analysis, usually linear and non-linear regression modeling, has been used in order to build prediction models [11] in a simulation environment of a mobile network. The quality of the predictions is affected by the ability of the considered models to represent reality as well as the accuracy of the data used to estimate their parameters.

Robust optimization has been increasingly used in mathematical optimization as an effective way to immunize solutions against data uncertainty. If the true value of a problem's data is not equal to its nominal value, the optimal solution computed using the nominal value might not be optimal or even feasible for the true value. Robust optimization considers uncertainty sets for the data of a problem and defines solutions that are immune to such uncertainty. In linear optimization problems, box-type uncertainty sets [10], and ellipsoidal uncertainty sets have been considered. In linear, as well as mixed integer optimization problems, robust counterparts with budgeted uncertainty sets that connect the uncertainties developed by the coefficients of the constraints have been efficiently solved [2,3]. The robust optimization paradigm has been successfully applied to regression and classification problems to deal with uncertainty in the statistical data and has been connected with regularized regression models [1,15].

In this paper, we apply robust optimization in order to refine the accuracy of quality prediction models. More specifically:

1. We use measurements data that acquired by a drive-test campaign of a live 3G network in Switzerland. Test mobile speech and video calls were performed by experimental equipment.
2. We focus on speech, video as well as on audio-visual quality prediction modelling based on live-network measurements.
3. We compute the linear and non-linear regression estimates that connect QoE with QoS, as well as the linear regression estimates that follow the robust optimization paradigm. The estimates depend on the size of the uncertainty set that is considered.

The robust estimates suggest a weaker dependence than the one suggested by linear regression between QoE and QoS. Our approach enables the choice of the strength of the dependence based on the accuracy of the data used to compute the estimates. The remaining of the paper is organized as follows: Section 2 is devoted to UMTS network architecture, radio KPIs as well as to QoE aspects regarding speech and video transmission. Followingly, we present regression analysis and robust optimization for speech and video quality prediction in Section 3. Finally, a general discussion of our contribution on QoE estimation is placed in Section 4.

2 Quality in Mobile Multimedia Networks

2.1 QoS in UMTS networks

A UMTS network consists of a radio access network (RAN) and a core network (CN). RAN basically consists of RNCs (Radio Network Controllers) and NodeBs (base stations) that are connected with UEs (User Equipments). The CN can be interconnected to various backbone networks like IP-networks (Internet) public fixed telephone networks (ISDN/PSTN). The report in [5] presents details for a UMTS coverage measurements methodology in order to characterize the quality of radio network coverage, specifically:

RSCP (Received Signal Code Power) is the received power on one code measured, in *dBm*, on the pilot bits of the P-CPICH (Primary Common Pilot Channel).

RSSI (Received Signal Strength Indicator) is the wideband received power, in *dBm*, within the relevant channel bandwidth.

E_c/N_0 is the ratio, in *dB*, of received pilot energy, E_c, to the total received energy or the total power spectral density, I_0. The received energy per chip, E_c, divided by the power density in the band. The E_c/N_0 is identical to $RSCP/RSSI$. We note that E_c/N_0 is the most crucial radio quality parameter.

2.2 QoE of Mobile Multimedia Services

Commercial UMTS networks provide both mobile speech and video telephony services. Specifically, a 64 kbit/s (speech is coded at 12.2 kbit/s and video at 50 kbit/s respectively) transport channel is dedicated for video communication. On the one hand, an objective method for end-to-end (E2E) narrow-band speech telephony quality is PESQ described in ITU-T P.862. Besides, a mapping function for transforming PESQ raw result scores to MOS (Mean Opinion Score) scale is presented in ITU-T P.862.1. On the other hand, video quality algorithms shall be eligible and applicable for end-to-end mobile applications and predict the perceived quality by the user-viewer according to the ETSI TR 102 493. J.247 is recommended by ITU-T for objective perceptual video assessment in the presence of a full reference. Audio-visual quality can be computed by speech, MOS_{SQ}, and video, MOS_{VQ}, quality parameters according to the formula of the ITU-T P.911: $MOS_{AVQ} = \lambda + \mu \cdot MOS_{SQ} \cdot MOS_{VQ}$, where $\lambda = 0.765$ and $\mu = 0.1925$.

2.2.1 QoE Evaluation

The assessment of a QoE indicator, either MOS_{SQ} or MOS_{VQ}, is made by comparing the original speech/video samples transmitted by the calling party, A-side, $S(\tau) =$

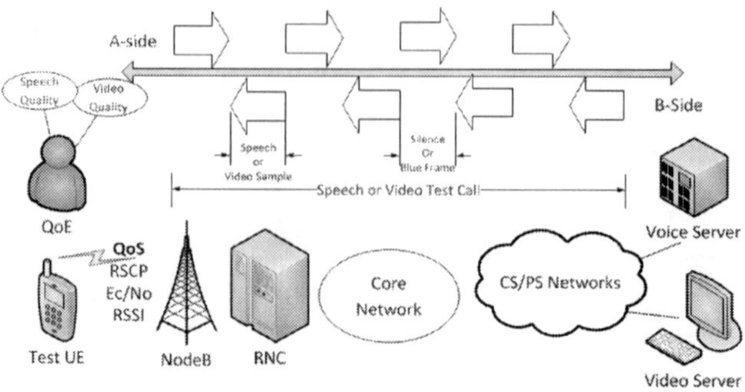

Fig. 1 UMTS architecture and Measurement Campaign configuration.

$\{s_1(\tau_1),\ s_2(\tau_2),\ \ldots,\ s_n(\tau_N)\}$, with the corresponding degraded received samples, $\tilde{S}(\tau) = \{\tilde{s}_1(\tau_1),\ \tilde{s}_2(\tau_2),\ \ldots,\ \tilde{s}_N(\tau_N)\}$, at the called party, B-side, by applying the objective assessment algorithm, φ. The objective QoE indicator, $MOS_{i \to j}$, varies between 1 (bad/very annoying) and 5 (excellent/imperceptible) and $(i, j) \in \{(A,B), (B,A)\}$:

$$MOS_{i \to j} = \frac{1}{N} \sum_{v=1}^{N} \varphi(s_{v,i}(\tau_v), \tilde{s}_{v,j}(\tau_v)) \tag{1}$$

Fig.1 presents a measurement campaign configuration for E2E QoE assessment with a *"Diversity Benchmarker"* [12]. Besides, we conducted post-processing analysis with *NetQual NQDI* on a measurements database managed by *MS SQL Server* and we used MATLAB for regression and optimization modelling.

3 Robust Optimization in Regression Analysis

In order to predict QoE from QoS, a non-linear regression method can be performed for speech and video quality estimation from E_c/N_0, $QoE = \exp(\alpha \cdot E_c/N_0 + \beta)$. A log-linear model can be derived, $ln(QoE) = \alpha \cdot E_c/N_0 + \beta$. The estimates for α and β can be computed through non-linear regression or linear regression and are almost the same. The estimates of linear regression are the optimal solution to

$$\min_{\alpha,\beta} \sqrt{\sum_{i=1}^{n}(ln(QoE_i) - \alpha \cdot (E_c/N_0)_i - \beta)^2}, \tag{2}$$

where QoE_i, $(E_c/N_0)_i$ denotes the i-th samples of QoE, E_c/N_0 respectively.

The accuracy of the data used to compute them is limited. To deal with this problem, we propose to follow a robust optimization approach. We assume that that data $(E_c/N_0)_i$ incorporate errors and that their true value resides in a ρ-sized interval centered at $(E_c/N_0)_i$, $\rho \geq 0$. The robust estimates are the optimal solution to the min-max problem

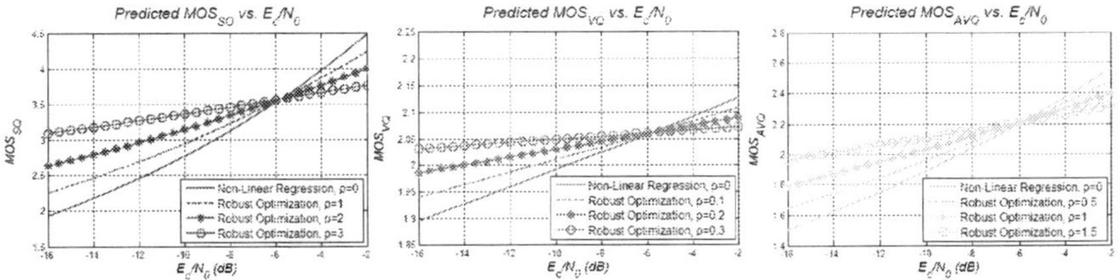

Fig. 2 Robust optimization in non-linear regression for QoE prediction models.

$$\min_{\alpha,\beta} \ \max_{\forall i:\ |x_i-(E_c/N_0)_i|\leq\rho} \sqrt{\sum_{i=1}^{n}(ln(QoE_i)-\alpha x_i-\beta)^2}. \tag{3}$$

The constraint $\forall i : |x_i - (E_c/N_0)_i| \leq \rho$ is equivalent to $\|\mathbf{x} - (\mathbf{E_c/N_0})\|_\infty \leq \rho$, where $\|\cdot\|_\infty$ is the infinity norm and $\mathbf{x}$, $(\mathbf{E_c/N_0})$ are the vectors which contain the x_i, $(E_c/N_0)_i$ respectively. Problem (3) is equivalent [15] to

$$\min_{\alpha,\beta} \ \sqrt{\sum_{i=1}^{n}(ln(QoE_i)-\alpha(E_c/N_0)_i-\beta)^2}+\rho|\alpha|, \tag{4}$$

which defines an l_1-regularized regression estimator [13]. Using our data and Se-DuMi [9], we computed the robust estimates for α and β for various values of the size ρ of the uncertainty set. The results can be seen in Fig. 2.

We observe that as the size of uncertainty ρ increases, α drops, namely the robust estimates suggests a weaker dependence between QoE and E_c/N_0. This weaker dependence is compensated by a smaller fixed term β, as seen in Fig. 3. Thus, in the presence of errors, one should be more conservative in the prediction of QoE for a given E_c/N_0. The rate of change for QoE with respect to E_c/N_0 estimated by linear or non-linear regression is too optimistic in the presence of errors. The regularized estimator of Eq. (4) takes this phenomenon into consideration by adding the trade-off term $\rho\,|\alpha|$. Our method connects the impact ρ of this trade-off with the size of the uncertainty sets for the data $(E_c/N_0)_i$. In this way, we can use the information

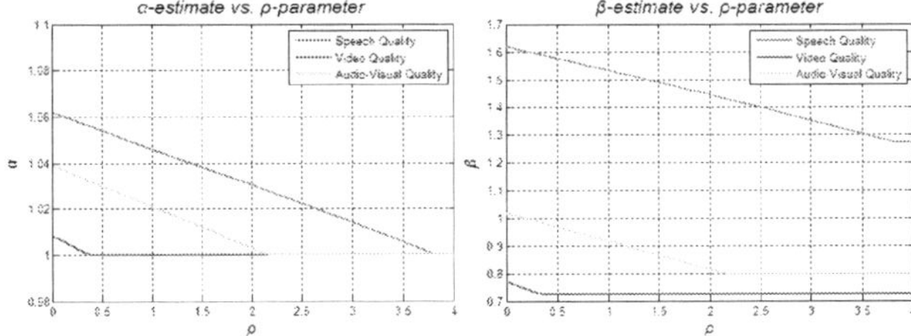

Fig. 3 Robust estimates of α and β parameters for various values of the size ρ of the uncertainty set.

on the accuracy of the data to assess the strength of the dependence between QoE and E_c/N_0.

4 Conclusions

The use of robust optimization to deal with uncertain data in regression-based speech, video, and audio-visual quality prediction was addressed in our paper. Indeed, our method suggested a weaker dependence between QoS and QoE parameters than the one suggested by linear regression estimates. In particular, it explicitly connected the strength of this dependence with the accuracy of the measurements data used to compute the estimates. The used QoE empirical models for mobile video telephony have been extracted from live 3G multimedia network measurements. Our models can be applied in quality-centric network planning and optimization processes to tackle the effect of errors in a measurement campaign.

Acknowledgements We would like to thank Mr. Arthur Tollenaar from SwissQual AG, Switzerland, for providing to us measurements files regarding mobile video-telephony.

References

1. Ben-Tal, A., El Ghaoui, L., Nemirovski, A.: Robust Optimization. Princeton University Press (2009).
2. Bertsimas, D., Sim, M.: Robust Discrete Optimization and Network Flows, Mathematical Programming **98**(1-3), 49-71 (2003).
3. Bertsimas, D., Sim, M.: The price of robustness. Operations Research **52**(1), 35-53 (2004).
4. Boyd, S., Vandenberghe, L.: Convex Optimization. Cambridge University Press (2004).
5. ECC Report 103: UMTS Coverage Measurements. Electronic Comm. Committee (ECC), European Conf. of Postal and Telecomm. Administrations (CEPT), Nice (2007).
6. Malkowski, M., Claßen, D.: Performance of Video Telephony Services in UMTS using Live Measurements and Network Emulation. Wireless Personal Communications, Springer, **46**(1), 19-32 (2008).
7. Goudarzi, M., Sun, L., Ifeachor, E.: PESQ and 3SQM measurement of voice quality over live 3G networks. 8th Int'l Conf. on Measurement of Speech, Audio and Video Quality in Networks (MESAQIN), Prague (2009).
8. Holma, H., Toskala, A.: WCDMA for UMTS - HSPA Evolution and LTE. 4th Ed., John Wiley & Sons Ltd. (2007).
9. SeDuMi 1.3: A Matlab toolbox for optimization over symmetric cones. Available [On Line] http://sedumi.ie.lehigh.edu/
10. Soyster, A. L.: Convex Programming with Set-Inclusive Constraints and Applications to Inexact Linear Programming. Operations Research, **21**(5), 1154-1157 (1973).
11. Sun, L., Ifeachor, E.: Voice Quality Prediction Models and their Applications in VoIP Networks. IEEE Transactions on Multimedia **8**(4), 809-820 (2006).
12. SwissQual AG: Diversity Benchmarker. [On Line]: http://www.swissqual.com/
13. Tibshirani, R.: Regression shrinkage and selection via the lasso. Journal of the Royal Statistical Society, Series B (Methodological) **57**(1), 267-288 (1995).
14. Vlachodimitropoulos, K., Katsaros, E.: Monitoring the end user perceived speech quality using the derivative mean opinion score (MOS) key performance indicator. 18th Annual IEEE Int'l Symp. on Personal, Indoor and Mobile Radio Comm. (PIMRC'07), Athens (2007).
15. Xu, H., Caramanis, C., Mannor, S.: Robust Regression and Lasso. IEEE Transactions on Information Theory **56**(7), 3561-3574 (2010).

Stream XI
OR in Industry, Software Applications, Modeling Languages

Ulrich Dorndorf, INFORM GmbH, Aachen, GOR Chair
Eleni Pratsini, IBM, Zurich, SVOR Chair
Alfred Taudes, Vienna University of Economics and Business, ÖGOR Chair

Scheduling steel plates on a roller furnace

Eric Ebermann and Stefan Nickel

Abstract We introduce a single machine scheduling problem arising in the heat treatment of steel plates. To the best of our knowledge, there is no study on this problem in the literature up to now. We refer to this problem as the *Heat Treatment Furnace Scheduling Problem with Distance Constraints* (HTFSPD) and propose a mixed integer linear program (MILP) formulation. Since the problem itself is NP-hard and computational times for real world instances are too high, a genetic algorithm is developed in order to provide heuristic solutions. Computational results for some real data sets[1] demonstrate the performance of the algorithm compared to the current solution method used in practice.

1 Introduction and problem description

Given a single machine (a roller heat furnace) with finite capacity L and a set of products (steel plates) $i = 1,...,n$, each with a processing time p_i, a length l_i and a furnace temperature f_i, the objective is to find a sequence which minimizes the total completion time. The process can be described more in detail by means of Figure 1. The steel plates were deposited by a bridge crane onto the roller conveyor on the left side. If there is no reason to wait, a plate enters the furnace and drives to the next possible position. When the heat treatment is completed, it leaves the furnace on the right side where another bridge crane removes it from the roller conveyor.

Steel plates of the *Dillinger Hütte GTS* have a length of 2 to 28 meters, a width of 1 to 5.2 meters and a thickness of 5 to 450 millimeters. The length of the furnace spans ca. 80 meters and the both conveyors (before the entrance door and behind the exit door) have a length of ca. 30 meters. As a consequence, the furnace can process

Eric Ebermann
Institute of Operations Research (IOR), Karlsruhe Institute of Technology (KIT),
e-mail: eric.ebermann@kit.edu

Stefan Nickel
Institute of Operations Research (IOR), Karlsruhe Institute of Technology (KIT),
e-mail: stefan.nickel@kit.edu

[1] received from *Dillinger Hütte GTS* (http://www.dillinger.de/dh/index.shtml.en)

D. Klatte et al. (eds.), *Operations Research Proceedings 2011*, Operations Research Proceedings,   389
DOI 10.1007/978-3-642-29210-1_62, © Springer-Verlag Berlin Heidelberg 2012

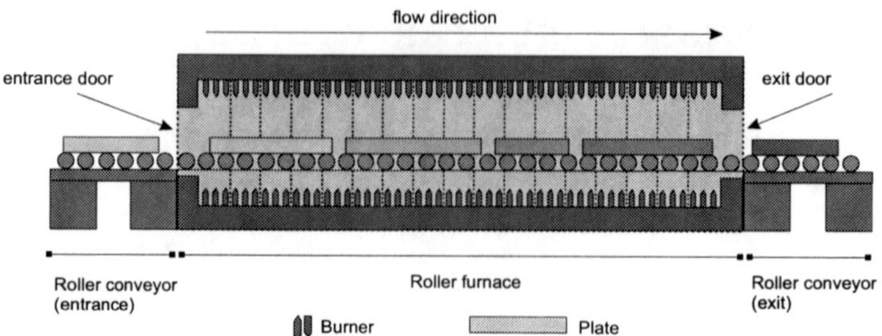

Fig. 1 Drawing of a continuous roller furnace

multiple plates at the same time. The plates are passing the furnace continously and
one by one. In addition, they have to leave the furnace in the same order as they
entered it because inside the furnace overtaking is not possible. Due to technical,
physical and logistical restrictions, a plate cannot always enter the furnace directly
after its predecessor. The minimal time a plate has to wait is equal to the duration its
predecessor needs to completely enter into the furnace. In the following, we assume
this duration to be constant ($t^{min} = 8$ minutes).

One reason for additional waiting times is the finite furnace capacity. A plate
cannot enter the furnace if there is not enough space. It has to wait until one or
more plates leave the furnace, so that the free furnace capacity is sufficiently large.
A further complication results from a special kind of setup procedure in cases when
two consecutive plates need different heat treatment temperatures. Depending on
the temperature deviation, these plates have to keep a specific distance from each
other in the furnace. This distance is necessary in order to avoid an overheating or
inhomogeneous heating of the plates. Apart from this, a minimal safety distance
$s^{rc,min}$ of two meters has always to be kept between two consecutive plates.

Another additional waiting time occurs if the processing time of the next plate
needed to enter the furnace is smaller than the processing time of its predecessor.
We already mentioned that overtaking is not possible inside the furnace. In addition, each plate has to keep its predefined holding time, so that processing times are
never exceeded. As a consequence, the waiting time of a plate whose processing
time is smaller than that of its predecessor is at least as long as the processing time
deviation.

To the best of our knowledge, there is no study on this problem in the literature
up to now. However, there are some studies having some parts in common with the
HTFSPD (see [3], [10], [9]).

2 Problem formulation

Koné et al. [7] demonstrate different MILP models for the *Resource-Constrained
Project Scheduling Problem* (RCPSP), two discrete-time formulations, a flow-based

continuous-time formulation and two event-based formulations. The advantage of the last method is that it involves limited complexity in terms of number of binary variables. In the following, we present an event-based MILP formulation for the HTFSPD. We define the decision variables:

$t_e =$ start time of event e, $e = 0, ..., k$

$r_e =$ used furnace capacity after event e, $e = 0, ..., k$

$$x_{ije} = \begin{cases} 1, & \text{if plate } i \text{ enters the furnace at event } e \text{ directly succeeding plate } j \\ 0, & \text{else;} \end{cases}$$

$$y_{ije} = \begin{cases} 1, & \text{if plate } i \text{ leaves the furnace at event } e \text{ directly preceding plate } j \\ 0, & \text{else;} \end{cases}$$

Notice that $k = 2n'$ and $n' = n + 1$ because we have to introduce two Dummy-plates (0 and $n + 1$). Then, the MILP formulation is as follows:

$$\min \quad t_k \tag{1}$$

$$\text{s.t.} \quad t_0 = 0 \tag{2}$$

$$t_f \geq t_e + \left(\sum_{j=0}^{n'} p_i + t^{min} \right) \cdot x_{ije}$$

$$- (p_i + t^{min}) \cdot \left(1 - \sum_{j=0}^{n'} y_{ijf} \right) \qquad \begin{aligned} &\forall i = 1, ..., n, \\ &e, f = 0, ..., k, \ e < f \end{aligned} \tag{3}$$

$$t_{e+1} \geq t_e + t^{min} \qquad \forall e = 0, ..., k-1 \tag{4}$$

$$\sum_{i=1}^{n} \sum_{j=0}^{n'} x_{ije} \leq 1 \qquad \forall e = 0, ..., k \tag{5}$$

$$\sum_{i=1}^{n} \sum_{j=0}^{n'} y_{ije} \leq 1 \qquad \forall e = 0, ..., k \tag{6}$$

$$x_{0n'0} = 1 \tag{7}$$

$$\sum_{e=0}^{k} \sum_{j=0}^{n'} x_{ije} = 1 \qquad \forall i = 0, ..., n' \tag{8}$$

$$\sum_{e=0}^{k} \sum_{i=0}^{n'} x_{ije} = 1 \qquad \forall j = 0, ..., n' \tag{9}$$

$$y_{n'0k} = 1 \tag{10}$$

$$\sum_{e=0}^{k} \sum_{j=0}^{n'} y_{ije} = 1 \qquad \forall i = 0, ..., n' \tag{11}$$

$$\sum_{e=0}^{k} \sum_{i=0}^{n'} y_{ije} = 1 \qquad \forall j = 0, ..., n' \tag{12}$$

$$\sum_{j=0}^{n'} x_{ije} + \sum_{j=0}^{n'} y_{ije} \leq 1 \qquad \begin{aligned} &\forall i = 1, ..., n, \\ &e = 0, ..., k \end{aligned} \tag{13}$$

$$x_{0n'0} + \sum_{j=0}^{n'} y_{0j0} = 2 \tag{14}$$

$$\sum_{j=0}^{n'} x_{n'jk} + y_{n'0k} = 2 \tag{15}$$

$$\sum_{e=0}^{k} x_{ije} + \sum_{e=0}^{k} x_{jie} \leq 1 \qquad \forall i,j = 0,...,n' \tag{16}$$

$$\sum_{e=0}^{k} y_{ije} + \sum_{e=0}^{k} y_{jie} \leq 1 \qquad \forall i,j = 0,...,n' \tag{17}$$

$$\sum_{e'=e}^{k} \sum_{j=0}^{n'} x_{ije'} + \sum_{e''=0}^{e-1} \sum_{j=0}^{n'} y_{ije''} \leq 1 \qquad \begin{array}{l} \forall i = 0,...,n', \\ e = 0,...,k \end{array} \tag{18}$$

$$\sum_{e=0}^{k} y_{ije} \cdot e \leq \sum_{e=0}^{k} \sum_{i'=0}^{n'} y_{ji'e'} \cdot e \qquad \begin{array}{l} \forall i = 0,...,n, \\ j = 1,...,n' \end{array} \tag{19}$$

$$\sum_{e=0}^{k} x_{ije} = \sum_{e=0}^{k} y_{jie} \qquad \forall i,j = 0,...,n' \tag{20}$$

$$\sum_{e=0}^{k} x_{ije} \cdot e \geq \sum_{e=0}^{k} \sum_{j'=0}^{n'} x_{jj'e} \cdot e + \left(\sum_{e=0}^{k} x_{ije} - 1 \right) \cdot M \qquad \begin{array}{l} \forall i = 1,...,n', \\ j = 0,...,n' \end{array} \tag{21}$$

$$r_0 = L - \sum_{i=0}^{n'} \sum_{j=0}^{n'} x_{ij0} \cdot (l_i + s_{ij}^{rc}) \tag{22}$$

$$r_e = r_{e-1} - \sum_{i=0}^{n'} \sum_{j=0}^{n'} (l_i + s_{ij}^{rc}) \cdot x_{ije} + \sum_{i=0}^{n'} \sum_{j=0}^{n'} (l_i + s_{ji}^{rc}) \cdot y_{ije} \qquad \forall e = 1,...,k \tag{23}$$

$$t_e, r_e \geq 0 \qquad \forall e = 1,...,k \tag{24}$$

$$x_{ije}, y_{ije} \in \{0,1\} \qquad \begin{array}{l} \forall i,j = 0,...,n', \\ e = 0,...,k \end{array} \tag{25}$$

In the above formulation, the objective function (1) minimizes the makespan. Constraints (2)-(21) assure that a feasible solution conforms with the continuous process described in Section 1. Constraints (22)-(23) make sure that the furnace capacity is never exceeded.

3 A genetic algorithm

A genetic algorithm (GA) is developed in order to solve real world problems. Its procedure is shown in Figure 2.

The most individuals of the initial population are generated randomly. In order to increase the performance of the algorithm, the others are generated by some simple sorting methods. One of them corresponds to the method the *Dillinger Hütte GTS* applies in practice. First, the plates are sorted by their furnace temperatures. Then,

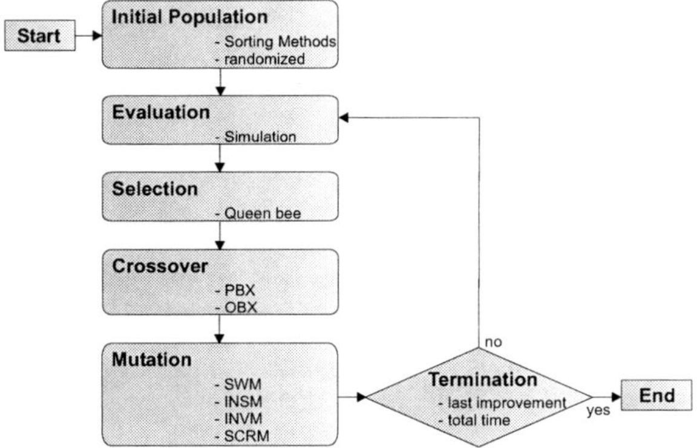

Fig. 2 A genetic algorithm

all plates with identical furnace temperatures are sorted by its processing time in ascending order. We refer to this heuristic in the following as the *rule-based sorting* (RBS).

The next step of the algorithm is to evaluate the indiviuums. A simulation of the furnace process serves as evaluation function.

The Selection scheme is similar to a proposal made by Karci [5]. It is based on the intelligence in bee swarms. The individual with the best fitness value of the current generation is the queen bee. All the other parents (selected by tournament selection) are crossed over with this fixed parent. For a survey about algorithms simulating bee swarm intelligence, we refer to Karaboga and Akay [4].

For crossover, we use the *Position based crossover operator* (PBX) with probability p_{pbx} and the *Order based crossover operator* (OBX) with probability p_{obx}. Kellegöz et al. [6] compare eleven crossover operators for the one machine n-jobs problem. Experimental results demonstrate the effectiveness of the OBX and PBX crossover operators. With probability p_{nox}, the selected individual does not cross with the queen bee.

As last step of each iteration, the individuals are mutated by the use of the *swap operator* (SWM) with probability p_{swm}, the *insert operator* (INSM) with probability p_{insm}, the *inverse operator* (INVM) with probability p_{invm} and *the scramble operator* (SCRM) with probability p_{scrm}. These operators can be found in [2].

There are two stopping criterions to terminate the algorithm. The first is if there has been no improvement since $stop_{limp}$ iterations whereas the second is a total run time limit of $stop_{time}$ seconds.

4 Computational results

ILOG CPLEX is used to solved the MILP formulation presented in Section 2. The genetic algorithm described in Section 3 has been implemented in C++. All computational studies have been performed on real data sets[2] on a PC with a *Pentium(R) Dual Core* processor with 2.8 GHz and 2 GB of RAM. Some of the results are given in Table 1.

Table 1 Sample of computational results

		RBS	GAQB		MILP		
data set	n	solution	solution	time	solution	time	GAP
1	6	459	459	34.41	459	4186	72.11%
2	8	624	528	39.92	806	6402	80.15%
3	12	938	860	56.86	-	7200	100.00%
4	20	1057	973	70.61	-	7200	100.00%
5	40	1816	1590	62.71	-	7200	100.00%
6	60	2564	2233	217.44	-	7200	100.00%

For the MILP model, a time limit of 7200 seconds of CPU time was imposed. For the GA, we set $stop_{limp}$ to 1000 iterations and $stop_{time}$ to 300 seconds.

It can be observed that the *MILP* model cannot be solved to optimality in reasonable time. However, the GA could improve the solutions of RBS on average by 9.5%.

References

1. Baker, K.R., Trietsch, D.: Principles of Sequencing and Scheduling. John Wiley & Sons, New York, (2009)
2. Eiben, A., Smith, J.: Introduction to Evolutionary Computation. Springer, Berlin, (2003)
3. Hartmann, S., Briskorn, D.: A survey of variants and extensions of the resource-constrained project scheduling problem. European Journal Of Operational Research **207**, 1–14 (2010) doi: 10.1016/j.ejor.2009.11.005
4. Karaboga, D., Akay, B.: A survey: algorithms simulating bee swarm intelligence. Artificial Intelligence Review **31**, 61–85 (2009) doi: 10.1007/s10462-009-9127-4
5. Karci, A.: Imitation of Bee Reproduction as a Crossover Operator in Genetic Algorithms. Lecture Notes in Computer Science **3157**, 1015-1016 (2004)
6. Kellegöz, T., Toklu, B., Wilson, J.: Comparing efficiencies of genetic crossover operators for one machine total weighted tardiness problem. Applied Mathematics and Computation **199**, 590-598 (2008)
7. Koné, O., Artigues, C., Lopez, P., Mongeau, M.: Event-based MILP models for resource-constrained project scheduling problems. Computers & Operations Research **38**, 3–13 (2011) doi: 10.1016/j.cor.2009.12.011
8. Pinedo, M.L.: Scheduling - Theory, Algorithms, and Systems (3rd edition). Springer, New York (2008)
9. Wang, X., Tang, L.: A simplified scatter search for a special single machine scheduling problem to minimize total weighted tardiness. CDC 2009, 6250–6255 (2009)
10. Zhang, G., Cai, X., Wong, C.K.: Some results on resource constrained scheduling. IIE Transactions **36**, 1–9 (2004) doi: 10.1080/07408170490257862

---

[2] the data can be obtained from the authors upon request.

Check-In Counter Planning via a Network Flow Approach

Torsten Fahle

Abstract Finding an appropriated number of open check-in counters is a typical planning problem on an airport. The problem has to take into account the flight plan, the arrival patterns of passengers and a given service level (max. waiting time of a passenger before being served). The goal is to generate a schedule for the check-in counters that uses as few as possible check-in counters per time.
We present a network flow formulation for this problem and discuss additional constraints relevant for real-world scenarios. We show how to solve the model and present numerical results on artificial as well as on real-world data. The model is part of the Inform GroundStar suite that successful is in use on more than 200 airports world-wide.

1 Introduction

Operating an airport efficiently is nowadays unimaginable without sophisticated resources management systems that optimally utilize the limited resources at reasonable costs. Besides scheduling infrastructure and equipment, personnel staffing is a major domain here. We consider the problem of finding an appropriated number of open check-in counters over the week. The problem has to take into account the flight plan, the arrival patterns of passengers and a given service level (max. waiting time of a passenger before being served). The goal is to generate a schedule for the check-in counters that uses as few as possible check-in counters per time. In a later step that schedule is the basis for the check-in allocation, see e.g. [5, 6].

Check-in counter schedules are typically highly irregular. Passengers arrive at the airport some time ahead of take off. That forerun depends on service class (business vs. tourists), day (week vs. weekend), or destination (long haul vs. domestic flights). Train connections can imply passenger waves arriving e.g. every 20 minutes, etc. Flight operations, on the other hand, typically conglomerate at some peak times (typically morning and evening). All in all, the number of passengers and thus the

Inform GmbH, Airport System Division, Aachen, Germany, e-mail: Torsten.Fahle@inform-software.com

D. Klatte et al. (eds.), *Operations Research Proceedings 2011*, Operations Research Proceedings, 395
DOI 10.1007/978-3-642-29210-1_63, © Springer-Verlag Berlin Heidelberg 2012

time and personnel needed to serve these passengers is highly volatile, both over the day and over the week. Figure 1 shows a typical arrival profile with 4 peaks over the day and 2 different daily patterns over the week. In order to reduce costs it is thus desirable to smooth the demand by making use of the allowed service level.

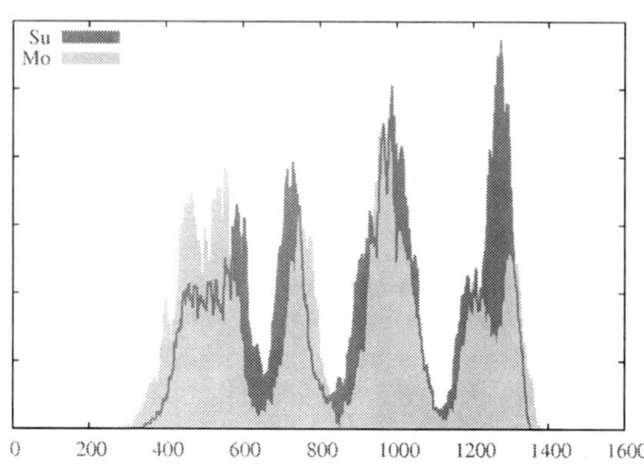

Fig. 1 Two days of the same week. Check-In demand over the day (minutes 0 – 1440) for a smaller carrier on a major European airport. 4 large peaks, but different layout: strong morning peak over the week, strong evening peak on Sunday

Our goal is to smooth the irregular demand by making some passengers wait, i.e. we try to move peaks into valleys. We present a network flow formulation for this problem and discuss additional constraints relevant for real-world scenarios. We show how to solve the model and present numerical results on real-world data.

The model is part of the Inform GroundStar suite. GroundStar is an integrated resources management system to optimize all planning and control processes in aircraft handling at passenger airports. GroundStar is in successful use in various handling areas of more than 200 airports of every size world-wide.

There is not much literature on check-in counter planning (consisting of calculating the demand and allocating the counters). Check-In counter allocation problems have been studied by [5]. They use a simulation for generating the appropriated input. [6] work with a number of passengers for a flight without assumptions on their distribution or on waiting times. To our knowledge, this problem has also not been considered in queuing theory.

2 A Network Model

We consider a planning horizon of T minutes. Typically, we plan for 10 days ($T = 14400$). Let n_t denote the demand at time t. The queue length, Q, is the maximum time a demand is allowed to wait before being served, i.e. demand at time t should be served at time t' with $t \leq t' \leq t + Q$. As this is not always possible we allow violating queue length Q. However, demand at time t must being served before $t + L$, where $L \geq Q$ is a limit of e.g. 6 hours.

An upper bound $rm(t)$ on the number of available resources at time t is given. Finally, we may have K intervals, $[s_1,e_1],\ldots,[s_K,e_K]$ representing start and end of check-in for flight k, at which $d_1,\ldots,d_K$ demand must be served.

The outcome of the planning is $p = (p_1,\ldots,p_T)$, the number of planned resources for the planning horizon.

We construct a leveled network consisting of T nodes in the input level $I(t)$, output level $O(t)$, resource limit level $M(t)$, and a cost level $C(t)$, respectively, $t = 1,\ldots,T$. A last level $Z(k)$, $k = 1,\ldots,K$, is devoted to the dates at which a certain amount of demand must be served. (see figure 2)

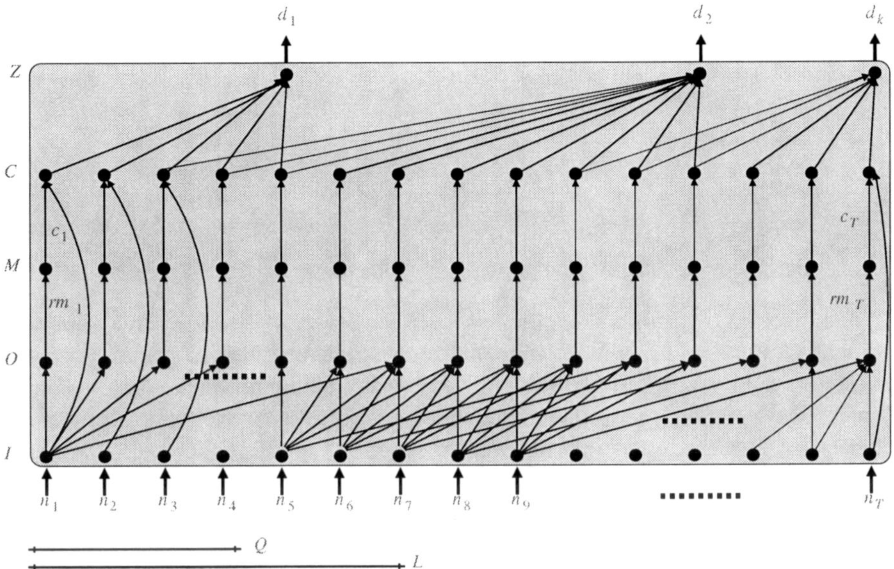

Fig. 2 The network model

1. Edges between I and O form the distribution network. Thus, a node $I(t)$ is connected to all nodes $O(t')$ that are reachable within the given queue length Q: $t \le t' \le t+Q$. Additional edges from $I(t)$ to nodes $O(t'')$ with $t+Q < t'' \le L$ allow to violate queue length at higher costs. $I(t)$ is also connected to $C(t)$. This edge serves as an overflow edge if not all demand can be distributed. The flow over the edge between $I(t)$ and $C(t)$ is basically the demand that cannot be served even if we violate the allowed queue length.
2. A node $O(t)$ has just one outgoing edge to $M(t)$. The purpose of this edge is to introduce a capacity limit if the number of resources is limited at some time intervals. Notice, that the flow over an edge between $O(t)$ and $M(t)$ is basically the number of resources needed at time t.
3. A node $M(t)$ is connected to $C(t)$. The edges here are used for cost calculation.

4. In order to count demand for resources with deadlines, a node $Z(k)$ is connected to all nodes $C(t)$ that represents a point in time that is not later than the deadline of resource k, i.e. $s_{kl} \leq t \leq e_k$.

Using that notation, we are ready to formulate a quadratic network flow model:

$$\min \quad \sum_{t=0}^{T} x^2_{M(t),C(t)} + \beta \sum_{t=0}^{T} x_{I(t),C(t)} + \gamma \sum_{t=0}^{T} \sum_{j=t+1+Q}^{t+L} x_{I(t),O(j)} \tag{1}$$

$$\text{s.t.} \quad x_{I(t),C(t)} + \sum_{j=t}^{t+L} x_{I(t),O(j)} = n_t \qquad \forall t \in \{1,\ldots,T\} \tag{2}$$

$$x_{O(t),M(t)} - \sum_{j=t-L}^{t} x_{I(j),O(t)} = 0 \qquad \forall t \in \{1,\ldots,T\} \tag{3}$$

$$x_{O(t),M(t)} \leq rm(t) \qquad \forall t \in \{1,\ldots,T\} \tag{4}$$

$$\sum_{t=s_k}^{e_k} x_{C(t),Z(e_k)} = d_k \qquad \forall k \in \{1,\ldots,K\} \tag{5}$$

$$x_{I(\cdot),C(\cdot)}, \; x_{I(\cdot),O(\cdot)}, \; x_{C(\cdot),Z(\cdot)}, \; x_{O(\cdot),M(\cdot)}, \; x_{C(\cdot),Z(\cdot)} \geq 0 \tag{6}$$

The objective (1) minimizes the flow between levels M and C. Notice, that we use a quadratic term here. Also, we penalize using the overflow edges or the edges leading to more waiting time than allowed by the maximum queue length. Thus, β and γ have to be large numbers and since overflow is worse than having some demand in the queue longer than allowed we require $\beta \gg \gamma$. Equations (2), (3) define flow conservation for the input and output level, respectively. A resource limitation is defined by (4), and (5) ensures that all demand required in the respective interval is collected. Finally, all edges must have non-negative flows (6).

Before generating the LP-model we try to identify useless edges in the graph. E.g. overflow edges $x_{I(t),C(t)}$ are only needed if a resource limitation for time t is given: $rm(t) = \infty \Rightarrow x_{I(t),C(t)} = 0$. If no relevant resource limit is given edges $x_{I(t),O(j)}$ with $j > t + Q$ can also be removed.

In order to use an LP solver, we have to linearize the quadratic term in the objective (see e.g. [1]). We simulate variable $x_{M(t),C(t)}$ by $y_t^1 + \cdots + y_t^{h(t)}$, where $h(t)$ is an upper bound on the expected flow over $x_{M(t),C(t)}$. We require $0 \leq y_t^i \leq 1$, $i = 1,\ldots,h(t)$ and use cost coefficients $c_1,\ldots,c_{h(t)}$ such that $\forall j = 0,\ldots,h(t): \sum_{i=1}^{j} c_i = j^2$. We estimate $h'(t)$ as the sum of all input demand that could flow into node $C(t)$: Let $\Gamma(t) := \{j \mid \exists \text{ path } I(j) \rightsquigarrow M(t), t - L \leq j \leq t\}$, then $h'(t) := \sum_{j \in \Gamma(t)} n_j$ and $h(t) := \min\{h'(t), rm(t)\}$.

After solving the model, variables $x_{O(t),M(t)}$ give the (fractional) number of resources needed at time t. We use a simple rounding scheme: tiny fractionalities smaller than 0.2 are rounded down, all others are rounded up

The model used within GroundStar is more complex. E.g. penalty terms for missing the queue length are defined as slightly increasing to serve demand as early as possible, etc. A better estimation of $h(t)$ is used if the estimation above yields large numbers. And there are additional rules, e.g. as soon as demand is there at least one check-in counter must be opened.

3 Numerical Results

The above model is implemented in C++ using the COIN-OR LP solver Coin CLP 1.11 [2].[1] Run time is measured on an Intel Core2, 2.8GHz, 3GB Ram on Windows XP using Visual Studio 2008. We start with an artificial test case, Fig. 3. The demand is given as a double sinus wave over 1001 minutes ($n_t := 12 + 10\sin(t\frac{4\pi}{1000})$, $t = 0\ldots1000$). The two lines show the resource requirement for a maximum waiting time of 20 and 100 minutes, respectively[2]. No departure flights are assumed. As can be seen in the statistics (Table 1), enlarging the queue length increases the number of edges between I and O and thus the number of LP variables. Nevertheless, run times of less than 2 seconds are no problem for the overall planning process. Quality-wise, the result is good as the number of resources only required for short time intervals is small and only up to 1.6% additional demand is induced by rounding (i.e. the demand of $\approx 12,000$ minutes is only increased by 172 or 192 minutes, resp. when rounding).

The figure on the right shows one day of a real-world scenario originally consisting of 9 days, 149 departure flights. The queue length is $Q = 24$. Whereas the number of variables scales with the number of days when comparing artificial and real model, the latter introduces new edges between C and Z. These variables ensure that all demand is served before departure. Despite their large number, they do barely impact running time here.

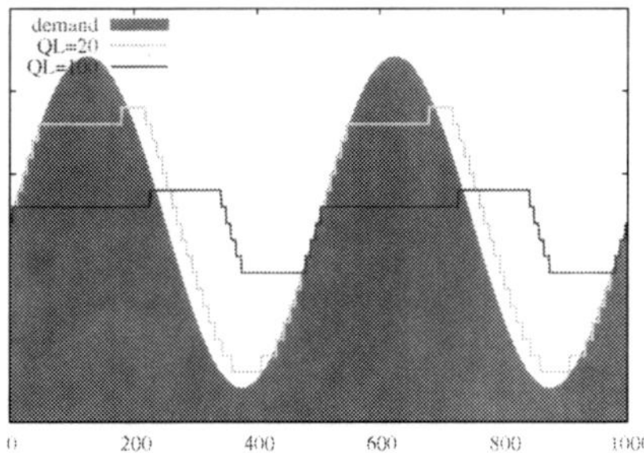

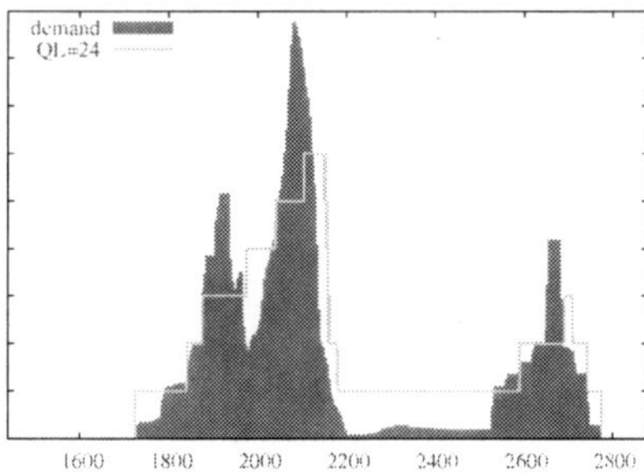

Fig. 3 Artificial example (right), 1 day from a real-world scenario (left)

---

[1] The model above could be adequately solved by a pure network solvers (e.g. [3, 4]). We prefer the more generic LP solver as that gives us some more freedom for future enhancements and constraints.

[2] waiting time is typically ≤ 1 hour on real airports. $Q = 100$ is chosen for illustration here.

	Sinus, $Q = 20$	Sinus, $Q = 100$	real-word, $Q = 24$
time horizon	1001m	1001m	9d
edges $I - O$	9907	37,195	103,503
edges $I - C / O - M$	1001	1001	9,070
edges $M - C$	27,707	27,707	75,742
edges $C - Z$	–	–	128,420
departures	–	–	149
LP vars	34,616	61,904	218,461
LP constr.	5,003	5,003	408,029
additional demand	1.4%	1.6%	0.2%
run time	0.6 sec.	1.4sec.	18.3sec

Table 1 Statistics for the test cases in figure 3

4 Conclusions

We have described a network model that calculates resource usage over time. Given a demand, the model makes use of a predefined service level in order to move demand from peaks to valleys, thus utilizing infrastructure and personnel more efficiently. Our discussion above was motivated by check-in counter allocations. The model, however, can be used as well in other areas, e.g. planning for security or loading personnel and equipment. Solutions found by the model have been in use for several years now on many airports world-wide.

Especially in recent years, the model is used for strategic management decisions as well. By inspecting the variables in the model it is possible to report key performance indicators like waiting time figures or average demand enqueued. Also it is possible to detect how often and where the contracted service level is not granted. For negotiations with airport customers the latter is a helpful tool as it allows the airport to check beforehand if a certain service level requested by the customer is in reach.

References

1. Ahuja, R.K., Magnati, T.L., Orlin, J.B.: Network Flows. Prentice Hall (1993)
2. CLP. Available from COIN-OR at http://www.coin-or.org/projects/Clp.xml. Cited 27 Jun 2011.
3. Goldberg, A.: An Efficient Implementation Of A Scaling Minimum-Cost Flow Algorithm. J. of Algorithms. **22**, 1–29 (1992)
4. Loebl, A.: MCF Version 1.3 - A network simplex implementation (2004). Available for academic use free of charge http://www.zib.de. Cited 27 Jun 2011.
5. van Dijk, van Sluis: Check-In computation and Optimization by simulation and IP in combination, EJOR **171**, 1152–1168 (2006)
6. Yan, S., Tang, C., Chen, M.: A model and a solution algorithm for airport common use check-in counter assignment, Transp. Res. Part A **38**, 101–125 (2004)

OpenSolver - An Open Source Add-in to Solve Linear and Integer Progammes in Excel

Andrew J Mason

Abstract OpenSolver is an open source Excel add-in that allows spreadsheet users to solve their LP/IP models using the COIN-OR CBC solver. OpenSolver is largely compatible with the built-in Excel Solver, allowing most existing LP and IP models to be solved without change. However, OpenSolver has none of the size limitations found in Solver, and thus can solve larger models. Further, the CBC solver is often faster than the built-in Solver, and OpenSolver provides novel model construction and on-sheet visualisation capabilities. This paper describes OpenSolver's development and features. OpenSolver can be downloaded free of charge at http://www.opensolver.org.

1 Introduction

Microsoft Excel for Windows [5], and its inbuilt Solver optimiser [3], form an ideal tool for delivering optimisation systems to end users. However, Solver imposes limitations on the maximum size of problem that can be solved. OpenSolver has been developed as a freely available open-source Excel add-in for Microsoft Windows that uses the COIN-OR (Computational Infrastructure for OR) [2] CBC optimiser [1] to solve large instances of linear and integer programs.

We start by briefly describing how optimisation models are built using Solver. Readers familiar with Solver may wish to skip to the next section. A typical linear programming model and its Solver data entry are shown in Figure 1. The decision variables are given in cells C2:E2; of these, C2 must be binary and C3 must be integer. Cell G4 defines the objective function using a 'sumproduct' formula defined in terms of the decision variables and the objective coefficients in cells C4:E4. Each of the constraints are similarly defined in terms of a constraint left hand side (LHS) in G6:G9 and right hand side (RHS) in I6:I9. Note that the 'min', 'binary', 'integer'

Andrew J Mason
Department of Engineering Science, University of Auckland, Private Bag 92019, Auckland, e-mail: a.mason@auckland.ac.nz

D. Klatte et al. (eds.), *Operations Research Proceedings 2011*, Operations Research Proceedings, 401
DOI 10.1007/978-3-642-29210-1_64, © Springer-Verlag Berlin Heidelberg 2012

and constraint relations shown on the spreadsheet are for display purposes only, and
are not used by Solver.

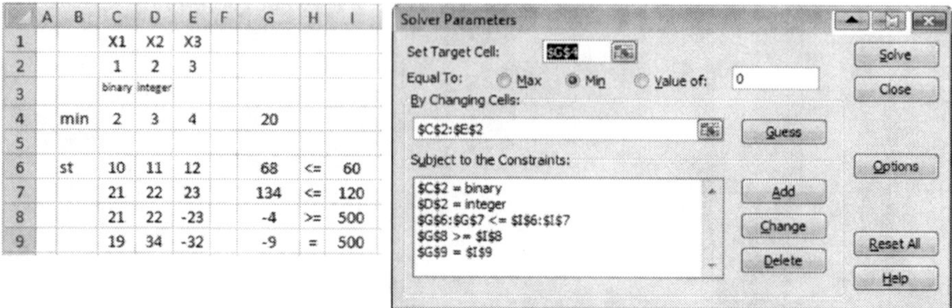

Fig. 1 A typical spreadsheet optimization model (left) and the Solver entry for this. Many models
do not follow this layout, but instead 'hide' the model inside the spreadsheet formulae.

2 Constructing a Mathematical Model from a Solver Model

The means by which Solver stores a model does not appear to be documented. However, using the Excel Name Manager add-in [7], developed by Jan Karel Pieterse
of Decision Models UK, shows that Solver uses hidden 'names' to contain all the
model's details. OpenSolver reads these values to determine the cells that define the
model and Solver options.

Excel's representation of the optimisation model is given in terms of cells that
contain constants and formulae. Because OpenSolver restricts itself to linear models, we wish to analyse the spreadsheet data to build an optimisation model with
equations of the form:

$$\text{Min/max} \quad c_1x_1 + c_2x_2 + \dots + c_nx_n$$
$$\text{Subject to } a_{i1}x_1 + a_{i2}x_2 + \dots + a_{in}x_n \geq/=/\leq \quad b_i, \ i = 1, 2, \dots, m$$

where $\geq/=/\leq$ denotes either $\geq$, $=$ or $\leq$. Assuming the model is linear, then the
Excel data can be thought of as defining an objective function given by

$$\text{Obj}(\mathbf{x}) = c_0 + c_1x_1 + c_2x_2 + \dots c_nx_n$$

where $\mathbf{x} = (x_1, x_2, \dots, x_n)$ is the vector of n decision variables, $\text{Obj}(\mathbf{x})$ is the objective
function cell value, and c_0 is a constant. Similarly, each constraint equation i is
defined in Excel by

$$\text{LHS}_i(\mathbf{x}) \geq/=/\leq \text{RHS}_i(\mathbf{x}) \Rightarrow \text{LHS}_i(\mathbf{x}) - \text{RHS}_i(\mathbf{x}) \geq/=/\leq 0, \quad i = 1, 2, \dots, m$$

where $\mathrm{LHS}_i(\mathbf{x})$ and $\mathrm{RHS}_i(\mathbf{x})$ are the cell values for the left hand side and right hand side of constraint i respectively given decision variable values $\mathbf{x}$. Because Solver allows both $\mathrm{LHS}_i(\mathbf{x})$ and $\mathrm{RHS}_i(\mathbf{x})$ to be expressions, we assume that both of these are linear functions of the decision variables. Thus, assuming the model is linear, we have

$$a_{i1}x_1 + a_{i2}x_2 + ...a_{in}x_n - b_i = \mathrm{LHS}_i(\mathbf{x}) - \mathrm{RHS}_i(\mathbf{x}), \quad i = 1,2,...,m.$$

OpenSolver determines the coefficients for the objective function and constraints through numerical differentiation. First, all the decision variables are set to zero, $\mathbf{x} = \mathbf{x}_0 = (0,0,...,0)$ giving:

$$c_0 = \mathrm{Obj}(\mathbf{x}_0)$$
$$b_i = \mathrm{RHS}_i(\mathbf{x}_0) - \mathrm{LHS}_i(\mathbf{x}_0), i = 1,2,...,m$$

Then, each variable x_j is set to 1 in turn on the spreadsheet, giving a sequence of decision variable values $\mathbf{x} = \mathbf{x}_j$, $j = 1,2,...,n$ where $\mathbf{x}_j = (x_{1j},x_{2j},...,x_{nj})$ is a unit vector with the single non-zero element $x_{jj} = 1$. The spreadsheet is recalculated for each of these $\mathbf{x}_j$ values and the value of the objective and the left and right hand sides of each constraint recorded. This allows us to calculate the following coefficients:

$$c_j = Obj(\mathbf{x}_j) - c_0$$
$$a_{ij} = \mathrm{LHS}(\mathbf{x}_j) - \mathrm{RHS}(\mathbf{x}_j) + b_i, i = 1,2,...,m$$

The speed of this process depends on both the speed at which OpenSolver can read and write data to the spreadsheet, and the time Excel takes to re-calculate the spreadsheet. Newer versions of OpenSolver have been optimised to access larger groups of cells in each operation, reducing run times for large models. As an example, a staff scheduling model with 532 variables and 1909 constraints takes 36s to build (and just 1 second to solve) on an Intel Core-2 Duo 2.66GHz laptop.

3 Excel Integration and User Interface

OpenSolver is coded in Visual Basic for Applications and runs as an Excel add-in. The add-in presents the user with new OpenSolver controls within the standard ribbon interface that provide buttons for common operations and a menu for more advanced operations. (In earlier versions of Excel without a ribbon, OpenSolver appears as a menu item.) These OpenSolver controls are shown in Figure 2.

OpenSolver is downloaded as a single .zip file which when expanded gives a folder containing the CBC files and the OpenSolver.xlam add-in. Double clicking OpenSolver.xlam loads OpenSolver and adds the new OpenSolver buttons and menu to Excel. OpenSolver then remains available until Excel is quit. If required, the

Fig. 2 OpenSolver's buttons and menu appear in Excel's Data ribbon.

OpenSolver and CBC files can be copied to the appropriate Microsoft Office folder to ensure OpenSolver is available every time Excel is launched.

Users can construct their models either using the standard Solver interface or using a new OpenSolver dialog. The new dialog provides a number of advantages, including highlighting of selected constraints on the sheet, and easier editing of constraints.

We have found OpenSolver's performance to be similar or better than Solver's. CBC appears to be a more modern optimizer than Solver's, and so gives much improved performance on some difficult problems. For example, large knapsack problems which take hours with the Excel 2007 Solver are solved instantly using OpenSolver, thanks to the newer techniques such as problem strengthening and pre-processing used by CBC [1].

To review an optimisation model developed using the built-in Solver, the user needs to check both the equations on the spreadsheet and the model formulation as entered into Solver. This separation between the equations and the model form makes checking and debugging difficult. OpenSolver provides a novel solution to this in the form of direct model visualisation on the spreadsheet. As Figure 3 shows, OpenSolver can annotate a spreadsheet to display a model in which the objective cell is highlighted and labeled min or max, the adjustable cells are shaded with any binary and integer decision variable cells being labeled 'b' and 'i' respectively, and each constraint is highlighted and its sense shown. We have found this model visualisation to be very useful for checking large models.

4 Automatic Model Construction

We have developed additional functionality that allows OpenSolver to build Solver-compatible models itself without requiring the usual step-by-step construction process. Our approach builds on the philosophy that the model should be fully documented on the spreadsheet. Thus, we require that the spreadsheet identifies the objective sense (using the keyword 'min' or 'max' or variants of these), and gives the sense ($\leq$, $=$, or $\geq$) of each constraint. Our example spreadsheet shown in Figure 1 satisfies these layout requirements.

To identify the model, OpenSolver starts by searching for a cell containing the text 'min' or 'max' (or variants of these terms). It then searches the cells in the vicinity of this min/max cell to find a cell containing a formula (giving preference to any cell containing a 'sumproduct' formula); if one is found, this is assumed to define the objective function. The left and/or right hand side formulae for the constraints are then located in a similar fashion by searching for occurrences of $<=$ (or '$<$'), $=$ and, $>=$ (or '$>$'). The predecessor cells of all these formulae are then found, and the decision variables are then taken as those predecessors cells that have as successors either (1) at least two constraints or (2) the objective and at least one constraint.

The final step is to identify any binary or integer restrictions on the decision variables. These are assumed to be indicated in the spreadsheet by the text 'binary' or 'integer' (and variants of these) entered in the cells beneath any restricted decision variables of these types.

5 Advanced Features

OpenSolver offers a number of features for advanced users, including:

- The ability to easily solve an LP relaxation with a single menu click,
- Interaction with the COIN-OR CBC solver via the command line,
- Faster running using 'Quick Solve' when repeatedly solving the same problem with a succession of different right hand sides,
- Viewing of the CBC .lp input file showing the model's equations, and,
- Detection and display of non-linearities in the model.

	A	B	C	D	E	F	G	H	I
1			X1	X2	X3				
2			b 1	i 2	3				
3			binary	integer					
4		min	2	3	4		min 20		
5									
6		st	10	11	12		68	<=	60
7			21	22	23		134	<=	120
8			21	22	-23		-4	>=	>500
9			19	34	-32		-9	=	=500

	X1	X2	X3
8	X1	X2	X3
9	1	0.923	1
10	< 19	20	21
11			
12	20	1.846	< 50
13			
14			
15		12> 1	
16			
17		2< 1	
18		0.923	

Fig. 3 OpenSolver can display an optimisation model directly on the spreadsheet. The screenshot on the left shows OpenSolver's highlighting for the model given earlier, while the screenshot on the right illustrates OpenSolver's highlighting for several other common model representations.

6 User Feedback

It is difficult to determine how OpenSolver is being used, but a comment by Joel
Sokol from Georgia Tech on the OpenSolver web site [6] describes one use of Open-
Solver as follows:

> Thanks, Andrew! I'm using OpenSolver on a project I'm doing for a Major League Baseball
> team, and it's exactly what I needed. It lets the team use their preferred platform, creates and
> solves their LPs very quickly, and doesn't constrain them with any variable limits. Thanks
> again! - Joel Sokol, August 11, 2010

OpenSolver has been used to successfully solve problems which have as many as
70,000 variables and 76,000 constraints [6]. Further, users report that they appreciate
being able to view the algebraic form of the Excel model given in the .lp file [4].

7 Conclusions

We have shown that it is possible to leverage the COIN-OR software to provide
users with a new open source option for delivering spreadsheet-based operations
research solutions. While our software is compatible with Solver, it also provides
new innovative tools for visualising and building spreadsheet models that we hope
will benefit both students and practitioners of spreadsheet optimisation.

Acknowledgements The author wishes to thank Iain Dunning (Department of Engineering Science, University of Auckland) who has made a huge contribution in developing the AutoModel and
Model dialog code. The author also wishes to thank Kathleen Gilbert (Dept of Engineering Science) for the many improvements she has made to OpenSolver, including the non-linearity checks.
The author gratefully acknowledges the contribution of Paul Becker of Eclipse Engineering for
developing the Excel 2003 interface code in OpenSolver. Finally, OpenSolver would not have been
possible without the huge software contributions made by the COIN-OR team.

References

1. CBC: A COIN-OR Integer Programming Solver, https://projects.coin-or.org/Cbc. Cited
   November 2010
2. COIN-OR: Computational Infrastructure for Operations Research, http://www.coin-or.org.
   Cited July 2011
3. Frontline Systems: http://www.solver.org. Cited July 2011
4. Martin, K.: 'COIN-OR: Software for the OR Community,' Interfaces 40(6), pp. 465-476,
   INFORMS (2011)
5. Microsoft Excel. In: Wikipedia, http://en.wikipedia.org/wiki/Microsoft_Excel. Cited July
   2011
6. OpenSolver web site, http://opensolver.org. Cited July 2011
7. Pieterse, J.K.: Name Manager for Excel, http://www.jkp-ads.com/officemarketplacenm-
   en.asp. Cited November 2010

Producing Routing Systems Flexibly Using a VRP Metamodel and a Software Product Line

Tuukka Puranen

Abstract Routing problems occur in a wide variety of situations. Due to the heterogeneity of cases we do not yet know how to manage the complexity of addressing all the relevant aspects in logistic planning and solving the variety of different problem types in a cost-efficient way. In the last decade, we have witnessed an emergence of systematic approach into managing variation within a set of related software systems. This paper presents an application of these advances from software engineering into vehicle routing: we suggest the construction of a higher-level (meta-) model of routing problems and the application of a software product line approach. The proposed approach results in a flexible product line for constructing a family of routing systems cost-efficiently.

1 Introduction

A well-known combinatorial optimization problem, the vehicle routing problem (VRP) is at the center of designing effective transportation. In VRP, a set of vehicles is to be routed to visit a set of customers with known demands [15]. VRP has a number of extensions that consider for instance constraints on delivery times or vehicles of different sizes. Recently attention has been devoted to complex VRP variants which combine several of these extensions [4]. Solving these "rich" VRPs has proved to be a viable approach for real-life routing.

Vehicle routing has been studied for decades and the typical observed benefits of route optimization include cost savings, more effective operations, environmental benefits, improvements in quality of service, and easier planning. Not surprisingly, these benefits have caught the attention of software industry. In February 2010, OR/MS Today [9] surveyed sixteen software vendors providing 22 products for route optimization.

Tuukka Puranen
Department of Mathematical Information Technology, University of Jyväskylä, Finland e-mail:
tuukka.p.puranen@jyu.fi

D. Klatte et al. (eds.), *Operations Research Proceedings 2011*, Operations Research Proceedings,   407
DOI 10.1007/978-3-642-29210-1_65, © Springer-Verlag Berlin Heidelberg 2012

Despite the advances in theory and in practice, the routing problem has not yet been solved; a gap between the academic research and the industry-strength implementations exists. Different problems have similarities but differ in subtle ways due to differences in the processes, the needs of operators, the needs of the customers, the legislation, in the dependencies to other operations, and other context-dependent factors.

As producers of routing systems, Sörensen *et al.* [13] observe that "implementation of a commercial routing package typically requires a lot of manual work to be done". They note that since the set of problems addressed by the single software is quite heterogeneous, especially the optimization methodology requires a lot of manual tuning, and that [one of the challenges is] "inability to develop completely new methods each time a new problem is encountered... [which] would generally require rewriting large portions of the code base of the algorithms used". This imposes requirements, for instance, for the optimization methodology: it has to be flexible enough for being customized. Currently, only few approaches meet this criteria and software vendors have usually resorted to these techniques. These methods, however, do not represent the state of the art, and this makes customization an attractive alternative. And although modularization provides means for adaptation, the vendors have not really been able to address the heterogeneity of the domain efficiently. Mass customization has to give in for case-by-case customization.

To address a wider set of routing problems, researchers have begun an effort to unify models, solution methods, and implementation techniques in a generic setting. From the algorithm viewpoint, this can be seen, e.g., in the works of Funke *et al.* [2], who studied local search in routing and scheduling and provided an integrated view on the search procedure; and Gendreau and Potvin [3], who identified a set of common elements in the metaheuristic search. In addition, Ropke and Pisinger [14] have developed a unified heuristic search procedure for a wide set of problems. Modeling has been addressed, e.g., by Irnich [7], who provided a unified view on routing using resource extension functions. The implementation of routing systems has been addressed recently by, e.g., Sörensen *et al.* [13], who noted the need for a generic and adaptive solution approach for different problems; Hasle and Kloster [6], who based a solver on a single, rich and generic VRP model to reduce the efforts in development and maintenance, and employed a unified, metaheuristic approach to achieve flexibility and robustness; and Groër *et al.* [5] who implemented an reusable object-oriented library for solving routing problems.

The recent studies in modeling, solving, and implementing generic routing problem represent the state of the art in producing routing systems. Despite these advances it is, however, unlikely that a single system will be able to address the heterogeneity of the routing domain as a whole: the complexity of implementation will probably render the approach infeasible. And although frameworks and libraries do help, building a new system for each individual case, as noted, is not viable either. Thus something "in between" these approaches would be needed. The VRP domain introduces variation, and in the last decade, software engineers have developed successful techniques for managing variation within a set of related systems [1]. It turns out that routing systems are prime candidates for utilizing these advances.

2 Product Line Architecture for Routing Systems

The key to achieving cost-efficiency in implementation is efficient use of existing assets, that is, *reuse*. The key to reuse in turn, is to identify commonalities and variation between different elements. Now, managing variation is probably best achieved by making it explicit, and the so-called software product line engineering (PLE) attempts to aid in this by separating the development into two processes: building the common elements, the core assets, and the differing elements, the individual applications. These two processes are referred to as *domain engineering* and *application engineering*, respectively [10]. All the artifacts from design and implementation to documentation and tests are divided between the two processes. The central artifact in PLE is the product line architecture (PLA) defined in the domain engineering process. PLA is the structure on which the architectures of applications are built.

Especially w.r.t. PLA, the variation is made explicit by defining a variability model. The variability model expresses the points in where variation occurs and describes how it should occur. The former are referred to as "variation points" and the latter, "variants". A variation point, for example, "VRP model" may have a number of variants, e.g., "OVRP", "VRPB", or "FSMVRP"[1]. In fact, there are a number of potential variation points in routing systems. Apart from the optimization model, these include solution methodology, domain model, presentation model, data connections, dispatching, reporting and tracking functionality, and data generation process. The two major variation points, optimization model and solution methodology, largely define the optimization capabilities of the system.

However, the optimization methods depend on the optimization model as they need to operate on the data structures of the model during the search. In this case changing the optimization model would result in writing both the model (data structures, feasibility and objective evaluation) and the algorithms for each case, which largely undermines the reusability within the system. To address this, we developed an approach which breaks this dependency. This is illustrated w.r.t. the two development processes in Fig. 1. The elements on the upper half of the figure depict the

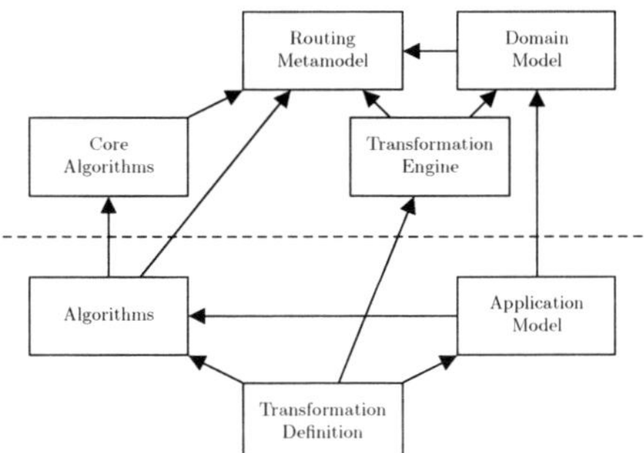

Fig. 1 A subset of the developed product line architecture for routing systems. The structure allows algorithms to operate independent of the case details as the algorithms are not dependent on the transformation defining the case. All the upper layer elements can thus be flexibly reused between the different cases.

---

[1] Open Vehicle Routing Problem, Vehicle Routing Problem with Backhauls, and Fleet Size and Mix Vehicle Routing Problem, respectively.

domain layer of PLA, the lower, the application-specific layer. The domain model defines the conceptual model of routing in object-oriented terms, whereas the routing metamodel is a generic encoding suitable for algorithmic manipulation. The transformation engine transforms the former to the latter using a case-specific transformation definition. Refinements to the conceptual model and the algorithms can also be defined on the application layer. The system is described in detail in [11].

Thus instead of defining different optimization models on the domain layer explicitly, we have defined a routing metamodel and a model transformation from the conceptual (domain) model into the case-specific optimization model. The algorithms operate on the metamodel and are independent of the specifics of the case.

3 Metamodel of Routing Problems

The routing metamodel is a model of routing models. It thus defines a modeling language for describing individual routing cases and exposes an interface for the individual applications for generating their models. It is notable that this metamodel is not a *surrogate model* in a sense that it is not used as a target of the optimization *per se*; instead, it defines a unified interface for algorithms, and the decision variables in the optimization models are themselves being manipulated during the search.

The developed metamodel needs to be generic enough for being able to describe different routing variants. The main construct is based on the General Pickup and Delivery Problem (GPDP) by Savelsbergh and Sol [12], and the main constraints are built using resource constrained paths and resource extension functions. In addition a set of additional constraints have been defined; these include mutual exclusion of tasks and compatibilities between resources. Central concepts in the metamodel are *activities* which represent events during the traversal of a route, and *actors* which represent any entity capable of traversing the paths of activities. The traversal is constrained by defining multidimensional resource (time, distance, *etc.*) accumulation between each transition from an activity to another. One of these values is used as an objective, others are constrained with given bounds, e.g., time windows.

In practice, a domain-specific transformation language has been defined for creating the metamodel elements, and the case-specific transformation script dictates which elements are generated from each domain model element. Thus, in different cases different elements are generated from the same conceptual model. The interface for generating the metamodel elements includes operations for generating an activity, generating an actor, setting a resource accumulation for traversal, defining bounds on resources, defining mutually exclusive groups of activities, and setting up dependencies between activities and resource accumulation functions to consider the dynamics of the state of the actor during the route traversal.

For example, if we need to define a fleet size and mix VRP instance with time windows, we create *an actor for each driver* object in the domain model, create *a driver pickup activity for each vehicle*, and *a delivery activity for each task* in the domain model. Each activity is then given an appropriate resource accumulation function and the necessary bounds on capacities (unique to each actor). In contrast, to define a PDP instance, we create *an actor for each vehicle* and *two activities for each task* in the domain model, add a function for capacity and (identical) bounds on the actors, and add a function for time and appropriate bounds on each activity.

Optimization using the metamodel is performed trough a unified interface which allows the algorithms to operate on the elements of the metamodel. Algorithms can, for example, move activities between routes, and a common feasibility evaluation mechanism ensures that all the constraints generated for the specified case are respected. This process is transparent to the algorithms.

The metamodel is largely compatible with the techniques examined in [8] and a constant time evaluation independent of segment length is achieved in most cases. Not all resource accumulation functions adhere to the assumptions needed for constant time evaluation and these need additional computation (which is the case with time-dependent travel times, for example). The additional constraints on, e.g., mutually exclusive activities introduce additional dimensions for the search operators, not unlike in cases where a depot or compartment selection is a part of the problem.

4 Results and Analysis

The proposed approach provides flexibility into modeling. We used 5 real-life cases and 3 benchmark types to evaluate the modeling capabilities of the developed system. The real-life cases were a pickup and delivery (PDP) case with, e.g., continuous planning, multidimensional capacities, dynamic incompatibilities, and soft-time windows; a multi-depot PDP case; VRP with incompatibilities between tasks; routing with heterogeneous fleet; and a two-trip VRP case with constraints for the time used in between the tasks of the trips. The benchmark cases were VRP, VRP with time windows, and PDP. The approach seems to be flexible enough for real-life use.

Performance-wise, the real-life cases are difficult to evaluate, but the initial results are encouraging. Only preliminary numerical results are available from the implementation of the developed metamodel. In mid-sized (400 customer) PDP instances, a simple variable neighborhood search scheme with both intensification and diversification components was able to find results on average at around 10% of the best known route length (with 15% more vehicles) on average in approx. 5 minutes on 3.00GHz CPU. We note that more research and measuring is needed for drawing conclusions on the impact of the approach to the performance of the system.

The cost-efficient implementation is achieved by, e.g., reuse of the existing assets, and with the proposed approach, we were able to achieve code reuse[2] of approx. 63%–99.5%. In addition, PLA allows customization of the individual applications flexibly: if heavy customization is required, case-specific refinements (customized algorithms, preprosessing steps, domain model extensions) may be used. Although up-front investments are required when building a product line, the approach is

---

[2] lines of reused code in system / lines of code in system

likely to be viable in cases where a wide range of heterogeneous set of problems need to be addressed. This hypothesis should be critically examined in future.

5 Conclusions

This paper presented an application of the recent advances from software engineering into vehicle routing. We suggested the construction of a metamodel of routing problems and the application of a software product line approach. The approach allows generation of multiple different models from the same conceptual model and abstracts the search space traversal from the algorithms. This offers reuse potential, but introduces overhead to the traversal. Multiple cases were solved with the approach, and although up-front investments are required, initial results indicate that the approach is suitable for a wide set of problems and yields less costs per case.

References

1. Clements, P., Northrop, L.: Software Product Lines: Practices and Patterns. Addison-Wesley (2001)
2. Funke, B., Grünert, T., Irnich, S.: Local Search for Vehicle Routing and Scheduling Problems: Review and Conceptual Integration. Journal of Heuristics 11(4), 267–306 (2005)
3. Gendreau, M., Potvin, J.-Y.: Metaheuristics in Combinatorial Optimization. Annals of Operations Research 140(1), 189–213 (2005)
4. Golden, B. L., Raghavan S., Wasil, E. A. (eds.) The Vehicle Routing Problem: Latest Advances and New Challenges. Springer (2008)
5. Groër, C., Golden, B., Wasil, E.: A library of local search heuristics for the vehicle routing problem. Mathematical Programming Computation 2 (2010)
6. Hasle, G., Kloster, O.: Industrial Vehicle Routing. In: Hasle G., Lie K.-A., Quak, E. (eds.) Geometric Modelling, Numerical Simulation, and Optimization, pp. 397–435. Springer Berlin Heidelberg (2007)
7. Irnich, S.: A Unified Modeling and Solution Framework for Vehicle Routing and Local Search-Based Metaheuristics. INFORMS Journal on Computing 20(2), 270–287 (2008)
8. Irnich, S.: Resource Extension Functions: Properties, Inversion, and Generalization to Segments. OR Spectrum 30, 133–148 (2008)
9. Partyka, J., Hall, R.: On the Road to Connectivity. OR/MS Today 37(1), 42–49 (2010)
10. Pohl, K., Böckle, G., van der Linden, F. J.: Software Product Line Engineering: Foundations, Principles and Techniques. Springer (2005)
11. Puranen, T.: Metaheuristics Meet Metamodels – A Modeling Language and a Product Line Architecture for Route Optimization Systems. PhD Thesis, University of Jyväskylä (2011)
12. Savelsbergh, M. W. P., Sol, M.: The General Pickup and Delivery Problem. Transportation Science 29, 17–29 (1995)
13. Sörensen, K., Sevaux, M., Schittekat, P.: "Multiple Neighbourhood" Search in Commercial VRP Packages: Evolving Towards Self-Adaptive Methods. Adaptive and Multilevel Metaheuristics, 239–253 (2008)
14. Ropke, S., Pisinger, D.: A unified heuristic for a large class of Vehicle Routing Problems with Backhauls. European Journal of Operational Research 171(3), 750–775 (2006)
15. Toth, P., Vigo, D. (eds.) The Vehicle Routing Problem. Society for Industrial and Applied Mathematics (2001)

Stream XII
Production Management, Supply Chain Management

Herbert Meyr, TU Darmstadt, GOR Chair
Stefan Minner, University of Vienna, ÖGOR Chair
Stephan Wagner, ETH Zurich, SVOR Chair

Fill Time, Inventory and Capacity in a Multi-Item Production Line under Heijunka Control

Andreas Tegel, Bernhard Fleischmann

Abstract The concept of production levelling (or Heijunka), which is part of the Toyota Production System, aims at reducing the variance of upstream material requirements and of workload utilization. When using Heijunka control, however, the fluctuation of the external demand has to be compensated by the fill time (the time from demand arrival until fulfilment) and/or end product inventory and/or reserve capacity. The inventory is usually controlled by Kanban loops. But precise statements on the necessary number of Kanban to guarantee a certain fill time are missing in literature. We consider a multi-item production line with a given Heijunka schedule and Poisson demand. We derive approximate analytic relationships between fill time, inventory and capacity for three different planning situations.

1 Introduction

During the past years the Toyota Production System (TPS) gained more and more attention in industry. Numerous companies are interested in the production system of the world's biggest automobile manufacturer, that seems to promise long lasting success by following easy to understand rules. After having spent a lot of effort in building up huge IT systems, it now appears to be much easier to control every day business in production and logistics by sending Kanban from one point in the supply chain to another. Nowadays, not only all German automotive manufacturers have adapted the TPS to their specific needs, but the TPS is also widespread at their

Dr. Andreas Tegel
Robert Bosch GmbH, Bosch Green Logistics, Max-Lang-Strasse 40-46, 70771 Leinfelden-Echterdingen, e-mail: andreas.tegel@de.bosch.com

Prof. em. Dr. Bernhard Fleischmann
Chair of Production and Logistics, University of Augsburg, Universitaetsstrasse 16, 86135 Augsburg e-mail: bernhard.fleischmann@wiwi.uni-augsburg.de

D. Klatte et al. (eds.), *Operations Research Proceedings 2011*, Operations Research Proceedings, DOI 10.1007/978-3-642-29210-1_66, © Springer-Verlag Berlin Heidelberg 2012

suppliers and even used in the food, aircraft or power tools industry.

A major building block of the TPS is the pull control of production. But in its pure form, pull control leads to fluctuating production following the fluctuating market demand. As a consequence, the utilization of manpower and material requirements vary as well and the latter variation is still aggravated by the bullwhip effect[1]. As a countermeasure, the TPS recommends Heijunka, that is production levelling at the last production stage in order to filter the short-term demand variation out. In a multi-product environment, Heijunka consists in fixing a schedule for one day or one week, which is composed of small production lots, ideally of one piece. The allocation of the products to the lots is fixed in such a way that, for any product, the intervals between consecutive lots are as even as possible and the allocated capacity is sufficient for at least the average demand. The fixed schedule is repeated daily or weekly, respectively, during a medium-term planning interval, as long as the external demand can be considered as stationary. The schedule is adapted several times a year.

When using Heijunka control, however, the deviation of the smooth production from the fluctuating demand has to be compensated by the fill time (the time from demand arrival until fulfilment) and/or end product inventory and/or reserve capacity. But precise statements on the relationship between these three factors are missing in literature. The purpose of this paper is to derive approximate analytic relationships between fill time, inventory and reserve capacity for a production line under Heijunka control.

The next section explains properties and use of a Heijunka schedule, notations and assumptions. Section 4 develops models and results.

2 Heijunka control

We consider a single production line where several products $p = 1,\ldots,P$ can be produced. The processing time for one unit, say one piece, of any product is constant and assumed to equal one time unit, without loss of generality. There is a given Heijunka schedule for a short planning interval of T time units, say one day, which consists of periods ("production slots") $t = 1,\ldots,T$ of length 1, each of which is dedicated to a certain product. Denote

d_p average demand of product p (pieces per day)
c_p number of periods in the schedule dedicated to product p.

Then $\sum_p c_p = T$ and c_p is the daily capacity available for product p. A feasible schedule must satisfy $d_p < c_p$ which implies $c_p < T$. The difference $c_p - d_p$ represents the reserve capacity for product p which can be used to compensate demand

---

[1] see [3], p. 612-616

variation.

When establishing a Heijunka schedule for given c_p and d_p, the products are allocated to the periods in such a way that, for every product, the production periods are distributed over the planning interval as evenly as possible. The ideal distance between the start of two consecutive production periods (a production cycle) of product p would be T/c_p, but it cannot be reached exactly due to the discrete periods and the competition of several products for the free periods. Minimizing the deviation of the actual lengths of the production cycles from the ideal one can be expressed in several ways as objective function for Heijunka scheduling. Various models and algorithms can be found in literature, see the recent survey in [4]. New exact and heuristic algorithms are proposed in [6]. But this is not the subject of this paper.

In the following we start from a given Heijunka schedule and analyze the resulting fill times as functions of the inventory and the reserve capacity. The following assumptions are made: Materials are always sufficiently available. The demand is a Poisson process with the orders for one piece of product p arriving with rate $\lambda = d_p/T$. The production rate is $\mu = c_p/T$, hence the capacity utilization $\rho = \lambda/\mu = d_p/c_p$.
For further analysis we will consider an example with six products $1,\ldots,6$. One production day consists of 16 hours working time and one production period is assumed to last two minutes. In total, $\sum_p c_p = T = 480$ production periods are available. The demand d_p, the daily capacity c_p, ρ and the cycle time T/c_p measured in production periods for the six products are listed in the table below. Note that in this example the total reserve capacity of 60 has been allocated equally to the six products, hence 10 for each product. But any other allocation with $c_p > d_p$ would be feasible.

p	1	2	3	4	5	6	Σ
d_p	220	60	70	30	25	15	420
c_p	230	70	80	40	35	25	480
ρ	0.957	0.857	0.875	0.750	0.714	0.600	0.875
T/c_p	2.087	6.857	6.000	12.00	13.71	18.400	

The following analysis is based on three essential considerations: First, given a schedule with dedicated periods, the decisions when to produce a certain product p are completely independent of the decisions on the other products. Therefore, the analysis can be done separately for single products. Second, the schedule for a single product p can be approximated by the ideal schedule with equal production cycles of length T/c_p as the construction of the schedule has minimized the deviation of the true cycles from this length. Third, a reserved period can only be used for production if a free Kanban is available at the production line. In case no customer order arrived until the reserved period, there is no production and the period remains unused.

3 Approximation of the average fill time of a customer order

In the following we analyze the performance of production levelling for a given Heijunka schedule in three different situations: without inventory, with inventory and backorders, with inventory and lost sales. As argued before, the analysis can be done separately for each product. Therefore, the product subscript is omitted. The derivation of the results can only be sketched here, details are presented in [6]. In that dissertation, Heijunka schedules have been calculated and the fill times or fill rates have been simulated for various schedules. The results of these simulations are shown below for the example in section 2 and compared to the derived formulae.

First, we consider the case without inventory of finished products. An arriving order has to wait until the beginning of the next production period for the desired product, and, if this period is occupied by previous orders, additionally until the next free production period. Let w_1, w_2 be these two waiting times. Then, adding the operation time of 1, the fill time is $FT = w_1 + w_2 + 1$. Following [5], w_1 equals one half cycle time and w_2 is according to [1] the waiting time in an $M/D/1$-system, hence

$$FT = \frac{1}{2\mu} + \frac{\rho}{2\mu(1-\rho)} + 1 = \frac{1}{2(\mu - \lambda)} + 1 = \frac{T}{2(c-d)} + 1. \tag{1}$$

Note that the fill time does not depend on the relative capacity utilization ρ of the product, but on its absolute reserve capacity $c - d$. Thus, allocating the total reserve capacity equally to the products leads to equal fill times. In our example $FT = 480/(2 \cdot 10) + 1 = 25$, as shown in the first line of Table 1. On the other hand, (1) allows to allocate the reserve capacity so as to optimize a given objective function, for instance a weighted sum of the fill times of the products as shown in [6].

In the second case, there is inventory after the production line which is Kanban-controlled. The K Kanbans are either attached to the units in stock or waiting before the line, so that the inventory level is at most K. When an order arrives, it can be fulfilled with fill time 0, if there is stock available, otherwise it becomes a backorder and has to wait. In [2] this situation is treated using an $M/G/1$-model, where the number of orders in the system equals the waiting Kanbans plus the backorders. For the $M/M/1$ case, they can calculate the expected value of the backorders $E(B)$ exactly, in the general case they give the approximation

$$E(B) = \frac{\rho \alpha^K}{1 - \alpha}$$

with some expression α. But this model is not immediately applicable to the considered situation. It is valid only for a single-product line and does not consider the waiting times for the next production period. Moreover, the present case has again deterministic operation times. But for the $M/D/1$ case, using results from [1], α can be calculated as

$$\alpha = \frac{\rho}{2-\rho}$$

and therefore, due to Little's law, the fill time is a function of K and equals

$$FT(K) = \frac{\alpha^K}{\mu(1-\alpha)}.$$

In order to adapt this result to the multi-product case, we replace $FT(0)$ by (1) and get

$$FT(K) = \left[\frac{T}{2(c-d)} + 1\right] \cdot \left[\frac{\rho}{2-\rho}\right]^K. \tag{2}$$

This equation quantifies the consequences of production levelling: waiting times for the customers (FT), inventory (K) and/or reserve capacity $(c-d)$. The fill time decreases exponentially with increasing inventory, and the decrease is the stronger the lower the capacity utilization is. This can also be seen in the simulation results shown in Table 1. The fill times for products $1,\dots,6$ are simulated for $K = 0,\dots,3$. The approximated fill times are calculated by (2). As can be seen in Table 1 the simulated and approximated fill times show a very good congruence.

K	1		2		3		4		5		6	
0	25.17	25.00	25.14	25.00	25.36	25.00	25.08	25.00	25.32	25.00	25.34	25.00
1	22.67	22.92	18.55	18.75	19.38	19.44	14.63	15.00	13.61	13.89	10.23	10.71
2	20.85	21.01	13.88	14.06	15.19	15.12	8.490	9.000	7.236	7.716	3.983	4.592
3	19.27	19.26	10.28	10.55	11.69	11.76	4.890	5.400	3.817	4.287	1.538	1.968

Table 1 Simulated (left column) and approximated (right column) fill times $FT(K)$

In the third case, customers are not willing to wait. Sales are lost, if there is no stock available. Then, the fill time is always zero for the accepted orders, but the performance of the system can now be expressed by the fill rate β, the rate of accepted orders. This situation can be modeled as a $M/D/1/K$-model with the limit K on the number of waiting orders. Adapting the model in [2] in an analogous way as before leads to the following result which now includes also the possibility of $\rho \geq 1$:

$$\beta \approx \begin{cases} \dfrac{1-\alpha^K}{1-\rho\alpha^K}, & \rho < 1 \\[2ex] \dfrac{2K}{2K+1}, & \rho = 1 \\[2ex] \dfrac{1-\alpha'^K}{1-\rho\alpha'^K}, & \rho > 1 \end{cases} \tag{3}$$

with $\alpha = \frac{\rho}{2-\rho}$ and $\alpha' = 2\rho - 1$. Here, the fill rate only depends on the capacity utilization ρ of the particular product.

This equation quantifies the impact of production levelling in case of lost sales. For $\rho < 1$ the fill rate increases with increasing inventory and the increase is the stronger the lower the capacity utilization is. For $\rho = 1$ the fill rate only depends on the inventory K and β increases with increasing inventory. For $\rho > 1$ the fill rate β converges to $1/\rho$ for high inventory K.

The simulated fill rates for products $1, \ldots, 6$ and $K = 1, \ldots, 3$ are shown in table 2. The approximated fill rates were calculated by (3).

K	1	2	3	4	5	6
1	56.4% 67.6%	62.5% 70.0%	60.6% 69.6%	67.9% 72.7%	68.7% 73.7%	73.1% 76.9%
2	75.6% 81.4%	81.8% 84.5%	80.3% 83.9%	86.3% 87.7%	87.3% 88.7%	91.1% 91.7%
3	84.3% 87.3%	89.4% 90.6%	88.3% 90.0%	93.2% 93.6%	94.1% 94.4%	96.8% 96.7%

Table 2 Simulated (left column) and approximated (right column) fill rates β

Table 2 shows that the accuracy of the approximation increases with increasing inventory K and the accuracy is higher for low capacity utilization ρ. In [6] further formulae to improve the exactness of the approximation are derived.

4 Conclusions

In this paper we analyzed relationships between fill time or fill rate, inventory and reserve capacity for a multi-item production line with Heijunka control. With the new approximation formulae the effects of capacity and inventory planning on fill time and fill rate can be estimated and the allocation of reserve capacity can be optimized. Additionally, a minimum number of Kanban needed to guarantee a certain fill time or fill rate can be calculated. Further research effort is required for analyzing the effects of production levelling on the variation of raw material demand as well, in order to fully evaluate the cost effectiveness of Heijunka control.

References

1. Arnold, D., Furmans, K.: Materialfluss in Logistiksystemen. Springer, Berlin, 5. ed. (2007)
2. Buzacott, J. A., Shanthikumar, J. G.: Stochastic models of manufacturing systems. Prentice Hall, Englewood Cliffs, NJ (1993)
3. Hopp, W. J., Spearman, M. L.: Factory physics. Irwin/McGraw-Hill, Boston, 2. ed. (2000)
4. Miltenburg, J.: Level schedules for mixed-model JIT production lines. International Journal of Production Research, **45**, 3555 - 3577 (2007)
5. Ross, S. M.: Introduction to probability models. Academic Press, Amsterdam, 9. ed. (2007)
6. Tegel, A.: Analyse und Optimierung der Produktionsglaettung fuer Mehrprodukt-Fliesslinien - Eine Studie zum Lean-Production-Konzept. Dissertation, Universitaet Augsburg (2010)

Tight lower bounds by semidefinite relaxations for the discrete lot-sizing and scheduling problem with sequence-dependent changeover costs

Celine Gicquel and Abdel Lisser

Abstract We study a production planning problem known as the discrete lot-sizing and scheduling problem with sequence-dependent changeover costs. We discuss two alternative QIP formulations for this problem and propose to compute lower bounds using a semidefinite relaxation of the problem rather than a standard linear relaxation. Our computational results show that for a certain class of instances, the proposed solution approach provides lower bounds of significantly better quality.

1 Introduction

We study a production planning problem known as the discrete lot-sizing and scheduling problem or DLSP. In this variant of lot-sizing problems, several key assumptions are used to model the production planning problem (see [5]):
- Demand for products is deterministically known and time-varying.
- A finite time horizon subdivided into discrete periods is used to plan production.
- At most one item can be produced per period and the facility processes either one product at full capacity or is completely idle ("discrete" production policy).
- Costs to be minimized are the inventory holding costs and the changeover costs.
We consider here the complicating case where the changeover costs to be incurred when the production of a new lot begins are sequence-dependent, i.e. depend on both the item produced before and the item produced after the changeover .
A wide variety of solution techniques from the Operations Research field have been proposed to solve lot-sizing problems (see e.g. [8]). Among them, most existing exact solution approaches are based on tight mixed-integer linear programming (MILP) formulations which are solved by standard Branch & Bound procedures (see

C. Gicquel, A. Lisser
Laboratoire de Recherche en Informatique, Universite Paris Sud, 91405 Orsay Cedex, France e-mail: celine.gicquel@lri.fr,abdel.lisser@lri.fr

D. Klatte et al. (eds.), *Operations Research Proceedings 2011*, Operations Research Proceedings,   421
DOI 10.1007/978-3-642-29210-1_67, © Springer-Verlag Berlin Heidelberg 2012

e.g. [9]). In particular, valid inequalities and extended reformulations have been proposed to obtain tight linear relaxations of the DLSP (see [4], [10]) . However, even if substantial improvements of the lower bounds can be obtained by strengthening the MILP formulation, there are still cases where the linear relaxation is of rather poor quality (see e.g. [6]). These difficulties motivate the investigation of stronger relaxation methods such as semidefinite relaxations which have proved interesting to solve difficult combinatorial optimization problems (see e.g. [7]).

To the best of our knowledge, there is no previous attempt at using semidefinite relaxations to solve lot-sizing problems. The main contributions of the present paper thus consist in proposing to compute lower bounds for the DLSP with sequence-dependent changeover costs (DLSPSD) using a semidefinite relaxation rather than a standard linear relaxation and in presenting a cutting-plane generation algorithm based on a semidefinite programming (SDP) solver to obtain tight SDP relaxations. The paper is organized as follows. We first discuss in section 2 two alternative QIP formulations for the DLSPSD. We then explain in section 3 how lower bounds can be obtained by computing SDP relaxations of both formulations. Some computational results are then presented in section 4.

2 QIP formulations

We first consider two alternative QIP formulations for the DLSPSD: an "aggregate" formulation where decision variables are related to the assignment of products to production periods and an "extended" formulation of the problem where decision variables are related to the assignment of demand units to production periods.

Aggregate formulation

We wish to plan production for a set of products denoted $p = 1...P$ to be processed on a single production machine over a planning horizon involving $t = 1...T$ periods. Product $p = 0$ represents the idle state of the machine and period $t = 0$ is used to describe the initial state of the production system.

Production capacity is assumed to be constant throughout the planning horizon. We can thus w.l.o.g. normalize the production capacity to one unit per period and express the demands to be satisfied as integer numbers of units (see e.g. [5]). We denote D_{pt} the demand for product p in period t. We assume that, for each product, the total cumulated production over the planning horizon cannot exceed the total cumulated demand. We denote h_p the inventory holding cost per unit per period for product p and SDC_{pq} the sequence-dependent changeover cost to be incurred whenever the resource setup state is changed from product p to product q.

Using this notation, the DLSPSD can be seen as the problem of assigning a single product to each period of the planning horizon while ensuring demand satisfaction. We thus introduce the binary decision variables x_{pt} where $x_{pt} = 1$ if product p is assigned to period t and 0 otherwise, and obtain the following QIP1 formulation.

$$Z_{QIP1} = min \sum_{p=1}^{P} \sum_{t=1}^{T} \sum_{\tau=1}^{t} h_p(x_{p\tau} - D_{p\tau}) + \sum_{p,q=0}^{P} \sum_{t=0}^{T-1} SDC_{p,q} x_{pt} x_{qt+1} \qquad (1)$$

$$\sum_{\tau=1..t} x_{p\tau} \geq \sum_{\tau=1..t} D_{p\tau} \qquad \forall p, \forall t \quad (2)$$

$$\sum_{\tau=1..T} x_{p\tau} = \sum_{\tau=1..T} D_{p\tau} \qquad \forall p \quad (3)$$

$$\sum_{p=0...P} x_{pt} = 1 \qquad \forall t \quad (4)$$

$$x_{pt} \in \{0,1\} \qquad \forall p, \forall t \quad (5)$$

The objective function (1) corresponds to the minimization of the inventory holding and changeover costs over the planning horizon. Constraints (2) impose that the cumulated demand over interval $[1,t]$ is sastified by the cumulated production over the same time interval. Constraints (3) ensure that the total cumulated production over the planning horizon cannot be greater than the total cumulated demand. Constraints (4) ensure that a single product is assigned to each period of the planning horizon.

Extended formulation

We now consider an alternative extended formulation for the DLSPSD involving a larger number of decision variables. More precisely, we exploit an idea proposed by [2] for single-item lot sizing problems and propose to replace the aggregate representation of the demand used in formulation (QIP1) with a more detailed representation where each individual unit demand $d = 1..\Delta$ is considered. Demand d corresponds to a unit of product $\pi_d \in [1,P]$ to be delivered to the customer on due-date $dd_d \in [1,T]$. In case $\Delta < T$, we add $T - \Delta$ demands for product 0 in period T, corresponding to $T - \Delta$ idle periods in the production plan. We use a demand $d = 0$ with $\pi_0 \in [0,P]$ and $dd_0 = 0$ to describe the system initial state. Using this notation, the DLSPSD can be seen as the problem of assigning a single unit demand to be processed to each period of the planning horizon. We thus introduce the binary decision variables y_{dt} where $y_{dt} = 1$ if demand d is assigned to period t and 0 otherwise, and obtain the following QIP2 formulation.

$$Z_{QIP2} = min \sum_{d=1}^{T} \sum_{t=1}^{dd_d} h_{\pi_d}(dd_d - t) y_{dt} + \sum_{d,d'=0}^{T} \sum_{t=0}^{T-1} SDC_{\pi_d, \pi_{d'}} y_{d,t} y_{d',t+1} \qquad (6)$$

$$\sum_{t=1..dd_d} y_{dt} = 1 \qquad \forall d \quad (7)$$

$$\sum_{d=1..T s.t. dd_d \geq t} y_{dt} = 1 \qquad \forall t \quad (8)$$

$$y_{dt} \in \{0,1\} \qquad \forall d, \forall t \quad (9)$$

The objective function (6) corresponds to the minimization of the inventory holding and changeover costs over the planning horizon. Constraints (7) ensure that each demand d is assigned to a single time period before its delivery due-date dd_d. Constraints (8) ensure that a single demand is assigned to each period.

3 Semidefinite programming relaxation

We now consider computing a lower bound on the value of Z_{QIP1} and Z_{QIP2} by reformulating the corresponding quadratic integer programs as semidefinite programs.

Semidefinite programming formulations

Semidefinite programming is a recent development of convex optimization which deals with optimizing over the set of symmetric positive semidefinite matrices. We denote S_n the set of symmetric matrices of size n, $tr(A \bullet B)$ the scalar product between two matrices A and B and $A \succeq 0$ a positive semidefinite matrix.
The first step of the reformulation consists in introducing:
- vector $x^t = [x_{00}, .., x_{pt}, .., x_{NT}]$ of size $n_1 = NT$ and matrix $X = xx^t \in S_{n_1}$,
- vector $y^t = [y_{00}, .., y_{dt}, .., y_{T,dd_T}]$ of size $n_2 = \sum_{d=0..T} dd_d$ and matrix $Y = yy^t \in S_{n_2}$.
We then define:
- two cost matrices $C_1 \in S_{n_1}$ and $C_2 \in S_{n_2}$ s.t.:
$Z_{QIP1} = tr(C_1 \bullet X)$ and $Z_{QIP2} = tr(C_2 \bullet Y)$,
- demand satisfaction constraint matrices $A_1^{pt} \in S_{n_1}$ and $A_2^d \in S_{n_2}$ s.t.:
$tr(A_1^{pt} \bullet X) = \sum_{\tau=1..t} x_{p\tau}$ and $tr(A_2^d \bullet Y) = \sum_{t=1..dd_d} y_{dt}$,
- limited capacity constraint matrices $B_1^t \in S_{n_1}$ and $B_2^t \in S_{n_2}$ s.t.:
$tr(B_1^t \bullet X) = \sum_{p=0..P} x_{pt}$ and $tr(B_2^t \bullet Y) = \sum_{d=1..T \, s.t. \, dd_d \geq t} y_{dt}$.
With this notation, the SDP relaxation of (QIP1) and (QIP2) can be written as:

$$Z_{SDP1} = min \, tr(C_1 \bullet X) \tag{10}$$

$$tr(A_{pt}^1 \bullet X) \geq \sum_{\tau=1}^{t} D_{p\tau} \qquad \forall p, \forall t \tag{11}$$

$$tr(A_{pt}^1 \bullet X) = \sum_{\tau=1}^{t} D_{p\tau} \qquad \forall p \tag{12}$$

$$tr(B_t^1 \bullet X) = 1 \qquad \forall t \tag{13}$$

$$Diag(X) = x \tag{14}$$

$$X - xx^t \succeq 0 \tag{15}$$

$$Z_{SDP2} = min \, tr(C_2 \bullet Y) \tag{16}$$

$$tr(A_d^2 \bullet Y) = 1 \qquad \forall d \tag{17}$$

$$tr(B_t^2 \bullet Y) = 1 \qquad \forall t \tag{18}$$

$$Diag(Y) = y \tag{19}$$

$$Y - yy' \succeq 0 \tag{20}$$

Formulation strengthening

We first add the valid inequalities proposed in [10] for the single-item DLSP to the aggregate formulation (QIP1) to obtain an initial strenghtening of the formulation. Moreover, formulations (QIP1) and (QIP2) can be strengthened by expliciting the binary exclusion constraints derived from assignment constraints (4) and (7)-(8). These quadratic constraints provide valid inequalities of the form $X_{ij} = 0$ and $Y_{ij} = 0$ for formulations (QIP1) and (QIP2). Simple valid inequalities $X_{ij} \geq 0$ and $Y_{ij} \geq 0$ can also be used to further strengthen both SDP formulations.

The number of binary exclusion and non-negativity valid inequalities grows very fast with the problem size. Hence it is not possible to include all of them directly in formulations (QIP1) and (QIP2). In the computational experiments to be presented in section 4, they are added to the formulations using the cutting-plane generation strategy described in [1] for the quadratic assignment problem.

4 Computational results

We now discuss the results of some computational experiments carried out to evaluate the quality of the lower bounds provided by the linear and semidefinite relaxations of formulations (QIP1) and (QIP2).

We created 6 sets of 10 randomly generated instances involving 4 or 6 products, 15 or 20 periods, with a production capacity utilization of 95%. The instances mainly differ with respect to the structure of the changeover cost matrix SDC. Set A instances have a general cost structure. Set B instances correspond to the frequently encountered case where products can be grouped into product families: there is a high changeover cost between products of different families and a smaller changeover cost between products belonging to the same family.

For each instance, we first used the solver CPLEX 12.1 to compute the linear relaxation of formulations (QIP1) and (QIP2). This was achieved by using the flow-conservation constraints discussed in [9] to link the linearization variables with the binary variables. We then used the SDP solver CSDP based on an interior-point type algorithm (see [3]) to compute the semidefinite relaxation of formulations (QIP1) and (QIP2). All tests were run on an Intel Core i5 (2,7 GHz) with 4 Go of RAM, running under Windows 7.

Table 4 displays the computational results. We provide for each set of instances:

- Gap_{lp1} and Gap_{sdp1}: the mean gap between the lower bounds provided by the linear / semidefinite relaxations of formulation (QIP1) and the optimal integer value,
- Gap_{lp2} and Gap_{sdp2}: the mean gap between the lower bounds provided by the linear / semidefinite relaxations of formulation (QIP2) and the optimal integer value,
- CPU_{lp1}, CPU_{sdp1}, CPU_{lp2}, CPU_{sdp2}: the mean computation time.

P T	QIP1 form.				QIP2 form.			
	Gap_{lp1}	CPU_{lp1}	Gap_{sdp1}	CPU_{sdp1}	Gap_{lp2}	CPU_{lp2}	Gap_{sdp2}	CPU_{sdp2}
A 4 15	2.8%	0.05s	2.8%	60s	53%	0.05s	0%	814s
6 15	0.9%	0.05s	0.9%	36s	46%	0.05s	0%	691s
6 20	3.6%	0.05s	3.6%	804s	56%	0.05s	2.1%	6600s
B 4 15	11.4%	0.05s	11.4%	65s	38%	0.05s	0%	684s
6 15	5.3%	0.05s	5.3%	85s	39%	0.05s	0%	666s
6 20	12.5%	0.05s	12.5%	1365s	49%	0.05s	3.5%	7200s

Table 1 Computational results for set A and B instances

The results displayed in table 1 show that the SDP relaxation of formulation (QIP2) provides lower bounds of excellent quality. Namely, $Gap_{sdp2} = 0\%$ for all 15-period instances whereas the other studied relaxations lead to mean positive gaps between 0.9% and 53%. For the larger 20-period instances, Gap_{sdp2} is also smaller than the gap obtained with the other relaxations.

We note that the observed gap reduction is particularly significant for the instances corresponding to the case where there are product families. This might be explained by the fact that the valid inequalities proposed by [10] are less efficient at strengthening formulation (QIP1) for this class of instances than for general cost stucture instances (see [6]).

References

1. Benajam, W, 2005. Relaxations semidéfinies pour les problemes d'affectation de fréquences dans les réseaux mobiles et de l'affectation quadratique, PhD thesis, Laboratoire de Recherche en Informatique, Orsay, France.
2. Bilde O. and Krarup J., 1977. Sharp lower bounds and efficient algorithms for the simple plant location problem. *Annals of Discrete Mathematics*, 1, 79-97, 1977.
3. Borchers B, 1999. CSDP, A C Library for Semidefinite Programming. *Optimization Methods and Software*, 11(1), 613-623.
4. Eppen G.D. and Martin R.K, 1987. Solving multi-item capacitated lot-sizing problems using variable redefinition. *Operations Research*, 35(6), 832-848.
5. Fleischmann B., 1990. The discrete lot sizing and scheduling problem. *European Journal of Operational Research*, 44, 337-348.
6. Gicquel C., Miegeville N., Minoux M. and Dallery Y., 2009. Discrete lot-sizing and scheduling using product decomposition into attributes. *Computers and Operations Research*,36, 2690-2698.
7. Helmberg C, 2002. Semidefinite programming. *European Journal of Operational Research*, 137 (3), 461-482.
8. Jans R. and Degraeve Z., 2007. Meta-heuristics for dynamic lot sizing: a review and comparison of solution approaches, *European Journal of Operational Research*, 177, 1855-1875.
9. Pochet Y. and Wolsey L.A., 2006. *Production planning by mixed integer programming*, Springer Science.
10. van Eijl C.A. and van Hoesel C.P.M., 1997. On the discrete lot-sizing and scheduling problem with Wagner-Whitin costs, *Operations Research Letters* 20, 7-13..

Condition-based Maintenance Policy for decentralised Production/Maintenance-Control

Ralf Gössinger and Michael Kaluzny

Abstract A condition-based release of maintenance jobs is analysed for a production/maintenance-system where the conditions of the machines deteriorate as a function of multiple production parameters and maintenance is performed by a separate department. If a predefined operational availability is to keep up, the machine condition that triggers the release of maintenance jobs at minimum costs has to be determined. In the paper a specific continuous condition monitoring and an information exchange protocol are developed, the components of waiting time between release and execution are operationalised and relevant effects of choosing the triggering condition are presented. Stochastic approaches to calculate the triggering condition are formulated and analysed by simulations.

1 Problem

A condition-based release of preventive maintenance jobs is analysed for a production/maintenance-system (*PMS*) where maintenance is performed by a separate department. Machine conditions deteriorate as a function of multiple production parameters. After reaching a critical level of deterioration the machine is designated to be defective, and reactive maintenance is required. Preventive maintenance can be carried out before reaching that level. In both cases an as-good-as-new-maintenance is assumed. The task of the maintenance department is to keep up a predefined operational availability α (probability of no machine failure in the time between an as-good-as-new-status and the next preventive maintenance job). In this context the problem of determining the optimal machine condition α' that triggers the release of a preventive maintenance job arises. Due to the division of labour this problem

Ralf Gössinger
Dept of Business Administration, Production Management and Logistics, University of Dortmund, Otto-Hahn Str. 6, 44227 Dortmund, e-mail: ralf.goessinger@udo.edu

Michael Kaluzny
Dept of Business Administration, Production Management and Logistics, University of Dortmund, Otto-Hahn Str. 6, 44227 Dortmund, e-mail: michael.kaluzny@tu-dortmund.de

D. Klatte et al. (eds.), *Operations Research Proceedings 2011*, Operations Research Proceedings, 427
DOI 10.1007/978-3-642-29210-1_68, © Springer-Verlag Berlin Heidelberg 2012

features the following: (1) Information exchange is required to diminish the information asymmetry between production and maintenance department. (2) The waiting time W between triggering the release and carrying out a maintenance job is determined by different stochastic processes. (3) During W production can be continued with a modified production parameter setting.

In the literature approaches of decentralised heterarchical[1] and hierarchical[2] production/maintenance-coordination are proposed that build on the central assumptions of instantaneous information exchange and immediate availability of maintenance resources. These assumptions are not always fulfilled in decentralised systems. Therefore in the present paper an information exchange protocol and a problem-specific continuous condition monitoring (*CM*) are proposed as well as the factors determining W are operationalised. To solve the coordination problem, decision relevant effects of choosing α' are presented and alternative stochastic approaches to calculate α' are formulated and tested.

2 Model

To coordinate the activities of production and maintenance department an exchange of local data about machine conditions, initiation of preventive maintenance jobs, chosen production parameters as well as required and realisable starting times for preventive maintenance jobs is necessary. The following structure of goods and information flow is one possibility to fulfill the information requirements (Fig.1)[3]. In order to realise *CM* a trigger (*TR*) is installed at the machine. It releases pre-

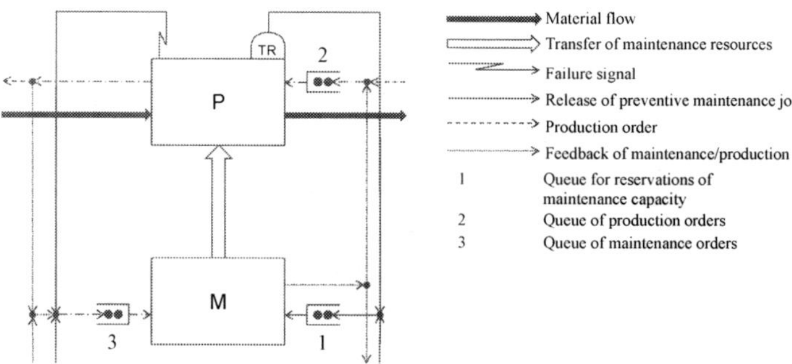

Fig. 1 Structure of goods and information flow for decentralised production/maintenance-control

---

[1] Cf. Berrichi et al. (2010), pp. 1584; Coudert/Grabot/Archiméde (2002), pp. 4611; Feichtinger (1982), pp. 168; Jørgensen (1984), pp. 77.

[2] Cf. Cho/Abad/Parlar (1993), pp. 155; Dehayem Nodem/Kenne/Gharbi (2009), pp. 173.

[3] According to the PAC-System proposed by Buzacott/Shanthikumar (1992), pp. 34.

ventive maintenance jobs, when defined events[4] occur and sends them to queue 1 (reservation of maintenance capacity). If capacity is available (waiting time W_1), the maintenance department enqueues a preparedness tag in queue 2 (production orders). After a waiting time W_2 this tag is forwarded to the machine and a requisition tag is sent to queue 3 (maintenance orders). If time span $W_{3.p}$ is elapsed the transfer of maintenance resources needed is carried out. After a transfer time W_4 the execution of the maintenance job starts. Thus W as the sum of W_i is a random variable. In case of a machine failure a requisition tag is enqueued in queue 3, and the execution of the reactive maintenance job starts after the time span $W_{3.r} + W_4$.

To keep up a predefined operational availability α, a preventive maintenance job has to be triggered in the way that the probability of reaching the critical deterioration level during W is not above $1 - \alpha$. Assuming that an availability higher than α results in higher maintenance costs, the aim is to realise α as accurate as possible by choosing the triggering condition α'. With π as the ratio of preventive maintenance jobs this leads to the objective:

$$min\ e(\alpha') := |\alpha - \pi(\alpha')|.$$

In addition to the machine runtime CM allows to consider multiple production parameters $v^P \in \mathbb{R}^n$ as influencing factors. Assuming a conditional failure time distribution $F_{T|v^P}$ for each constellation v^P, the family $(F_{T|v^P})_{v^P}$ of these distributions forms a hypersurface. In this case the availability restriction is the isosurface to the value $1 - \alpha$ and the development of the failure probability $F_T(t, v^P(t))$ is an utilisation path on $(F_{T|v^P})_{v^P}$ (Fig.2). To solve the problem at hand, at time t TR has to

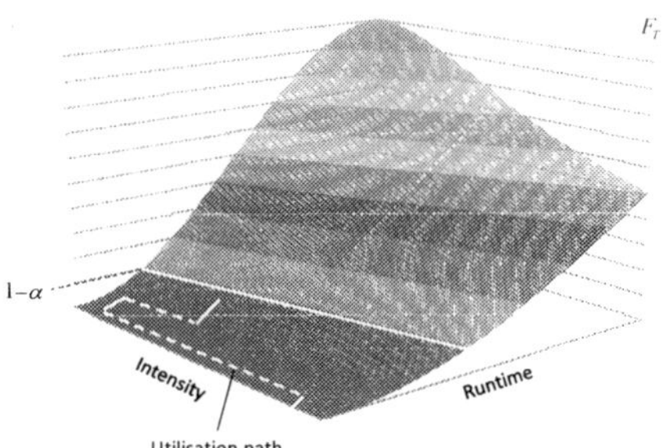

Fig. 2 Family of failure time distributions, availability restriction and utilisation path of a machine considering the influencing factors runtime and intensity

calculate two machine conditions: (1) Failure probability $\rho(t)$ based on the chosen production parameter settings after the as-good-as-new-status. (2) Failure probability $\rho(t+W)$ that would be reached after an estimated W, if a maintenance job would

---

[4] Cf. e.g. Abdulnour/Dudek/Smith (1995), pp. 565; Albino/Carella/Okogbaa (1992), pp. 369; Kuo/Chang (2007), pp. 602; Rishel/Christy (1996), pp. 421.

be triggered at t and production after t would be performed with an estimated production parameter setting:

$$\rho(t) = F_T(t, v^P(t))$$

$$\rho(t+W) = \hat{F}_T(t+W, v^P(t+W))$$

If F_W and F_{v^P} are given distributions of W and v^P respectively, the estimations $\hat{W}$ and $\hat{v}^P$ can simply be expected values (EV) or, more sophisticated, values determined with the chance constrained approach (CC):

$$\hat{W}_{EV} := E(W), \hat{W}_{CC} := F_W^{-1}(p_W)$$

$$\hat{v}^P_{EV} := E(v^P), \hat{v}^P_{CC} \in \{v^{P*} \in \mathbb{R}^n | P(F_T(t, v^P(t)) \leq F_T(t, v^{P*}(t))) = p_{v^P}, \forall t \in [0,\infty[\}^5$$

The estimation of the failure probability after W can be written as a function of $\hat{W}$ and $\hat{v}^P$ by using the conditional failure time distribution $F_{t|\hat{v}^P}$:

$$\rho(t+W) := F_{T|\hat{v}^P}(\tau + \hat{W}).$$

The time equivalent τ of reaching $\rho(t)$ with the estimated parameter constellation $\hat{v}^P$ is:

$$\tau := F_{T|\hat{v}^P}^{-1}(\rho(t))$$

The requirement $F_{T|\hat{v}^P}(F_{T|\hat{v}^P}^{-1}(\alpha') + \hat{W}) = 1 - \alpha$ implies:

Theorem 1. $\alpha'(\hat{W}, \hat{v}^P) = 1 - F_{T|\hat{v}^P}(F_{T|\hat{v}^P}^{-1}(1-\alpha) - \hat{W})\square$

As a result of the search necessary for approximating the parameters p_W and p_{v^P}, CC induces a higher computational effort than EV. Therefore the question is, whether better values of the objective function may legitimate this complexity.

3 Numerical analysis

In order to get first insights regarding the suitability of the two estimation types, their influence on the objective fulfillment is analysed for $\hat{W}^6$. In a single-stage PMS the required availability α and the strength a of the v^P-influence on deterioration is variied systematically ($\alpha \in \{0.79, 0.85, 0.9, 0.94, 0.97, 0.98, 0.99\}$, $a \in \{0.25, 0.5, 1, 1.5, 2, 2.5\}$). Per constellation (α, a) 30 simulations are carried out[7]. The analysed production system is defined as follows:

---

[5] For application it has to be proven, that this set is not empty. If this set includes more than one element, then any element can be chosen without changing the estimation of the deterioration.

[6] Assuming that v^P is constant during W the estimation $\hat{v}^P$ equals the last chosen parameter setting.

[7] The number of maintenance jobs is chosen in a way, that for each constellation the condition $P(\textit{preventive maintenance}) = \pi$ is accepted with an error probability $\varepsilon < 0,05$ and accuracy $\delta < 0,01$.

- Queue 2: Exponentially distributed interarrival time of production orders $E(D) = 1800$, maintenance-first/production-FCFS-rule and production rate v^P as a function of the queue length L:

L	0-10	11-20	21-30	31-40	41-50	51-60	61-70	71-80	81-90	91-
$\dfrac{1}{v^P}$	2280	2160	2040	1920	1800	1680	1560	1440	1320	1200

- Machine: Exponentially distributed deterioration $T|v^P$ conditioned by production rate v^P $E(T|v^P) = 1800 + a \cdot (\frac{1}{v^P} - 1800)$ and duration of reactive/preventive maintenance $m_r = 18000, m_p = 3600$.
- Maintenance orders: Exponentially distributed waiting times at the relevant queues: $E(W_1) = 3600, E(W_{3.p}) = 3600, E(W_{3.r}) = 600, E(W_4) = 3600$.

The observed relative deviations $\frac{e}{1-\alpha}$ induced by the application of $\hat{W}_{EV}$ und $\hat{W}_{CC}$ are shown as isoquants in Fig.3. The results show that CC enables an appropiate

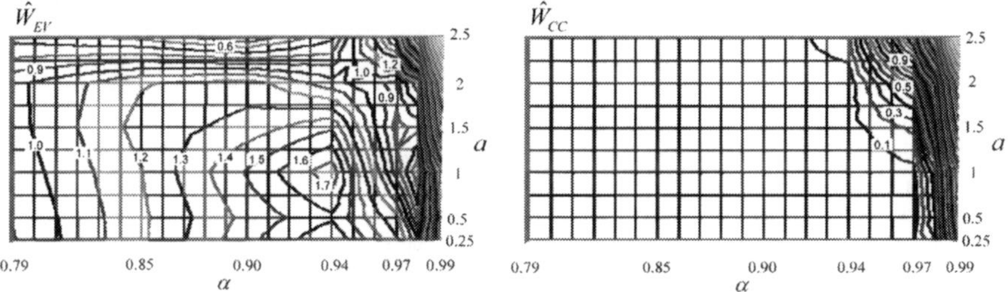

Fig. 3 Isoquants of the relative deviation from required availability

fulfillment of the required operational availability in a wide area of the (α,a)-constellations. Thereby the value of the objective function worsens with increasing α and a. Applying EV a best relative deviation of 42% is realised, and the influence of α and a is multivalued. So EV is not applicable for the analysed PMS.

4 Concluding remarks

Starting point of this paper was a structure of a goods and information flow that ensures the nessesary exchange of required information between the production and the maintenance department. On this basis a decision calculus to determine the triggering condition in a single-stage PMS was formulated for the case of fulfilling a defined operational availability. The estimations of the time between triggering the release and carrying out a preventive maintenance job and the expected deterioration, depending on the production parameter settings expected to be chosen, within the estimated waiting time were specified with the expected value and the value determined by the chance constrained approach. The numerical analysis shows, that an estimation of waiting time by using the chance constrained approach enables an appropriate fulfillment of the availability condition for a wide range of parameter values, whereas the expected value approach was inapplicable for the simulated system. Next steps of our future research are: (1) Numerical analysis of (a) the suitability of both methods to estimate parameter settings and of (b) the interdependencies between the estimation parameters of CC; (2) modelling and numerical analysis of multi-stage PMS.

References

1. Abdulnour, G., Dudek, R.A., Smith, M.L.: Effect of Maintenance Policies on the Just-in-Time Production System. International Journal of Production Research **33**, 565–583 (1995)
2. Albino, V., Carella, G., Okogbaa, O.G.: Maintenance Policies in Just-in-Time Manufacturing Lines. International Journal of Production Research **30**, 369–382 (1992)
3. Berrichi A. et al.: Bi-Objective Ant Colony Optimization Approach to Optimize Production and Maintenance Scheduling. Computers & Operations Research **37**, 1584–1596 (2010)
4. Buzacott, J.A., Shanthikumar J.G.: A General Approach for Coordinating Production in Multiple-Cell Manufacturing Systems. Production and Operations Management **1**, 34–52 (1992)
5. Cho, D.I., Abad, P.L., Parlar, M.: Optimal Production and Maintenance Decisions When a System Experience (sic!) Age-Dependent Deterioration. Optimal Control Applications & Methods **14**, 153–167 (1993)
6. Coudert, T., Grabot, B., Archiméde, B.: Production/Maintenance Cooperative Scheduling Using Multi-Agents and Fuzzy Logic. International Journal of Production Research **40**, 4611–4632 (2002)
7. Dehayem Nodem, F.I., Kenne, J.P., Gharbi, A.: Hierarchical Decision Making in Production and Repair/Replacement Planning With Imperfect Repairs Under Uncertainty. European Journal of Operational Research **198**, 173–189 (2009)
8. Feichtinger, G.: The Nash Solution of a Maintenance-Production Differential Game. European Journal of Operational Research **10**, 165–172 (1982)
9. Jørgensen, S.: A Pareto-Optimal Solution of a Maintenance-Production Differential Game. European Journal of Operational Research **18**, 76–80 (1984)
10. Jorgenson, D.W., McCall, J.J., Radner, R.: Optimal Replacement Policy. Amsterdam (1967)
11. Kuo, Y., Chang, Z.: Integrated Production Scheduling and Preventive Maintenance Planning for a Single Machine Under a Cumulative Damage Failure Process. Naval Research Logistics **54**, 602–614 (2007)
12. Rishel, T.D., Christy, D.P.: Incorporating Maintenance Activities Into Production Planning: Integration at the Master Schedule Versus Material Requirements Level. International Journal of Production Research **34**, 421–446 (1996)

Mid-Term Model-Plant Allocation for Flexible Production Networks in the Automotive Industry

Kai Wittek, Achim Koberstein, and Thomas S. Spengler

Abstract In order to match capacity with volatile demand, original equipment manufacturers in the automotive industry increase plant flexibility. This allows for car models to be re-allocated between plants over time, to optimally use available capacity. The planning problem moves from the strategic to the tactical level. Decision support in the form of mathematical planning models accounting for the increased flexibility and characteristics of the mid-term planning situation is required. Two different modeling approaches are considered in this work and compared for their applicability to the planning situation: An approach based on a single time index model formulation and an approach based on a double time index formulation. They differ in their computational time and ability to integrate lost customer rates.

1 Introduction and Motivation

Automotive markets are increasingly heterogeneous and manufacturers respond with an explosion of the number of car models (e.g. Polo, Golf) offered, to fill niches. In addition, high demand volatility can be observed. At the same time, technological advances standardize car models from a production point of view, resulting in allocation flexibility by multi-model lines. To capture the benefits of increased flexibility, decision support is required [14]. Having flexibility to re-allocate over time, the model-plant allocation decision moves from the strategic long-term to the mid-term planning level. Compared to strategic planning models e.g. [2] or [5], new requirements apply and have to be integrated into the models [14]. As in the strategic

Kai Wittek · Thomas S. Spengler
Institute of Automotive Management and Industrial Production, Technische Universität Braunschweig, Katharinenstr. 3, 38106 Braunschweig, e-mail: k.wittek | t.spengler@tu-braunschweig.de

Achim Koberstein
Decision Support & Operations Research Lab, Universität Paderborn, Warburger Str. 100, 33098 Paderborn, e-mail: koberstein@dsor.de

D. Klatte et al. (eds.), *Operations Research Proceedings 2011*, Operations Research Proceedings, 433
DOI 10.1007/978-3-642-29210-1_69, © Springer-Verlag Berlin Heidelberg 2012

models, a multi-period problem is considered, requiring a time indexed formulation. Similarly to the models of dynamic freight transportation in [9] two different time indexed formulations are feasible, a single or a double time index. Where in [9] a double time index allows for an explicit model of the information process, in the context of model-plant and volume allocation, it allows for an explicit model of the production process. The single index represents an implicit model. Another work, where model formulations based on different index structures were developed is [6].

For overviews concerning flexibility and integration into the planning context in the automotive industry refer to [10] or [7] and the mid-term model-plant allocation problem for flexible networks in the automotive industry was introduced in [14].

The goal of this paper is to compare both formulations in the context of mid-term model-plant allocation in flexible networks of the automotive industry, since network planning problems of realistic sizes reach limits in computation time [5]. The paper is structured as follows: First, both model formulations are introduced in section 2. In section 3 the models are applied to generic test networks, exhibiting characteristics of the automotive industry and different sizes, to analyze the scalability. The paper finishes with conclusions and outlook in section 4.

2 Comparison of Modeling Approaches

The models for mid-term model-plant allocation in flexible networks of the automotive industry consider a time horizon with smaller granularity (e.g. months not years) than current strategic models. Further, different revenues, the option to re-allocate as well as shifting or deferring production volume need to be considered. This results in integrated models of allocation planning (where to produce) and master production planning (when and what product to produce). A corresponding model based on a double time index formulation was presented in [14], having a discounted cash flow oriented objective function (maximizing) covering revenues, production costs, inventory costs and re-allocation costs. Here, costs refer to actual costs i.e. actual cash flows. Constraints include: Allowable allocations, capacity restrictions, upper bound on inventory, upper bound on number of models in one plant in parallel, upper bound on number of plants where a model is produced in parallel, lower bound on capacity utilization, lower bound on volume per allocated model, upper bound on re-allocations per period and lower bound on minimal allocation length. This results in a mixed integer linear programming model, with the following characteristics, using the classification criteria in [8]: Multi item, multi machine, constant capacity, lower bound on production levels, start up costs (re-allocation costs), lost sales and Wagner-Whitin condition (given set ups, it pays to produce later).

Having introduced the general model, the single and double time index formulation can be compared. The following notation will be used throughout the paper: p plants, m car models, $D_{m,t}$ demand, LR_m loss rate (rate of deferred customers cancelling) and t periods $[1,\ldots,T]$. Single time index specific: $I_{m,t}$ inventory, $N_{m,t}$ unfulfilled demand and $P_{m,p,t}$ production. Double time index specific: t^{prod} period

Continuation approach (single time index)   **Fragmentation approach** (double time index)

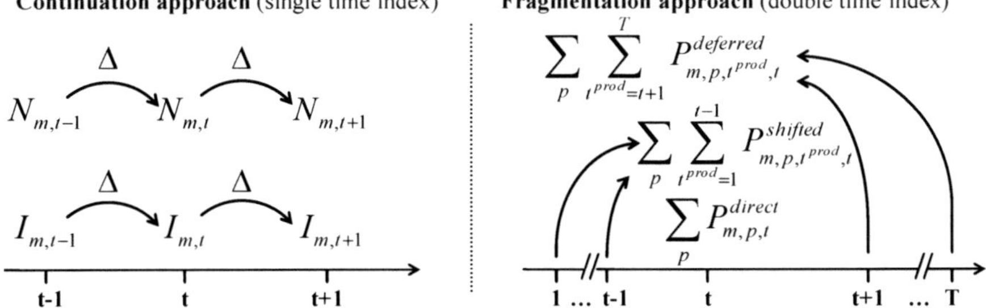

Fig. 1 Handling of shifted and deferred demand

of production, $P^{direct}_{m,p,t}$ production in t for demand in t, $P^{shifted}_{m,p,t^{prod},t}$ early production in t^{prod} for demand in t and $P^{deferred}_{m,p,t^{prod},t}$ late production in t^{prod} for demand in t.

A major difference lies in the way, shifted and deferred (resulting in some customers cancelling) demand are handled. As Figure 1 exhibits, the single time index resembles a continuation approach, where only differences in inventory $I_{m,t}$ and unfulfilled demand $N_{m,t}$ are tracked between consecutive periods, like in classic master production planning. The double time index is different by tracking, which volume was produced when for demand in which period. Hence, it offers more information and the option to address each 'type' of production volume individually.

Continuation approach (single time index)   **Fragmentation approach** (double time index)

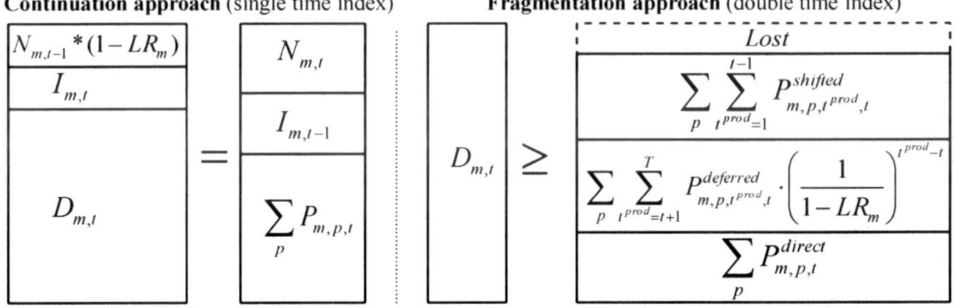

Fig. 2 Demand fulfillment restriction

The most interesting difference between these two formulations is the demand fulfillment restriction visualized in Figure 2 and given in (1) for the single time index as well as in (2) for the double time index. Both, for all m,t without initial inventory and starting/ending conditions. Due to space, all other constraints are omitted.

$$D_{m,t} = \sum_p P_{m,p,t} + I_{m,t-1} - I_{m,t} - N_{m,t-1} * (1 - LR_m) + N_{m,t} \qquad (1)$$

A separate constraint limiting $N_{m,t}$ is required to prevent 'free' inventory built up.

$$D_{m,t} \geq \sum_p P_{m,p,t}^{direct} + \sum_p \sum_{t^{prod}=1}^{t-1} P_{m,p,t^{prod},t}^{shifted} + \sum_p \sum_{t^{prod}=t+1}^{T} P_{m,p,t^{prod},t}^{deferred} * \left(\frac{1}{1-LR_m} \right)^{t^{prod}-t} \quad (2)$$

The lost customer rate is defined as $1/(1-LR_m)$ for the double index, to achieve identical results for both formulations with the same LR_m value. The double time index allows to include different customer loss rates, since the overall delay can be derived from the indices. Further, the detailed production tracking of the double time index allows for further analysis. It can be concluded that the double time index offers increased capabilities. However, it needs to be analyzed if this comes at a cost by affecting scalability.

3 Numerical Comparison

Based on the model formulations, scalability can be studied. First, the number of variables is considered for two network sizes. Sizes considered in the literature are in the range of 6 plants and 36 car models [2] or 6 to 18 plants with 16 to 186 car models in [5], which however could not all be solved. Table 1 presents figures for the single and double time index models based on the constraints listed in section 2.

Table 1 Number of required variables of modeling approaches

Example network: 10 plants, 20 car models, 24 time periods		
Single time index	Binary variables: 4,800	Total number of variables: 20,640
Double time index	Binary variables: 4,800	Total number of variables: 255,360

Example network: 20 plants, 40 car models, 24 time periods		
Single time index	Binary variables: 19,200	Total number of variables: 79,680
Double time index	Binary variables: 19,200	Total number of variables: 1,019,520

The number of variables increases rapidly with network size, a classic example of the variable explosion. The double time index formulation exceeds the single time index by magnitudes, potentially limiting the applicability to large networks.

For the computational study, a generic data set based on characteristics of the auto industry was used. Within the network, plants are connected analogous to the 'chaining' principle developed by [4]. Models were solved using branch and bound in ILOG OPL IDE 6.3 using CPLEX 12.1.0 on a 64 Bit; Intel Xeon E5430 @2.66 GHz (quadcore) with 8GB RAM. Runs exceeding 24 h computation time found integer solutions, but were terminated at 24 h before the optimal solution was found.

The notation in the tables is: 'x out of y' refers to the allowable number of x car models in parallel at a plant and the number of different models y for which allocation is feasible in general. 'In max. z plants in parallel' defines to how many plants a car model can be allocated to simultaneously. Units are h hours and s seconds.

Table 2 depicts the computation times for different flexibility settings. It exhibits a steep step in computation time, when flexibility is limited to '2 out of 5' and all plants receive the maximum feasible assignment, limiting cuts to the solution space.

Table 2 Impact of flexibility on computation time

	2 out of 5	3 out of 5	4 out of 5	5 out of 5
single time index	> 24 h	16 s	2 s	2 s
double time index	> 24 h	36 s	8 s	7 s

20 plants, 40 car models, 24 periods - in max. 2 plants in parallel

Table 3 exhibits similar results. When production costs for a car model are equal at all feasible plants, the computation time rises. Interestingly, removing the chaining concept and clustering the plants increases the computation time significantly.

Table 3 Impact of various parameters on computation time

	Inventory at no charge	Inventory prohibitive	Equal production costs	no chaining	customer loss rate = 0
single time index	1 s	4 s	191 s	7 h	2 s
double time index	6 s	20 s	1726 s	> 24 h	24 s

20 plants, 40 car models, 24 periods - 5 out of 5 - in max. 5 plants in parallel

In Table 4 the scalability for network size is provided. Computation times vary little for the exemplary data set. Fluctuations induced by changing the data set exceeded the differences by network size. However, for large networks, the double index formulation reaches memory limitations even for state of the art workstations.

Table 4 Impact of network size on computation time

Number of car models	5	10	20	40	80
Number of plants	3	5	10	20	40
Number of periods	24	24	24	24	24
single time index	< 1 s	< 1 s	< 1 s	< 1 s	7.1 s
double time index	1.0 s	2.6 s	3.3 s	8.1 s	out of memory*

5 out of 5 - in max. 5 plants in parallel   * upon initialization

4 Conclusion and Outlook

The comparison showed that the double time index offers superior information on shifted and deferred demand, as well as extended possibilities to manage customer loss rates. This comes at increased computation times. A well known solution from lot sizing to facilitate the LP relaxation and reduce the number of nodes in branch and bound by using the double time index to sharpen the bounds on big M in the indicator restriction for production volume [8] does not promise improvements, because applying available capacity as upper bound already provides a tight bound. Additional indicator restrictions might however further reduce the solution space.

Where both formulations can be applied to small and medium sized networks, the double time index turns infeasible due to memory limitations, not computation time, for large networks, e.g. the Volkswagen AG with ca. 40 final assembly plants. Future work includes the integration of stochastic elements into the model, where it can be drawn on approaches adapted from various works, mostly from the strategic level e.g. [1] (two stage stochastic), [11] (multi stage stochastic), [13] (approximate dynamic programming) or [3] (sample average approximation). Further, decomposition approaches like Benders decomposition (e.g. [1]), a temporal decomposition as in [12] or a decomposition into regions as proposed in [14] shall be applied to the mid-term planning model, to improve applicability to large scale problems.

References

1. Bihlmaier, R., Koberstein, A., Obst, R.: Modeling and optimizing of strategic and tactical production planning in the automotive industry. OR Spectrum **31**, 311–336 (2009)
2. Fleischmann, B., Ferber, S.; Henrich, P.: Strategic planning of BMW's global production network. Interfaces **36**, 194–208 (2006)
3. Francas, D., Kremer, M., Minner, S., Friese, M.: Strategic process flexibility under lifecycle demand. Int. J. Prod. Econ. **121**, 427–440 (2009)
4. Jordan, W.C., Graves, S.C.: Principles on the benefits of manufacturing process flexibility. Manage. Sci. **41**, 577–594 (1995)
5. Kauder, S., Meyr, H.: Strategic network planning for an international automotive manufacturer. OR Spectrum **31**, 507–532 (2009)
6. Mairs, T.G., Wakefield, G.W., Johnson, E.L., Spielberg, K.: On a production allocation and distribution problem. Manage. Sci. **24**, 1622–1630 (1978)
7. Meyr, H.: Supply chain planning in the German automotive industry. OR Spectrum **26**, 447–470 (2004)
8. Pochet, Y., Wolsey, L.A.: Production planning by mixed integer programming. Springer, New York (2006)
9. Powell, W.B., Bouzaïene-Ayari, B., Simao, H.P.: Dynamic models for freight transportation. In: Barnhart, C., Laporte, G. (eds.) Handbook in OR & MS, Vol. 14, pp. 285-365. Elsevier, Amsterdam (2007)
10. Sethi, A.K., Sethi, S.P.: Flexibility in manufacturing: A survey. Int. J. Flex. Manuf. Sys. **2**, 289–328 (1990)
11. Stephan, H.A., Gschwind, T., Minner, S.: Manufacturing capacity planning and the value of multi-stage stochastic programming under Markovian demand. Flex. Serv. Manuf. J. **22**, 143–162 (2010)
12. Timpe, C.H., Kallrath, J.: Optimal planning in large multi-site production networks. Eur. J. Opr. Res. **126**, 422–435 (2000)
13. Walter, M., Sommer-Dittrich, T., Zimmermann, J.: Evaluating volume flexibility instruments by design-of-experiments methods. Int. J. Prod. Res. **49**, 1731–1752 (2011)
14. Wittek, K., Volling, T., Spengler, T.S., Gundlach, F.-W.: Tactical planning in flexible production networks in the automotive industry. In: Hu, B., Morasch, K., Pickl, S., Siegle, M.(eds.) Operations Research Proceedings 2010, pp. 429-434. Springer, Berlin (2011)

Optimal (R, nQ) Policies for Serial Inventory Systems with Guaranteed Service

Peng Li and Haoxun Chen

Abstract In this paper, serial inventory systems with Poisson external demand and fixed costs at each stock are considered and the guaranteed-service approach (GSA) is used to design their optimal echelon (R, nQ) inventory policies. The problem is decomposed into two independent sub-problems, batch size decision sub-problem and reorder point decision sub-problem, which are solved optimally by using a dynamic programming algorithm and a branch and bound algorithm, respectively. Numerical experiments on randomly generated instances demonstrate the efficiency of these algorithms.

1 Introduction

In a supply chain modeled as a multi-echelon inventory system, the effective management of its inventory at each stock is critical to assure a high service level to customers at the minimal cost. Two approaches are often used in the optimization of inventory policies for multi-echelon inventory systems: stochastic-service approach (SSA) and guaranteed-service approach (GSA). In the GSA, each stock sets a service time to its downstream stocks and guarantees that their demands can always be satisfied in the given service time, under the assumption that final customer demand is bounded. For the excessive customer demand superior to a pre-specified bound, it is considered lost or to be treated by some extraordinary measures such as overtime and expediting in the approach.

We consider serial inventory systems with stochastic final demand and fixed costs at each stage. For such a system, its optimal inventory policy is not known, and previous studies only proposed heuristic approaches for finding a near-optimal in-

Peng Li
Laboratoire d'Optimisation des Système Industriels, Institut Charles Delaunay and UMR CNRS STMR 6279, Université de Technologie de Troyes, France, e-mail: peng.li@utt.fr

Haoxun Chen
Laboratoire d'Optimisation des Système Industriels, Institut Charles Delaunay and UMR CNRS STMR 6279, Université de Technologie de Troyes, France, e-mail: haoxun.chen@utt.fr

D. Klatte et al. (eds.), *Operations Research Proceedings 2011*, Operations Research Proceedings, 439
DOI 10.1007/978-3-642-29210-1_70, © Springer-Verlag Berlin Heidelberg 2012

ventory policy (Chen Fangruo, 1998). In this paper, we use an echelon (R, nQ) policy to control the inventory replenishment of each stock in the system. We use the guaranteed-service approach (GSA) to find an optimal echelon (R, nQ) policy for the system. After obtaining an analytical expression for the total cost of the system and decomposing its inventory policy optimization problem into two independent sub-problems, batch size decision sub-problem and reorder point decision sub-problem, we develop a dynamic programming algorithm and a branch and bound algorithm for the two sub-problems, respectively. Numerical experiments on randomly generated instances demonstrate that the two algorithms together are able to obtain optimal inventory policies of the system quickly.

2 Problem Description

Consider a serial inventory system with N (> 2) stages where stage N orders from an external supplier with unlimited stock, stage N-1 orders from stage N, stage N-2 orders from stage N-1, and so on. Finally, at the lowest stage, stage 1, customer demand occurs. The inventory replenishment of each stage is controlled by an echelon (R_i, nQ_i) policy, $i = 1, 2, \ldots, N$.

We assume the customer demand is a Poisson process with average demand rate λ. According to Graves and Willems(1996), the lead time demand over τ periods is bounded by $D(\tau)$ which can be described as follows:

$$D(\tau) \geq d[t - \tau, t)$$

where $d[t - \tau, t) = d_{t-\tau} + d_{t-\tau+1} + \ldots + d_{t-1}$, d_t is the customer demand in period t and $D(0) = 0$.

In this paper, we determine the bound $D(\tau)$ according to the expected service level to customer, denoted by α. Consider stage 1 at which only customer demand occurs, the lead time demand bound can be determined as the minimum number of $D(\tau)$ satisfying the following condition:

$$P\{d[t - \tau, t) \leq D(\tau)\} \geq \alpha$$

Then, for the Poisson demand, we have:

$$\sum_{k=0}^{D(\tau)} \frac{(\lambda \tau)^k e^{-\lambda \tau}}{k!} \geq \alpha$$

In GSA, each stage i is assumed to have an outbound service time S_i, a production time T_i, and an inbound service time $SI_i, i = 1, 2, \ldots, N$. The outbound service time S_i is the time required by stage i to satisfy the demand of its immediate downstream stage i-1. The inbound service time SI_i is the time required by stage i to receive the shipment of its order from its immediate upstream stage $i + 1$ after the placement of the order. The production time T_i is given, which includes processing time, waiting time and transportation time. The net lead time of stage i is $SI_i + T_i - S_i$.

For the system considered, we assume a setup cost C_i incurred at each stage i for each batch of Q_i units ordered. Since the average demand rate is λ, the average fixed cost per period at stage i in the long-run is thus $\frac{c_i \lambda}{Q_i}$.

3 Problem Formulation

Define $I_i(t), I_i^e(t), IL_i^e(t)$ and $IP_i^e(t)$ as on-hand inventory, echelon on-hand inventory, echelon inventory level and echelon inventory position after order decision and before demand occurrence in period t at stage i for $i=1,2,\ldots,N$. The objective of the serial system is to minimize the average total cost per period, which includes fixed costs and inventory costs and is given by

$$\sum_{i=1}^{N} \left(\frac{c_i \lambda}{Q_i} + h_i^e * E[I_i^e - d] \right) \tag{1}$$

where I_i^e is $I_i^e(t)$ at the steady state, d is a random variable which has the same stationary probability distribution of d_t. And h_i^e is the echelon holding cost per unit of product and per period at stage i.

The following inventory balance equations can be derived for each stage i:

$$IL_i^e(t) = IP_i^e(t - SI_i - T_i + S_i) - d[t - SI_i - T_i + S_i, t] \tag{2}$$

$$I_i(t) = I_i^e(t) - IP_{i-1}^e(t)$$

Under the GSA, $IL_i^e(t) = I_i^e(t) \geq 0$ and $I_i(t) \geq 0$, so we have:

$$IP_i^e(t - SI_i - T_i + S_i) \geq d[t - SI_i - T_i + S_i, t] + IP_{i-1}^e(t) \tag{3}$$

If we assume randomize initial conditions, $IP^e(t) = (IP_i^e(t), i \in N)$ is uniformly distributed in S^e, where $S^e = \{(s_1, s_2, \ldots, s_N) \mid s_i \in R_i + 1, R_i + 2, \ldots, R_i + Q_i, i \in N\}$, so in order to ensure that inequality (3) holds for any demand realization, we must have:

$$R_i + 1 \geq D(SI_i + T_i - S_i) + R_{i-1} + Q_{i-1}, \quad i = 1, 2, \ldots, N \text{ where } Q_0 = 0$$

Then, there exists an optimal solution such that $R_i, i = 1, 2, \ldots, N$ satisfy

$$R_i = \sum_{j=1}^{i} D(SI_j + T_j - S_j) + \sum_{j=0}^{i-1} Q_j - i, \quad i = 1, 2, \ldots, N \tag{4}$$

According to Hadley and Whitin (1961), the echelon inventory position $IP_i^e(t)$ is uniformly distributed in $\{R_i + 1, \ldots, R_i + Q_i\}$. So we can derive $E[I_i^e - d]$ as follows:

$$E[I_i^e - d] = \lim_{t \to +\infty} E[IP_i^e(t - SI_i - T_i + S_i) - d[t - SI_i - T_i + S_i, t]] - \lambda$$

$$= R_i + \frac{1+Q_i}{2} - \lambda(SI_i + T_i - S_i) - \lambda$$

Substitute (4) into the above equation of $E[I_i^e - d]$ for $i = 1,2,\ldots,N$, we can derive:

$$E[I_i^e - d] = \sum_{j=1}^{i} D(SI_j + T_j - S_j) - \lambda(SI_i + T_i - S_i) - \lambda + \sum_{j=0}^{i-1} Q_j + \frac{1+Q_i}{2} - i \quad (5)$$

Then, with equation (1) and (5), the optimization problem we study can be formulated as the following nonlinear programming problem:

P: Minimize

$$\sum_{i=1}^{N} \{ \frac{C_i\lambda}{Q_i} + h_i^e * [\sum_{j=1}^{i} D(SI_j + T_j - S_j) - \lambda(SI_i + T_i - S_i) - \lambda + \frac{1+Q_i}{2} - i] + \sum_{j=i}^{N} h_j^e * Q_{i-1} \}$$

s.t.

$$Q_i = m_{i-1}Q_{i-1}, \quad i = 1,2,\ldots,N \qquad (6)$$
$$SI_i + T_i - S_i \geq 0, \quad i = 1,2,\ldots,N \qquad (7)$$
$$SI_i \geq S_{i+1}, \quad i = 1,2,\ldots,N \qquad (8)$$
$$0 \leq S_1 \leq s_1 \qquad (9)$$

Constraint (6) is integer ratio constraints among the batch sizes. Constraint (7) assures that the net lead time at each stage is nonnegative. Constraint (8) ensures that the inbound service time of each stage is no less than the out bound service time of its upstream stage. Constraint (9) imposes a given upper bound on the outbound service time of stage1.

The model P can be divided into two independent sub-problems, batch size decision sub-problem (Q-problem) and reorder point decision sub-problem (R-problem). The Q-problem has a convex objective function composed of all Q-dependent cost terms and constraint (6), whereas the R-problem has a nonlinear objective function composed of all R-dependent cost terms and linear constraints(7),(8) and (9).

4 Dynamic Programming for Q-problem

A dynamic programming (DP) algorithm is used to solve Q-problem. The notations to be used in the DP algorithm are first defined:

i: stage index, $i = 0,1,2,\ldots,N+1$, stage 0 and stage $N+1$ are two additional stages representing the starting state and ending state of the DP algorithm, respectively.

Q_i: state variable of stage i.

m_{i-1}: decision variable of stage i, $i = 1,2,\ldots,N$.

$M_{i-1}(Q_{i-1})$: the set of permissible values of m_{i-1} given state Q_{i-1} of stage i-1.

$d_i(Q_{i-1},m_{i-1})$: the cost of stage i when its decision is m_{i-1} and the state of stage

i-1 is Q_{i-1}.

$f_i(Q_i)$: the minimal total cost from stage 0 to stage i when the state of stage i is Q_i.
The state transition functions and the recursion equations can be written as:

$$Q_i = m_{i-1}Q_{i-1}, \quad i = 1, 2, \dots, N$$

$$\begin{cases} d_i(Q_{i-1}, m_{i-1}) = \dfrac{C_i \lambda}{Q_i} + h_i^e * [\dfrac{1+Q_i}{2} - i] + \sum\limits_{j=i}^{N} h_j^e * Q_{i-1}, \ i = 1, 2, \dots, N \\ d_{N+1}(Q_N, m_N) = 0 \end{cases}$$

$$\begin{cases} f_i(Q_i) = \min\limits_{m_{i-1} \in M_{i-1}(Q_{i-1})} \{d_i(Q_{i-1}, m_{i-1}) + f_{i-1}(Q_{i-1})\}, \quad i = 1, 2, \dots, N \\ f_0(Q_0) = 0 \end{cases}$$

In order to apply above recursion equations to calculate $f_i(Q_i)$ for each stage, the
upper bound of Q_1 and m_i for $i = 1, 2, \dots, N-1$ can be specified as follows:

$$\overline{Q_1} = \sqrt{\dfrac{2\lambda \sum\limits_{i=1}^{N} (C_i)}{\sum\limits_{i=1}^{N} (2i-1)h_i^e}}, \quad \overline{m_i} = \dfrac{1}{Q_i} * \sqrt{\dfrac{2\lambda \sum\limits_{j=i+1}^{N} C_i}{\sum\limits_{j=1}^{N-i} (2j-1)*h_{i+j}^e}}, \quad i = 1, 2, \dots, N-1$$

5 Branch and Bound Algorithm for *R*-problem

Lower bound

Numerical experiments show that the cost function of the R-problem is approximately linear. For this reason, we use a linear function $\widehat{D}_i = a_i(SI_i + T_i - S_i) + b_i$ to
approximate from below the nonlinear function $D_i = D(SI_i + T_i - S_i)$ for each stage
i. Then the following constrained least square error problem can be derived.

$$Minimize \sum_{j=1}^{M} (D_{ij} - \widehat{D}_{ij})^2$$

s.t.

$$a_i((SI_i + T_i - S_i)_j) + b_i \le D_{ij} \ \ i = 1, 2, \dots, N, j = 1, 2, \dots, M$$

where M is the number of all integer values in the domain of variable $SI_i + T_i - S_i$,
D_{ij} is the value of function D_i and $\widehat{D}_{ij}$ is the value of function $\widehat{D}_i$ at the j-th integer
value $(SI_i + T_i - S_i)_j$ of the variable.

Then, we can obtain a linear programming problem RL which is derived from
the R-problem by replacing its cost function D_i with $\widehat{D}_i$. The optimal value of RL
provides a lower bound for the R-problem.

Upper Bound

By solving the problem *RL*, we can also obtain a feasible solution for the original *R*-problem. This feasible solution provides an upper bound for the *R*-problem.

The Branching scheme

We choose the best-first-search as the branching scheme. Suppose that initially the domain of variable $SI_i + T_i - S_i$ at the root node of the branch and bound tree is given by $[0, K_i], i = 1, 2, \ldots, N$. For each node (sub-problem) being examined with $[K_{i0}, K_{i1}]$ as the domain of variable $SI_i + T_i - S_i$, the B&B algorithm solves its corresponding problem *RL*. The algorithm then chooses a stage i and create two son nodes (also sub-problems). The sub-problems associated with the two son nodes are the same as that of their parent node except that the domain of variable $SI_i + T_i - S_i$ becomes $[K_{i0}, K_i']$ and $[K_i', K_{i1}]$, respectively, where $K_i', i = 1, 2, \ldots, N$ is the value of the variable at the optimal solution of the problem *RL*.

6 Experiment Results

The performances of the proposed algorithms are evaluated by randomly generated instances for serial systems with 2, 3 and 4 stages, respectively, and with different parameter settings. Totally, 22113 instances are tested for *Q*-problem and 7600 instances are tested for *R*-problem. For the *Q*-problem, the average computation time is no more than 0.1 seconds. For the *R*-problem, the upper bound obtained at the root node of the branch and bound tree is equal to its optimal value for 90.83% instances. This implies that the optimal solution of the problem can be found mostly at the root node. For the remaining 9.17% instances, the gap between upper bound at the root node and the optimal value is also small with the mean gap no more than 14.4% for each set of instances. The average computation time of the B&B algorithm is less than 1 seconds.

References

1. Chen, F., Y. -S. Zheng.:Near-optimal echelon-stock (R,nq) policiers in multistage serial systems. Operations Research. 46, 592-602.(1998)
2. Hadley, G., T. M. Whitin.:A family of inventory model. Management Science, 7, 351-371.(1961)
3. Graves, S. C., Willems, S. P.: Strategic safety stock placement in supply chain. Proceedings of the 1996 MSOM Conference, Hanover, NH.(1996)

Channel coordination in a HMMS-type supply chain with profit sharing contract

Imre Dobos, Barbara Gobsch, Nadezhda Pakhomova, Grigory Pishchulov and Knut Richter

Abstract We investigate a supply chain with a single supplier and a single manufacturer. The manufacturer is supposed to know the demand for the final product which is produced from a raw material ordered from the supplier just in time — i.e., the manufacturer holds no raw material inventory. Her costs consist of the linear purchasing cost, quadratic production cost and the final product quadratic holding costs. It is assumed that the market price of the final product is known as well, hence the sales of the manufacturer are known in advance. Her goal is to maximize her cumulated profits. The supplier's costs are the quadratic manufacturing and inventory holding costs; his goal is to maximize the revenues minus the relevant costs. We will not examine the bargaining process that determines the adequate price and quantity. The situation is modeled as a differential game. The decision variables of the supplier are the sales price and the production quantity, while the manufacturer chooses a production plan that minimizes her costs, so maximizing the cumulated profits. The basic problem is a Holt-Modigliani-Muth-Simon (HMMS) problem extended to linear purchasing costs. We examine two cases: the decentralized Nash-solution and a centralized Pareto-solution to optimize the behavior of the players of the game.

Imre Dobos
Corvinus University of Budapest, Institute of Business Economics, H-1093 Budapest, Fövám tér 8., Hungary, e-mail: imre.dobos@uni-corvinus.hu
Imre Dobos gratefully acknowledges the financial supports by the TÁMOP-4.2.1.B-09/KMR-2010-0005 research program and the Deutscher Akademischer Austauschdienst (DAAD).

Barbara Gobsch
European University Viadrina, Faculty of Business Administration and Economics, D-15230 Frankfurt (Oder), Germany, e-mail: gobsch@europa-uni.de

Nadezhda Pakhomova
St. Petersburg State University, Faculty of Economics, Russia, e-mail: pakhomova@europa-uni.de

Grigory Pishchulov
European University Viadrina, Frankfurt (Oder), Germany, e-mail: pishchulov@europa-uni.de

Knut Richter
European University Viadrina, Faculty of Business Administration and Economics, Germany &
St. Petersburg State University, Faculty of Economics, Russia, e-mail: richter@europa-uni.de

D. Klatte et al. (eds.), *Operations Research Proceedings 2011*, Operations Research Proceedings,   445
DOI 10.1007/978-3-642-29210-1_71, © Springer-Verlag Berlin Heidelberg 2012

1 Introduction

Most of inter-organizational supply chains are decentralized, that is they consist of multiple, legally independent parties having different information, objectives and power [1, 7]. To avoid extra costs of the participants in such supply chains in contrast to a centralized supply chain with system-wide optimization and to reduce the well-known bullwhip effect, supply chain coordination is necessary. There are a number of tools which lead to coordination – the most important one is the contract [2, 13]. Contracts are said to coordinate a supply chain if a Nash equilibrium is reached and "no firm has a profitable unilateral deviation from the set of supply chain optimal actions" [2] respectively with weaker requirements, if they improve the performance of the chain compared to the default solution without coordination characterized by double marginalization [1, 12]. There is a large body of literature on different kinds of contracts, in particular wholesale price contracts, buy-back contracts, revenue-sharing contracts, in a one-period newsvendor framework or multi-period EOQ framework with deterministic data. Reviews are presented in [1, 2, 9, 13], among others. In this paper we examine the profit sharing contract in a multi-period dynamic environment with continuous time. The literature on problems of this type is rather limited. Based on the basic problem by Holt et al. [8], Dobos analyzed a reverse logistics problem in a Holt-Modigliani-Muth-Simon (HMMS-) environment [3] and examined the bullwhip effect in a HMMS-supply chain [4], among others.

We investigate a two-stage supply chain with a single supplier, single manufacturer, and symmetric information. The manufacturer (she) knows the market price and the demand for the final product, which is produced with a raw material purchased from the supplier (him), and wants to maximize her cumulated profit over the planning horizon while taking into account the target inventory level and production rate, so that holding, production and purchasing costs are decision relevant. The supplier's objective is similar, while the purchasing costs of the manufacturer are his revenues. The parties seek to determine the purchase price of the raw material, the fraction of profit shared and the quantity to be ordered by the manufacturer. The situation is modeled (Section 2) as a differential game and represents a HMMS-type model extended to linear purchasing costs. We will examine two cases: the decentralized Nash solution (Section 2) and the centralized Pareto solution (Section 3).

2 The decentralized system: The Nash solution

We consider a supply chain consisting of a supplier and a manufacturer acting independently of each other to minimize their own costs. The final product is produced with a raw material purchased by the manufacturer from the supplier. Because of a negligibly low fixed ordering cost the manufacturer can realize a just in time procurement of the raw material and holds no raw material stock. The production processes have a known, constant lead time. The market price and the demand for the final product are deterministic, too. The following parameters are used in the model:

T length of the planning horizon ($t \in [0,T]$)
$S(t)$ demand rate, continuous differentiable
$\overline{I}_m(t), \overline{I}_s(t)$ inventory target level for the final product and the raw material, resp.
$\overline{P}_m(t), \overline{P}_s(t)$ manufacturer's and supplier's production target level, resp.
h_m, h_s inventory holding cost coefficients for the product & the raw material
c_m, c_s $2\times$ the production cost coefficient of the manufacturer & the supplier
p market price of the final product

In the HMMS-model it is assumed that the managers of both firms have fixed a production-inventory pattern, i.e., the production plans $\overline{P}_m(t), \overline{P}_s(t)$ and planned inventory levels $\overline{I}_m(t), \overline{I}_s(t)$ known at the beginning of the planning horizon. The managers' objective is to minimize the deviations from the target levels, which are defined as quadratic functionals with known parameters. This phenomenon was empirically tested in [8]. The decision variables are then (all non-negative, $t \in [0,T]$):

$I_m(t), I_s(t)$ inventory level of the final product and the raw material, resp.
$P_m(t), P_s(t)$ production rate of the manufacturer and the supplier, resp.
w wholesale price for the raw material
α fraction of the supply chain profit the manufacturer keeps

The decentralized model describes the situation where the supplier and the manufacturer act independently, i.e. the manufacturer determines her individual optimal production-inventory strategy first and orders the necessary quantity of the raw material. Then the supplier accepts the order and minimizes his own costs. The transfer payment of the manufacturer to the supplier is

$$\int_0^T w P_m(t)dt + (1-\alpha)\left[\int_0^T pS(t)dt - \int_0^T \left\{\tfrac{h_m}{2}[I_m(t)-\overline{I}_m(t)]^2 + \tfrac{c_m}{2}[P_m(t)-\overline{P}_m(t)]^2\right\}dt\right],$$

i.e., the manufacturer pays the supplier for the supplied items and a fraction of the profit she earns [10]. For the given purchase price w^N she then solves the problem:

$$J_m\left(P_m(\cdot); w^N\right) = \alpha \cdot \left[\int_0^T pS(t)dt - \right.$$
$$\left. \int_0^T \left\{\tfrac{h_m}{2}\left[I_m(t)-\overline{I}_m(t)\right]^2 + \tfrac{c_m}{2}\left[P_m(t)-\overline{P}_m(t)\right]^2\right\}dt\right] - \int_0^T w^N P_m(t)dt \to \max \tag{1}$$

$$\text{s.t.} \qquad \dot{I}_m(t) = P_m(t) - S(t), \quad I_m(0) = I_{m0}, \quad 0 \le t \le T. \tag{2}$$

Functional (1) maximizes the manufacturer's cumulated profit over the planning horizon. The problem can obviously be reformulated as a cost minimization one:

$$J'_m\left(P_m(\cdot); w^N\right) = \alpha \cdot \int_0^T \left\{\tfrac{h_m}{2}\left[I_m(t)-\overline{I}_m(t)\right]^2 + \tfrac{c_m}{2}\left[P_m(t)-\overline{P}_m(t)\right]^2\right\}dt +$$
$$+ \int_0^T w^N \cdot P_m(t)dt \to \min. \tag{1'}$$

Assuming that the optimal production-inventory policy of the manufacturer for the given purchase price w is $\left(I_m^N(\cdot), P_m^N(\cdot)\right)$ in model (1')–(2) and she orders $P_m^N(\cdot)$, the supplier solves the following problem:

$$J_s\left(P_s(\cdot), w; P_m^N(\cdot)\right) = w \int_0^T P_m^N(t)\, dt - \int_0^T \left\{ \tfrac{h_s}{2}[I_s(t) - \bar{I}_s(t)]^2 + \tfrac{c_s}{2}[P_s(t) - \bar{P}_s(t)]^2 \right\} dt$$

$$+ (1 - \alpha)\left[\int_0^T pS(t)\, dt - \int_0^T \left\{ \tfrac{h_m}{2}[I_m(t) - \bar{I}_m(t)]^2 + \tfrac{c_m}{2}[P_m(t) - \bar{P}_m(t)]^2 \right\} dt \right] \to \min \tag{3}$$

$$\text{s.t.} \qquad \dot{I}_s(t) = P_s(t) - P_m^N(t), \quad I_s(0) = I_{s0}, \quad 0 \le t \le T \tag{4}$$

We apply the Pontryagin's Maximum Principle [6, 11] to solve (1')–(2) which is an optimal control problem with pure state variable constraints. The optimal production plan of the manufacturer can be obtained as follows:

Lemma 1. *Assume that production–inventory strategy $\left(I_m^N(\cdot), P_m^N(\cdot)\right)$ is an optimal solution to model (1')–(2). Then it must satisfy the following differential equation:*

$$\begin{pmatrix} \dot{I}_m^N(t) \\ \dot{P}_m^N(t) \end{pmatrix} = \begin{pmatrix} 0 & 1 \\ \frac{h_m}{c_m} & 0 \end{pmatrix} \cdot \begin{pmatrix} I_m^N(t) \\ P_m^N(t) \end{pmatrix} + \begin{pmatrix} -S(t) \\ \dot{\bar{P}}_m(t) - \frac{h_m}{c_m} \cdot \bar{I}_m(t) \end{pmatrix},$$

with the initial and terminal condition: $\quad I_m^N(0) = I_{m0}, \quad P_m^N(T) = \bar{P}_m(T) - \frac{w^N}{\alpha \cdot c_m}.$

The proof of this lemma can be found in [3].

After optimal production strategy $P_m^N(\cdot)$ is given we can solve (3)–(4) which is an optimal control problem with pure state variable constraints, too.

Value $\bar{w}$ denotes the upper bound for the wholesale price. The manufacturer will only order from the supplier if the cumulated profit is nonnegative in the planning horizon in dependence of the purchase price, i.e.

$$\int_0^T pS(t)\, dt - \int_0^T \left\{ \tfrac{h_m}{2}\left[I_m^N(t) - \bar{I}_m(t)\right]^2 + \tfrac{c_m}{2}\left[P_m^N(t) - \bar{P}_m(t)\right]^2 + w^N P_m(t) \right\} dt \ge 0.$$

By successively increasing the value of the purchase price w^N and re-solving (1)–(2), the upper bound $\bar{w}$ at which the manufacturer's optimal cumulated profit turns to zero can accordingly be determined.

For the case of a positive inventory level and production rate there holds

Lemma 2. *Assume that production-inventory strategy $\left(I_s^N(\cdot), P_s^N(\cdot)\right)$ is an optimal solution to (3)–(4). Then it must satisfy the following differential equation:*

$$\begin{pmatrix} \dot{I}_s^N(t) \\ \dot{P}_s^N(t) \end{pmatrix} = \begin{pmatrix} 0 & 1 \\ \frac{h_s}{c_s} & 0 \end{pmatrix} \cdot \begin{pmatrix} I_s^N(t) \\ P_s^N(t) \end{pmatrix} + \begin{pmatrix} -P_m^N(t) \\ \dot{\bar{P}}_s(t) - \frac{h_s}{c_s} \cdot \bar{I}_s(t) \end{pmatrix},$$

with the initial and terminal conditions: $\quad I_s^N(0) = I_{s0}, \quad P_s^N(T) = \bar{P}_s(T).$

Let J_m^N and J_s^N be the optimal values of the objective functions (1') and (3) (here the latter is expressed as a loss function) respectively, that is, let

$$J_m^N = \int_0^T \left\{ \frac{h_m}{2} \left[I_m^N(t) - \bar{I}_m(t) \right]^2 + \frac{c_m}{2} \left[P_m^N(t) - \bar{P}_m(t) \right]^2 + w^N \cdot P_m^N(t) \right\} dt,$$

$$J_s^N = \int_0^T \left\{ \frac{h_s}{2} \left[I_s^N(t) - \bar{I}_s(t) \right]^2 + \frac{c_s}{2} \left[P_s^N(t) - \bar{P}_s(t) \right]^2 - w^N \cdot P_m^N(t) \right\} dt.$$

The solutions of the two models are the well-known Nash solution of this game model. This relation between them can be written as:

$$J_m^N = J_m \left(P_m^N(\cdot); w^N \right) \le J_m \left(P_m(\cdot); w^N \right), J_s^N = J_s \left(P_s^N(\cdot), w^N; P_s^N(\cdot) \right) \le J_s \left(P_s(\cdot), w; P_s^N(\cdot) \right).$$

As the Nash equilibrium has the property that the profit of the manufacturer is zero, all of the profits are realized at the supplier in our decentralized model.

3 The centralized system: The Pareto optimal solution

In this section we investigate the coordinated solution of the examined model. We assume that the supplier and the manufacturer act as an integrated firm and jointly minimize the system-wide costs. As the purchase price is an internal accounting tool of the gains, the cost functional does not depend on it. The model is then as follows:

$$J_{ms}\left(P_m(\cdot), P_s(\cdot)\right) = \int_0^T \left\{ \frac{h_m}{2} \left[I_m(t) - \bar{I}_m(t) \right]^2 + \frac{c_m}{2} \left[P_m(t) - \bar{P}_m(t) \right]^2 + \right.$$
$$\left. + \frac{h_s}{2} \left[I_s(t) - \bar{I}_s(t) \right]^2 + \frac{c_s}{2} \left[P_s(t) - \bar{P}_s(t) \right]^2 \right\} dt \to \min \quad (5)$$

$$\text{s.t. } \dot{I}_m(t) = P_m(t) - S(t), \; \dot{I}_s(t) = P_s(t) - P_m(t), \; 0 \le t \le T, \; \begin{pmatrix} I_m(0) \\ I_s(0) \end{pmatrix} = \begin{pmatrix} I_{m0} \\ I_{s0} \end{pmatrix} \quad (6)$$

The necessary and sufficient conditions become a system of linear differential equations:

$$\begin{pmatrix} \dot{I}_m^P(t) \\ \dot{I}_s^P(t) \\ \dot{P}_m^P(t) \\ \dot{P}_s^P(t) \end{pmatrix} = \begin{pmatrix} 0 & 0 & 1 & 0 \\ 0 & 0 & -1 & 1 \\ \frac{h_m}{c_m} & -\frac{h_s}{c_m} & 0 & 0 \\ 0 & \frac{h_s}{c_s} & 0 & 0 \end{pmatrix} \cdot \begin{pmatrix} I_m^P(t) \\ I_s^P(t) \\ P_m^P(t) \\ P_s^P(t) \end{pmatrix} + \begin{pmatrix} -S(t) \\ 0 \\ \dot{\bar{P}}_m(t) - \frac{h_m}{c_m}\bar{I}_m(t) + \frac{h_s}{c_m}\bar{I}_s(t) \\ \dot{\bar{P}}_s(t) - \frac{h_s}{c_s}\bar{I}_s(t) \end{pmatrix} \quad (7)$$

with the initial and terminal conditions

$$\begin{pmatrix} I_m^P(0) \\ I_s^P(0) \end{pmatrix} = \begin{pmatrix} I_{m0} \\ I_{s0} \end{pmatrix} \quad \text{and} \quad \begin{pmatrix} P_m^P(T) \\ P_s^P(T) \end{pmatrix} = \begin{pmatrix} \bar{P}_m(T) \\ \bar{P}_s(T) \end{pmatrix}.$$

Finally, let $J_{ms}^P = J_{ms}\left(P_m^P(.), P_s^P(.)\right)$ denote the optimal value of cost function (5). It is easy to see that $J_{ms}^P \le J_m^N + J_s^N$, i.e. the Pareto solution of the problem has lower costs and a higher profit compared to the Nash solution of Section 2.

4 Conclusion and further research

In this paper we have solved a two-stage HMMS-type supply chain model in a decentralized and a centralized setting. By comparing the Nash solution of the decentralized system with the Pareto solution of the centralized system, we have shown that a cooperation of the two players induces savings in costs.

The presented study may be a starting point for studying a number of further research topics, e.g.: 1) analyzing a supply chain with different bargaining power of the manufacturer and the supplier, 2) analyzing the performance of other contracts in the given multi-period deterministic environment with continuous time, and 3) extending the model to a system with reverse logistics as recently approached in [4].

References

1. Albrecht, M.: Supply chain coordination mechanisms: New approaches for collaborative planning. Springer, Berlin et al. (2010)
2. Cachon, G.P.: Supply chain coordination with contracts. In: de Kok, A.G., Graves, S.C. (eds.) Supply Chain Management: Design, Coordination and Operation, pp. 229–339. Elsevier (2003)
3. Dobos, I.: Optimal production-inventory strategies for a HMMS-type reverse logistics system. International Journal of Production Economics **81–82**, 351–360 (2003)
4. Dobos, I.: The analysis of bullwhip effect in a HMMS-type supply chain. International Journal of Production Economics **131**, 250–256 (2011)
5. Dobos, I., Gobsch, B., Pakhomova, N., Richter, K.: Remanufacturing of used products in a closed-loop supply chain. In: Csutora, M., Kerekes, S. (eds.) Accounting for climate change – What and how to measure, pp. 130–146. Proceedings of the EMAN-EU 2011 Conference, 24–25 January 2011, Budapest, Hungary (2011)
6. Feichtinger, G., Hartl, R.F.: Optimale Kontrolle ökonomischer Prozesse: Anwendungen des Maximumprinzips in den Wirtschaftswissenschaften. De Gruyter, Berlin (1986)
7. Höhn, M.I.: Relational supply contracts: Optimal Concessions in Return Policies for Continuous Quality Improvements. Springer, Berlin et al. (2010)
8. Holt, C.C., Modigliani, F., Muth, J.F., Simon, H.A.: Planning Production, Inventories, and Work Forces. Prentice-Hall, Englewood Cliffs, N.J. (1960)
9. Lariviere, M.A.: Supply chain contracting and coordination with stochastic demand. In: Tayur, S., Ganeshan, R., Magazine, M. (eds.) Quantitative Models in Supply Chain Management, pp. 233–268. Kluwer Academic Publishers, Boston (1999)
10. Leng, M., Zhu, A.: Side-payment contracts in two-person non-zero supply chain games: Review, discussion and applications. European J. of Operational Research **196**, 600–618 (2009)
11. Seierstad, A., Sydsaeter, K.: Optimal control theory with economic applications. North-Holland, Amsterdam (1987)
12. Spengler, J.J.: Vertical integration and antitrust policy. Journal of Political Economy **58**, 347-352 (1950)
13. Tsay, A.A., Nahmias, S., Agrawal, N.: Modeling Supply Chain Contracts: A Review. In: Tayur, S., Ganeshan, R., Magazine, M. (eds.) Quantitative Models in Supply Chain Management, pp. 299–336. Kluwer Academic Publishers, Boston (1999)

Enhancing Aggregate Production Planning with an Integrated Stochastic Queuing Model

Gerd J. Hahn, Chris Kaiser, Heinrich Kuhn, Lien Perdu and Nico J. Vandaele

Abstract Mathematical models for Aggregate Production Planning (APP) typically omit the dynamics of the underlying production system due to variable workload levels since they assume fixed capacity buffers and predetermined lead times. Pertinent approaches to overcome these drawbacks are either restrictive in their modeling capabilities or prohibitive in their computational effort. In this paper, we introduce an Aggregate Stochastic Queuing (ASQ) model to anticipate capacity buffers and lead time offsets for each time bucket of the APP model. The ASQ model allows for flexible modeling of the underlying production system and the corresponding optimization algorithm is computationally very well tractable. The APP and the ASQ model are integrated into a hierarchical framework and are solved iteratively. A numerical example is used to highlight the benefits of this novel approach.

1 Introduction

Aggregate Production Planning (APP) covers the mid-term level of rough-cut capacity planning to determine the master production schedule [2]. Mathematical programming models for APP are widely applied deriving cost-optimal overtime and inventory levels to serve customer demand for a given planning period [7]. However, APP models typically make two simplifying assumptions omitting the dynamics of the underlying production system due to variable workload levels in a complex job shop environment [1]. First, fixed capacity buffers are applied to accommodate for capacity losses due to setups and machine breakdowns [9]. Second, APP models use predetermined lead times [2] that are multiples of the time bucket length.

Gerd J. Hahn, Chris Kaiser, and Heinrich Kuhn
Catholic University of Eichstaett-Ingolstadt, Auf der Schanz 49, 85049 Ingolstadt/Germany, e-mail: gerd.hahn@kuei.de, chris.kaiser@kuei.de, heinrich.kuhn@kuei.de

Lien Perdu and Nico J. Vandaele
Catholic University of Leuven, Naamsestraat 69, 3000 Leuven/Belgium, e-mail: lien.perdu@kuleuven-kortrijk.be, nico.vandaele@econ.kuleuven.be

D. Klatte et al. (eds.), *Operations Research Proceedings 2011*, Operations Research Proceedings,   451
DOI 10.1007/978-3-642-29210-1_72, © Springer-Verlag Berlin Heidelberg 2012

Two major research streams can be distinguished in the literature providing different approaches to overcome the aforementioned drawbacks. In the first stream, clearing functions typically based on queuing theory are integrated into APP models to anticipate workload-dependent lead times [1]. However, clearing functions impose restrictive assumptions on the queuing model to obtain a closed-form expression that can be implemented in mathematical programming models [8]. Consequently, a comprehensive clearing function approach to approximate a complex multi-product multi-machine job shop environment with significant setups and general distributions for the stochastic parameters has not yet been presented.

Iterative approaches constitute the second research stream applying less restrictive analytical and descriptive models within the framework of APP to better approximate the underlying production system [4]. However, the analytical approaches are still quite limited in flexibility and simulation-based approaches typically result in prohibitive computing times provided that an optimization is performed based on the results of different simulation runs [3]. We therefore propose an iterative framework for APP integrating an Aggregate Stochastic Queuing (ASQ) model that is considerably less restrictive compared to pertinent analytical approaches. Nonetheless, the ASQ model can be solved optimally and is computationally very well tractable [6].

The remainder of the paper is structured as follows: the decision framework with the conceptual approach and the modified APP model is outlined in section 2. Section 3 highlights the benefits of the integrated approach presented in this paper using a case-oriented numerical example. We conclude the paper in section 4 with a summary of the findings and an outlook for further research.

2 Decision Framework

In a two-level hierarchical framework for production planning and scheduling, the top level typically includes an anticipated base level model to approximate detailed decisions that influence parameters of the top level [9]. We use a classical APP model in our approach for rough-cut planning at the top level and integrate an ASQ model as the anticipated base level to determine capacity supply and demand in the production system more accurately. The bucketized APP model derives a rough-cut capacity plan with respect to overtime and seasonal inventory levels [7]. The continuous ASQ model for optimal order batching minimizes lead times in a multi-machine multi-product job shop environment with stochastic interarrival times, machine breakdowns, and setup and processing times [6].

The hierarchical planning framework pursues an iterative solution approach in three steps. First, the APP model provides total capacity for each time bucket and aggregate production quantities being converted into average interarrival times of the orders for the ASQ model. Second, the ASQ model is solved separately for each time bucket of the APP model and determines average lead times per product and capacity buffer per work center due to setups and machine breakdowns. The results of the ASQ model are converted into lead time offsets to allocate capacity demand

and capacity supply buffers per work center in the APP model. Third, the updated parameters are fed back into the APP model and a new iteration is conducted.

Only the APP model is presented in the following in order not to go beyond the scope of this paper. A classical LP formulation for APP is applied according to [2]. The objective function in (1) minimizes total cost of overtime O and inventory I at a unit cost of overtime co and inventory ci for product p. (2) represents the mass balance equation taking into account production quantity X of product p in period t to fulfill demand d of product p in period t. P denotes the set of end products p.

$$\min \quad \sum_{t=1}^{T}(co \cdot O_t) + \sum_{p \in P}\sum_{t=1}^{T}(ci_p \cdot I_{pt}) \tag{1}$$

$$I_{pt-1} + X_{pt} - I_{pt} = d_{pt} \quad \forall p \in P; t = 1 \ldots T \tag{2}$$

The capacity restriction in (3) is modified in two ways compared to a classical formulation. First, capacity demand on the LHS of the equation is allocated to the respective work center and period according to the work plan and the lead time offset. Lead time offset lt for work center w and product p represents the fraction of the capacity need kx for work center w that is allocated to offset period t for the finishing period z. Lead time offset coefficients are derived from the processing and queuing times in the ASQ model using backward scheduling from the finishing period z in the APP model. Since the ASQ model minimizes lead times, WIP is also minimized according to Little's law as throughput remains constant [5]. Consequently, WIP does not need to be considered separately in the APP model. X denotes the production quantity of product p in period z; W denotes the set of work centers.

$$\sum_{p \in P}\sum_{z=t}^{T} lt_{wpzt} \cdot kx_{wp} \cdot X_{pz} \leq (1 - \alpha_{wt}) \cdot nm_w \cdot (capX + O_t) \quad \forall w \in W; t = 1 \ldots T \tag{3}$$

Second, a capacity buffer α for work center w and period t is introduced on the RHS of (3) to account for setups and machine breakdowns. Production capacity per work center is derived from the number of parallel machines nm and available working time $capX$ that can be extend using overtime O in period t. (4) ensures that production quantity X for product p in period t does not exceed a maximum production quantity $X^{\max}$ with corresponding preproduction levels determined in an earlier planning iteration of the rolling horizons approach.

$$X_{pt} \leq X_{pt}^{\max} \quad \forall p \in P; t = 1 \ldots T \tag{4}$$

A maximum level of overtime $O^{\max}$ must be complied with in (5), and (6) determines initial and target inventories (I^0, I^T) for product p. All decision variables are restricted to the non-negative domain.

$$O_t \leq O^{max} \quad t = 1 \ldots T \tag{5}$$

$$I_{p0} = I_p^0, \quad I_{pT} = I_p^T \quad \forall p \in P \tag{6}$$

A standard LP solver is applied to the APP model. We use a steepest descent algorithm [6] to optimally solve the ASQ model given a minimum and maximum batch size. This interval is gradually reduced in each iteration to ensure convergence.

3 A Case-oriented Numerical Example

We investigate a numerical example to highlight the benefits of the approach presented in this paper. A job shop with 4 work centers each consisting of 11 machines is considered that produces 3 different product types: nuts, bolts, and screws. The planning horizon covers a full seasonal cycle of 13 standard months (each equal to 4 weeks). Available working hours are 320 hrs per month, and can be extended up to 480 hrs using overtime. Cost of overtime is 4,500 monetary units (mu) per hour. Storage cost, the base level of customer demand, and initial and target inventories per product are provided in Table 1.

	Nuts	**Bolts**	**Screws**
Storage cost [mu/bu]	5.5	7.2	8.4
Base demand [bu]	10,000	5,000	15,000
Inventory [bu]	3,000	1,500	4,500

Table 1 Storage cost, base level of customer demand, inventories

Demand seasonality is modeled using a harmonic oscillation around the base level of customer demand with an amplitude of 35% and a peak in period 7. Initial lead time offsets correspond to 0, i.e., that all preproduction steps are scheduled in the period before the finishing production step is performed. Initial capacity buffer α for all work centers and periods corresponds to 30%. Figure 1 depicts product routings for each product, work center availability, average setup times ($\overline{st}$) and average processing times ($\overline{kx}$) per product bundle unit (bu). Furthermore, we assume coefficients of variation for setup and processing times of 0.2 and 0.35.

The iterative solution approach is applied as described in section 2 until the objective function value of the APP model changes within a range of 0.1%. The data repository and the control flow to manage the solution procedure including data preparation and conversion are implemented in Microsoft Access and Visual Basic for Applications (VBA). The APP model consists of 94 continuous decision variables and 149 contraints, and is solved using IBM ILOG OPL with CPLEX v12.1.0. The steepest descent algorithm for the ASQ model is implemented in C++. For the given numerical example, the iterative solution procedure terminates after 20 iterations with a total computation time of 25 seconds on a computer with an Intel I3 Dual Core with 2.53 GHz per core and 4 GB RAM.

The results of our integrated framework are compared to a conventional approach that corresponds to iteration 0 of the integrated framework. Total costs are reduced

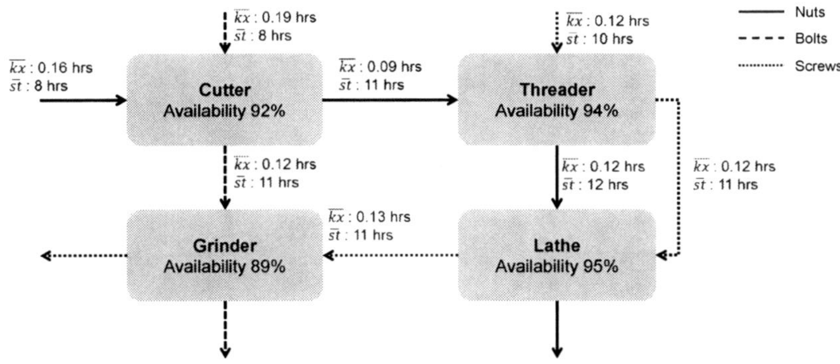

Fig. 1 Product routings, average setup and processing times

by 33% in the integrated approach since average capacity utilization can be improved substantially. Capacity buffers are adjusted according to the workload in the production system, and seasonal inventories are used to a greater extent for production leveling to avoid overly long lead times. In general, the integrated approach leads to better results since suboptimal or infeasible solutions as the consequence of pessimistic or optimistic assumptions in a conventional approach are prevented.

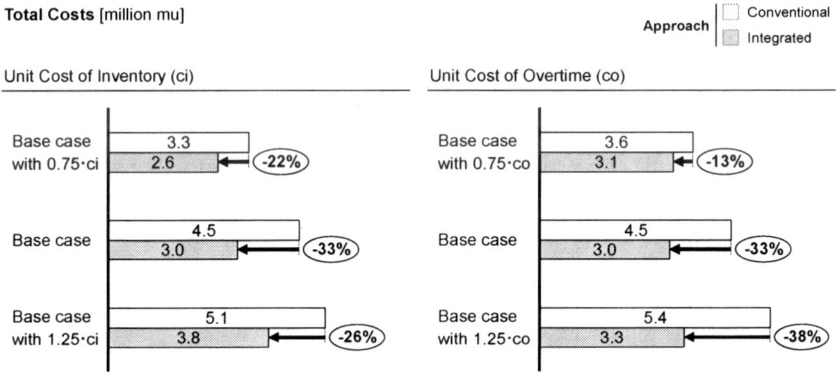

Fig. 2 Results of the sensitivity analyses

Further analyzing the results shows that the product mix is adjusted in each period of the integrated approach to account for changing workload levels per work center and their implications on lead times and capacity utilization. Shifting bottlenecks in a job shop can thus be anticipated more accurately in rough-cut capacity planning. Sensitivity analyses for the unit cost of inventory and overtime confirm the improvement potential of the integrated approach within a range of 22–33% and

13–38%, respectively (see Figure 2). In summary, total costs react more sensitive to changes in the unit cost of overtime since the unit cost of overtime are relatively higher and overtime costs account for approx. 68–74% of total costs.

4 Conclusion and Outlook

This paper presents an integrated decision framework for Aggregate Production Planning (APP) using an Aggregate Stochastic Queuing (ASQ) model to approximate capacity supply and demand in the production system more accurately. In contrast to pertinent analytical and simulation-based approaches, the ASQ model allows for flexible modeling and optimization of the underlying production system. Although the integrated framework pursues an iterative approach, the computational effort is very well tractable. A case-oriented numerical example highlights the improvement potential of this approach compared to conventional APP.

With respect to future research, three major directions should be pursued from our point of view. First, further numerical investigations should be conducted and the framework should be applied to a large-scale industry case study. Second, the scope of the APP model should be extended towards sales and operations planning (S&OP) including backlogs and lost sales to analyze the interdependencies between a profitable sales product mix and reliable order lead times for due date quoting. Furthermore, robust optimization methods could be introduced to analyze the impact of data uncertainty. Third, adapting the framework to the supply chain context would allow investigation into load balancing within a production network.

References

1. Asmundsson, J., Rardin, R.L., Turkseven, C.H., Uzsoy, R.: Production planning with resources subject to congestion. Nav. Res. Logist. **56**, 142–157 (2009)
2. Hopp, W.J., Spearman, M.L.: Factory physics. McGraw-Hill/Irwin, New York (2008)
3. Hung, Y.-F., Hou, M.-C.: A production planning approach based on iterations of linear programming optimization and flow time prediction. J. Chin. Inst. Ind. Eng. **18**, 55–67 (2001)
4. Jansen, M.M., de Kok, T.G., Fransoo, J.C.: Lead time anticipation in supply chain operations planning. OR Spectrum (2011) doi: 10.1007/s00291-011-0267-y
5. Karmarkar, U.S.: Lot sizes, lead times and in-process inventories. Manag. Sci. **33**, 409–418 (1987)
6. Lambrecht, M.R, Ivens, P.L., Vandaele, N.J.: ACLIPS – a capacity and lead time integrated procedure for scheduling. Manag. Sci. **44**, 1548–1561 (1998)
7. Nam, S.-J., Logendran, R.: Aggregate production planning – a survey of models and methodologies. Eur. J. Oper. Res. **61**, 255–272 (1992)
8. Pahl, J., Voß, S., Woodruff, D.L.: Production planning with load dependent lead times – an update of research. Ann. Oper. Res. **153**, 297–345 (2007)
9. Söhner, V., Schneeweiss, C.: Hierarchically integrated lot size optimization. Eur. J. Oper. Res. **86**, 73–90 (1995)

Remanufacturing of used products in a closed-loop supply chain with quantity discount

Grigory Pishchulov, Imre Dobos, Barbara Gobsch, Nadezhda Pakhomova and Knut Richter

Abstract An extended joint economic lot size problem is studied which incorporates the return flow of repairable (remanufacturable) used products. The supply chain under consideration consists of a single supplier and a single buyer. The buyer orders a single product from the supplier, uses it for her own needs, and collects a certain fraction of used items for the subsequent remanufacturing. The product is shipped to the buyer in the lot-for-lot fashion by a vehicle which also returns the collected used items to the supplier for remanufacturing and subsequent service of the buyer's demand in the next order cycle. To satisfy the buyer's demand, the supplier can remanufacture used items as well as manufacture new ones. For the given demand, productivity, disposal, setup, ordering and holding costs for serviceable and nonserviceable items at the supplier and the buyer, the optimal lot sizes and collection rates are determined which minimize the joint as well as individual total costs. Further, a problem of coordinating this supply chain by a type of quantity discount is addressed.

Grigory Pishchulov
European University Viadrina, Faculty of Business Administration and Economics, D-15230 Frankfurt (Oder), Germany, e-mail: pishchulov@europa-uni.de

Imre Dobos
Corvinus University of Budapest, Institute of Business Economics, H-1093 Budapest, Fővám tér 8., Hungary, e-mail: imre.dobos@uni-corvinus.hu

Imre Dobos gratefully acknowledges the financial supports by the TÁMOP-4.2.1.B-09/KMR-2010-0005 research program and the Deutscher Akademischer Austauschdienst (DAAD).

Barbara Gobsch
European University Viadrina, Faculty of Business Administration and Economics, D-15230 Frankfurt (Oder), Germany, e-mail: gobsch@europa-uni.de

Nadezhda Pakhomova
St. Petersburg State University, Faculty of Economics, Russia, e-mail: pakhomova@europa-uni.de

Knut Richter
European University Viadrina, Faculty of Business Administration and Economics, D-15230 Frankfurt (Oder), Germany & St. Petersburg State University, Faculty of Economics, Russia, e-mail: richter@europa-uni.de

D. Klatte et al. (eds.), *Operations Research Proceedings 2011*, Operations Research Proceedings,   457
DOI 10.1007/978-3-642-29210-1_73, © Springer-Verlag Berlin Heidelberg 2012

1 Introduction

Consider a supply chain in which a single supplier manufactures and ships a particular product per order placed by a single buyer. The buyer is able to collect used items for taking them back to the supplier at the end of each order cycle; otherwise the used items are disposed of. The supplier manufactures new items and can also produce as-good-as-new items by remanufacturing the used ones. Both kinds of items (shortly called *serviceables*) serve the buyer's demand — which is assumed to be deterministic constant and known to both parties. We are interested in the analysis of coordination within such supply chain.

A similar coordination problem has been addressed, among others, by Banerjee [1] and Goyal [6]; however, they do not consider the opportunity of returning the used items (*nonservicables*) for remanufacturing. Therefore we combine in this paper the basic model of [1] with the reverse logistics model studied by Richter [8,9].

It is known [1] that the order size that is optimal to one supply chain member is unlikely to be optimal to the other as well as for the whole supply chain. Coordination may, however, be achieved if an acceptable contract between the parties can be designed that would guide their decision making towards the supply chain optimum.

In the present work we study optimal lot sizes and reuse rates in the above closed-loop supply chain and address the question whether it can be coordinated by one particular form of the quantity discount. In Section 2 we present the basic model and construct the relevant cost functions. Section 3 addresses the system-optimal reuse rate and the associated lot sizes. Section 4 addresses the coordination problem by means of a quantity discount. Section 5 concludes with a summary and an outlook.

2 The basic model

Let the assumptions of the joint economic lot size model [1] apply. Additionally to that, we assume that the buyer collects and sends a certain fraction of used items back to the vendor for remanufacturing, while the rest is disposed of. The portion of the buyer's demand that cannot be satisfied by remanufacturing is covered by manufacturing. Both are run in the JIT fashion — i.e., there is neither raw material nor component stock held at the supplier; this simplifying assumption allows us to neglect the respective inventory holding costs. We furthermore assume that the sequence of manufacturing and remanufacturing operations has no impact on the supplier's setup cost per order cycle. Let the problem data be as follows:

D	buyer's demand for the product per time unit
P_M, P_R	supplier's manufacturing and remanufacturing productivities ($P_M, P_R > D$)
s_b, s_v	buyer's fixed cost and supplier's setup cost per order, resp.
h_b, u_b	buyer's holding cost rate per serviceable and nonserviceable unit, resp.
h_v, u_v	that of the supplier, resp. ($h_v > u_v$)
d_b	buyer's disposal cost per nonserviceable unit

The buyer chooses the fraction β of used items to be collected and remanufactured (*collection rate*, $0 \leq \beta \leq 1$) as well as the order size q.

The relevant costs of the supplier and the buyer consist of fixed costs per order, stock holding costs and, in the buyer's case, disposal costs. The evolution of the inventory levels at the buyer is shown in Figure 1, where $t^c = q/D$ is the order cycle length. It is straightforward to see that the buyer's cost function expresses then as

$$TC_b(q,\beta) = s_b \cdot \frac{D}{q} + \frac{q}{2} \cdot (h_b + \beta \cdot u_b) + d_b \cdot (1 - \beta) \cdot D.$$

The structure of the inventory levels at the supplier depends on what operation is run first within each order cycle. Figure 2 illustrates the inventory levels during an order cycle with manufacturing executed prior to remanufacturing; t_2 and t_3 represent there the respective durations, and $t_1 = t^c - t_2 - t_3$ is the slack time. Due to the assumption $P_M, P_R > D$, it holds $t_1 > 0$. The expression of the supplier's costs accordingly depends on the sequence of operations during an order cycle [4]:

$$TC_v(q,\beta) = \begin{cases} s_v \cdot \frac{D}{q} + \frac{q}{2} \cdot h_{MR}(\beta), & \text{if manufacturing is run first} \\ s_v \cdot \frac{D}{q} + \frac{q}{2} \cdot h_{RM}(\beta), & \text{otherwise} \end{cases}$$

where

$$h_{MR}(\beta) = \beta^2 \cdot \left[h_v \cdot \left(\frac{D}{P_M} - \frac{D}{P_R} \right) - u_v \frac{D}{P_R} \right] - 2\beta \cdot \left[h_v \cdot \left(\frac{D}{P_M} - \frac{D}{P_R} \right) - u_v \right] + h_v \frac{D}{P_M},$$

$$h_{RM}(\beta) = \beta^2 \cdot \left[(h_v - u_v) \cdot \left(\frac{D}{P_R} - \frac{D}{P_M} \right) + u_v \frac{D}{P_M} \right] + 2\beta u_v \cdot \left(1 - \frac{D}{P_M} \right) + h_v \frac{D}{P_M}.$$

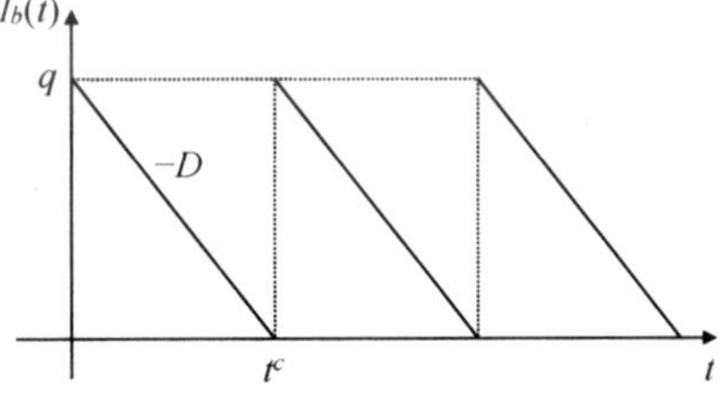
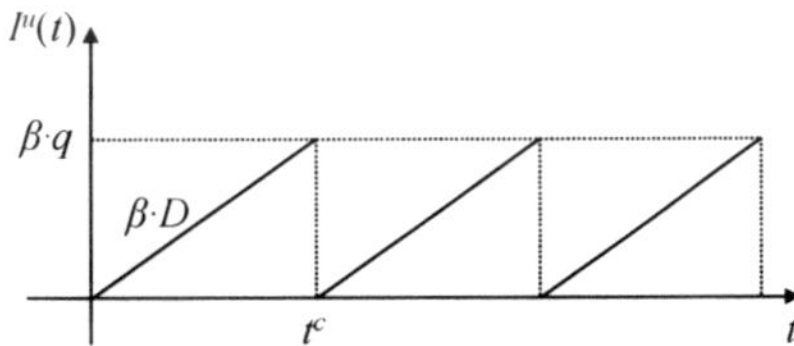

Fig. 1 Inventory levels of serviceables (left) and nonserviceables (right) at the buyer over time

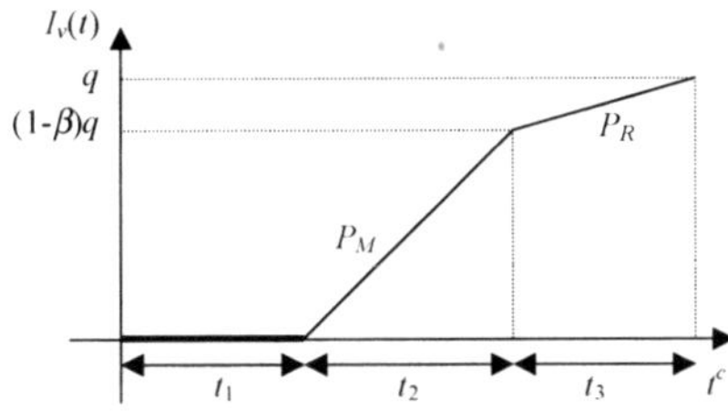
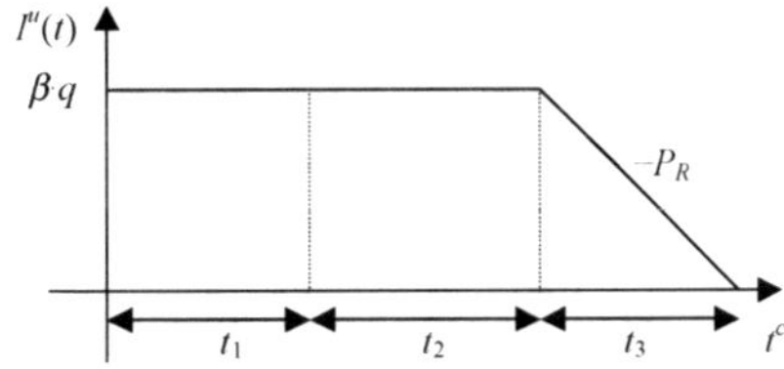

Fig. 2 Inventory levels of serviceables (left) and nonserviceables (right) at the supplier; manufacturing precedes remanufacturing

The optimal manufacturing-remanufacturing strategies are discussed in the context of this model in [4]. It can be shown that running manufacturing before remanufacturing is more economical whenever the following holds:

$$\frac{P_R}{P_M} > 1 + \frac{u_v}{h_v - u_v},$$

i.e., when the remanufacturing productivity sufficiently exceeds the manufacturing one. Due to the space limitations, we restrict the analysis below to this setting only.

3 The joint economic lot size and collection rate

In this section we address the system-wide optimal lot size and collection rate that would minimize the partners' joint total (relevant) costs

$$TC_{MR}(q,\beta) = (s_b + s_v) \cdot \frac{D}{q} + \frac{q}{2} \cdot [h_b + \beta u_b + h_{MR}(\beta)] + d_b \cdot (1 - \beta) \cdot D$$

in the setting where manufacturing precedes remanufacturing in each order cycle. The above cost function is a convex EOQ-type function in q and a quadratic one in β which must not necessarily be convex. The joint economic lot size for the given collection rate can then be obtained in the usual way:

$$q_{MR}^*(\beta) = \sqrt{\frac{2 \cdot (s_b + s_v) \cdot D}{h_b + \beta u_b + h_{MR}(\beta)}},$$

what yields the following minimum total costs for the given collection rate:

$$TC_{MR}^*(\beta) = \sqrt{2(s_b + s_v) \cdot D \cdot [h_b + \beta u_b + h_{MR}(\beta)]} + d_b \cdot (1 - \beta) \cdot D.$$

It is straightforward to verify that $TC_{MR}^*(\beta)$ is either convex or concave (or both) on its entire domain $[0, 1]$, therefore one can figure out whether the function has its minimum in the interior of its domain by referring to the sign of the function's derivative at the borders. The extremal analysis of the function allows to express the joint economic collection rate in the closed form, yielding the following

Proposition 1. *A mixed strategy represented by the collection rate*

$$\beta^* = \frac{\Omega_M}{\Delta_M} + \frac{d_b D}{\Delta_M} \sqrt{\frac{(V + h_b)\Delta_M - (\Omega_M - u_b/2)^2}{2D(s_v + s_b)\Delta_M - d_b^2 D^2}} - \frac{u_b}{2\Delta_M} \quad \in (0,1)$$

where

$$\Delta_M = h_v \left(\frac{D}{P_M} - \frac{D}{P_R} \right) - u_v \frac{D}{P_R}, \qquad \Omega_M = h_v \left(\frac{D}{P_M} - \frac{D}{P_R} \right) - u_v, \qquad and \qquad V = h_v \frac{D}{P_M},$$

is optimal for the entire system if the following conditions simultaneously hold:

$$\Delta_M > \max \left\{ \frac{(\Delta_M - u_b/2)^2}{V + h_b}, \frac{d_b^2 D}{2(s_v + s_b)} \right\}$$

$$\frac{u_v \cdot (D/P_R - 1) - u_b/2}{\sqrt{(h_v - u_v) \cdot D/P_R + 2u_v + h_b + u_b}} < \frac{-d_b\sqrt{D}}{\sqrt{2(s_v + s_b)}} < \frac{\Omega_M - u_b/2}{\sqrt{V + h_b}}$$

otherwise a pure *strategy with* $\beta^* = 0$ *or* $\beta^* = 1$ *is optimal.*

We refer the reader to Corollary 3 in [3] (with $c_M = c_R$ and $c := d_b$) for the proof.

4 Individual optimization and a quantity discount

Analysis conducted in [3] as well as Proposition 1 allow to obtain the following

Proposition 2. *A* mixed *strategy represented by the collection rate* $\beta_v^* = \frac{\Omega_M}{\Delta_M} \in (0, 1)$ *is preferred by the vendor if* $\Delta_M, \Omega_M > 0$, *otherwise a* pure *strategy with* $\beta_v^* \in \{0, 1\}$.

At the same time, it is straightforward to verify that the buyer's minimum total costs for the given collection rate express as

$$TC_b^*(\beta) = \sqrt{2Ds_b(h_b + \beta u_b)} + (1 - \beta)d_b D$$

what represents a function concave in β and provides us with the following

Proposition 3. *The buyer's optimal choice is always a* pure *strategy:* $\beta_b^* \in \{0, 1\}$.

We below refer to a quantity discount type of contract [2] in the form of a two-part tariff [5] which is known to coordinate the serial supply chain [7]; we consider namely an "all unit" type of discount specified by two entities: a fixed fee w_0 and a per-unit charge w_1, so that the amount payable is $w(q) = w_0 + w_1 q$, where q is the purchase quantity. The unit purchase price is accordingly $\frac{w(q)}{q} = \frac{w_0}{q} + w_1$. Thus by adopting the order size q, the buyer pays an amount of $\frac{D}{q} \cdot (w_0 + w_1 \cdot q)$ per time unit, where $\frac{D}{q}$ is the respective number of orders. The buyer's cost function is accordingly

$$TC_b(q, \beta) = (s_b + w_0) \cdot \frac{D}{q} + \frac{q}{2} \cdot (h_b + \beta u_b) + d_b \cdot (1 - \beta) \cdot D + w_1 D$$

and the supplier's cost function $TC_v(q, \beta)$ is adjusted in a similar way.

It is straightforward to see that the concavity property of the minimum cost function $TC_b^*(\beta)$ retains, what in turn makes Proposition 3 hold also in the case of the above form of quantity discount — which therefore fails to coordinate the closed-loop supply chain under consideration whenever the system-optimal collection rate does not represent a pure strategy. To determine a most favourable combination of the quantity discount parameters w_0, w_1, the supplier can still resort to the following non-linear optimization problem (cf. [10]):

$$\min \ TC_v^*(\beta_b^*(w_0, w_1)) \quad \text{s.t.} \quad TC_b^*(\beta_b^*(w_0, w_1)) \le TC_b^*(\beta_b^*(0, \overline{w})), \quad w_0, w_1 \ge 0$$

where $\beta_b^*(w_0, w_1)$ represents the buyer's optimal collection rate for the given w_0, w_1, and $\overline{w}$ represents the initial (wholesale) price.

5 Conclusions and further research

We considered a generalization of the joint economic lot size model [1] and the reverse logistics model [8, 9] and derived the system-wide as well as individually optimal policies with regard to the collection rate and the order size. We further addressed the problem of coordination by means of a quantity discount whose further properties w.r.t. the coordination deficit need to be addressed in the future research. It is known that the quantity discounts may lose the coordination capability in other settings, as well, but still can be augmented by further parameters to restore coordination [11]. A similar insight needs to be developed in the future research with regard to the coordination problem under study. A further interesting extension can be a bargaining study in the context of private information and one party's absolute bargaining power as in [10] as well as other distributions of bargaining power [7].

References

1. Banerjee, A.: A joint economic lot-size model for purchaser and vendor. Decision Sciences **17**, 292–311 (1986)
2. Cachon, G.P.: Supply chain coordination with contracts. In: de Kok, A.G., Graves, S.C. (eds.) Supply Chain Management: Design, Coordination and Operation, pp. 229–339. Elsevier (2003)
3. Dobos, I., Gobsch, B., Pakhomova, N., Pishchulov, G., Richter, K.: A vendor-purchaser economic lot size problem with remanufacturing and deposit. Discussion paper no. 304, Faculty of Economics and Business Administration, European University Viadrina, Germany (2011)
4. Dobos, I., Gobsch, B., Pakhomova, N., Richter, K.: Remanufacturing of used products in a closed-loop supply chain. In: Csutora, M., Kerekes, S. (eds.) Accounting for climate change – What and how to measure, pp. 130–146. Proceedings of the EMAN-EU 2011 Conference, 24–25 January 2011, Budapest, Hungary (2011)
5. Dolan, R.J.: Quantity discounts: Managerial issues and research opportunities. Marketing Science **6**, 1–22 (1987)
6. Goyal, S.K.: An integrated inventory model for a single supplier single customer problem. International Journal of Production Research **15**, 107–111 (1976)
7. Kohli, R., Park, H.: A cooperative game theory model of quantity discounts. Management Science **35**, 693–707 (1989)
8. Richter, K.: An EOQ repair and waste disposal model. In: Proceedings of the Eighth International Working Seminar on Production Economics, Innsbruck, Vol. 3, pp. 83–91 (1994)
9. Richter, K.: Pure and mixed strategies for the EOQ repair and waste disposal problem. OR-Spektrum **19**, 123–129 (1997)
10. Sucky, E.: A bargaining model with asymmetric information for a single supplier–single buyer problem. European Journal of Operational Research **171**, 516–535 (2006)
11. Weng, Z.K.: Channel coordination and quantity discounts. Man. Sci. **41**, 1509–1522 (1995)

Integrated capacity and inventory decisions

N.P. Dellaert, S.D.P. Flapper, T. Tan, J. Jeunet

Abstract This paper deals with the simultaneous acquisition of capacity and material in a situation with uncertain demand, with non-zero lead-times for the supply of both material and capacity. Although there is a lot of literature on the time-phased acquisition of capacity and material, most of this literature focuses on one of the two decisions. By using a dynamic programming formulation, we describe the optimal balance between using safety stocks and contingent workforce for various lead-time situations. We compare the cost ingredients of the optimal strategy with the standard inventory approach that neglects capacity restrictions in the decision. The experimental study shows that co-ordination of both decisions in the optimal strategy leads to cost reductions of around 10%. We also derive characteristics of the optimal strategy that we expect to provide a thorough basis for operational decision making.

1 Introduction

The performance of a production system is determined by decisions about inventory levels of raw material, subassemblies and finished goods as well as by decisions about the production capacity in terms of machine and labour availability. Production managers face fluctuations in demand and supply, which they try to manage not only by keeping safety stock, but also by the acquisition of temporary capacity. In this paper we investigate the use of contingent workforce for coping with fluctuations of both demand and supply as contingent employment allows for a reduction in labour costs, an improvement of labour productivity and an adaptation of the production to a variable demand. We deal with the integral acquisition of capacity

N.P. Dellaert, S.D.P. Flapper, T. Tan
Dept. Industrial Engineering & Innovation Sciences, Operations, Planning, Accounting and Control, Technische Universiteit Eindhoven, P.O. Box 513, 5600 MB Eindhoven, The Netherlands
e-mail: n.p.dellaert@tue.nl; s.d.p.flapper@tue.nl, t.tan@tue.nl

J. Jeunet
CNRS, Lamsade, University Paris-Dauphine, Place du Marchal de Lattre de Tassigny, 75775 Paris Cedex 16, France e-mail: jully.jeunet@dauphine.fr

D. Klatte et al. (eds.), *Operations Research Proceedings 2011*, Operations Research Proceedings,   463
DOI 10.1007/978-3-642-29210-1_74, © Springer-Verlag Berlin Heidelberg 2012

and material in a situation with uncertain demand, with non zero lead-times for the supply of both materials and capacity.

One of the few papers dealing with both decisions simultaneously is that of Dellaert and De Kok (2004) [4] but the lead-time for capacity is assumed to be zero. Bradley and Arntzen (1999) [1] simultaneously plan and control production, inventory and capacity with a number of limiting assumptions that are not made in this paper: deterministic demand, always enough materials available, instantaneous supply of extra capacity. The last two assumptions are also made in Hu et al. (2004) [5] and Tan and Gershwin (2004) [7].

In order to focus on the essentials of the integration of decisions on production, inventory and workforce, we consider the simplest typical situation for which these combined decisions are relevant. Such a situation is characterized by the process of one raw material, using both permanent and contingent workforce. Many examples of this situation can be found in practice, including companies where raw materials like metal or wood are cut for other companies or department where these products are packed so specific package materials have to be ordered. From this simple model, we derive useful insights for more complex production systems.

2 Situation description and strategies

We consider a simple production-inventory system with two stock points: one for the input material and one for the single end product. The end product is produced to stock and the demand for this product is assumed to be stationary with a given distribution. The input material is ordered from an external supplier and is delivered after a constant replenishment lead-time. The production process uses labour capacity that is acquired with a lead-time. We assume the required capacity is proportional to the production quantity and the capacity is needed during the production lead-time of the end item. Added value at a stage consists of material, labour and equipment costs. Inventory holding costs are assumed to be proportional to the product value at a stage.

Our objective is to minimize the expected total relevant cost per period which is the sum of the inventory holding costs, the costs for permanent and contingent capacity as well as the shortage costs. In every period we have to take decisions about the order size for the raw material, the production quantity for the end-item, and the flexible labour capacity. The decision about the permanent labour capacity is not taken every period, but only once, beforehand. We consider two strategies: the optimal strategy and the inventory strategy.

In the optimal strategy we use a dynamic programming method to minimize the average long-term costs per period. As Markovian states, we consider the elements: inventory levels; goods in pipeline; flexible workforce in pipeline. The decisions taken every period are: amount taken into production; new raw material ordered; new flexible workforce ordered. Such an optimal strategy can easily be formulated, but can only be solved for simple situations, for instance with a discrete demand with a limited maximum quantity and very short lead-times, because of the size of the state space. The state space can be further limited by excluding states that are never reached, because other choices are always more attractive.

The inventory strategy aims at reaching lowest possible costs for inventories and shortages, disregarding the capacity costs. This is the basic strategy developed in the multi-echelon inventory area, starting off with Clark and Scarf (1960) [2] to more recent results like Rosling (1989) [6] or De Kok and Visschers (1999) [3]. This strategy can be useful in two ways: if the capacity costs are very small then minimizing inventory costs will result in a good strategy and furthermore, it can serve as a lower bound for the integrated inventory and shortage costs. In cases of shortages of raw material, end product or labour capacity, orders are backlogged. Depending upon the lead-times, the ordered quantity of contingent labour is determined by the expected available amount of raw material or by the backlog quantity.

3 Numerical results

Under several unit cost parameter values for the contingent and permanent capacity, shortages, and inventory, we compare the performance of the 2 strategies for all combinations of lead time between 0 and 2 for both the raw material (L_R) and the contingent workforce (L_M). For each lead time situation, the best value of the permanent workforce P is obtained by considering its cost for a relevant set of P values. From the results we learn that the differences in total costs are depending upon the lead times and vary from 0 to 12%, with an average of 7%. Two levels of unit cost for the fixed workforce c_P are considered: $c_P = 10$ and $c_P = 20$. For both strategies, the flexible capacity costs are obviously higher when the unit cost of the fixed workforce is higher: as the permanent workforce becomes more expensive, the difference with the cost of flexible workforce gets smaller, we thus hire a higher number of flexible workers. The inventory strategy shows much more variation in flexible capacity costs, perhaps because the simple decision rules do not always give sufficient power to control the costs in other ways. We also observe that the holding costs are obviously lower in the Inventory Strategy (IS), compared to the Optimal Strategy (OS). The shortage costs are much higher in the IS. It seems therefore that in the OS, we have more stocks that prevent from shortages than in the IS. Total costs are higher in the IS than in the OS because of higher shortage costs that are not balanced by lower inventory costs. Having longer lead-times always leads to higher costs for both strategies. The OS performs better than the IS in terms of shortages. In the OS, shortage costs increase with both lead times but the lead time for labour has a stronger impact than the lead-time for the raw material.

In order to better characterize the optimal solution, we examine three lead time situations in further details: $(L_M, L_R) = \{(2,0),(0,2),(2,2)\}$. For each situation, we consider reasonable values for the permanent capacity, as this capacity can not be optimized in the short run which is our framework of analysis. In this way we expect to get more intuition in the costs and decisions as a function of the permanent capacity.

In the first situation $(L_M, L_R) = (2, 0)$, with an average demand of 5 units per period, the optimal value for the total available capacity (permanent and flexible) is also 5 units. The Optimal Strategy hardly wastes any flexible capacity for a fixed capacity level less than 5 units, whereas the Inventory Strategy has additional shortage costs of about 19, which corresponds approximately to 0.76 units of unused capacity per period. In this example, for some given values of the fixed capacity P, the additional cost of the Inventory Strategy may be almost 50% higher than those of the Optimal Strategy. When we consider the decisions in some states more carefully, for instance in the situation with a permanent capacity $P = 5$, we find differences between the Inventory Strategy and the Optimal Strategy in various aspects:

- the inventory level of raw material is never below 3 in the Optimal Strategy, always allowing us to use at least 3 units of capacity when necessary, whereas in the Inventory Strategy this level is sometimes even 0. Thus, due to a lack of raw material, capacity has to be wasted;
- the maximum quantity produced ahead is 6 in the Optimal Strategy, whereas production ahead does not take place in the Inventory Strategy; not producing ahead also means a waste of capacity.

In the second situation $(L_M, L_R) = (0, 2)$, for high and low levels of fixed capacity, the strategies are almost alike in terms of costs. However, for intermediate values of the fixed capacity around the average demand (P varying from 3 to 7) differences are suddenly up to 20%. For instance with $P = 5$, the minimum level of raw material in the Optimal Strategy is exactly equal to this fixed capacity level whereas in the IS, in about 40% of the situations, some fixed capacity has to be wasted because of a lack of raw material. Clearly, due to its myopic behavior, the Inventory Strategy is not very good in foreseeing sufficient material availability.

Table 1 presents the results of the third situation where both lead times equal 2 periods $(L_M, L_R) = (2, 2)$. Again, the differences between the strategies are small when the fixed capacity is high, but now for both low and medium sized fixed capacity we find considerable differences in performance, partly due to additional shortage costs (not enough capacity to produce) and partly due to higher flexible capacity costs (unused available capacity and need for additional capacity later). Thus, for most levels of fixed capacity the Inventory Strategy does not properly use the available fixed capacity and therefore is performing quite poor compared to the Optimal Strategy.

4 Characteristics of the optimal solution

From our numerical experiments, we have found that usually the structure of the optimal solution is quite complex unless we have extreme capacity levels. In order to get some intuition for the structure of the optimal solution, we consider the situation in which both lead times are equal to 2 periods (see Table 1) and we try to

	Holding cost		Shortage cost		Flexible capa. cost		Total cost	
P	OS	IS	OS	IS	OS	IS	OS	IS
0	26.03	24.83	10.28	18.62	125.00	137.89	161.31	181.34
1	26.06	24.83	10.25	18.61	100.00	112.89	136.31	156.33
2	26.29	24.83	10.07	18.53	75.00	87.89	111.35	131.25
3	27.28	24.82	9.37	18.18	50.00	62.93	86.65	105.93
4	29.41	24.81	9.99	17.44	26.11	38.29	65.52	80.54
5	28.48	24.78	11.10	14.65	8.10	16.60	47.68	56.03
6	29.22	24.88	9.07	15.47	1.54	4.65	39.83	44.99
7	27.50	24.88	9.11	13.62	0.21	0.73	36.82	39.23
8	28.24	24.85	7.60	11.85	0.02	0.13	35.85	36.82
9	28.61	24.83	6.99	11.12	0.00	0.01	35.60	35.95
10	28.76	24.82	6.78	10.82	0.00	0.00	35.54	35.65
11	24.80	24.82	10.72	10.72	0.00	0.00	35.52	35.54
12	24.82	24.82	10.68	10.68	0.00	0.00	35.50	35.50

Table 1 Costs of the Optimal Strategy (OS) and the Inventory Strategy (IS)

characterize the optimal solution for and for . In the following, we let $Q(t)$ be the advanced production quantity at the beginning of period t, after production decisions are made. The optimal solution looks a lot like an order-up-to policy for raw material and a synchronized policy for the raw material, with some specific elements listed hereafter.

- The maximum amount produced ahead is equal to 3 for $P = 5$, and 12 when $P = 3$;
- The ordered contingent workforce is always fully used, otherwise it would lead to a situation where $Q(t) \geq 4$ (with $P = 5$). For $P = 3$, the contingent workforce is always fully used, and only the fixed workforce is occasionally not used, otherwise it would lead to $Q(t) \geq 13$;
- The ordered amount of contingent workforce is always limited in two ways: we do not order more than can be used for production, which is the foreseen inventory of raw material minus the fixed capacity, and we do not order more than the production backlog minus one (four);
- As an observation of the results of the decision model, the quantity of raw material is such that the amount in pipeline is always equal to 10, except when there is a production backlog in which case the amount in pipeline becomes $(9 - Q(t)$. When $P = 3$, the pipeline inventory is reduced when we have produced ahead and the total stock becomes $\max(6, 10 - Q(t))$.
- The amount produced is always such that in the next period the inventory of raw material is at least equal to the fixed capacity P.

One of the most typical elements of the optimal strategy turns out to be the maximum quantity that is produced in advance and the maximum amount of production backlog that is allowed. These quantities obviously depend on the fixed capacity level P; when $P = 12$ it is not interesting to produce in advance, for low P-values hardly any capacity will be wasted.

5 Conclusion

In this paper, we have proposed a mathematical model to support the integrated acquisition and allocation of input materials and temporary workforce for the simplest relevant situation with one material and one type of labour.

The analysis based on this model shows that separating inventory decisions from capacity decisions results to higher costs. In most production environments, capacity plays some role and by co-ordination of both decisions the optimal strategy shows that cost reductions of around 10% are quite achievable. Looking closer at the structure of the optimal solution, we find that it has a lot of similarities with the order-up-to policy, but that co-ordination is essential, by ensuring that always sufficient raw material is available to avoid the waste of flexible capacity. One simple way to avoid capacity waste is to produce ahead, creating more stock of finished goods than strictly necessary. The additional holding costs are usually compensated by reduced capacity costs. There are however always limits for the maximum quantities to be produced ahead. Building upon these characteristics we would now be able to define some heuristics that could perform well in situations that cannot be analysed exactly.

References

1. Bradley, J.R., and Arntzen, B.C.: The simultaneous planning of production, capacity, and inventory in seasonal demand environments. Operations Research. **47**, 795–806 (1999)
2. Clark, A.J., and Scarf, H.: Optimal policies for a multi-echelon inventory problem. Management Science. **6**, 475–490 (1960)
3. De Kok, A.G., and Visschers, J.W.C.H.: Analysis of assembly systems with service level constraints. International Journal of Production Economics. **59**, 313–326 (1999)
4. Dellaert, N.P., and De Kok, A.G.:Integrated resource and production decisions in a simple multi-stage assembly system. International Journal of Production Economics. **90**, 281–294 (2004)
5. Hu, J.Q., and Vakili, P., and Huang, L.:Capacity and production management in a single product manufacturing system. Annals of Operations Research. **125**, 191–204 (2004)
6. Rosling, K.:Optimal inventory policies for assembly systems under random demands. Annals of Operations Research. **37**, 565–579 (1989)
7. Tan, B., and Gershwin, S.B..:Production and subcontracting strategies for manufacturers with limited capacity and volatile demand. Annals of Operations Research. **125**, 205–232 (2004)

Stream XIII
Scheduling, Time Tabling and Project Management

Erwin Pesch, University of Siegen, GOR Chair
Ulrich Pferschy, University of Graz, ÖGOR Chair
Norbert Trautmann, University of Bern, SVOR Chair

Lower bounds for a multi-skill project scheduling problem

Cheikh Dhib, Anis Kooli, Ameur Soukhal and Emmanuel Néron

Abstract The aim of this paper is to present lower bounds for a Multi-Skill Project Scheduling Problem, including some classical lower bounds and more enhanced ones. Starting from the lower bounds found in the literature, we focus on the particularity of our problem and on the adaptation of these lower bounds to our problem. We present preliminary results obtained with these new lower bounds. The adaptation of the time window adjustments used with energetic reasoning, shows the specificity of this problem, especially slack computation, which requires solving a max-flow with minimum cost problem.

1 Introduction

The problem addressed in this paper is a project scheduling problem defined in an industrial context. The aims is to integrate a project scheduling module to the open source framework *OFBiz/Neogia* (open source ERP, www.neogia.org). Indeed firms hold an increasing share of their activities in project mode and wish that this part of their activities could be linked to other ERP modules, eg, billing passed time. Therefore, we aim at providing a generic model which can integrate an open source platform, but also closely match the needs of the customers of Nereide (a service

Cheikh Dhib
Laboratory of Computer Sciences (LI), University of Tours, 64 Avenue Jean Portalis, 37200 Tours, Neride Company,3bis, les Isles, 37270 VERETZ, e-mail: cheikh.mohameddhib@etu.univ-tours.fr

Anis Kooli
Combinatorial Optimization Research Group - ROI, Ecole Polytechnique de Tunisie, BP 743, 2078, La Marsa, Tunisia, e-mail: anis.kooli@gmail.com

Ameur Soukhal
Laboratory of Computer Sciences (LI), University of Tours, 64 Avenue Jean Portalis, 37200 Tours, e-mail: ameur.soukhal@univ-tours.fr

Emmanuel Néron
Laboratory of Computer Sciences (LI), University of Tours, 64 Avenue Jean Portalis, 37200 Tours, e-mail: emmanuel.neron@univ-tours.fr

D. Klatte et al. (eds.), *Operations Research Proceedings 2011*, Operations Research Proceedings, 471
DOI 10.1007/978-3-642-29210-1_75, © Springer-Verlag Berlin Heidelberg 2012

company specialized in open source software, at the initiative of this project [1]).
The model selected may be seen as a multi-skill project scheduling problem with
preemption [6,7]. Indeed, in our case, some tasks can be preempted without penalty.
The tasks are subject to classical precedence constraints. Multi skilled staff members are considered as resources. The tasks to be executed require one or more skills.
The need in skills is expressed as man-day work. For example the same task may
require 15 days of development and 5 days of analysis. A task cannot begin before the end of all its predecessors. At each time, a person performs one skill and
is asigned to a single task. In our example we must assign a person to development
and a responsible person to analysis. These two individuals are necessarily different.
The estimated workload is thus converted in duration. The person associated to one
of the skills of a task must remain the same, even if the task is preempted, i.e., that
person will resume its contribution to this task later on. An additional constraint occurs on the beginning of tasks: different contributions on all skills must begin at the
same time for practical reasons (synchronization and coordination in carrying out
the task). A task cannot start until all people assigned to each skill it requires, are
available. We also include the fact that people may not be available entirely on the
planning horizon (downtime, holidays, etc.). The objective is to propose a schedule
of minimum duration.
Recently, many studies have taken into account the notion of skills in project
scheduling, e.g., [2,10]. The specificity of the problem addressed here is the simultaneous consideration of both preemption and the expression of resource requirements
in terms of skills. This is a generalization of the multi-skill project scheduling problem proposed by Néron and Bellenguez-Morineau, and for which exact methods
and lower bounds have been proposed.

2 Notation

- $\mathscr{A} = \{A_0, A_1, \ldots, A_n, A_{n+1}\}$ set of activities. A_0 and A_{n+1} are the starting and
  ending dummy activities.
- $\mathscr{A}_{\overline{p}} \subset \mathscr{A}$ set of non preemptable activities
- $\mathscr{G} = (\mathscr{A}, E)$ the precedence graph. $E = \{(i,j) \in \mathscr{A} \times \mathscr{A}\}, (i,j) \in E$ then A_j
  cannot start before the end of A_i
- $\mathscr{P} = \{P_1, \ldots, P_M\}$ set of staff members;
- $\mathscr{S} = \{S_1, \ldots, S_K\}$ set of skills
- $\forall i \in \{1, \ldots, n\}, k \in \{1, \ldots, K\}, p_{i,k}$ workload of skill S_k for activity A_i
- $\forall m \in \{1, \ldots, M\}, k \in \{1, \ldots, K\}, S_{m,k} = 1$ if P_m is able to do S_k and 0 otherwise
- $\mathscr{T} = \{0, \ldots, T\}$: the set of relevant time-points
- $\forall m \in \{1, \ldots, M\}, t \in \{1, \ldots, T\}\ Dispo(m,t) = 1$ if p_m is available on the interval
  $[t, t+1[$ and 0 otherwise.

---

[1] This work is supported by the *ANRT* governmental agency in the context of a *CIFRE* convention

3 Classical lower Bounds

The proposed lower bounds developed in this section have been presented in [3].

Critical Path lower bound: based on the work of Stinson *et al.* [4], for each activity we associate the maximal workload to its duration, and then we construct a precedence graph from which we compute the release dates r_i. Thus r_{n+1} (the longest path from A_0 to A_{n+1}) is a valid lower bound denoted by LB_{cp_0}. This lower bound can be enhanced when we fix a project due date D. From the value D we compute a deadline for each task $d_i(D)$ according to the precedence graph. So we have for each task i an execution time window $[r_i, d_i(D)]$. For each task i' in the critical path we look for potential parallel tasks,i.e., tasks that can be executed in parallel with it, and we compute the lack of resources. If this lack exists, it can be added to the trivial value. We denote it by LB_{cp_1}. For more details readers can refer to [4].

Capacity Lower Bound: here we relax all constraints except for the availability of resources. For each skill we compute the total needed workload then we divide it by the number of staff member mastering this skill, the result is a valid lower bound. It can be formulated as the following: $LB_{cap} = \max_k \frac{\sum_i p_{ik}}{\sum_m S_{m,k}}$.

4 Destructive lower bounds

In this section we present three destructive lower bounds which are based on feasibility tests, as well as a binary search which leads to detect an infeasability if we consider a given trial value D of project duration.

Flow model based lower bound: in this lower bound we relax the precedence constraints, and we choose an integer number D such that $LB_{cp_0} < D < T$, then we compute for each task i its time-window $[r_i, d_i(D)]$ where $d_i(D)$ is the deadline in the case of D considered as the whole project deadline. We construct a flow model per skill, to assign skill workload to staff members and a global model considering all skills as the same, which allows to detect an infeasibility. If an infeasibility occurs then $D + 1$ is a valid lower bound, we denote it by LB_{FM} [3].

Preemptive based LP Model: this lower bound has been presented in [3]. It is based on discretizing the time-horizon into successive intervals using release dates, due dates and resources availability intervals. We use an LP formulation based on continuous variables $x_{i,k,l,m}$ which represent the amount of time spent by person m for skill k of task i during the interval l. We denote this lower bound by LB_{LM}.

Adapted energetic reasoning: Based on the energetic reasoning approach proposed for the classical RCPSP [8], we develop new lower bounds. The main difference being the synchronization of the skills of each task, and the preemption for given tasks. In fact regarding preemption the mandatory part for a given skill of a task corresponds to the difference between the needed workload and the part that can be processed outside the interval. For synchronization we can have a mandatory part equal to 1, despite that it can be processed totally outside the interval. For example, in Fig. 2(right side), a preemptive task i needs both skills k_1 and k_2. We can see that skill k_2 can be executed entirely outside the interval, but because skill k_1 cannot start outside the interval and due to synchronization constraints the mandatory part of skill k_2 becomes 1 instead of zero. We denote this lower bound by LB_{ER}.

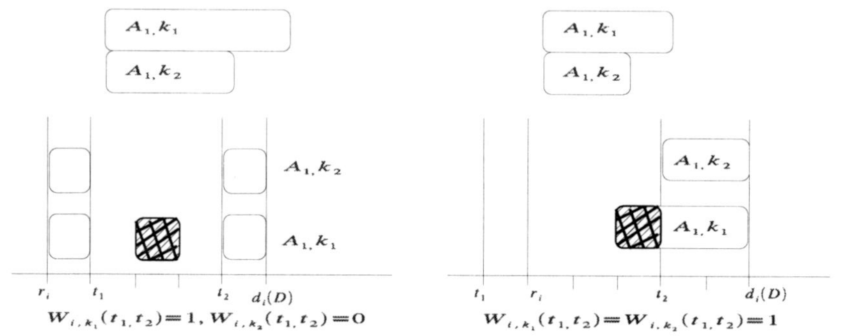

Fig. 1 Mandatory part computation with preemption and synchronization constraints

5 Time window adjustments and shaving

In order to enhance the energetic reasoning lower bound we use the adjustments procedure as in [8]. The main specificity of our problem is the manner to compute a slack for a given skill and task. In fact the skills are interdependent because they share the resources, so to compute a slack for a skill of a given task on an interval we have to solve a max flow with minimum cost problem as illustrated in Fig. 2. Let i be a non preempted task which needs a p_{ik} man-day of skill k, $W_{i,k}(t_1,t_2)$ be the mandatory part of skill k and $W_{i,k}^l(t_1,t_2)$, $W_{i,k}^r(t_1,t_2)$ be respectively the left and right part of skill k of task i. We denote by $\mathscr{P}_k \subseteq \mathscr{P}$ the set of staff members mastering skill k. We denote by $Slk_{i,k}(t_1,t_2)$ the slack of skill k of task i, which is the time remaining for executing this skill after satisfying all other skills mandatory parts on the interval $[t_1,t_2]$. Due to the multi-skill constraints, a person $m \in \mathscr{P}_k$ can do other skills or the same skill for other tasks, so to find out the time remaining for skill k, after executing all mandatory parts, we solve the assignment problem as in

Fig. 2. For each $m \in \mathscr{P}_k$ the cost (value of c in the figure) is very high, and it is

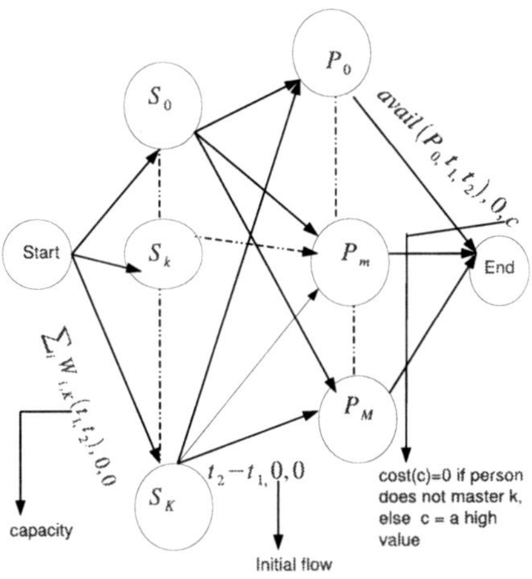

Fig. 2 Slack computation network model for a task i and skill k

null for all the other ones, which implies that we do not use the people mastering this skill for other skills, except if nobody else can perform them. We denote by $flw(m,t_1,t_2)$ the flow passed on the edge from a node P_m to the final node, which represents the duration person m works on $[t_1,t_2]$, and $avl(m,t_1,t_2)$ the availability of person m on the interval $[t_1,t_2]$. so the slack, $Slk_{i,k}(t_1,t_2) = \sum_{m \in \mathscr{P}_k} avail(m,t_1,t_2) - \sum_{m \in \mathscr{P}_k} flw(m,t_1,t_2) + W_{ik}(t_1,t_2)$. This slack is then compared with left and right parts. If it is less than one of them, a time window is adjusted. So we can follow the same approach as presented in [8] to compute a lower bound which we denote by LB_{ERA}.

To strengthen this lower bound we apply a shaving procedure as described in [5], in this study the shaving is done only on the non preempted tasks, we denote this bound by LB_{SHA}.

6 Results

Experiments have been conducted to study the performance of the proposed lower bounds using 160 instances adapted from the PSPLIB KSD30 set of data [9]. The results show that LB_{LM} and LB_{ER} are almost equivalent in terms of quality of solutions. Moreover, LB_{ER} runs within a negligible time, while LB_{LM} needs more than

1 second to compute a solution. The lower bound LB_{FM} is dominated by both LB_{ER} and LB_{LM}. The lower bound LB_{ERA} was able to enhance the lower bound based on the energetic reasoning without adjustments (LB_{ER}) on 11 instances, in an average time of 18 sec. Contrary to the shaving procedure LB_{SHA} which enhanced only 1 instance compared to LB_{ERA} and over a longer period. We denote also that both shaving procedure (LB_{SHA}) and the energetic reasoning with adjustments (LB_{ERA}) did the same adjustments (15 instances on the 160).

7 Conclusion

In this paper we presented destructive lower bounds based on energetic reasoning concepts. The most particular aspect is the notion of skill synchronization and preemption which makes the mandatory parts of resources(skills) dependent to each other for the same task. Also, due to resources being shared by skills, a max flow min cost assignment procedure was used to compute resource slacks. We show that time window adjustments leads to enhance the energetic reasoning lower bound which was not the case for the shaving procedure(LB_{SHA}) which did not enchance over the energetic reasonning with adjustments (LB_{ERA}).

References

1. Montoya, C., Bellenguez-Morineau, O., Rivreau, D.: New models for the multi-skill project scheduling problem. In EURO 2010. Lisbonne, July 2010.
2. Heimerl, Ch., Kolisch, R.: Scheduling and staffing multiple projects with a multi- skilled workforce. OR Spectrum. **32**, 343–368 (2010)
3. Dhib, Ch., Roumdani, T., Soukhal, A., Néron, E.: Bornes infrieures pour un problme dordonnancement de projets multi-comptence premptif. ROADEF 2011, Saint Etienne, France, 2011.
4. Stinson,J., Davis, E., Khumawala, B.: Multiple resource constraine d scheduling using branch and bound, AIEE Transactions 10 (1978) 252-259.
5. Haouari, M., Kooli, A., Néron, E.: Enhanced energetic reasoning-based lower bounds for the resource constrained project scheduling problem. Computers & Operations Research. (2011) doi:10.1016/j.cor.2011.05.022
6. Bellenguez-Morineau, O., Néron, E.: Lower bounds for the multi-skill project scheduling problem with hierarchical levels of skills. In Practice and Theory of Automated Timetabling (PATAT2004), pages 429432, Pittsburgh, PA, USA, 2004.
7. Bellenguez-Morineau, O., Néron, E.: Methods for solving the multi-skill project scheduling problem. In 9th International Workshop on Project Management and Scheduling (PMS2004), pages 6669, Nancy, France, 2004.
8. Baptiste, P., Le Pape, C., Nuijten, W.: Constrained project scheduling problem, Satisfiability tests and time bound adjustments for cumulative scheduling problems. Annals of Operations Research . **92**, 305–333 (1999)
9. Kolisch, R., Sprecher, A., Drexl, A.: PSPLIB-a project scheduling library. European Journal of Operational Research. **96**, 205–216 (1997).
10. Hartmann, S., Briskorn, D.: A survey of variants and extensions of the resource- constrained project scheduling problem. European Journal of Operational Research. **207(1)**, 1–14 (2010)

Synchronized Two-Stage Lot Sizing and Scheduling Problem in Automotive Industry

Fadime Uney-Yuksektepe, Rifat Gurcan Ozdemir

Abstract This study proposes a mixed-integer linear programming model to solve the synchronized two-stage lot sizing and scheduling problem with setup times and costs. The main motivation behind this study is based on the need of improvement in the inventory control and management system of a supplier that serves to automotive industry. The supplier considered in this study has synchronized two-stage production system which involves three parallel injection machines in the first stage and a single spray dying station in the second stage. In order to achieve synchronization of the production between these stages and improve the inventory levels, a mixed-integer linear programming model is proposed. By the help of developed model, synchronization between the parallel injection machines and the single spray dying station with minimum total cost is obtained. Optimal lot sizes for each stage and optimal schedules of these lot sizes are determined by the developed model as well.

1 Introduction

The automotive industry is one of the most important sectors by revenue and has been significantly affected by the global financial crisis. Hence, the overall efficiency of the automotive supply chains becomes very important. As the suppliers are the first stage of automotive supply chain, the sufficiency and quality of semi-products produced by them has an important effect on the overall efficiency of the supply chain. The main motivation behind this study is based on the need of improvement in the inventory control and management system of one of the main suppliers of an automotive manufacturer.

Fadime Uney-Yuksektepe
Department of Industrial Engineering, Istanbul Kultur University, Istanbul, Turkey, e-mail:
f.yuksektepe@iku.edu.tr

Rifat Gurcan Ozdemir
Department of Industrial Engineering, Istanbul Kultur University, Istanbul, Turkey, e-mail:
rg.ozdemir@iku.edu.tr

D. Klatte et al. (eds.), *Operations Research Proceedings 2011*, Operations Research Proceedings, 477
DOI 10.1007/978-3-642-29210-1_76, © Springer-Verlag Berlin Heidelberg 2012

The supplier considered in this study is producing bumpers for different models and colors of a passenger car based on the demands realized from a car manufacturer. This supplier has a synchronized two-stage production system which involves parallel injection machines in the first stage and a single spray dying station in the second stage. There exist three injection machines in the first step whose production rates are different according to models produced. Furthermore, in order to produce different models of car distinct moulds have to be used by these injection machines. During the production, changing the mould types is costly and time consuming. On the other hand, as each model has only one mould to be used, the production of a specific model can not be done on different injection machines simultaneously. The uncolored bumpers that exit the injection machines enter to the spray dying station. According the demand characteristics of the passenger car models and colors, bumpers are painted in a synchronized fashion. Here the synchronization means that front and back bumpers of a car should enter to the dying station together. As dying station has a slow rate and should not be stopped, there exists a buffer stock between the first and second stages. After this dying step, bumpers are stocked in the inventory and sent to the assembly line of the automotive manufacturer based on the demand. As a bumper supplier, the problem is to synchronize the injection machines and dying station as they have different rate of production.

In this study, we proposed a mixed integer programming model that integrates the synchronized two-stage lot sizing and scheduling decisions. The developed model is solved for a real life automotive supplier's synchronized two-stage lot sizing and scheduling problem. The remaining part of this paper is organized as follows. In the next section, proposed optimization model is described in detail. Data related to the considered automotive supplier and the results for the computational tests are given in Section 3. Paper concludes with conclusion and suggestions in Section 4.

2 Proposed Model

The mixed-integer linear programming (MILP) model presented in this section considers the synchronization between the production stages and integrates the lot sizing and scheduling decisions. The problem is considered in weekly manner and for each week the scheduling and lot sizing decision could be obtained by the running the model repeatedly. Each day is divided into T micro periods with fixed-length. Hence, the following indices are used to develop the proposed model: i represents the model type of the car $(1, \ldots, I)$. The model type of the bumper is represented by j (1:front, 2: back). Furthermore, t is used for micro periods $(1, \ldots, T)$, r is used for type of injection machines $(1, \ldots, R)$, k represents the color of the bumpers $(1, \ldots, K)$, and g represents the day of the week $(1, \ldots, 6)$.

In order to solve this problem, the following parameters are assumed to be known:
d_{ikg}: demand of model i in color k for day g
s_{ijr}: setup time needed for injection machine r to produce model i of type j

at: available time of injection machine r in micro period t
atd: available time of dying station in micro period t
td_{ik}: dying time of model i with color k
tp_{ijr}: processing time needed for unit product of model i of type j by injection machine r
ic: inventory cost per unit product per day
sc: Setup cost of injection machines
bc: costs of backlogging per product per day
idc: cost of idleness of dying station
$MaxINX$: maximum number of products that can be kept in available inventory

Given these parameters and sets, the following decision variables are sufficient to model the synchronized two-stage lot sizing and scheduling problem:
X_{ijrt}: number of products of model i of type j produced by injection machine r in micro period t
Y_{ikt}: number of products of model i dyed by dying station with color k in micro period t
INY_{ikg}: number of products of model i with color k held in buffer stock II at the end of day g
INX_{ijt}: number of products of model i of type j held in buffer stock I at the end of in micro period t
BL_{ikg}: number of backlogged products of color k of model i in day g
IDL_t: idle time of the dying station in micro period t
S_{ijrt}: 1 if model i of type j is produced by machine r in micro period t; 0 otherwise
SI_{ijrt}:1 setup incurred for model i of type j by machine r in micro period t; 0 otherwise

The following mixed integer programming model is developed to solve the synchronized two-stage lot sizing and scheduling problem:

$$\min z = \sum_i \sum_j \sum_t \frac{6ic}{T} INX_{ijt} + \sum_i \sum_j \sum_r \sum_t sc \cdot SI_{ijrt} + \sum_i \sum_k \sum_g bc \cdot BL_{ikg}$$

$$+ \sum_t idc \cdot IDL_t + \sum_i \sum_k \sum_g ic \cdot INY_{ikg} \qquad (1)$$

subject to

$$X_{ijrt} \leq M \cdot S_{ijrt} \qquad \forall ijrt \qquad (2)$$

$$S_{ijrt} - S_{ijr(t-1)} \leq SI_{ijrt} \qquad \forall ijrt \qquad (3)$$

$$\sum_r S_{ijrt} \leq 1 \qquad \forall ijt \qquad (4)$$

$$\sum_i \sum_j S_{ijrt} \leq 1 \qquad \forall rt \qquad (5)$$

$$\sum_i \sum_j (tp_{ijr} \cdot X_{ijrt} + st_{ijr} \cdot SI_{ijrt}) \leq at \qquad \forall rt \qquad (6)$$

$$\sum_i \sum_j INX_{ijt} \leq MaxINX \quad \forall t \quad (7)$$

$$INX_{ij(t-1)} + \sum_r X_{ijrt} - INX_{ijt} = \sum_k Y_{ikt} \quad \forall ijt \quad (8)$$

$$INY_{ik(g-1)} + \sum_{t:(T(g-1)+6)/6}^{Tg/6} Y_{ikt} + BL_{ikg} = d_{ikg} + INY_{ikg} + BL_{ik(g-1)} \quad \forall ikg \quad (9)$$

$$\sum_i \sum_k td_{ik} \cdot Y_{ikt} + IDL_t \leq atd \quad \forall t \quad (10)$$

$$X_{ijrt}, Y_{ikt} \geq 0, \text{integer} \quad \forall i,j,r,k,t \quad (11)$$

$$INX_{ijt}, INY_{ikg}, BL_{ikg}, IDL_t \geq 0 \quad \forall i,k,g,t \quad (12)$$

$$S_{ijrt}, SI_{ijrt} \in 0,1 \quad \forall i,j,r,t \quad (13)$$

The objective function (1) is to minimize the total sum of semi-product and finished product inventory, demand backorder, machine setup and dying station idleness costs. Constraint (2) ensures that if there is a production of model i of type j on injection machine r at micro period t, then decision variable S_{ijrt} will be 1 which indicates whether model i of type j on injection machine r at micro period t is produced or not. In order to count the setups incurred at each micro period, constraint (3) is necessary. Since each model has only one mould to be used, the production of a specific model can not be done on different injection machines simultaneously. This restriction is given by constraint (4). Constraint (5) is necessary to ensure that exactly one type of a model is produced in each micro period at each injection machine. Constraint (6) represents the injection machines capacity constraint and provides the total production and setup times do not to exceed available time for injection machine r at micro period t. For semi-products, available buffer stock field is limited and at any micro period total inventory level should not exceed $MaxINX$ value (7). The synchronization of injection machines and dying station is attained by constraint (8). As constraint (8) is valid for all j, the synchronization at the entering of back and front bumpers to the dying station is established. For each day, demand should be satisfied by balancing the total production, inventory and backlogging amounts (9). Available time of the dying station should not be exceeding as it is given in constraint (10). Moreover, as the idleness of dying station is not preferable, the idle time of dying machine for each micro period is counted by the help of that constraint. Finally, constraints (11), (12) and (13) give the non-negativity and integrality of the decision variables.

3 Computational Tests

In this section, computational results of the developed MILP model to solve real instances of the production planning problem of automotive supplier are presented. Before analyzing the results, some information related to the real problem is given.

Supplier considered in this study produces five different bumper models with front and back options. There exist three injection machines with different production rates (45, 57, 56, 45,45 unit/hour) for distinct models. For the injection machines, setup time is approximately 1 hour and setup cost is 600 TL. Setup time and cost is independent of the model, type and injection machines. In the dying station, the rate of production is 70 bumper pairs (front and back) per hour. As the bumpers are put on the hangers and dyed on the hangers, while a color change occurs leaving a single hanger empty will be sufficient. Hence, the setup time at the dying station could be ignored since the production rate of dying station is much higher than that setup time. Similarly, no setup cost related to the color change at the dying station will be considered. As the idleness of dying station is not preferable, idleness cost is taken as 80 TL. Due to the restricted plant area, the semi-product inventory has a capacity of 2472 unit products. Therefore, this restriction should be considered while determining the lot sizes and schedules. For the inventory holding cost of both semi-product and final product is taken as 10 TL/day. On the other hand, cost of backlogging is assumed to be 1000 TL/day. All of these parameter values are obtained by the help of production manager of that automotive supplier. Thus, each of the values reflects the real problem efficiently. For the bumpers, there exist nineteen different colors. For each of the model-color pairs, there exists a daily demand value which is obtained from the automotive manufacturer just before the start of a week.

Based on the data obtained from the supplier, proposed model is formulated in GAMS 23.6[1] and solved by using CPLEX 12.0[2] solver in order to obtain the optimal results. The runs were executed on a 1.66GHz Intel Pentium 4 notebook with 2 GB of RAM. A number of different runs are obtained by changing the micro period lengths to determine the optimal micro period length for this problem.

Table 1 Objective function values for distinct micro period lengths.

Micro Period Lengths (hours)	Objective Function
2	23694800
3	18141686
4	8126983
6	11282400
8	15333400
12	18959000

As it is shown in Table 1, the best (minimum total cost) objective function value is obtained with the micro period length of 4 hours. The ones with micro period length of 2, 3, 6, 8 and 12 hours are the worst from the perspective of objective function. Hence, the results for the micro period length of 4 hours will be examined in detail.

---

[1] Brooke, A., Kendrick, D., Meeraus, A., Raman, R., 1998. GAMS:A User's Guide. GAMS Development Co., Washington, DC.

[2] Ilog, 2010. CPLEX 12.0 User's Manual , ILOG S. A. See website www.cplex.com

At the optimal solution of the model with micro period of 4 hours, lot sizes in the injection machines are found as 100 units for each period. For instance; for the first micro period, front and back bumpers for model I5 are produced on injection machines R2 and R1, respectively. Furthermore, injection machine R3 is used to produce the back bumper of model I2. Hence, at the end of first micro period, 100 units of back bumper of model I2 will be kept at semi-product inventory as the dying station will dye the model of I5 bumpers at the first micro period. Moreover, in order to dye a specific model, each of the front and back bumpers should be ready and enter to the dying station together. Therefore, at semi-product inventory exists at the end of first micro period.

4 Conclusions

In this paper, we studied the lot sizing and scheduling problem in an automotive supplier. A mixed integer programming model is developed to solve this two stage problem optimally. Moreover, the sensitivity of optimum solution with respect to micro period length is analyzed and efficient micro period length is determined by considering the objective function and computational times. By the help of developed model, synchronization between the parallel injection machines and the single spray dying station with minimum total cost for all types of demand patterns is attained for the determined micro period length. Optimal lot sizes for each stage and optimal schedules of these lot sizes are determined by the developed model as well. Furthermore, the characteristics of the proposed model are given. Possible future work could be trying flexible micro period lengths rather than fixed lengths. This can also be achieved by permitting more than one type of product produced in a single micro period and determining the orders of these products.

References

1. Brooke, A., Kendrick, D., Meeraus, A., Raman, R., 1998. GAMS:A User's Guide. GAMS De-velopment Co., Washington, DC.
2. Ilog, 2010. CPLEX 12.0 User's Manual , ILOG S. A. See website www.cplex.com

A branch-and-price algorithm for the long-term home care scheduling problem

M. Gamst and T. Sejr Jensen

Abstract In several countries, home care is provided for certain citizens living at home. The long-term home care scheduling problem is to generate work plans such that a high quality of service is maintained, the work hours of the employees are respected, and the overall cost is kept as low as possible. We propose a branch-and-price algorithm for the long-term home care scheduling problem. The pricing problem generates a one-day plan for an employee, and the master problem merges the plans with respect to regularity constraints. The method is capable of generating plans with up to 44 visits during one week.

1 Introduction

In several countries, home care is provided for certain citizens living at home. Home care offers cleaning, grocery shopping, help with personal hygiene and medicine, etc. The long-term home care scheduling problem is to generate work plans spanning a longer period of time, such that a high quality of service is maintained, the work hours of the employees are respected, and the overall cost is kept as low as possible.

Quality of service consists of the following. *Regularity*: all visits at a citizen should be conducted at the same time of the day and by the same (small group of) employee(s) in order for the citizen to feel safe. *Skill set requirements*: certain tasks can only be performed by a subset of the employees due to skill requirements.

All time windows are soft, i.e., preferred visit times and employee work hours can be violated at a cost. Such violations are denoted "busyness". The overall cost of a solution consists of a linear combination of travel time between visits, quality of service, and busyness.

The long-term home care scheduling problem is *NP*-hard and is typically approached in one of two ways in the literature: (1) plans for employees for a single day are generated. This corresponds to a modified VRPTW and is denoted the *daily*

University of Southern Denmark, DK-5230 Odense M, e-mail: gamst@man.dtu.dk and thomassejr@gmail.com

D. Klatte et al. (eds.), *Operations Research Proceedings 2011*, Operations Research Proceedings, 483
DOI 10.1007/978-3-642-29210-1_77, © Springer-Verlag Berlin Heidelberg 2012

planning problem [1, 4, 5]. (2) Otherwise a Periodic VRPTW is solved, i.e., plans are made for several days, but regularity constraints are ignored [2, 6, 7].

We propose a branch-and-price (BP) algorithm for the full long-term home care scheduling problem. The pricing problem generates a one-day plan for an employee, and the master problem merges the plans with respect to regularity constraints. The method is capable of generating plans with up to 44 visits during one week. This truly illustrates the complexity of the problem.

2 Exact solution algorithm

The problem considers a given period of time consisting of $L \in \mathbb{N}$ days. Time is discretized into time steps, which together span L. The set of employees is denoted E. For each employee $j \in E$ is given a set of days H_j and time windows $[a_{jh}, b_{jh}], h \in H_j$, that specify the work hours of j.

The set of visits is denoted V, the set of citizens is C, and the set of visits at citizen c is $V_c \subseteq V$. Each visit $j \in V$ is repeated after p_j amount of time. The visit repetitions are scheduled independently of each other and are denoted *activities*. Let A be the set of all activities and $A_j \subseteq A$ the set of activities for visit $j \in V$. Two consecutive activities from visit j are denoted $(i, k) \in A_j$. An activity $i \in A$ has attached a prioritized list of employees to conduct the activity, denoted pr_i. The prioritized list represents the *skill set requirement*. The duration of $i \in A$ is d_i and the time window is $[a_i, b_i]$. The travel time between two activities $i, j \in A$ is $c_{ij} \geq 0$. Finally, busyness, i.e., the amount of time employee j is late for conducting visit i, is denoted $b_{ij} \geq 0$.

Dantzig-Wolfe decomposing the problem results in a pricing problem, which generates a daily schedule for a given employee on a given day, and a master problem, which merges the daily schedules into an overall feasible solution.

The daily schedule for an employee is denoted a *path* and contains an ordered list of activities with attached starting times. The overall solution consists of paths for appropriate employees on appropriate days, covering all activities. Let $s_i \in \mathbb{N}$ denote the starting time of activity $i \in A$. Recall that the objective function consists of travel time, quality of service, and busyness. We aggregate these into a single, weighted objective function. Let **w** be the non-negative weight vector. The objective function which is to be minimized consists of:

1. Travel time (TT) between activities.
2. Busyness (B), i.e., how late an employee j is for conducting an activity i with respect to travel times and time windows.
3. Employee priority (EP).
4. Employee regularity (ER), i.e., the number of different employees at a citizen.
5. Visit regularity (VR), i.e., if the time between two consecutive activities $(i, k) \in A_j$ differs from p_j.

Let p be a path and P the set of all generated paths. Let $x_p \in \{0,1\}$ denote whether or not path p is part of the solution, and let $u_{ik} \geq 0$ denote the difference in the starting times between two consecutive activities $(i,k) \in A_j$. Each path p has a number of constants attached: $\delta_{Bp}^{ij} \geq 0$ denotes the amount of busyness for employee j and activity i, $\delta_p^i \in \{0,1\}$ denotes if activity i is visited, $\delta_{sp}^i \geq 0$ denotes the start time at activity i, $\delta_p^{ij} \in \{0,1\}$ denotes if employee j visits activity i in the path, and $\delta_p^{jh} \in \{0,1\}$ denotes if the path is generated for employee j on day $h \in H_j$. The master problem is formulated as:

$$\min \sum_{p \in P} \sum_{i \in A} \left(\sum_{k \in A} w^{TT} c_{ik} \delta_p^i \delta_p^k x_p + \sum_{j \in E} \left(w^B \delta_{Bp}^{ij} x_p + w^{EP} \, pr_i(j) \, \delta_p^{ij} x_p \right) \right) +$$

$$\sum_{c \in C} \sum_{j \in E} w^{ER} y_c^j + \sum_{j \in V} \sum_{(i,k) \in A_j} w^{VR} u_{ik} \tag{1}$$

$$\text{s. t.} \qquad \sum_{p \in P} \delta_p^i x_p = 1 \qquad\qquad \forall i \in A \tag{2}$$

$$\sum_{p \in P} \delta_{sp}^i x_p + p_j - \sum_{p \in P} \delta_{sp}^k x_p \leq u_{ik} \qquad \forall j \in V, \, \forall (i,k) \in A_j \tag{3}$$

$$\sum_{p \in P} \delta_{sp}^k x_p - \left(\sum_{p \in P} \delta_{sp}^i x_p + p_j \right) \leq u_{ik} \qquad \forall j \in V, \, \forall (i,k) \in A_j \tag{4}$$

$$\sum_{p \in P} \delta_p^{ij} x_p \leq y_c^j \qquad\qquad \forall c \in C, \, \forall v \in V_c,$$

$$\forall i \in A_v, \, \forall j \in E \tag{5}$$

$$\sum_{p \in P} \delta_p^{jh} x_p \leq 1 \qquad\qquad \forall j \in E, \, \forall h \in H_j \tag{6}$$

$$x_p \in \{0,1\} \qquad\qquad \forall p \in P \tag{7}$$

$$u_{ik} \geq 0 \qquad\qquad \forall j \in V, \, \forall (i,k) \in A_j \tag{8}$$

$$y_c^j \in \{0,1\} \qquad\qquad \forall c \in C, \forall j \in E \tag{9}$$

The objective function (1) minimizes a weighted sum of travel times, busyness, employee priorities, employee regularity and visit regularity. Constraints (2) ensure that every activity is visited. Constraints (3) and (4) measure visit regularity. Constraints (5) measure employee regularity. Constraints (6) ensure that at most one path per employee per day is part of a solution. The number of columns in the master problem is reduced by fixing daily visits to days: if a visit must be repeated every day, then the corresponding activities are fixed to day 1, 2, 3, etc., respectively. The pricing problem only allows such activities to be part of paths on appropriate days.

Let $\pi_i^{(2)} \in \mathbb{R}$ be the dual of constraints (2), $\pi_{ik}^{(3)} \leq 0$ the dual of (3), $\pi_{ik}^{(4)} \leq 0$ the dual of (4), $\pi_{icj}^{(5)} \leq 0$ the dual of (5), and $\pi_{jh}^{(6)} \leq 0$ the dual of (6). The pricing problem is solved for each employee $j \in E$ on each day $h \in H_j$. The reduced cost of visiting activity $i \in A_v, v \in V_c, c \in C$ is:

$$\bar{c}_{jh}^i = w^{EP} \, pr_i(j) - \pi_i^{(2)} - \pi_{icj}^{(5)} - \begin{cases} s_i(\pi_{ik}^{(3)} - \pi_{ik}^{(4)}) & \exists k \in A : (i,k) \in A_v \\ 0 & \text{otherwise} \end{cases}$$

Activity $i \in A$ is visited exactly once, hence the reduced cost for employee $j \in E$ on day $h \in H_j$ is defined as:

$$\bar{c}_{jh} = \sum_{i \in A} \left(\bar{c}^i_{jh} + w^B b_{ij} + \sum_{k \in A} w^{TT} c_{ik} \right) \leq \pi^{(6)}_{jh} \qquad (10)$$

Recall that $b_{ij} \geq 0$ is the amount of busyness for employee j at activity i and that $c_{ik} \geq 0$ is the travel time between activities $i, k \in A$. Now, $\pi^{(6)}_{jh}$ is a constant, so if the pricing problem generates a path where $\bar{c}_{jh} \leq \pi^{(6)}_{jh}$, then the path has negative reduced cost and the corresponding column is added to the master problem. The pricing problem is recognized as a shortest path problem with time constraints and potentially negative edge weights. This is also denoted the Elementary Shortest Path Problem with Resource Constrained (ESPPRC), which is *NP*-hard. We solve the problem to optimality using the labeling algorithm in [3]. Initially, we try to solve the pricing problem heuristically using the labeling algorithm, where only a single label is stored at each activity.

Branching is necessary when the optimal solution in a branch node is fractional. The following strategy is finite and eventually ensures a feasible solution. Fractional solutions occur when:

An employee j visits a citizen c fractionally. Two branching children are generated with added cut: $\left(y^j_c = 0 \right)$ resp. $\left(y^j_c = 1 \right)$.

An activity i is visited by several employees j, j' or on several days h, h'. Two branching children are generated with rules: $\left(\sum_{p \in P} \delta^i_p \delta^{jh}_p x_p = 0 \right)$ resp. $\left(\sum_{p \in P} \delta^i_p \delta^{j'h'}_p x_p = 0 \right)$. The pricing problem ensures that employee j (resp. j') never visits activity i on day h (resp. h').

An employee j travels on edges $(ik), (ik')$ on a given day h a fractional number of times. Let i be the first activity, from which employee j travels on different edges or at different times. Let constant $\delta^{s_{ik}}_p$ denote whether or not path p travels from i to k at time s_{ik}. Two branching children are generated with the following rules: $\left(\sum_{p \in P} \delta^{jh}_p \delta^{s_{ik}}_p x_p = 0 \right)$ resp. $\left(\sum_{p \in P} \delta^{jh}_p \delta^{s_{ik'}}_p x_p = 0 \right)$. The pricing problem ensures that employee j never travels from i to k (resp. k') at time s_{ik} (resp. $s_{ik'}$) on day h.

The branching rules do not complicate the pricing problem, because they either consider different columns (y^j_c) or consist of rules, which can trivially be handled by forbidding appropriate extensions in the labeling algorithm.

3 Computational Results

The BP algorithm is implemented using the framework COIN Bcp [8] and tested on an Intel 2.13GHz Xeon CPU with 4 cores and 8 GB RAM. Note that test results stem from using a single core. CPLEX 12.1 is used as standard MIP solver.

The BP algorithm is tested on a number of real-life benchmark instances provided by Papirgarden, a home care center in Funen, Denmark. Time is discretized into either 5 or 10 minute time steps. The objective weights are as follows: $W^{TT} = 500$, $w^B = 750$, $w^{EP} = 0$, $w^{ER} = 750 \cdot 5/\tau$ and $w^{VR} = 50$, where τ denotes the size of a time step. We have tested two algorithms: (1) The exact BP algorithm as described and (2) a heuristic BP algorithm where columns are only generated heuristically.

| Instance | Master Problem | | Tree | Gap | Sol. | Time | Master Problem | | Tree | Sol. | Tim |
	Cols	Rows	Size		Value	Sec.	Cols	Rows	Size	Value	Sec.
1-20-5	6327	199	1	0.00	27000.0	3.97	1794	199	1	38750.0	0.75
1-25-5	8671	250	4687	0.79	29750.0	1801.33*	608	250	1	44500.0	0.18
1-30-5	12883	295	2079	3.17	38000.0	1801.97*	2852	295	1	86000.0	1.45
1-40-5	28365	397	171	0.00	43000.0	812.94	903	397	1	67500.0	0.55
1-50-5	31480	493	27	110.91	87000.0	1803.35*	1235	493	1	97500.0	1.45
1-80-5	8737	787	1	333.89	259250.0	1906.50*	2954	787	1	259250.0	10.96
2-20-5	8917	346	5	0.00	27000.0	8.58	1917	346	1	31750.0	0.66
2-25-5	10089	432	2405	0.94	29750.0	1803.39*	1416	432	3	34750.0	0.27
2-30-5	16383	512	695	1.41	38400.0	1803.29*	5637	512	481	73500.0	117.37
2-80-5	8261	1354	1	297.49	237500.0	1908.73*	12677	1354	185	237500.0	388.09
1-20-10	4877	199	59	0.00	17250.0	6.37	1286	199	1	24750.0	0.29
1-25-10	5507	250	277	0.00	18125.0	27.36	324	250	1	29625.0	0.11
1-30-10	10934	295	2449	0.32	25300.0	1801.07*	1845	295	1	54250.0	0.71
1-40-10	14037	397	63	0.00	27750.0	90.16	733	397	1	48750.0	0.32
1-50-10	19907	493	3	0.00	36500.0	337.04	1002	493	1	67500.0	0.71
1-55-10	19912	544	3	0.00	60375.0	123.64	1409	544	1	86875.0	0.69
1-58-10	40434	571	111	73.60	64450.0	1803.82*	2270	571	1	106650.0	1.89
1-80-10	19711	787	1	224.63	161100.0	1847.78*	2255	787	1	161100.0	4.44
2-20-10	5913	346	39	0.00	17250.0	9.19	1770	346	1	20750.0	0.43
2-25-10	6594	432	81	0.00	18125.0	16.30	714	432	3	23125.0	0.15
2-30-10	10109	512	1245	0.20	25200.0	1802.52*	4585	512	751	48000.0	115.49
2-40-10	15937	684	77	0.00	27250.0	162.93	3267	684	7	37500.0	1.00
2-50-10	25457	850	21	0.00	35750.0	654.53	6914	850	191	61250.0	58.38
2-55-10	29331	936	49	133.70	80625.0	1841.08*	6090	936	67	80625.0	21.94
2-58-10	21593	984	31	160.61	96750.0	1826.54*	7242	984	143	96750.0	76.81
2-80-10	22136	1354	3	201.76	149750.0	2475.74*	8552	1354	221	149500.0	265.50

Table 1 Results for the exact BP (left hand side) and heuristic BP (right hand side) for instances named "$|E| - |A| - \tau$", $\tau =$ time step size. Tests marked with * exceeded the 30 minute time bound.

Test results are summarized in Table 1. As can be seen from the left hand size of the Table, only few instances with 5 minute time steps can be solved to optimality within half an hour. A coarser discretization helps, but the BP algorithm still suffers from a large time usage. The number of columns is large, which is caused partly by large time windows and partly by busyness, i. e., that time windows may be violated. The tree size grows very large for some instances not solved to optimality, hence branching also constitutes a bottleneck. As can be seen from the right hand side of the Table, the heuristic BP algorithm is generally faster. Few instances suffer from large tree sizes and many columns, but the far majority of instances are solved in seconds. Unfortunately, the objective values generally suffer from the heuristic approach.

Improving the exact BP approach would require methods for reducing the number of columns and for improving the bounds to prune larger parts of the tree. The authors attempted stabilizing the value of dual variables using an interior point method [9], but with no avail. Other stabilization methods could be investigated, as better values for the dual variables could reduce the number of generated columns. The authors also tried different primal and incumbent heuristics for improving the bounds with little luck. Future work could continue on such heuristics or on changing the branching strategy.

4 Conclusion

In this paper, we presented a BP algorithm for the long-term home care scheduling problem. The NP-hard pricing problem consisted of calculating a work plan on a given day for a given employee and the master problem merged the plans into an overall optimal solution. The BP algorithm was tested on a number of real-life instances and was capable of only solving smaller instances due to the large number of combinations of visits, visit times and employees. This truly illustrates the complexity of the problem.

Acknowledgements We would like to thank Papirgaden for sharing knowledge and real-life data for home care scheduling, and Villum-Kann-Rasmussen foundation for supporting this work.

References

1. Begur S. V., Miller D. M., Weaver J. R.: An Integrated Spatial DSS for Scheduling and Routing Home-Health-Care Nurses. Interfaces **27**, 35–48 (1997)
2. Claassen G.D.H., Hendriks Th.H.B.: An application of Special Ordered Sets to a periodic milk collection problem. European Journal of Operational Research **180**, 754 – 769, (2007)
3. Dumitrescu I., Boland N.: Improved preprocessing, labelling and scaling algorithms for the weight-constrained shortest path problem. Networks **42**, 135–153 (2004)
4. Eveborn P., Flisberg P., Rönnqvist M.: Laps Care - an operational system for staff planning of home care. European Journal of Operational Research **171**, 962–976 (2006)
5. Hansen A. D., Rasmussen M. S., Justensen T., Larsen J.: The Home Care Crew Scheduling Problem. In 1st International Conference on Applied Operational Research volume **1** of Lecture Notes in Management Science, 1–9 (2008)
6. Hemmelmayr V., Doerner K., Hartl R., Savelsbergh M.: Delivery strategies for blood products supplies. OR Spectrum **31**, 707–725 (2008)
7. Jang W., Lim H. H., Crowe T. J., Raskin G., Perkins T. E.: The Missouri Lottery Optimizes Its Scheduling and Routing to Improve Efficiency and Balance. Interfaces **36**, 302–313 (2006)
8. Lougee-Heimer R.: The Common Optimization INterface for Operations Research. IBM Journal of Research and Development **47**, 57–66 (2003)
9. Rousseau L.-M., Gendreau M., Feillet D.: Interior point stabilization for column generation. Operations Research Letters **35**, 660–668 (2007)

On a multi-project staffing problem with heterogeneously skilled workers

Matthias Walter and Jürgen Zimmermann

Abstract We examine a workforce assignment problem arising in the multi-project environment of companies. For a given set of workers and for a given set of projects, we want to find one team of workers for each project that is able to satisfy the project workload during each period of project execution. The aim is to minimize average team size. We outline two mixed integer linear programming (MIP) formulations and introduce two heuristic approaches to this problem. A numerical study indicates the performance of the approaches and provides managerial insights into flexibility issues.

1 Introduction

We investigate a workforce assignment problem, which tpyically arises in companies where a significant share of work is carried out in projects and where workers master various skills at differing levels of experience. In these companies workers are usually uniquely assigned to one department, but join cross-departmental teams to accomplish projects that ask for different skills. The team of workers assigned to each project must provide the required skills and be able to execute and complete the project on time, taking into account the limited working time of each team member. Since team coordination and communication within a team is easier for a small team than for a large one, it is advisable to keep teams as small as possible [4, p. 144]. Furthermore, avoiding large project teams prevents social loafing, i.e. the decrease in a group member's productivity with increasing group size [7]. From the viewpoint of a single worker it is more efficient to work for a small number of projects instead of being scattered over many projects [6]. Hence, it is our aim to minimize the number of assignments of workers to projects, i.e. the average team size. In literature, only variable costs of project staffing have been minimized

Matthias Walter, Jürgen Zimmermann
Department for Operations Research, Clausthal University of Technology
38678 Clausthal-Zellerfeld, Germany
e-mail: matthias.walter@tu-clausthal.de, juergen.zimmermann@tu-clausthal.de

D. Klatte et al. (eds.), *Operations Research Proceedings 2011*, Operations Research Proceedings,   489
DOI 10.1007/978-3-642-29210-1_78, © Springer-Verlag Berlin Heidelberg 2012

so far, which depend on the amount of time a worker contributes to a project [5]. In contrast, fixed costs have not been considered yet. Since the number of assignments can be perceived as the sum of uniform fixed costs, we contribute to close this gap.

In order to tackle our problem, we describe it formally by an optimization model in Section 2 and outline two heuristic solution approaches in Section 3. Based on preliminary test results we evaluate the performance of the proposed approaches in Section 4, where we also sketch implications of a flexible workforce for planning. Section 5 concludes our contribution with final remarks.

2 Optimization models

We assume that a set P of projects was selected by a company for execution in the forthcoming year and that for each project $p \in P$ a schedule was constructed. Let the schedule of project p span the interval $T_p \subseteq T$ between start and finish time and let the schedule state the requirements r_{pst} for skill $s \in S_p \subseteq S$ required by project p in period $t \in T_p$. We presume that a period $t \in T$ represents one month. The requirements are expressed in man hours. For example, a project requires 80 man hours of the skill *actuarial mathematics* in February. Let K denote the company's workforce. Each worker $k \in K$ has a limited capacity R_{kt} in period $t \in T$, e.g. 160 hours in February, and belongs to one department $d \in D$. He masters a subset $S_k \subseteq S$ of all skills and for skill $s \in S_k$ his level of experience $l_{ks} \in \{0.5, 1, 1.5, 2\}$ indicates his efficiency. A skill level of 2 means that he can accomplish one hour of workload r_{pst} within half an hour, whereas 0.5 means that it takes him two hours to complete a requirement of one hour [5]. Assuming static skill levels we do not consider learning effects. To model a worker's contribution to a project p for skill s in period t we introduce continuous variables y_{kpst} which can only be positive if worker k is assigned to project p. Such an assignment is captured by a binary decision variable x_{kp} which is equal to 1 if worker k is assigned to project p and 0 otherwise. Not only projects but also departments ask for man hours. The requirement rd_{dt} of a department d in period t can be met by workers $k \in K_d$, i.e. by department members only.

Having introduced all identifiers we can formulate a MIP model to state our multi-project staffing problem with heterogeneously skilled workers where objective function (1) minimizes the number of assignments of workers to projects. Constraints (2) ensure that the skill requirements of the projects are satisfied, while Constraints (3) guarantee that an assignment is established if a worker contributes to a project. Capacity limits of workers are taken into account in Constraints (4) while Constraints (5) assure that in every period the remaining total capacity of workers who belong to the same department is large enough to cover the workload of their department.

$$\text{Min.} \quad \sum_k \sum_p x_{kp} \tag{1}$$

$$\text{s. t.} \quad \sum_k \left(y_{kpst} \cdot l_{ks} \right) \geq r_{pst} \qquad \forall p, s \in S_p, t \in T_p \tag{2}$$

$$y_{kpst} \leq x_{kp} \cdot R_{kt} \qquad \forall k, p, s \in S_k \cap S_p, t \in T_p \tag{3}$$

$$\sum_p \sum_{s \in S_k \cap S_p} y_{kpst} \leq R_{kt} \qquad \forall k, t \tag{4}$$

$$\sum_{k \in K_d} \left(R_{kt} - \sum_p \sum_{s \in S_k \cap S_p} y_{kpst} \right) \geq rd_{dt} \qquad \forall d, t \tag{5}$$

$$x_{kp} \in \{0, 1\} \qquad \forall k, p \tag{6}$$

$$y_{kpst} \geq 0 \qquad \forall k, p, s \in S_k \cap S_p, t \in T_p \tag{7}$$

In our numerical study we additionally consider a network flow based formulation for our problem described in [8]. In that formulation capacities R_{kt} represent the sources of a network. In that network man hours flow to satisfy the demands r_{pst} of the projects which represent sinks of the network.

3 Heuristic solution approaches

Since an exact solution of our problem using MIP solvers is quite time consuming for large-scale instances we propose two heuristic methods. The first one is a DROP procedure, which is similar to the DROP method known from the warehouse location problem [2]. Our DROP procedure is sketched in Algorithm 1. Each time we solve problem (1)–(7) during this procedure, we solve a linear program (LP) because all binary variables x_{kp} are fixed to either 1 or 0. Such a solution for the binary variables is denoted by X. For the choice of a pair (k', p') in line 4 we apply a roulette wheel selection with a probability for a selection of (k, p) proportional to $\frac{1}{\sum_{s \in S_k \cap S_p} l_{ks}}$. Thus, we can embed the DROP procedure in a randomized multistart process to generate several solutions from which we pick the best one.

The second heuristic approach is a construction method that is described in Algorithm 2. It is a greedy randomized assignment procedure (GRAP) inspired by [1,3]. Here projects are randomly chosen one after the other and staffed by repeatedly selecting workers mastering required skills and assigning as much workload to a selected worker as possible. The term *remaining matching skills between worker k and project p* in line 6 refers to skills $s \in S_k \cap S_p$ for which not all the workload r_{pst}, $t \in T_p$, has been assigned to workers yet. To select a worker in line 5 we apply a roulette wheel selection where the selection probability for a worker k is proportional to the sum of the skill levels of the remaining matching skills. In line 7 the skill is randomly selected from the remaining matching skills with highest level l_{ks} for worker k. Again we incorporate the procedure in a multistart process.

Algorithm 1 DROP procedure

1: $V := \{(k,p)|S_k \cap S_p \neq \emptyset\}$
2: Obtain a start solution X by setting $x_{kp} := 1$ for all $(k,p) \in V$
3: **while** $V \neq \emptyset$ **do**
4:     Randomly select one pair $(k',p') \in V$
5:     Set $x_{k'p'} := 0$ and call the corresponding solution X'
6:     **if** X' is not a feasible solution **then**
7:         $x_{k'p'} := 1, V := V \setminus \{(k',p')\}$
8:     **else**
9:         $V := V \setminus \{(k',p')\}, X := X'$
10:     **end if**
11: **end while**
12: **return** X

Algorithm 2 Greedy randomized assignment procedure (GRAP)

1: **while** Projects exist which have not been staffed yet **do**
2:      Randomly select a project p which has not been staffed
3:      **while** Project p has remaining requirements for at least one skill **do**
4:          **if** A worker with remaining capacity exists who masters a skill still required by p **then**
5:              Select a worker k with remaining capacity who masters a skill still required by p
6:              **while** Remaining matching skills between worker k and project p exist **do**
7:                  Randomly select a matching skill s
8:                  Assign as much workload of $r_{pst}, t \in T_p$, to worker k as possible
9:              **end while**
10:         **else** Abort GRAP as a feasible solution could not be constructed
11:         **end if**
12:     **end while**
13: **end while**
14: **return** A feasible solution

Remark 1. While for Algorithm 1 there always exists a drop sequence that leads to an optimal solution, there are instances for which feasible solutions exist, but for which Algorithm 2 cannot construct a feasible solution.

4 Numerical study

To evaluate the efficiency of the two mentioned MIP formulations and of our heuristic approaches we solved instances of different size, which were artificially created. The size was characterized by parameter values such as the number of workers, projects, and skills and the minimum and maximum capacity of workers. For each size ten instances were randomly generated. Note that we chose $|S_k| \in \{1,2,3,4\}$ for all $k \in K$ and that $|T|$ was always set to twelve periods and $|D| := |S|$. Algorithms were coded in Microsoft Visual C++ 9.0 and Cplex 12.2 was called to solve MIPs and LPs. The time limit was set to one hour (3600 s) for each approach. Additionally, we set the limit on the number of passes for the multistart DROP and GRAP approach to 10 and 1,000 passes, respectively. Test runs were executed on computers operating under Windows XP with a Pentium 4 CPU on 3.2 GHz and 2 GB RAM. Results are displayed in Table 1 where we report the number of instances that were solved to optimality (column *opt.*), the average solution time per instance (*time*), the average gap for the MIP models (*gap*) and the average relative deviation of the heuristic solutions to the solutions found by the network model (*dev.*).

Table 1 Results of the four approaches for several sets of ten instances

			Standard model			Network model			DROP			GRAP		
$|K|$	$|P|$	$|S|$	opt.	time	gap	opt.	time	gap	opt.	time	dev.	opt.	time	dev.
10	5	3	10	1	-	10	1	-	6	1	3.0	1	0.1	11.2
20	10	4	10	79	-	10	17	-	0	5	13.2	0	0.4	31.6
30	15	6	7	192	5.2	7	162	4.3	0	20	16.6	0	0.6	37.3
40	20	8	1	lim.	6.1	1	lim.	4.1	0	56	21.9	0	1.1	46.4
100	25	10	0	lim.	10.2	0	lim.	7.9	0	427	24.8	0	4.2	25.0
200	50	20	0	lim.	38.7	0	lim.	29.5	0	lim.[a]	−3.7	0	14.9	20.6

[time] = s, [gap] = [dev.] = %, lim. indicates that the time limit was hit, [a] Time limit was hit after 8 to 10 passes.

On average the network model clearly outperforms the standard model in terms of both solution time and quality. The DROP procedure provides good solutions and outperforms Cplex applied to the network model on instances with 200 workers and more, whereas solutions of the GRAP procedure deviate considerably from solutions of the network model. Though, GRAP is much faster in comparison with the DROP procedure because the latter requires to repeatedly solve LPs. For example, for a specific instance with 800 workers and 140 projects 42,099 LPs have to be solved, one for each possible assignment. Thus, the time required for one pass of the DROP procedure increases drastically for large-scale instances as can be seen from Table 2. Those instances were solved in reasonable time only by GRAP.

Table 2 Results for sets of large-scale instances

				Network model			DROP			GRAP		
$|K|$	$|P|$	$|S|$	time limit	time	obj.	gap	time	passes	obj.	time	passes	obj.
800	140	50	1 h	lim.	no sol.	-	lim.	0	35,767	208	1,000	3,213
800	140	50	8.5 h	lim.	3,552	54.5	lim.	1	2,593	2,049	10,000	3,195
1,500	250	100	1 h	-	-	-	-	-	-	627	1,000	7,037

[time] = s, [gap] = %, lim. indicates that the time limit was hit, colum *obj.* states the average objective function value, *passes* states the number of passes completed, no sol. indicates that no solution was returnd by the solver.

To shed light on the effect of workforce flexibility on team size we considered a set of ten instances with 30 workers, 30 projects, and 6 skills. Each instance was solved for six different scenarios of workforce flexibility applying a time limit of 1 h. In scenario 1 each worker masters only one skill. In scenario 2, each worker masters an additional skill, and so on. Finally in scenario 6, workers master all six skills, ceteris paribus. The corresponding average number of assignments is depicted in Figure 1. Only instances of scenario 1 and 2 could be solved to optimality. As expected, team size decreases with increasing flexibility. Though, to exploit the growing potential high computational effort is necessary for the exact approach.

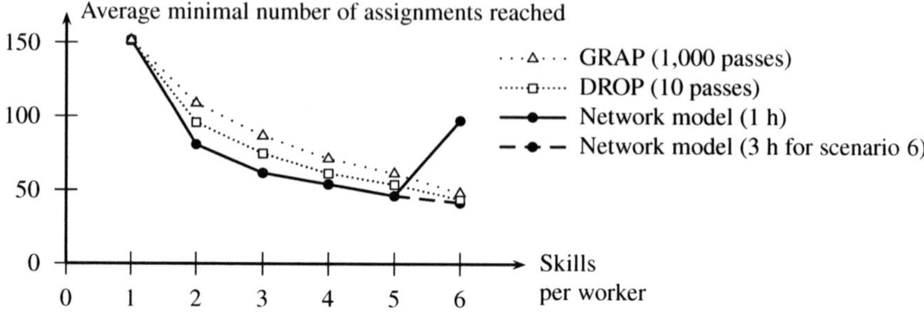

Fig. 1 The curse of flexibility: For the exact approach objective function values deteriorate for the scenario with highest workforce flexibility due to computational complexity.

5 Conclusion

We presented solution approaches to a workforce assignment problem faced by companies in multi-project environments and we pointed to pitfalls of planning when flexibility is high. To obtain good solutions for large-scale instances in less time it seems beneficial to combine both heuristics. GRAP can quickly construct start solutions that can be improved by DROP.

References

1. Drexl, A.: Scheduling of project networks by job assignment. *Management Science*, 37(12), 1590–1602 (1991)
2. Feldman, E., Lehrer, F. A., Ray, T. L.: Warehouse location under continuous economies of scale. *Management Science*, 12(9), 670–684 (1966)
3. Gutjahr, W. J., Katzensteiner, S., Reiter, P., Stummer, C., Denk, M.: Competence-driven project portfolio selection, scheduling and staff assignment. *Central European Journal of Operations Research*, 16(3), 281–306 (2008)
4. Hammer, M., Champy, J.: *Reengineering the corporation*. HarperCollins, New York (1993)
5. Heimerl, C., Kolisch, R.: Scheduling and staffing multiple projects with a multi-skilled workforce. *OR Spectrum*, 32(2), 369–394 (2010)
6. Hendriks, M. H. A., Voeten, B., Kroep, L.: Human resource allocation in a multi-project R&D environment. *International Journal of Project Management*, 17(3), 181–188 (1999)
7. Karau, S. J., Williams, K. D.: Social loafing: A meta-analytic review and theoretical integration. *Journal of Personality and Social Psychology*, 65(4), 681–706 (1993)
8. Walter, M., Zimmermann, J.: A heuristic approach to project staffing. *Electronic Notes in Discrete Mathematics*, 36, 775–782 (2010)

Integrated schedule planning with supply-demand interactions for a new generation of aircrafts

Bilge Atasoy, Matteo Salani and Michel Bierlaire

Abstract We present an integrated schedule planning model where the decisions of schedule design, fleeting and pricing are made simultaneously. Pricing is integrated through a logit demand model where itinerary choice is modeled by defining the utilities of the alternative itineraries. Spill and recapture effects are appropriately incorporated in the model by using a logit formulation similar to the demand model. Furthermore class segmentation is considered so that the model decides the allocation of the seats to each cabin class. We propose a heuristic algorithm based on Lagrangian relaxation to deal with the high complexity of the resulting mixed integer nonlinear problem.

1 Introduction and Motivation

Transportation demand is constantly increasing in the last decades for both passenger and freight transportation. According to the statistics provided by the Association of European Airlines (AEA), air travel traffic has grown at an average rate over 5% per year over the last three decades. Given the trends in the air transportation, actions need to be taken both in supply operations and the demand management to have a demand responsive transportation capacity for the sustainability of transportation.

This study is in the context of a project regarding the design of a new air transportation concept, Clip-Air, at Ecole Polytechnique Fédérale de Lausanne. Clip-Air is designed to introduce flexibility in various aspects including multi-modality, mod-

Bilge Atasoy
Transport and Mobility Laboratory, ENAC, EPFL, e-mail: bilge.kucuk@epfl.ch

Matteo Salani
Dalle Molle Institute for Artificial Intelligence (IDSIA), Lugano, e-mail: matteo.salani@idsia.ch

Michel Bierlaire
Transport and Mobility Laboratory, ENAC, EPFL, e-mail: michel.bierlaire@epfl.ch

D. Klatte et al. (eds.), *Operations Research Proceedings 2011*, Operations Research Proceedings, 495
DOI 10.1007/978-3-642-29210-1_79, © Springer-Verlag Berlin Heidelberg 2012

ularity, demand management and robustness. The main characteristic of Clip-Air is that it simplifies the fleet management by allowing to decouple the carrying (wing) and the load (capsule) units. The wing carries the flight crew and the engines, and the capsules carry the passengers/freight. The separation of passengers from the pilot and the fuel has several advantages in terms of security. Furthermore, capsules are modular detachable units such that the transportation capacity can be modified according to the demand. Clip-Air is already tested in a simulation environment and the aircraft design is shown to be feasible. From the operations viewpoint, a preliminary analysis on the potential advantages of Clip-Air is presented in [1]. An integrated schedule design and fleet assignment model is developed to compare the performance of Clip-Air with standard aircraft. In this paper we focus on modeling aspects and solution methods. Therefore we present the model for standard aircraft where we further integrate supply-demand interactions into the scheduling model.

2 Related Literature

Supply-demand interactions are studied in the context of fleet assignment with different perspectives. In an itinerary-based setting, [2] consider the spill and recapture effects separately for each class resulting from insufficient capacity. Similarly, [3] study the network effects including the demand adjustment in case of flight cancellations.

Advanced supply and demand interactions can be modeled by letting the model to optimize itinerary's attributes (e.g., the price). [5] integrate discrete choice modeling into the single-leg, multiple-fare-class revenue management model that determines the subset of fare products to offer at each point in time. [4] develops a market-oriented integrated schedule design and fleet assignment model with integrated pricing decisions. Several specifications are proposed for the pricing model including discrete choice models such as logit model and nested logit model where the explanatory variable is taken as the fare price.

3 Integrated schedule planning model

We develop an integrated schedule design, fleet assignment and pricing model. The pricing decision is integrated through a logit model. The utility of the itineraries are defined as below:

$$V_i^h = \beta_{fare}^h p_i^h + \beta_{time}^h time_i + \beta_{stops}^h nonstop_i, \tag{1}$$

where p_i^h is the price of itinerary $i \in I$ for class $h \in H$, $time_i$ is a dummy variable which is 1 if departure time is between 07:00-11:00, and $nonstop_i$ is a dummy variable which is 1 if i is a non-stop itinerary. The coefficients of these variables, which are specific to each class h, are estimated with maximum likelihood estimation. Total expected demand for each market segment $s \in S$ (D_s^h) is attracted by the itineraries in the segment ($I_s \in I$) according to the logit formula in equation (2).

$$\tilde{d}_i^h = D_s^h \frac{\exp(V_i^h)}{\sum\limits_{j \in I_s} \exp(V_j^h)} \quad \forall s \in S, h \in H, i \in I_s. \tag{2}$$

In case of capacity shortage some passengers, who can not fly on their desired itineraries, may accept to fly on other available itineraries in the same market segment. This effect is referred as spill and recapture effect. In this paper we model accurately the spill and recapture in order to better represent the demand. We assume that the spilled passengers are recaptured by the other itineraries with a recapture ratio based on a logit choice model as given by the equation (3). We include no-purchase options ($I_s' \in I_s$), which are the itineraries offered by competitive airlines, in order to represent the lost passengers in a more realistic way.

$$b_{i,j}^h = \frac{\exp(V_j^h)}{\sum\limits_{k \in I_s \setminus i} \exp(V_k^h)} \quad \forall s \in S, h \in H, i \in (I_s \setminus I_s'), j \in I_s. \tag{3}$$

The integrated model, which considers a single airline, is provided in Figure 1. Let F be the set of flight legs, there are two subsets of flights one being mandatory flights (F_M), which should be flown, and the other being the optional flights (F_O) which can be canceled in the context of schedule design decision. A represents the set of airports and K is for the set of fleet. The schedule is represented by time-space network such that $N(k,a,t)$ is the set of nodes in the time-line network for plane type k, airport a and time $t \in T$. $In(k,a,t)$ and $Out(k,a,t)$ are the sets of inbound and outbound flight legs for node (k,a,t).

Objective (4) is to maximize the profit which is calculated with revenue for business and economy demand, that takes into account to lost revenue due to spill, minus operating costs. Operating cost for flight f when using fleet type k is represented by $C_{k,f}$ which is associated with a binary variable of $x_{k,f}$ that is one if a plane of type k is assigned to flight f. Constraints (5) ensure the coverage of mandatory flights which must be served according to the schedule development. Constraints (6) are for the optional flights that have the possibility to be canceled. Constraints (7) are for the flow conservation of fleet, where $y_{k,a,t-}$ and $y_{k,a,t+}$ are the variables representing the number of type k planes at airport a just before and just after time t. Constraints (8) limit the usage of fleet by R_k for fleet type k. It is assumed that the network configuration at the beginning and at the end of the day is the same in terms of the number of planes at each airport, which is ensured by the constraints (9).

The assigned capacity for a flight should satisfy the demand for the corresponding itineraries considering the spill effects as maintained by the constraints (10). Similarly when a flight is canceled, all the related itineraries should not realize any demand. $\delta_{i,f}$ is a binary parameter which is 1 if itinerary i uses flight f and enables us to have itinerary-based demand. We let the allocation of business and economy

$$Max \sum_{s\in S}\sum_{h\in H}\sum_{i\in(I_s\setminus I'_s)} \left(d_i^h - \sum_{\substack{j\in I_s\\ i\neq j}} t_{i,j}^h + \sum_{\substack{j\in(I_s\setminus I'_s)\\ i\neq j}} t_{j,i}^h b_{j,i}^h\right) p_i^h - \sum_{\substack{k\in K\\ f\in F}} C_{k,f} x_{k,f} \tag{4}$$

$$s.t. \sum_{k\in K} x_{k,f} = 1 \qquad \forall f \in F^M \tag{5}$$

$$\sum_{k\in K} x_{k,f} \leq 1 \qquad \forall f \in F^O \tag{6}$$

$$y_{k,a,t^-} + \sum_{f\in In(k,a,t)} x_{k,f} = y_{k,a,t^+} + \sum_{f\in Out(k,a,t)} x_{k,f} \qquad \forall [k,a,t] \in N \tag{7}$$

$$\sum_{a\in A} y_{k,a,t_n} + \sum_{f\in CT} x_{k,f} \leq R_k \qquad \forall k \in K \tag{8}$$

$$y_{k,a,minE_a^-} = y_{k,a,maxE_a^+} \qquad \forall k \in K, a \in A \tag{9}$$

$$\sum_{s\in S}\sum_{i\in(I_s\setminus I'_s)} \delta_{i,f} d_i^h - \sum_{\substack{j\in I_s\\ i\neq j}} \delta_{i,f} t_{i,j}^h + \sum_{\substack{j\in(I_s\setminus I'_s)\\ i\neq j}} \delta_{i,f} t_{j,i}^h b_{j,i}^h \leq \sum_{k\in K} \pi_{k,f}^h \qquad \forall h \in H, f \in F \tag{10}$$

$$\sum_{h\in H} \pi_{k,f}^h = Q_k x_{k,f} \qquad \forall f \in F, k \in K \tag{11}$$

$$\sum_{\substack{j\in I_s\\ i\neq j}} t_{i,j}^h \leq d_i^h \qquad \forall s \in S, h \in H, i \in (I_s\setminus I'_s) \tag{12}$$

$$\tilde{d}_i^h = D_s^h \frac{\exp(V_i^h)}{\sum_{j\in I_s}\exp(V_j^h)} \qquad \forall s \in S, h \in H, i \in I_s \tag{13}$$

$$b_{i,j}^h = \frac{\exp(V_j^h)}{\sum_{k\in I_s\setminus i}\exp(V_k^h)} \qquad \forall s \in S, h \in H, i \in (I_s\setminus I'_s), j \in I_s \tag{14}$$

$$x_{k,f} \in \{0,1\} \qquad \forall k \in K, f \in F \tag{15}$$

$$y_{k,a,t} \geq 0 \qquad \forall [k,a,t] \in N \tag{16}$$

$$\pi_{k,f}^h \geq 0 \qquad \forall h \in H, k \in K, f \in F \tag{17}$$

$$d_i^h \leq \tilde{d}_i^h \qquad \forall h \in H, i \in I \tag{18}$$

$$0 \leq p_i^h \leq UB_i^h \qquad \forall h \in H, i \in I \tag{19}$$

$$t_{i,j}^h, b_{i,j}^h \geq 0 \qquad \forall s \in S, h \in H, i \in (I_s\setminus I'_s), j \in I_s \tag{20}$$

Fig. 1 Integrated schedule planning model

seats to be decided by the model and we need to make sure that the total does not exceed the capacity (11) Q_k being the capacity of plane type k. Constraints (12) maintain that total redirected passengers from itinerary i to all other itineraries including the no-purchase options do not exceed its realized demand. We have already explained the constraints (13) and (14) in the beginning of the section. Constraints (15)-(20) specify the decision variables. Demand value provided by the logit model, $\tilde{d}_i^h$, serves as an upper bound for the realized demand, d_i^h. The price of each itinerary has an upper bound UB_i, which is assumed to be the average market price plus the standard deviation.

A dataset from a major European airline company is used. The model is implemented in AMPL and solved with BONMIN. For large instances computational time becomes unmanageable. For example, for an instance with 78 flights no feasible solution is found in 12 hours. Therefore we propose a Lagrangian relaxation based heuristic in the next section.

4 Heuristic approach

We relax the constraints (10) introducing the Lagrangian multipliers λ_f^h for each flight f and class h. With this relaxation the model can be decomposed into two subproblems. The first is a revenue maximization model with fare prices modified by the Lagrangian multipliers. The objective function is given by equation (21) subject to constraints (12)-(14) and (18)-(20).

$$z_{REV}(\lambda) = Max \sum_{h \in H} \sum_{f \in F} \sum_{s \in S} \sum_{i \in (I_s \setminus I_s')} \delta_{i,f}(p_i^h - \lambda_f^h) \left(d_i^h - \sum_{\substack{j \in I_s \\ i \neq j}} t_{i,j}^h + \sum_{\substack{j \in (I_s \setminus I_s') \\ i \neq j}} t_{j,i}^h b_{j,i}^h \right) \tag{21}$$

The second is a fleet assignment model with class-fleet seat prizes. The objective function is given by equation (22) subject to constraints (5)-(9), (11) and (15)-(17).

$$z_{FAM}(\lambda) = Min \sum_{k \in K} \sum_{f \in F} \left(C_{k,f} x_{k,f} - \sum_{h \in H} \lambda_f^h \pi_{k,f}^h \right) \tag{22}$$

The overall procedure is given by algorithm 4. Upper bound to the problem is obtained by separately solving the problems of $z_{REV}(\lambda)$ and $z_{FAM}(\lambda)$. Since the capacity constraint is relaxed, we may have infeasibility. Therefore we need to obtain a primal feasible solution which serves as a lower bound. To achieve that we devise a Lagrangian heuristic that uses the optimal solution to $z_{FAM}(\lambda)$ to fix the fleet assignment and class capacity variables. Then it solves a revenue optimization problem with these fixed values dropping the constraints (5)-(9) and (11). This Lagrangian heuristic is included in a neighborhood search loop where neighborhood solutions are explored by changing the fleet assignment of a subset of flights at each iteration. Information provided by the Lagrangian multipliers is used such that flights with high multipliers have higher probability to be selected for the neighborhood search.

The study is in progress and the preliminary results regarding the performance of the heuristic are presented in Table 1. For these small and medium instances, it is seen that we gain computational time and the deviation from the optimal solution is less than 2%. The ongoing work is on the improvement of the heuristic via neighborhood search methods. The heuristic will then be tested for larger data instances.

Algorithm 1 Lagrangian procedure

Require: z_{LB}, $\bar{k}$, $\bar{j}$, ε

$\quad \lambda^0 := 0$, $k := 0$, $z_{UB} := \infty$, $\eta := 0.5$, $j := 0$

$\quad$ **repeat**

$\qquad \{\bar{d}, \bar{\iota}, \bar{b}\} :=$ solve $z_{REV}(\lambda^k)$

$\qquad \{\bar{x}, \bar{y}, \bar{\pi}\} :=$ solve $z_{FAM}(\lambda^k)$

$\qquad z_{UB}(\lambda^k) := z_{REV}(\lambda^k) - z_{FAM}(\lambda^k)$

$\qquad z_{UB} := \min(z_{UB}, z_{UB}(\lambda^k))$

$\qquad$ **if** no improvement(z_{UB}) **then**

$\qquad\quad \eta := \eta \div 2$

$\qquad$ **end if**

$\qquad$ **repeat**

$\qquad\quad \{\bar{x}, \bar{\pi}\} :=$ Neighborhood search($\{\bar{x}, \bar{\pi}\}$)

$\qquad\quad lb :=$ Lagrangian heuristic ($\{\bar{x}, \bar{\pi}\}$)

$\qquad\quad j := j + 1$

$\qquad$ **until** $j \geq \bar{j}$

$\qquad z_{LB} := \max(z_{LB}, lb)$

$\qquad G :=$ compute sub-gradient($z_{UB}, z_{LB}, \{\bar{d}, \bar{\iota}, \bar{b}, \bar{x}, \bar{y}, \bar{\pi}\}$)

$\qquad T :=$ compute step($z_{UB}, z_{LB}, \{\bar{d}, \bar{\iota}, \bar{b}, \bar{x}, \bar{y}, \bar{\pi}\}$)

$\qquad \lambda^{k+1} := \max(0, \lambda^k - TG)$

$\qquad k := k + 1$

$\quad$ **until** $||TG||^2 \leq \varepsilon$ or $k \geq \bar{k}$

Table 1 Performance of the heuristic

	BONMIN solver		Heuristic method		
Instances	opt solution	time(min)	best solution	deviation	time(min)
9 flights - 800 pax.	52,876	0.24	52,876	0%	0.07
18 flights - 1096 pax.	78,275	41.04	77,224	1.34%	16.09
26 flights - 2329 pax.	176,995	204.56	173,509	1.97%	41.24

Acknowledgements The authors would like to thank EPFL Middle East for funding the project.

References

1. Atasoy, B., Salani, M., Leonardi, C., Bierlaire, M.: Clip-Air, a flexible air transportation system. Technical report TRANSP-OR 110929. Transport and Mobility Laboratory, ENAC, EPFL (2011)

2. Barnhart, C., Kniker, T.S., Lohatepanont, M.: Itinerary-based airline fleet assignment. Transportation Science **36**, 199–217 (2002)

3. Lohatepanont, M., Barnhart, C.: Airline schedule planning: Integrated models and algorithms for the schedule design and fleet assignment. Transportation Science **38**, 19–32 (2004)

4. Schön, C.: Market-oriented airline service design. Operations Research Proceedings, 361–366 (2006)

5. Talluri, K.T., van Ryzin, G.J.: Revenue management under a general discrete choice model of customer behavior. Management Science **50**, 15–33 (2004)

Stream XIV
Stochastic Programming, Stochastic Modeling and Simulation

Karl Frauendorfer, University of St.Gallen, SVOR Chair
Georg Pflug , University of Vienna, ÖGOR Chair
Rüdiger Schultz, University of Duisburg-Essen, GOR Chair

Approximate Formula of Delay-Time Variance in Renewal-Input General-Service-Time Single-Server Queueing System

Yoshitaka Takahashi, Yoshiaki Shikata and Andreas Frey

Abstract Approximate formulas of the variance of the waiting-time (also called as delay-time variance) in a renewal-input general-service-time single-server (GI/GI/1) system play an important role in practical applications of the queueing theory. However, there exists almost no literature on the approximate formulas of the delay-time variance in the GI/GI/1 system. The goal of this paper is to present an approximate formula for the delay-time variance. Our approach is based on the combination of a higher-moment relationship between the unfinished work and the waiting time, and the diffusion process approximation for the unfinished work. To derive the former relationship, we apply Miyazawa's rate conservation law for the stationary point process. Our approximate formula is shown to converge to the exact result for the Poisson-input system as traffic intensity goes to the unity. The accuracy of our approximation is validated by simulation results.

1 Introduction

We consider an iid-general (renewal) input, iid-general service-time stationary GI/GI/1 queueing system with infinite-capacity waiting room under the First-Come-First-Served (FCFS) service discipline. Bounds and approximations for the GI/GI/1 system are well-developed, but most of them are for the mean delay time (mean waiting time). Here, we focus on the variance of the delay time to propose a new

Yoshitaka Takahashi
Faculty of Commerce, Waseda University, 169-8050 Tokyo, Japan, e-mail: yoshitak@waseda.jp

Yoshiaki Shikata
Faculty of Informatics for Arts, Shobi University, 350-1153 Saitama, Japan, e-mail: y-shikata@shobi-u.ac.jp

Andreas Frey
Faculty of Business Management and Social Sciences, University of Applied Sciences Osnabrueck, 49076 Osnabrueck, Germany, e-mail: a.frey@hs-osnabrueck.de

D. Klatte et al. (eds.), *Operations Research Proceedings 2011*, Operations Research Proceedings, 503
DOI 10.1007/978-3-642-29210-1_80, © Springer-Verlag Berlin Heidelberg 2012

approximation, since the second-moment of the delay time is practically of importance.

2 Relationship between Unfinished Work and Waiting Time

We denote by $V(t)$ the unfinished work at time t in our queueing system and N be the arrival point process with intensity $\lambda := E\{N((0,1])\}$. We denote by P_0 the Palm probability measure with respect to the arrival point process N, and E_0 the expectation with P_0. Let ρ be the traffic intensity, i.e.,

$$\rho = \lambda E_0(B) = \lambda E(B) \tag{1}$$

It should be noted that $E_0(B^n) = E(B^n)$ for any non-negative integer n and an iid service time random variable B; see Takahashi and Miyazawa [7] for the proof. Note also that the sample-path discontinuous points of $V(t)$ are included in the point process N.

If we let $X(t) := [V(t)]^2$, we have

$$X'(t) = 2V(t)V'(t) = -2V(t) \qquad \text{a.s.-}P$$

$$\begin{aligned}
X(0-) - X(0+) &= V(0-)^2 - V(0+)^2 \\
&= [V(0-) - V(0+)][V(0-) + V(0)] \\
&= (-B_0)[2V(0-) + B_0] \qquad \text{under } P_0
\end{aligned}$$

Here, B_0 is the service time of the arriving customer at time origin.

Applying Miyazawa's rate conservation law (RCL, see [5] and [6]):

$$E[X'(0)] = \lambda E_0[X(0-) - X(0+)], \tag{2}$$

we then have

$$2E[V(0)] = 2\lambda E_0(B_0)E_0[V(0-)] + \lambda E_0(B_0^2)$$

from which we obtain the well-known Brumelle's formula [1]:

$$E[V(0)] = \rho E_0[W_0] + \lambda E_0(B_0^2)/2 \tag{3}$$

Indeed, Eq.(3) comes from the fact that $V(0-) = W_0$ the waiting time of the customer arriving at time origin under P_0.

In a similar way to derive Eq.(3), if we let $X(t) := [V(t)]^3$, we have

$$X'(t) = 3[V(t)]^2 V'(t) = -3[V(t)]^2 \qquad \text{a.s.-}P$$

$$\begin{aligned}
X(0-) - X(0+) &= V(0-)^3 - V(0+)^3 \\
&= (V(0-) - V(0+))\{[V(0-)]^2 + V(0-)V(0+) + [V(0+)]^2\} \\
&= (-B_0)\{[V(0-)]^2 + V(0-)[V(0-)+B_0] + [V(0-)+B_0]^2\} \\
&= (-B_0)\{3[V(0-)]^2 + 3V(0-)B_0 + B_0^2\} \qquad \text{under } P_0
\end{aligned}$$

Applying Miyazawa's RCL [Eq.(2)] again enables us to obtain the second-moment relationship between the unfinished work and the waiting time (using the first three moments of the service time):

$$E\{[V(0)]^2\} = \rho E_0(W^2) + \lambda E_0(W)E_0(B_0^2) + \lambda E_0(B_0^3)/3 \tag{4}$$

We finally obtain the following formula. For any non-negative integer m, we have

$$E\{[V(0)]^m\} = \frac{\lambda}{m+1} \sum_{k=0}^{m} \sum_{j=0}^{k} E_0(W^{m-j})E_0(B^{j+1}) \tag{5}$$

by mathematical induction.

3 Diffusion Approximations for Unfinished Work

Let $V(t)$ represent the unfinished work at time t in the single-server system. We approximate the unfinished work process by a diffusion process with a negative drift α and diffusion coefficient β. These two parameters are obtained in Gelenbe [2].

$$\alpha = \rho - 1 \tag{6}$$

$$\beta = \rho \frac{C_a^2 + C_b^2}{\mu} \tag{7}$$

Here, $\rho = \lambda/\mu$, λ: arrival rate, μ: service rate. C_a^2 (or C_b^2): the squared coefficient of variation of the inter-arrival time (service time).

We further need an appropriate boundary condition at space origin ($x = 0$). One is the reflecting barrier (RB) boundary. Another one is the elementary return (ER) boundary. Let f_{RB} (or f_{ER}) be the probability density function (pdf) of the diffusion process with RB (ER) boundary. Assuming the steady state for which ρ is necessarily less than unity ($\rho < 1$), these pdfs f_{RB} and f_{ER} are explicitly obtained in e.g. Heyman [3] and Gelenbe [2].

$$f_{RB}(x) = -\frac{2\alpha}{\beta} e^{\frac{2\alpha x}{\beta}} \tag{8}$$

$$f_{ER}(x) = -\frac{2\alpha}{\beta} \lambda e^{\frac{2\alpha x}{\beta}} \int_0^\infty (1 - B(y)) e^{\frac{-2\alpha y}{\beta}} dy \tag{9}$$

where $B(y)$ is the cumulative distribution of the service time.

From these pdfs, any moment of the unfinished work can be derived. Using Brumelle's formula [1] which links the mean unfinished work and the mean waiting time, we can derive the mean waiting (delay) time approximations. The mean waiting time approximations are respectively obtained as

$$E_0(W_{RB}) = \frac{(C_a^2 - 1) + \rho(C_b^2 + 1)}{2(1 - \rho)\mu} \tag{10}$$

$$E_0(W_{ER}) = \frac{\rho(C_a^2 + C_b^2)}{2(1 - \rho)\mu} \tag{11}$$

The index RB (or ER) denotes the result for the reflecting barrier (or elementary return) boundary condition. We will use the same notation in the subsequent section. For the accuracy for the mean waiting (delay) time approximation, see Hoshi et al. [4] and Takahashi et al. [8], showing ER is better than RB for several situations.

4 Variance Approximations for Delay Time

Our relationship between the unfinished work and the waiting time (5) together with the diffusion process approximation with RB or ER boundary now enables us to propose any higher moment approximation. However, due to the lack of higher moment information than two in the diffusion parameters (α and β), we will restrict ourselves to the second moment, i.e., variance of the waiting (delay) time. Here are our final approximations after some calculations.

$$Var(W_{RB}) = E_0([W_{RB}]^2) - \{E_0(W_{RB})\}^2$$
$$= \{E(V_{RB}^2) - \lambda E_0(W_{RB})E_0(B^2) - \frac{\lambda}{3}E_0(B^3)\}/\rho - \{E_0(W_{RB})\}^2 \tag{12}$$

where

$$E(V_{RB}^2) = \frac{1}{2}\{\frac{\rho(C_a^2 + C_b^2)}{(1 - \rho)\mu}\}^2 \tag{13}$$

$$Var(W_{ER}) = [\frac{\rho(C_a^2 + C_b^2)}{2(1 - \rho)\mu}]^2 \tag{14}$$

For the Poisson-input ($C_a^2 = 1$) system it should be noted that the mean delay approximations [$E_0(W_{RB})$ and $E_0(W_{ER})$] are consistent with exact (P-K) formula, however, the delay-time variance approximations are no longer consistent but converge to the exact result as traffic intensity goes to the unity (in heavy traffic).

Comparing this delay-time variance exact result and $Var(W_{ER})$ in the Poisson-input system, we heuristically suggest the following formula for the GI/GI/1 queueing system.

$$Var(W_{HR}) = [\frac{\rho(C_a^2 + C_b^2)}{2(1-\rho)\mu}]^2 + \frac{\lambda(C_a^2 + 1)b^{(3)}}{6(1-\rho)} \tag{15}$$

Here, $b^{(3)} = E_0(B^3)$ is the third moment of the service time. Eq.15 is reduced to the exact result for the Poisson-input system.

In Figs 1 and 2, we show numerical and simulated results for a two-stage hyper exponential (H_2, $C_a^2 = 2.0$) or two-stage Erlang input (E_2, $C_a^2 = 0.5$), two-stage Erlang service (E_2, $C_b^2 = 0.5$) queueing system, assuming $\mu = 1$. The accuracy of $Var(W_{HR})$ is seen to be best as we can expect, while the accuracy of $Var(W_{ER})$ is better than that of $Var(W_{RB})$.

5 Conclusion

We have presented simple delay-time variance approximations for the GI/GI/1 queueing system. Unlike the mean delay time, the diffusion approximations for the unfinished work with reflecting barrier (RB) and elementary return (ER) boundary conditions have not seen to be consistent with the exact result for delay-time variance in the Poisson-input system. We have therefore proposed an approximation to refine the diffusion approximation with elementary return boundary so that it is consistent with the exact result for the delay-time variance in the Poisson-input system.

References

1. Brumelle, S.L.: On the relationship between customer and time averages in queues. J. Appl. Prob. **8(3)**, 508–520 (1971)
2. Gelenbe, E.: Probabilistic models of computer systems, Part II, Diffusion approximations, waiting times and batch arrivals. Acta Informatica **12**, 285–303 (1979)
3. Heyman, D.P.: A diffusion model approximation for the GI/G/1 queue in heavy traffic. Bell System Tech. J. **54**, 1637–1646 (1976)
4. Hoshi, K., Takahashi, Y., Komatsu, N.: A further remark on diffusion approximations with elementary return and reflecting barrier boundaries. In: Proceedings of OR 2010 Munich, pp. I.6.TA-15-2:1–6. Springer (2010)
5. Miyazawa, M.: The derivation of invariance relations in complex queueing systems with stationary inputs. Adv. Appl. Prob. **15**, 875–885 (1983)
6. Miyazawa, M.: Rate conservation laws: a survey. Queueing Systems **15(1-4)**, 1–58 (1994)
7. Takahashi, Y., Miyazawa, M.: Relationship between queue-length and waiting time distributions in a priority queue with batch arrivals. J. Operations Res. Soc. Japan **37(1)**, 48–63 (1994)
8. Takahashi, Y., Shikata, Y., Frey, A.: Diffusion approximation for a web-server system with proxy servers. In: Proceedings of OR 2010 Munich, pp. I.6.TC-15-3:1–6. Springer (2010)

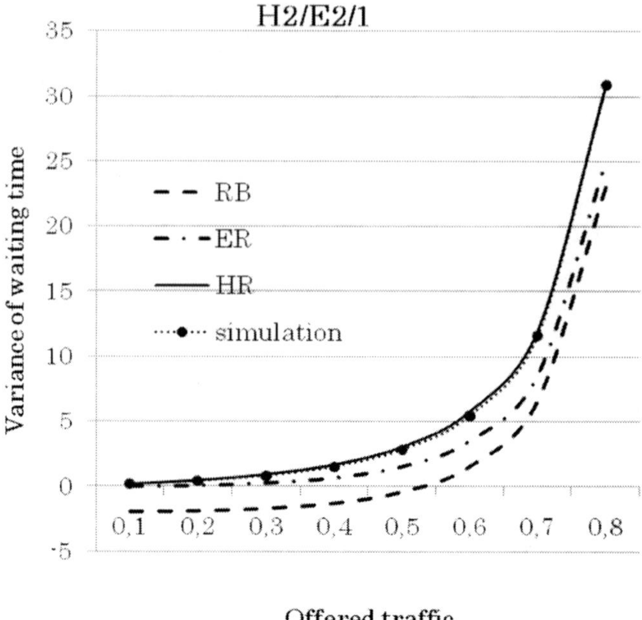

Fig. 1 Variance of waiting time in the $H_2/E_2/1$ system. (Dots: simulation results with 95% confidence intervals).

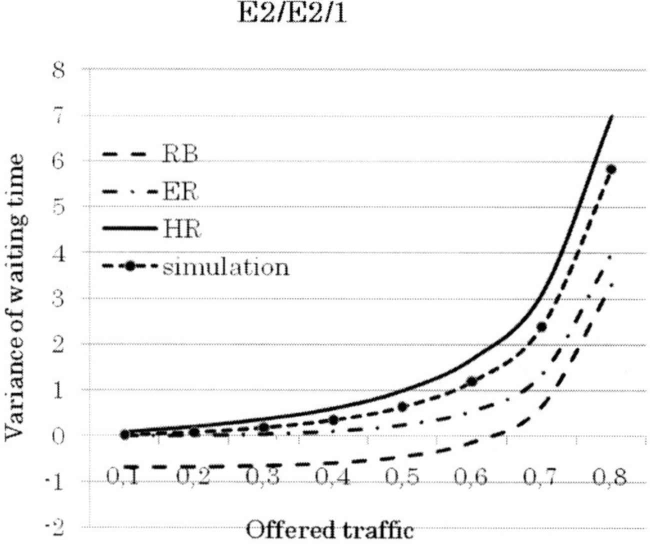

Fig. 2 Variance of waiting time in the $E_2/E_2/1$ system. (Dots: simulation results with 95% confidence intervals).

On solving strong multistage nonsymmetric stochastic mixed 0-1 problems

Laureano F. Escudero, M. Araceli Garín, María Merino and Gloria Pérez

Abstract In this note we present the basic ideas of a parallelizable Branch-and-Fix Coordination algorithm for solving medium and large-scale multistage mixed 0-1 optimization problems under uncertainty, where this one is represented by nonsymmetric scenario trees. An assignment of the constraint matrix blocks into independent scenario cluster MIP submodels is performed by a named cluster splitting variable - compact representation (explicit non anticipativity constraints between the cluster submodels until break stage t^*). Some computational experience using CPLEX within COIN-OR is reported while solving a large-scale real-life problem.

1 Introduction

Stochastic Optimization is currently one of the most robust tools for decision making. It can be used in real-world applications in a wide range of problems from different areas, see e.g. [8]. The mixed integer problems under uncertainty have been studied in [1,6], just for citing a few references. An extended bibliography of Stochastic Integer Programming (SIP) has been collected in [7].

It is well known that a multistage mixed 0-1 problem under uncertainty with a finite number of possible future scenarios has a mixed 0-1 Deterministic Equivalent Model (DEM), where the risk of providing a wrong solution is included in the model via a set of representative scenarios. We should point out that the scenario tree in real-life problems is very frequently nonsymmetric and then, the traditional splitting

Laureano F. Escudero
Dpto. de Estadística e Investigación Operativa, Universidad Rey Juan Carlos, Calle Tulipán, s/n. 28933 Móstoles (Madrid), Spain. e-mail: laureano.escudero@urjc.es

M. Araceli Garín
Dpto. de Economía Aplicada III, Universidad del País Vasco, Avenida Lehendakari Aguirre, 83, 48015 Bilbao (Vizcaya), Spain. e-mail: mariaaraceli.garin@ehu.es

María Merino · Gloria Pérez
Dpto. de Matemática Aplicada y Estadística e Investigación Operativa, Universidad del País Vasco, Barrio Sarriena s/n Leioa (Vizcaya), Spain. e-mail: {maria.merino@ehu.es gloria.perez@ehu.es}

D. Klatte et al. (eds.), *Operations Research Proceedings 2011*, Operations Research Proceedings, 509
DOI 10.1007/978-3-642-29210-1_81, © Springer-Verlag Berlin Heidelberg 2012

variable representation for the nonanticipativity constraints (for short, NAC) on the 0-1 and continuous variables does not appear readily accessible for manipulations that are required by the decomposition strategies for problem solving. The NAC have been stated in [9]. So, a new type of strategies is necessary for solving medium and large scale instances of the problem. The decomposition approaches appear as the unique way to deal with these large-scale problems.

In this note we present the basic ideas of both the stochastic mixed 0-1 optimization modeling approach and the parallelizable Branch-and-Fix Coordination MultiStage algorithm, named BFC-MS, for solving general mixed 0-1 optimization problems under uncertainty, where this one is represented by nonsymmetric scenario trees. In the full paper, see [4], we present the reasons for considering scenario cluster submodels in a mixture of the splitting variable and compact representations. Given the structuring of the clusters, independent cluster submodels are generated, then, allowing parallel computation for obtaining lower bounds to the optimal solution value as well as feasible solutions for the problem until getting the optimal one. (Tighter lower bounds can be obtained by following the lines presented in [5] by using Lagrangean decomposition approaches in a risk aversion environment). A splitting variable representation with explicit NAC is presented for linking the submodels together, as well as a compact representation for each submodel to treat the implicit NAC related to each of the scenario clusters. Then, the algorithm that we propose uses the Twin Node Family (TNF) concept introduced in [2, 3], and it is specially designed for coordinating and reinforcing the branching nodes and the branching 0-1 variable selection strategies at each Branch-and-Fix tree, such that the nonsymmetric scenario tree which will be partitioned into smaller scenario cluster subtrees. Some computational experience is reported while solving up to optimality a large-scale real-life problem.

The remainder of the note is organized as follows. Section 2 presents the multistage mixed 0-1 problem under uncertainty in a splitting variable representation. Section 3 presents the basic ideas for generating the required information in order to know up to what stage the cluster submodels have common information. Section 4 reports some computational experience.

2 Splitting variable representation in stochastic optimization

Let us consider the following multistage deterministic mixed 0-1 model

$$\min \sum_{t \in \mathcal{T}} a_t x_t + c_t y_t$$
$$s.t. \ A'_t x_{t-1} + A_t x_t + B'_t y_{t-1} + B_t y_t = b_t \ \forall t \in \mathcal{T} \tag{1}$$
$$x_t \in \{0, 1\}^{nx_t}, \ y_t \in \mathbb{R}^{+ny_t}, \qquad \forall t \in \mathcal{T}$$

where $\mathcal{T}$ is the set of stages, such that $T = |\mathcal{T}|$, x_t and y_t are the nx_t and ny_t dimensional vectors of the 0-1 and continuous variables, respectively, a_t and c_t are the vectors of the objective function coefficients, A_t and B_t are the constraint matrices, and b_t is the right-hand-side vector (*rhs*), for stage t.

This model can be extended to consider uncertainty in some of the main parameters, in our case, the objective function, the right hand side and the constraint matrix coefficients. To introduce the uncertainty in the parameters, we will use a scenario analysis approach. A scenario consists of a realization of all random variables in all stages, that is, a path through the scenario tree. Let Ω denote the set of scenarios, $\omega \in \Omega$ will represent a specific scenario and w^ω will denote the likelihood or probability assigned by the modeler to scenario ω, such that $\sum_{\omega \in \Omega} w^\omega = 1$. Let also $\mathcal{G}$ denote the set of scenario groups (i.e., nodes in the underlying scenario tree), and $\mathcal{G}_t$ denote the subset of scenario groups that belong to stage $t \in \mathcal{T}$, such that $\mathcal{G} = \cup_{t \in \mathcal{T}} \mathcal{G}_t$. Ω_g denotes the set of scenarios in group g, for $g \in \mathcal{G}$. Notice that the scenario group concept corresponds to the node concept in the underlying scenario tree.

A symmetric scenario tree assumes that the number of branches is the same for all conditional distributions in the same stage, that is, the number of branches arising from any scenario group at each stage t to the next one is the same for all groups in $\mathcal{G}_t, t \leq T - 1$. A *nonsymmetric* tree assumes that the number of branches is not the same for at least one stage $t \in \mathcal{T}$. Moreover, without lost of generality, we consider nonsymmetric trees in the full paper.

We say that two scenarios belong to the same group in a given stage provided that they have the same realizations of the uncertain parameters up to the stage. Following the *nonanticipativity* principle stated in [9], both scenarios should have the same value for the related variables with the time index up to the given stage.

The *splitting variable* representation of the DEM of the full recourse stochastic version related to the multistage deterministic problem (1) can be expressed as follows,

$$z_{MIP} = \min \sum_{\omega \in \Omega} \sum_{t \in \mathcal{T}} w^\omega \left(a_t^\omega x_t^\omega + c_t^\omega y_t^\omega \right)$$

$$s.t. \ A_t'^\omega x_{t-1}^\omega + A_t^\omega x_t^\omega + B_t'^\omega y_{t-1}^\omega + B_t^\omega y_t^\omega = b_t^\omega, \ \forall \omega \in \Omega, \ t \in \mathcal{T}$$

$$x_t^\omega - x_t^{\omega'} = 0, \ \forall \omega, \omega' \in \Omega_g : \omega \neq \omega', \ g \in \mathcal{G}_t, \ t \in \mathcal{T} - T \qquad (2)$$

$$y_t^\omega - y_t^{\omega'} = 0, \ \forall \omega, \omega' \in \Omega_g : \omega \neq \omega', \ g \in \mathcal{G}_t, \ t \in \mathcal{T} - T$$

$$x_t^\omega \in \{0,1\}^{nx_t^\omega}, \ y_t^\omega \in \mathbb{R}^{+ny_t^\omega}, \ \forall \omega \in \Omega, \ t \in \mathcal{T}.$$

According to the nonanticipativity principle cited above, the equalities of the constraint matrices and *rhs* for the scenarios that belong the same scenario group must be satisfied for each stage t. Observe that for a given stage t, $A_t'^\omega$ and A_t^ω are the technology and recourse matrices for the x_t variables and $B_t'^\omega$ and B_t^ω are the corresponding ones for the y_t variables. Notice that $x_t^\omega - x_t^{\omega'} = 0$ and $y_t^\omega - y_t^{\omega'} = 0$ are the NAC. Finally, nx_t^ω and ny_t^ω denote the dimensions of the vectors of the variables x and y, respectively, related to stage t under scenario ω.

3 Scenario clustering in nonsymmetric scenario trees

It is clear that the explicit representation of the NAC is not desirable for all pairs of scenarios in order to reduce the dimensions of the model. In fact, we can represent implicitly the NAC for some pairs of scenarios in order to gain computational efficiency.

We will decompose the scenario tree into a subset of scenario clusters denoted by $\mathscr{P} = \{1, ..., q\}$ with $q = |\mathscr{P}|$. Let Ω^p denote the set of scenarios that belongs to a generic cluster p, where $p \in \mathscr{P}$ and $\sum_{p=1}^{q} |\Omega^p| = |\Omega|$. It is clear that the criterion for scenario clustering is instance dependent. In any case, notice that $\Omega^p \bigcap \Omega^{p'} = \emptyset$, $p, p' = 1, \ldots, q : p \neq p'$ and $\Omega = \cup_{p=1}^{q} \Omega^p$. We propose to choose the number of scenario clusters q as any value from the subset $\mathscr{Q} = \{|\mathscr{G}_1|, |\mathscr{G}_2|, \ldots, |\mathscr{G}_T|\}$. As we will see in the definitions below, the value q will be associated with the number of stages with explicit NAC between scenario clusters.

Definition 1. A *break stage* t^* is the stage t such that the number of scenario clusters is $q = |\mathscr{G}_{t^*+1}|$, where $t^* + 1 \in \mathscr{T}$. In this case, each cluster $p \in \mathscr{P}$ is induced by one scenario group $g \in \mathscr{G}_{t^*+1}$ and contains all scenarios belonging to that group, i.e., $\Omega^p = \Omega_g$.

Definition 2. The *scenario cluster* models are those that result from the relaxation of NAC until some break stage t^* in model (2), called t^*-*decomposition*.

The decomposition of the scenario tree into scenario clusters is due to the way in which our algorithm BFC-MS works. It considers the explicit NAC related to the stages from $t = 1$ until the break stage t^*, since the NAC from its next stage are implicitly considered while solving the scenario cluster MIP submodels.

4 Computational experience

The algorithm BFC-MS has been implemented in a C++ experimental code. It alternatively uses the open source optimization engine COIN-OR and the state-of-the-art commercial optimization engine CPLEX v12.2 for solving the LP relaxation and mixed 0-1 submodels. See in the full paper [4] an extensive computational experience for non-symmetric multistage scenario trees. Moreover, in this note we report the main results obtained while solving a large-scale real-life case by using the SW/HW platform given by a workstation Dell Precision T7500 under Windows 7, operating system 64 bits, 2.4 GHz and 12 GB RAM, such that the scenario cluster MIP submodels have been solved by using CPLEX.

Our pilot case is a large-scale problem for supporting the decision making process on a system of production, distribution, processing and transportation of natural gas by considering a highly volatile product demand along a time horizon. In this case, the multistage scenario tree is a symmetric one. The objective consists of maximizing the expected profit along a horizon of $|\mathscr{T}| = 4$ years over a set of $|\Omega|{=}1000$

scenarios with $|\mathscr{G}| = 1111$ scenario groups in a risk neutral environment. The dimensions of the DEM in compact representation of the stochastic problem are m=98456 constraints, nx=34680 0-1 variables, ny=22221 continuous variables, nel =253340 nonzero elements in the constraint matrix and $dens$=0.0045% constraint matrix density. These are big dimensions for plain using of today state-of-the-art optimization engines such as XPRESS and CPLEX, since those engines could not solve the problem in an affordable computing time for medium size hardware. In particular, CPLEX could not solve the DEM for the RAM available in the experiment, at least (see above).

The time period $t = 1$ has been chosen as the t^*-*decomposition*. So, we have considered $q = 10$ scenario clusters Ω^p for $p = 1, ..., q$, such that they are constructed by favoring the approach that shows higher scenario clustering for greater number of scenario groups in common. Notice that, in this case, the 0-1 and continuous variables implicitly satisfy the NAC while solving the compact scenario cluster MIP submodels, see [2]. Those submodels have the same dimensions, due to the symmetric scenario subtrees to represent the uncertain parameters. The dimensions of the MIP submodels to be solved at the root Twin Nodes Family (TNF) in the Branch-and-Fix tree generated by the algorithm BFC-MS are m=9950 constraints, nx=3540 0-1 variables, ny=2241 continuous variables, nel =25568 nonzero elements in the constraint matrix and $dens$=0.04% constraint matrix density.

It is worthy to point out that only 10 twin nodes are explored for obtaining the optimal solution for the original stochastic problem. The other main results of our computational experience are as follows: z_0 =79234.6, initial upper bound computed as the expected solution value at the root node obtained by solving independently the scenario cluster MIP submodels; z_{DEM} =78995.9, solution value of the original DEM; GAP_0=0.3%, optimality gap defined as $\frac{z_{DEM} - z_0}{z_0}$ (in %); t_0 =31, elapsed time (in secs.) to obtain the z_0 solution; and $t_{DEM} = 182$, elapsed time (in secs.) to obtain the z_{DEM} solution, including the $t_0 = 31$ secs.

As the first observation, notice the very small value for GAP_0. So, z_0 is a strong upper bound of the solution value of the original stochastic problem that is obtained in a very affordable elapsed time. Secondly, the total elapsed time required by the algorithm BFC-MS is very small. These astonishing results are confirmed by the similar results reported in the full paper for large-scale testbed of randomly generated instances, whose LP solution value could be calculated and its related gap, say GAP_{LP}, is not as small as its GAP_0 that itself is also very small.

5 Conclusions

A modeling approach and an exact Branch-and-Fix Coordination algorithmic framework, named BFC-MS, have been proposed for solving multistage mixed 0-1 & combinatorial problems under uncertainty in the parameters. The 0-1 and continuous variables can also appear at any stage. The approach treats the uncertainty by scenario cluster analysis, allowing the scenario tree to be a nonsymmetric one.

We have reported the astonishing small computing time required by the proposed algorithm for solving a large-scale real-life case, using CPLEX as a solver of the required MIP scenario cluster submodels.

Acknowledgements This research has been partially supported by the projects ECO2008-00777 ECON from the Ministry of Education and Science, Grupo de Investigación IT-347-10 from the Basque Government, URJC-CM-2008-CET-3703 and RIESGOS CM from Comunidad de Madrid, and PLANIN MTM2009-14087-C04-01 from the Ministry of Science and Innovation, Spain. We would like to express our gratefulness to Prof. Dr. Fernando Tusell for making easier the access to the Laboratory of Quantitative Economics, University of the Basque Country (UPV/EHU, Bilbao, Spain) to perform and check some of our computational experience. The authors would also like to thank to Prof. Dr. Asgeir Tomasgard and his PhD student Marte Fodstad from Norges Teknisk Naturvitenskapelige Universitet (NTNU, Trondheim, Norway) for providing the instance on natural gas transportation infrastructures decision making that has been used for testing of the proposed approach.

References

1. Birge JR, Louveaux FV. Introduction to Stochastic Programming. Springer (1997).
2. Escudero LF, Garín A, Merino M, Pérez G. BFC-MSMIP: an exact Branch-and-Fix Coordination approach for solving multistage stochastic mixed 0-1 problems. TOP **17**, 96-122 (2009).
3. Escudero LF, Garín A, Merino M, Pérez G. On BFC-MSMIP strategies for scenario cluster partitioning, and twin node family branching selection and bounding for multistage stochastic mixed integer programming. Computers & Operations Research. **37**, 738-753 (2010).
4. Escudero LF, Garín A, Merino M, Pérez G. An algorithmic framework for solving large scale multistage stochastic mixed 0-1 problems with nonsymmetric scenario trees. Computers & Operations Research. **39**, 1133-1144 (2012).
5. Escudero LF, Landete M, Rodriguez-Chia A. Stochastic Set Packing Problem. European Journal of Operational Research. **211**, 232-240 (2011).
6. Schultz R. Stochastic programming with integer variables. Mathematical Programming, Ser. B. **97**, 285-309 (2003).
7. van der Vlerk MH. Stochastic Integer Programming Bibliography. http://www.eco.rug.nl/mally/biblio/sip.html (2007).
8. Wallace SW, Ziemba WT, editors. Applications of Stochastic Programming. MPS-SIAM Series on Optimization (2005).
9. Wets RJ-B. Stochastic programs with fixed recourse: The equivalent deterministic program. *SIAM Review.* **16**, 309-339 (1974).

A Cost-Equation Analysis of General-Input General-Service Processor Sharing System

Kentaro Hoshi, Naohisa Komatsu, and Yoshitaka Takahashi

Abstract In the computer-communication field, we frequently encounter a situation in which the processor sharing (PS) rule is adopted for a time-shared server next to the first-come-first-serve (FCFS) rule. There has been much work on the Poisson-input general-service M/GI/1 (PS) system. However, there have been few results for a general-input general-service GI/GI/1 (PS) system. We deal with this general GI/GI/1 (PS) system. We show that the cost-equation analysis enables us to derive the relationship between the mean (time-average) unfinished work and the mean (customer-average) sojourn time. Our relationship is then applied to extend and generalize the previous results, e.g., Brandt et al.'s relationship between the mean (customer-average) sojourn times under the FCFS and PS rules, and Klein-rock's conservation law for the M/GI/1 (PS) system.

1 Introduction

We consider a stochastic service system, assuming the followings. (1) There is single server. Customers are served under the processor-sharing (PS) rule. Arriving customers do not have to wait for service under the PS rule, because customers are served promptly unlike the first-come-first-serve (FCFS) rule, although the service rate for an individual customer becomes slow. Indeed, each customer receives $1/n$ of the service capacity if there are $n(n > 0)$ customers in the single-server system. (2) Customers arrive at the single-server system. The customer inter-arrival times, A, are identically distributed with an arrival rate $\lambda : \lambda = 1/E_0(A)$. Here, E_0 denotes the expectation with respect to the Palm measure P_0 for the arrival point process; Our system is customer-stationary under P_0; see Refs [3], [6], [8]. (3) The requested service times, B, are independent, and identically distributed (iid) with a mean $E(B) = 1/\mu$, and second moment $E(B^2)$. Here, E denotes the expectation with respect to the probability measure P under which our system is stationary. The traffic intensity ρ is then given by $\rho = \lambda/\mu$ which is assumed to be less than

Kentaro Hoshi
Faculty of Science and Engineering, Waseda University, Japan, e-mail: sizer@aoni.waseda.jp

D. Klatte et al. (eds.), *Operations Research Proceedings 2011*, Operations Research Proceedings, 515
DOI 10.1007/978-3-642-29210-1_82, © Springer-Verlag Berlin Heidelberg 2012

unity ($\rho < 1$) for stability. Using the Kendall's notation our system will be denoted by GI/GI/1 (PS). It should be noted that the iid requested service time sequence yields that a stochastic behavior of the requested service time B under the probability P and the corresponding stochastic behavior of B under the Palm measure P_0 are identical. Thus, we have $E(B^n) = E_0(B^n)$ *for any natural number* $n(n \in N)$ and $P(B > x) = P_0(B > x)$ *for any real number* $x(x \in R)$ See Takahashi et al. [16] for the proof. Next to the first-come-first-serve (FCFS) rule, the PS rule arises as a natural paradigm in a variety of practical situations, including a time-shared computer-communication system. There has been much work on the PS rule, starting with the pioneering work of Kleinrock [9]. However, most of work assumed a Poisson-input general service M/GI/1 (PS) system, or a renewal-input exponential-service GI/M/1 (PS) system [11]. Sengupta [15] proposed an approximation for the sojourn time distribution for a renewal-input general-service GI/GI/1 (PS) system, but his approximation required a numerical solution for the transcendental equation appearing in the GI/M/1 queueing analysis; see also Refs [1], [2] and references therein. For the general-input general-service GI/GI/1 (PS) system there is only one exception to the best of our knowledge. Brandt et al. [5] applied a sample-path analysis to derive the relationship between the mean sojourn times under FCFS and PS rules. However, they required additional assumptions on the service time distribution. The main purpose of the paper is to show that the cost-equation approach enables us to derive the relationship between the mean unfinished work and sojourn time. The cost-equation approach is different from the sample-path approach taken in Brandt et al [5].

2 Relationship Between Mean Unfinished Work and Mean Sojourn Time

Imagine that entering customers are forced to pay money to the system. According to Ross [12], we have the following cost equation for a wide class of stationary queueing systems.

2.1 Cost Equation

$$\textit{Average rate at which the system earns}$$

$$= \lambda (\textit{Average amount an entering customer pays}), \tag{1}$$

where λ is the arrival rate of entering customers. To be more exact, Average on the left-hand side of Eq. (1) corresponds to the time-average (E), while Average on the right-hand side of Eq. (1) corresponds to the customer-average (E_0). Cost equation links the time-average and customer-average. We have to consider an appropriate cost mechanism. For our GI/GI/1 (PS) system, we consider the following cost mechanism: Each customer pays at a rate of y (per unit time) when his remaining requested service-time is y. Thus, the rate at which the system earns is simply the unfinished work in the system. The cost equation (1) yields

$$E(U) = \lambda E_0[amount\ paid\ by\ a\ customer], \tag{2}$$

where E(U) denotes the mean (time-average) unfinished work in the GI/GI/1 system.

2.2 Relationship between E(U) and Conditional Mean Sojourn Time $E_0(T|B = x)$

In this subsection we derive the relationship between mean (time-average) unfinished work E(U) and $E_0(T|B = x)$, where $E_0(T|B = x)$ is the conditional mean (customer-average) sojourn time given that the customer's requested service time is equal to x (sec), and B denotes the requested service time random variable. We start with the following lemma. [Lemma-1] Suppose that each customer pays at a rate of y (per unit time) when his remaining requested service time is y. For discrete random variables, we then have eq.(3). For continuous random variables, we then have eq.(4)

$$E_0[amount\ paid\ by\ a\ customer] = \sum_{j=1}^{\infty} E_0(T|B = x_j)P(B \geq x_j). \tag{3}$$

$$E_0[amount\ paid\ by\ a\ customer] = \int_0^{\infty} E_0(T|B = x_j)P(B > x)dx. \tag{4}$$

[Proof of Lemma 1] We will prove Eq. (3) for a discrete-random variable case. Without loss of generality, we can assume that time and space are slotted as $x_0 = 0 < x_1 < \cdots < x_{j-1} < x_j < \cdots < x_n < \cdots$, where the adjacent distance is Δ, i.e.,

$$x_j - x_{j-1} = \Delta \quad (j = 1, 2, \cdots). \tag{5}$$

Given that $B = x_n = n\Delta$, the average amount paid by a customer is equal to: $\sum_{j=1}^{\infty} E_0(T|B = x_j)$ By removing the conditions $(B = x_n)$ above, we get $E_0[amount\ A\ paid\ by\ a\ customer] = \sum_{n=1}^{\infty} \sum_{j=1}^{n} E_0(T|B = x_j)P(B = x_n)$ (changing the orders of these two summations)

$$\sum_{j=1}^{\infty} E_0(T|B = x_j)P(B \geqq x_j), \tag{6}$$

which completes the proof of Eq. (3). By taking the limit as $\Delta \to 0$ in Eq. (6), we have Eq. (4), assuming the existence of the Riemann integral. This completes the proof of Lemma-1. (q.e.d.) Substituting Lemma-1 [Eqs (3) and (4)] into the cost equation result [Eq. (2)] leads the following theorem. [Theorem-1] Consider a general-input general service GI/GI/1 (PS) system with iid requested service time

B. Let E(U) be the mean unfinished work and $E_0(T|B = x)$ be the conditional mean sojourn time given that the service time requested by a customer is equal to x. We then have the following relationship between E(U) and $E_0(T|B = x)$: for discrete random variables eq.(7), and for continuous random variables eq.(8),

$$E(U) = \lambda \sum_{j=1}^{\infty} E_0(T|B = x_j)P(B \geq x_j); \tag{7}$$

$$E(U) = \lambda \int_0^{\infty} E_0(T|B = x)P(B > x)dx; \tag{8}$$

[Remark-1] Note that the relationship developed here treats both discrete- and continuous-random variables, unlike the previous work [3, p.165, Eq.(2.3.12)] that treated only continuous random variables. Our proof is different from the previous work, since they used Campbell-Little-Mecke formula. On the other hand, Kleinrock [10, p.168, Eq.(4.16)] has shown that the conditional mean (customer-average) sojourn time has linear function property even for the general-input general-service GI/GI/1 (PS) system with iid requested service time B, i.e.,

$$E_0(T|B = x) = Const \ x. \tag{9}$$

by using Little's law [3], [12]. Substituting Eq. (11) into Theorem-1 [Eqs. (7) and (8)], we get eq.(10), from which we get *Const* as eq.(11)

$$E(U) = \lambda Const \ E(B^2)/2. \tag{10}$$

$$Const = 2E(U)/\lambda E(B^2). \tag{11}$$

From Eqs.(11) and (13) we obtain the next theorem [Eq.(14)]. [Theorem-2] Consider a general-input general service GI/GI/1 (PS) system with iid requested service time B. Let E(U) be the mean unfinished work and $E_0(T|B = x)$ be the conditional mean sojourn time given that the service time requested by a customer is equal to x. We then have the following relationship between E(U) and $E_0(T|B = x)$:

$$E_0(T|B = x) = 2E(U)x/\lambda E(B^2). \tag{12}$$

2.3 *Relationship between E(U) and Mean Sojourn Time E(T)*

By removing the condition $(B = x)$ in Eq.(14), we have the following theorem. [Theorem-3] Consider a general-input general service GI/GI/1 (PS) system with iid requested service time B. Let E(U) be the mean unfinished work and $E_0(T)$ be the mean sojourn time. We then have the following relationship:

$$E_0(T) = 2E(U)E(B)/\lambda E(B^2). \tag{13}$$

3 Mean Performance Comparison between FCFS and PS System

So far, we have considered the general-input iid-service GI/GI/1 (PS) system. In this section we compare the mean performance measures between the FCFS and PS systems. In the corresponding GI/GI/1 (FCFS) system we assume infinite-capacity waiting room. Since the PS and FCFS rules are work-conserving, their resulting unfinished work processes, busy periods, and idle periods are identical. However, this identical property is not valid for the sojourn times of customers, which are sensitive with respect to the service rules. For our comparing purpose, let W_{FCFS} and T_{FCFS} be respectively the waiting time and sojourn time for the corresponding GI/GI/1 (FCFS) system. Schrage's conservation law [14] or Brumelle's formula [6] reads for the GI/GI/1 (FCFS) system:

$$E(U) = \rho E_0(W_{FCFS}) + \rho E(B^2)/2E(B).\tag{14}$$

Here, note that their argument [6], [14] for the renewal-input GI/GI/1 system is seen to be still valid for our general-input GI/GI/1 system. Substituting Eq.(17) into our Theorem-3 [Eq.(16)], we have

$$E_0(T) = 2E(B)^2 E_0(W_{FCFS})/E(B^2) + E(B).\tag{15}$$

From the definitions of the sojourn time and waiting time, it follows that

$$E_0(W_{FCFS}) = E_0(T_{FCFS}) - E(B).\tag{16}$$

Substituting Eq.(19) into Eq. (18), we obtain the next theorem [Eq.(20)].

[Theorem-4] Consider a general-input general-service GI/GI/1 (PS) system with iid requested service time B. Let T be the sojourn time in the GI/GI/1 (PS) system. Let T_{FCFS} be the sojourn time in the corresponding GI/GI/1 (FCFS) system. We then have the following relationship:

$$E_0(T) - E(B) = 2E(B)^2 \left[E_0(T_{FCFS}) - E(B)\right]/E(B^2).\tag{17}$$

[Remark-2] Using the sample-path analysis, Brandt et al. [5] proved Eq.(20) for the Poisson-input general-service M/GI/1 system, and general-input, a special service-time (mixture of a zero and deterministic services, or mixture of a zero and exponential service times) system; see [4, Theorems 3.1 and 3.2]. Our Theorem-4 states that the additional assumption on the service time distribution in Brandt et al. [5] is not necessary. Eq. (20) is valid for any generally distributed requested service time.

4 Concluding Remarks

Applying the cost-equation analysis [12], we have derived the relationship between the mean (time-average) unfinished work and the mean (customer-average) sojourn time for the GI/GI/1 (PS) system. Note that the cost-equation is closely related to $H = \lambda G$ formula [6], but we cannot straightforwardly apply the $H = \lambda G$ formula. This is because under the PS rule, it is hard to evaluate the integral for the kernel function $f_n(t)$ associated with the n-th arriving customer due to the complicated PS-service sample-path influenced by arriving customers after the n-th arriving customer unlike the FCFS rule. Our relationship has been subsequently applied to find the relationship between the mean (customer-average) sojourn times under the FCFS and PS rules, generalizing the recent results obtained by Brandt et al. [5] via the sample-path analysis. Our cost-equation analysis developed here is fairly simple, but it links only the mean (time-average and customer-average) performance measures, as in Takahashi [17]. Therefore, it would be worthwhile as a future study to derive a higher-order moment relationship, e.g. how the second moment of the unfinished work is related to the sojourn time distribution for the GI/GI/1 (PS) system, by using the point-process theory[6,8,21].

References

1. Aalto, S., Ayesta, U., Borst, S., Misra, V., Nunez-Queija, R.: Beyond processor sharing. ACM SIGMETRICS Performance Evaluation Review, vol.34, pp.36–43 (2007)
2. Berg, J.L., Boxma, O.J.: The M/G/1 queue with processor sharing and its relation to a feedback queue. Queueing Systems, vol.9, pp.365–402 (1991)
3. Brandt, A., Franken, P., Lisek, B.: Stationary Stochastic Models. Wiley, Chichester (1990)
4. Baccelli, F., Bremaud, P.: Elements of Queueing Theory: Palm-Martingale Calculus and Stochastic Recurrences, Springer, New York (1994)
5. Brandt, A., Brandt, M.: A sample path relation for the sojourn times in G/G/1-PS systems and its applications. Queuing Systems, vol.52, pp.281–286 (2006)
6. Brumelle, S.L.: On the relation between customer and time averages in queues. J. Applied Probability, vol.8, pp.508–520 (1971)
7. Franken, P., Koenig, D., Arndt, U., Schmidt, V.: Queues and Point Process. Akademie-Verlag, Berlin (1982)
8. Hoshi, K., Shikata, Y., Takahashi, Y., Komatsu, N.: Mean approximate formulas for GI/G/1 processor-sharing system. Proc. International Conference on Operations Research (OR 2010 Munich), I.6, WE-15-1 (2010)
9. Kleinrock, L.: Time-shared systems: a theoretical treatment. J. of the ACM, vol.14, pp.242–261 (1967)
10. Kleinrock, L.: Queueing Systems : Computer Applications. vol.II, Wiley, New York (1976)
11. Ramaswami, V.: The sojourn time in the GI/M/1 queue with processor sharing. J. Applied Probability, vol.21, pp.437–442 (1984)
12. Ross, S.: Introduction to Probability Models, Academic Press, San Diego (1993)
13. Sakata, M., Noguchi, S., Oizumi, J.: An analysis of the M/G/1 queue under round-robin scheduling. Operations Research, vol.19, no.2, pp.371–385 (1971)
14. Schrage, L.: An alternative proof of a conservation law for the queue G/G/1. Operations Research, vol.18, pp.185–187 (1970)
15. Sengupta, B.: An approximation for the sojourn-time distribution for the GI/G/1 processor-sharing queue. Stochastic Models, vol.8(1), pp.3–57 (1992)
16. Takahashi, Y., Miyazawa, M.: Relationship between queue length and waiting time in a stationary discrete-time queue. Stochastic Models, 11(2), pp.249–271, (1995)
17. Takahashi, Y.: Shikata, Y., Frey, A.: A single-server queueing system with modified service mechanism. Proceedings of IEEE International Conference on Ultra Modern Telecommunications (ICUMT 2009), IEEE, pp.1–6 (2009)

Multiperiod Stochastic Optimization Problems with Time-Consistent Risk Constraints

M. Densing and J. Mayer

Abstract Coherent risk measures play an important role in building and solving optimization models for decision problems under uncertainty. We consider an extension to multiple time periods, where a risk-adjusted value for a stochastic process is recursively defined over the time steps, which ensures time consistency. A prominent example of a single-period coherent risk measure that is widely used in applications is Conditional-Value-at-Risk (CVaR). We show that a recursive calculation of CVaR leads to stochastic linear programming formulations. For the special case of the risk-adjusted value of a random variable at the time horizon, a lower bound is given. The possible integration of the risk-adjusted value into multi-stage mean-risk optimization problems is outlined.

1 Introduction

Coherent risk measures are defined in the seminal work of Artzner et al. [2] by a set of axioms, which should be fulfilled by any acceptance threshold for the risk of a financial stochastic loss. For an overview of the many developments since then, see the monographs [10, 17, 20] and the survey [16]. In particular, the example of the coherent risk measure Conditional-Value-At-Risk[1](CVaR) is suitable for applications, mainly because of its reformulation as a linear optimization problem [21], leading to efficient implementations for mean-risk problems [12, 13].

Based on the success of single-period coherent risk measures, there is a considerable effort to extend the concepts to multiple periods (or continuous time), see [1]

Martin Densing
Energy Economics Group, Paul Scherrer Institute, 5232 Villigen PSI, Switzerland, e-mail: martin.densing@psi.ch

János Mayer
Institute for Business Administration, University of Zurich, 8044 Zürich, Switzerland, e-mail: janos.mayer@business.uzh.ch

[1] Closely related notions by slightly differing definitions are *Tail Value-at-Risk, Tail Conditional Expectation* [2], *(Expected) Shortfall* [4], and *Average Value-at-Risk* [10, 17].

D. Klatte et al. (eds.), *Operations Research Proceedings 2011*, Operations Research Proceedings, 521
DOI 10.1007/978-3-642-29210-1_83, © Springer-Verlag Berlin Heidelberg 2012

for an overview. Multi-period risk measurement can be categorized in: (i) risk measurement for a final random variable at a time horizon, (ii) risk measurement for a stochastic (value) process. In both cases, the risk is assessed from the viewpoint of different times and states.

Straightforward dynamic (non-recursive) extensions of CVaR are not time consistent (in the sense of [3]). In discrete time, time consistency is equivalent to a recursive calculation of risk [3, 15]. A recursively defined CVaR is known under different names with slightly varying definitions, and mostly defined for final random variables: *Nested average value-at-risk* (nAV@R) [17, p. 156], *Conditional risk mapping* CV@R$_{\mathcal{X}_{t+1}|\mathcal{X}_t}$ from time $t+1$ to t, concatenated over time [18, Def. (5.8, 6.18)], *Dynamically consistent Tail Value-at-Risk* (DTVaR) [19, Def. 8.2], *Dynamic AVaR* [5, Ex. 2.3.1], and *Composed AVaR* (ComAVaR) [6, Sect. 5]. Note that there are also multi-period extensions of CVaR without ensuring time consistency [9, 17, 19]. Applications of multi-period coherent risk measurement in optimization problems can be found in electric energy production [7, 8, 11] and pension funds management [14] models.

In this paper, the recursive calculation of CVaR for a stochastic process is considered. The corresponding risk-adjusted value can be calculated by stochastic linear programming. In a special case, a lower bound is given, and the integration in mean-risk optimization problems is shortly illustrated.

2 Coherent and Recursive Risk-Adjusted Values

Let a probability space $(\Omega, \mathcal{F}, \mathbb{P})$ be given. To avoid technicalities in this application-oriented work, the set Ω is assumed to be finite; for simplicity, we use the more concise general notation if applicable.

A stochastic future value shall be represented by a bounded random variable $X \in L^\infty(\Omega; \mathbb{R})$. A coherent risk-adjusted value $\pi[X]$ can be represented by (see [3])

$$\pi[X] = \inf_{\mathbb{Q} \in \mathcal{P}} \mathbb{E}_\mathbb{Q}[X], \tag{1}$$

where $\mathbb{E}_\mathbb{Q}[\cdot]$ denotes the integration with respect to ('test-scenario') $\mathbb{Q}$, and $\mathcal{P}$ is a set of probability measures on $(\Omega, \mathcal{F})$. The *risk measure* is the sign-reversed $\rho[X] = -\pi[X]$, in comparison with [2]. A prominent example is (in our sign-convention)

Definition 1. The risk-adjusted value *Conditional-Value-at-Risk (CVaR)* is

$$\text{CVaR}^{X,\alpha} := \inf_{\mathbb{Q} \in \mathcal{Q}} \mathbb{E}_\mathbb{Q}[X], \quad \mathcal{Q} := \left\{ \mathbb{Q} \left| \frac{d\mathbb{Q}}{d\mathbb{P}} \le \frac{1}{\alpha} \; a.s. \right. \right\}, \tag{2}$$

where $\frac{d\mathbb{Q}}{d\mathbb{P}}$ denotes the Radon-Nikodym probability density of the probability measure $\mathbb{Q}$ with respect to $\mathbb{P}$, and where the level $\alpha \in (0, 1)$.

For multi-period risk measurement, we consider a finite number of time-steps $t = 0, \ldots, T$. The gain of information over time is modeled by a filtration of σ-

algebras $(\mathscr{F}_t)_{t=0,\ldots,T}$, $\mathscr{F}_t \subseteq \mathscr{F}_{t+1}$, $t = 0,\ldots,T-1$, from $\mathscr{F}_0 := \{\emptyset, \Omega\}$ to $\mathscr{F}_T \equiv \mathscr{F}$. In our finite setting, the filtration corresponds to a scenario tree: Each atom of $\mathscr{F}_t$ is identified with a node $n \in \mathscr{N}_t$ of the tree at stage t, where $\mathscr{N}_t$ is the set of nodes at stage t. The set Ω is identified with the root node, and the probability $\mathbb{P}[n]$ for a node n is the probability of the corresponding atom.

An uncertain sequence of financial values over time is described by an $(\mathscr{F}_t)$-adapted stochastic process $(X_t)_{t=0,\ldots,T}$ (i.e. the values of X_t realize on the nodes $\mathscr{N}_t$ at stage t on the scenario tree). In the multi-period setting, the risk can be considered from the viewpoint of different states (nodes) at different times, such that the risk-adjusted values over time form also an $(\mathscr{F}_t)$-adapted stochastic process, denoted by $(R_t^{(X)})_{t=0,\ldots,T}$, where the dependence on the process of values is denoted by the subscript $(X) := (X_t)_{t=0,\ldots,T}$.

A special case of a risk-adjusted value process can be motivated as follows [3]: (i) The risk-adjusted value process has lower values than the value process, (ii) in each 'test-scenario' $\mathbb{Q}$ (cf. (1)) the riskiness decreases with increasing time because information is gained over time, (iii) the decision maker uses the most favorable (largest possible) risk-adjusted values fulfilling (i) and (ii). This leads to (see [3])

Definition 2. Let $(X) := (X_t)_{t=0,\ldots,T}$ be an $(\mathscr{F}_t)$-adapted stochastic process. Let $\mathscr{P}$ be a set of probability measures on $(\Omega, \mathscr{F}_T)$. The *risk-adjusted value process* $(R_t^{(X)})_{t=0,\ldots,T}$ is recursively defined by

$$R_t^{(X)} := \begin{cases} X_T, & \text{if } t = T, \\ \min\left(X_t, \; \min_{\mathbb{Q} \in \mathscr{P}} \mathbb{E}_{\mathbb{Q}}[R_{t+1}^{(X)} | \mathscr{F}_t]\right), & \text{if } t = 0,\ldots,T-1, \end{cases} \tag{3}$$

where the minimization of the conditional expectation is pointwise, that is, separately on each node $n \in \mathscr{N}_t$.

Note that (3) is a simplified formulation requiring that $\mathscr{P}$ and (X) are such that the pointwise minimum of the conditional expectation is attained (this will hold for the given example later on). In our finite setting, minimization preserves measurability, so the risk-adjusted value process is $(\mathscr{F}_t)$-adapted; $R_0^{(X)}$ is $\mathscr{F}_0$-measurable and therefore deterministic.

The single period property of coherency can be extended to multi-period risk-adjusted values (see [3] for an in-depth discussion). With the short-hand notation $(\lambda X + Y + c) := (\lambda X_t + Y_t + c)_{t=0,\ldots,T}$ the coherency is given in

Lemma 1. *Let (X) and (Y) be $(\mathscr{F}_t)$-adapted stochastic processes. Then*
(i) $R_0^{(X+Y)} \geq R_0^{(X)} + R_0^{(Y)}$,   (ii) $R_0^{(\lambda X)} = \lambda R_0^{(X)} \; \forall \lambda \geq 0, \, \lambda \in \mathbb{R}$,
(iii) $X_0 \leq Y_0, \ldots, X_T \leq Y_T$ a.s. $\Rightarrow R_0^{(X)} \leq R_0^{(Y)}$,   (iv) $R_0^{(X+c)} = R_0^{(X)} + c \; \forall c \in \mathbb{R}$.

See also [7] for the simple proof in our notation. The recursive risk adjusted value is strongly *time consistent* (see [3,7]). A short-hand definition of time consistency is as follows. For every pair of processes (X) and (Y) and every time $t \geq 0$ such that the processes are identical in $0,\ldots,t$: $R_t^{(X)} \leq R_t^{(Y)}$ a.s. $\Rightarrow R_0^{(X)} \leq R_0^{(Y)}$. Hence, 'preferences' (orderings) are preserved backward in time. Time-consistency as well as coherency require in our setting no particular assumptions on the set $\mathscr{P}$.

3 Stepwise-CVaR Sets and Stochastic Linear Formulation

Next, an example of the set $\mathscr{P}$ (cf. (3)) will be given such that feasible conditional single-step probabilities are those of CVaR.

Let a probability measure $\mathbb{Q}$ be given. In our finite setting on a scenario tree, $\mathbb{Q}$ is determined by its values on the terminal nodes. Consider a non-terminal node n. If $\mathbb{Q}[n] > 0$, then the transition probabilities from n to its children nodes can be combined into a vector $\mathbf{q}_n \in \mathbb{R}_+^{k(n)}$, where $k(n)$ is the cardinality of the children. Note that $\mathbf{e}_n^T \mathbf{q}_n = 1$ with $\mathbf{e}_n := (1,\ldots,1) \in \mathbb{R}^{k(n)}$. If $\mathbb{Q}[n] = 0$, then a vector of transition probabilities from such an unreachable n is arbitrarily chosen (e.g. as $\frac{1}{k(n)}(1,\ldots,1) \in \mathbb{R}^{k(n)}$). On the other hand, given a tree topology, a set of vectors of transition probabilities for each non-terminal node determines a probability measure:

$$(\mathbf{q}_n)_{n \in \mathscr{N}_t,\, t=0,\ldots,T-1} \mapsto \mathbb{Q}. \tag{4}$$

The components of $\mathbf{q}_n$ can be viewed as the values of a probability measure on a finite sample space of cardinality $k(n)$. In each such space, we choose the set of feasible probability measures as for CVaR (see (2)) with (for simplicity) constant level α.

Definition 3. Let $\alpha \in (0,1)$. The *stepwise-CVaR set* $\mathscr{P}^\alpha$ of probability measures is given with the representation (4) by

$$\mathscr{P}^\alpha := \prod_{\substack{n \in \mathscr{N}_t, \\ t=0,\ldots,T-1}} \left\{ \mathbf{q}_n \in \mathbb{R}_+^{k(n)} \,\middle|\, \mathbf{q}_n \le \frac{1}{\alpha}\mathbf{p}_n,\ \mathbf{e}_n^T \mathbf{q}_n = 1 \right\},$$

where $\mathbf{p}_n$ is the vector of transition probabilities going from node n to its children given by the probability measure $\mathbb{P}$. The recursive risk-adjusted value (Def. 2) with set $\mathscr{P}^\alpha$ for process (X) at time t is denoted by $R_t^{(X),\alpha}$.

The following proposition is a multi-period generalization of that in [21].

Proposition 1. *Let a finite filtration $(\mathscr{F}_t)_{t=0,\ldots,T}$ be given (i.e. a scenario tree), and let $(X) := (X_t)_{t=0,\ldots,T}$ be an $(\mathscr{F}_t)$-adapted stochastic process. The recursive risk-adjusted value $R_0^{(X),\alpha}$ with set $\mathscr{P}^\alpha$ is given by the optimal objective value of the following stochastic linear optimization problem:*

$$\max R_0,$$

$$\text{s.t.} \begin{cases} R_t \le X_t, & t = 0,\ldots,T, \\ R_t \le Q_t - \dfrac{1}{\alpha}\mathbb{E}[Z_{t+1}|\mathscr{F}_t], & t = 0,\ldots,T-1, \\ Z_t \ge Q_{t-1} - R_t, & t = 1,\ldots,T, \\ Z_t \ge 0, & t = 1,\ldots,T, \end{cases} \tag{5}$$

where R_t, Q_t and Z_t are $\mathscr{F}_t$-measurable random variables. In addition, we have for feasible $(R_t)_{t=0,\ldots,T}$ and for the risk-adjusted value process $(R_t^{(X),\alpha})_{t=0,\ldots,T}$ that $R_t \le R_t^{(X),\alpha}$ a.s. holds, for all t.

Proof (sketch; for the details see [7]). The conditional minimization in (3) is for $\mathscr{P} := \mathscr{P}^\alpha$ of the same form as in (2) for CVaR. By CVaR-duality [7, 10, 17], the minimization is replaced by a maximization problem. It rests to check equivalence of the modified (3) with (5).

Consider the case where all intermediate values of a stochastic value process are sufficiently large (e.g. $X_t \geq \|X_T\|_\infty, t < T$). Then only the final values X_T are relevant in the recursive Def. 2. In this case, a lower bound can be given.

Proposition 2. *Let X_T be an $\mathscr{F}_T$-measurable random variable and let α^T be the T-th power of $\alpha \in (0,1)$. Then*

$$\mathrm{CVaR}^{X_T,\alpha^T} \leq R_0^{(\|X_T\|_\infty,\ldots,\|X_T\|_\infty,X_T),\alpha}. \tag{6}$$

Given an appropriate α of a single-period model, eq. (6) suggests that a reasonable multi-period model choice is $\sqrt[T]{\alpha}$. This multiplicative effect is discussed in [19, Ex. 8.8]. Note that the given bound is different from that in [17, Prop. 3.36].

Proof (sketch; for the details see [7]). Problem (5) is simplified for the case of a final random variable. After dualizing, it is shown (by multiplication of constraints) that a feasible solution is in the set $\mathscr{Q}$ of CVaR (2) for level α^T.

4 Application in Mean-Risk Optimization

The value process $(X) := (X_t)_{t=0,\ldots,T}$ is called *accepted* at time $t = 0$ if $R_0^{(X)} \geq 0$ (see [3]). Because of the translation equivariance (Lem. 1, (iv)), shifting the value process by a deterministic cash amount shifts the risk-adjusted value by the same amount. Thus, the acceptability of a non-normalized process (X) can be defined by $R_0^{(X)} \geq \rho_{\min}$ for a suitable $\rho_{\min} \in \mathbb{R}$. Such an acceptability may be used in mean-risk optimization, for example as a constraint:

$$\sup \mathbb{E}[g(X_0,\ldots,X_T)],$$
$$\mathrm{s.t.} \begin{cases} R_0^{(X),\alpha} \geq \rho_{\min}, \\ (X_t)_{t=0,\ldots,T} \in \mathscr{X}, \end{cases} \tag{7}$$

where $g \colon \mathbb{R}^{T+1} \to \mathbb{R}$ is a profit (utility) function of the value process (X), and the set $\mathscr{X}$ are budget and technical constraints. Prop. 1 permits us to write the constraint on risk in problem (7) as a system of linear inequalities in the following equivalent problem (see [7] for details):

$$\sup \mathbb{E}[g(X_0, \ldots, X_T)],$$

$$\text{s.t.} \begin{cases} R_0 \geq \rho_{\min}, \quad R_t \leq X_t, & t = 0, \ldots, T, \\[2mm] R_t \leq Q_t - \dfrac{1}{\alpha}\mathbb{E}[Z_{t+1} | \mathscr{F}_t], & t = 0, \ldots, T-1, \\[2mm] Z_t \geq Q_{t-1} - R_t, \quad Z_t \geq 0 & t = 1, \ldots, T, \quad (X_t)_{t=0,\ldots,T} \in \mathscr{X}, \end{cases}$$

where X_t, R_t, Q_t and Z_t are $\mathscr{F}_t$-measurable random variables.

References

1. Acciaio, B., and Penner, I.: Dynamic Risk Measures, In: Di Nunno, G., Oksendal, B. (eds.) Adv. Math. Meth. for Finance, pp. 1-34. Springer, Berlin Heidelberg (2011)
2. Artzner, P., Delbaen, F., Eber, J.-M., Heath, D.: Coherent measures of risk. Math. Finance, **9**, 203–228 (1999)
3. Artzner, P., Delbaen, F., Eber, J.-M., Heath, D., Ku, H.: Coherent multiperiod risk adjusted values and Bellman's principle. Ann. Oper. Res., **152**, 5–22 (2007)
4. Bertsimas, D., Lauprete, G.J., Samarov, A.: Shortfall as a risk measure: properties, optimization and applications. J. Econ. Dyn. Control, **28**, 1353–1381 (2004)
5. Cheridito, P., Kupper, M.: Composition of time-consistent dynamic monetary risk measures in discrete time. Int. J. Theoretical Appl. Finance, **14**, 137–162 (2011)
6. Cheridito, P., Stadje, M.: Time-inconsistency of var and time-consistent alternatives. Finance, **6**, 40–46 (2009)
7. Densing, M.: Multi-Stage Stochastic Optimization of Hydro-Energy Plant with Time-Consistent Constraints on Risk. PhD, ETH Zurich (2007) doi: 10.3929/ethz-a-005464814
8. Eichhorn, A., Heitsch, H., Römisch, W.: Stochastic optimization of electricity portfolios: Scenario tree modeling and risk management. In: Pardalos, P.M. et al. (eds.) Handbook of Power Systems II, pp. 405-432. Springer, Berlin Heidelberg (2010)
9. Eichhorn, A., Römisch, W.: Polyhedral risk measures in stochastic programming. SIAM J. Optimiz., **16**, 69–95 (2005)
10. Föllmer, H., Schied, A.: Stochastic Finance. de Gruyter Studies in Mathematics (2002)
11. Geman, H., Ohana, S.: Time-consistency in managing a commodity portfolio: A dynamic risk measure approach. J. of Bank. Financ., **32**, 1991–2005 (2008)
12. Kall, P., Mayer, J.: Stochastic Linear Programming: Models, Theory, and Computation. 2nd Edition, Intern. Series in Oper. Res. & Manag. Sci., 156. Springer, New York (2010)
13. Künzi-Bay, A., Mayer, J.: Computational aspects of minimizing conditional value-at-risk. Comput. Manag. Sci., **3**, 3–27 (2006)
14. Kilianova, S., Pflug, G.C.: Optimal pension fund management under multi-period risk minimization. Ann. Oper. Res., **166**, 261–270 (2009)
15. Kydland, F.E., Prescott, E.C.: Rules rather than discretion: The inconsistency of optimal plans. J. Polit. Econ., **85**, 473–492 (1977)
16. Krokhmal, P., Zabarankin, M., Uryasev, S.: Modeling and optimization of risk. Surveys in Oper. Res. and Manag. Sci., **16**, 49–66 (2011)
17. Pflug, G.C., Römisch, W.: Modeling, measuring, and managing risk. World Scientific (2007)
18. Ruszczyński, A., Shapiro, A.: Conditional risk mappings. Math. Oper. Res., **31**, 544–561 (2006)
19. Roorda, B., Schumacher, J.M.: Time consistency conditions for acceptability measures, with an application to tail value at risk. Ins.: Mathematics Econ., **40**, 209–230 (2007)
20. Szegö, G. (ed.): Risk Measures for the 21st Century. Wiley (2004)
21. Uryasev, S., Rockafellar, R.T.: Optimization of conditional value-at-risk. J. Risk, **2**, 21–41 (2000)

Robustness in SSD portfolio efficiency testing

Miloš Kopa

Abstract This paper deals with portfolio efficiency testing with respect to second-order stochastic dominance (SSD) criterion. These tests classify a given portfolio as efficient or inefficient and give some additional information for possible improvements. Unfortunately, these tests usually assumes discrete probability distribution of returns with equiprobable scenarios. Moreover, they can be very sensitive to the probability distribution. Therefore, [1] derived a SSD portfolio efficiency test for a general discrete distribution and its robust version allowing for changes in probabilities of the scenarios. However, these formulations are not suitable for numerical computations. Therefore, in this paper, the numerically tractable reformulations of these SSD efficiency tests are derived.

1 Introduction

Consider N assets and a random vector of their returns $\mathbf{r}$. Since all existing portfolio efficiency tests are derived for a discrete probability distribution of returns we assume that $\mathbf{r}$ takes S values $\mathbf{x}^s = (x_1^s, x_2^s, \ldots, x_N^s)$, called scenarios, with probabilities $\mathbf{p} = (p_1, p_2, \ldots, p_S)'$. The scenarios can be collected in matrix

$$X = \begin{pmatrix} \mathbf{x}^1 \\ \mathbf{x}^2 \\ \vdots \\ \mathbf{x}^S \end{pmatrix}.$$

We will use $\lambda = (\lambda_1, \lambda_2, \ldots, \lambda_N)'$ for a vector of portfolio weights and the portfolio possibilities are given by

$$\Lambda = \{\lambda \in R^N | \mathbf{1}'\lambda = 1, \ \lambda_n \geq 0, \ n = 1, 2, \ldots, N\}.$$

Miloš Kopa
Charles University in Prague, Faculty of Mathematics and Physics, Department of Probability and Mathematical Statistics, Sokolovská 83, 186 75 Prague 8, Czech Republic. e-mail: kopa@karlin.mff.cuni.cz

D. Klatte et al. (eds.), *Operations Research Proceedings 2011*, Operations Research Proceedings, 527
DOI 10.1007/978-3-642-29210-1_84, © Springer-Verlag Berlin Heidelberg 2012

Alternatively, one can consider any bounded polyhedron:

$$\Lambda' = \{\lambda \in R^N | A\lambda \geq \mathbf{b}\}.$$

The tested portfolio is denoted by $\tau = (\tau_1, \tau_2, ..., \tau_N)'$. For any portfolio $\lambda \in \Lambda$, let $(-X\lambda)^{[k]}$ be the k-th smallest element of $(-X\lambda)$, i.e. $(-X\lambda)^{[1]} \leq (-X\lambda)^{[2]} \leq ... \leq (-X\lambda)^{[S]}$ and let $I(\lambda)$ be a permutation of index set $I = \{1, 2, ..., S\}$ such that $-\mathbf{x}^{i(\lambda)}\lambda = (-X\lambda)^{[i]}$. Similarly, we can order the corresponding probabilities, that is, let $p_i^\lambda = p_{i(\lambda)}$. Hence, $p_i^\lambda = P(-\mathbf{r}\lambda = (-X\lambda)^{[i]})$. The same notation is applied for portfolio τ.

Following [9], [4], [3], [2] and [1], we define second-order stochastic dominance relation in the strict form in the context of SSD portfolio efficiency.

Let $F_{\mathbf{r}'\lambda}(x)$ denote the cumulative probability distribution function of returns of portfolio λ. The twice cumulative probability distribution function of returns of portfolio λ is defined as:

$$F_{\mathbf{r}'\lambda}^{(2)}(y) = \int_{-\infty}^{y} F_{\mathbf{r}'\lambda}(x)dx. \tag{1}$$

Definition 1. Portfolio $\lambda \in \Lambda$ dominates portfolio $\tau \in \Lambda$ by second-order stochastic dominance ($\mathbf{r}'\lambda \succ_{SSD} \mathbf{r}'\tau$) if and only if

$$F_{\mathbf{r}'\lambda}^{(2)}(y) \leq F_{\mathbf{r}'\tau}^{(2)}(y) \qquad \forall y \in \mathbb{R}$$

with strict inequality[1] for at least one $y \in \mathbb{R}$.

Following [6], [3] and [1] we will express the SSD relation using conditional value at risk (CVaR) of portfolio λ and τ. Conditional value at risk (CVaR) of portfolio λ at level $\alpha \in [0, 1)$ can be obtained as a solution of the following optimization problem:

$$\text{CVaR}_\alpha(-\mathbf{r}'\lambda) = \min_{v \in \mathbb{R}, z_t \in \mathbb{R}^+} \quad v + \frac{1}{1-\alpha} \sum_{t=1}^{S} p_t z_t \tag{2}$$

$$\text{s.t.} \quad z_t \geq -\mathbf{x}^t\lambda - v, \ t = 1, 2, ..., S$$

and the similar expression can be considered for portfolio τ. However, for portfolio τ, we will rather use the dual formulation:

---

[1] This type of SSD relation is sometimes referred to as the strict second-order stochastic dominance. If no strict inequality is required then the relation can be called the weak second-order stochastic dominance.

$$\text{CVaR}_\alpha(-\mathbf{r}'\tau) = \max_{\kappa_t \in \mathbb{R}^+} \frac{1}{1-\alpha} \sum_{t=1}^{S} \kappa_t(-\mathbf{x}^t\tau) \tag{3}$$

$$\text{s.t.} \quad \kappa_t \leq p_t, \quad t = 1,2,...,S$$

$$\sum_{t=1}^{S} \kappa_t = 1 - \alpha$$

See Rockafellar & Uryasev [8] for more details.

Since we limit our attention to a discrete probability distribution of returns, the inequality of CVaRs need not to be verified in all $\alpha \in \langle 0,1 \rangle$, but only in at most S particular points and the following result was shown in [1].

Theorem 1. *Let* $q_s^\lambda = \sum_{i=1}^{s} p_i^\lambda$, $s = 1,2,...,S-1$. *Let* $q_0^\lambda = 0$. *Then* $\mathbf{r}'\lambda \succ_{SSD} \mathbf{r}'\tau$ *if and only if* $\text{CVaR}_{q_s^\lambda}(-\mathbf{r}'\lambda) \leq \text{CVaR}_{q_s^\lambda}(-\mathbf{r}'\tau)$ *for all* $s = 0,1,2,...,S-1$ *with strict inequality for at least one* q_s^λ.

Following [7], [4], [3], [2] and [1] we define efficiency of portfolio τ with respect to the second-order stochastic dominance.

Definition 2. A given portfolio $\tau \in \Lambda$ is SSD inefficient if there exists portfolio $\lambda \in \Lambda$ such that $\mathbf{r}'\lambda \succ_{SSD} \mathbf{r}'\tau$. Otherwise, portfolio τ is SSD efficient.

This definition classifies portfolio $\tau \in \Lambda$ as SSD efficient if and only if no other portfolio is better (in the sense of the SSD relation) for all risk averse and risk neutral decision makers. See e.g. [5]

2 SSD portfolio efficiency test

Using Theorem 1, [1] introduced a new measure of SSD portfolio efficiency for a given portfolio τ. This measure depends also on the underlying probability distribution of returns represented by scenario matrix X and corresponding probabilities vector $\mathbf{p}$. The measure is defined as the optimal solution of mathematical programming problem:

$$\xi(\tau,X,\mathbf{p}) = \min_{a_s,\lambda} \sum_{s=0}^{S-1} a_s \tag{4}$$

$$\text{s.t.} \quad \text{CVaR}_{q_s^\lambda}(-\mathbf{r}'\lambda) - \text{CVaR}_{q_s^\lambda}(-\mathbf{r}'\tau) \leq a_s, \quad s = 0,1,...,S-1$$

$$a_s \leq 0, \quad s = 0,1,...,S-1$$

$$\lambda \in \Lambda.$$

The objective function of (4) represents the sum of differences between CVaRs of a portfolio λ and CVaRs of the tested portfolio τ where CVaRs are considered at levels q_s^λ, $s = 0,1,...,S-1$. Similarly as in [7], [4] or [3], the following SSD efficiency test based on measure $\xi(\tau,X,\mathbf{p})$ was derived in [1].

Theorem 2. *A given portfolio* τ *is SSD efficient if and only if* $\xi(\tau,X,\mathbf{p}) = 0$. *If* $\xi(\tau,X,\mathbf{p}) < 0$ *then the optimal portfolio* λ^* *in (4) is SSD efficient and it dominates portfolio* τ *by SSD.*

Unfortunately, problem (4) can not be solved directly, because each CVaR can be seen as a separate optimization problem and, moreover, the levels of CVaRs depend on chosen portfolio λ. The problem (4) can be easily handled, see [3], only in the case of equally probable scenarios. If the probabilities are not equal then the order of probabilities plays a key role in CVaRs expressions. Therefore, for each portfolio λ, we consider a permutation matrix $\Pi = \{\pi_{i,j}\}_{i,j=1}^{S}$ that arranges the losses of portfolio λ in ascending order, that is, $\sum_{j=1}^{S} \pi_{1,j}(-\mathbf{x}^j\lambda) \le \sum_{j=1}^{S} \pi_{2,j}(-\mathbf{x}^j\lambda) \le \ldots \le \sum_{j=1}^{S} \pi_{S,j}(-\mathbf{x}^j\lambda)$ and $\sum_{j=1}^{S} \pi_{s,j}(-\mathbf{x}^j\lambda) = (-X\lambda)^{[s]}$. Using the same permutation matrix, we can reorder also the vector of probabilities: $\sum_{j=1}^{S} \pi_{s,j}p_j = p_s^{\lambda}$. Hence $q_s^{\lambda} = \sum_{i=1}^{s} \sum_{j=1}^{S} \pi_{i,j}p_j$. Finally, we apply (2) for each $\text{CVaR}_{q_s^{\lambda}}(-\mathbf{r}'\lambda)$ and (3) for each $\text{CVaR}_{q_s^{\lambda}}(-\mathbf{r}'\tau)$. Summarizing, we can rewrite (4) in the following non-linear integer program:

$$\xi(\tau,X,\mathbf{p}) = \min_{a_s,v_s,z_{s,t},\kappa_{s,t},q_s^{\lambda},\lambda} \sum_{s=1}^{S} a_s \tag{5}$$

$$\text{s.t.} \quad v_s + \frac{1}{1-q_{s-1}^{\lambda}} \sum_{t=1}^{S} p_t z_{s,t} - \frac{1}{1-q_{s-1}^{\lambda}} \sum_{t=1}^{S} \kappa_{s,t}(-\mathbf{x}^t\tau) \le a_s, \quad s = 1,2,\ldots,S$$

$$z_{s,t} \ge -\mathbf{x}^t\lambda - v_s, \quad s,t = 1,2,\ldots,S$$

$$\kappa_{s,t} \le p_t, \quad s,t = 1,2,\ldots,S$$

$$\sum_{t=1}^{S} \kappa_{s,t} = 1 - q_{s-1}^{\lambda}, \quad s = 1,2,\ldots,S$$

$$q_0^{\lambda} = 0, \quad q_s^{\lambda} = \sum_{i=1}^{s} \sum_{j=1}^{S} \pi_{i,j}p_j, \quad s = 1,2,\ldots,S-1$$

$$\sum_{j=1}^{S} \pi_{s,j}(-\mathbf{x}^j\lambda) \le \sum_{j=1}^{S} \pi_{s+1,j}(-\mathbf{x}^j\lambda), \quad s = 1,2,\ldots,S-1$$

$$\sum_{s=1}^{S} \pi_{s,j} = \sum_{j=1}^{S} \pi_{s,j} = 1, \quad j,s = 1,2,\ldots,S$$

$$a_s \le 0, \quad s = 1,2,\ldots,S$$

$$z_{s,t}, \kappa_{s,t} \ge 0, \quad s,t = 1,2,\ldots,S$$

$$\pi_{i,j} \in \{0,1\}, \quad i,j = 1,2,\ldots,S$$

$$\lambda \in \Lambda.$$

3 Portfolio efficiency with respect to ε-SSD relation

Following [1], we consider ε-SSD efficiency approach as a robustification of the classical SSD portfolio efficiency. It guarantees stability of the SSD efficiency classification with respect to small changes (prescribed by parameter $\varepsilon > 0$) in probability vector $\mathbf{p}$. Assume that the probability distribution $\bar{P}$ of random returns $\bar{\mathbf{r}}$ takes again values $\mathbf{x}^s$, $s = 1, 2, ..., S$ but with other probabilities $\bar{\mathbf{p}} = (\bar{p}_1, \bar{p}_2, ..., \bar{p}_S)$. We define the distance between P and $\bar{P}$ as $d(\bar{P}, P) = \max_i |\bar{p}_i - p_i|$.

Definition 3. A given portfolio $\tau \in \Lambda$ is ε-SSD inefficient if there exists portfolio $\lambda \in \Lambda$ and $\bar{P}$ such that $d(\bar{P}, P) \leq \varepsilon$ with $\bar{\mathbf{r}}'\lambda \succ_{SSD} \bar{\mathbf{r}}'\tau$. Otherwise, portfolio τ is ε-SSD efficient.

For testing ε-SSD efficiency of a given portfolio τ we modify (4) in order to introduce a new measure of ε-SSD efficiency:

$$\xi_\varepsilon(\tau, X, \mathbf{p}) = \min_{a_s, v_s, z_{s,t}, \kappa_{s,t}, \bar{p}_i \bar{q}_s^\lambda, \lambda} \sum_{s=1}^{S} a_s \tag{6}$$

$$\text{s.t.} \quad v_s + \frac{1}{1 - \bar{q}_{s-1}^\lambda} \sum_{t=1}^{S} \bar{p}_t z_{s,t} - \frac{1}{1 - \bar{q}_{s-1}^\lambda} \sum_{t=1}^{S} \kappa_{s,t}(-\mathbf{x}^t \tau) \leq a_s, \quad s = 1, 2, ..., S$$

$$z_{s,t} \geq -\mathbf{x}^t \lambda - v_s, \quad s, t = 1, 2, ..., S$$

$$\kappa_{s,t} \leq \bar{p}_t, \quad s, t = 1, 2, ..., S$$

$$\sum_{t=1}^{S} \kappa_{s,t} = 1 - \bar{q}_{s-1}^\lambda, \quad s = 1, 2, ..., S$$

$$\bar{q}_0^\lambda = 0, \quad \bar{q}_s^\lambda = \sum_{i=1}^{s} \sum_{j=1}^{S} \pi_{i,j} \bar{p}_j, \quad s = 1, 2, ..., S-1$$

$$\sum_{j=1}^{S} \pi_{s,j}(-\mathbf{x}^j \lambda) \leq \sum_{j=1}^{S} \pi_{s+1,j}(-\mathbf{x}^j \lambda), \quad s = 1, 2, ..., S-1$$

$$\sum_{s=1}^{S} \pi_{s,j} = \sum_{j=1}^{S} \pi_{s,j} = 1, \quad j, s = 1, 2, ..., S$$

$$a_s \leq 0, \quad s = 1, 2, ..., S$$

$$z_{s,t}, \kappa_{s,t} \geq 0, \quad s, t = 1, 2, ..., S$$

$$\pi_{i,j} \in \{0, 1\}, \quad i, j = 1, 2, ..., S$$

$$\lambda \in \Lambda$$

$$\sum_{i=1}^{S} \bar{p}_i = 1$$

$$-\varepsilon \leq \bar{p}_i - p_i \leq \varepsilon, \quad i = 1, 2, ..., S$$

$$\bar{p}_i \geq 0, \quad i = 1, 2, ..., S.$$

Since (6) is obtained from (5) by an additional minimization over $\bar{\mathbf{p}}$ from ε-neighborhood of the original probability vector $\mathbf{p}$, we may modify Theorem 2 into an ε-SSD portfolio efficiency test, see [1] for more details.

Theorem 3. *Portfolio $\tau \in \Lambda$ is ε-SSD efficient if and only if $\xi_\varepsilon(\tau, X, \mathbf{p})$ given by (6) is equal to zero.*

The test classifies τ as ε-SSD efficient if and only if no portfolio λ SSD dominates τ neither for the original probabilities $\mathbf{p}$ nor for arbitrary probabilities $\bar{\mathbf{p}}$ from ε-neighborhood of the original vector $\mathbf{p}$.

Acknowledgements The research was partly supported by the project "Methods of modern mathematics and their applications" – MSM 0021620839 and by the Czech Science Foundation (grant 402/10/1610).

References

1. Dupačová, J., Kopa, M.: Robustness in stochastic programs with risk constraints. Annals of Operations Research, (2011) doi: 10.1007/s10479-010-0824-9.
2. Kopa, M.: Measuring of second-order stochastic dominance portfolio efficiency. Kybernetika **46**, 488–500 (2010)
3. Kopa, M., Chovanec, P.: A second-order stochastic dominance portfolio efficiency measure. Kybernetika **44**, 243–258 (2008)
4. Kuosmanen, T.: Efficient diversification according to stochastic dominance criteria. Management Science **50**, 1390–1406 (2004)
5. Levy, H.: Stochastic Dominance: Investment Decision Making Under Uncertainty. Second edition. Springer Science, New York (2006)
6. Ogryczak, W., Ruszczyński, A.: Dual stochastic dominance and related mean-risk models. SIAM Journal on Optimization **13**, 60–78 (2002)
7. Post, T.: Empirical tests for stochastic dominance efficiency. Journal of Finance **58**, 1905–1932 (2003)
8. Rockafellar, R. T. and Uryasev, S.: Conditional Value-at-Risk for General Loss Distributions. Journal of Banking and Finance **26** (7), 1443–1471 (2002)
9. Ruszczyński, A., Vanderbei, R. J.: Frontiers of Stochastically Nondominated Portfolios. Econometrica **71**, 1287–1297 (2003)

Order Fulfillment and Replenishment Policies for Fully Substitutable Products

Beyazit Ocaktan and Ufuk Kula

Abstract We consider a fully substitutable multi product inventory system in which customer demand may be satisfied by delivering any combination of products. A customer requesting a certain quantity, D may accept different combinations of products as long as its order of size D is fulfilled. We model the system as a Markov decision process and develop a approximate dynamic programming algorithm to determine order fulfillment and replenishment policies. In order to reduce the run time of the algorithm, we use two-layer neural network that iteratively fits a function to the state values and finds an approximately optimal order fulfillment and can order type replenishment policy.

1 Introduction

Consider a multi-product inventory system in which all the products are fully substitutable. One example for such inventory systems is Automated Teller Machines (ATMs): A customer arriving to an ATM to withdraw money requests a certain amount and the software that controls the ATM decides on the combination of banknotes given to the customer. A similar situation arises in manufacturing settings too. Consider a manufacturer producing a single product which are packed together in different quantities. A customer requesting a certain quantity, D may accept different combinations of packages as long as its order of size D is fulfilled.

Our study is motivated by an ATM software company which approached us enquiring if it was possible to develop a sensible algorithm which decides on the combination of banknotes to dispense, when to replenish the ATMs, and how many should be delivered from each banknote type. Since each banknote type may be seen as a different product which are fully substitutable as long as the total quan-

Beyazit Ocaktan
Sakarya University, Adapazari, 54187, Turkey e-mail: ocaktan@sakarya.edu.tr

Ufuk Kula
Sakarya University, Adapazari, 54187, Turkey, e-mail: ukula@sakarya.edu.tr

D. Klatte et al. (eds.), *Operations Research Proceedings 2011*, Operations Research Proceedings,   533
DOI 10.1007/978-3-642-29210-1_85, © Springer-Verlag Berlin Heidelberg 2012

tity delivered is equal to the total quantity requested, this ATM inventory system is a multi-product system in which a multiple number of one product substitutes a single quantity of an other product. Even tough our work is motivated by the ATM software company's customer order fulfillment and inventory replenishment problem, we consider the generalized problem in which an arriving customer requests a total quantity D, and the manufacturer or the distributor decides on the product combination to deliver to the customer to satisfy the ordered quantity D. A product combination may consist of multiple units of a single product or several product types. The objective of the manufacturer is to determine how to fulfill a customer's order (i.e, to determine the combination of products) and to determine a replenishment policy for the products. We assume that the manufacturer uses a can order type policy, which is determined by three parameters S, c, and s, which represents the up-to level, can order, and must order levels of a product.

We model the problem as a markov decision process (MDP). However, since the state space of the problem is prohibitively large to solve the problem to the optimality, we use Q-learning, a reinforcement learning algorithm to find an approximately optimal order fulfillment and can order policy, (S, c, s) for each product type. In order to shorten the run time of the algorithm we use a function approximation scheme developed by using a two-layer neural network.

Section 2 reviews the relevant work related to ours and points out to the differences between our work and the existing literature. Section 3 defines and models the problem, and gives the MDP formulation. Section 4 describes the Q-learning algorithm with function approximation and provides a numerical example. Finally, section 5 concludes our work and points out to future work.

2 Literature Review

For brevity, we give a very short review of the related literature on joint replenishment that includes can order policies, on substitution, and point out the similarities and the differences between our work and the literature.

First study on can order policies is [2]. It suggests can order type policies as an alternative to optimal joint replenishment policies. Silver in [6] finds the parameters S, c, and s when demand arrives according to a Poisson process, and consider the cases in which instantaneous replenishment and the lead time L is deterministic. [7] extends [6] to the Compound Poisson case. [3] models a multiproduct inventory system as an MDP to find the best policy within the class of (S, c, s) policies. For a recent review on can order type policies see [5]. Substitution literature, on the other hand, is divided into two categories as consumer-driven and decision-maker substitution. In our model, since the decision- maker (distributer) decides how to substitute the most desired products in an order. To the best of our knowledge ours is the first to consider the problem in which customer order D is satisfied fully substitutable product combinations. For a review on substitutable products see [1].

3 Problem Definition and the MDP Model

In this section, we will first define the problem, describe how the considered multi-product inventory system works, state the problem assumptions, and give the Markov decision process (MDP) model that we developed.

Problem Definition Consider a multi-product inventory system in which customers arrive to the system according to a Poisson process with rate λ. An arriving customer demands a certain quantity D according to a probability distribution, $f_D()$. In other words, customer demands arrive to the system according to a Compound Poisson distribution with rate λ. Upon arrival, the customer declares the quantity D that she demands. Requested demand has a discrete distribution $f_D()$. Upon arrival, the distributer (for example, the ATM) is informed on the demanded quantity, D, and he decides how to fulfill this order of size D by deciding on a combination of available product types in the inventory. When the amount of products' inventory is insufficient to fulfill the quantity, D, we assume that all demanded quantity D is lost.

We assume that the customer accepts the combination offered to her as long as the quantity offered equals to D. However, when a customer demand is satisfied by a product combination which is less than desired, the utility that the customer receives decreases. Assume that the customers have an identical and discrete utility function, which is given by $U(n_1, n_2, ..., n_l)$, where n_i denotes the number of product i used to fulfill a customers demand and k denotes the number of products available in the inventory. Note that a customer's utility function depends on the number of product i, $i = 1, 2, ..., l$ delivered. Let U_{max} denote the maximum utility that a customer gets when she receives the demanded amount, Q as the most desired product combination $n_1^*, n_2^*, ..., n_l^*$, i.e., let $U_{max} = U\left(n_1^*, n_2^*, ..., n_l^*\right)$. When an arriving customer does not get the most desired product combination she wants the difference between U_{max}, and $U(n_1, n_2, ..., n_l)$ represents the utility that the customer looses by accepting the product combination $n_1, n_2, ..., n_l$. Let U_a denote the utility difference when a product combination, i.e, an action a is chosen. If the customer demand is not satisfied by using the most desirable combination, we penalize this by multiplying U_a by a penalty cost p. After a customer's order D is satisfied, the distributer checks the inventory level x_i of each product i and decides whether to place an order according to a (S, c, s) policy.

Let h, w, and p denote the holding cost, the lost sales cost per customer, and penalty cost per customer, respectively. Also let K and k denote the major and minor replenishment costs, which are incurred when an inventory replenishment order is given. A major replenishment cost, K is incurred whenever an order is given form a product type. A minor replenishment cost is incurred whenever one or more product types are included into the replenishment order of a product.

MDP Model We model the problem as a Markov decision process. We denote the state of the system by $\mathbf{x} = (x_1, x_2, ..., x_l, d)$, where x_i, $i = 1, 2, ..., l$ represents the inventory level of each product i, and d represents the value that the random variable D takes. In other words, d is the amount that an arriving customer requests. Admissible action set in a given state $\mathbf{x}$ is $A(\mathbf{x})$ consists of all the possible combinations of product types such that $x_i \leq n_i$ for all $i = 1, 2, ..., l$, and $n_1 + n_2 +, ..., + n_l = D$. In other words, when a feasible action is chosen upon arrival of customer request, D, the quantity of a product i given to the customer cannot exceed the available inventory of the product, and the total amount of products used to satisfy the customer order should be equal to the ordered amount, D. When the following customer arrives to the system and her demand, D is revealed, the process makes a transition from a state $\mathbf{x}$ to a state $\mathbf{x}'$ according to the transition function

$$(x_i - n_i + Q_i) I_i + (x_i - n_i)(1 - I_i), \text{ for all } i = 1, 2, ..., l \tag{1}$$

where the indicator variable I_i is defined as,

$$I_i = \begin{cases} 0, & \text{if product } i \text{ is not included into the replenishment order;} \\ 1, & \text{otherwise,} \end{cases} \tag{2}$$

Our objective is to minimize the average cost per unit time over an infinite horizon. The cost of being in state $\mathbf{x}$, and an action a is taken the current state cost is given as

$$g(x, a) = \sum_{i=1}^{l} h_i . t_i + (K + m.k).I_1 + p.U_a(n_1, n_2, ..., n_l) + w.I_2 \quad , \tag{3}$$

where h_i, and t_i denotes the inventory holding cost per unit per unit time for product i respectively. Therefore, the first term in equation (3) represents the total holding cost incurred when the process is in state $\mathbf{x}$. The second term is the joint replenishment cost for ordering m products together, and the third one is the penalty cost of not delivering the most desired combination to the customer. Finally, the last term represents the lost sales cost incurred when an order is not fulfilled. I_2 in (3) is a 0-1 indicator function, where it takes a value of 0 when order is fulfilled.

The optimality equation corresponding to the objective function given in equation (4) is

$$v(\mathbf{x}) = \min_{a \in A(\mathbf{x})} \left[g(x, a) + \sum_{\mathbf{x}' \in X} p(\mathbf{x}' | \mathbf{x}, a) v(\mathbf{x}') \right] \tag{4}$$

where $v(\mathbf{x})$ is the value of being in state $\mathbf{x}$, and $p(\mathbf{x}' | \mathbf{x}, a)$ is the probability that action a takes the process from state $\mathbf{x}$ to state $\mathbf{x}'$. Because the probability transition matrix that governs the evaluation of the system is too large, and because to calculate $v(\mathbf{x})$ for each state $\mathbf{x}$ is computationally impossible we use Q-learning, a reinforcement learning algorithm to learn state values and hence the optimal policy π^*.

4 The Algorithm

In this section we briefly describe the steps of the algorithm that determines the order fulfillment and the can order policies for each product. Figure 1 gives the details of the algorithm. The algorithm starts by determining an initial demand rate for each product for an arbitrary order fulfillment policy, π. As the figure 1 shows the algorithm starts by determining an approximate demand rate for each product i. In the second step, the parameters S_i, and c_i are found for instantaneous replenishment by following [6] and [7] . The resulting can order parameters are adjusted to find S_i, c_i, s_i for positive lead time, L. If the resulting product i fill rate, β_i for policy (S_i, c_i, s_i) is less than the required service levels of any of l product types, the can order parameters are increased by a unit, and the parameters are recalculated for positive lead time, L. This step is repeated until $\beta_i \geq \beta_i^*$, for each $i = 1, 2, \ldots, l$. Once the desired fill rates are achieved, it means that can order policy parameters are found for the arbitrary order fulfillment policy, π used in step 1. Once, the can order policy is decided Q-learning algorithm with function fitting is used to find an improved policy. If the can order policy found by using the improved policy is identical to the policy from the previous step the algorithm stops.

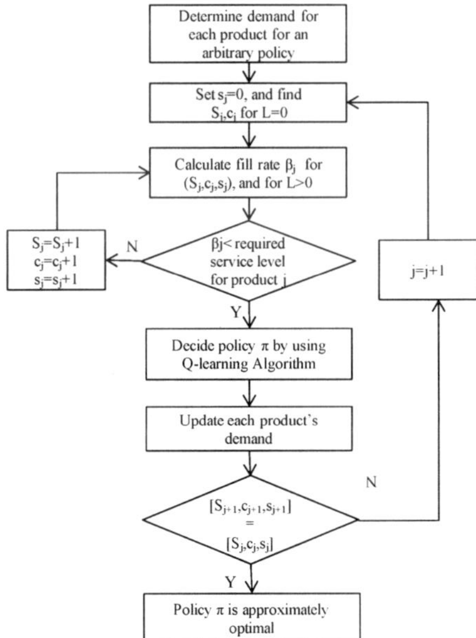

Fig. 1 Flowchart of the algorithm

5 Numerical study

We run the described algorithm and decided on an approximate optimal policy, π^* and compared it with a myopic policy, π^m. Figure 2 plots the average cost per unit time. As the figure 2 shows, both policies converge as the number of iterations increase.

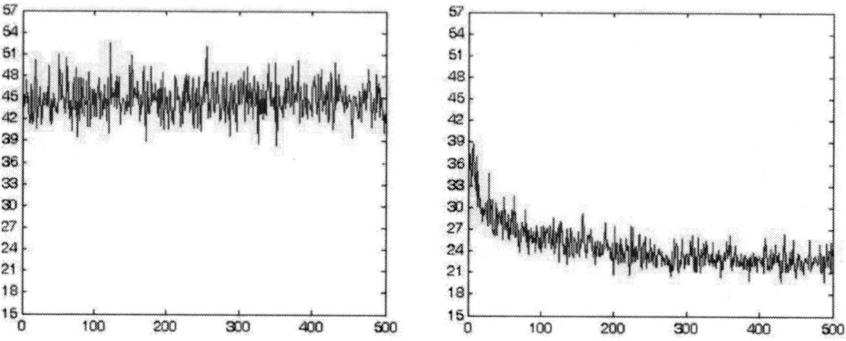

Fig. 2 Convergence of policies π^m (left), and π^* (right)

Note that the approximate optimal policy, π^* outperforms the myopic policy, π^m by nearly 50 %. Other numerical examples we performed also provided savings of similar magnitude.

References

1. Axsater, S.: A new decision rule for lateral transshipments in inventory systems. Management Science **49**, 1168–1179 (2003)
2. Balintfy, J.L.:On a basic class of multi-item inventory problems. Management Science **10**, 287–297 (1964)
3. Federgruen,A., Groenevelt,H., Tijms, H.C.: Coordinated replenishments in a multi-item inventory system with compound poisson demands. Management Science **30**, 344–357 (1984)
4. Ignall, E.: Optimal continous review policies for two product inventory systems with joint setup costs. Management Science **15**, 278–283 (1969)
5. Melchiors, P.:Calculating can-order policies for the joint replenishment problem by the compensation approach. European Journal of Operational Research **14**, 587–595 (2002)
6. Silver,E.A.: A control system for coordinated inventory replenishment. Int.J.Prod.Res. **12**, 647–671 (1974)
7. Thompstone,R.M. and Silver,E.A.:A coordinated inventory control system for compound poisson demand and zero lead time. Int.J.Prod.Res. **13**, 581–602 (1975)

Stochastic Programming DEA Model of Fundamental Analysis of Public Firms for Portfolio Selection

N.C.P. Edirisinghe

Abstract A stochastic programming (SP) extension to the traditional Data Envelopment Analysis (DEA) is developed when input/output parameters are random variables. The SPDEA framework yields a robust performance metric for the underlying firms by controlling for outliers and data uncertainty. Using accounting data, SPDEA determines a relative financial strength (RFS) metric that is strongly correlated with stock returns of public firms. In contrast, the traditional DEA model overestimates actual firm strengths. The methodology is applied to public firms covering all major U.S. market sectors using their quarterly financial statement data. RFS-based portfolios yield superior out-of-sample performance relative to sector-based ETF portfolios or broader market index.

1 Introduction

Based on the economic notion of Pareto optimality, Data Envelopment Analysis (DEA) employs a non-parametric approach for estimating the production frontier based on the specification of the production possibility set rather than a production functional form itself. The so-called CCR-DEA optimization model, first developed in [3], constructs this efficient frontier from the observed data on multi-input and multi-output combinations that arise from available production technologies pertaining to various firms (or Decision Units). Those firms with production input-output mixes that lie on the efficient frontier are termed operationally efficient (relative to a pool firms). Therefore, DEA can be used as a tool to identify "best-practice" firms that can be used as benchmarks of strong performance.

A relative evaluation of firm strength is fundamental in identifying public firms whose equities may be considered in investment portfolios. As suggested by the

N.C.P. Edirisinghe
College of Business, University of Tennessee, Knoxville, TN, U.S.A. e-mail: chanaka@utk.edu
October 25, 2011 (*To appear in OR2011 Proceedings, Springer.*)

D. Klatte et al. (eds.), *Operations Research Proceedings 2011*, Operations Research Proceedings,   539
DOI 10.1007/978-3-642-29210-1_86, © Springer-Verlag Berlin Heidelberg 2012

Efficient Market Hypothesis (EMH), see [6], future expectations of firm strength are strongly correlated with (the direction of) stock returns. Expectations of firm strength are generally formed on the basis of (historical) quarterly or annual financial data on firm's operation in competitive markets for supply and demand, see [5]. Since past observations exhibit variability, future benchmarks of firm strength under DEA must be able to accommodate randomness in the data. However, (standard) DEA models treat data in inputs and outputs (for a firm) as non-random, i.e., just one sample point. Thus, the efficient frontier so-computed is just an estimate of the true frontier, which may lead to picking dominated firms as benchmarks or excluding stochastically nondominated firms, not to mention the increased bias and variance in computed DEA efficiencies.

Data uncertainty in DEA has been recognized in the early literature, [10]. Sampling errors stemming from having an insufficient number of firms to describe the production possibility set are addressed and statistical foundations are developed in [1], [2]. In contrast, in the presence of stochastic noise, along with stochastic technology in production process leading to random outcomes, chance-constrained (CC) DEA formulations are used to develop efficient frontiers that envelop the stochastic input/output data with specified probability, see [7], [8].

Deterministic DEA models are quite sensitive to outlier data because the efficient frontier is determined by extreme points in the data, thus, adversely biasing the identification of efficient firms. The existing literature focuses on identifying outlier firms rather than outlier observations of inputs/outputs for a given firm, see [4], [11]. Removal of firms in this manner only adds to the difficulties of small sample biases.

Relative firm strength evaluation using DEA under stochastic data must also take into account of risk attitudes of the decision maker. A utility-based approach is offered in [9] allowing for risk preferences to enter into determining the DEA-frontier. Relying on the theory of second-order stochastic dominance (SSD), the author develops a mean-variance model that trades off the (estimated) frontier location with the variance of the data from a fixed set of values.

This paper presents a stochastic programming (SP) approach to DEA, termed SPDEA, where risk attitude toward variation of firm performance is incorporated via expected utility maximization. While risk is measured in [9] around an already accepted set of (estimated) values for the inputs/outputs, in SPDEA each firm is allowed to have a domain of uncertainty for their input/output values, which may be correlated with those of other firms. Moreover, the SPDEA model allows a firm to treat certain outcomes at a lower importance level if such an outcome is deemed extreme or possibly an outlier.

Using quarterly financial information of public firms as inputs/outputs, SPDEA is employed to determine future expectations of relative financial strength (RFS) of a public firm. RFS is then used to determine long/short pools of stocks for further analysis within portfolio mean-variance optimization. When applied on S&P 500 stocks in the U.S., the proposed methodology yields superior returns, easily exceeding the out-of-sample performances of either sector-based ETF (Exchange Traded Funds) portfolios or broader market index.

2 DEA and Stochastic Extensions

Consider n (financial) parameters chosen to evaluate the fundamental performance of J firms. Let $\mathscr{I}$ and $\mathscr{O}$ denote disjoint subsets (of indices) of the n parameters that identify input and output data, respectively. For some period (quarter), the value of parameter P_i for firm j is denoted by ξ_{ij}. The vector of data elements $(\xi_{ij}, i = 1, \ldots, n, j = 1, \ldots, J)$ is denoted by $\xi \in \mathfrak{R}^{nJ}$. Then, DEA-based efficiency for firm k, relative to the remaining $J - 1$ firms, is determined by solving the following linear programming (LP) model for each firm k:

$$\eta_k(\xi) := \max_{y_k} \sum_{i \in \mathscr{O}} \xi_{ik} y_{ik} \tag{1}$$
$$\text{s.t.} \ \sum_{i \in \mathscr{I}} \xi_{ik} y_{ik} \leq 1$$
$$- \sum_{i \in \mathscr{I}} \xi_{ij} y_{ik} + \sum_{i \in \mathscr{O}} \xi_{ij} y_{ik} \leq 0, \ j = 1, \ldots, J$$
$$y_{ik} \geq 0, \ i \in \mathscr{I} \cup \mathscr{O}.$$

In (1), y_{ik} is the nonnegative multiplier (or weight allocation) for input/output parameter i by firm k. When the data ξ is random, with known domain Ξ, $\eta_k(\xi)$ is a random variable in the interval $[0, 1]$. Furthermore, the optimal multipliers $y_k(\xi) \in \mathfrak{R}^n$ form a random weighting vector. Ξ is assumed to be a compact set in $\mathfrak{R}^{nJ}$.

Usually, after a particular outcome vector $\hat{\xi} \in \Xi$ is observed, (1) is solved with $\xi = \hat{\xi}$ to determine $\eta_k(\hat{\xi})$. The associated multiplier vector $y_k(\hat{\xi})$ is a optimal weight allocation to inputs and outputs by firm k considering the operation of the remaining firms. Thus, $y_k(\hat{\xi})$ is a "wait-and-see" decision assigning importance to certain aspects of the business, and $\eta_k(\hat{\xi})$ is an ex-poste measure of relative firm performance. But, what the firm management needs for the future is to identify the levels of importance to be placed on various financial parameters so that resources and efforts can be committed pro-actively in order to achieve a stronger competitive position for the firm. An ideal tool to determine such a *nonanticipative* weight allocation is the "here-and-now" modeling paradigm in stochastic programming.

SPDEA Model. A *well-hedged* weighting vector y_k^0 for firm k must be evaluated considering all possible outcomes in Ξ according to their probabilities of occurrence, but independent of any particular realization in Ξ. The focus here is on determining y_k^0 for the immediate future period (quarter) of firm performance under a stochastic (and robust) estimation of the frontier using SP modeling.

Denote the feasible set of multipliers $y_k(\xi)$ in (1), for some $\xi \in \Xi$, by $\mathscr{Y}_k(\xi) \in \mathfrak{R}^n$, which is a convex compact polyhedron. Then, $\eta_k(\xi)$ is the maximum of $\sum_{i \in \mathscr{O}} \xi_{ik} y_{ik}$ over $y_k(\xi) \in \mathscr{Y}_k(\xi)$. Given an ex-ante weighting decision y_k^0, and for a future (random) $\xi \in \Xi$, define $f_k(y_k^0, \xi) := \sum_{i \in \mathscr{O}} \xi_{ik} y_{ik}^0$, which is the aggregate output of firm k under ξ provided $y_k^0 \in \mathscr{Y}_k(\xi)$. Since it is unknown a priori which ξ might be observed, the (hedging) multipliers must be feasible for all $\xi \in \Xi$, i.e.,

$$y_k^0 \in \mathscr{Y}_{(k)} := \bigcap_{\xi \in \Xi} \mathscr{Y}_k(\xi). \tag{2}$$

Let firm k be endowed with a non-decreasing utility function $U_k(.)$. Then, the utility of firm k's aggregate output is $U_k[f_k(y_k^0, \xi)]$, expectation of which is maximized to determine y_k^0. Thus, firm k's SPDEA relative strength score is given by,

$$\text{(SPDEA):} \qquad s_k := \max_{y_k^0} \left\{ E_\xi \left[U_k \left(\sum_{i \in \mathcal{O}} \xi_{ik} y_{ik}^0 \right) \right] : y_k^0 \in \mathcal{Y}_{(k)} \right\}. \tag{3}$$

Note that $U_k(0) \leq s_k \leq U_k(1)$. It can be shown that:

Proposition 1. *Suppose firm k is risk-neutral and the first moment of ξ is denoted by $\bar{\xi}$. Then, $s_k \leq \eta_k(\bar{\xi})$ holds, implying that DEA model (1) specified with average outcomes over-estimates (true) firm strength under stochastic data. Moreover, $s_k \leq E_\xi[\eta_k(\xi)]$, implying that average DEA efficiency is an upper bound on (true) firm strength.*

Worst-case Bounds on Firm Performance. Given the domain Ξ, when only partial information is available on ξ, one needs to compute bounds on firm performance. If the first moment information is known, then the upper bounds in Proposition 1 are applicable. When probabilistic information is not available on ξ, the worst-case scenarios in Ξ are considered to develop lower (upper) bounds by considering the minimum (maximum) of the objective utility in (3) over the outcomes in Ξ. These bounds can be reformulated for efficient computation as:

$$s_k \geq s_k^{\text{LB}} = \max_{y_k^0} \left\{ B_L : B_L - U_k \left(\sum_{i \in \mathcal{O}} \xi_{ik} y_{ik}^0 \right) \leq 0, y_k^0 \in \mathcal{Y}_{(k)}, \xi \in \Xi \right\} \tag{4}$$

and, $$s_k \leq s_k^{\text{UB}} = \max_{\beta, y_k^0} \left\{ B_U \mid B_U - U_k \left(\sum_{i \in \mathcal{O}} \xi_{ik} y_{ik}^0 \right) + U_k(1)\beta(\xi) \leq U_k(1), \right.$$
$$\left. y_k^0 \in \mathcal{Y}_{(k)}, \int_\Xi \beta(d\xi) = 1, \beta(\xi) = 0 \text{ or } 1, \xi \in \Xi \right\}. \tag{5}$$

Handling Outlier-data. This paper subscribes to the notion that outliers might represent a significant part of the data representing important phenomena that should not be overlooked completely. Thus, SPDEA in (3) is extended to allow a firm to treat certain outcomes at a lower importance level rather than eliminating completely. Recall that the weighting vector y_k^0 in SPDEA is a "hedge" against all future outcomes; however, in the event Ξ contains extreme or outlier data, then hedging against all outcomes will unduly bias y_k^0. Such biases caused by "outliers" often weaken the predictive power of the underlying stochastic models. In other words, when Ξ contains outliers, the constraint $y_k^0 \in \mathcal{Y}_{(k)}$ may render SPDEA lose the predictive accuracy of the performance strength s_k for firm k's future period.

Accordingly, $y_k^0 \notin \mathcal{Y}_k(\xi)$ will be allowed for certain outcomes, and in such cases, the magnitude of violation is measured and penalty is imposed. Consider the following *relaxed* version of $\mathcal{Y}_k(\xi) \in \mathfrak{R}^n$, labeled $\mathcal{Z}_k(\xi) \in \mathfrak{R}^{n+J+1}$, for $\xi \in \Xi$:

$$\mathscr{Z}_k(\xi) := \left\{ \begin{array}{l} (y_k^0, v_k(\xi)) \in \mathfrak{R}^{n+J+1} \mid \sum_{i \in \mathscr{I}} \xi_{ik} y_{ik} - v_{0k}(\xi) \le 1, \\[2mm] -\sum_{i \in \mathscr{I}} \xi_{ij} y_{ik} + \sum_{i \in \mathscr{O}} \xi_{ij} y_{ik} - v_{jk}(\xi) \le 0, \quad j = 1, \ldots, J \\[2mm] y_{ik} \ge 0, \; i = 1, \ldots, n, \; v_{jk}(\xi) \ge 0, \; j = 0, 1, \ldots, J. \end{array} \right. \tag{6}$$

Note that $y_k^0 \in \mathscr{Y}_k(\xi)$ implies $(y_k^0, 0) \in \mathscr{Z}_k(\xi)$. Also, if $y_k^0 \notin \mathscr{Y}_k(\xi)$ then $(y_k^0, v_k(\xi)) \in \mathscr{Z}_k(\xi)$ only if $v_{jk}(\xi) > 0$ for some j. Using linear penalty, the SPDEA model under outlier correction seeks a trade-off between the expected utility of relative performance and the expected total violation of extreme outcomes, as follows:

$$\text{(SPDEA-C):} \quad S_k(\lambda) := \max_{y_k^0, v_k} E_\xi \left[U_k \left(\sum_{i \in \mathscr{O}} \xi_{ik} y_{ik}^0 \right) \right] - \lambda E_\xi \left[\sum_{j=0}^{J} v_{jk}(\xi) \right] \tag{7}$$
$$\text{s.t.} \;\; (y_k^0, v_k(\xi)) \in \mathscr{Z}_k(\xi), \; \xi \in \Xi,$$

where $\lambda \ge 0$ is a trade-off scalar. For sufficiently-large λ, observe that $S_k(\lambda) = s_k$.

3 SPDEA-based Portfolio Selection

We consider the S&P 500 index stocks individually, as well as the 9 separate sector-ETFs (spanning the index) with ticker symbols: XLK (Technology), XLV (Health-Care), XLB (Basic Materials), XLI (Industrial Goods), XLE (Energy), XLY (Consumer Discretionary), XLP (Consumer Staples), XLU (Utilities), and SLF (Financials). The number of firms in each sector for performance analysis is 86, 57, 31, 53, 29, 90, 38, 32, and 84, respectively. The chosen financial parameters for the SPDEA-C model are: inputs $\mathscr{I} = \{$Acct receivables, Long-term debt, Capital expenditure, Cost of goods sold$\}$, and outputs $\mathscr{O} = \{$Revenue, Earnings, Price/book, Net income growth$\}$. Historical quarterly financial statements for the period 1984-2004 are used to obtain the data for the 8 parameters. Financial sector was dropped due to incomplete data. The (out-of-sample) investment period is taken to be Jan-Jun, 2005. The expected fundamental strength of each firm is computed for the first quarter of 2005 using (7) under linear utility for two levels of λ: *large* (no control for outliers) and *small* (controlling for outliers). These RFS scores are presented in Figure 1.

Stocks for long (short) investments are identified via firm-RFS with at least 0.8 (at most 0.2). For *large-λ* case, this results in 53 long-stocks and 74 short-stocks, see (a) in Figure 1, and for *small-λ* case, 48 long-stocks and 40 short-stocks are obtained, see (b) in Figure 1. These two sets of 127 and 88 stocks are used to form two separate portfolios under monthly-rebalanced mean-variance portfolio optimization for Jan-Jun, 2005. A third portfolio is also formed using the 9 sector ETFs directly. Historical returns from 2003-04 are used to estimate distribution parameters for all optimizations. Risk tolerance levels of all portfolios are adjusted to yield the same 10.5% annualized volatility as the market index. It is evident from Figure 2 that the RFS-based portfolio with outlier control exceeds the performance of all other portfolios, including that of the S&P500 index.

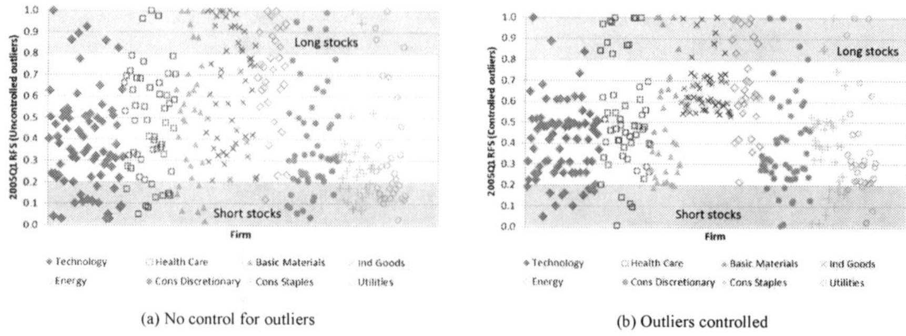

(a) No control for outliers (b) Outliers controlled

Fig. 1 2005Q1 expected firm-RFS scores

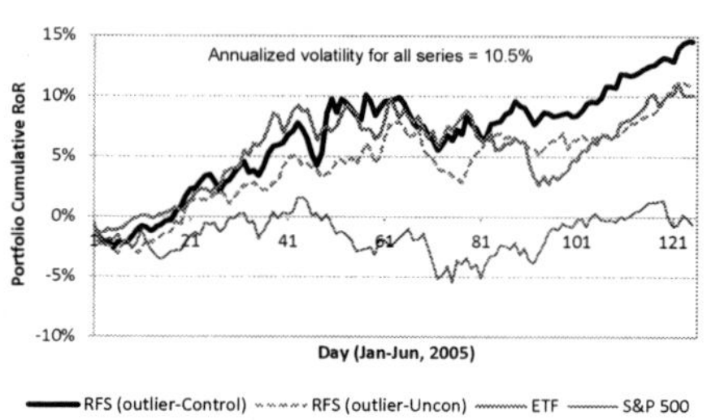

Fig. 2 Out-of-sample performance of RFS-based and ETF portfolios

References

1. Banker, R.D.: Maximum likelihood, consistency and DEA: Statistical foundations. Management Science **39** 1265–1273 (1993)
2. Banker, R.D., Natarajan, R.: Evaluating contextual variables affecting productivity using Data Envelopment Analysis. Operations Research **56** 48–58 (2008)
3. Charnes, A., Cooper, W.W., Rhodes, E.: Measuring the efficiency of decision-making units. European Journal of Operational Research **2**, 429–444 (1978)
4. Dusansky, R., Wilson, P.W.: On the relative efficiency of alternative modes of producing public sector output: The case of the developmentally disabled. European Journal of Operational Research **80** 608–628 (1995)
5. Edirisinghe, N.C.P., Zhang, X.: Portfolio selection under DEA-based relative financial strength indicators: Case of US industries. Journal of the Operational Research Society **59** 842–56 (2008)
6. Fama, E.F.: Efficient capital markets: A review of theory and empirical work. Journal of Finance **25** 383-417 (1970)
7. Huang, M., Li, S.X,: Stochastic DEA models with different types of input-output disturbances. Journal of Productivity Analysis **15** 95–113 (2001)
8. Olesen, O.B., Petersen, N.C.: Chance constrained efficiency evaluation. Management Science **41** 442–457 (1995)
9. Post,T.: Performance evaluation in stochastic environments using mean-variance Data Envelopment Analysis. Operations Research **49** 281–292 (2001)
10. Sengupta, J.K.: Efficiency measurement in stochastic input-output systems. International Journal of Systems Science **13** 273–287 (1982)
11. Wilson, G.W., Jadlow, J.M.: Competition, profit incentives, and technical efficiency in the provision of nuclear medicine services. The Bell Journal of Economics **13** 472–482 (1982)

Stream XV
Accounting and Revenue Management

Matthias Amen, University of Bielefeld, GOR Chair
Michaela Schaffhauser-Linzatti, University of Vienna, ÖGOR Chair
Claudius Steinhardt, University of Augsburg, GOR Chair

Consumer Choice Modelling in Product Line Pricing: Reservation Prices and Discrete Choice Theory

Stefan Mayer and Jochen Gönsch

Abstract In the literature on product line pricing, consumer choice is often modelled using the max-surplus choice rule. In terms of this concept, consumer preferences are represented by so-called reservation prices and the deterministic decision rule is to choose the product that provides the highest positive surplus. However, the distribution of the reservation prices often seems somewhat arbitrary. In this paper, we demonstrate how reservation prices can be obtained using discrete choice analysis and that these two concepts are not as different as often perceived in the literature. A small example illustrates this approach, using data from a discrete choice model taken from the literature.

1 Introduction and Problem Description

In the literature, the problem of determining optimal prices for a product line has been discussed from multiple points of view (see, e.g., the literature reviews in [2] and [3]) for several decades. A monopolist is usually confronted with the task of choosing static product-specific prices r_j for all his products $j \in \mathscr{J} = \{1,\ldots,J\}$ at the beginning of the selling period, such that his total revenue is maximised. The products are usually assumed to be substitutes. Depending on the analysed industry sector and the research focus, additional modifications, for example, capacity constraints, are considered. During the selling period, I consumers, $i \in \mathscr{I} = \{1\ldots,I\}$, arrive and evaluate the available products. Each consumer i chooses the alternative

Stefan Mayer
Department of Analytics & Optimization, University of Augsburg, 86135 Augsburg, Germany,
e-mail: Stefan.Mayer@wiwi.uni-augsburg.de

Jochen Gönsch
Department of Analytics & Optimization, University of Augsburg, 86135 Augsburg, Germany,
e-mail: Jochen.Goensch@wiwi.uni-augsburg.de

D. Klatte et al. (eds.), *Operations Research Proceedings 2011*, Operations Research Proceedings, 547
DOI 10.1007/978-3-642-29210-1_87, © Springer-Verlag Berlin Heidelberg 2012

that maximises her utility; that is, she either self-selects one of the products j or leaves without buying anything.

Three major streams of research can be distinguished with regard to consumer choice behaviour. The first stream generally assumes that the consumer population consists of a number of segments populated by identical individuals. Their preferences are represented by so-called reservation prices, which denote their willingness-to-pay for the products and are known to the seller. The consumers then select a product according to the deterministic max-surplus choice rule (MSCR). Another research stream adopts a stochastic approach and assumes that all consumers' preferences for a given product come from a single probability distribution, implying that the consumer population consists of a single segment. Obviously, this can also be extended to multiple segments. Here, choice behaviour is usually described using discrete choice models – like the Multinomial Logit (MNL) model –, which are widely accepted in theory as well as in practice due to their deep theoretical foundations and the availability of appropriate software packages. The third stream is somewhat in between. Again, non-identical consumers from a single segment are considered, their reservation prices are stochastic, but the choice decision follows the MSCR. After reservation prices have been generated, the resulting problem of optimising prices for a single or a set of scenarios is formulated with similar MILPs as those used in the first stream of research. Among others, the possibility to explore more realistic settings and varying information conditions are cited as advantages of this numerical approach, which considers demand in a disaggregated form.

The aim of this article is to show that the MSCR and discrete choice modelling concepts are not as different as often perceived in academic discussions. Specifically, we derive a distribution for reservation prices, which allows consumers following MSCR to behave according to a simple MNL model. Our results can be used directly in the third literature stream outlined above, where the distribution of reservation prices often seems arbitrary. Moreover, the results can help to design more realistic test cases for the models and algorithms developed in the first stream. In the following, an overview is given of consumer choice with MSCR (Section 2) and MNL (Section 3). In Section 4, an MNL model is taken as starting point in order to show how reservation prices have to be drawn to obtain an equivalent MSCR. This approach is illustrated in Section 5, using an MNL model taken from the literature.

2 MSCR and Reservation Prices in Product Line Pricing

This section describes the basic idea of the max-surplus choice rule in the context of product line pricing. In this concept, the willingness-to-pay of consumer i for product j is represented by a reservation price, which is denoted by v_{ij}. The surplus s_{ij} (also referred to as utility in the literature) is then defined as the difference between her reservation price and the price of the product: $s_{ij} = v_{ij} - r_j$. To account for rational consumer behaviour, the max-surplus choice rule incorporates two well-known assumptions (see, e.g., [5]): the participation constraint (PC) and the incentive compatibility constraint (IC). Evaluating all alternatives, each consumer i self-selects at most one available product j', which provides her with the highest positive surplus; that is, she buys product j' if the following constraints hold:

$$s_{ij'} = v_{ij'} - r_{j'} \geq 0 \tag{PC}$$

$$s_{ij'} = v_{ij'} - r_{j'} \geq s_{ij} = v_{ij} - r_j \qquad \text{for all } j \in \mathcal{J} \tag{IC}$$

According to the PC, a consumer i only considers products that provide a non-negative utility. Hence, if $s_{ij} < 0$ for all $j \in \mathcal{J}$, she decides on the no-purchase option 0 with $s_{i0} = 0$. Note that this alternative is modelled implicitly in this concept. Otherwise, the IC ensures that consumer i chooses the product j' that maximises her utility.

3 Multinomial Logit

Similar to the max-surplus choice rule, utility-maximising consumers are assumed in discrete choice theory and its most famous model, the MNL. (See, e.g., the classical textbooks by [1] or [6] for comprehensive references.) Consumers have to choose exactly one alternative from a choice set $\mathcal{C}$, which is defined as $\mathcal{C} = \mathcal{J} \cup \{0\}$ in the context of product line pricing. According to the random utility theory, consumers' utilities for alternatives are random variables. This can reflect either heterogeneity in preference among individual consumers or measurement errors. In either case, the choice outcome of any given consumer is uncertain due to this uncertainty in the utilities. Formally, the utility U_{ij} of consumer i for alternative j consists of two parts: a deterministic component V_{ij} and a zero-mean random component ε_{ij}

$$U_{ij} = V_{ij} + \varepsilon_{ij} \qquad \text{for all } i \in \mathcal{I}, j \in \mathcal{C} \tag{1}$$

The deterministic component V_{ij} is usually modelled as a linear-in-parameter combination of observable attributes

$$V_{ij} = \beta^T x_{ij}, \tag{2}$$

where x_{ij} is a vector of known attributes of alternative j for consumer i (e.g., product characteristics and price) and β is an unknown vector of weights (parameters) on these attributes. The weights β are estimated from historical data on choice outcomes via maximum likelihood estimation. Since utility is ordinal, we normalise the deterministic utility of the no-purchase alternative to zero ($V_{i0} = 0$). Thus, the probability that consumer i chooses alternative j' from the choice set $\mathcal{C}$ is given by

$$P_{ij'} = Prob(U_{ij'} \geq U_{ij}, \forall j \in \mathcal{C}) \tag{3}$$

While these basic ideas are shared by all random utility models, they differ in the distribution assumed for the ε_{ij}. The MNL model is derived by assuming that the ε_{ij} are i.i.d. random variables following a Gumbel distribution with mean zero and scale parameter one for all i and j. This distribution can be viewed as an approximation of the normal distribution and leads to analytically tractable choice probabilities. In respect of the MNL model, one can show that the choice probability that consumer i chooses alternative $j' \in \mathscr{C}$ is given by

$$P_{ij'} = e^{V_{ij'}} / \sum_{j \in \mathscr{C}} e^{V_{ij}} \tag{4}$$

The existence of this closed form for the choice probabilities is an important advantage of the MNL model and, together with its robustness in practice, the reason for its broad application in economics and marketing.

4 Transformation of MNL to MSCR

In order to illustrate how to obtain an MSCR model that is equivalent to an MNL model, we start with a model where utility is specified according to (1) and (2). The deterministic component V_{ij} of the utility is split into two parts, namely one part influenced by the price r_j and a second part that reflects the other product characteristics, which is denoted by $\bar{x}_{ij}$. Then, the utility U_{ij} of consumer i for product j can be rewritten as

$$U_{ij} = \beta^T \bar{x}_{ij} - \beta' r_j + \varepsilon_{ij}, \tag{5}$$

where $\beta' > 0$ denotes the price parameter. Maximising her utility, consumer i selects product j' if no other product yields a higher utility; that is, if $U_{ij'} \geq U_{ij}$ for all $j \in \mathscr{J}$. Substituting (5) for utility, we obtain

$$\left(\frac{1}{\beta'} \beta^T \bar{x}_{ij'} + \frac{\varepsilon_{ij'}}{\beta'} \right) - r_{j'} \geq \left(\frac{1}{\beta'} \beta^T \bar{x}_{ij} + \frac{\varepsilon_{ij}}{\beta'} \right) - r_j \qquad \text{for all } j \in \mathscr{J} \tag{6}$$

This closely resembles the IC in the MSCR, with the terms in brackets reflecting the reservation prices. But the constraints only reflect the choice between the products that are offered. In the MNL model, product j' is chosen if and only if $U_{ij'} \geq U_{i0}$ holds in addition to (6). Thus, using (5) and $U_{i0} = \varepsilon_{i0}$ since $V_{i0} = 0$, we obtain the following constraint after some minor rearrangement of terms:

$$\left(\frac{1}{\beta'} \beta^T \bar{x}_{ij'} + \frac{\varepsilon_{ij'} - \varepsilon_{i0}}{\beta'} \right) - r_{j'} \geq 0 \tag{7}$$

When reservation prices are calculated according to the term in brackets, formula (7) corresponds to the PC in the MSCR. By subtracting $\frac{\varepsilon_{i0}}{\beta'}$ from (6), it is easy to see that this choice of reservation prices also reflects the choice among products when these reservation prices are used in the IC. Hence, MNL and MSCR with

$$v_{ij} = \frac{1}{\beta'}\beta^T \bar{x}_{ij} + \frac{\varepsilon_{ij} - \varepsilon_{i0}}{\beta'} \tag{8}$$

lead to the same decision for every consumer. Similar to utility in the MNL model, this reservation price consists of two parts, a deterministic component $\frac{1}{\beta'}\beta^T \bar{x}_{ij}$ and a stochastic component $\frac{\varepsilon_{ij}-\varepsilon_{i0}}{\beta'}$, where all ε_{ij} and ε_{i0} are i.i.d. Gumbel distributed with mean zero. Interestingly, consumer i's reservation prices for the products are not independent as is widely assumed in the literature. There are consumers with generally higher or lower reservation prices, reflecting different valuations for the alternative of not buying any of the products. To the best of our knowledge, this correlation of reservation prices induced by the implicit incorporation of the no-purchase alternative in MSCR has not yet been considered in the literature on product line pricing.

5 Illustrative Example

We apply the approach outlined above to data from a study describing consumers' choice of orange juice in a major North American city. According to the MNL model in [4], the four products that we created for this example are associated with non-price valuations of 0.773, 0.853, 0.559, and 0.143, respectively. The value of β' is 1.480. To keep the example simple, we consider the problem of a seller selecting prices from a set of discrete price points, having increments of 0.1, to maximise his revenue without taking any further restrictions or costs into account. We first generate reservation prices according to (8) for 30 scenarios, each consisting of 1000 consumers. For each scenario, optimal prices are selected and the average revenue is described as Rev1 (see Table 1). We then evaluate these price vectors using another 30 scenarios (Rev2). Thus, Rev1 indicates the optimal average revenue when prices are set with knowledge of the exact reservation prices, whereas Rev2 indicates the revenue that is realised when only the distribution of the reservation prices is known.

Since reservation prices are often considered independent in the literature, we also draw reservation prices without considering ε_{i0} in (8). The resulting optimal price vectors are then evaluated analogously (Rev3), using the same scenarios as for Rev2. The tests reveal that if reservations prices are (wrongly) considered uncorrelated, then not only are the optimal prices for all products significantly lower than in the correlated case, but so are the resulting revenues if these prices are used in the evaluation scenarios (Rev3 – Rev2). This difference in revenue is significant at the 99 % level. Furthermore, it becomes apparent – at least in this example – that using the correct demand model is more important than knowing the exact consumer valuations.

Table 1. Revenues (averaged over 30 scenarios)

	mean	confidence interval (99 % level)
Rev1 (valuations known)	797.55	[788.65 ; 806.45]
Rev2 (distr. of valuations known)	780.65	[771.90 ; 789.40]
Rev3 (assuming uncorrelated valuations)	744.63	[735.44 ; 753.81]
Rev3 – Rev2	–36.02	[–44.39 ; –27.65]

6 Conclusion

If the basic model of the product line pricing problem is extended by additional aspects, such as capacity restrictions, it is often necessary to analyse the consumer behaviour in a disaggregated form. In this case, simulation-based optimisation (e.g., sample path optimisation) is often applied to account for the stochastic nature of the problem. Assuming that consumer behaviour follows a discrete choice model, reservation prices should be generated accordingly in order to obtain realistic scenarios. Specifically, this involves implicitly considering the valuation of the no-purchase option in the reservation prices for all other alternatives. Computational tests by means of an example based on a published study show that the effects resulting from this approach can be significant.

References

1. Ben-Akiva, M., Lerman, S.R.: Discrete choice analysis – Theory and application to travel demand. MIT Press, Cambridge (1985)
2. Day, J.M., Venkataramanan, M.A.: Profitability in product line pricing and composition with manufacturing commonalities. European Journal of Operational Research **175**, 1782–1797 (2006)
3. Kraus, U.G., Yano, C.A.: Product line selection and pricing under a share-of-surplus choice model. European Journal of Operational Research **150**, 653–671 (2003)
4. Louviere, J.J., Hensher, D.A., Swait, J.D.: Stated choice methods – Analysis and application. Cambridge University Press, Cambridge, 283–292 (2000)
5. Talluri, K.T., van Ryzin, G.J.: The theory and practice of revenue management. Springer, New York, 357–359 (2004)
6. Train, K.E.: Discrete choice methods with simulation. 2nd edn. Cambridge University Press, Cambridge (2009)

Incentives for Risk Reporting with Potential Market Entrants

Anne Chwolka, Nicole Kusemitsch

Abstract We analyze a situation in which an incumbent firm, endowed with private information about a risk factor in its market, can credibly disclose quantitative risk information to the market and, thus, also to its opponents. Favorable information increases the market price of the firm, but it may also induce the opponent to enter the market, which imposes a proprietary cost on the firm. We show that there exist partial-disclosure equilibria with two distinct nondisclosure intervals. If the firm does not disclose, the opponent will not enter the market. We conclude that one reason for the empirically observed lack of quantitative risk disclosure is the fact that firms are required to explain the underlying assumptions and models that they use to measure the risk, which can lead to important competitive disadvantages.

1 Introduction

Empirical studies on risk reporting of German companies reveal that most firms do not disclose quantitative risk information, while other studies find evidence for the successful implementation of risk management-systems.[1] The latter observation implies that firms possess quantitative risk information, but do not disclose it. This behavior is surprising, because many benefits of risk management can only be realized, if the presence of an appropriate and effective risk management system is credibly communicated, and the addressees of financial information improve their assessments of the riskiness of the firm's cash flows, resulting in a possible increased market price of the firm. Indeed, as the disclosure of externally procured credit rat-

Prof. Dr. Anne Chwolka
Department of Economics and Management, Otto-von-Guericke University Magdeburg, e-mail: chwolka@ovgu.de

Dipl.-Kff. Nicole Kusemitsch
Department of Economics and Management, Otto-von-Guericke University Magdeburg, e-mail: nicole.kusemitsch@ovgu.de

[1] See for a survey of empirical studies on risk reporting and implementation of risk management-systems for German companies Chwolka/Kusemitsch (2010).

D. Klatte et al. (eds.), *Operations Research Proceedings 2011*, Operations Research Proceedings, 553
DOI 10.1007/978-3-642-29210-1_88, © Springer-Verlag Berlin Heidelberg 2012

ings documents, many German firms have the desire to publish risk information, even if the provision of the information is costly. This raises the question why firms buy external credit ratings, instead of publishing similar information, which they can access for free with their risk management-systems.

The aim of our study is to find reasons for this seemingly paradoxical disclosure behavior. We argue that the legal requirements are too strict and prevent informative risk reports. In general, established firms in a market have a competitive advantage, because they can base their decisions on information that is not available to other people. In particular, they are better in assessing and handling the risks in their markets than potential competitors. However, according to DRS 5, when disclosing quantitative risk information, the reporting firm also has to disclose additional, detailed information, i.e., the assumptions underlying the quantitative risk information. Consequently, the firm is forced to give up its informational advantage. The publicly available information in turn may induce a competing firm to enter the market with corresponding negative consequences, presumably a loss of revenues, for the reporting firm.

We analyze disclosure strategies of an incumbent firm endowed with private information about a risk factor, e.g., the variance of its cash flows (CF), which influences its market price. The firm can credibly disclose this information to outsiders, i.e., its opponents and investors in the market. Favorable information, i.e., a smaller variance, increases the market price of the firm, but it may also induce potential opponents to enter the market, which then imposes a proprietary cost on the firm. Without disclosure the opponent will not enter the market. We show that there exists a partial-disclosure equilibrium with two distinct nondisclosure intervals.

Our analysis builds on existing models. Wagenhofer (1990), Darrough/Stoughton (1990), Feltham/Xie (1992), Newman/Sansing (1993) and Suijs (2005) all analyze disclosure strategies of firms faced with the threat of opponents' market entrance. However, they focus on situations in which the market price or the CF of the firm is increasing in the information, whereas in our situation, it is decreasing. In contrast to Wagenhofer (1990), we model the proprietary costs as variable and dependent on the disclosed information, because these costs capture the loss of market share resulting in a loss of CF. In contrast to Suijs (2005), we exclude direct costs in our analysis, because we see the risk information as a byproduct of effective risk management. Direct costs are relevant, if the firm can disclose credit ratings, whereas in that case no proprietary costs arise.

The remainder of the paper is organized as follows. In section 2 the basic model is described, section 3 derives the equilibria, and section 4 concludes.

2 The Model

In the following we modify the model of Wagenhofer (1990) in order to adopt it to a situation in which an incumbent firm holds private information about its CF-variance. We consider a setting with three players, i.e., the incumbent firm, one potential competitor, and a group of investors in the financial market. The firm wants to maximize its market price, and has to decide whether to disclose its private information. The disclosure of the information is assumed to be truthful, there do not occur direct costs if the information is disclosed, and the competitor and the financial market know that the firm possesses the information. Acting as a monopolist, the incumbent firm gains a risky CF at the sales market, $\tilde{y} = x + \tilde{\varepsilon}$, at the end of the period, where $\tilde{\varepsilon}$ is a normally distributed random variable with $\tilde{\varepsilon} \sim N\left(0; \sigma^2\right)$. Thus, the expected CF is $E\left(\tilde{y}\right) = x$ with variance $Var\left(\tilde{y}\right) = \sigma^2$.

In the beginning, the firm, the competitor, and the financial market have homogeneous prior beliefs about the variance $E(\sigma^2)$, where the variance is uniformly distributed in the interval $I = \left[\underline{\sigma}^2; \overline{\sigma}^2\right]$. The firm's risk management-system determines the CF-variance. After privately learning the variance, the firm decides whether to disclose it, taking into account that a potential competitor may enter the market. In case of market entrance, the firm has to share its sales market, and looses the fraction c of its CF. The remaining CF for the firm is $\tilde{y}^I = (1-c)(x+\tilde{\varepsilon})$. Market entrance causes entry costs K at the beginning of the period, and must be paid back at the end of the period with interest. Thus, the CF for the competitor is $\tilde{y}^E = c(x+\tilde{\varepsilon}) - K(1+r_f)$, where r_f is the risk-free interest rate. Observe that this sharing of CF reflects our assumption that the competitor is able to replicate the variance of CF, if the firm discloses this information σ^2 and the additional detailed information underlying the quantitative risk information. But if the firm withholds the information, the competitor is not able to replicate the volatility of the CF. Accordingly, in case of nondisclosure, the competitor will never enter the market. Note that our assumption here is different to the respective assumption of Wagenhofer (1990) and the other papers mentioned above.

For expositional convenience, it is assumed that investors and the competitor have the same utility function $U(\tilde{y}) = -\exp(-2a\tilde{y})$, with the same constant absolute risk aversion, $2a$, where $\tilde{y} \in \{\tilde{y}^I; \tilde{y}^E\}$ represents the risky CF of the incumbent and the entering firm, respectively. The market price of each firm is assessed by discounting the certainty equivalent of the corresponding CF of the firm with the risk-free discount rate $R_f = 1 + r_f$, i.e., $P = (E(\tilde{y}) - aVar(\tilde{y}))/R_f$ with $\tilde{y} \in \{\tilde{y}^I; \tilde{y}^E\}$.

If the firm discloses, the beliefs of investors and competitor equal the disclosed information, σ_D^2. If the firm does not disclose, the beliefs of investors about the variance (except for the case of skeptical beliefs) equal the expected value conditional on observing nondisclosure: $E\left(\tilde{\sigma}^2 \,\middle|\, \tilde{\sigma}^2 \in N\right) = \frac{1}{\int_N dF(\sigma^2)} \int_N \sigma^2 dF\left(\sigma^2\right) =: \hat{\sigma}_N^2$.

The competitor will only enter the market, if the risk-free discounted certainty equivalent of his CF is greater than zero. Therefore, the decision function of the competitor is:

$$b\left(\sigma_D^2\right) = \begin{cases} 0 \text{ for } \frac{cx - c^2 a\sigma_D^2}{R_f} < K\,, \\[2ex] 1 \text{ for } \frac{cx - c^2 a\sigma_D^2}{R_f} \geq K\,. \end{cases}$$

Dependent on the competitor's strategy, the market price of the incumbent firm in case of disclosure, P_D^I, and nondisclosure, P_N^I, is given by

$$P_D^I = \frac{x\left(1 - cb\left(\sigma_D^2\right)\right) - a\sigma_D^2\left(1 - cb\left(\sigma_D^2\right)\right)^2}{R_f}, \qquad P_N^I = \frac{x - a\hat{\sigma}_N^2}{R_f}.$$

3 Equilibrium

The solution concept used is that of a sequential equilibrium. We only consider pure strategies. Disclosing low variances will increase the market price, but increases the danger of market entrance. Thus, the disclosure strategy of the firm has to tradeoff between maximizing the market price through disclosing low variances and deterring market entrance.

Proposition 1. *There always exists a full-disclosure equilibrium.*

This result is driven by skeptical beliefs held by investors, as argued by Wagenhofer (1990) for his setting. Skeptical beliefs are the worst information of the firm, i.e., in our setting $\overline{\sigma}^2$. If the firm does not disclose, the competitor will not enter the market, and the investors assume that the undisclosed variance is $\overline{\sigma}^2$. Thus, the investors value the firm at the lowest possible price. Consequently, all firms, whose variance is smaller than $\overline{\sigma}^2$, will disclose their variance, and the competitor will only enter the market, if the risk-free discounted certainty equivalent of his CF is greater than zero.

If there are no skeptical beliefs, and if the firm faces the threat of market entrance, the firm has an incentive to withhold some variances. Given the decision strategy of the competitor, this is the case, if the market price is lower when the firm discloses than when it withholds the information, i.e., when

$$\frac{x\left(1 - cb\left(\sigma_D^2\right)\right) - a\sigma_D^2\left(1 - cb\left(\sigma_D^2\right)\right)^2}{R_f} \leq \frac{x - a\hat{\sigma}_N^2}{R_f},$$

which leads to the following proposition:[2]

Proposition 2. *There exists a partial-disclosure equilibrium with the nondisclosure set*

$$N = \left[\max\left\{\underline{\sigma}^2; \frac{a\hat{\sigma}_N^2 - cx}{a\left(1 - c\right)^2}\right\}; \frac{cx - R_f K}{ac^2}\right] \cup \left[\hat{\sigma}_N^2; \overline{\sigma}^2\right].$$

If the firm does not disclose, the competitor will not enter the market.

The partial-disclosure equilibrium consists of two nondisclosure intervals (see Fig. 1). The dotted line in Fig. 1 represents the market price of the firm, if the firm

---

[2] The formal proof of this proposition is mainly analogous to Wagenhofer (1990).

does not disclose and the investors estimate the market price according to the expected variance conditional on nondisclosure, $\hat{\sigma}_N^2$. The slope of this market price is zero because the price depends on the expected variance conditional on nondisclosure. In equilibrium, the investors do not change their expectations about the undisclosed variance and, therefore, the market price is constant. The solid line with a jump at σ_2^2 represents the market price of the firm, if the firm discloses the variance. The slope of the solid line is flatter for small variances, because for small variances the competitor would enter the market. In this case, the firm and the competitor share the risk of the market as well as the expected cash flow. If the competitor does not enter, the firm has to bear the risk by itself. Accordingly, the market price is less decreasing in the variance for small variances with market entrance than for higher variances with no market entrance.

In the lower interval, variances are not disclosed, because the market price of the firm assessed by the investors cannot compensate for the disadvantage of the proprietary costs due to market entrance. If the firm would disclose low variances, it must reveal additional, detailed information such as the assumptions underlying this quantitative information. Then, the competitor would be able to estimate the risk in the market and replicate the variance of CF. If the firm discloses variances that induce the competitor to enter, the firm looses part of its cash flow and, therefore, its market price is lower than in case of nondisclosure. Thus, the firm withholds the information to deter entrance. In the second interval of nondisclosure, the competitor would not enter if the firm discloses the variance. But the firm observes a variance which is greater than the variance expected by investors conditional on observing nondisclosure. Thus, the firm withholds the information in order to realize a higher market price.

In the partial-disclosure equilibrium depicted in Fig. 1, only average variances are disclosed. However, there might be cases, in which variances somewhat greater than $\underline{\sigma}^2$ are disclosed, too, implying the competitor's market entrance. This is the case, if the market price of the firm assessed by investors is high enough to compensate for the disadvantage of the proprietary costs caused by market entrance.

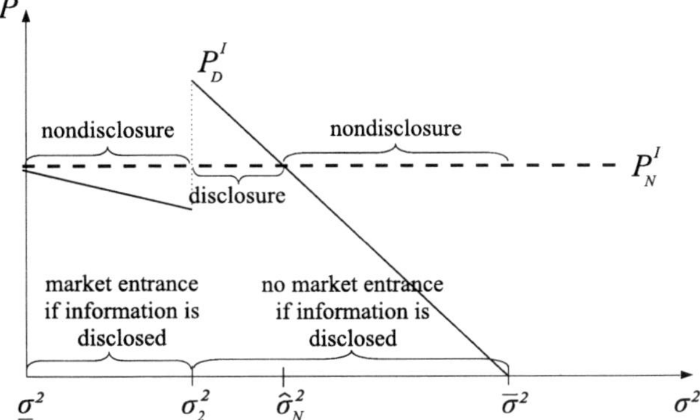

Fig. 1 A partial-disclosure equilibrium

4 Conclusion and Discussion

This paper analyzes risk disclosure strategies of privately informed firms, who face the threat of a market entrance of competitors. In our model, the variance of CF represents the relevant risk information. According to DRS 5, the firm has to reveal additional information when disclosing quantitative risk information, which can result in a market entrance and sales market loss. Because the firm has to disclose detailed information, the competitor might be able to deduce the risk situation of the market and can replicate it.

We showed that there exists a partial-disclosure equilibrium with two nondisclosure intervals. In the lower interval the firm does not disclose, because a good market valuation cannot outweigh the disadvantage of bearing the proprietary costs due to the market entrance of the competitor. In the upper nondisclosure interval, the firm does not disclose, due to the lower market valuation, which results from the fact that the observed variance is greater than the expected variance conditional on observing nondisclosure.

We conclude that one reason for the empirically observed unusual quantitative risk disclosure is the fact that firms are required to explain the underlying assumptions and models used to measure the risk, which can lead to significant competitive disadvantages. As Casson (2005) argues, a successful entrepreneur must be an effective information manager. If risk-reporting standards force the disclosing firm to give up its informational advantage, it is not surprising that we do not often observe firms publishing quantitative risk information. Hence, if standard setters actually want the disclosure of quantitative risk information, one should check whether the verification of the assumptions and models underlying the risk information through an auditor could be an alternative to publishing the detailed information.

References

1. Casson, M.: Entrepreneurship and the theory of the firm, J. Econ. Behav. Organ. **58**, 327–348 (2005)
2. Chwolka, A., Kusemitsch, N.: Auswirkungen einer zunehmenden Regulierung auf die Qualität der Risikoberichterstattung, 2010, FEMM Working Paper No. 15, University of Magdeburg, July (2010)
3. Darrough, M.N., Stoughton, N.M.: Financial Disclosure Policies in an Entry Game. J. Account. Econ. **12**, 219–243 (1990)
4. Feltham, G.A., Xie, J.Z.: Voluntary financial disclosure in an entry game with continua of types. Contemp. Account. Res. **9**, 46–80 (1992)
5. Newman, P., Sansing, R.: Disclosure Policies with Multiple Users. J. Account. Res. **31**, 92–112 (1993)
6. Suijs, J.: Voluntary Disclosure of Bad News. J. Bus. Finan. Account. **32**, 1423–1435 (2005)
7. Wagenhofer, A.: Voluntary disclosure with a strategic opponent. J. Account. Econ. **12**, 341–363 (1990)

On the Integration of Customer Lifetime Value into Revenue Management

Johannes Kolb

Abstract This paper provides insights into assessing customer relationships in a revenue management environment and depicts methodological problems that may arise in this context. For this purpose, we formulate a multi-period deterministic linear programing model, which accounts for customers' future demand behavior being dependent on the provider's previous availability decisions. Moreover, on the basis of an illustrative example, we investigate the impact on the control policy of a traditional deterministic linear program when both the product-related revenue and the previously developed long-term profitability measure are considered.

1 Introduction

In the field of customer relationship management (CRM), the customer lifetime value (CLV) – often defined as the present value of all future profits generated from a customer – serves as an important decision-making criterion. However, hardly any revenue management (RM) methodologies comprise such a long-term perspective. Even though a service provider's current pricing and availability decisions may affect customers' future purchase behavior, these implications are ignored. The up-to-date optimization approaches in capacity control are almost exclusively transaction-based, that is, only short-term attainable revenue is maximized. The incorporation of the expected CLV into RM systems is therefore identified as a current challenge. This requirement is often termed "customer relationship and revenue management" (CR^2M, see [2]) or "customer value-based revenue management" (see [7]). In the relevant literature, there are some RM approaches that incorporate a customer oriented view, for example, passenger name record-based no-show forecasts or inventory control models considering customer choice behavior (see [3], [1]). Furthermore, the first models were developed where capacity is allocated with respect to the customers' lifetime values (see [5], [7]).

Johannes Kolb
Department of Analytics & Optimization, University of Augsburg, 86135 Augsburg, Germany,
e-mail: johannes.kolb@wiwi.uni-augsburg.de

D. Klatte et al. (eds.), *Operations Research Proceedings 2011*, Operations Research Proceedings, 559
DOI 10.1007/978-3-642-29210-1_89, © Springer-Verlag Berlin Heidelberg 2012

In order to justify the incorporation of CLV – as a proxy for a customer's long-term profitability – into RM models, it is necessary to assume that current availability decisions affect the lifetime value. Otherwise, the traditional transaction-based approaches would suffice. In this paper, we regard the CLV as the present value of all revenues to be generated from a customer in future periods, that is, the revenue attainable in the present period is not included. In this context, for optimization purposes, it is generally not adequate to estimate the CLV, but rather its changes depending on whether a customer's current request is accepted or denied. There may be several reasons why the control policy induces changes in CLV. If a customer's present request for a certain product is met, her lifetime value may increase (compared to the denial of her request), since, in the next period(s), this customer will purchase:

- higher revenue products while not changing her purchase frequency;
- more frequently while still spending the same amount per transaction; or
- higher revenue products, thereby overcompensating a lower purchase frequency.

However, such potential reactions of the customer cause a change of the demand structure in subsequent periods. Given that the supplier already faces excess demand, he will, in turn, adapt his future availability decisions, which implies further CLV-changes that affect other customers. These interdependences are not considered in the currently available models. Moreover, these approaches do not address the above-listed issues that influence the CLV, but incorporate lifetime value measures as an exogenous parameter, which means that the implications of the present control policy on the future customer purchase behavior are not directly considered. In addition, the relevant literature neither offers approaches that allow for estimating the CLV in the context of scarce capacity nor are there any analyses evaluating the performance of traditional RM models that are adjusted for a CLV component.

The remainder of this paper is organized as follows: In Section 2, we introduce a multi-period model accounting for customer behavior dependent on the provider's control policy. Section 3 analyzes the CLV within an RM environment and a customer equity (CE)[1]-based measure for long-term profitability is developed. Section 4 contains closing remarks.

2 Problem Description and Mathematical Modeling

In each time period $t \in \mathscr{T} = \{1, ..., T\}$, a provider offers different products to heterogeneous classes of customers. We regard a customer segment $i \in \mathscr{I}$ as a combination of a customer class and an offered product, that is, each segment i is characterized as a customer class requesting a particular product.

We consider two sets of customer segments. First, there are segments $i \in \mathscr{I}_{stable}$ whose future demand is invariant towards the provider's previous control policy. On

---

[1] In the marketing literature, the customer equity is defined as the sum over all lifetime values of a firm's current and future customers.

the other hand, there are customer segments $i \in \mathscr{I}_{\text{policy}}$ that are sensitive to current availability decisions, where $\mathscr{I}_{\text{policy}} = \mathscr{I} \setminus \mathscr{I}_{\text{stable}}$. As stated above, we assume that current availability decisions can affect future demand. If the provider accepts a request by a customer of a sensitive segment $i \in \mathscr{I}_{\text{policy}}$, the respective customer will once more request the corresponding product in the subsequent period with probability $p_i(a)$. If the request is denied, the relevant repurchase probability is $p_i(d)$. For ease of notation, we assume that customers do not switch between segments over time so that availability decisions only affect the frequency of purchase, but not the requested product.

For each $i \in \mathscr{I}_{\text{stable}}$, the demand in period t is denoted as $D_{i,t}$. In respect of the segments $i \in \mathscr{I}_{\text{policy}}$, only the initial demand arriving in $t = 1$ is known to be $D_{i,1}$, whereas the demand in future periods $t = 2, ..., T$ is a function of the provider's control policy. In each period t the provider faces a fixed capacity C_h of resources $h \in \mathscr{H}$. Let $a_{h,i}$ be the number of capacity units of resource h required to fulfill one request of segment i. The revenue connected to each segment i is denoted r_i.

The provider's objective is to maximize the revenue attainable over all periods by allocating contingents to the different segments in each period where the decision variables $x_{i,t}$ denote the number of expected requests from segment i in period t to be accepted. Thus, it is necessary to optimize simultaneously over all T periods. The optimization problem can be stated as a variant of the traditional deterministic linear programming model (DLP, see [6], which we adapt to the here considered multi-period scenario as follows (MP-DLP):

$$V = \max \sum_{t \in \mathscr{T}} \sum_{i \in \mathscr{I}} x_{i,t} r_i \tag{1}$$

subject to

$$\sum_{i \in \mathscr{I}} a_{h,i} x_{i,t} \leq C_h \qquad\qquad \forall h \in \mathscr{H}, t \in \mathscr{T} \tag{2}$$

$$x_{i,t} \leq D_{i,t} \qquad\qquad \forall i \in \mathscr{I}_{\text{stable}}, t \in \mathscr{T} \tag{3}$$

$$x_{i,t} \leq D_{i,1} p_i(d)^{t-1} + [p_i(a) - p_i(d)] \sum_{\tau=1}^{t-1} x_{i,\tau} p_i(d)^{t-\tau-1} \qquad \forall i \in \mathscr{I}_{\text{policy}}, t \in \mathscr{T} \tag{4}$$

$$x_{i,t} \geq 0 \qquad\qquad \forall i \in \mathscr{I}, t \in \mathscr{T} \tag{5}$$

The objective function (1) maximizes the total revenue over all periods and segments, that is the customer equity (CE). Constraints (2) ensure that the capacity C_h is sufficient. Constraints (3) and (4) guarantee that accepted requests do not exceed the expected demand. Regarding (4), the expected demand of the segments $i \in \mathscr{I}_{\text{policy}}$ is a function of the variables $x_{i,t}$. The number of denied requests from a segment i in Period 1 weighted by $p_i(d)$ and the number of accepted requests weighted by $p_i(a)$ determine this segment's demand in Period 2, that is, $D_{i,2} = (D_{i,1} - x_{i,1}) p_i(d) + x_{i,1} p_i(a)$. By definition, the demand met in Period 2 is $x_{i,2}$. Thus, the denied requests in Period 2 are $D_{i,2} - x_{i,2} = (D_{i,1} - x_{i,1}) p_i(d) + x_{i,1} p_i(a) - x_{i,2}$. By multiplying the requests denied and accepted in Period 2 by $p_i(d)$ and $p_i(a)$ respectively, the expected demand in Period 3 is derived. This procedure leads to the right hand side of Constraints (4). Finally, the decision variables must be nonnegative (Constraints (5)). Note that the traditional DLP model can be attained by setting $T = 1$.

To illustrate the optimal solution of the MP-DLP and the analyses in Section 3, we use the following running example where we consider two segments: Segment A's demand is invariant to previous availability decisions whereas customers of Segment B terminate their relationship with the provider if their request is denied $(p_i(d) = 0)$. However, for Segment B customers, there is a repurchase probability of $p_i(a) = 1$, if their request was accepted. To fulfill a request of either Segment A or B, one unit of the sole resource is needed. The provider has a planning horizon of $T = 5$ periods and a capacity of $C = 10$. The following table shows the demand in each period, where the last column contains the prices r_i.

Segment i	$D_{i,1}$	$D_{i,2}$	$D_{i,3}$	$D_{i,4}$	$D_{i,5}$	r_i
A	10	7	9	8	6	100
B	8	$x_{B,1}$	$x_{B,2}$	$x_{B,3}$	$x_{B,4}$	70

Solving this scenario to optimality gives the following contingents with an overall revenue of $V = 4450$

Segment i	$x_{i,1}$	$x_{i,2}$	$x_{i,3}$	$x_{i,4}$	$x_{i,5}$	$\sum_{t \in \mathscr{T}} x_{i,t} r_i$
A	7	7	7	7	6	3400
B	3	3	3	3	3	1050
$\sum_{i \in \mathscr{I}} x_{i,t} r_i$	910		3540			4450

and optimal Period 1 contingents $(x^*_{A,1}, x^*_{B,1}) = (7,3)$. $V_2 = \sum_{t=2}^{T} \sum_{i \in \mathscr{I}} x^*_{i,t} r_i$ denotes the revenue that is obtained in Periods $2, ..., T$. According to this, the revenue of Period 1 is 910 and the long-term attainable revenue $V_2 = 3540$.

3 Customer Lifetime Value in the Context of Revenue Management

3.1 Deriving CLV measures from MP-DLP

Although the incorporation of CLV into RM systems has been identified as a challenge in the relevant literature, there are no approaches that allow for estimating the CLV while facing scarce capacity. The methods discussed in the CRM and marketing community for valuing customer relationships can be categorized as probabilistic, econometric, persistence, computer science, and diffusion models (see [4]). However, these approaches do not account for fixed capacity, but model a particular customer's lifetime value assuming that it is independent from the provider's relationship with other customers. In an RM environment, different customers are connected through capacity, so that the isolated CLV estimation for a customer belonging to a particular segment fails.

As stated above, we are interested in estimating the future monetary impact of currently accepting (instead of denying) a customer of a particular segment i. Therefore, we consider the effects of accepting a segment i customer with regard to CE in the parameter ΔCE_i, which is determined via an opportunity cost-based approach. In this context, the optimal Period 1 contingent $x_{i,1}^*$ derived from the MP-DLP is reduced by one unit. Subsequently, the MP-DLP is resolved with $\bar{x}_{i,1} = x_{i,1}^* - 1$ where $V(\bar{x}_{i,1})$ denotes this model's optimal objective function value and $V_2(\bar{x}_{i,1})$ describes the long-term attainable revenue from Period 2 onwards. Now, ΔCE_i is calculated as follows:[2]

$$\Delta CE_i = V_2 - V_2(\bar{x}_{i,1}) \ .$$

For estimating the changes in CE caused by accepting a Segment B customer, we apply this procedure to our running example and set $\bar{x}_{B,1} = 3 - 1 = 2$. Subsequently, the optimal solution is

segment i	$x_{i,1}$	$x_{i,2}$	$x_{i,3}$	$x_{i,4}$	$x_{i,5}$	$\sum_{t \in \mathcal{T}} x_{i,t} r_i$
A	8	7	8	8	6	3700
B	2	2	2	2	2	700
$\sum_{i \in \mathcal{I}} x_{i,t} r_i$	940		3460			4400

where the short-term revenue attainable in Period 1 is 940 and $V_2(\bar{x}_{B,1}) = 3460$. This yields to $\Delta CE_B = V_2 - V_2(\bar{x}_{B,1}) = 3540 - 3460 = 80$, which is consequently, based on the optimal solution of Section 2, the cost of serving one less Segment B customer. The effects of accepting a customer from segment A can be obtained in the same way. Resolving the MP-DLP with $\bar{x}_{A,1} = 7 - 1 = 6$ gives $\Delta CE_A = V_2 - V_2(\bar{x}_{A,1}) = 0$. This result may seem somewhat intuitive, since Segment A's future demand is independent of the current availability decisions. However, $i \in \mathcal{I}_{\text{stable}}$ does not necessarily imply $\Delta CE_i = 0$.

Modifying this scenario by setting Segment A's revenue ceteris paribus $r_A' \neq r_A$ leads to $\Delta CE_B' \neq \Delta CE_B$ which once more illustrates that an isolated consideration of a single segment is not appropriate for long-term profitability estimation in an RM environment.

3.2 CLV-adjusted revenues in a DLP model

The requirement of integrating the CLV into RM models is motivated by the idea of being able to secure long-term profitability. Thus, the following questions arise: How do traditional RM models perform if they are adjusted according to a CLV or

---

[2] Note that the value of ΔCE_i is solely the opportunity cost for denying the first of $x_{i,1}^*$ segment i customers. An average value would be obtained by setting $\bar{x}_{i,1} = 0$ and analyzing $[V_2(\bar{x}_{i,1}) - V_2]/x_{i,1}^*$.

CE component and is it thereby possible to reproduce the results of a multi-period optimization? In this context, we analyze a traditional DLP model ($T = 1$) where the coefficients of the objective function are composed as the sum of the ΔCE_i-measures and the short-term attainable revenues r_i. Hence, the objective function (1) can be written as

$$\max \sum_{i \in \mathscr{I}} x_{i,1} \left(r_i + \Delta CE_i \right) .$$

This leads to optimal contingents in Period 1 – the sole period considered – of $(x_{A,1}^*, x_{B,1}^*) = (2,8)$, which obviously differs from the optimal solution of the MP-DLP. This stands to reason because, in the DLP model, it is impossible for more than one segment to be served partially, which can however be the case in the optimal MP-DLP solution. Even if other lifetime value measures are added to the coefficients of the DLP's objective function, this circumstance persists.

4 Conclusion

In an RM context, where lifetime values of different customers are connected through scarce capacity, currently available methods of CLV estimation fall short due to the isolated consideration of only one customer segment. The lifetime value measures derived from the here developed MP-DLP overcome these difficulties by directly accounting for the interdependencies between the availability decisions and demand of all segments. Although the integration of the CLV into RM systems is identified as a current challenge, we showed that a lifetime value adjusted DLP model cannot necessarily reproduce the long-term optimal results derived from a multi-period approach.

References

1. Bront, J.J.M., Méndez-Díaz, I., Vulcano, G.: A column generation algorithm for choice-based network revenue management. Operations Research **57**(3), 769-784 (2009)
2. Gallego, G., Stefanescu, C.: Service engineering: The future of service feature design and pricing. Working Paper (2010)
3. Goring, T., Brunger, W.G., White, M.M.: No-show forecasting: A blended cost-based, PNR-adjusted approach. Journal of Revenue and Pricing Management **5**(3), 188-206 (2006)
4. Gupta, S., Hanssens, D., Hardie, B., Kahn, W., Kumar, V., Lin, N., Ravishanker, N., Driram, S.: Modeling customer lifetime value. Journal of Service Research **9**(2), 139-155 (2006)
5. Metters, M., Queenan, C., Ferguson, M., Harrison, L., Higbie, J., Ward, S., Barfield, B., Farley, T., Kuyumcu, H.A., Duggasani, A.: The killer application of revenue management: Harrah's Cherokee Casino & Hotel. Interfaces **38**(3), 161-175 (2008)
6. Talluri K.T., van Ryzin, G.J.: The theory and practice of revenue management. Kluwer, Boston (2004)
7. von Martens, T., Hilbert, A.: Customer-value-based revenue management. Journal of Revenue and Pricing Management **10**(1), 87-98 (2011)

Stream XVI
Forecasting, Neural Nets and Fuzzy Systems

Petros Koumoutsakos, ETH Zurich, SVOR Chair
Martin Kührer, Fin4Cast Data Research Services GmbH Vienna, ÖGOR Chair
Hans Georg Zimmermann, Siemens Germany, GOR Chair

Formulation of A Sale Price Prediction Model Based on Fuzzy Regression Analysis

Michihiro Amagasa

Abstract It is indispensable for companies to take some marketing strategy to predict and analyze the sale price met to customer satisfaction. In sale price prediction model, human factors are included in the elements composing the model, and it is getting more difficult to define and describe entire systems precisely. Therefore in case we obtain data from such a model, the data are accompanied by human subjective and experiential uncertainty. In this paper, we develop a theoretical formulation of sale price prediction model based on fuzzy regression with fuzzy input-output data (SPP-model). The solutions of SPP-model is found by solving two LP problems, both Min- and Max-problems, and each of them indicates upper and lower bounds of possibility for solutions (prices).

1 Introduction

It is indispensable for companies to take a marketing strategy to predict and analyze the sale price met to customer satisfaction. In sale price prediction model, human factors are included in critical function composing the model, and it is getting more difficult to define and describe entire systems precisely. Therefore in case we obtain data from such a model, the data are accompanied by human subjective and experiential uncertainty. Up to now regression analysis has been used as one of the methods to construct models from data. Regarding usual regression models, deviations between the observed data and the estimated values from models are taken as observation errors, and it is assumed that error fluctuation follows probability distribution. But in some cases, it is difficult to conceive that deviations to data obtained through models such as economic systems, management systems, etc. are only caused by observation errors. In this paper, it is assumed that discrepancy between estimated values and observed data by analysis models is caused by fuzziness structure of systems. In short, we express the regression model with fuzziness.

Michihiro Amagasa
Hokkai-Gakuen University, 1-40 Asahimachi, Toyohira-ku, Sapporo 062-8605 Japan, e-mail: amagasa@ba.hokkai-s-u.ac.jp

D. Klatte et al. (eds.), *Operations Research Proceedings 2011*, Operations Research Proceedings, 567
DOI 10.1007/978-3-642-29210-1_90, © Springer-Verlag Berlin Heidelberg 2012

Fuzzy regression analysis has been formulated by Tanaka, H.et al.[4], and the application of it has been carried out. Since then, various types of regression analysis have been proposed for usual fuzzy data[5]. The approach such as the input is non-fuzzy and the output is fuzzy has been also proposed by using feasible linear regression models[3]. But, in these studies, regarding data treated in fuzzy regression models, most of them are to ordinary non-fuzzy data. Regarding ones toward fuzzy data as well, only output is expressed with fuzzy number, and ones toward data, with which input and output are expressed with fuzzy numbers, have not been discussed. In case models are accompanied by human subjective fuzziness, it is considered that the vagueness is included into both data supplied to the models and data obtained from the models.

In this paper, we propose a sale price prediction model (SPP-model) based on fuzzy regression with fuzzy input-output data as an extension of former paper[2]. The SPP-model is found by solving two LP problems, both Min- and Max-problems, and each of them indicates upper and lower bounds of possibility for solutions. Even in a special case as fuzzy numbers degenerate into usual non-fuzzy data, the SPP-model is applicable.

2 Fuzzy input-output data

Fuzzy input-output data are data with which input to the system and output from the system are expressed with fuzzy numbers[1]. When the input X and the output Y are shown as symmetric L-R fuzzy numbers, X and Y are expressed as (1) and (2).

$$X = (x,d)_L, d > 0 \tag{1}$$

$$Y = (y,e)_L, e > 0 \tag{2}$$

,where x, and y represent centers for X and Y. d and e widths or extents for X and Y.

Membership functions $\mu_x(z)$ and $\mu_y(z)$, of fuzzy numbers X and Y are expressed as (3) and (4).

$$\mu_x(z) = L((z-x)/d), d > 0 \tag{3}$$

$$\mu_y(z) = L((z-y)/e), e > 0 \tag{4}$$

But $L(z)$ is a function which has a distinctive feature as below.
(1) $L(z) = L(-z)$
(2) $L(0) = 1$
(3) $L(z) =$ strictly decreases in $(0, \infty)$

3 Formulation of a sale price prediction model(SPP-Model)

The relationships between the standard prices Y_i and typical marketing variables X_{ij} is given by a linear combination (5) and called a sale price prediction model.

$$Y_i = \sum_{j=1}^{n} A_j X_{ij} \quad (i = 1, 2, ...m)$$

,where A_i indicates the coefficient (weight) for the marketing variable. The decision makers, in general, suppose from their experienced that sale price prediction will be linear, and has uncertainty. Especially since it predicts the price of matured products at stage of development planning, the sale price and the marketing variables contain the fuzziness.

In this study, we here develop the theoretical formulation of sale price prediction model by using fuzzy input-output data defined in the previous section.

Definition 1: Let fuzzy input, fuzzy output and coefficient be fuzzy numbers $X_j(j=1,2,...n)$, Y and $A_j(j=1,2,...n)$. Then the fuzzy linear regression model with fuzzy input-output data is expressed as (5).

$$Y_i = \sum_{j=1}^{n} A_j X_{ij} \quad (i = 1, 2, ...m) \tag{5}$$

,where A_j represents L-R fuzzy number and is shown in (6).

$$A_j = (a_j, c_j)_L, \quad c_j \geq 0 \tag{6}$$

a_j represents the center for A_j. c_j represents the width or the extent for A_j and is expressed as the symmetric type function $L(z)$.

Here the sale price prediction model is defined by (5), but in general, we can't directly find a membership function of fuzzy number produced by multiplication between fuzzy numbers. We don't discuss about it here.

Expressing h level sets, $[Y]_h$, $[X]_h$, and $[A]_h$ of fuzzy numbers, Y, X, A with closed interval, they are expressed as (7), (8) and (9).

$$[Y]_h = [y - |L^{-1}(h)|e, y + |L^{-1}(h)|e] \tag{7}$$

$$[X]_h = [x - |L^{-1}(h)|d, x + |L^{-1}(h)|d] \tag{8}$$

$$[A]_h = [a - |L^{-1}(h)|c, a + |L^{-1}(h)|c] \tag{9}$$

From the above, let every interval be positive. Then h level set $[AX]_h$ of AX is expressed as (10) by using (8) and (9).

$$[AX]_h = [(a - |L^{-1}(h)|c)(x - |L^{-1}(h)|d), (a + |L^{-1}(h)|c)(x + |L^{-1}(h)|d)] \tag{10}$$

Therefore, h level set $[\tilde{Y}]_h$ of estimated values $\tilde{Y}$ of fuzzy linear regression model is expressed as (11) transformed by (10).

$$[\tilde{Y}]_h = [\sum_{j=1}^{n}(a_j - |L^{-1}(h)|c_j)(x_j - |L^{-1}(h)|d_j), \sum_{j=1}^{n}(a_j + |L^{-1}(h)|c_j)(x_j + |L^{-1}(h)|d_j)] \tag{11}$$

Here we take two estimated fuzzy outputs $\bar{Y}$ and $\underline{Y}$ shown in (12) and (13) into considtration.

$$\bar{Y} = \bar{A}_1 X_1 + ... + \bar{A}_n X_n \tag{12}$$

$$\underline{Y} = \underline{A}_1 X_1 + ... + \underline{A}_n X_n \tag{13}$$

The fuzzy coefficients $\bar{A}_j$, $\underline{A}_j$ represent the symmetric L-R fuzzy numbers, that is, $\bar{A} = (\bar{a}, \bar{c})_L, \underline{A} = (\underline{a}, \underline{c})_L$. The inclusion relation between (5), (12) and (13) is as follows;

$$\underline{A}_1 X_{i1} + ... + \underline{A}_n X_{in} = \underline{Y}_i \subseteq Y_i \subseteq \bar{Y}_i = \bar{A}_1 X_{i1} + ... + \bar{A}_n X_{in}, \quad {}^{\forall}i \in 1, 2, ..., m \quad 0 \leq h < 1 \tag{14}$$

Therefore the fuzzy coefficients should just be determined so that (14) is satisfied.

However, since the estimated $\bar{Y}$ shows the possibility of upper bound of Y, that is, an upper bound of price, and $\underline{Y}$ shows the possibility of lower bounds of Y, that is, a lower bound of price, it is necessary both to make the uncertainty of $\bar{Y}$ as small as possible and to make the uncertainty of $\underline{Y}$ as large as possible.

The widths of estimated fuzzy outputs $\bar{Y}_i$, $\underline{Y}_i$, indicate the width of h level set at h=0. Using (11), h level set of each at $h = 0$ is expressed as (15) and (16).

$$[\bar{Y}_i]_0 = [\sum_{j=1}^{n}(\bar{a}_j - |L^{-1}(0)|\bar{c}_j)(x_{ij} - |L^{-1}(0)|d_{ij}), \sum_{j=1}^{n}(\bar{a}_j + |L^{-1}(0)|\bar{c}_j)(x_{ij} + |L^{-1}(0)|d_{ij})] \tag{15}$$

$$[\underline{Y}_i]_0 = [\sum_{j=1}^{n}(\underline{a}_j - |L^{-1}(0)|\underline{c}_j)(x_{ij} - |L^{-1}(0)|d_{ij}), \sum_{j=1}^{n}(\underline{a}_j + |L^{-1}(0)|\underline{c}_j)(x_{ij} + |L^{-1}(0)|d_{ij})] \tag{16}$$

Therefore, the widths of fuzzy outputs, that is, $[\bar{Y}_i]$ and $[\underline{Y}_i]$ are expressed as (17) and (18).

$$[\bar{Y}_i] = \sum_{j=1}^{n}(2|L^{-1}(0)|)(\bar{a}_j d_{ij} + x_{ij}\bar{c}_j) \tag{17}$$

$$[\underline{Y}_i] = \sum_{j=1}^{n}(2|L^{-1}(0)|)(\underline{a}_j d_{ij} + x_{ij}\underline{c}_j) \tag{18}$$

Since $|L^{-1}(0)|$ is constant, it doesn't affect the objective functions. the uncertainty of estimated fuzzy outputs is expressed as (19) and (20).

$$J(\bar{a}, \bar{c}) = \sum_{i=1}^{m}\sum_{j=1}^{n}(\bar{a}_j d_{ij} + x_{ij}\bar{c}_j) \tag{19}$$

$$J(\underline{a},\underline{c}) = \sum_{i=1}^{m} \sum_{j=1}^{n} (\underline{a}_j d_{ij} + x_{ij}\underline{c}_j) \tag{20}$$

Consequently, the problem is to determine the coefficients $(\bar{a},\bar{c})$ and $(\underline{a},\underline{c})$ so as to minimize (19) and maximize (20).

From the above, Min-problem and Max-problem shown in (21) and (23) are derived.

(1) Min-Problem

$$\min_{\bar{A}_j=(\bar{a}_j,\bar{c}_j)_L} J(\bar{a},\bar{c}) = \sum_{i=1}^{m} \sum_{j=1}^{n} (\bar{a}_j d_{ij} + X_{ij}\bar{c}_j)$$

$$Y_i \subseteq_h \bar{Y}_i = \bar{A}_1 X_1 + \ldots + \bar{A}_n X_n$$

$$\bar{a}_j,\bar{c}_j \geq 0, \quad 0 \leq h < 1, \quad {}^{\forall} i \in 1,2,\ldots,m \tag{21}$$

This Min-problem is to make the uncertainty of $\bar{Y}$ minimize under the constraint such that estimated fuzzy output $\bar{Y}_i$ includes fuzzy output Y_i with a certain degree h. In other words, the problem is to find the fuzzy coefficients $\bar{Y}_i$ so that h level set $[Y_i]_h$ of the estimated fuzzy output $\bar{Y}_i$ includes h level set $[Y_i]_h$ of the fuzzy output Y_i, that is, under the constraint of $[Y_i]_h \subseteq [\bar{Y}_i]_h$.

Further, (21) can be transformed into LP problem as shown in (22) by using (7) and (11).

$$\min_{\bar{A}_j=(\bar{a}_j,\bar{c}_j)_L} J(\bar{a},\bar{c}) = \sum_{i=1}^{m} \sum_{j=1}^{n} (\bar{a}_j d_{ij} + X_{ij}\bar{c}_j)$$

$$Y_i - |L^{-1}(h)|e_i \geq \sum_{j=1}^{n} (\bar{a}_j - |L^{-1}(h)|\bar{c}_j)(x_{ij} - |L^{-1}(h)|d_{ij})$$

$$Y_i + |L^{-1}(h)|e_i \leq \sum_{j=1}^{n} (\bar{a}_j + |L^{-1}(h)|\bar{c}_j)(x_{ij} + |L^{-1}(h)|d_{ij})$$

$$\bar{a}_j,\bar{c}_j \geq 0, \quad 0 \leq h < 1, \quad {}^{\forall} i \in 1,2,\ldots,m \tag{22}$$

(2) Max-problem

$$\max_{\underline{A}_j=(\underline{a}_j,\underline{c}_j)_L} J(\underline{a},\underline{c}) = \sum_{i=1}^{m} \sum_{j=1}^{n} (\underline{a}_j d_{ij} + X_{ij}\underline{c}_j)$$

$$Y_i \supseteq_h \underline{Y}_i = \underline{A}_1 X_1 + \ldots + \underline{A}_n X_n$$

$$\underline{a}_j,\underline{c}_j \geq 0, \quad 0 \leq h < 1, \quad {}^{\forall} i \in 1,2,\ldots,m \tag{23}$$

This problem is to find the fuzzy coefficients $\underline{A}_j$ so as to make the uncertainty of $\underline{Y}$ maximize under the constraint such that the estimated fuzzy output $\underline{Y}_j$ is included in fuzzy output Y_i with a certain degree h. In other words, under the constraint of $[Y_i]_h \subseteq [\underline{Y}_i]_h$, the estimated fuzzy output $\underline{Y}_i$ should be found so that makes the uncertainty of $\underline{Y}$ maximize.

In the same way as Min-problem, in this case, (23) can be transformed into LP problem shown as (24) by using (7) and (11).

$$\max_{\underline{A}_j=(\underline{a}_j,\underline{c}_j)_L} J(\underline{a},\underline{c}) = \sum_{i=1}^{m} \sum_{j=1}^{n} (\underline{a}_j d_{ij} + X_{ij}\underline{c}_j)$$

$$Y_i - |L^{-1}(h)|e_i \leq \sum_{j=1}^{n} (\underline{a}_j - |L^{-1}(h)|\underline{c}_j)(x_{ij} - |L^{-1}(h)|d_{ij})$$

$$Y_i + |L^{-1}(h)|e_i \geq \sum_{j=1}^{n} (\underline{a}_j + |L^{-1}(h)|\underline{c}_j)(x_{ij} + |L^{-1}(h)|d_{ij})$$

$$\underline{a}_j, \underline{c}_j \geq 0, \quad 0 \leq h < 1, \quad {}^{\forall}i \in 1,2,...,m \qquad (24)$$

4 Conclusion

In this paper, we developed a theoretical formulation of sale price prediction model (SPP-model) based on fuzzy regression with fuzzy input-output data. The solutions of SPP-model is found by solving two LP problems, and each of them indicates the possibilities of upper and lower bounds of price. The model we have developed in this paper can deal with fuzzy input and output data, and degenerate into the regular model with ordinary non-fuzzy data by letting the widths of fuzzy numbers be zero. This study can bring about the rational solution to actual problems such that include fuzzy data like human elements. This means that compared to the current regression models, the models are more practical ones. From this, by identifying the possibility linear regression model while taking fuzziness into account, we will be able to do more effectively the prediction and control of sale price.

References

1. Inoue,H., Amagasa,M.: Fundamentals of Fuzzy Theory. Asakurasyoten,p.30 (1997)
2. Inoue, Y., Uematsu, Y., Amagasa, M., Tomizawa, G.: The method of setting forecast selling price model by fuzzy regression analysis. Proceedings of Japan industrial Management Association, Autumn Congress, pp.194-195 (1991)
3. Tanaka, H.: Fuzzy analysis by possibility linear models: Fuzzy Sets and Systems, Vol.24, pp.363-375 (1987)
4. Tanaka, H., Uejima, S., Asai K.: Linear regression analysis with fuzzy model. IEEE Transactions on Systems, Man and Cybernetics, SMC-12, No.6, pp.903-907(1982).
5. Guo,P., Tanaka,H.: Dual Models for Possibilistic Regression Analysis. Computational Statistics & Data Analysis, Vol.51(1) 252-266 (2006).

Visualizing Forecasts of Neural Network Ensembles

Hans-Jörg von Metthenheim, Cornelius Köpp and Michael H. Breitner

Abstract Advanced neural network architectures like, e.g., Historically Consistent Neural Networks (HCNN) offer a host of information. HCNN produce distributions of multi step, multi asset forecasts. Exploiting the entire informational content of these forecasts is difficult for users because of the sheer amount of numbers. To alleviate this problem often some kind of aggregation, e.g., the ensemble mean is used. With a prototypical visualization environment we show that this might lead to loss of important information. It is common to simply plot every possible path. However, this approach does not scale well. It becomes unwieldy when the ensemble includes several hundred members. We use heat map style visualization to grasp distributional features and are able to visually extract forecast features. Heatmap style visualization shows clearly when ensembles split into different paths. This can make the forecast mean a bad representative of these multi modal forecast distributions. Our approach also allows to visualize forecast uncertainty. The results indicate that forecast uncertainty does not necessarily increase significantly for future time steps.

1 Introduction

The goal of the present paper is to provide new ideas how to visualize forecasts of neural network ensembles. As a prototypical workhorse we use the Historically Consistent Neural Network (HCNN), see [8]. An important aspect of this archi-

Hans-Jörg von Mettenheim
Institut für Wirtschaftsinformatik, Leibniz Universität Hannover, e-mail: mettenheim@iwi.uni-hannover.de

Cornelius Köpp
Institut für Wirtschaftsinformatik, Leibniz Universität Hannover, e-mail: koepp@iwi.uni-hannover.de

Michael H. Breitner
Institut für Wirtschaftsinformatik, Leibniz Universität Hannover, e-mail: breitner@iwi.uni-hannover.de

D. Klatte et al. (eds.), *Operations Research Proceedings 2011*, Operations Research Proceedings,   573
DOI 10.1007/978-3-642-29210-1_91, © Springer-Verlag Berlin Heidelberg 2012

tecture is that it allows multi step forecasts. Additionally, by training an ensemble of HCNNs, we get a distribution of possible outcomes. In this paper we will *not* primarily evaluate the merits of HCNNs as a forecasting model. See, e.g., [5, 6, 8] for detailed performance evaluations. Instead we focus on supporting the decision maker in interpreting the considerable amount of data that the forecast produces.

2 Basic Visualization Techniques for Ensemble Forecasts

Typically, we visualize time series ensembles by plotting summary data, see for example [1–4]. As a first step, we can simply plot the mean or median, see figure 1. The figure shows a 20-day ahead forecast for the price of natural gas in US-Dollar. The exact forecast asset is not central to the following discussion. We continue using the same 20-day ahead natural gas price forecast throughout the paper. Mean and median can, of course, differ. The general visual impression of figure 1 is that of strong downtrend. The figure does not convey any distributional information.

This changes slightly when we add representative percentiles, like the quartiles, see figure 2. Note an interesting feature: future uncertainty, as measured by the difference of maximum and minimum, does not necessarily increase. In fact, uncertainty decreases during the last five forecast days. The figure now only conveys the visual impression of a slight downtrend, due to the width of the distribution. Nevertheless, we still have no idea of the distribution of individual paths.

An additional step is to plot every path of individual ensemble members, see also [7]. This leads to figure 3. For clarity, the figure also shows mean and median in bold lines. It becomes apparent that we do not gain much by plotting every path. To the contrary, the information becomes less clear because the paths overlap. All we can see is that the distribution is dense in the middle and less dense to its borders. We might therefore visually conclude that the distribution is unimodal — a possibly dangerous conclusion as we will see later on. The approach of plotting every ensemble member does not scale well. It becomes unwieldy when the ensemble includes several hundred members. The figure is, e.g., a plot of 190 individual paths.

Fig. 1 Average and median of neural network ensemble forecast.

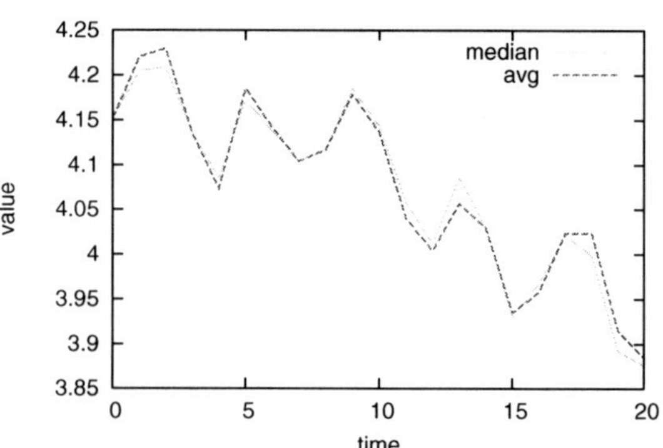

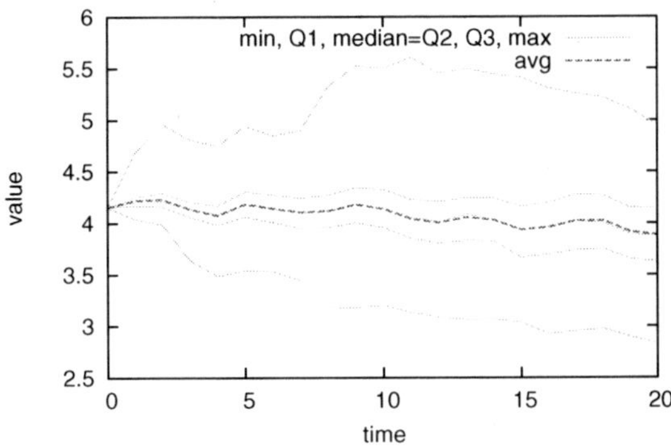

Fig. 2 Neural network ensemble forecast. The figure shows from the top: maximum, 75 percent quartile, mean and median, 25 percent percentile, minimum.

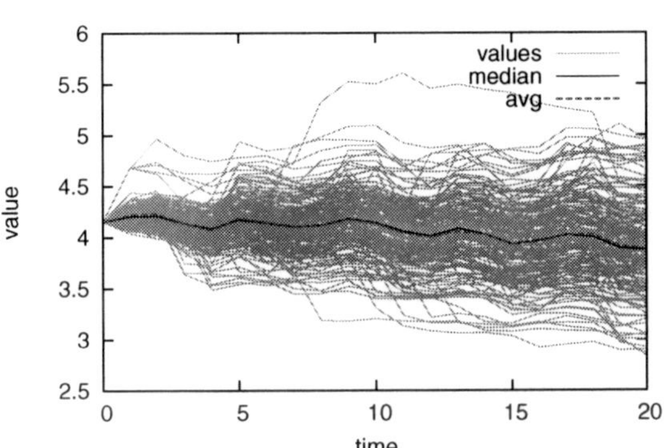

Fig. 3 All paths of neural network ensemble forecast. Plot of 190 ensemble members.

3 Prototypical Visualization Environment with Heatmaps

The question arises: how can we present the forecast information to a decision maker in an intuitive way but still exploit all distribution information? We propose a heatmap style visualization. A heatmap allows to differentiate between more active and less active regions of the forecast space by color coding, see figure 4. We obtain the color values as follows:

- At every timestep compute the maximum of ensemble members within an (adjustable) ε-environment.
- Optionally divide the number of ensemble members within an ε-environment by the maximum for the specific timestep.
- Assign the color red to a ratio of 1 (the maximum) and light blue to a ratio of almost 0.
- Values with no ensemble members at all in the given ε-environment are color coded in white.

Therefore, red regions indicate regions of high activity according to the forecast. For clarity, figure 4 also shows mean, median, maximum and minimum in thick lines. We note that often, but not always, the mean coincides with red regions. However, we also note that approximately from day 5 to day 15 the red region splits into several paths and becomes quite large. This becomes especially clear in a smoothed version of the heatmap, see figure 5. The conclusion is that according to the forecast the mean does not represent the distribution well at all, because (especially for days 6 to 8) the distribution is bimodal or multi modal. This significantly changes the interpretation of the decision maker. Looking only at figures 1–3 it was apparent that the forecast for days 6 to 8 is a slight downtrend. Looking at the heatmap we see that the correct answer actually is: the model doesn't know! This is a warning to the decision maker. On the other hand, during the last few forecast days (days 16-20) the uncertainty becomes smaller and the model clearly forecasts a downtrend. Remember, that we are not dealing with realized forecast accuracy. We are just exploiting the forecast information a priori.

We implement heatmap style visualization in a prototype, see the screenshot in figure 6. For plotting the open source software Gnuplot (version 4.2 patchlevel 6) is used. The right part of the program window controls several plot parameters. Especially, vertical and horizontal plot resolution can be configured. Higher parameters produce smoother heatmaps, see 4 and 5 for two extremes. The screenshot 6 shows an intermediate setting. At first sight no additional information seems to be gained by interpolating between different forecast steps. However, the smoothed result allows to follow intersecting paths. See, for example, the split path from timestep 8 to 9 in figures 4 and 5. Another important parameter is the class radius divider which allows to set the width of the mentionned ε-environment. Finally, different checkboxes allow the plot additional information, like extrema, percentiles, etc.

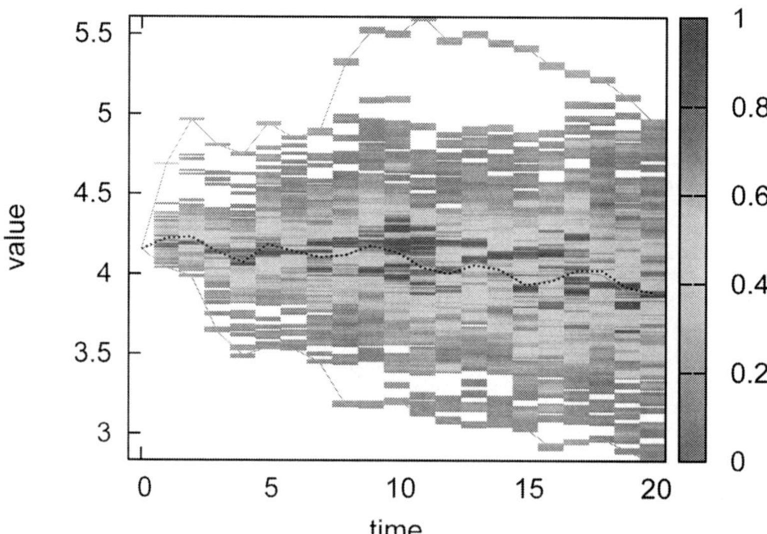

Fig. 4 Heatmap style forecast visualization.

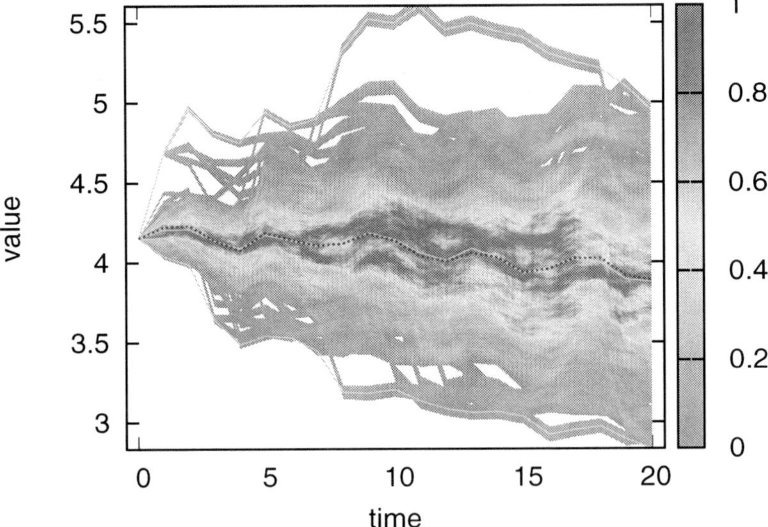

Fig. 5 Smoothed heatmap.

4 Conclusions and Outlook

This paper presents a visualization approach for distributional forecasts. Individual distributional forecasts often arise in the context of neural network ensembles. However, the same visualization style could be useful for other forecasts. We see, that mean and median do not necessarily confer the right information because the distribution may split. This makes mean and median bad representatives. We also see, that forecast uncertainty does not uniformly increase with future timesteps. A split distribution may actually become unimodal again.

An apparent alternative to heatmap style visualization would be a three dimensional plot. This would replace the color coded information by height information. However, the disadvantage of three dimensional plots is that often parts of the plot may hide other parts. Generally, getting adequate information out of a three dimensional plot is more difficult. Heatmap style visualization provides a good compromise between information density and visual interpretability.

The paper only focuses on *visual* representation. It would be interesting to analyze, if we could also *quantify* the advantage of better using the forecast information. This involves, e.g., identifying peaks of the distribution and benchmarking a tri-state model forecast (increases, decreases, don't know) against the realization. Other means of better quantifying the forecast information can, of course, be devised.

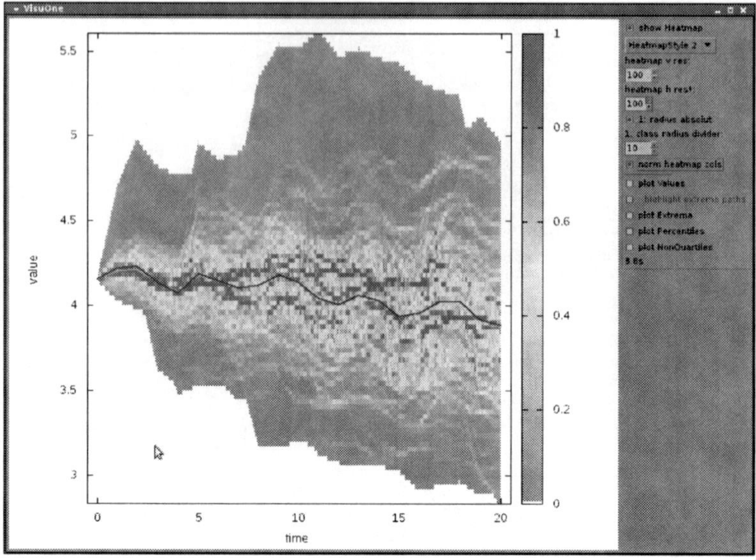

Fig. 6 Screenshot of the prototypical visualization environment (alternative heatmap style).

References

1. Andrienko, G., Andrienko, N.: Visual Exploration of the Spatial Distribution of Temporal Behaviors. In: Proceedings of the Ninth International Conference on Information Visualisation, pp. 799–806 (2005) doi: 10.1109/IV.2005.135
2. Andrienko, G., Andrienko, N., Mladenov, M., Mock, M., Poelitz, C.: Extracting Events from Spatial Time Series. In: Proceedings of the 14th International Conference on Information Visualisation, pp. 48–53 (2010) doi: 10.1109/IV.2010.17
3. Buono, P., Plaisant, C., Simeone, A., Aris, A., Shneiderman, B., Shmueli, G., Jank, W.: Similarity-Based Forecasting with Simultaneous Previews: A River Plot Interface for Time Series Forecasting. In: Proceedings of the 11th International Conference Information Visualization (2007) doi: 10.1109/IV.2007.101
4. Feng, D., Kwock, L., Lee, Y., Taylor II, R.M.: Matching Visual Saliency to Confidence in Plots of Uncertain Data. IEEE Transactions on Visualization and Computer Graphics (2010) doi: 10.1109/TVCG.2010.176
5. von Mettenheim, H.-J. Advanced Neural Networks: Finance, Forecast, and other Applications. Leibniz Universität Hannover (2009)
6. von Mettenheim, H.-J., Breitner, M.H., Robust Decision Support Systems with Matrix Forecasts and Shared Layer Perceptrons for Finance And other Applications. In: ICIS 2010 Proceedings (2010)
7. Potter, K., Wilson, A., Bremer, P.-T., Williams, D., Doutriaux, C., Pascucci, V., Johnson, C.R.: Ensemble-Vis: A Framework for the Statistical Visualization of Ensemble Data. In: International Conference on Data Mining Workshops, pp. 233–240 (2009) doi: 10.1109/ICDMW.2009.55
8. Zimmermann, H.G., Grothmann, R., Tietz, C., von Jouanne-Diedrich, H.: Market Modeling, Forecasting and Risk Analysis with Historical Consistent Neural Networks. In: Selected Papers of the Annual International Conference of the German Operations Research Society, pp. 531–536 (2010) doi: 10.1007/978-3-642-20009-0_84

Forecasting Market Prices with Causal-Retro-Causal Neural Networks

Hans-Georg Zimmermann, Ralph Grothmann and Christoph Tietz

1 Introduction

Forecasting of market prices is a basis of rational decision making [Zim94]. Especially recurrent neural networks (RNN) offer a framework for the computation of a complete temporal development. Our applications include short- (20 days) and long-term (52 weeks) forecast models. We describe neural networks (NN) along a correspondence principle, representing them in form of equations, architectures and embedded local algorithms.

By definition causality explains the present state of a system by other features, which are prior to the current state. Open RNNs offer a nonlinear modeling depended on hidden variables and past observations [Hay08]. Here, the major drawback is the assumption, that the environment variables are constant from present time on. This is questionable in our fast changing times.

Thus, we have developed a causal RNN framework for closed dynamical systems. To train such models we have invented a form of architectural teacher forcing [ZGTJ10]. Despite their universal approximation capabilities these models show only partly good forecasting results. Looking for alternative approaches, we might remember that markets are human made dynamical systems. Microeconomics states that human behavior can be described by utility functions. If we would know the human reward function, we could describe the behavior of the market participants and thus the market by retro causal equations (see also the Hamilton Jacobi necessary conditions in optimal control theory) [FH86].

This approach improves the forecasting accuracy in time periods for which the causal model fails and, interestingly, the retro-causal approach fails when the causal approach is appropriate. Therefore we work out an integrated model, which combines a causal and retro-causal information flow. Thus, we explain the present state of the system by other features, which not necessarily have to be in the past.

Siemens AG, Corporate Technology, Otto-Hahn-Ring 6, 81730 Munich, e-mail: Hans_Georg.Zimmermann@siemens.com

D. Klatte et al. (eds.), *Operations Research Proceedings 2011*, Operations Research Proceedings, 579
DOI 10.1007/978-3-642-29210-1_92, © Springer-Verlag Berlin Heidelberg 2012

2 Causal-Retro-Causal Recurrent Neural Networks

2.1 Causal Recurrent Neural Networks

To derive the CRCNN, we let us first introduce causal RNN for closed dynamical systems, a NN for modeling the dynamics of observables Y_τ in a causal formulation [ZGTJ10]. We formulate the RNN as a state space model in discrete time τ that describes the observables Y_τ by using a state transition and output equation [ZGTJ10]:

$$\text{state transition } s_\tau = f(s_{\tau-1}) = \tanh(As_{\tau-1}), \; s_0 \tag{1}$$

$$\text{output equation } \quad Y_\tau = g(s_\tau) \quad = [Id, 0]s_\tau \tag{2}$$

$$\text{system identification} \quad \text{Error} \quad = \sum_{\tau=t-m}^{t} \left(Y_\tau - Y_\tau^d \right)^2 \to \min_A \tag{3}$$

The time-recurring state transition equation $s_\tau = f(s_{\tau-1})$ describes the current state s_τ solely dependent on the previous system state $s_{\tau-1}$. The observables Y_τ are derived from the state s_τ using the output equation g. Without loss of generality we can approximate the functions f and g of the state space model with a NN, i. e. $\tanh(As_{\tau-1})$ and $[Id, 0]s_\tau$ [SZ06, ZGN02].

We use the technique of finite unfolding in time to transform the temporal equations into a spatial architecture. The RNN is unfolded across the entire time path, i. e. we learn the unique history of the system. Fig. 1 depicts the resulting RNN. The

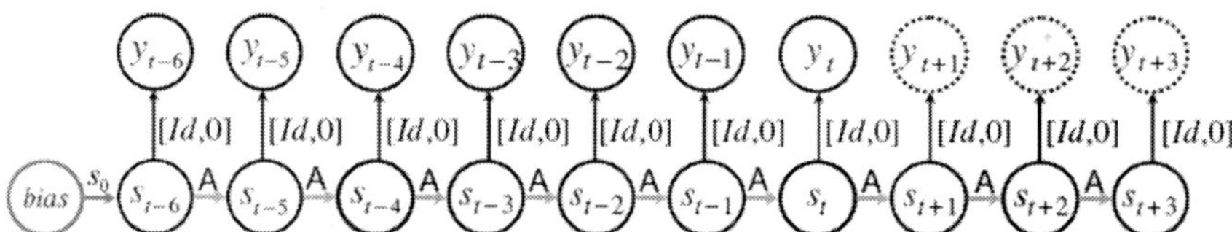

Fig. 1 Architecture of the Causal Recurrent Neural Network

RNN is trained using error back propagation through time (BPTT) [Wer74] together with an architecture based formulation of teacher forcing (TF) [ZGTJ10].

We applied the causal RNN to forecast prices of energy and copper futures traded on the European Energy Exchange (EEX) and the London Metal Exchange (LME) over 20 days. Fig. 2 compares the RNN forecasts with the actual prices for two forecast horizons in sequence.

As depicted in Fig. 2 the causal RNN gives a fairly good forecast of the EEX (LME) market dynamics in case of the first (second) forecast horizon, while it fails in case of the second (first) time horizon. Now the question arises, how we can improve the forecasting accuracy of the causal RNN.

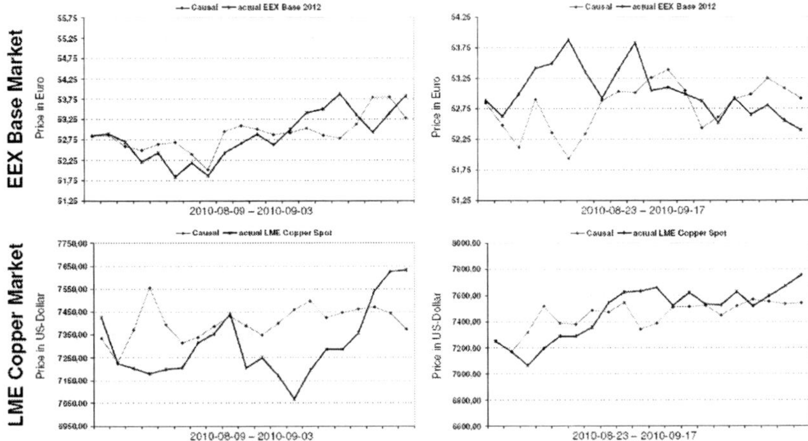

Fig. 2 Forecasting EEX and LME future contracts over 20 days with causal RNNs

2.2 Retro-Causal Recurrent Neural Networks

Causal models might fail due to the impact of utility maximization, which implies an information flow from the future backwards to make an optimal decision at present time. Thus we formulate a model with a retro-causal information flow:

$$\text{state transition } s'_\tau = \tanh(A's'_{\tau+1}), \ s'_T \tag{4}$$

$$\text{output equation } Y_\tau = [Id, 0]s'_\tau, \tag{5}$$

In the RCNN the internal state s'_τ depends on $s'_{\tau+1}$. The observables Y_τ are part of s'_τ. The final state s'_T and the transition matrix A' are trained such that the historical system behavior is described. Fig. 3 depicts the architecture of the RCNN.

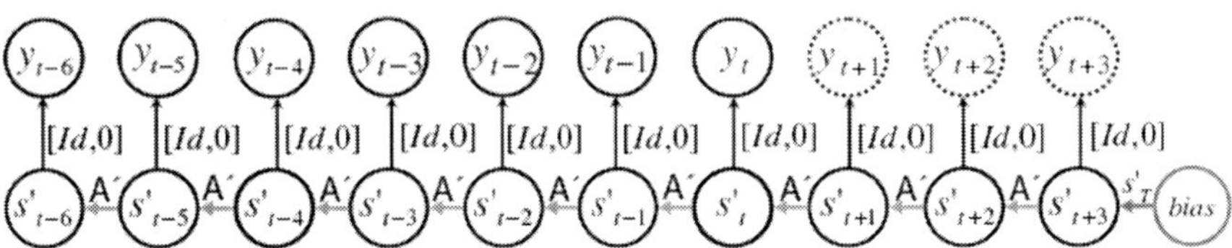

Fig. 3 Architecture of the Retro-Causal Recurrent Neural Network (RCNN)

Likewise to the causal RNN, the RCNN in Fig. 3 is also trained using the BPTT algorithm and an architecture based formulation of TF [ZGTJ10]. We applied the RCNN model on the same time periods as the causal RNN (Fig. 4).

In cases the causal RNN fails, the RCNN is able to give an accurate forecast and vice versa. However, *ex ante* we do not if we are in a causal or retro-causal regime.

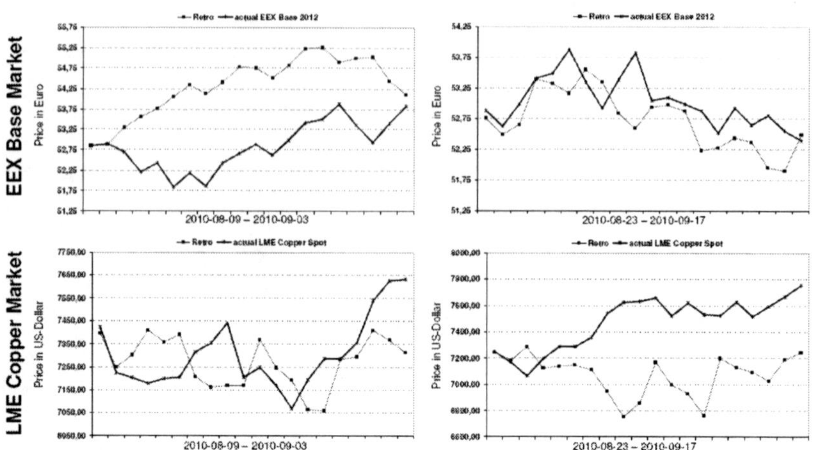

Fig. 4 Forecasting EEX and LME future contracts over 20 days with RCNNs

2.3 Causal-Retro-Causal Recurrent Neural Networks

Causal-retro-causal neural networks (CRCNN) combine a causal and a retro-causal information flow in an integrated NN model given by

$$\text{causal state transition } s_\tau = \tanh(As_{\tau-1}), \; s_0 \tag{6}$$

$$\text{retro-causal state transition } s'_\tau = \tanh(A's'_{\tau+1}), \; s'_T \tag{7}$$

$$\text{output equation } Y_\tau = [Id,0]s_\tau + [Id,0]s'_\tau. \tag{8}$$

The dynamics Y_τ is explained in the CRCNN by a sequence of causal s_τ and retro-causal states s'_τ using a transition matrices A and A' for the causal and retro-causal information flow (Eq. 6 and Eq. 7). Fig. 5 depicts the architecture of the CRCNN.

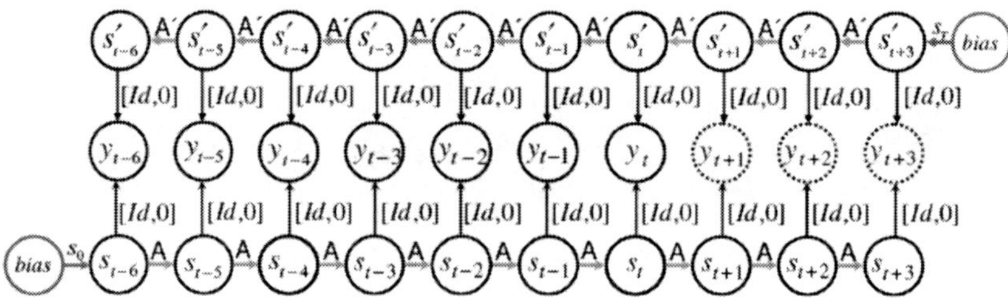

Fig. 5 Architecture of the Causal-Retro-Causal Recurrent Neural Network (CRCNN)

To train the CRCNN we use a combined teacher forcing approach on both branches, resulting in an extended CRCNN architecture (Fig. 6).

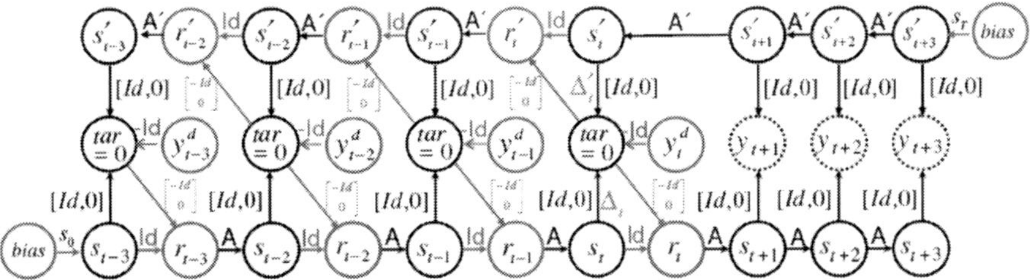

Fig. 6 Extended CRCNN incorporating a Teacher Forcing (TF) mechanism

The causal as well as the retro-causal branch have to explain only the incompleteness of the opposite side. Thus, we have a moving target problem, where the TF has to take into account the behavior of the opposite side. If the error approaches zero, TF is inactive and thus, the extended architecture converges to the original CRCNN (Fig. 5). The advantage of extended CRCNN (Fig. 6) is that it allows a fully dynamical superposition of the causal and retro-causal information flows. We have to state that the numerical solution is difficult, because the extended CRCNN contains many closed loops. Mathematically this means to identify a dynamical system on a manifold, where we have to find the dynamics and the manifold in parallel.

Fig. 7 depicts the application of the CRCNN to the two consecutive time periods.

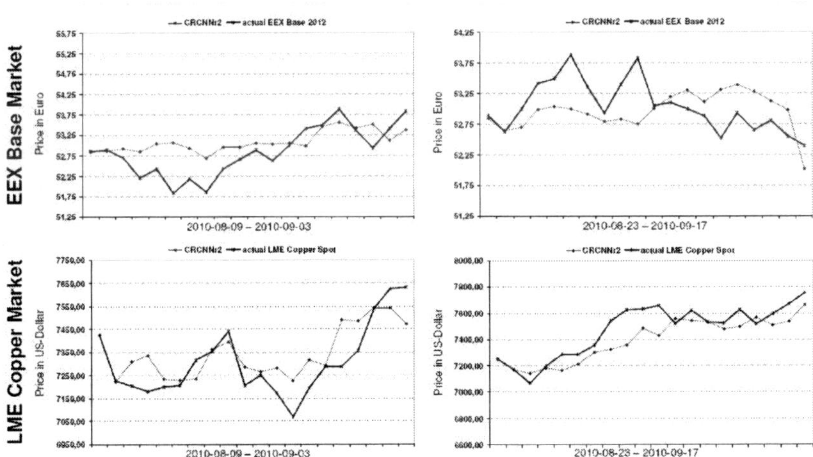

Fig. 7 Forecasting EEX and LME future contracts over 20 days with CRCNNs

The combination of the causal and retro-causal information flow within an integrated NN is able to give an accurate prediction for the future contracts under consideration. Remarkably, the CRCNN is able to detect beforehand if the market is in an causal or retro-causal regime as well as a mixture thereof, since the causal and retro-causal information flows are dynamically combined.

3 Conclusion and Outlook

An integrated model of a causal and retro-causal information flow significantly increases the forecast accuracy. The CRCNN dynamically combines causal and retro-causal information to describe the prevailing market regime.

We usually work with ensembles of CRCNN to predict market prices. All solutions have a model error of zero in the past, but show a different behavior in the future. The reason for this lies in different ways of reconstructing the hidden variables from the observations and is independent of different random sparse initializations. Since every model gives a perfect description of the observed data, we can use the simple average of the individual forecasts as the expected value, assuming that the distribution of the ensemble is unimodal [MB10]. The analysis of the ensemble spread opens up new perspectives on market risks. We claim that the model risk of a CRCNN is equal to the forecast risk [ZGTJ10].

Work currently in progress concerns the analysis of the forecast ensemble distribution and its practical application. Further on we focus on simplifying the training of the CRCNNs, since the extended architecture (Fig. 6) has many fix point loops, which makes the training extremely difficult.

All NN architectures and algorithms are implemented in the Simulation Environment for Neural Networks (SENN), a product of Siemens Corporate Technology.

References

[Hay08] Haykin S.: Neural Networks and Learning Machines, 3rd Edition, Prentice Hall, 2008.

[FH86] Feichtinger, G., Hartl, R. F.: Optimale Kontrolle ökonomischer Prozesse. Anwendungen des Maximumprinzips in den Wirtschaftswissenschaften. Berlin-New York, Walter de Gruyter 1986.

[MB10] Mettenheim, H.-J. and Breitner, M.: Forecasting Complex Systems with Shared Layer Perceptrons, in: Hu, B. et al. (Eds.): Operations Research Proceedings 2010, Munich, Sept. 2010, Berlin and Heidelberg, Springer 2011.

[SZ06] Schäfer, A. M. und Zimmermann, H.-G.: Recurrent Neural Networks Are Universal Approximators. ICANN, Vol. 1., 2006, pp. 632-640.

[ZGN02] Zimmermann, H. G., Grothmann, R. and Neuneier, R.: Modeling of Dynamical Systems by Error Correction Neural Networks. In: Soofi, A. und Cao, L. (Ed.): Modeling and Forecasting Financial Data, Techniques of Nonlinear Dynamics, Kluwer, 2002.

[Zim94] Zimmermann, H. G.: Neuronale Netze als Entscheidungskalkül. In: Rehkugler, H. und Zimmermann, H. G. (Ed.): Neuronale Netze in der Ökonomie, Grundlagen und wissenschaftliche Anwendungen, Vahlen, Munich 1994.

[ZGTJ10] Zimmermann H. G., Grothmann R., Tietz Ch. and v. Jouanne-Diedrich H.: Market Modeling, Forecasting and Risk Analysis with Historical Consistent Neural Networks, in: Hu, B. et al. (Eds.): Operations Research Proceedings 2010, Munich, Sept. 2010, Berlin and Heidelberg, Springer 2011.

[Wer74] Werbos P. J.: Beyond Regression: New Tools for Prediction and Analysis in the Behavioral Sciences, PhD Thesis, Harvard University, 1974.

Stream XVII
Award Winners

Leena Suhl, University of Paderborn, GOR Chair

The Gomory-Chvátal Closure of a Non-Rational Polytope is a Rational Polytope

Juliane Dunkel and Andreas S. Schulz

Abstract The question as to whether the Gomory-Chvátal closure of a non-rational polytope is a polytope has been a longstanding open problem in integer programming. In this paper, we answer this question in the affirmative, by combining ideas from polyhedral theory and the geometry of numbers.

1 Introduction

Cutting-plane methods, when combined with branch and bound, are among the most successful techniques for solving integer programming problems in practice; numerous types of cutting planes have been studied in the literature and several of them are used in commercial solvers (see, e.g., [2] and the references therein). Cutting planes also give rise to a rich theory (see again [2]). In general, a cutting plane for a polyhedron P is an inequality that is satisfied by all integer points in P, and, when added to the polyhedron P, it typically yields a stronger relaxation of its integer hull. A Gomory-Chvátal cutting plane [1, 8] is an inequality of the form $cx \leq \lfloor \delta \rfloor$, where c is an integral vector and $cx \leq \delta$ is valid for P. The Gomory-Chvátal closure of P is the intersection of all half-spaces defined by such inequalities; it is usually denoted by P'. Even though the Gomory-Chvátal closure is defined as the intersection of an infinite number of half-spaces, the Gomory-Chvátal closure of a rational polyhedron is again a rational polyhedron. Namely, Schrijver [10] showed that, for a rational polyhedron P, the Gomory-Chvátal cuts corresponding to a totally dual integral system of linear inequalities describing P specify its closure P' fully. For polyhedra that cannot be described by rational data the situation is different. It is well-known that the integer hull P_I of an unbounded non-rational polyhedron P may not be a polyhedron (see, e.g., [9]). In fact, the integer hull may not be a closed set, and the Gomory-Chvátal closure may not be rational polyhedron. On the other hand, in the case of a non-rational polytope, P_I is the convex hull of a finite set of

Juliane Dunkel, Andreas S. Schulz
Massachusetts Institute of Technology, 77 Massachusetts Avenue, Cambridge, MA 02139-4307, USA, e-mail: juliane@mit.edu, schulz@mit.edu

D. Klatte et al. (eds.), *Operations Research Proceedings 2011*, Operations Research Proceedings, 587
DOI 10.1007/978-3-642-29210-1_93, © Springer-Verlag Berlin Heidelberg 2012

integer points and, therefore, a rational polytope. Yet, there is no notion of total dual integrality for non-rational systems of linear inequalities. In fact, it was unknown whether the Gomory-Chvátal closure of an arbitrary polytope is a rational polytope. We show that this is indeed the case: also the Gomory-Chvátal closure of a non-rational polytope is again a rational polytope, that is, it can be described by a finite set of rational inequalities.

Even though Gomory-Chvátal cuts were originally introduced for polyhedra, they have lately been applied to other convex sets as well. Of particular relevance is the work by Dey and Vielma [5] who showed that the Gomory-Chvátal closure of a full-dimensional ellipsoid described by rational data is a polytope. They recently extended this result to strictly convex bodies and to the intersection of strictly convex bodies with rational polyhedra [3]. Since the original proof of Schrijver for rational polyhedra relies strongly on polyhedral properties, a new proof technique had to be developed in [3, 5], which can roughly be described as follows: One first shows that there exists a finite set of Gomory-Chvátal cuts that separate every non-integral point on the boundary of the strictly convex body. In a second step, one proves that if the intersection of the boundary of a convex body with a finite set of Gomory-Chvátal cuts is contained in the Gomory-Chvátal closure, only a finite set of additional inequalities is needed to fully describe the Gomory-Chvátal closure of the body.

Our general proof strategy for showing the polyhedrality of the Gomory-Chvátal closure of a non-rational polytope is inspired by the work in [3]. Yet, the key argument is very different and new, since the proof in [3] relies on properties of strictly convex bodies that do not extend to polytopes. More precisely, strictly convex bodies do not have any higher-dimensional "flat faces", and therein lies the main difficulty in establishing the polyhedrality of the elementary closure for non-rational polytopes. Our proof is geometrically motivated and uses ideas from convex analysis, polyhedral theory, and the geometry of numbers. In particular, the underlying geometric idea relies on properties of integer lattices and reduced lattice bases.

Simultaneously and independently from this work, it was proven in [4] that the Gomory-Chvátal closure of any compact convex set is a rational polytope.

2 Basics and Notations

For a closed and convex set $K \subseteq \mathbb{R}^n$ and any vector $a \in \mathbb{R}^n$, we define a_K as the minimal right-hand side such that $ax \leq a_K$ is a valid inequality for K, that is, $a_K := \max\{ax|x \in K\}$. We denote the hyperplane $\{x \in \mathbb{R}^n|ax = a_0\}$ by $(ax = a_0)$ and, similarly, $(ax \leq a_0)$ denotes the half-space of all points satisfying the inequality $ax \leq a_0$. If y is a real number, $\lfloor y \rfloor$ denotes the largest integer less than or equal to y. For a subset U of $\mathbb{R}^n$, the boundary of U is denoted by $\mathrm{bd}(U)$. The relative boundary and relative interior of U (that is, the boundary and interior of U considered as a subset of $\mathrm{aff}(U)$) are denoted by $\mathrm{rbd}(U)$ and $\mathrm{ri}(U)$, respectively. For any set $S \subseteq \mathbb{Z}^n$, we use $C_S(K) := \bigcap_{a \in S} (ax \leq \lfloor a_K \rfloor)$ to denote the intersection of all half-spaces corresponding to Gomory-Chvátal cuts for K with normal vector in S. Note that for $S = \mathbb{Z}^n$, one obtains the Gomory-Chvátal closure K' of the set K.

3 General Proof Idea

Due to space limitations, we can only provide a sketch of our proof. For the complete proof we refer the reader to [7] and [6].

Our general strategy for proving that for any polytope a finite number of Gomory-Chvátal cuts is sufficient to describe the polytope's closure is a modification of the two-step technique developed in [3] for a strictly convex body K: One first constructs a finite set $S \subseteq \mathbb{Z}^n$ such that

$$C_S(K) \subseteq K \ , \tag{K1}$$
$$C_S(K) \cap \mathrm{bd}(K) \subseteq \mathbb{Z}^n \ . \tag{K2}$$

Then one argues that S needs to be augmented by, at most, a finite set of vectors. In particular, property (K2) demonstrates that every fractional point on the boundary of the strictly convex body is separated by a Gomory-Chvátal cut. Obviously, the same cannot be true for polytopes, since this would otherwise imply that the Gomory-Chvátal procedure separates fractional points in the relative interior of the facets of an integral polytope. Furthermore, in the case of a general polytope P, we cannot assume full-dimensionality. This is because a unimodular transformation that maps P to a full-dimensional polytope in a lower-dimensional space is not guaranteed to exist if P is contained in some non-rational affine subspace. In particular, this observation forces us to consider the *relative* boundary of the polytope instead of its boundary. Hence, our general strategy for proving the polyhedrality of P' is as follows: First, we show that one can find a finite set S of integral vectors such that

$$C_S(P) \subseteq P \ , \tag{P1}$$
$$C_S(P) \cap \mathrm{rbd}(P) \subseteq P' \ . \tag{P2}$$

We then argue that, given the polytope $C_S(P)$, no more than a finite number of additional Gomory-Chvátal cuts are necessary to describe the closure P'.

The main challenge of this proof strategy lies in showing the existence of a set S satisfying property (P1). This is due to the presence of higher-dimensional faces with non-rational affine hulls. These require the development of completely new arguments compared to the proof for strictly convex bodies. The outlined general strategy is implemented in four main steps:

1. Show that there exists a finite set $S \subseteq \mathbb{Z}^n$ such that $C_S(P) \subseteq P$.
2. Show that for any face F of P, $F' = P' \cap F$. In particular, if $F = P \cap (ax = a_P)$, then for every Gomory-Chvátal cut for F there exists a Gomory-Chvátal cut for P that has the same impact on the maximal rational affine subspace of $(ax = a_P)$.

3. Show that if there exists a finite set S satisfying (P1) and (P2), then P' is a rational polytope.
4. Prove that P' is a rational polytope by induction on the dimension of $P \subseteq \mathbb{R}^n$.

In the remainder of this section, we describe the reasoning behind each step of the proof and sketch some of the applied techniques.

Step 1: Constructing a subset of P from a finite number of Gomory-Chvátal cuts. Suppose that we can find a set $S \subseteq \mathbb{Z}^n$ with $C_S(P) \subseteq P$ for some polytope $P \subseteq \mathbb{R}^n$ for which a non-rational inequality $ax \leq a_P$ is facet-defining. As this inequality cannot be facet-defining for the *rational* polytope $C_S(P)$, there must exist a finite set of Gomory-Chvátal cuts that dominate $ax \leq a_P$. More formally, there must exist a subset $S_a \subseteq S$ such that $C_{S_a}(P) \subseteq (ax \leq a_P)$. If V_R denotes the maximal rational affine subspace of $(ax = a_P)$, that is, the affine hull of all rational points in $(ax = a_P)$, then the Gomory-Chvátal cuts associated with the vectors in S_a have to separate every point in $(ax = a_P) \setminus V_R$.

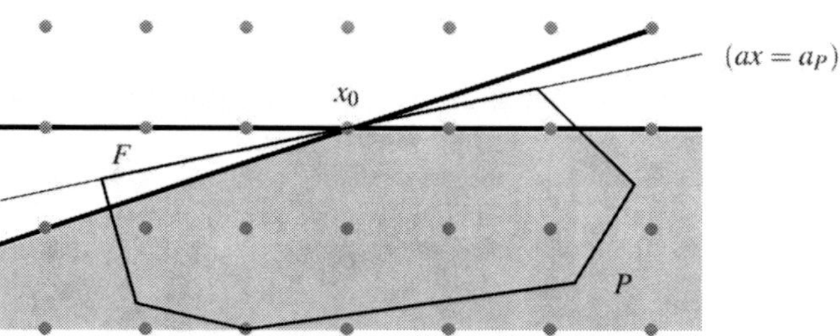

Fig. 1 Construction of a finite set of Gomory-Chvátal cuts that dominate a non-rational facet-defining inequality $ax \leq a_P$. Here, the hyperplane $(ax = a_P)$ contains only one rational, in fact, one integral point x_0 and has, therefore, one non-rational direction. That is, $V_R = \{x_0\}$ and $\dim(V_R) = 0$. Two Gomory-Chvátal cuts separate every point in the hyperplane $(ax = a_P)$ that is not in V_R.

Indeed, our strategy for the first step of the proof is to show that for each non-rational facet-defining inequality $ax \leq a_P$ for P there exists a finite set of integral vectors S_a that satisfies $C_{S_a}(P) \subseteq (ax \leq a_P)$. This fact is proven in a series of steps. First, we establish the existence of a sequence of integral vectors satisfying a specific list of properties. These vectors give rise to Gomory-Chvátal cuts that separate all points in the non-rational facet $F = P \cap (ax = a_P)$ that are not contained in the maximal rational affine subspace V_R of $(ax = a_P)$. The number of Gomory-Chvátal cuts needed in our construction for separating the points in $(ax = a_P) \setminus V_R$ depends only on the dimension of V_R. If $\dim(V_R) = n - 2$, that is, the hyperplane $(ax = a_P)$ has a single "non-rational direction", then only two cuts are necessary. One can visualize these cuts to form a kind of "tent" in the half-space $(ax \leq a_P)$, with the ridge being V_R (see Fig. 1 for an illustration). With each decrease in the dimension of V_R by 1, the number of necessary cuts is doubled. Hence, at most 2^{n-1} Gomory-

Chvátal cuts are required to separate the non-rational parts of a non-rational facet of the polytope.

The proof of Step 1 uses many classic results from convex and polyhedral theory, as well as from number theory. In particular, integral lattices and reduced lattice bases play a crucial role.

Step 2: A homogeneity property: $F' = P' \cap F$**.** As the second step of the proof, we show a property of the Gomory-Chvátal closure that has been well-known for rational polytopes (see, e.g., [11]): if one applies the closure operator to a face of a polytope, the result is the same as if one intersected the closure of the polytope with the face. As it turns out, the same is true for non-rational polytopes. The proof for the rational case is based on the observation that any Gomory-Chvátal cut for a face $F = P \cap (ax = a_P)$ can be "rotated" so that it becomes a valid Gomory-Chvátal cut for P. In particular, the rotated cut has the same impact on the hyperplane $(ax = a_P)$ as the original cut for F. While the exact same property does not hold in the non-rational case, we show that there is a rotation of any cut for F that results in a Gomory-Chvátal cut for P, which has the same impact on the maximal rational affine subspace V_R of $(ax = a_P)$. As Step 1 of our proof implies that the non-rational parts of a face are separated in the first round of the Gomory-Chvátal procedure in any event, this property suffices to show that $F' = P' \cap F$.

The insights gained in this second step will be useful for Step 4 of the proof, where we show the main result by induction on the dimension of the polytope. Knowing that the Gomory-Chvátal closure of a lower-dimensional facet F of P is a polytope, each of the finite number of cuts describing F' can be rotated in order to become a Gomory-Chvátal cut for P. We thereby establish the existence of a finite set $S_F \subseteq \mathbb{Z}^n$ with the property that $C_{S_F}(P) \cap F = F'$. Since the facets of P constitute the relative boundary of the polytope, the union of all these sets will give rise to a set $S \subseteq \mathbb{Z}^n$ that satisfies property (P2).

Step 3: Finite augmentation property. A statement similar to the one in Step 3 has been established in [3] for full-dimensional convex bodies. Since P can be contained in some non-rational affine subspace and, thus, a unimodular transformation of P to a full-dimensional polytope in a lower-dimensional space is not possible, we need the extension to lower-dimensional polytopes. However, the basic observation for proving this part is similar to [3]: every additional undominated Gomory-Chvátal cut has to separate a point that is contained in the *relative* interior of the polytope. Even though in the non-full-dimensional case there are infinitely many cuts with this property, we argue that only a finite number of them need to be considered.

Step 4: Proof of the main result. As the final step of the proof, we establish the main result, namely that the Gomory-Chvátal closure of any polytope can be described by a finite set of inequalities. The proof is by induction on the dimension of the polytope and uses the observations made in the steps above. Step 1 provides a finite set $S \subseteq \mathbb{Z}^n$ satisfying $C_S(P) \subseteq P$. Applying the induction assumption to the facets of P and using the homogeneity property of Step 2, we augment S for each facet F by a finite set $S_F \subseteq \mathbb{Z}^n$ such that the resulting set of integral vectors satisfies properties (P1) and (P2). From that it follows with the finite augmentation property proven in Step 3 that P' is a polytope.

References

1. Chvátal, V.: *Edmonds polytopes and a hierarchy of combinatorial problems,* Discrete Mathematics 4, 305–337 (1973)
2. Cornuéjols, G.: *Valid inequalities for mixed integer linear programs*, Mathematical Programming B 112, 3–44 (2008)
3. Dadush, D.; Dey, S. S.; Vielma, J. P.: *The Chvátal-Gomory closure of a strictly convex body*, http://www.optimization-online.org/DB/HTML/2010/05/2608.html (2010)
4. Dadush, D.; Dey, S. S.; Vielma, J. P.: *The Chvátal-Gomory closure of a compact convex set*, Proceedings of the 15th Conference on Integer Programming and Combinatorial Optimization (IPCO 2011), Lecture Notes in Computer Science Vol. 6655 (O. Günlük and G. J. Woeginger, editors), 130–142 (2011)
5. Dey, S. S.; Vielma, J. P.: *The Chvátal-Gomory closure of an ellipsoid is a polyhedron*, Proceedings of the 14th Conference on Integer Programming and Combinatorial Optimization (IPCO 2010), Lecture Notes in Computer Science Vol. 6080 (F. Eisenbrand and F. B. Shepherd, editors), 327–340 (2010)
6. Dunkel, J.: *The Gomory-Chvátal Closure: Polyhedrality, Complexity, and Extensions*, PhD Thesis, Massachusetts Institute of Technology, http://stuff.mit.edu/people/juliane/thesis/ (2011)
7. Dunkel, J., Schulz, A. S. *The Gomory-Chvátal Closure of a non-rational polytope is a rational polytope*, http://www.optimization-online.org/DB_HTML/2010/11/2803.html (2010)
8. Gomory, R. E.: *Outline of an algorithm for integer solutions to linear programs*, Bulletin of the American Mathematical Society 64, 275–278 (1958)
9. Halfin, S.: *Arbitrarily complex corner polyhedra are dense in $\mathbb{R}^n$*, SIAM Journal on Applied Mathematics 23, 157–163 (1972)
10. Schrijver, A.: *On cutting planes*, Annals of Discrete Mathematics 9, 291–296 (1980)
11. Schrijver, A.: *Theory of Linear and Integer Programming*, John Wiley & Sons, Chichester (1986)

Dynamic Fleet Management for International Truck Transportation

Steffen Schorpp

Abstract Recent developments in IC technologies have facilitated the practical application of dynamic vehicle routing systems. So far, most of the available dynamic approaches have dealt with local area environments. In this work, however, the focus is set to wide area dynamic freight transportation (tramp transportation mode) covering the entire European continent.

We develop and evaluate two new dynamic planning approaches and incorporate all important real-life restrictions, such as regulations on driving hours (EC 561), working hours and traffic bans. Extensive numerical tests are carried out with a five-week real-life data set from an international freight forwarding company. The results indicate that significant improvements, especially in terms of empty kilometers and service quality, can be achieved.

1 Motivation and problem setting

Vehicle routing has been a popular field in Operations Research since the 1960s. A large number of researchers have dealt with all kinds of static problems. In recent years, the *center of attention has* also *moved to the field of dynamic vehicle routing problems*, where new information evolves concurrently to the plan execution and has to be handled efficiently by a dynamic planning approach.

Many authors have developed dynamic procedures, basically extensions to already existing static ideas. In the majority of cases, the dynamic publications focused on *local area* problems with the goal of minimizing distance traveled. There are only a few works available so far which deal with dynamic *wide area* planning problems.

This finding may be due to the *predominance of line transportation* in wide area environments: A recurring medium-term plan – effective for several weeks or months – is generated and constitutes each vehicle's circulation between the nodes of a *fixed transportation network*. New requests are fed into the driving routes of the existing line operation schedule, which, for example, induces advantages in consolidation of

Production & Logistics, University of Augsburg, e-mail: steffen.schorpp@gmx.de

D. Klatte et al. (eds.), *Operations Research Proceedings 2011*, Operations Research Proceedings,   593
DOI 10.1007/978-3-642-29210-1_94, © Springer-Verlag Berlin Heidelberg 2012

less-than-truckload requests. Short-term dynamic planning is not required.

It can, however, be observed that some internationally operating freight forwarding companies – Willi Betz International, LKW Walter, Hindelang, etc. – have also successfully specialized in another type of *wide area* freight transportation: *tramp transportation, independent of predefined line networks.* Since there is no line schedule on a fixed network, there is the need for dynamic replanning to react in the short term to newly occurring requests and other changing information.

In literature, there is only a small group of publications available that cover the specified planning problem (e.g. [3, 4]). All these publications neglect most of the specific *European real-life requirements for long-haul transportation.*

Our problem can be classified as an extension to the dynamic Multi Load Pickup and Delivery Problem with Time Windows (MLPDPTW, see [5] chapter 2.3 for a detailed description). As given **data**, there are a limited number of *vehicles* (attributes: load-type, hazardous goods equipment, no. of drivers, initial geographical position and initial time of availability), a number of dynamic *orders* (attributes: call-in time, locations for Pickup and Delivery, required vehicle type, hazardous goods information, loading/unloading time, soft time windows) and a static *travel time matrix*. Due to real-life requirements, order bundles with more than one Pickup and/or Delivery location are also considered. The following **constraints** have to be ensured: (i) all orders must be served, (ii) compliance with general requirements for wide area truck transportation in Europe, (iii) feasible order to vehicle assignment, (iv) hazardous goods equipment on the vehicle, if required. The **objective** is the minimization of a weighted sum of empty travel time, delay and waiting time.

2 General requirements for wide area truck transportation in Europe

Basically, there are two categories of requirements: **driver-based regulations** (driving and rest periods, working time restrictions) and **general traffic bans** [5].

Driving and rest period regulations EC 561 and AETR turned out to be the most important requirements: EC 561 is primarily applied in Western Europe, AETR is used in Eastern Europe. For cross border transports between countries with EC 561 and countries having ratified the AETR, AETR must be applied for the whole journey. AETR regulation is very similar to EC 561, although in some points a little less restrictive. In order to reduce complexity while keeping the Europe-wide schedule feasible, we decided to apply the more restrictive EC 561 regulation for all European transportation tasks. In the new planning procedures, EC 561 rules for Single and for Team driver mode are included. In addition, we make use of possible exceptions to reduce potential delay (e.g. extended daily driving time, reduced daily rest period, reduced weekly rest period, etc.). Working time restrictions are handled in directive EC 15: those restrictions are included for team driver mode.

General traffic bans are very heterogeneous across Europe: nearly every country has different traffic bans (e.g. Germany: Sunday traffic ban from Saturday 10pm until Sunday 10pm). The inclusion of all different rules is impractical, therefore a general Sunday traffic ban for *all* European countries is included.

3 Dynamic planning procedures

The following approach was chosen: In a *first step*, two dynamic planning procedures were developed for the dynamic MLPDPTW standard problem. In a *second step*, both procedures' performance was evaluated with test data from the literature [2] and also with self-generated test data sets. *Finally*, one procedure was selected and adapted to the actual real-life planning problem with all its additional requirements.

The first procedure is called **Multiple Neighborhood Search (MNS)**. It uses a classical Best Insertion algorithm for the inclusion of new dynamic orders. For improvement, it applies a variety of different local search techniques based on various neighborhoods [5]. Re-planning is triggered by clock: Time intervals, e.g. with a duration of 10 minutes, are specified; a new re-planning run is started at the beginning of each time interval. The complete time interval is used for re-planning calculations.

The second procedure is called **Assignment** (an advancement to [1]): The problem of assigning all open orders to the available vehicles is optimally solved with the help of a sophisticated matrix. In contrast to MNS, re-planning is triggered by the occurrence of new dynamic events ("event based") and takes less than a second of computation time.

For performance evaluation, both procedures were applied on test data sets from literature [2] and on self-generated test data sets. While MNS performed better for the test data from literature, Assignment showed better performance for the self-generated test data. Due to this ambiguous finding, three additional performance criteria were introduced: robustness, parameterization effort and procedure's complexity. Since MNS showed advantages in all three extra criteria, we decided to adapt MNS to the actual real-life planning situation.

4 Computational results

Our cooperating freight forwarding company provided a five-week (17.08.2009 to 19.09.2009) **real-life test data set** with 953 vehicles and 14,025 requests. We performed an ex-post analysis of the actual manual planning and derived the subsequent performance indicators: delay = 159,647 h; total travel time = 229,661 h; empty travel time = 25,816 h (empty-to-all ratio = 11.2%); loading/unloading time = 24,000 h; break/waiting time = 496,471 h.

When parameterizing the adapted MNS procedure, we found a **contrary** behavior between the two **objectives** *minimization of delay* and *minimization of empty travel time*. If more relative weight is put on one of both objective criteria, this produces the desired improvement; however, simultaneously the solution quality of the other objective decreases by nearly the same magnitude.

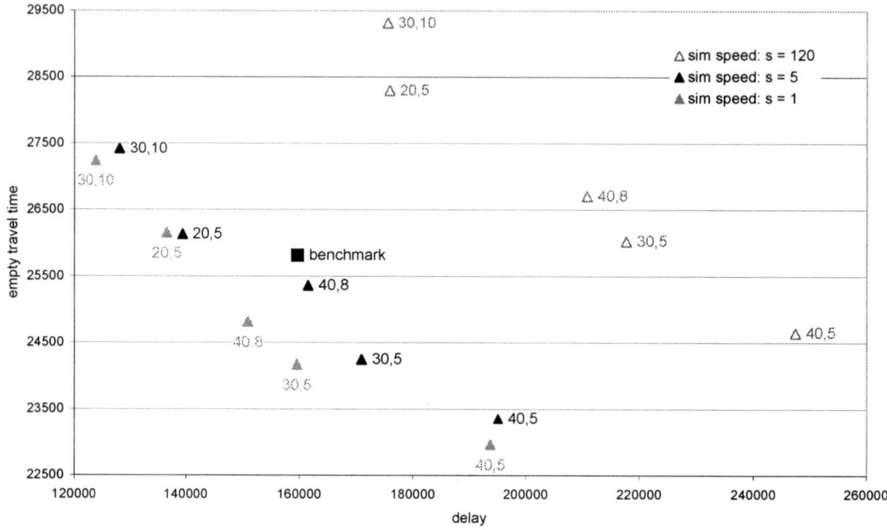

Fig. 1 Pareto curves for different simulation speeds

Figure 1 illustrates this finding: The black square indicates the benchmark value in terms of delay (x-axis) and empty travel time (y-axis). The triangles show the results of different penalty cost parameter settings and different simulation speeds. For a simulation speed of s = 5 (black triangles), it can be observed that cost parameters (30,10) and (20,5) result in reduced delay, while the cost parameters (40,8), (30,5) and (40,5) result in a reduced empty travel time. However, there is no black triangle solution that reaches the *preferred area* with simultaneous improvement of both objective criteria. The *preferred area* is only achieved with a simulation speed of s = 1 (real-time simulation, grey triangles).

It is interesting to observe the general **impact of simulation speed** on overall solution quality: The slower the simulation is run, the better the solution quality. This finding can be exemplarily investigated for cost parameter (40,8): From simulation speed s = 120 (blank triangles) to simulation speed s = 5 (black triangles) empty travel time is reduced by 5.1% and delay is reduced by 23.4%. From s = 5 (black triangles) to s = 1 (grey triangles) empty travel time and delay are reduced by an additional 2.2% and 6.6%, respectively. This finding is due to the MNS procedure's improvement framework: additional time is efficiently used to perform extra calculations resulting in an improved overall solution quality.

Figure 2 shows a typical planning result generated by the adapted MNS procedure. Such type of **graphical illustrations** were used to discuss and verify the planning results with the cooperating freight forwarding company. The *first loaded trip* directs the vehicle from Reutlingen (Germany) at point A, to Miskolc (Hungary) at point B. Afterwards, the vehicle has to perform an empty trip from Miskolc (Hungary) at point B to Mosonszolnok (Hungary) at point C. Here, the vehicle gets its *second loaded trip* from Mosonszolnok (Hungary) at point C, to Vienna (Austria) at point D. In Vienna (Austria) at point D, the *third loaded trip* is directly available: Vienna (Austria) at point D, to Mannheim (Germany) at point E. And so on...

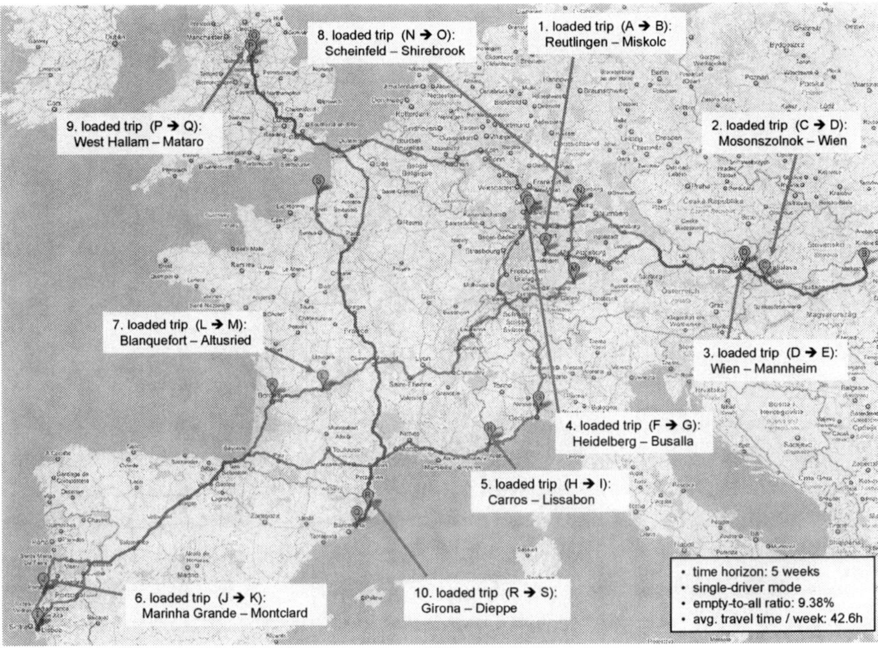

Fig. 2 Exemplary five-week vehicle tour – geographical illustration

Due to the contrary objectives, it is not possible to report *one* best result. Instead, the **best results** of three penalty cost combinations are presented in Figure 3:

For penalty cost combination (40,8) empty travel time is reduced by 3.9% and delay by 6.0%. More relative weight is put on the reduction of empty travel time with penalty cost combination (30,5): Consequently, a reduction of 7.3% in empty travel time is now achieved. However, the reduction in delay decreases to only 1.7%. The highest reduction of empty travel time is achieved with penalty cost combination (40,5): 11.9%. However, at the same time, delay increases by 16.6%.

Our cooperating freight forwarding company chose the results of penalty cost combination (30,5) as their "preferred ones".

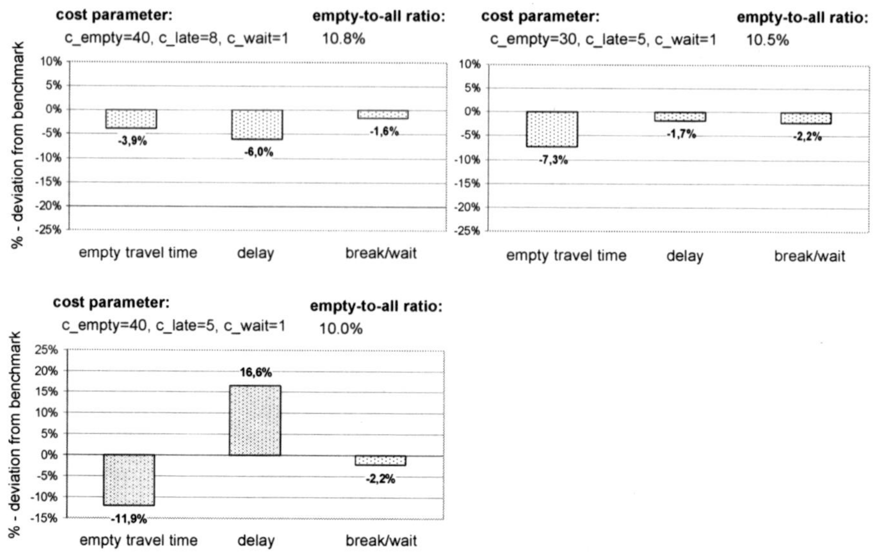

Fig. 3 Best results for penalty cost combinations (40,8), (30,5) and (40,5)

5 Conclusion and outlook

The study shows that computer-based dynamic vehicle routing generates a significant improvement potential. If we project the savings in empty travel time (results of penalty cost combination (30,5)) to a whole year, there are potential savings of up to 1.4 million empty kilometres. Transferred to an environmental context, these are yearly savings of up to 1,120 tons of CO_2.

Future work in this area of research should deal with (i) coordinated tour and driver scheduling, (ii) decision support for the order acquisition process and (iii) the inclusion of stochastic information on the future.

References

1. Fleischmann, B., Gnutzmann, S., Sandvoss, E.: Dynamic Vehicle Routing Based on Online Traffic Information. Transportation Science. 38 (4), 420–433 (2004)
2. Gendreau, M., Guertin, F., Potvin, J., Seguin, R.: Neighborhood Search Heuristics for a Dynamic Vehicle Dispatching Problem with Pickups and Deliveries. Transportation Research Part C. 14 (3), 157–174 (2006)
3. Powell, W.: A Stochastic Formulation of the Dynamic Assignment Problem with an Application to Truckload Motor Carriers. Transportation Science. 30 (3), 195–219 (1996)
4. Powell, W., Marar, A., Gelfand, J., Bowers, S.: Implementing Real-Time Optimization Models: A Case Application from the Motor Carrier Industry. Operations Research. 50 (4), 571–581 (2002)
5. Schorpp, S.: Dynamic Fleet Management for International Truck Transportation - Focusing on Occasional Transportation Tasks. Gabler Research, Wiesbaden (2011)

Algorithmic Cost Allocation Games: Theory and Applications

Nam-Dũng Hoàng

Abstract This article gives an overview on some of the results of the author's PhD thesis [7]. This thesis deals with the cost allocation problem, which arises when several participants share the costs of building or using a common infrastructure. We attempt to answer the question: What is a fair cost allocation among participants? By combining cooperative game theory and state-of-the-art algorithms from linear and integer programming, our work not only defines fair cost allocations but also calculates them numerically for large real-world applications.

1 Introduction

Due to economy of scale, it is suggested that individual users, in order to save costs, should join a cooperation rather than acting on their own. However, a challenge for individuals when cooperating with others is that every member of the cooperation has to agree on how to allocate the common costs among members, otherwise the cooperation cannot be realised. Taken this issue into account, we set the objective of the thesis in investigating the issue of fair allocations of common costs among users in a cooperation. This thesis combines cooperative game theory and state-of-the-art algorithms from linear and integer programming in order to define fair cost allocations and calculate them numerically for large real-world applications.

In the first, theoretical part of the thesis, summarized in Section 2 and Section 3, we present and discuss several game-theoretical concepts. These concepts consider not only different aspects of fairness but also practical requirements, which, to the best of our knowledge, have not been considered in previous research. In addition, this part also investigates the computational complexity by calculating allocations based on the game-theoretical concepts. If the cost function is submodular, then one can find them in oracle-polynomial time. However, the problem is NP-hard in general. The biggest challenge is that there is exponential number of possible

Nam-Dũng Hoàng
Faculty of Mathematics, Mechanics, and Informatics, Vietnam National University, 334 Nguyen Trai Str., Hanoi, Vietnam, e-mail: hoangnamdung@hus.edu.vn

D. Klatte et al. (eds.), *Operations Research Proceedings 2011*, Operations Research Proceedings, 599
DOI 10.1007/978-3-642-29210-1_95, © Springer-Verlag Berlin Heidelberg 2012

coalitions. To tackle this issue, we construct a constraint generation approach as well as primal and dual heuristics for its separation problem.

In the second part of the thesis, summarized in Section 4, we consider several applications based on the framework in the first part, one of which is the ticket pricing problem of the Dutch IC railway network. The current distance tariff results in a situation that some passengers in the central region of the country pay over 25% more than the costs they incur, and these excess payments subsidize operations elsewhere. In this case, it is obvious that the cost allocation is unfair. Using our method, we suggest new ticket prices which can reflect costs better and reduce the overpayments to less than 1.68%.

2 Game Theoretical Setting

Cooperative game theory analyzes the possible grouping of individuals to form their coalitions. It provides mathematical tools to understand fair prices in the sense that a fair price prevents the collapse of the grand coalition and increases the stability of the cooperation. The current definition of cost allocation game does not allow us to restrict the set of possible coalitions of players and to set conditions on the output prices, which often occur in real-world applications. Our generalization bring the cost allocation game model a step closer to practice.

In the literature, the cost allocation game is commonly defined as a pair of a set N of players and a cost function $c : 2^N \to \mathbb{R}$. The goal is to find a fair allocation of the total cost $c(N)$ among players. The allocation is required to fulfil only two conditions. The first one is that the allocation must be non-negative. The second one is that the total price must cover exactly the total cost $c(N)$. We generalize in [7] this classical definition of the cost allocation game in order to handle constraints on allocations and on the formation of coalitions. Such constraints arise in real-world applications of cost allocation games. Indeed, it is commonplace that some players do not want to cooperate with certain other players, or only if third players also enter the coalition, some coalitions may be prohibited because of political or technical reasons, etc. The output prices are also often subject to constraints, e.g., to be "logical", to avoid large fluctuations, to be similar to a price system computed in another way, etc. For instance, consider the ticket pricing problem in public or rail transport. It is natural to stipulate that the ticket price p_{AC} for a trip from station A to station C via station B should fulfil the monotonicity conditions

$$0 \le p_{AB}, p_{BC} \le p_{AC} \le p_{AB} + p_{BC},$$

where p_{AB} and p_{BC} are ticket prices from A to B, and B to C, respectively, see [7] for a thorough discussion and further examples.

Instead of two, our constrained cost allocation game contains now four components: a finite set of *players* $N = \{1, 2, \ldots, n\}$, a non-empty family of possible *coalitions* $\Sigma \subseteq 2^N$, a *cost function* $c : \Sigma \to \mathbb{R}$ that associates with each coalition a price, and a polyhedron $P = \{x \in \mathbb{R}^n_+ \,|\, Ax \leq b\}$, that encodes conditions on the prices x that the players are asked to pay. We denote by $\Sigma^+ := \Sigma \setminus \{\emptyset\}$ the *set of non-empty coalitions*, and we assume $\Sigma \ni N$ (the *grand coalition* N is possible), $\Sigma \setminus \{\emptyset, N\} \neq \emptyset$ (there are non-empty coalitions other than the grand coalition), $c(S) > 0$ for all $S \in \Sigma^+$ (non-empty coalitions have positive cost), and the imputation set $\mathscr{X}(N, c, P, \Sigma) := \{x \in P \,|\, x(N) = c(N)\}$ is non-empty. Note that the imputation set is then a polytope, i.e., bounded. The tuple $\Gamma := (N, c, P, \Sigma)$ is called a *constrained cost allocation game*.

Based on this new definition, we consider several game theoretical concepts, which model fairness. One of the concept that one can come up is the Shapley value. However, it is unclear how can we apply it for our constrained cost allocation game. The efficiency, i.e., the property that the sum of the price of each player must equal the total cost $c(N)$, is only guaranteed if the family of coalitions Σ is the power set 2^N of N. The Shapley value may also not satisfy the conditions on the prices, i.e., it does not belong to P. Moreover, Shapley value is not a core-allocation in general, i.e., it may lie outside the core in the case that the core is non-empty. The core is defined as the set of all price vector x in the imputation set which satisfy that for each coalition S in Σ^+ the price $x(S) = \sum_{i \in S} x_i$ that S is asked to pay is not larger than its own cost $c(S)$, i.e, $x(S) \leq c(S)$. That means, a price vector x in the core allocates the total cost $c(N)$ among players in such a way that every coalition S will have a profit by accepting the asked price $x(S)$. It is hard to convince the players that a price vector, which does not belong to the core, is fair if the core is non-empty. Therefore we only consider core cost allocation concepts.

Let $f : \Sigma^+ \to \mathbb{R}_{>0}$ be a given weight function and define for each vector $x \in \mathbb{R}^n$ and each coalition $S \in \Sigma^+$ the *f-excess* of S at x as $e_f(S, x) := \frac{c(S) - x(S)}{f(S)}$. The f-excess represents the f-weighted gain (or loss, if it is negative) of coalition S, if its members accept to pay $x(S)$ instead of operating some service themselves at cost $c(S)$. The excess measures price acceptability: the smaller $e_f(S, x)$, the less favorable is price x for coalition S. Therefore, keeping the minimal f-excess as large as possible increases the incentive to stay in the grand coalition N of the players and is fair from the point of view of coalitions. The so-called *f-least core* $\mathscr{LC}_f(\Gamma)$ is the set of all price vectors which maximize the minimal f-excess, i.e., the set of all price vectors x^* which satisfy that (x^*, ε^*) is an optimal solution of the following linear program

$$\max \ \varepsilon$$
$$\text{s.t.} \ \ x(S) + \varepsilon f(S) \leq c(S), \ \forall S \in \Sigma^+ \setminus \{N\} \tag{1}$$
$$x \in \mathscr{X}(\Gamma),$$

where ε^* is the optimal value of (1). Under the assumption that the imputation set $\mathscr{X}(\Gamma)$ is non-empty, one can easily show that (1) has optimal solutions and the f-least core belongs to the core if the core is non-empty.

By imposing a lexicographic order we obtain the so-called *f-nucleolus*, which belongs to the f-least core. The f-nucleolus can be defined as follows. For each

$x \in \mathbb{R}^n$, let $\theta_f(x)$ be the vector in $\mathbb{R}^{|\Sigma^+|-1}$ whose components are the f-excesses $e_f(S,x)$ of $S \in \Sigma^+ \setminus \{N\}$ arranged in increasing order, i.e., $\theta_f^i(x) \leq \theta_f^j(x)$ for each $1 \leq i < j \leq |\Sigma^+| - 1$. The f-nucleolus $\mathcal{N}_f(\Gamma)$ is the set of all vectors in $\mathcal{X}(\Gamma)$ that maximizes θ_f in $\mathcal{X}(\Gamma)$ with respect to the lexicographic ordering.

These two concepts are extensions of the ones which come from the game theory and may not appropriate to define prices, which are considered fair by each player. The reason for this is that the opinion about fairness of different players can be quite different. The quotation of H.L. Mencken, "a wealthy man is one who earns \$100 a year more than his wife's sister's husband", shows the relativity of human notion. Human notions of fairness may not be absolute, but relative to the opinions of the other. The concept (f,r)-*least core* was introduced as a compromise between the coalitional fairness and individual satisfaction, where r is a chosen reference price vector. The (f,r)-least core contains the price vector x^* in the f-least core $\mathcal{LC}(\Gamma)$ such that $\frac{x^*}{r}$ is lexicographically smallest, where for each $x \in \mathbb{R}^n$ the vector $\frac{x}{r}$ denote the one in $\mathbb{R}^n$ whose components are $\frac{x_i}{r_i}$, $i \in N$, arranged in decreasing order. That means x^* is the closest one to r in $\mathcal{LC}(\Gamma)$ in a special sense.

This thesis also considers the question of whether there exists a "best" cost allocation, which people naturally like to have. We can prove that there is no cost allocation which can satisfy all of our desired properties, which are coherent and seem to be reasonable or even indispensable. Our game theoretical concepts try to minimize the degree of axiomatic violation while the validity of some most important properties is kept.

3 Computational Aspects

The best concepts are useless unless we can solve the corresponding models. From the complexity point of view, it is NP-hard to calculate allocations based on the game theoretical concepts in general [4]. The hardest challenge is that there are exponentially many possible coalitions and their costs are a-priori unknown. However, this difficulty can be overcome by using constraint generation approaches [6]. We can prove that if the cost function c is submodular and the weight function f and the family Σ of possible coalitions satisfy certain properties then one can calculate the allocations in oracle-polynomial time. Our result is an extension of the one in [3] for submodular constrained cost allocation games and at the same time to a larger class of weight functions f (not only for $f = 1$ as in [3]), that is, our result is more general in two ways. Our proof is based on the well known theorems of Grötschel et al. on the equivalence between separation and optimization and the oracle-polynomial solvability of submodular function minimization problems on certain sets [5].

But since this submodularity is rarely fulfilled in reality, we consider therefore a class of combinatorial cost allocation games, whose coalition costs are given as objective values of combinatorial optimization problems. The separation of the coalition constraint is also a combinatorial optimization problem. Several primal and dual heuristics are constructed in order to decrease the solving time of the separation problem. These heuristics use informations obtained in the previous separation steps to fix the variables in a reliable way and to evaluate the quality of founded cutting planes quickly. Based on these techniques, we are able to solve our applications, whose sizes vary from small with 4 players, to medium with 18 players, and even large with 85 players and $2^{85} - 1$ possible coalitions.

4 Applications

Based on the theoretical part of the thesis, we can solve three real world cost allocation problems numerically. The first example deals with the distribution of costs for the construction of a sewage network for four households. The second example considers the distribution of the costs of a water network in Sweden with 18 players (municipalities). The cost function of the problem is based on a multi-commodity flow problem with a nonlinear objective function. With piecewise linear approximation and SOS conditions, we can approximate it by a mixed-integer program [2]. The resulting approximation error of the cost allocation is controlled. In this paper we only present the results of the last example in details. This example considers the ticket pricing problem of the Dutch InterCity railway network, whose data comes from the dissertation [1]. Through computational results, we show the unfairness of traditional cost allocations. The current distance tariff asks for almost the same amount of money for one travelling kilometer independent on travelling route. This system does not distinguish between economically efficient and inefficient routes and the ones have to subsidize the others thereby. These unfair prices result in a situation where the passengers in the central region of the country pay over 25% more than the costs they incur. Table 1 lists several major coalitions, which would earn a substantial benefit from shrinking the network. This table demonstrates the unfairness and instability of the distance price vector. In contrast, our game theory based prices, i.e., the (f, r)-least core prices with $f = c$ and r is the distance price vector, decrease this unfairness and increase the incentive to stay in the grand coalition for players. The maximum relative loss of all coalitions with our prices is a mere 1.68%. Simultaneously, the difference in the prices for one travelling kilometer between two arbitrary routes is bounded by a chosen factor of 3. The so-called fairness distribution diagram in Figure 1 presents a comparison between the dis-

Coalition ID	Relative loss	Percentage of all passengers
0	25.67%	15.34%
26	18.39%	18.35%
56	16.36%	30.81%
133	11.90%	57.34%
191	10.43%	78.14%

Table 1 Unfairness of the distance price vector

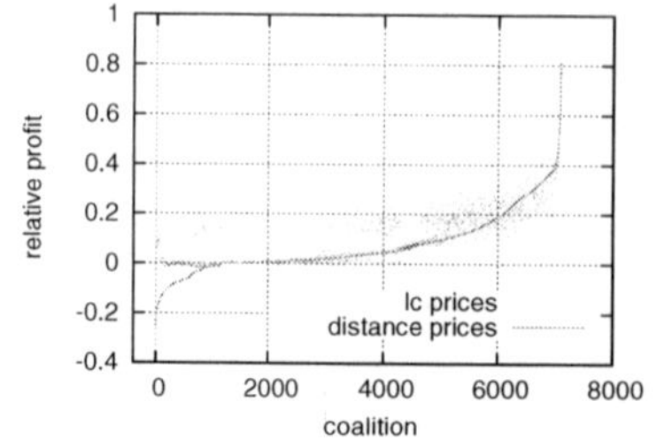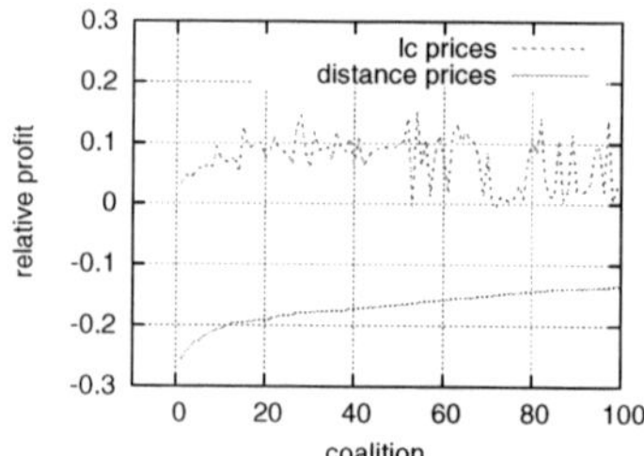

Fig. 1 Distance vs. game theoretical prices

tance and our game theoretical prices. Obviously, it is impossible to consider all of the $2^{85} - 1$ coalitions. The best thing that we can do is to sample and create a pool of some "essential coalitions" among them. Based on this pool one can show graphically a representative comparison between the two price vectors. The picture on the left side represents the relative profits of 7084 essential coalitions in the pool with the two price vectors, while the picture on the right side is just a zoom of the first one. They show that our prices for all of the 100 coalitions which have the worst relative profit with the distance price vector increase significantly. Many of them have even a relative profit of more than 10% with our prices. This example shows clearly that using our game theoretical approach one can come up with price systems that constitute a good compromise between fairness and enforceability and are better (fairer) than the ad-hoc allocations.

Acknowledgements I would like to express my sincere gratitude to my advisors, Prof. Dr. Dr. h.c. mult. Martin Grötschel and Dr. habil. Ralf Borndörfer, for the interesting research theme and for their valuable supports and suggestions. I am grateful to the Zuse Institute Berlin for providing me a Konrad-Zuse Scholarship.

References

1. Bussieck, M. R.: Optimal Lines in Public Rail Transport. PhD Thesis, TU Braunschweig (1998)
2. Dantzig, G. B.: On the Significance of Solving Linear Programming Problems with Some Integer Variables. Econometrica **28**, 30–44 (1960)
3. Faigle, U., Kern, W., Kuipers, J.: On the computation of the nucleolus of a cooperative game. Internat. J. Game Theory **30**, 79–98 (2001)
4. Faigle, U., Kern, W., Paulusma, D.: Note on the Computational Complexity of Least Core Concepts for min-cost Spanning Tree Games. Math. Methods of Operations Research **52**, 23–38 (2000)
5. Grötschel, M., Lovász, L., Schrijver, A.: Geometric algorithms and combinatorial optimization, 2. corr. ed., Springer (1993)
6. Hallefjord, A., Helming, R., Jørnsten, K.: Computing the Nucleolus when the Characteristic Function is Given Implicitly: A Constraint Generation Approach. International Journal of Game Theory **24**, 357–372 (1995)
7. Hoang, N. D.: Algorithmic Cost Allocation Games: Theory and Applications. PhD Thesis, TU Berlin (2010)

Author Index

D. Klatte et al. (eds.), *Operations Research Proceedings 2011*, Operations Research Proceedings, 605
DOI 10.1007/978-3-642-29210-1, © Springer-Verlag Berlin Heidelberg 2012